MRI in Psychiatry

Christoph Mulert • Martha E. Shenton
Editors

MRI in Psychiatry

Springer

Editors
Christoph Mulert
Psychiatry Neuroimaging Branch
UKE, Department of Psychiatry
Hamburg
Germany

Martha E. Shenton
Department of Psychiatry and Radiology
Harvard Medical School
Boston, Massachusetts
USA

ISBN 978-3-642-54541-2 ISBN 978-3-642-54542-9 (eBook)
DOI 10.1007/978-3-642-54542-9
Springer Heidelberg New York Dordrecht London

Library of Congress Control Number: 2014942985

Printed on acid-free paper

Springer is part of Springer Science+Business Media (www.springer.com)

Preface

Magnetic resonance imaging (MRI) is a technique that is widely used in many fields of modern medicine today for the investigation of structure and function of the human body. MRI is a routine procedure for medical diagnosis, for staging of disease, and for follow-up without exposure to ionizing radiation. It has also become an essential tool for understanding the brain as it offers an important window for viewing both brain structure and function in living humans, where previously most information came from postmortem studies. Neuroscience, and a focus on the brain, most particularly in neuropsychiatric disorders, has been an area of intense scientific inquiry. For understanding these conditions – with unknown etiologies and many open questions concerning pathophysiology and therapy – insights into brain structure and function have been greatly enhanced with the advent of advanced MRI. MRI can be safely done at many imaging facilities all over the world and has become very attractive for young researchers who are thinking about a career in the area of neuroscience that involves neuroimaging techniques. This book is intended to provide an overview and to introduce this fascinating area of scientific inquiry to the Ph.D. student in psychology or neuroscience, to the medical student, and also to residents in psychiatry or clinical psychiatrists. The chapters are written by leading experts in their respective fields. Given the fast-paced development of new and distinctive MRI techniques, this book will also be of interest to experienced researchers who want an overview about topics they are not specialized in themselves.

This book is divided into three sections. With respect to the background of the three sections, we thought it would be important to provide a comprehensive textbook that includes information about methodology and concepts, as well as brain systems and psychiatric disorders, all in one place. Accordingly, the book is divided into three sections with the first part of the book focused on relevant methods. It introduces the basic aspects of fMRI statistics and advanced statistical procedures for effective connectivity analyses, as well as technology and applications of real-time fMRI. Moreover, different MRI techniques are described such as MR spectroscopy, diffusion tensor Imaging (DTI), or simultaneous EEG-fMRI. In addition, we have the impression that there is a need to go beyond MRI technology itself and shift to more hypothesis- and model-driven approaches. Accordingly, combinations of MRI with other approaches that have been successfully used are introduced, such as the combination of MRI with brain stimulation procedures, or imaging genetics, a scientific field that is very influential in the understanding of psychiatric disorders today. The second part of the book introduces the reader to major brain systems. Here, possibilities, including recent findings and limitations, are reviewed for MRI imaging in the context of perception, cognition, emotion, and reward. Finally, the third part of the book covers most of the major psychiatric diseases. Here, findings with different MR techniques are summarized in individual chapters. As the reader will come to understand, the number of studies is different among the different psychiatric conditions. We nonetheless decided to include disorders that have a smaller number of studies completed thus far given their clinical relevance, such as personality disorders. This book can thus be read in different ways: from the very beginning to the end for obtaining a comprehensive overview – or as a reference for certain topics both in advanced MRI techniques and in MR findings that are relevant to specific psychiatric disorders.

We note that while MRI techniques are widely used clinically for medical diagnosis, and, as noted previously, for staging of disease as well as for follow-up, MRI in psychiatry is still more in its infancy and is used more as a research tool where groups of clinical populations are compared to healthy controls rather than as a clinical tool that provides information for an individual patient. The latter, however, is changing and the potential is quite high for neuroimaging to be used diagnostically in the near future where individual patients will benefit from the tools that are now used solely for research purposes. Some of these latter ideas are also expressed in this book, particularly with respect to the early diagnosis of Alzheimer's dementia, as well as therapeutic applications with real-time fMRI neurofeedback.

Finally, no project of this scope is ever done alone. First and foremost, we want to thank all of the authors who contributed their time and effort to make this book a reality. Second, we wish to thank Meike Stoeck from Springer for her assistance on all aspects of this book. We also thank Marius Mußmann who was of great help in making this book come to fruition. And we also wish to thank our spouses, Marianne and George, who have supported our endeavors.

Hamburg, Germany — Christoph Mulert
Boston, MA, USA — Martha E. Shenton

Contents

Contributors

Chris I. Baker, PhD Laboratory of Brain and Cognition, National Institute of Mental Health, National Institutes of Health, Bethesda, MD, USA

Anne Beck Department of Psychiatry and Psychotherapy, Charité Universitätsmedizin Berlin, Berlin, Germany

Andreas Böhringer, Dipl-Psych Department of Psychiatry and Psychotherapy, Medical Faculty Mannheim, Central Institute of Mental Health, University of Heidelberg, Mannheim, Germany

Arun L.W. Bokde, PhD Discipline of Psychiatry, School of Medicine and Trinity College Institute of Neuroscience, Trinity College Dublin, Adelaide and Meath Hospital incorporating National Children's Hospital, Dublin, Ireland

Valentina Cardi, PhD Department of Psychological Medicine,, Institute of Psychiatry, King's College London, London, UK

Katrin Charlet Department of Psychiatry and Psychotherapy, Charité Universitätsmedizin Berlin, Berlin, Germany

Michael Czisch Max Planck Institute of Psychiatry, Munich, Germany

Birgit Derntl Department of Psychiatry, Psychotherapy and Psychosomatics, RWTH Aachen University, Aachen, Germany

Irina Falkenberg Department of Psychiatry and Psychotherapy, Philipps-University Marburg, Marburg, Germany

Sophia Frangou, MD, MSc, PhD, FRCPsych Department of Psychiatry, Icahn School of Medicine at Mount Sinai, New York, NY, USA

Karl Friston, FRS Wellcome Trust Centre for Neuroimaging, University College London, London, UK

Thomas Frodl, MD, MA Department of Psychiatry, University of Dublin, Trinity College, Dublin, Ireland

Department of Psychiatry, University of Regensburg, Regensburg, Germany

Rainer Goebel Department of Cognitive Neuroscience, Faculty of Psychology and Neuroscience, Maastricht University, Maastricht, The Netherlands

Department of Neuroimaging and Neuromodelling, Netherlands Institute for Neuroscience, Amsterdam, The Netherlands

Roberto Goya-Maldonado Department of Psychiatry and Psychotherapy, Center for Translational Research in Systems Neuroscience and Psychiatry, Georg August University, Goettingen, Germany

Michel J. Grothe German Center for Neurodegenerative Diseases (DZNE), Rostock, Germany

Oliver Gruber Department of Psychiatry and Psychotherapy, Center for Translational Research in Systems Neuroscience and Psychiatry, Georg August University, Goettingen, Germany

Ute Habel Department of Psychiatry, Psychotherapy and Psychosomatics, RWTH Aachen University, Aachen, Germany

Assaf Harel Laboratory of Brain and Cognition, National Institute of Mental Health, National Institutes of Health, Bethesda, MD, USA

Andreas Heinz Department of Psychiatry and Psychotherapy, Charité Universitätsmedizin Berlin, Berlin, Germany

Georg Juckel Department of Psychiatry, Ruhr University Bochum, LWL University Hospital, Bochum, Germany

Walker Keenan Department of Radiology, Brigham and Women's Hospital, Harvard Medical School, Boston, MA, USA

Tilo Kircher Department of Psychiatry and Psychotherapy, Philipps-University Marburg, Marburg, Germany

Inga Katharina Koerte, MD Division of Radiology, Ludwig-Maximilians-University, Munich, Germany

Axel Krug Department of Psychiatry and Psychotherapy, Philipps-University Marburg, Marburg, Germany

Gregor Leicht Psychiatry Neuroimaging Branch, Department of Psychiatry and Psychotherapy, University Medical Center Hamburg-Eppendorf, Hamburg, Germany

Huijun Liao, BS Department of Radiology, Brigham and Women's Hospital, Harvard Medical School, Boston, MA, USA

Alexander Lin, PhD Department of Radiology, Brigham and Women's Hospital, Harvard Medical School, Boston, MA, USA

Psychiatric Neuroimaging LaboratoryBrigham and Women's Hospital, Harvard Medical School, Boston, MA, USA

David Linden Institute of Psychological Medicine and Clinical Neurosciences, Cardiff University School of Medicine, Cardiff, UK

Sai Merugumala, MS Department of Radiology, Brigham and Women's Hospital, Harvard Medical School, Boston, MA, USA

Andreas Meyer-Lindenberg Department of Psychiatry and Psychotherapy, Medical Faculty Mannheim, Central Institute of Mental Health, University of Heidelberg, Mannheim, Germany

Marc Muehlmann, MD Division of Radiology, Ludwig-Maximilians-University, Munich, Germany

Christoph Mulert Psychiatry Neuroimaging Branch, Department of Psychiatry and Psychotherapy, University Medical Center Hamburg-Eppendorf, Hamburg, Germany

William Penny Wellcome Trust Centre for Neuroimaging, University College London, London, UK

Saadalah Ramadan, PhD Faculty of Health, School of Health Sciences, University of Newcastle, Callaghan, NSW, Australia

Stefan Roepke Department of Psychiatry and Psychotherapy, Charité Universitätsmedizin Berlin, Berlin, Germany

Alexander T. Sack Faculty of Psychology and Neuroscience, Maastricht University, Maastricht, The Netherlands

Frank Schneider Department of Psychiatry, Psychotherapy and Psychosomatics, RWTH Aachen University, Aachen, Germany

Teresa Schuhmann Faculty of Psychology and Neuroscience, Maastricht University, Maastricht, The Netherlands

Lars Schulze Department of Clinical Psychology and Psychotherapy, Freie Universität Berlin, Berlin, Germany

Victor I. Spoormaker, PhD Max Planck Institute of Psychiatry, Munich, Germany

Masashi Suda, MD, PhD Department of Psychological Medicine,, Institute of Psychiatry, King's College London, London, UK

Stefan J. Teipel German Center for Neurodegenerative Diseases (DZNE), Rostock, Germany

Department of Psychosomatic Medicine, University Rostock, Rostock, Germany

Heike Tost, Dipl-Psych Department of Psychiatry and Psychotherapy, Medical Faculty Mannheim, Central Institute of Mental Health, University of Heidelberg, Mannheim, Germany

Janet Treasure, PhD, FRCP FRCPsych Department of Psychological Medicine, Institute of Psychiatry, King's College London, London, UK

Eric Vermetten, MD, PhD Department of Psychiatry, Leiden University Medical Center, Leiden, The Netherlands

Luke Y-J. Wang, MD Department of Radiology, Brigham and Women's Hospital, Harvard Medical School, Boston, MA, USA

Department of Anesthesiology, Perioperative and Pain Medicine, Boston Children's Hospital, Harvard Medical School, Boston, MA, USA

Frank H. Wilhelm, PhD Clinical Psychology, Psychotherapy & Health Psychology, Institute for Psychology, University of Salzburg, Salzburg, Austria

Yen Yu Wellcome Trust Centre for Neuroimaging, University College London, London, UK

Part I

Methods

1 Statistical Analysis and Modeling of Functional MRI Data

Rainer Goebel

Abbreviations

AC	Anterior commissure
ANCOVA	Covariance analysis
ANOVA	Analysis of variance
AR	Autoregressive
ASL	Arterial spin labeling
BOLD	Blood oxygenation level dependent
CBF	Cerebral blood flow
EPI	Echoplanar imaging
ERP	Event-related potential
FDR	False discovery rate
FFA	Fusiform face area
FFX	Fixed-effects analysis
FT	Fourier transform
FWE	Family-wise error
FWHM	Full width at half maximum
GE	Gradient echo
GLM	General linear model
GLS	Generalized least squares
ITI	Intertrial interval
LTI	Linear time invariant
MANOVA	Multivariate analysis of variance
MFX	Mixed-effects analysis
MNI	Montréal Neurological Institute
MVPA	Multi-voxel pattern analysis
OLS	Ordinary least squares
PC	Posterior commissure
PET	Positron emission tomography
RFX	Random-effects analysis
ROI	Region of interest
SVM	Support vector machine
TR	Volume time to repeat

R. Goebel
Department of Cognitive Neuroscience, Faculty of Psychology and Neuroscience, Maastricht University, Maastricht, The Netherlands

Neuroimaging and Neuromodelling, Netherlands Institute for Neuroscience, Amsterdam, The Netherlands
e-mail: r.goebel@maastrichtuniversity.nl

1.1 Introduction

Since its invention in the early 1990s, functional magnetic resonance imaging (fMRI) has rapidly assumed a leading role among the techniques used to localize brain activity. The spatial and temporal resolution provided by state-of-the-art MR technology and its noninvasive character, which allows multiple studies of the same subject, are some of the main advantages of fMRI over the other functional neuroimaging techniques that are based on changes in blood flow and cortical metabolism (e.g., positron emission tomography, PET). fMRI is based on the discovery of Ogawa et al. (1990) that magnetic resonance imaging (MRI, also called nuclear magnetic resonance imaging) can be used in a way that allows obtaining signals depending on the level of blood oxygenation. The measured signal is therefore also called "BOLD" signal (BOLD = blood oxygenation level dependent). Since locally increased neuronal activity leads to increased local blood flow, which again changes local blood oxygenation (in combination with changes in blood volume and oxygen

C. Mulert, M.E. Shenton (eds.), *MRI in Psychiatry*,
DOI 10.1007/978-3-642-54542-9_1, © Springer Berlin Heidelberg 2014

metabolism), fMRI allows indirect measurements of neuronal activity changes.

A major goal of functional MRI (fMRI) measurements is the localization of the neural correlates of sensory, motor, and cognitive processes in healthy subjects and patients. The term "brain mapping" is often used to refer to this goal of relating operations of the mind to specific areas and networks in the brain. As a well-known example, fMRI has been used to localize the fusiform face area (FFA, Kanwisher et al. 1997), a specific region in the ventral posterior temporal cortex that responds preferentially to faces, i.e., it responds stronger to images depicting faces than to images depicting other objects. Another major goal of fMRI studies is the detailed characterization of the response profile for known regions of interest (ROIs) across experimental conditions. In this context, the aim of conducted studies is often not to map new functional brain regions (whole-brain analysis) but to characterize further how known specialized brain areas respond to (subtle) differences in experimental conditions (ROI-based analysis).

While the location of FFA is, for example, well established, many additional studies have been conducted "on this area" to learn how FFA responds to various face-related and face-unrelated stimuli or to internally generated mental processes such as face imagery.

Furthermore, it is often of interest to estimate the shape of the response and how it varies across different conditions and brain areas. This chapter describes analysis steps that are usually conducted to reach these goals. The described analysis steps include data preprocessing such as head motion correction and filtering and, most importantly, statistical single-subject and group analyses. Specific advantages and problems will be discussed when conducting whole-brain and ROI-based analyses. This chapter focuses on univariate statistical analysis where time series are analyzed independently for each *voxel* ("voxel" = "volume element") or for the average signal time course in a region of interest.

1.1.1 Typical Data Sets for fMRI Analysis

Most fMRI experiments collect functional time series using the gradient echo EPI sequence that allows acquisition of a 64×64 matrix in 50–100 ms. A typical functional scan of the whole brain with about 30 slices lasts only about 2 s on state-of-the-art 3 Tesla MRI scanners. The data obtained from scanning all slices located at different brain positions once (e.g., 30 slices covering the whole brain) is subsequently referred to as a *functional volume* or a *functional (3D) image*. The measurement of an uninterrupted series of functional volumes is referred to as a *run* or *functional scan* in this chapter. A run, thus, consists of *repeated measurements* of a functional volume at regularly spaced intervals. The sampling interval – the time until the same brain region is measured again – is called *volume TR*. The volume TR specifies the temporal sampling resolution of the functional measurements since all slices comprising one functional volume are obtained once during that time. Note, however, that the slices of a functional volume are not recorded simultaneously, i.e., the data from different regions of the brain are recorded at different moments in time (see Sect. 1.2.2). During a functional MRI experiment, a participant usually performs active or passive tasks following a *stimulation protocol* that specifies the sequence of experimental conditions. A short experiment can be completed in a single run, which typically consists of 100–1,000 functional volumes. Assuming a run with 500 volumes, each consisting of 30 slices of 64×64 voxels and that 2 bytes are needed to store one measurement at each voxel, the amount of raw data acquired per run would be 500 × 30 × 64 × 64 × 2 = 122,880,000 bytes or roughly 117 MB. In more complex experiments, a subject typically performs multiple runs in one or more scanning sessions resulting in about 500 MB of functional raw image data per subject.

The raw data collected for a typical multi-subject experiment with 10–20 subjects will, thus, easily consist of several gigabytes (GB). Using parallel imaging and high-resolution scanning, several tenth of GB will be collected for a single group study.

Given the small amplitude of task-related BOLD signal changes of typically 1–5 % and the presence of many confounding effects, such as signal drifts and head motion, the localization and characterization of brain regions responding to experimental conditions of the stimulation protocol is a nontrivial task. The major analysis steps of functional – as well as associated anatomical – data will be described in the following sections including spatial and temporal *preprocessing*, *coregistration* of functional and anatomical data sets, individual *statistical data analysis*, *spatial normalization*, and *group analyses*. Although these essential data analysis steps are performed in a rather standardized way in all major software packages, including AFNI (http://afni.nimh.nih.gov/afni/), BrainVoyager (http://www.brainvoyager.com/), FSL (http://www.fmrib.ox.ac.uk/fsl/), and SPM (http://www.fil.ion.ucl.ac.uk/spm/), there is still room for improvements as will be discussed below. For the visualization of functional data, high-resolution anatomical data sets with a resolution of (or close to) 1 mm in all dimensions are often collected in a recording session. In most cases, these anatomical volumes are scanned using slow T1 weighted MR sequences that are optimized to produce high-quality images with very good contrast to separate gray and white matter. In some analysis packages, anatomical data sets do not only serve as a structural reference for the visualization of functional information but are often also used to improve the functional analysis itself, for example, by restricting statistical data analysis to cortical voxels, i.e., voxels located within gray matter, or to analyze topological representations on extracted cortex meshes. The preprocessing of high-resolution anatomical data sets and their role in functional data analysis will be described in Sect. 1.2.

Since some data analysis steps depend on the details of the experimental paradigm, the next section shortly describes the two most frequently used experimental designs.

1.1.2 Experimental Design

In the first years of fMRI measurements, experimental designs were adapted from positron emission tomography (PET) studies. In a typical PET design, several *trials* (individual stimuli, or more generally, cognitive events) were clustered in *blocks*, each of which contained trials of the same condition. As an example, one block may consist of a series of different pictures showing happy faces, and another block may consist of pictures showing sad faces. The statistical analysis of such *block designs* compares the mean activity obtained in the different experimental blocks. Block designs were necessary in PET studies because of the limited temporal resolution of this imaging technique requiring about a minute to obtain a single whole-brain image. Since the temporal resolution of fMRI is much higher than that of PET, it has been proposed to use also *event-related designs* (Blamire et al. 1992; Buckner et al. 1996; Dale and Buckner 1997). The characteristics of these designs follow closely those used in event-related potential (ERP) studies. In event-related designs, individual trials of different conditions are not clustered in blocks but are presented in a random sequence. Responses to trials belonging to the same condition are selectively averaged, and the calculated mean responses are statistically compared with each other. While block designs are well suited for many experiments, event-related designs offer several advantages over block designs, especially for cognitive tasks. An important advantage of event-related designs is the possibility to present stimuli in a randomized order avoiding cognitive adaptation or expectation strategies. Such cognitive adaptations are likely to occur in block designs since a subject knows what type of stimuli to expect within a block after having experienced the first few trials. Another important advantage of event-related designs is that the *response profile* for different trial types (and even single trials) can be *estimated* by event-related averaging.

Furthermore, event-related designs allow *post hoc sorting* of individual brain responses. One important example of post hoc sorting is the separation of brain responses for correctly vs. incorrectly performed trials.

The possibilities of event-related fMRI designs are comparable to standard behavioral and ERP analyses. Note, however, that the hemodynamic response extends over about 20–30 s after presentation of a short stimulus; if only the positive BOLD response is considered, the signal response to short events extends over 10–15 s. The easiest way to conduct event-related fMRI designs is to temporally separate individual trials far enough to avoid overlapping responses of successive trials. Event-related designs with long temporal intervals between individual trials are termed *slow event-related designs*. For stimuli of duration of 1–2 s, the optimal *intertrial interval* (ITI) has been identified as 12 s (Bandettini and Cox 2000; Maus et al. 2010). Since it has been shown that the fMRI signals of closely spaced trials add up approximately linearly (Boynton et al. 1996; Dale and Buckner 1997), it is also possible to run experiments with inter trial intervals of 2–6 s. Designs with short temporal intervals between trials are called *rapid event-related designs*. While the measured response of rapid event-related designs will contain a mix of overlapping responses from closely spaced trials at each time point, condition-specific event-related time courses can be isolated using *deconvolution analysis*. Deconvolution analysis works correctly only under the assumption of a linear system (see Sect. 1.3.4) and requires randomized intertrial intervals ("jitter"), which can be easily obtained by adding "null" (baseline) trials when trial sequences are created for an experiment. While simple permutation approaches produce already good event sequences for rapid event-related designs, statistical power can be maximized by using more advanced randomization procedures (Wager and Nichols 2003; Maus et al. 2010). In general, block and event-related designs can be statistically analyzed using the same mathematical principles (Sect. 1.3.3).

BOLD-related signal fluctuations do have neither a defined origin nor a unit. The signal strength in main experimental conditions can, thus, not be interpreted absolutely but needs to be assessed relative to the signal strength in other main or control conditions at the same voxels. As a general control condition, many fMRI experiments contain a baseline ("rest," "fixation") condition with "no task" for the subject. Such simple control conditions allow analyzing brain activity that is common in multiple main conditions that would not be detectable when only contrasts between main conditions could be performed. More complex experimental control conditions are often included that differ from the main condition(s) only in a subtle cognitive component allowing isolating brain responses specific to that component. Responses to main conditions are often expressed as percent signal change relative to a baseline condition. Furthermore, it is recommended to vary conditions *within subjects* – and even within runs – since the lack of an absolute signal level increases variability when comparing effects across runs, sessions, or subjects. Many experiments, however, require a *between-subject* design, including comparisons of responses between different groups, e.g., treatment group vs. control group. Note that the BOLD signal measured with conventional fMRI may be affected by medication that modifies the neurovascular coupling, e.g., by increasing or decreasing baseline cerebral blood flow (CBF). In order to obtain more quantitative evaluation of activation responses, it is, thus, recommended for patient studies to combine standard BOLD measurements with CBF measurements using arterial spin labeling (ASL) techniques (e.g., Buxton et al. 2004).

1.2 Preprocessing

In order to reduce artifact and noise-related signal components, a series of *preprocessing* operations are typically performed prior to statistical data analysis. The most essential preprocessing steps are (1) head motion detection and correction, (2) slice scan timing correction, (3) removal of linear and nonlinear trends in voxel time courses, and (4) spatial and temporal smoothing of the data.

1.2.1 Detection and Correction of Head Motion

The quality of fMRI data is strongly hampered in the presence of substantial head movements. Data sets are often rejected for further analysis if head motion of 5 or more millimeters is detected. Although head motion can be corrected in image space, displacements of the head reduce the homogeneity of the magnetic field, which is fine-tuned ("shimmed") ahead of functional scans with respect to the head position at that time. If head movements are small, motion correction is a useful and important step to improve the data quality for subsequent statistical data analysis. Motion correction operates by selecting a functional volume of a run (or a volume from another run of the same scanning session) as a reference to which all other functional volumes are aligned. Most head motion algorithms describe head movements by six parameters describing translation (displacement) and rotation at each time point. These six parameters are sufficient to characterize motion of *rigid bodies* since rigid objects do not show scale or sheer transformations. Any spatial displacement of the brain can, thus, be described by translation along the x-, y-, and z-axes and rotation around these axes. The values of these six parameters are estimated by analyzing how a source volume should be translated and rotated in order to better align with the reference volume; after applying a first estimate of the parameters, the procedure is repeated to improve the "fit" between the transformed (motion-corrected) and target (reference) volume. A *similarity* or *error measure* quantifies the approximation quality of the transformed volume with respect to the reference volume. An often-used error measure is the sum of the squared intensity differences at corresponding positions in the reference volume and the transformed volume. The iterative adjustment of the parameter estimates stops if no further improvement can be achieved, i.e., when the error measure reaches a minimum. After the final motion parameters have been *detected* by the iterative procedure, they can be applied to the source volume to produce a *motion-corrected* volume that replaces the original volume in the output data set. For visual inspection, fMRI software packages are usually presenting line plots of the translation and rotation parameters across time showing how the estimated values change from volume to volume. The obtained parameter time courses may also be integrated in subsequent statistical data analysis with the aim to remove residual motion artifacts.

1.2.2 Slice Scan Time Correction

For statistical analysis, a functional volume is usually considered as measured at the same time point. Slices of a functional volume are, however, scanned sequentially in standard functional (EPI) measurements, i.e., each slice is obtained at a different time point within a functional volume. For a functional volume of 30 slices and a volume TR of 3 s, for example, the data of the last slice is measured almost 3 s later than the data of the first slice. Despite the sluggishness of the hemodynamic response, an imprecise specification of time in the order of 3 s will lead to suboptimal statistical analyses, especially in event-related designs. It is, thus, desirable to preprocess the data in such a way that the resulting processed data appears as if all slices of a functional volume were measured at the same moment in time. Only then would it be, for example, possible to compare and integrate event-related responses from different brain regions correctly with respect to temporal parameters such as onset latency. In order to correct for different slice scan timings, the time series of individual slices are temporally "shifted" to match a reference time point, e.g., the first or middle slice of a functional volume. The appropriate temporal shift of the time courses of the other slices is then performed by resampling the original data accordingly. Since this process involves sampling at time points that fall between measurement time points, the new values need to be estimated by interpolation of values from past and future time points. The most often used interpolation methods are linear and sync interpolation. Note that the time points of slice scanning depend on the *acquisition order* specified during preparation of functional scanning at the scanner console. Besides an ascending or descending order, slices are often scanned interleaved, i.e., the odd slice numbers are recorded first followed by the even slice numbers. After appropriate temporal resampling, all slices within a functional volume of the new data set represent the same time point and can be analyzed with the same hemodynamic response function.

1.2.3 Temporal Filtering

1.2.3.1 Removal of Linear and Nonlinear Drifts

Due to physical and physiological noise, voxel time courses are often nonstationary exhibiting signal drifts. If the signal rises or falls with a constant slope from beginning to end, the drift is called a *linear trend*. If the signal level slowly varies over time with a nonconstant slope, the drift is called a *nonlinear trend*. Since drifts describe slow signal changes, they can be removed by Fourier analysis using a *temporal high-pass filter*. The original signal in the time domain is transformed in the frequency domain using the Fourier transform (FT). In this domain, drifts can be easily removed because low-frequency components, underlying drifts, are isolated from higher-frequency components underlying task-related signal changes. The filtered frequency representation (suppressing low frequencies) is then transformed back into the time domain by the inverse Fourier transform. As an alternative approach, drifts can be modeled and removed in the time domain using appropriate basis sets in a general linear model analysis. This approach can be performed either as a preprocessing step or as part of statistical data analysis.

1.2.3.2 Temporal Smoothing

Another temporal preprocessing step consists in temporal smoothing of voxel time courses removing *high-frequency signal fluctuations*, which are considered as noise. While this step increases the signal-to-noise ratio, (strong) temporal smoothing is not recommended when analyzing event-related designs since it may distort estimates of temporally relevant parameters, such as the onset or width of (average) event-related responses. Temporal smoothing also increases serial correlations between values of successive time points that need to be corrected (Sect. 1.3.5).

1.2.4 Spatial Smoothing

To further enhance the signal-to-noise ratio, the data is often spatially smoothed by convolution with a 3D Gaussian kernel. In this process, each voxel is replaced by a weighted average value calculated across neighboring voxels. The shape and width of the Gaussian kernel determines the weights used to include neighboring voxels in the average, i.e., voxels further apart will receive smaller weights and will influence the average less than voxels closer to the center. Note that smoothing reduces the spatial resolution of the data and should be therefore applied with care. Many studies, however, aim to detect regions larger than a few voxels, i.e., specialized brain areas in the order of 1 cm^3 or larger. Under these conditions, spatial smoothing with an appropriate kernel width of 4–8 mm is very useful since it suppresses noise and enhances the task-related signal. Furthermore, spatial smoothing increases the extent of activated brain regions, which is exploited in the context of group analyses (see Sect. 1.6) facilitating the integration of signals from corresponding but not perfectly aligned brain regions.

From the description and discussion of standard preprocessing steps, it should have become clear that there are no universally correct criteria to choose preprocessing steps and parameters because choices depend to some extent on the goal of data analysis. Some steps depend also on the experimental design of a study. If, for example, a high-pass temporal filter is used with a cut-off point that is too high, interesting task-related signal fluctuations could easily be removed from the data.

1.2.5 Distortion Correction

The gradient echo (GE) echoplanar imaging (EPI) sequence is used for most fMRI studies because of its speed, but it has the disadvantage that images suffer from signal dropouts and geometric distortions, especially in brain regions close to other tissue types such as air and liquor (susceptibility artifacts). These artifacts can be reduced substantially by using optimized EPI sequence parameters (e.g., Weiskopf et al. 2006) and parallel imaging techniques. A complete removal of dropouts and geometric distortions is, however, not possible. Further improvements may be obtained by distortion correction routines, which may benefit from special scans measuring magnetic field distortions (e.g., field maps). The distortion corrected images may improve coregistration results between functional and anatomical data sets enabling a more precise localization of brain function.

1.3 Statistical Analysis of Single-Subject Data

Statistical data analysis aims at identifying brain regions exhibiting increased or decreased responses in specific experimental conditions as compared to other (e.g., control) conditions. Due to the presence of physiological and physical noise fluctuations, observed differences between conditions may occur simply by chance. Note that measurements provide only a *sample* of data, but we are interested in true effects in the underlying *population*. Statistical data analysis protects from wrongly accepting effects in small sample data sets by explicitly assessing the effect of measurement variability (noise fluctuations) on estimated condition effects: If it is very unlikely that an observed effect is the result of noise fluctuations, it is assumed that the observed effect reflects a true difference between conditions in the population. In standard fMRI analyses, this assessment is performed independently for the time course of each voxel. Since independent testing at each voxel increases the chance to find some voxels with strong differences between conditions simply due to noise fluctuations, further adjustments for *multiple comparisons* need to be made. In order to localize activation effects, the estimated statistical values are integrated in three-dimensional *statistical maps*.

1.3.1 Basic Statistical Concepts

Figure 1.1 shows two fMRI time courses obtained from two different brain areas of an experiment with two conditions, a control condition ("Rest") and a main condition ("Stim"). Each condition has been measured several times. How can we assess whether the response values are higher in the main condition than in the control condition? One approach consists in subtracting the mean value of the "Rest" condition, $\bar{X}_1$, from the mean value of the "Stim" condition, $\bar{X}_2$: $d = \bar{X}_2 - \bar{X}_1$. Note that one would obtain in this example the same mean values in both conditions and, thus, the same difference in cases (a) and (b). Despite the fact that the means are identical in both cases, the difference in case b) seems to be more "trustworthy" than the difference in case (a) because the measured values exhibit less fluctuations, i.e., they *vary less* in case (b) than in case (a). Statistical data analysis goes beyond simple subtraction by taking into account the amount of variability of the measured data points. Statistical analysis essentially asks how likely it is to obtain a certain effect (e.g., difference of condition means) in a data sample if there is no effect at the population level, i.e., how likely it is that an observed sample effect is solely the result of noise fluctuations. This is formalized by the *null hypothesis* that states that there is no effect, i.e., no true difference between conditions in the population. In the case of comparing the two means μ_1 and μ_2, the null hypothesis can be formulated as H_0: $\mu_1 = \mu_2$. Assuming the null hypothesis, it can be calculated how likely it is that an observed sample effect would have occurred simply by chance. This requires knowledge about the amount of noise fluctuations (and its distribution), which can be estimated from the data. By incorporating the number of data points and the variability of measurements, statistical data analysis allows to *estimate the uncertainty of effects* (e.g., of mean differences) in data samples. If an observed effect is large enough so that it is very unlikely that it has occurred simply by chance (e.g., the probability is less than $p = 0.05$), one *rejects* the null hypothesis and *accepts* the *alternative hypothesis* stating that there exists a true effect in the population. Note that the decision to accept or

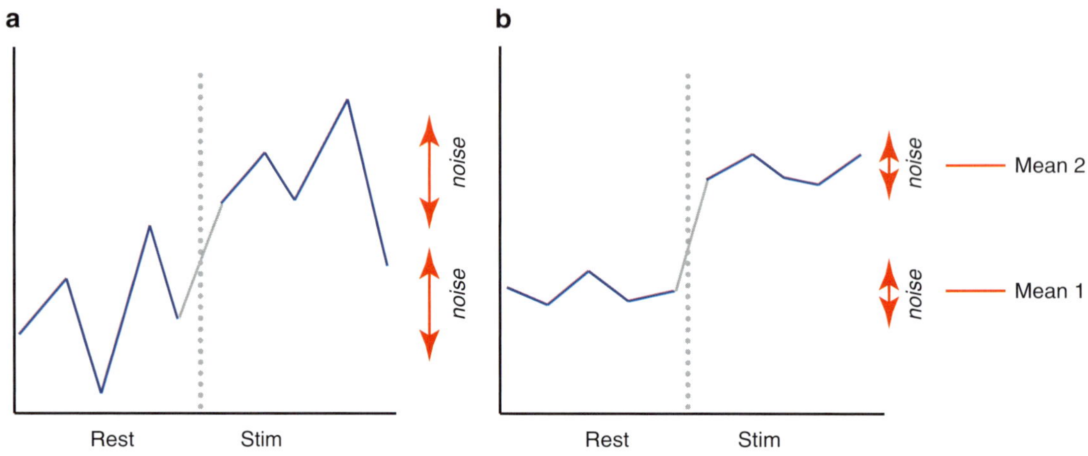

Fig. 1.1 Principle of statistical data analysis. An experiment with two conditions ("Stim" and "Rest") has been performed. (**a**) Time course obtained in area 1. (**b**) Time course obtained in area 2. Calculation and subtraction of Mean 1 ("Rest" condition) from Mean 2 ("Stim" condition) leads to the same result in cases (a) and (b). In a statistical analysis, the estimated effect (mean difference) is related to its uncertainty, which is estimated by the variability of the measured values within conditions. Since the variance within the two conditions is smaller in case (b), the estimated effect is more likely to correspond to a true difference in case (b) than in case (a)

reject the null hypothesis is based on a probability value of $p<0.05$ that has been accepted by the scientific community. A statistical analysis, thus, does never *prove* the existence of an effect, it only suggests "to believe in an effect" if it is very unlikely that the observed effect would have occurred by chance. Note that a probability of $p=0.05$ means that if we would repeat the experiment 100 times, we would accept the alternative hypothesis in about five cases even if there would be no real effect in the population. Since the chosen probability value thus reflects the likelihood of wrongly rejecting the null hypothesis, it is also called *error probability*. The error probability is also referred to as the *significance level* and denoted with the Greek letter α. If one would know that there is no effect in the population but one would incorrectly reject the null hypothesis in a particular data sample, a "false-positive" decision would be made (type 1 error). Since a false-positive error depends on the chosen error probability, it is also referred to as alpha error. If one would know that there is a true effect in the population but one would fail to reject the null hypothesis in a sample, a "false-negative" decision would be made, i.e., one would miss a true effect (type 2 error or beta error).

1.3.2 T-Test and Correlation Analysis

The uncertainty of an effect is estimated by calculating the variance of the noise fluctuations from the data. For the case of comparing two mean values, the observed difference of the means is related to the variability of that difference resulting in a t statistic:

$$t = \frac{\bar{X}_2 - \bar{X}_1}{\hat{\sigma}_{\bar{X}_2 - \bar{X}_1}}$$

The numerator contains the calculated mean difference while the denominator contains the estimate of the expected variability, the *standard error* of the mean difference. Estimation of the standard error $\hat{\sigma}_{\bar{X}_2 - \bar{X}_1}$ involves pooling of the variances obtained within both conditions. Since we observe a high variability in case (a) in the data of Fig. 1.1, we will obtain a small t value. Due to the small variability of the data points, we will obtain a larger t value in case (b). The higher the t value, the less likely it is that the observed mean difference is just the result of noise fluctuations. It is obvious that measurement of many data points allows a more robust estimation of this probability than the measurement of only a few data points. The error probability p can be calculated exactly from the obtained t value (using the so-called incomplete beta function) and the number of measured data points N. If the computed error probability falls below the standard value ($p<0.05$), the alternative hypothesis is accepted stating that the observed mean difference exists in the population from which the data points have been drawn (i.e., measured). In that case, one also says that the two means differ *significantly*. Assuming that in our example the obtained p value falls below 0.05 in case (b) but not in case (a), we would only infer for brain area 2 that the "Stim" condition differs significantly from the "Rest" condition.

The described mean comparison method is not the ideal approach to compare responses between different conditions since this approach is unable to capture the gradual rising and falling profile of fMRI responses. As long as the temporal resolution is low (volume TR > 4 s), the mean of different conditions can be calculated easily because transitions of expected responses from different conditions occur mainly within a single time point. If the temporal resolution is high, the expected fMRI responses change gradually from one condition to the next due to the sluggishness of the hemodynamic response. In this case, time points in the transitional zone cannot be assigned easily to different conditions. Without special treatment, the mean response can no longer be easily computed for each condition. As a consequence, the statistical power to detect mean differences may be substantially reduced, especially for short blocks and events.

This problem does not occur when *correlation analysis* is used since this method allows explicitly incorporating the gradual increase and decrease of the expected BOLD signal. Predicted (noise-free) time courses exhibiting the gradual increase and decrease of the fMRI signal can be used as a *reference function* in a correlation analysis. At each voxel, the time course of the reference function is compared with the time course of the measured data from a voxel by calculating the *correlation coefficient* r, indicating the strength of covariation:

$$r = \frac{\sum_{t=1}^{N}\left(X_t - \bar{X}\right)\left(Y_t - \bar{Y}\right)}{\sqrt{\sum_{t=1}^{N}\left(X_t - \bar{X}\right)^2 \sum_{t=1}^{N}\left(Y_t - \bar{Y}\right)^2}}$$

Index runs over time points (t for "time") identifying pairs of temporally corresponding values from the reference (X_t) and data (Y_t) time courses. In the numerator, the mean of the reference and data time course is subtracted from the respective value of each data pair, and the two differences are multiplied. The resulting value is the sum of these cross products, which will be high if the two time courses *covary*, i.e., if the values of a pair are both often above and below their respective mean. The term in the denominator normalizes the covariation term in the numerator so that the correlation coefficient lies in a range of −1 and +1. A value of +1 indicates that the reference time course and the data time course go up and down in exactly the same way, while a value of −1 indicates that the two time courses run in opposite direction. A correlation value of 0 indicates that the two time courses do not covary, i.e., the value in one time course cannot be used to predict the corresponding value in the other time course.

While the statistical principles are the same in correlation analysis as described for mean comparisons, the null hypothesis now corresponds to the statement that the population correlation coefficient ρ equals zero (H0: $\rho=0$). By including the number of data points N, the error probability can be computed assessing how likely it is that an observed correlation coefficient would occur solely due to noise fluctuations in the signal time course. If this probability falls below 0.05, the alternative hypothesis is accepted stating that there is indeed significant covariation between the reference function and the data time course. Since the reference function is the result of a model assuming different response strengths in the two conditions (e.g., "Rest" and "Stim"), a significant correlation coefficient indicates that the two conditions lead indeed to different mean activation levels in the respective voxel or brain area. The statistical assessment can be performed also by converting an observed r value into a corresponding t value using the equation $t = r\sqrt{N-2} / \sqrt{1-r^2}$.

1.3.3 The General Linear Model

The described t-test for assessing the difference of two mean values is a special case of an analysis of a *qualitative* (*categorical*) *independent variable*. A qualitative variable is defined by discrete levels, e.g., "stimulus on" vs. "stimulus off," "male" vs. "female." If a design contains more than two levels, a more general method such as analysis of variance (ANOVA) needs to be used, which can be considered as an extension of the t-test to more than two levels and to more than one experimental factor. The described correlation coefficient on the other hand is suited for the analysis of *quantitative independent variables*. A quantitative variable may be defined by any gradual time course. If more than one reference time course has to be considered, *multiple regression analysis* can be performed, which can be considered as an extension of the described linear correlation analysis. The *general linear model* (GLM) is mathematically identical to a multiple regression analysis but stresses its suitability for both multiple qualitative and multiple quantitative variables. Note that in the fMRI literature, the term "general linear model" refers to its univariate version where "univariate" refers to the number of dependent variables; in its general form, the general linear model has been defined for multiple dependent variables. The univariate GLM is suited to implement any parametric statistical test with one dependent variable, including any factorial ANOVA design as well as designs with a mixture of qualitative and quantitative variables (covariance analysis, ANCOVA). Because of its flexibility to incorporate multiple quantitative and qualitative independent variables, the GLM has become the core tool for univariate fMRI data analysis after its introduction to the neuroimaging community by Friston and colleagues (1994, 1995). The following sections briefly describe the mathematical background of the GLM in the context of fMRI data analysis; a comprehensive treatment of the GLM can be found in the standard statistical literature, e.g., Draper and Smith (1998) and Kutner et al. (2005).

From the perspective of multiple regression analysis, the GLM aims to "explain" or "predict" the variation of a dependent variable in terms of a *linear combination* (weighted sum) of several reference functions. The dependent variable corresponds to the observed fMRI time course of a voxel, and the reference functions correspond to time courses of expected (idealized) fMRI responses for different conditions of the experimental paradigm. The reference functions are also called *predictors*, *regressors*, *explanatory variables*, *covariates*, or *basis functions*. A set of specified predictors forms the *design matrix*, also called the *model*. A predictor time course is typically obtained by convolution of a "boxcar" time course with a standard hemodynamic response function. A boxcar time course is usually defined by setting values to 1 at time points at which the modeled condition is defined ("on") and 0 at all other time points.

Each predictor time course x_i gets an associated coefficient or *beta weight* b_i that quantifies the contribution of a predictor in explaining variance in the voxel time course y. The voxel time course y is modeled as the sum of the defined predictors, each multiplied with the associated beta weight b. Since this linear combination will not perfectly explain the data due to noise fluctuations, an error value e is added to the GLM system of equations with n data points and p predictors:

$$\begin{aligned} y_1 &= b_0 + b_1 X_{11} + \quad \ldots \quad + b_p X_{1p} + e_1 \\ y_2 &= b_0 + b_1 X_{21} + \quad \ldots \quad + b_p X_{2p} + e_2 \\ &\vdots \qquad\qquad\qquad \ddots \qquad\qquad \vdots \\ y_n &= b_0 + b_1 X_{n1} + \quad \ldots \quad + b_p X_{np} + e_n \end{aligned}$$

The y variable on the left side corresponds to the data, i.e., the measured time course of a single voxel. Time runs from top to bottom, i.e., y_1 is the measured value at time point 1, y_2 the measured value at time point 2, and so on. The voxel time course (left column) is "explained" by the terms on the right side of the equation. The first column on the right side corresponds to the first beta weight b_0. The corresponding predictor time course X_0 has a value of 1 for each time point and is, thus, also called "constant." Since multiplication with 1 does not change the value of b_0, this predictor time course (X_0) does not explicitly appear in the equation. After estimation (see below), the value of b_0 typically represents the signal level of the baseline condition and is also called intercept. While its absolute value is not very informative in the context of fMRI data, it is important to include the constant predictor in a design matrix since it allows the other predictors to model small condition-related fluctuations as increases or decreases relative to the baseline signal level. The other predictors on the right side model the expected time courses of different conditions. For multifactorial designs, predictors may be defined that code combinations of condition levels allowing to estimate main and interaction effects. The beta weight of a predictor scales the associated predictor time course and reflects the unique contribution of that predictor in explaining the variance of the voxel time course. While the exact interpretation of beta values depends on the details of the design matrix, a large positive (negative) beta weight typically indicates that the voxel exhibits strong activation (deactivation) during the modeled experimental condition relative to baseline. All beta values together characterize a voxels "preference" for one or more experimental conditions. The last column in the system of equations contains error values, also called *residuals*, *prediction errors*, or *noise*. These error values quantify the deviation of the measured voxel time course from the predicted time course.

The GLM system of equations may be expressed elegantly using matrix notation. For this purpose, the voxel time course, the beta values, and the residuals are represented as vectors, and the set of predictors as a matrix:

$$\begin{bmatrix} y_1 \\ \vdots \\ \vdots \\ y_n \end{bmatrix} = \begin{bmatrix} 1 & X_{11} & \cdots & \cdots & X_{1p} \\ \vdots & \vdots & \vdots & \vdots & \vdots \\ \vdots & \vdots & \vdots & \vdots & \vdots \\ 1 & X_{n1} & \cdots & \cdots & X_{np} \end{bmatrix} \begin{bmatrix} b_0 \\ b_1 \\ \vdots \\ b_p \end{bmatrix} + \begin{bmatrix} e_1 \\ e_2 \\ \vdots \\ e_n \end{bmatrix}$$

Representing the indicated vectors and matrix with single letters, we obtain this simple form of the GLM system of equations:

$$\mathbf{y} = \mathbf{X}\mathbf{b} + \mathbf{e}$$

In this notation, the matrix **X** represents the *design matrix* containing the predictor time courses as column vectors. The beta values now appear in a separate vector **b**. The term **Xb** indicates matrix-vector multiplication. Figure 1.2 shows a graphical representation of the GLM. Time courses of the signal, predictors, and residuals have been arranged in column form with time running from top to bottom as in the system of equations.

Given the data **y** and the design matrix **X**, the GLM fitting procedure has to find a set of beta values explaining the data as good as possible. The time course values $\hat{\mathbf{y}}$ predicted by the model are obtained by the linear combination of the predictors:

$$\hat{\mathbf{y}} = \mathbf{X}\,\mathbf{a}$$

A good fit would be achieved with beta values leading to predicted values $\hat{\mathbf{y}}$ that are as close as possible to the measured values **y**. By rearranging the system of equations, it is evident that a good prediction of the data implies small error values:

$$\begin{aligned} \mathbf{e} &= \mathbf{y} - \mathbf{X}\mathbf{b} \\ &= \mathbf{y} - \hat{\mathbf{y}} \end{aligned}$$

An intuitive idea would be to find those beta values minimizing the sum of error values. Since the error values contain both positive and negative values (and because of additional statistical considerations), the GLM procedure does not estimate beta values minimizing the sum of error values, but finds those beta values that *minimize the sum of squared error values*:

$$\mathbf{e}'\mathbf{e} = (\mathbf{y} - \mathbf{X}\mathbf{b})'(\mathbf{y} - \mathbf{X}\mathbf{b}) \rightarrow \min$$

The term $\mathbf{e}'\mathbf{e}$ is the vector notation for the sum of squares $\left(\sum_{t=1}^{N} e_t^2\right)$. The apostrophe symbol denotes transposition of a vector (or matrix), i.e., a row vector version of **e** is multiplied by a column vector version of **e** resulting in the sum of squared values e_t. The optimal beta weights minimizing the

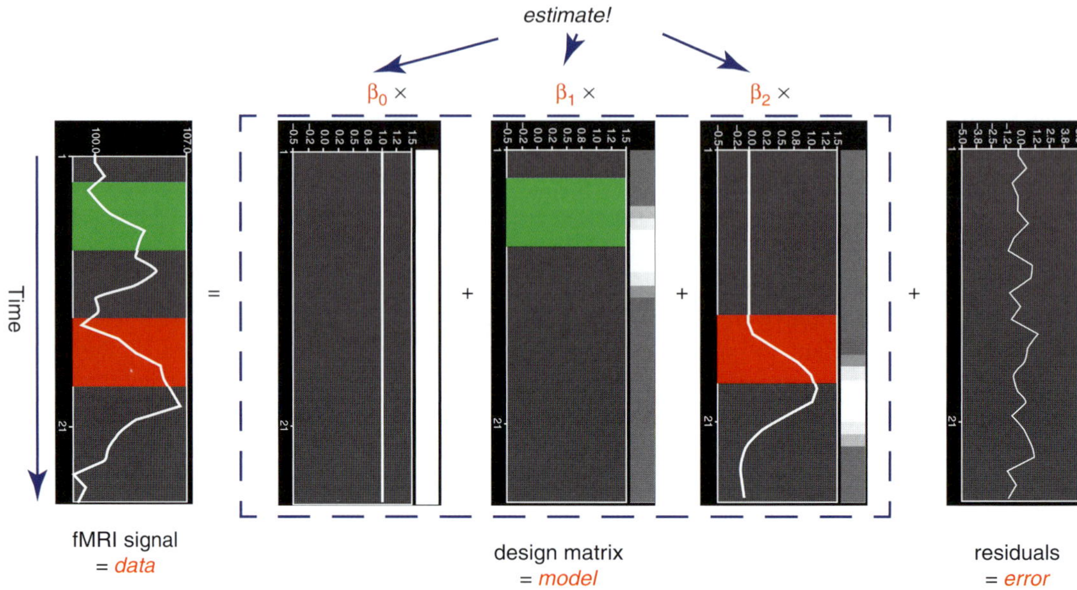

Fig. 1.2 Graphical display of a general linear model. Time is running from *top* to *bottom*. *Left side* shows observed voxel time course (data). The model (design matrix) consists of three predictors, the constant and two main predictors (*middle part*). *Filled green* and *red rectangles* depict stimulation time, while the *white curves* depict expected BOLD responses obtained by convolution of the stimulus on periods with the two-gamma function. Expected responses are also shown using a *black-to-white color range* (*right side* of each predictor plot). Beta values have to be estimated (*top*) to scale the expected responses (predictors) in such a way that their weighted sum predicts the measured data values as good as possible (in the least squares sense, see text). Unexplained fluctuations (residuals, error) are shown on the *right side*

squared error values (the "least squares estimates") are obtained non-iteratively by the following equation:

$$\mathbf{b} = \left(\mathbf{X}'\mathbf{X}\right)^{-1} \mathbf{X}'\mathbf{y}$$

The term in brackets contains a matrix-matrix multiplication of the transposed, $\mathbf{X}'$, and non-transposed, $\mathbf{X}$, design matrix. This term results in a square matrix with a number of rows and columns corresponding to the number of predictors. Each cell of the $\mathbf{X}'\mathbf{X}$ matrix contains the scalar product of two predictor vectors. The scalar product is obtained by summing all products of corresponding entries of two vectors corresponding to the (non-mean normalized) calculation of covariance. This $\mathbf{X}'\mathbf{X}$ matrix, thus, corresponds to the (non-mean normalized) predictor variance-covariance matrix.

The non-normalized variance-covariance matrix is inverted as denoted by the "−1" symbol. The resulting matrix $(\mathbf{X}'\mathbf{X})^{-1}$ plays an essential role not only for the calculation of beta values but also for testing the significance of contrasts (see below). The remaining term on the right side, $\mathbf{X}'\mathbf{y}$, evaluates to a vector containing as many elements as predictors. Each element of this vector is the scalar product (non-mean normalized covariance term) of a predictor time course with the observed voxel time course.

An important property of the least squares estimation method (following from the independence assumption of the errors, see below) is that the variance of the measured time course can be decomposed into the sum of the variance of the predicted values (model-related variance) and the variance of the residuals:

$$\mathrm{Var}(\mathbf{y}) = \mathrm{Var}(\hat{\mathbf{y}}) + \mathrm{Var}(\mathbf{e}) \cdot$$

Since the variance of the voxel time course is fixed, minimizing the error variance by least squares corresponds to maximizing the variance of the values explained by a model. The square of the *multiple correlation coefficient R* provides a measure of the proportion of the variance of the data which can be explained by a specified model:

$$R^2 = \frac{\mathrm{Var}(\hat{\mathbf{y}})}{\mathrm{Var}(\hat{\mathbf{y}})} = \frac{\mathrm{Var}(\hat{\mathbf{y}})}{\mathrm{Var}(\hat{\mathbf{y}}) + \mathrm{Var}(\mathbf{e})}$$

The values of the multiple correlation coefficient vary between 0 (no variance explained) and 1 (all variance explained by the model). A coefficient of $R=0.7$, for example, corresponds to an explained variance of 49 % (0.7·0.07). An alternative way to calculate the multiple correlation coefficient consists in computing a standard correlation coefficient between the predicted values and the observed values: $R = r_{\hat{\mathbf{y}}\mathbf{y}}$. This equation provides another view on the meaning of the multiple correlation coefficient quantifying the interrelationship (correlation) of the *combined* set of optimally weighted predictor variables with the observed time course.

1.3.3.1 GLM Diagnostics

Note that if the design matrix (model) does not contain all relevant predictors, condition-related increases or decreases in the voxel time course will be explained by the error values instead of by the model. It is, therefore, important that the design matrix is constructed with all expected effects, which may also include reference functions not related to experimental conditions, for example, estimated motion parameters or drift predictors (if not removed during preprocessing, see Sect. 1.2.3). In case that all expected effects are properly modeled, the residuals should reflect only "pure" noise fluctuations. If some effects are not (correctly) modeled, a plot of the residuals may show low-frequency fluctuations instead of a stationary noisy time course. A visualization of the residuals (for selected voxels or regions of interest) is, thus, a good diagnostic to assess whether the design matrix has been specified properly.

1.3.3.2 GLM Significance Tests

The multiple correlation coefficient is an important measure of the "goodness of fit" of a GLM. In order to test whether a specified model significantly explains variance in a voxel time course, an F statistic can be calculated for a R value with $p-1$ degrees of freedom in the numerator and $n-p$ degrees of freedom in the denominator:

$$F_{n-1,n-p} = \frac{R^2(n-p)}{(1-R^2)(p-1)}$$

An error probability value p can then be obtained for the calculated F statistics. A high F value (p value smaller than 0.05) indicates that the experimental conditions as a whole have a significant modulatory effect on the data time course (omnibus effect).

While the overall F statistic may tell us whether the specified model significantly explains a voxel's time course, it does not allow assessing which individual conditions differ significantly from each other. Comparisons between conditions can be formulated as *contrasts*, which are linear combinations of beta values corresponding to specific null hypotheses. To test, for example, whether activation in a single condition 1 deviates significantly from baseline, the null hypothesis would be that there is no effect in the population, i.e., $H_0: b_1 = 0$. To test whether activation in condition 1 is significantly different from activation in condition 2, the null hypothesis would state that the beta values of the two conditions would not differ, $H_0: \cdot b_1 = b_2$ or $H_0: (+1)b_1 + (-1)b_2 = 0$. To test whether the mean of condition 1 and condition 2 differs from condition 3, the following contrast could be specified: $H_0: (b_1 + b_2)/2 = b_3$ or $H_0: (+1)b_1 + (+1)b_2 + (-2)b_3 = 0$. The values used to multiply the respective beta values are often written as a *contrast vector* **c**. In the latter example,[1] the contrast

[1] Note that the constant term is treated as confound and it is not included in contrast vectors, i.e., it is implicitly assumed that b_0 is multiplied by 0 in all contrasts. To include the constant explicitly, each contrast vector must be expanded by one entry, e.g., at the beginning or end.

vector would be written as $\mathbf{c} = [+1 + 1 - 2]$. Using matrix notation, the linear combination defining a contrast can be written as the scalar product of contrast vector **c** and beta vector **b**. The null hypothesis can then be simply described as $\mathbf{c}'\mathbf{b} = 0$. For any number of predictors k, such a contrast can be tested with the following t statistic with $n-p$ degrees of freedom:

$$t = \frac{\mathbf{c}'\mathbf{b}}{\sqrt{\mathrm{Var}(\mathbf{e})\mathbf{c}'(\mathbf{X}'\mathbf{X})^{-1}\mathbf{c}}}$$

The numerator of this equation contains the described scalar product of the contrast and beta vector. The denominator defines the standard error of $\mathbf{c}'\mathbf{b}$, i.e., the variability of the contrast estimate due solely to noise fluctuations. The standard error depends on the variance of the residuals Var(**e**) as well as on the design matrix **X**. With the known degrees of freedom, a t value for a specific contrast can be converted in an error probability value p using the incomplete beta function mentioned earlier. Note that the null hypotheses above were formulated as $\mathbf{c}'\mathbf{b} = 0$ implying a *two-sided* alternative hypothesis, i.e., $H_a: \mathbf{c}'\mathbf{b} \neq 0$. For one-sided alternative hypotheses, e.g., $H_a: b_1 > b_2$, the obtained p value from a two-sided test can be simply divided by 2 to get the p value for the one-sided test. If this p value is smaller than 0.05 and if the t value is positive, the null hypothesis may be rejected.

1.3.3.3 Conjunction Analysis

Experimental research questions often lead to specific hypotheses, which can best be tested by the *conjunction* of two or more contrasts. As an example, it might be interesting to test with contrast $\mathbf{c}_1$ whether condition 2 leads to significantly higher activity than condition 1 *and* with contrast $\mathbf{c}_2$ whether condition 3 leads to significantly higher activity than condition 2. This question could be tested with the following conjunction contrast: $\mathbf{c}_1 \wedge \mathbf{c}_2 = [-1 + 1 \ \ 0] \wedge [0 - 1 + 1]$. Note that a logical "and" operation is defined for Boolean values (true/false) but that t values associated with individual contrasts can assume any real value. An appropriate way to implement a logical "and" operation for conjunctions of contrasts with continuous statistical values is to use a minimum operation, i.e., the significance level of the conjunction contrast is identical to the significance level of the contrast with the smallest t value: $t_{\mathbf{c}_1 \wedge \mathbf{c}_2} = \min\left(t_{\mathbf{c}_1}, t_{\mathbf{c}_2}\right)$. For more details about conjunction testing, see Nichols et al. (2006).

1.3.3.4 Event-Related Averaging

As mentioned earlier, event-related designs cannot only be used to *detect* activation effects but also to *estimate* the time course of task-related responses. Visualization of mean response profiles can be achieved by averaging all responses of the same condition across corresponding time points with respect to stimulus onset. Averaged (or even single trial) responses can be used to characterize the temporal dynamics of brain activity within and across brain areas by comparing estimated features such as response latency, duration, and amplitude (e.g., Kruggel and von Cramon 1999; Formisano and Goebel 2003). In more complex, temporally extended tasks, responses to subprocesses may be identified. In working memory paradigms, for example, encoding, delay and response phases of a trial may be separated. Note that event-related selective averaging works well only for slow event-related designs. In rapid event-related designs, responses from different conditions lead to substantial overlap, and event-related averages are often meaningless. In this case, deconvolution analysis is recommended (see below).

In order to avoid circularity, event-related averages should only be used descriptively if they are selected from significant clusters identified from a whole-brain statistical analysis of the same data. Even a merely descriptive analysis visualizing averaged condition responses is, however, helpful in order to ensure that significant effects are caused by "BOLD-like" response shapes and not by, e.g., signal drifts or measurement artifacts. If ROIs are determined using independent (localizer) data, event-related averages extracted from these regions in a subsequent (main) experiment can be statistically analyzed. For a more general discussion of ROI vs. whole-brain analyses, see Friston and Henson (2006),

Friston et al. (2006), Saxe et al. (2006), and Frost and Goebel (2013).

1.3.4 Deconvolution Analysis

While standard design matrix construction (convolution of boxcar with two-gamma function) can be used to estimate condition amplitudes (beta values) in rapid event-related designs, results critically depend on the appropriateness of the assumed standard BOLD response shape: Due to variability in different brain areas within and across subjects, a *static* model of the response shape might lead to nonoptimal fits. Furthermore, the isolated responses to different conditions cannot be visualized due to overlap of condition responses over time. To model the shape of the hemodynamic response more flexibly, multiple basis functions (predictors) may be defined for each condition instead of one. Two often-used additional basis function sets are derivatives of the two-gamma function with respect to two of its parameters, delay and dispersion. If added to the design matrix for each condition, these basis functions allow capturing small variations in response latency and width of the response. Other sets of basis functions (i.e., gamma basis set, Fourier basis set) are much more flexible, but obtained results are often more difficult to interpret. *Deconvolution analysis* is a general approach to estimate condition-related response profiles using a flexible and interpretable set of basis functions. It can be easily implemented as a GLM by defining an appropriate design matrix that models each bin after stimulus onset by a separate condition predictor (delta or "stick" functions). This is also called a finite impulse response (FIR) model because it allows estimating any response shape evoked by a short stimulus (impulse). In order to capture the BOLD response for short events, about 20 s are typically modeled after stimulus onset. This would require, for each condition, 20 predictors in case of a TR of 1 s, or ten predictors in case of a TR of 2 s. Despite overlapping responses, fitting such a GLM "recovers" the underlying condition-specific response profiles in a series of beta values, which appear in plots as if event-related averages have been computed in a slow event-related design. Since each condition is modeled by a series of temporally shifted predictors, hypothesis tests can be performed that compare response amplitudes *at different moments in time* within and between conditions. Note, however, that deconvolution analysis assumes a linear time-invariant (LTI) system, i.e., it is assumed that overlapping responses can be modeled as the sum of individual responses. Furthermore, in order to uniquely estimate the large number of beta values from overlapping responses, variable ITIs must be used in the experimental design. The deconvolution model is very flexible allowing to capture any response shape. This implies that also non-BOLD-like time courses will be detected easily since the trial responses are not "filtered" by the ideal response shape as in conventional analysis.

1.3.5 Serial Correlations and Generalized Least Squares

Given a correct model (design matrix), the standard estimation procedure of the GLM, ordinary least squares (OLS), operates correctly only under the following assumptions.

1.3.5.1 GLM Assumptions

The population error values ε must have an expected value of zero and constant variance at each time point i:

$$E[\varepsilon_i]=0$$

$$\mathrm{Var}[\varepsilon_i]=\sigma^2$$

Furthermore, the error values are assumed to be uncorrelated, i.e., $\mathrm{Cov}(\varepsilon_i, \varepsilon_j)=0$ for all $i \neq j$. To justify the use of t and F distributions in hypothesis tests, errors are further assumed to be normally distributed, i.e., $\varepsilon_i \sim N(0, \sigma^2)$. In summary, errors are assumed to be normal, independent, and identically distributed (often abbreviated as "normal i.i.d."). Under these assumptions, the solution obtained by the least squares method is optimal in the sense that it provides the most efficient unbiased estimation of the beta values. While the OLS approach is robust with respect to small violations, assumptions should be checked. In the context of fMRI measurements, the assumption of uncorrelated error values requires special attention.

1.3.5.2 Correction for Serial Correlations

In fMRI data, one typically observes *serial correlations*, i.e., high values are followed more likely by high values than by low values and vice versa. Assessment of these serial correlations is not performed on the original voxel time course but on the time course of the residuals since serial correlations in the recorded signal are induced by (and therefore expected to some extent from) slow task-related fluctuations. Task-unrelated serial correlations most likely occur because data points are measured in rapid succession, i.e., they are also observed to some extent when scanning phantoms. Likely sources of temporal correlations are physical and physiological noise components such as hardware-related low-frequency drifts, oscillatory fluctuations related to respiration and cardiac pulsation, and residual head motion artifacts. Serial correlations violate the assumption of uncorrelated population errors (see above). Fortunately, the beta values estimated by the GLM are correct (unbiased) estimates even in case of serial correlations. The standard errors of the betas are biased, however, leading to "inflated" test statistics, i.e., the resulting t or F values are higher than they should be. This can be explained by considering that the presence of serial correlations (serial dependence) reduces the true number of independent observations (effective degrees of freedom) that will be lower than the number of observations. Without correction, the degrees of freedom are systematically overestimated leading to an underestimation of the error variance resulting in inflated statistical values, i.e., t or F values are too high. It is, thus, necessary to correct for serial correlations in order to obtain valid error probabilities. Serial correlations can be corrected using several approaches. In *pre-whitening* approaches, autocorrelation is first estimated and removed from the data; the pre-whitened data can then be analyzed with a standard OLS GLM solution. In *pre-coloring* approaches (e.g., Friston et al. 1995), a strong autocorrelation structure is imposed on the data by temporal smoothing, and degrees of freedom are adjusted according to the imposed (known) autocorrelation. The pre-coloring (temporal smoothing) operation acts, however, as a low-pass filter and may weaken experimentally induced signals of interest and is thus not the preferred method. The pre-whitening approach can be expressed in terms of a more powerful estimation procedure than OLS called *generalized least squares* (GLS, Searle et al. 1992). As opposed to the OLS method, GLS works correctly also in case that error values exhibit correlations or when error variances are not homogeneous. Note, however, that this more powerful estimation approach only provides correct results in case that the true (population) variances and covariances of the error values are known. With the known error covariance matrix $\mathbf{V}$, the betas and their (co-)variances can be calculated with GLS as follows:

$$\mathbf{b} = \left(\mathbf{X}'\mathbf{V}^{-1}\mathbf{X}\right)^{-1}\mathbf{X}'\mathbf{V}^{-1}\mathbf{y}$$

$$\mathrm{Cov}(\mathbf{b}) = \left(\mathbf{X}'\mathbf{V}^{-1}\mathbf{X}\right)^{-1}$$

With the obtained b values and their covariances, any contrast can then be assessed statistically as described above for the OLS method. When comparing the GLS solution with the OLS solution, it is evident that the inverse of the population error covariance matrix $\mathbf{V}^{-1}$ is needed to properly treat the effect of covariance of the errors on the parameter estimates (betas and their covariances). Note also that when setting $\mathbf{V}$ as a diagonal matrix (entries outside the main diagonal are zero, i.e., no covariation of errors) with equal variance values (all values of the main diagonal are the same, e.g., 1), the GLS equation reduces to the OLS solution, i.e., the $\mathbf{V}^{-1}$ term vanishes.

Since the population covariance matrix of the error values $\mathbf{V}$ is usually not known, it needs to be estimated from the data itself. Since there are too many (n^2) degrees of freedom, **V** cannot be estimated for the general case of arbitrary covariance matrices. It is, however, often possible to estimate **V** for special cases where only some parameters need to be estimated. The two most important special cases in the context of fMRI data analysis are the treatment of serial correlations (see below) and the treatment of unequal variances when integrating data from different subjects in the context of mixed-effects group analyses (see Sect. 1.6.1).

A simple example of a pre-whitening method was developed independently from the GLS approach (Cochrane and Orcutt 1949; Bullmore et al. 1996) but can be shown to be identical to a GLS solution. The procedure assumes that the errors follow a first-order autoregressive, or $AR(1)$, process. After calculation of a GLM using OLS, the amount of serial correlation a_1 is estimated using pairs of successive residual values (e_t, e_{t+1}) i.e., the residual time course is correlated with itself shifted by one time point (lag = 1). In the second step, the estimated serial correlation is removed from the measured voxel time course by calculating the transformed time course $y_t^n = y_{t+1} - a_1 \cdot y_t$. The superscript "$n$" indicates the values of the new, adjusted time course. The same calculation is also applied to each predictor time course resulting in an adjusted design matrix $\mathbf{X}^n$. In the third step, the GLM is recomputed using the adjusted voxel time course and adjusted design matrix resulting in correct standard errors for beta estimates and, thus, correct significance levels for contrasts (of course under the assumption that the AR(1) model is correct). If autocorrelation is not sufficiently reduced in the new residuals, the procedure can be repeated. If performed using the GLS approach, the first step is identical to the Cochrane-Orcutt method, i.e., OLS is used to fit the GLM, and the obtained residuals are used to estimate the value of the AR(1) term. The adjustment of the time course y_t and the design matrix described above need, however, not be performed explicitly since these adjustments are handled implicitly in the next step by using a $\mathbf{V}^{-1}$ term in the GLS equations that contains values in the off-diagonal elements derived from the estimated serial correlation term.

While an AR(1) autocorrelation model substantially reduces serial correlations in fMRI data, better results are obtained when using an AR(2) model, i.e., both first-order and second-order autocorrelation terms should be estimated and used to construct the error covariance matrix **V** for GLS estimation. Since serial correlations differ across voxels, serial correlation correction should be performed separately for each voxel time course as opposed to the estimation of serial correlation values from multiple averaged (neighboring) voxel time courses. An AR(2) serial correlation model applied separately for each voxel time course has been recently shown to be the most accurate approach to treat serial correlations when compared to other models (Lenoski et al. 2008).

1.3.6 From Single Voxels to Statistical Maps

The statistical analysis steps above were described for a single voxel's time course since standard statistical methods are performed independently for each voxel. Since a typical fMRI data set contains several hundred thousand voxels, a statistical analysis is performed independently hundred thousands of times. Running a GLM, for example, results in a set of estimated beta values attached to each voxel. A specified contrast $\mathbf{c}'\mathbf{b}_v$ will be performed using the same contrast vector $\mathbf{c}$ for each voxel v, but it will use a voxel's vector of beta values $\mathbf{b}_v$ to obtain voxel-specific t and p values. Statistical test results for individual voxels are integrated in a 3D data set called a *statistical map*. To visualize a statistical map, the obtained values, e.g., contrast t values, can be shown at the location of each voxel replacing anatomical intensity values shown as default. A more useful approach shows the statistical values only for those voxels that exceed a specified *statistical threshold*. This allows visualizing anatomical information in large parts of the brain, while statistical information is shown (overlaid) only in those regions exhibiting suprathreshold (usually statistically significant) signal modulations (see Fig. 1.3). While anatomical

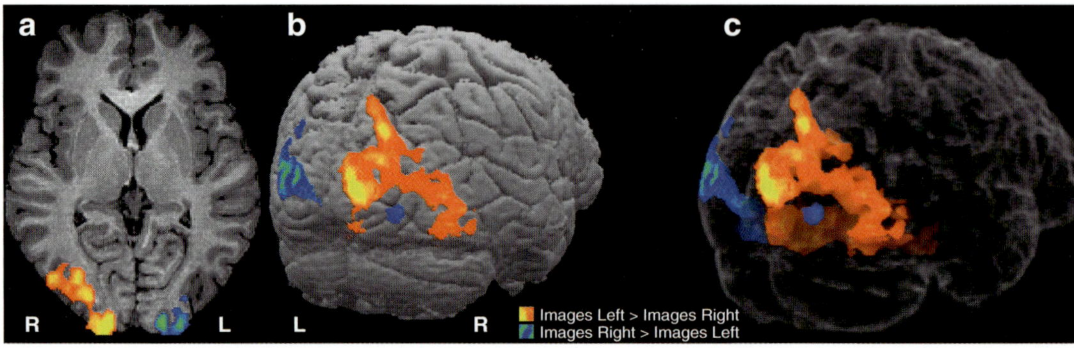

Fig. 1.3 Visualization of statistical maps in an experiment showing visual stimuli (images of objects) in either the *left* or *right* visual field. After application of the threshold, the result of a two-sided hypothesis test (null hypothesis: there is no difference between the two conditions) is visualized on an anatomical slice (**a**), on the surface of the brain (**b**), and within a transparently rendered 3D brain view. Significant voxels where visual stimuli presented on the *left side* exhibited larger activation than stimuli presented on the *right side* are color-coded using an *orange-to-yellow* color gradient, while a *blue-to-green* color gradient is used for voxels for the inverse contrast

information is normally visualized using a range of gray values, suprathreshold statistical test values are typically visualized using multiple colors, for example, a red-to-yellow range for positive values and a green-to-blue range for negative statistical values (see Fig. 1.3). With these colors, a positive (negative) t value just passing a specified threshold would be colored in red (green), while a very high positive (negative) t value would be colored in yellow (blue).

1.3.7 Multiple Comparison Correction

An important issue in fMRI data analysis is the specification of an appropriate threshold for statistical maps. If there would be only a single voxel's data, a conventional threshold of $p<0.05$ (or $p<0.01$) could be used to assess significance of an observed effect quantified by an R, t, or F statistic. Running the statistical analysis separately for each voxel creates, however, a *massive multiple comparison problem*. If a *single test* is performed, the conventional threshold protects from wrongly declaring a voxel as significantly modulated with a probability of $p<0.05$ when there is no effect in the population (α error). Note that in case that the null hypothesis (no effect) holds, an adopted error probability of $p=0.05$ implies that if the same test would be repeated 100 times, the alternative hypothesis would be accepted wrongly on average in five cases, i.e., we would expect 5 % of false positives. If we assume that there is no real effect in any voxel time course, running a statistical test spatially in parallel is statistically identical to repeating the test 100,000 times at a single voxel (each time with new measured data). It is evident that this would lead to about 5,000 false positives, i.e., about 5,000 voxels would be labeled "significant" although these voxels would reach the 0.05 threshold purely due to chance.

Several methods have been suggested to control this massive multiple comparison problem. The *Bonferroni correction* method is a simple multiple comparison correction that controls the α error *across all voxels*, and it is therefore called a *family-wise error* (FWE) correction approach. The method calculates single-voxel threshold values in such a way that an error probability of 0.05 is obtained at the global level. With N-independent tests, this is achieved by using a statistical significance level which is N times smaller than usual. The Bonferroni correction can be derived mathematically as follows. Under the assumption of independent tests, the probability that all of N performed tests lead to a subthreshold result is $(1-p)^N$ and the probability to obtain one or more false-positive results is $1-(1-p)^N$. In order to guarantee a family-wise (global) error probability of $p_{\mathrm{FWE}}=1-(1-p)^N$, the threshold for a single test, p, has to be adjusted as follows: $p = 1-(1-p_{\mathrm{FWE}})^{1/N}$. For small p_{FWE} values (e.g., 0.05), this equation can be approximated by $p = p_{\mathrm{FWE}}/N$. This means that to obtain a global error probability of $p_{\mathrm{FWE}} < 0.05$, the significance level for a single test is obtained by dividing the family-wise error probability by the number of independent tests. Given 100,000 voxels, we would obtain an adjusted single-voxel threshold of $p_v = p_{\mathrm{FWE}}/N = 0.05/100000 = 0.0000005$. The Bonferroni correction method ensures that we do not declare even a single voxel wrongly as significantly activated with an error probability of 0.05. For fMRI data, the Bonferroni method would be a valid approach to correct the α error if the data at neighboring voxels would be truly independent from each other. Neighboring voxels, however, show similar response patterns within functionally defined brain regions, such as the fusiform face area (FFA). In the presence of such spatial correlations, the Bonferroni correction method operates too conservative, i.e., it corrects the error probability more strongly than necessary. As a result of a too strict control of the α error, the sensitivity (power) to detect truly active voxels is reduced: Many voxels will be labeled as "not significant" although they likely reflect true effects. Wrongly accepting (rejecting) a null (alternative) hypothesis is called β error or type II error.

Worsley et al. (1992) suggested a less conservative approach to correct for multiple comparisons taking explicitly the observation into account that neighboring voxels are not activated independently from each other but are more likely to activate together in clusters. In order to incorporate spatial neighborhood relationships in the calculation of global error probabilities, the method describes a statistical map as a *Gaussian random field* (for details, see Worsley et al. 1992). Unfortunately, application of this correction method requires that the fMRI data are spatially smoothed substantially reducing one of its most attractive properties, namely, its high spatial resolution.

Another correction method incorporating the observation that neighboring voxels often activate in clusters is based on Monte Carlo simulations that generate many random images (maps) using the spatial correlation structure of the original map; the generated maps are used to calculate the likelihood to obtain different sizes of functional clusters by chance for specific (less conservative) single-voxel thresholds (Forman et al. 1995). The calculated cluster extent threshold combined with a less strict single-voxel threshold is finally applied to the statistical map ensuring that a global error probability of $p < 0.05$ is met. This approach does not require spatial smoothing and appears highly appropriate for fMRI data. A disadvantage is that the method is quite compute intensive and that small functional clusters might not be discovered.

While the described multiple comparison correction methods aim to control the family-wise error rate, the *false discovery rate (FDR) approach* (Benjamini and Hochberg 1995) uses a different statistical logic and has been proposed for fMRI analysis by Genovese and colleagues (2002). This approach does not control the overall number of false-positive voxels but the number of false-positive voxels among the subset of voxels labeled as significant. Given a specific threshold, suprathreshold voxels are called "discovered" voxels or "voxels declared as active." With a specified false discovery rate of $q < 0.05$, one would accept that 5 % of the discovered (suprathreshold) voxels would be false positives. Given a desired false discovery rate, the FDR algorithm calculates a single-voxel threshold, which ensures that the voxels beyond that threshold contain on average not more than the specified proportion of false positives. With a q value of 0.05, this also means that one can "trust" 95 % of the suprathreshold (i.e., color-coded) voxels since the null hypothesis has been rejected correctly. Since the FDR logic relates the number of false positives to the amount of truly active voxels, the FDR method adapts to the amount of activity in the data: The method is very strict if there is not much evoked activity in the data, but assumes less conservative thresholds if larger regions of the brain show task-related effects. In the extreme case that not a single voxel is truly active, the calculated single-voxel threshold is identical to the one computed with the Bonferroni method. The FDR method appears ideal for fMRI data because it does not require spatial smoothing and it detects voxels with a high sensitivity (low β error) if there are true effects in the data.

Another simple approach to the multiple comparisons problem is to reduce the number of tests by using anatomical masking. Most correction methods, including Bonferroni and FDR, can be combined with this approach since a smaller number of tests leads to a less strict control of the α error and thus a smaller β error is made as compared to inclusion of all voxels. In a simple version of an anatomical mask, an intensity threshold for the basic signal level can be used to remove voxels outside the head. The number of voxels can be further reduced by masking the brain, e.g., after performing a brain extraction step. These simple steps typically reduce the number of voxels from about 100,000 to about 50,000 voxels. In a more advanced version, statistical data analysis may be restricted to gray matter voxels, which may be identified by standard cortex segmentation procedures (e.g., Kriegeskorte and Goebel 2001). This approach not only removes voxels outside the brain but also excludes voxels in the white matter and ventricles. Note that anatomically informed correction methods do not require spatial smoothing of the data and not only reduce the multiple comparisons problem but also reduce computation time since fewer tests (e.g., GLM calculations) have to be performed.

1.4 Integration of Anatomical and Functional Data

The localization of the neural correlates of sensory, motor, and cognitive functions requires a precise relationship between voxels in calculated statistical maps with voxels in high-resolution anatomical data sets. While it is recommended to also view statistical maps overlaid on a volume of the functional data itself, EPI data sets often do not contain sufficient anatomical detail to specify the precise location of an active cluster in a subject's brain. 3D renderings of high-resolution anatomical data sets may greatly aid in visualizing activated brain regions (see Fig. 1.3). Advanced visualization requires that a high-resolution 3D data set is recorded for a subject and that the functional data is *coregistered* to the 3D data set as precisely as possible. Anatomical data sets are also important for most brain normalization methods, which is a prerequisite of whole-brain group studies (see Sect. 1.5). High-resolution anatomical data sets are typically recorded with T_1-weighted MRI sequences. A typical structural scan covering the whole brain with a resolution of 1 mm in all three dimensions (e.g., 180 sagittal slices) lasts between 5 and 20 min on current 1.5 and 3 Tesla scanners.

1.4.1 Coregistration of Functional and Anatomical Data

If functional images are superimposed on coplanar high-resolution anatomical images, spatial transformations (translations and rotations) to align the two data sets are not necessary (except maybe the correction of small head movements and small geometric distortions), since the respective slices are measured at the same 3D positions. Since coplanar anatomical images are usually recorded with a higher resolution (typically with a 256×256 matrix) than the functional images (typically with 64×64 or 128×128 matrices), only a scaling factor has to be applied. To allow high-quality visualization of the functional data in arbitrary resliced anatomical planes, the functional data must be coregistered with 3D data sets with a high resolution (e.g., 1 mm) in all three dimensions. These high-resolution 3D data sets are usually recorded with different slice orientation and position than the functional data, and the coregistration step, thus, requires an affine spatial transformation including translation, rotation, and scaling. These three elementary spatial transformations can be integrated in a single transformation step expressed in a standard 4×4 spatial transformation matrix. If the high-resolution 3D data set has been recorded in the same scanning session as the functional data, the coregistration matrix can be constructed simply by using the scanning parameters (slice positions, pixel resolution, slice thickness) from both recordings. The alignment based on this

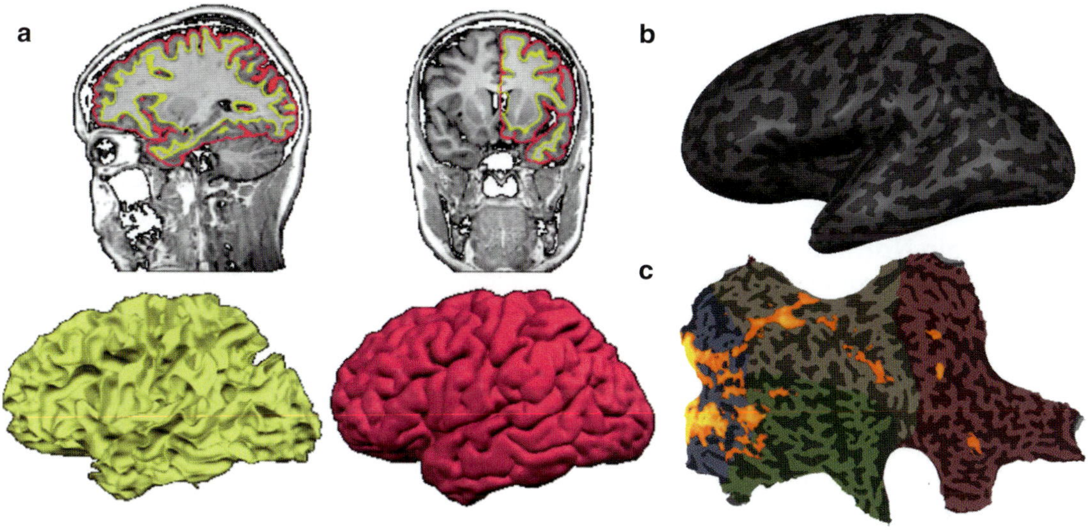

Fig. 1.4 Cortex representations used for advanced visualization of statistical maps. (**a**) Segmentation and surface reconstruction of the inner (white/gray matter, *yellow*) and outer (gray matter/CSF, *magenta*) boundary of gray matter. (**b**) "Inflated" cortex representation of the left hemisphere obtained by iterative morphing process. (**c**) "Flat map" of the right cortical hemisphere with superimposed functional data

information would be perfect if there would be no head movement between the anatomical and functional images. To further improve coregistration results, an additional intensity- or gradient-driven alignment step is usually performed after the initial (mathematical) alignment correcting for head displacements (and eventually geometric distortions) between the functional and anatomical recordings.

1.4.2 Visualizing Statistical Maps in Volume and Cortical Surface Space

Besides aiding in visualizing functional data (e.g., statistical maps, see Fig. 1.3), high-resolution anatomical data sets can be used to create 3D volume or surface renderings of the brain, which allow additional helpful visualizations of functional data on a subject's brain. These visualizations require segmentation of the brain, which can be performed automatically with most available software packages. For more advanced visualizations, segmentation of cortical voxels allows to construct topologically correct mesh representations of the cortical sheet, one for the left and one for the right hemisphere (e.g., Fischl et al. 1999; Kriegeskorte and Goebel 2001). The obtained meshes (Fig. 1.4a) may be further transformed into inflated (Fig. 1.4b) and flattened (Fig. 1.4c) cortex representations. Functional data can then be superimposed on folded, inflated, and flattened representations (Fig. 1.4c), which is particularly useful for topologically organized functional information, for example, in the context of retinotopic mapping experiments. To help in orientation, inflated and flattened cortex representations indicate gyral and sulcal regions by color-coding local curvature; concave regions, indicating sulci, may be depicted, e.g., with a dark gray color, while convex regions, indicating gyri, may be depicted, e.g., with a light gray color (Fig. 1.4b). A general advantage of visualizing functional data on flat maps is that all cortical activation foci from different experiments can be visualized at once at their correct anatomical location in a canonical view. In contrast, visualizing several activated regions using a multi-slice representation depends on the chosen slice orientation and number of slices. Note that anatomical data is not only important to visualize functional data. Anatomical information may also be used to constrain statistical data analysis as has been described in Sect. 1.3.7. Furthermore, the explicit segmentation of cortical voxels is also the prerequisite for advanced anatomical analyses, including cortical morphometry.

1.5 Brain Normalization for Whole-Brain Group Studies

The described analysis steps allow analyzing and visualizing the functional data of individual subjects. The goal of many imaging studies is, however, to integrate and compare data from many subjects. Such *group studies* allow generalizing findings from a sample of subjects to the population from which patient groups or healthy subjects have been drawn. Group analysis of functional data sets is of clinical relevance when the effects of various brain pathologies or different therapies (e.g., behavioral or pharmacological effects) on brain function are subject to study.

1.5.1 The Correspondence Problem

The integration of fMRI data from multiple subjects is challenging because of the *spatial correspondence problem* between different brains. This problem manifests itself already at a purely anatomical level, but presents a fundamental problem of neuroscience when considered as a question of the consistency of structure-function relationships. At the anatomical level, the correspondence problem refers to the differences in brain shape and, more specifically, to differences in the gyral and sulcal pattern varying substantially across subjects. At this macroanatomical level, the correspondence problem would be solved, if brains could be matched in such a way that for each macroanatomical structure in one brain, the corresponding region in the other brain would be known. In neuroimaging, the matching of brains is usually performed by a process called *brain normalization*, which involves warping each brain into a common space. After brain normalization, a point in the common space identified by its x, y, and z coordinates is assumed to refer to a similar region in any other normalized brain. The most commonly used target space for normalization is the *Talairach space* (see below) and the closely related *MNI template space*. Unfortunately warping brains in a common space does not solve the anatomical correspondence problem very well, i.e., macroanatomical structures, such as banks of prominent sulci are often still misaligned deviating from brain to brain in the order of 0.5–1 cm. In order to increase the chance that corresponding regions overlap, functional data is therefore often smoothed with a Gaussian kernel with a width of about 1 cm. More advanced anatomical matching schemes attempt to directly align macroanatomical structures such as gyri and sulci (see below).

1.5.2 Spatial Normalization in Volume Space

The most often used standard space for brain normalization is the *Talairach space* (Talairach and Tournaux 1988) or the closely related *MNI template space*.

1.5.2.1 Talairach Transformation

Talairach transformation is controlled either by the (automatic) specification of a few prominent landmarks or by a data-driven alignment of a subject's brain to a target (average) brain that has been previously already brought into Talairach space. In the explicit landmark-based approach (Talairach and Tournaux 1988), the midpoint of the anterior commissure (AC) is located first, serving as the origin of Talairach space. The brain is then rotated around the new origin (AC) so that the posterior commissure (PC) appears in the same axial plane as the anterior commissure (see Fig. 1.5). The connection of AC and PC in the middle of the brain forms the y-axis of the Talairach coordinate system. The x-axis runs from the left to the right hemisphere through AC. The z-axis runs from the inferior part of the brain to the superior part through AC. In order to further specify the x- and z-axes, a y-z plane is rotated around the y (AC-PC)-axis until it separates the left and right hemispheres (midsagittal

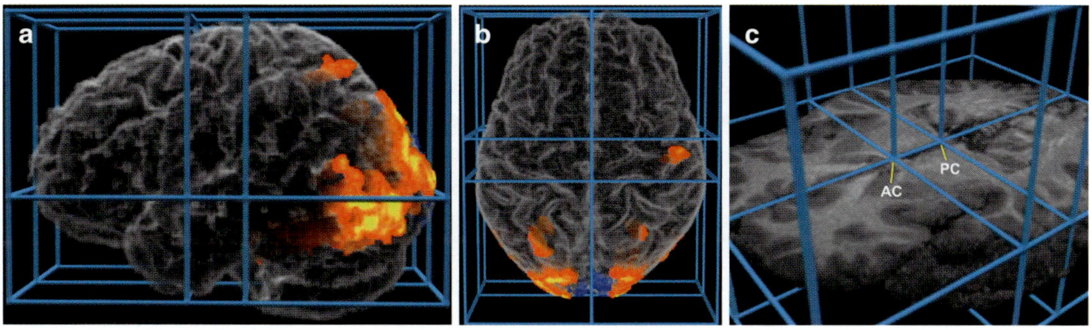

Fig. 1.5 Definition of Talairach space. (**a**) View from *left side*. (**b**) View from *top*. (**c**) View from *front-left side* with the *upper* part of the brain removed. Talairach space is defined by three orthogonal axes pointing from left to right (*x*-axis), posterior to anterior (*y*-axis), and inferior to superior (*z*-axis). The origin of the coordinate system is defined by the anterior commissure (*AC*). Coordinates are in millimeters. The posterior commissure (*PC*) is located on the y-axis ($y = -23$ mm). The borders of the Talairach grid (**a**) correspond to the borders of the cerebrum. The most right point of the brain corresponds to $x = 68$ mm, the most left one to $x = -68$ mm, the most anterior one to $y = 70$, the most posterior one to $y = -102$, the most upper one to $z = 74$, and the most lower one to $z = -42$

plane). The obtained AC-PC space is attractive for individual clinical applications, especially presurgical mapping and neuronavigation since it keeps the original size of the subject's brain intact while providing a common orientation for each brain. For a full Talairach transformation, a cuboid is defined running parallel to the three axes enclosing precisely the cortex. This cuboid or bounding box requires specification of additional landmarks that identify the borders of the cerebrum. The bounding box is subdivided by several sub-planes. The midsagittal *y-z* plane separates two sub-cuboids containing the left and right hemispheres, respectively. An axial (*x-y*) plane through the origin separates two sub-cuboids containing the space below and above the AC-PC plane. Two coronal (*x-z*) planes, one running through AC and one running through PC, separate three sub-cuboids; the first contains the anterior portion of the brain anterior to the AC, the second contains the space between AC and PC, and the third contains the space posterior to PC. These planes together divide the brain in 12 sub-cuboids. In a final Talairach transformation step, each of the 12 sub-cuboids is expanded or shrunken linearly to match the size of the corresponding sub-cuboid of the standard Talairach brain. To reference any point in the brain, *x*, *y*, and *z* coordinates are specified in millimeters of Talairach space. Talairach and Tournaux (1988) also defined the "proportional grid," to reference points within the defined cuboids.

In summary, Talairach normalization ensures that the anterior and posterior commissures obtain the same coordinates in each brain and that the sub-cuboids defined by the AC-PC points and the borders of the cortex will have the same size. Note that the specific distances between landmarks in the original post-mortem brain are not important for establishing the described spatial relationship between brains. The important aspect of Talairach transformation is that correspondence is established across brains by linearly interpolating the space between important landmarks.

While Talairach transformation provides a recipe to normalize brains, regions at the same coordinates in different individuals do not necessarily point to corresponding brain areas. This holds especially true for cortical regions. For subcortical structures around the AC-PC landmarks, however, the established correspondence is remarkably good, even when analyzing high-resolution fMRI data (e.g., De Martino et al. 2013a).

1.5.2.2 MNI Template Space

As an alternative to specify crucial landmarks, a direct approach of stereotactic normalization has been proposed (e.g., Evans et al. 1993; Ashburner and Friston 1999) that attempts to align each individual brain as good as possible to an average target brain, called *template brain*. The most often used template brain is provided by the Montréal Neurological Institute (MNI) and has been created by averaging many (>100) single brains after manual Talairach transformation. Although automatic alignment to a template brain has the potential to result in a better correspondence between brain regions, comparisons have shown that the achieved results are not substantially improved as compared to the explicit landmark specification approach, even when using nonlinear spatial transformation techniques. This can be explained by noting that the template brain has lost anatomical details due to extensive averaging. In order to bring functional data of a subject into Talairach space, the obtained spatial transformation for the anatomical data may be applied to the functional data if it has been coregistered with the unnormalized anatomical data set. Using the intensity-driven matching approach, functional data sets may also be directly normalized (without the help of anatomical data sets) because versions of the MNI template brain for functional (EPI) scans are also available. If possible, it is, however, recommended to apply the transformation obtained for the anatomical data also to the functional data because this approach guarantees that the precision of functional-anatomical alignment achieved during coregistration is not changed during the normalization step. More advanced volume-based normalization schemes have been proposed that replace the presented simple approaches (e.g., DARTEL, Ashburner 2007).

1.5.3 Spatial Normalization in Cortex Space

In recent years, more advanced brain normalization techniques have been proposed going beyond simple volume space alignment approaches. A particular interesting method attempts to explicitly align the cortical folding pattern (macroanatomy) across subjects (Fischl et al. 1999; Goebel et al. 2004, 2006; Frost and Goebel 2012). These methods start with topologically correct cortex mesh representations that are first morphed to spherical representations since the restricted space of a sphere allows alignment using only two dimensions (longitude and latitude) instead of three dimensions as needed in volume space. Since the inflation of cortex hemispheres to spheres removes information of the gyral/sulcal folds, the respective information is retained by calculating curvature maps prior to inflation that are projected on the spherical representations. Cortex meshes from different subjects are then aligned on the sphere by *increasing the overlap of curvature information*. Since the curvature of the cortex reflects the gyral/sulcal folding pattern of the brain, this brain matching approach essentially aligns corresponding gyri and sulci across brains. It has been shown that cortex-based alignment substantially increases the statistical power and spatial specificity of group analyses by increasing not only the overlap of macroanatomical regions but also the overlap of corresponding functionally defined specialized brain areas (Frost and Goebel 2012).

1.5.4 Establishing Correspondence with Functional Localizers

An interesting approach to establish correspondence between brains is to use functional information directly. Using standardized stimuli, a specific region of interest (ROI) may be functionally identified in each subject. The ROIs identified in such *functional localizer experiments* are then used to extract time courses in subsequent main experiments. The extracted time courses of individual subjects are then integrated in group analyses. If the assumption is correct that localizer experiments reveal corresponding brain regions in different subjects, the approach provides an optimal solution to the correspondence problem and will allow detection of subtle differences in fMRI responses at the group level with high statistical power. Statistical sensitivity is further enhanced by avoiding the massive multiple comparison correction problem. Instead of ca. 100,000 voxel-wise tests, only a few tests have to be performed in parallel, one for each ROI. The approach is statistically sound (no circularity) because the considered regions have been determined *independently* from the main data using special localizer runs. It may also be acceptable to use the same functional data for both localizer and main analysis as long as the contrast to localize ROIs is orthogonal to any contrast used to statistically test more subtle differences. The localizer approach has been applied successfully in many experiments, most notably in studies of the ventral visual cortex (e.g., O'Craven and Kanwisher 2000).

Unfortunately, it is often difficult to define experiments localizing the same pattern of activated brain areas in all subjects, especially in studies of higher cognitive functions, such as attention, mental imagery, working memory, and planning. If at all possible, the selection of corresponding functional brain areas in these experiments is very difficult and depends on the investigator's choice of thresholding statistical maps and often on additional decisions such as grouping sub-clusters to obtain the same number of major clusters for each subject. Note that the increased variability of activated regions in more complex experiments could be explained by at least two factors. On the one hand, the location of functionally corresponding brain regions may vary substantially across subjects with respect to aligned macroanatomical structures. On the other hand, subjects may engage in different cognitive strategies to solve the same task leading to partially different set of activated brain areas. Most likely, the observed variability is caused by a mixture of both sources of variability. Another problem of the localizer approach is the tendency to focus only on a few brain areas, namely, those which can be mapped consistently in different subjects. This tendency bears the danger to overlook other important brain regions. This can be avoided by a recently proposed approach, functionally informed cortex-based alignment (Frost and Goebel 2013), that integrates ROI-based and whole-cortex analysis using a modified version of cortex-based alignment that uses corresponding pre-mapped ROIs as alignment targets in addition to curvature information.

1.6 Statistical Group Analysis

After brain normalization, the whole-brain data from multiple subjects can be statistically analyzed simply by concatenating time courses at corresponding locations. The corresponding locations can be voxel coordinates in Talairach/MNI space or corresponding mesh vertices in cortex space. Note that the power of statistical analysis depends on the quality of brain normalization. If the achieved alignment of corresponding functional brain areas is poor, suboptimal group results may be obtained since active voxels of some subjects will be averaged with nonactive voxels (or active voxels from a non-corresponding brain area) from other subjects. In order to increase the overlap of activated brain areas across subjects in volume space, the functional data of each subject is often smoothed, typically using rather large Gaussian kernels with a full width at half maximum (FWHM) of 8–12 mm. While such an extensive spatial smoothing increases the overlap of active regions, it introduces other problems including potential averaging of non-corresponding functional areas within and across brains; furthermore, functional clusters smaller than the smoothing kernel will be suppressed. While spatial smoothing may be beneficial to reduce noise, it may also reduce detection sensitivity of truly active but small functional clusters. Extensive spatial smoothing may not be necessary when using advanced volumetric normalization schemes (e.g., Ashburner 2007), cortex-based alignment (e.g., Frost and Goebel 2012), or functional localizers.

1.6.1 Fixed Effects, Random Effects, and Mixed Effects

After concatenating the data, the same statistical analysis described for single-subject data can be applied to group data. In fact, in the context of the GLM, the multi-subject voxel time courses as well as the multi-subject predictors may be obtained by appending the data and design matrices of all subjects. After estimating the beta values, contrasts can be tested in the same way as described for single-subject data. Note that this approach leads to a high statistical power since a large number of events are used to estimate the beta values, i.e., the number of analyzed data points is the sum of the data points from all subjects; in case that all N subjects have the same number of data points n, the total number of data points N_T is $N_T = N \times n$. Note, however, that the obtained statistical results (e.g., contrast t maps) from concatenated data and design matrices cannot be generalized to the population level since the data is analyzed as if it originates from a single individual that is scanned for a (very) long time. Inferences drawn from obtained results are, thus, only valid for the included group of subjects since the group data is treated like a virtual single-case study. In order to test whether the obtained results are valid at the population level, the statistical procedure needs to consider that subjects constitute a randomly drawn sample from a large population. Subjects are thus random quantities, and the statistical analysis must assess the variability of observed effects *between subjects* (σ_b^2) in a *random-effects* (*RFX*) *analysis*. In contrast, the simple concatenation approach constitutes a *fixed-effects* (*FFX*) *analysis* assessing observed activation effects with respect to the scan-to-scan measurement error, i.e., with respect to the precision with which the fMRI signal can measure. The source of variability used in a FFX analysis, thus, represents within-subject variance (σ_w^2).

The optimal approach to perform valid population inferences is to run a single GLM using the full time course data from all subjects as in the simple concatenation approach described above. Instead of simple concatenation, however, a special multi-subject design matrix is created that allows modeling explicitly both within-subject and between-subject variance components (ref; Beckmann et al. 2003). Because this "all-in-one" approach leads to large concatenated data sets and huge design matrices that are difficult to handle on standard computer platforms, many authors have proposed to separate the group analysis in (at least) two successive stages by performing a *multilevel analysis* (e.g., Holmes and Friston 1998; Worsley et al. 2002; Beckmann et al. 2003; Friston et al. 2005). The suggested analysis strategies are related to the *multilevel summary statistics approach* (e.g., Kirby 1993). In the first analysis stage, parameters (*summary statistics*) are estimated for each subject independently (level 1, fixed effects). Instead of the full time courses, only the resulting first-level parameter estimates (betas) from each subject are carried forward to the second analysis stage where they serve as the *dependent variables*. The second-level analysis assesses the consistency of effects within or between groups based on the variability of the first-level estimates across subjects (level 2, random effects). This hierarchical analysis approach reduces the data for the second-stage analysis enormously since the time course data of each subject has been "collapsed" to only one or a few parameter estimates per subject. Since the summarized data at the second level reflects the variability of the estimated parameters across subjects, obtained significant results can be generalized to the population from which the subjects were drawn as a random sample.

To summarize the data at the first level, standard linear modeling (GLM) is used to estimate parameters (beta values) separately for each subject k.

$$\mathbf{y}_1 = \mathbf{X}_1\mathbf{b}_1 + \mathbf{e}_1$$

$$\mathbf{y}_k = \mathbf{X}_k\mathbf{b}_k + \mathbf{e}_k$$

$$\mathbf{y}_N = \mathbf{X}_N\mathbf{b}_N + \mathbf{e}_N$$

Instead of one set of beta values as in the concatenation approach, this step will provide a *separate set of beta values* $\mathbf{b}_k$ *for each subject*. The obtained beta values serve as the dependent variable at the second level (i.e., they appear on the left side) and can be analyzed again with a second-level general linear model:

$$\mathbf{b} = \mathbf{X}_G\mathbf{b}_G + \mathbf{e}_G$$

Note that the error term $\mathbf{e}_G$ is a random variable that specifies the deviation of individual subjects from group parameter estimates, i.e., it models the random-effects variability (actually a combination of within-subject and between-subject variance, see below). The group design

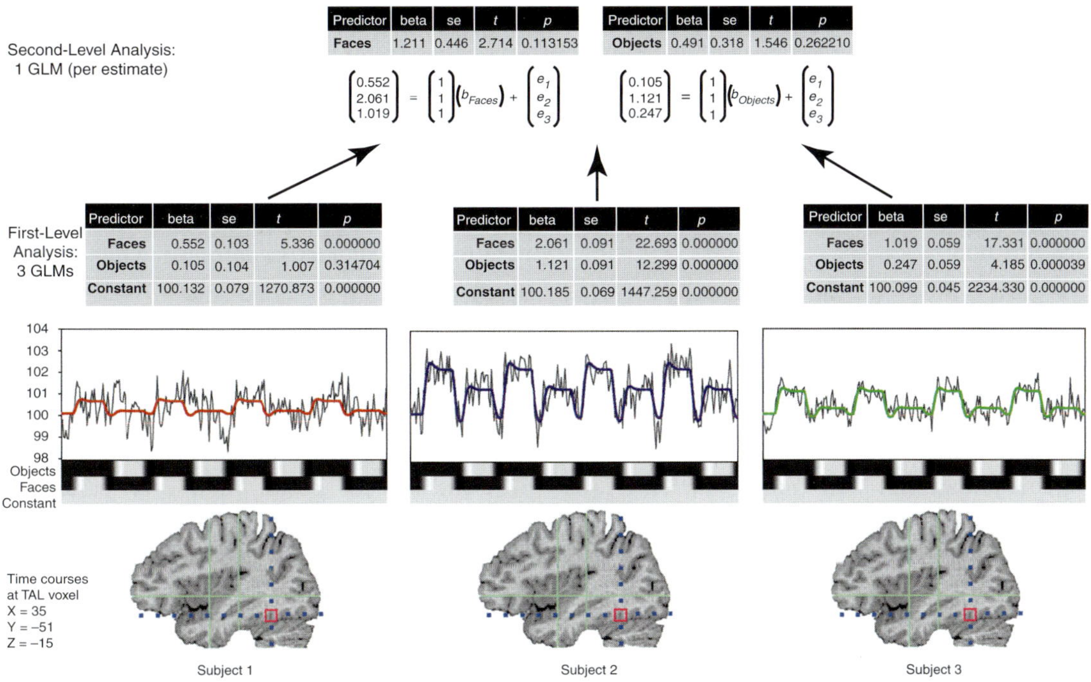

Fig. 1.6 Principle of multilevel statistical group analysis for a simple experiment where images of faces and objects were shown in separate blocks that were separated by fixation blocks. *Bottom row* shows the brains of three scanned subjects and the corresponding voxel in normalized space from which time courses have been extracted (see time course plots). The design matrices for the first-level analysis are shown below each time course plot. The results of the first-level (single-subject) GLMs are shown above the time course plots. For each of the two estimated main conditions (faces, objects), a second-level GLM is performed to assess the obtained effect at the population level (random-effects analysis). Note that the estimated beta values from the first-level analysis now serve as the dependent variable for the second-level analysis. While all three subjects show consistent activation increases with respect to baseline, results for this small sample of subjects do not reach significance

matrix $\mathbf{X}_G$ may be used to model group mean effects of first-level beta estimates of a single condition (one per subject), implementing, e.g., a one-sample *t*-test (i.e., H0: mean(b_G)=0) with a simple design matrix containing just a constant ("1") predictor (see Fig. 1.6). More generally, any mixed design with one or more within-subject factors and one or more between-subject factors can be analyzed using a GLM/ANOVA formulation at the second level.

1.6.1.1 Mixed-Effects Analysis

The random-effects analysis at the second level described above does not differ from the usual statistical approach in behavioral and medical sciences: The units of observation are measurements from randomly and independently drawn individuals with experimental group factors (e.g., specific patient groups as levels), repeated measures factors (e.g., specific tasks as levels), and one or more covariates (e.g., age or IQ). We know, however, from the first-level analysis that our "measurements" (beta values) are parameter estimates and not the true (unobservable) population values. When using parameter estimates as the dependent variable in a GLM (instead of error-free true values), the second-level analysis can no longer be treated as a pure random-effects model since the entered values are "contaminated" by variance from the first-level analysis. The second-level analysis, thus, constitutes a *mixed-effects* (MFX) model containing variance components from both the first (fixed-effects) and second (random-effects) level. It can be shown (see also below) that the two-level summary statistics approach nonetheless leads to valid inferences, but it requires that the variances of first-level parameter estimates are homogeneous across subjects. A safe way to correctly estimate the fixed- and random-effects variance components is to avoid the summary statistics approach and to construct the "all-in-one" model mentioned earlier that combines the two levels in one equation:

$$\mathbf{y} = \mathbf{X}\mathbf{b} + \mathbf{e}$$

$$\mathbf{b} = \mathbf{X}_G \mathbf{b}_G + \mathbf{e}_G$$

$$\rightarrow \mathbf{y} = \mathbf{X}\mathbf{X}_G \mathbf{b}_G + \mathrm{X}\mathbf{e}_G + \mathbf{e}$$

The resulting well-known mixed-effects model (e.g., Verbeke and Molenberghs 2000) can be fitted with several methods, including the previously mentioned generalized least squares (GLS) approach (see Sect. 1.3.5). This all-in-one model does not use the summary statistics approach since the group betas are directly estimated from the full time series data of all subjects. The resulting betas $\mathbf{b}_G$ are used to test contrasts at the group level with standard deviation values that take into account (inhomogeneous) fixed- and random-effects variance components. This approach is mathematically the most adequate solution providing valid population inferences even in case of unbalanced designs (see below). The all-in-one approach leads, however, to large design matrices and large concatenated data sets, and it may be challenging to perform the analysis on standard computer platforms.

It would, thus, be advantageous to use the multilevel summary statistics approach described above to perform mixed-effects analysis. Since the first-level betas are only estimates of the true (population) betas, the two-level model is, however, not strictly identical to the all-in-one model. Nonetheless it has been shown that by incorporating first-level information about the precision (variance) of the estimated betas (as well as their covariances) into the second-level model, a two-level summary statistics approach can be performed that is identical to the all-in-one model for typical hypothesis testing scenarios (Beckmann et al. 2003; Friston et al. 2005). Since in this case both first-level and second-level variance components are incorporated at the second level, the analysis extends a random-effects model to a mixed-effects model allowing for non-homogeneous variances at the first level. Solutions of these modified two-stage mixed-effects models require, however, time-consuming iterative estimation procedures using either a frequentist (e.g., Worsley et al. 2002) or Bayesian (e.g., Woolrich et al. 2004; Friston et al. 2005) framework. A major advantage of mixed-effects (generalized two-level or all-in-one) models as compared to the simple two-level summary statistics model is that they can take into account different subject-specific first-level parameter variances. If, for example, a subject has a very high first-level beta value but it enters the second level also with a very high parameter variance, the mixed-effects model is able to "down-weight" its effect on the group variance estimates resulting in an effective treatment of outliers.

1.6.1.2 Balanced Designs

While the modified mixed-effects model does not require the homogeneous variance assumption, the simpler multilevel summary statistics approach (using non-iterative ordinary least squares (OLS) estimation) has been shown to be robust with respect to modest violations of the homogeneity-of-variance assumption (Friston et al. 2005; Mumford and Nichols 2009). More specifically, when calculating one-sample *t-test* contrasts of first-level parameter estimates, group inferences based on OLS estimation are fully valid and even achieve nearly optimal sensitivity under modest violations of the variance homogeneity assumption (Penny and Holmes 2007; Mumford and Nichols 2009). For realistic scenarios, the power differences between advanced mixed-effects models and the simple two-stage summary statistics approach are modest (Friston et al. 2005; Penny and Holmes 2007; but see Beckmann et al. 2003). As a general advice, *balanced designs* should be employed whenever possible since such designs lead to valid estimates and inferences without complex iterative fitting routines. Furthermore, standard time course normalization routines (percent signal change or z transformation) further reduce the issue of inhomogeneous error variance terms across subjects. Since the two-level summary statistics approach is robust with respect to modest violations of the homogeneous variance assumption (e.g., Mumford and Nichols 2009), its application has become common practice (Friston et al. 2005). For designs deviating substantially from the equality of variance assumption (e.g., in designs requiring post hoc classification of events), one of the more complex mixed-effects analysis approaches should be employed.

1.6.2 Whole-Brain Group Analysis

Figure 1.6 shows the essential steps required for whole-brain multi-subject analysis in normalized (Talairach or MNI) space. The data is analyzed independently for all voxels of the brain (or vertices in aligned cortex space). For a specific voxel, the data is extracted at the respective coordinates from the normalized time course data of each subject and analyzed using first-level GLM modeling. In the example shown in Fig. 1.6, time courses from three subjects are extracted for a voxel with Talairach coordinates $x=35$, $y=-51$, $z=-15$ (black curves). The three time courses are analyzed separately using a GLM with two main predictors, "Faces" and "Houses," as indicated in the design matrices plotted below the time course data. While in this example the experimental conditions (blocks) were presented in the same order (same design matrix), it is possible (and recommended) to use different condition sequences for different subjects. Above the time course data, results of the three conducted single-subject GLM analyses are shown including beta values for the two main conditions and their standard errors. Using the estimated beta values, the predicted time course data is calculated and plotted on top of the extracted time course data (red, blue, green curves for subjects 1, 2, 3, respectively). The estimated beta values (two per subject) serve themselves as the dependent data for the second-level analysis. In case that only the estimated beta values are used at the second level (e.g., measurement errors at the first level are ignored), a random-effects analysis is performed. For a mixed-effects analysis, also the variances (standard errors) of the beta estimates need to be taken to the second level. In case that contrasts are calculated at the first level, the first-level variances and covariances of the parameter estimates need to be taken to the second level for mixed-effects analysis.

1.6.3 ROI-Based Group Analysis

Using functional localizers to identify corresponding brain regions is one way to solve the spatial correspondence problem. This approach can, however, only be used for those specialized brain areas that can be robustly mapped with a standard experimental protocol. To avoid circularity in data analysis, it is important to use independent data for localizing a specialized brain area. To ensure independence, localizer runs are usually performed prior to a main experiment. The mean time course from the main experiment originating from the previously localized region can then be used for further data analysis avoiding the massive multiple comparison problem of whole-brain analysis and benefitting from a nearly perfect correspondence across brains. Note that this approach requires some extra effort since the corresponding brain area has to be identified in the data of each subject. One way to reduce this effort is to run a fixed-effects group analysis on the localizer data and to define a single mean ROI from the resulting data. This *group ROI analysis approach* requires less work, but the identified mean ROI may fit better to some subjects than to others, i.e., the correspondence problem is not solved. Only a true *subject-specific ROI analysis approach* largely solves the correspondence problem by extracting the time course data from ROIs defined separately for each subject.

1.7 Multivariate Pattern Analysis

While this chapter mainly describes *univariate* statistical analysis, various tools for the *multivariate* (statistical) analysis of distributed patterns have gained increasing popularity in the last 10 years. While univariate analyses process the time course of each voxel independently, several new analysis approaches focus on local or global, multivariate pattern analyses (e.g., Haxby et al. 2001; Kriegeskorte et al. 2006). In these multi-voxel pattern analysis (MVPA) approaches, time courses from voxels within a small region or the whole brain are *jointly analyzed*, and distributed activity patterns corresponding to different conditions are compared using multivariate statistical approaches (e.g., multivariate analysis of variance (MANOVA)) or by using machine learning tools (see below). In the multivariate searchlight mapping approach (Kriegeskorte et al. 2006), for example, a small sphere is moved across the brain, and the time courses of the voxels within the sphere are statistically analyzed jointly using MANOVA. Since the sphere is centered at each voxel, this multivariate analysis approach opens the possibility to localize spatially distributed but local effects, which would be potentially too weak to be discovered by single-voxel analysis. Note that in multivariate analysis neighboring voxels do not need to respond in the same way to different conditions, since the amount of information within the distributed pattern is extracted in a given region. If, for example, some voxel time courses in a local neighborhood show a weak increase and other voxel time courses show a weak decrease, these effects would potentially cancel out in an ROI average, but would well contribute to a measure of multivariate information. MVPA tools are, thus, able to detect differences between conditions with higher sensitivity than conventional univariate analysis. MVPA is often presented in the context of "brain reading" applications reporting that specific mental states or representational content can be decoded from fMRI activity patterns. These applications often use data from the whole brain, and they do usually not use statistical tools such as MANOVA but rely on classifiers developed in machine learning such as support vector machines (SVMs). In a training phase, classifiers learn to distinguish between brain activity patterns that are evoked by different mental states that are usually linked to specific experimental conditions. The performance of trained classifiers is then tested on new data and includes the assessment whether obtained accuracy values reach statistical significance.

Conclusions

In the last 20 years, functional brain imaging has been successfully employed in many psychiatric neuroimaging studies and fMRI measurements have the potential to further elucidate neural correlates of specific diseases in the future. This chapter has described the principles of statistical localization of brain function. It has been also emphasized that a number of preprocessing steps are required in order to protect the calculation and visualization of functional maps from potential signal artifacts that may be caused by head motion, signal drifts, and other physical and physiological noise sources. It is also recommended in the context of patient studies to consider the use of arterial spin labeling (ASL) MR pulse sequences in order to minimize the effect of medication on estimated effect sizes in different experimental groups.

While multivariate voxel pattern analysis and functional/effective connectivity analysis have gained increased popularity in recent years, univariate statistical analysis remains an important fMRI analysis tool since it allows to exploit the high spatial resolution of fMRI at the level of individual voxels. This is especially important in the context of very high (submillimeter) spatial resolution scans at ultrahigh magnetic field scanners (7 Tesla and higher) that aim to unravel the functional organization of the cortex at the level of cortical columns and cortical layers (e.g., Zimmermann et al. 2011; De Martino et al. 2013b).

References

Ashburner J (2007) A fast diffeomorphic image registration algorithm. Neuroimage 38(1):95–113

Ashburner J, Friston KJ (1999) Nonlinear spatial normalization using basis functions. Hum Brain Mapp 7:254–266

Bandettini PA, Cox RW (2000) Event-related fMRI contrast when using constant interstimulus interval: theory and experiment. Magn Reson Med 43:540–548

Beckmann CF, Jenkinson M, Smith SM (2003) General multilevel linear modeling for group analysis in FMRI. Neuroimage 20:1052–1063

Benjamini Y, Hochberg Y (1995) Controlling the false discovery rate: a practical and powerful approach to multiple testing. J R Stat Soc Ser B 57:289–300

Blamire AM, Ogawa S, Ugurbil K et al (1992) Dynamic mapping of the human visual cortex by high-speed magnetic resonance imaging. Proc Natl Acad Sci U S A 89:11069–11073

Boynton GM, Engel SA, Glover GH, Heeger DJ (1996) Linear systems analysis of functional magnetic resonance imaging in human V1. J Neurosci 16: 4207–4221

Buckner RL, Bandettini PA, O'Craven KM, Savoy RL, Petersen SE, Raichle ME, Rosen BR (1996) Detection of cortical activation during averaged single trials of a cognitive task using functional magnetic resonance imaging. Proc Natl Acad Sci U S A 93:14878–14883

Bullmore E, Brammer M, Williams SC et al (1996) Statistical methods of estimation and inference for functional MR image analysis. Magn Reson Med 35:261–277

Buxton RB, Uludag K, Dubowitz DJ, Liu TT (2004) Modeling the hemodynamic response to brain activation. Neuroimage 23:S220–S233

Cochrane D, Orcutt GH (1949) Application of least squares regression to relationships containing autocorrelated error terms. J Am Stat Assoc 44:32–61

Dale AM, Buckner RL (1997) Selective averaging of rapidly presented individual trials using fMRI. Hum Brain Mapp 5:329–340

De Martino F, Moerel M, van de Moortele PF, Ugurbil K, Goebel R, Yacoub E, Formisano E (2013a) Spatial organization of frequency preference and selectivity in the human inferior colliculus. Nat Commun 4: 1386

De Martino F, Zimmermann J, Muckli L, Ugurbil K, Yacoub E, Goebel R (2013b) Cortical depth dependent functional responses in humans at 7 T: improved specificity with 3D GRASE. PLoS One 8(3):e60514

Draper NR, Smith H (1998) Applied regression analysis, 3rd edn. Wiley, New York

Evans AC, Collins DL, MIlls SR, Brown ED, Kelly RL, Peters TM (1993) 3D statistical neuroanatomical models from 305 MRI volumes. In: Proceedings of IEEE-Nuclear Science Symposium and Medical Imaging Conference, Institute of Electrical and Electronics Engineers, Piscataway, NJ, pp 1813–1817

Fischl B, Sereno MI, Tootell RBH, Dale AM (1999) High-resolution inter-subject averaging and a coordinate system for the cortical surface. Hum Brain Mapp 8:272–284

Forman SD, Cohen JD, Fitzgerald M, Eddy WF, Mintun MA, Noll DC (1995) Improved assessment of significant activation in functional magnetic resonance imaging (fMRI): use of a clustersize threshold. Magn Reson Med 3(5):636–647

Formisano E, Goebel R (2003) Tracking cognitive processes with functional MRI mental chronometry. Curr Opin Neurobiol 13:174–181

Friston KJ, Henson RN (2006) Commentary on: divide and conquer; a defence of functional localisers. Neuroimage 30:1097–1099

Friston KJ, Jezzard P, Turner R (1994) The analysis of functional MRI time-series. Hum Brain Mapp 1:153–171

Friston KJ, Holmes AP, Worsley KJ, Poline JP, Frith CD, Frackowiak RSJ (1995) Statistical parametric maps in functional imaging: a general linear approach. Hum Brain Mapp 2:189–210

Friston KJ, Stephan KE, Lund TE, Morcom A, Kiebel S (2005) Mixed-effects and fMRI studies. Neuroimage 24:244–252

Friston KJ, Rotshtein P, Geng JJ, Sterzer P, Henson RN (2006) A critique of functional localisers. Neuroimage 30:1077–1087

Frost M, Goebel R (2012) Measuring structural-functional correspondence: spatial variability of specialised brain regions after macro-anatomical alignment. Neuroimage 59:1369–1381

Frost M, Goebel R (2013) Functionally informed cortex based alignment: an integrated approach for whole-cortex macro-anatomical and ROI-based functional alignment. Neuroimage 83:1002–1010

Genovese CR, Lazar NA, Nichols T (2002) Thresholding of statistical maps in functional neuroimaging using the false discovery rate. Neuroimage 15:870–878

Goebel R, Hasson U, Lefi I, Malach R (2004) Statistical analyses across aligned cortical hemispheres reveal high-resolution population maps of human visual cortex. Neuroimage 22(Suppl 2)

Goebel R, Esposito F, Formisano E (2006) Analysis of functional image analysis contest (FIAC) data with brainvoyager QX: from single-subject to cortically aligned group general linear model analysis and self-organizing group independent component analysis. Hum Brain Mapp 27:392–401

Haxby JV, Gobbini MI, Furey ML, Ishai A, Schouten JL, Pietrini P (2001) Distributed and overlapping representations of faces and objects in ventral temporal cortex. Science 293:2425–2430

Holmes AP, Friston KJ (1998) Generalisability, random effects & population inference. In: Fourth international conference on Mapping of the Human Brain. NeuroImage 7:S754

Kanwisher N, McDermott J, Chun MM (1997) The fusiform face area: a module in human extrastriate cortex specialized for face perception. J Neurosci 17: 4302–4311

Kirby KN (1993) Advanced data analysis with SYSTAT. Van Nostrand Reinhold, New York

Kriegeskorte N, Goebel R (2001) An efficient algorithm for topologically correct segmentation of the cortical sheet in anatomical MR volumes. Neuroimage 14: 329–346

Kriegeskorte N, Goebel R, Bandettini P (2006) Information-based functional brain mapping. Proc Natl Acad Sci U S A 103:3863–3868

Kruggel F, von Cramon DY (1999) Temporal properties of the hemodynamic response in functional MRI. Hum Brain Mapp 8:259–271

Kutner MH, Nachtsheim CJ, Neter J, Li W (2005) Applied linear statistical models, 5th edn. McGraw-Hill, Boston

Lenoski B, Baxter LC, Karam LJ, Maisog J, Debbins J (2008) On the performance of autocorrelation estimation algorithms for fMRI analysis. IEEE J Sel Top Signal Proc 2:828–838

Maus B, van Breukelen GJ, Goebel R, Berger MP (2010) Optimization of blocked designs in fMRI studies. Psychometrica 75:373–390

Mumford JA, Nichols T (2009) Simple group fMRI modeling and inference. Neuroimage 47:1469–1475

Nichols T, Brett M, Andersson J, Wager T, Poline JB (2006) Valid conjunction inference with the minimum statistic. Neuroimage 25(3):653–660

O'Craven KM, Kanwisher N (2000) Mental imagery of faces and places activates corresponding stimulus-specific brain regions. J Cogn Neurosci 12:1013–1023

Ogawa S, Lee TM, Kay AR, Tank DW (1990) Brain magnetic resonance imaging with contrast dependent on blood oxygenation. Proc Natl Acad Sci U S A 87:9868–9872

Penny WD, Holmes AP (2007) In: Friston K, Ashburner K, Kiebel S, Nichols T, Penny W (eds) Random-effects analysis, in statistical parametric mapping: the analysis of functional brain images. Academic, London, pp 156–165

Saxe R, Brett M, Kanwisher N (2006) Divide and conquer: a defense of functional localizers. Neuroimage 30:1088–1096

Searle SR, Casella G, McCulloch CE (1992) Variance components. Wiley, New York

Talairach G, Tournaux P (1988) Co-planar stereotaxic atlas of the human brain. Thieme, New York

Verbeke G, Molenberghs G (2000) Linear mixed models for longitudinal data. Springer, New York

Wager TD, Nichols TE (2003) Optimization of experimental design in fMRI: a general framework using a genetic algorithm. Neuroimage 18:293–309

Weiskopf N, Hutton C, Josephs O, Deichmann R (2006) Optimal EPI parameters for reduction of susceptibility-induced BOLD sensitivity losses: a whole-brain analysis at 3 T and 1.5 T. Neuroimage 33(2): 493–504

Woolrich MW, Behrens TEJ, Beckmann CF, Jenkinson M, Smith SM (2004) Multilevel linear modelling for FMRI group analysis using Bayesian inference. Neuroimage 21(4):1732–1747

Worsley KJ, Marrett S, Neelin P, Evans AC (1992) A three-dimensional statistical analysis for CBF activation studies in human brain. J Cereb Blood Flow Metab 12:900–918

Worsley KJ, Liao CH, Aston J, Petre V, Duncan GH, Morales F, Evans AC (2002) A general statistical analysis for fMRI data. Neuroimage 15(1):1–15

Zimmermann J, Goebel R, De Martino F, van de Moortele PF, Feinberg D, Adriany G, Chaimov D, Shmuel A, Uğurbil K, Yacoub E (2011) Mapping the organization of axis of motion selective features in human area MT using high-field fMRI. PLoS ONE 6:e28716

2 Neurofeedback with Real-Time Functional MRI

Rainer Goebel and David Linden

Abbreviations

ACPC	Anterior commissure-posterior commissure
ADHD	Attention deficit/hyperactivity disorder
DBS	Deep-brain stimulation
DCT	Discrete cosine transform
ETH	Swiss Federal Institute of Technology
fMRI-/EEG-NF	fMRI-/EEG-based neurofeedback
GLM	General linear model
GP-GPUs	General purpose graphic processing units
ICA	Independent component analysis
MVPA	Multi-voxel pattern analysis
PD	Parkinson's disease
SMA	Supplementary motor area
SVM	Support vector machine
tDCS	Transcranial direct current stimulation
TMS	Transcranial magnetic stimulation
TR	Volume time to repeat

R. Goebel (✉)
Department of Cognitive Neuroscience, Faculty of Psychology and Neuroscience, Maastricht University, Maastricht, The Netherlands

Department of Neuroimaging and Neuromodelling, Netherlands Institute for Neuroscience, Amsterdam, The Netherlands
e-mail: r.goebel@maastrichtuniversity.nl

D. Linden
Institute of Psychological Medicine and Clinical Neurosciences, Cardiff University School of Medicine, Cardiff, UK

2.1 Introduction

Since its invention 20 years ago, functional magnetic resonance imaging (fMRI) has become one of the most widely used and probably the publicly most visible noninvasive technique to measure brain activation. fMRI has played a central role in the development of cognitive neuroscience, and several new fields, including social neuroscience, neuroeconomics, and genetic imaging, may not have developed had it not been for the unique opportunities afforded by fMRI. The particular strengths of this technique are in its spatial resolution and fidelity, ability to reach deep subcortical structures, and whole-brain coverage, enabling the mapping of functionally connected networks and the extraction of information from activation patterns that are distributed across different brain areas. In the psychiatric domain, fMRI has made major contributions to the understanding of psychopathology and the effects of risk genes on cognitive and affective networks (Linden 2012a), and in neurology fMRI has become a central technique for mapping neuroplasticity, for example, in recovery from stroke (Seitz 2010), and for presurgical mapping. However, fMRI has not yet fulfilled its

C. Mulert, M.E. Shenton (eds.), *MRI in Psychiatry*,
DOI 10.1007/978-3-642-54542-9_2, © Springer Berlin Heidelberg 2014

translational potential, and there is as of today no established diagnostic, prognostic, or therapeutic use of this technique for any of the neuropsychiatric disorders.

fMRI-based neurofeedback (fMRI-NF) has the potential to open up radically new paths to translation. During fMRI-NF training, participants receive feedback on their brain activity in real time and are instructed to change this activation (see Fig. 2.1 for a schematic overview of an fMRI neurofeedback setup). This change is normally a simple up- or downregulation but could also entail changing the activation difference between two areas, their correlation, or the output of a multivariate pattern classification algorithm. The "hemodynamic" delay between neural activity and the vascular response that contributes to the fMRI signal, which is approximately 5 s, does not pose an obstacle when participants are informed of this delay (Weiskopf et al. 2004a). The technical requirements include direct network access to the scanner hardware, software that processes the incoming data in real time and computes incremental online statistics, and software for the conversion of the real-time activation data into sensory stimulation (e.g., visual, tactile) for the participant. The relevant workflows (see below) have been established on systems from all major MRI scanner manufacturers, and several custom-made and commercial software packages have been developed for online preprocessing and statistical analysis of fMRI data. Over the last 10 years, fMRI-NF has been used to train healthy participants in the self-regulation of motor, sensory, language, and emotion areas (Weiskopf 2012). In analogy to the effects of other noninvasive or invasive brain stimulation techniques, one should expect that such self-regulation also has consequences at the behavioral, cognitive, and affective level. However, clinical applications have been sparse and had initially focused on chronic pain (deCharms et al. 2005). Compared to other neurofeedback techniques (with EEG or MEG) and to noninvasive physical stimulation techniques (tDCS and TMS), fMRI-NF has the advantage of higher spatial resolution and localization accuracy and better access to deep limbic and subcortical structures. Compared to deep-brain stimulation, fMRI-NF has the advantage of noninvasiveness and spatial flexibility (although it is not intended to replace established DBS indications, e.g., in Parkinson's disease [PD]). Finally, compared to all external stimulation techniques, neurofeedback has the attractive characteristic of enabling the patients themselves to control their brain activity and thus contributing to their experience of self-efficacy, which is an important therapeutic factor in many neuropsychiatric disorders (Bandura 1997).

There are, in principle, at least two ways in which self-regulation of brain activity through fMRI-NF may be beneficial in neurological and psychiatric disorders. It may help rectify pathological hyper- or hypoactivation of specific brain areas or network, and/or it may boost the recruitment of compensatory circuits for particular tasks (e.g., motor control) or cognitive processes (e.g., emotion regulation). The latter possibility is particularly attractive in cases where brain tissue may have been damaged by the disease, for example, in stroke, or where no clear abnormalities have been established, as in most psychiatric disorders (Linden 2012b). Here, the knowledge about the canonical neural circuits underlying emotion and cognition derived from 20 years of fMRI can provide the basis for neurofeedback protocols that help patients recruit the functional systems needed to overcome their symptoms (Linden 2013).

fMRI-NF has been boosted by developments in MR physics, allowing for fast acquisition of high-quality data sets, and data analysis, allowing for online calculation of univariate and multivariate statistics. While improvements in hard- and software are still being made, fMRI-NF is now a mature experimental technique, and the exciting question for the next decade will be whether it can make a true translational contribution in medicine. The central challenges are identifying the symptoms and disorders that will respond to fMRI-NF, adapting the treatment protocols to the neural networks involved in each of them, evaluating the underlying neuroplastic mechanisms, and devising training strategies that enable sustainable long-term effects (Sulzer et al. 2013).

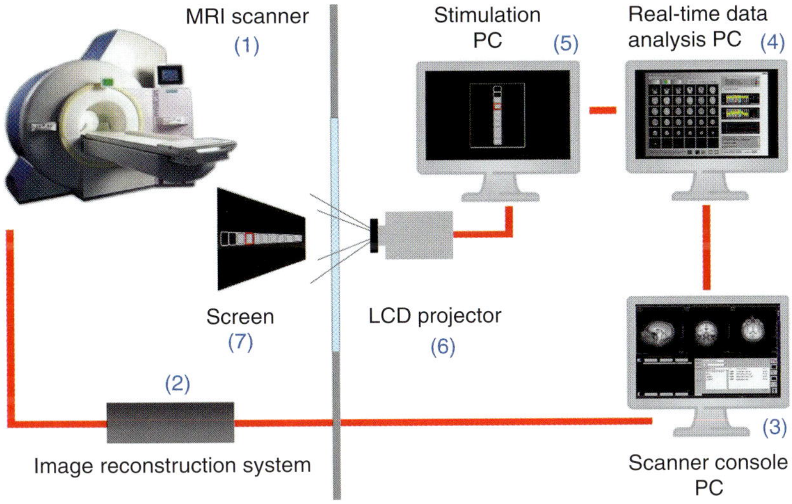

Fig. 2.1 General technical setup and information flow during real-time fMRI experiments used for neurofeedback and other brain-computer interface (*BCI*) applications (Adapted from Goebel et al. 2010). Following acquisition of functional data (*1*), the images are reconstructed (*2*) and sent to the scanner console's hard disk (*3*). The real-time fMRI analysis PC (*4*) accesses the reconstructed images as soon as possible. Calculated feedback signals (or visualizations) are transferred to the stimulation PC (*5*), which sends the final feedback visualization (here a simple thermometer display) via a projector (*6*) onto the screen (*7*) and is then visible to the participant

2.2 Principles and Methods

2.2.1 Learning Theory

Attempts to train humans and animals to regulate their own brain activity, feeding back signals from noninvasive (electroencephalography [EEG]) or invasive recordings, go back to the 1960s (reviewed by Birbaumer and Cohen 2007). The theoretical principles were largely derived from operant conditioning, whereby the participant learns the optimal strategy through the contingencies between their actions and a reward. In an animal, the reward would have to be a primary reward. For example, the desired neural response would be reinforced by the delivery of food or drink. In humans, the information about the achievement of a particular neural target (e.g., increasing the ratio between theta and alpha power of the EEG by 20 %) could itself be the reward. Whether this reward reinforces a particular mental strategy that leads to the desired physiological response or the physiological response itself is a futile question because the two cannot be separated in any meaningful way under the assumptions of psychophysical unity. It has been reported that humans can achieve reliable self-control over parameters of their EEG, for example, the topography of slow cortical potentials, the alpha-theta ratio, or the ratio between sensorimotor rhythm and theta activity, through such operant conditioning.

FMRI-based neurofeedback (fMRI-NF) can be conducted in a similar fashion, where participants are blind to the functional relevance of the targeted activation and essentially aim to achieve self-regulation by trial and error (Weiskopf et al. 2004b; Birbaumer et al. 2013). However, fMRI-NF can also harness the considerable knowledge about the neural basis of particular mental and cognitive processes that the last 20 years of functional brain mapping have achieved and introduce a cognitive component into the learning strategy. For example, in our study on upregulation of the supplementary motor area in Parkinson's disease (Subramanian et al. 2011), patients were told about the motor planning functions of the target area and informed that motor imagery might be one viable strategy to upregulate it. Giving patients initial hints about potential strategies for the upregulation training, which they can then refine through an operant conditioning protocol with fMRI-NF, can considerably reduce scanning time and provide patients with tangible results within one scanning session (Subramanian et al. 2011). Using this cognitive hypothesis testing strategy for neurofeedback training, the temporal relationship between mental actions and reward (information about achievement) may be delayed for several seconds without problems since participants use cognitive systems such as working memory to evaluate performed mental tasks with respect to feedback information obtained with a few seconds of delay.

2.2.2 Real-Time Processing and Analysis of fMRI Data

In order to enable neurofeedback applications, the measured fMRI data needs to be processed online, that is, during functional scanning; data analysis should preferentially operate in real time, that is, analysis for a newly measured functional volume should be completed before the next functional volume becomes available. Real-time processing (as opposed to near real-time fMRI), thus, restricts processing time to a maximum duration (volume time to repeat (TR)) that is defined by the temporal interval between successive functional volumes, which typically assumes values between 1 and 3 s. The requirement for incremental analysis in limited time windows is in contrast to conventional fMRI analyses processing data offline, that is, data analysis only starts after the fMRI scanning session has actually ended without specific restrictions in calculation time.

A basic prerequisite of real-time fMRI (rt-fMRI) analysis is that images are reconstructed from k-space to image space as fast as possible. Incremental image reconstruction is usually performed directly on the scanner using fast image reconstruction computers, that is, data is stored to disk volume by volume (or slice by slice) in image space, usually in DICOM format. Besides image reconstruction, most scanner manufacturers also provide basic real-time analysis tools, which are sufficient for simple applications such as quality assurance. For advanced real-time applications such as neurofeedback, specialized software packages are usually employed that provide comprehensive preprocessing options, advanced (statistical) analysis tools, ROI extraction, and specialized visualization routines. Data acquisition techniques and analysis software have been considerably improved since the introduction of real-time fMRI (Cox et al. 1995). The first real-time fMRI setups provided limited processing capabilities, lacking, for example, motion correction, or moment-to-moment statistical analysis. Recent real-time fMRI studies employ analysis pipelines that include almost all preprocessing and analysis steps used in conventional offline analysis (see chapter fMRI analysis). Here we provide a short overview of typical analysis workflows; more detailed reviews are provided, among others, by Weiskopf et al. (Weiskopf et al. 2004b), deCharms (2007, 2008), LaConte (2011), and Caria et al. (2012).

Most real-time processing pipelines include rigid-body 3D motion correction that detects head movements in the same way as used in offline analysis; in order to allow motion correction from the second time point onwards, the first functional volume (instead of, e.g., the middle one) is used as the reference to which subsequent functional volumes are aligned. In order to reduce calculation time, realignment of images to the reference volume is usually performed using trilinear interpolation, but since interpolation is nowadays also possible in real time when using parallel implementations that exploit the power of general purpose graphics processing units (GP-GPUs). To enhance signal contrast, spatial Gaussian smoothing may be optionally performed. Removal of linear and nonlinear trends is typically not performed as a preprocessing step, but it is incorporated in the statistical analysis using low-frequency drift predictors.

In order to guarantee constant (and fast) processing time, univariate statistical data analysis is usually performed recursively, that is, estimated statistical parameters are updated by the information arriving with the next available functional volume instead of estimating them using the whole available time course from the first volume up to the volume of the current time point (Bagarinao et al. 2006). If the whole time course is used, calculation time of conventional algorithms (e.g., correlation analysis) increases with growing data sets bearing the risk of lagging behind the incoming data at some point. Instead, incremental algorithms provide constant calculation time per data point (volume) and enable real-time processing even for very long functional scans.

While correlation analysis was used in early real-time fMRI studies, full (incremental) general linear model (GLM) analyses are now usually performed (Goebel 2001; Smyser et al. 2001; Bagarinao et al. 2003; Weiskopf et al. 2004b; Hinds et al. 2011). While the design matrix for main experimental conditions may be built in advance for a planned experiment, sophisticated implementations allow building the design matrix incrementally allowing to incorporate real-time imaging and behavioral data as it becomes available. This allows, for example, that trials are assigned "on the fly" to specific experimental conditions with respect to trial-by-trial performance of the participant in the scanner. Incrementally built design matrices also allow incorporation of parameters obtained volume by volume from a 3D motion correction routine that may help to reduce residual motion artifacts. The real-time GLM design matrix may also contain confound predictors to model drifts in voxel time courses. Basic low-frequency drift removal can be achieved by adding a linear trend predictor; for nonlinear trends, discrete cosine transform (DCT) confound predictors may be incrementally added to the design matrix. Removing drifts is especially important for neurofeedback experiments in order to ensure that visualized increases or decreases of activity are caused by mental tasks and are not the result of unrelated signal drifts.

Note that the described (incremental) statistical analyses are usually operating at the single voxel level providing dynamic whole-brain statistical maps that integrate information of the whole voxel time course data from the beginning of the functional scan up to the currently processed volume. It is also possible to restrict the calculation of whole-brain voxel-wise statistical maps to a sliding window; depending on the specified size of the sliding window, obtained maps reflect more dynamic changes (short sliding window) or more stable effects (large sliding window). While these whole-brain voxel-wise maps are not strictly necessary to calculate region-specific neurofeedback signals (see below), whole-brain maps are also very useful during neurofeedback runs serving as a quality assurance tool; whole-brain maps allow, for example, to inspect activity in emotion, attention, and control networks indicating whether the participant is engaging in mental tasks.

2.2.3 Feedback and Stimulus Delivery

Neurofeedback signals are often calculated from restricted temporal windows that start with a rest period from which an up-to-date baseline signal level is calculated for a subsequent active mental task period. The windowed analysis approach with a rest and mental task period of 15–30 s has the advantage that slow drifts have minimal impact on the neurofeedback signal since all relevant time points are in close temporal proximity. Given a baseline level ybl, the feedback value yfb for subsequent time points can then be calculated simply as

$$\text{yfb} = \frac{(\text{y} - \text{ybl})}{\text{ybl} * 100}$$

The value yfb expresses activity changes with respect to the baseline level as a percent signal change value. In order to remove local drifts within a single neurofeedback window, they can be estimated in the baseline period and removed from subsequent time points, for example, by using a GLM with a linear confound predictor.

Calculation of (window-based) neurofeedback signals does usually not operate at the single voxel level but uses the mean signal time course from selected neurofeedback target regions or networks. The regions of interest (ROIs) serving as target regions are defined anatomically or by running functional localizer scans prior to the actual neurofeedback session. The averaged signal time course from a functionally defined ROI usually provides higher signal modulation values and lower noise fluctuations than when using anatomical ROIs.

The calculated neurofeedback signal can be visualized in various ways to the participant in the scanner, for example, as the level in a thermometer display (see Fig. 2.1). In order to convert the calculated percent signal change value into a corresponding level of a thermometer display, the percent value is usually related to a maximum percent signal value; the maximum modulation value is either specified based on general experience with BOLD signal amplitudes in various regions, or it is derived from measured responses in the same individual using prior localizer or neurofeedback runs. If, for example, the calculated feedback value is 1.5 % and the maximum percent value has been set to 3 %, the thermometer will be shown as half filled to the participant. This visualization approach assumes that only increases of activity relative to the baseline are considered. In case one wants to also visualize deactivations, the (thermometer) display needs to be extended for negative values accordingly. Besides simple activation displays (thermometer, continuous scrolling time course, numerical value), more expressive visualizations have been used in the past such as images of a virtual fire that increased in size with increasing activation levels (deCharms et al. 2005), level of a racket in a "brain pong" computer game (Goebel et al. 2004), and virtual reality (Sitaram et al. 2005). Furthermore, feedback may also be provided via haptic or auditory stimuli, but not much experience is currently available. While simple feedback representations may be generated directly from the used real-time analysis software, more complex visualizations usually require custom-made software accessing feedback signals calculated by the real-time analysis software. Note that it is also possible to present feedback from multiple areas simultaneously. In a recent study (Zilverstand et al. 2014), two thermometer displays have been used where one thermometer represented activity in the control network, while the other thermometer represented the activity level of areas related to fear responses.

Since neurofeedback studies often span over multiple sessions, it is important to transfer defined ROIs from one session to the next. This can be done by aligning 3D anatomical data sets across sessions and projecting voxel coordinates marked on anatomical data sets to the aligned functional data measured in subsequent sessions. While not strictly necessary, across-session alignment may be facilitated by normalizing the brain of a participant to a standard (e.g., ACPC or Talairach) space that keeps generic representation of the location of ROIs and allows to project them in any scan-specific space using spatial transformations obtained from alignment procedures. If brain normalization is performed prior to functional runs, real-time (statistical) data can be projected on standard orthogonal brain slices as opposed to visualizations on arbitrarily positioned slices in the native scanning space in a particular session. If a high-quality anatomical data set is available prior to functional scanning, it is even possible to visualize results, for example, dynamic statistical maps, on reconstructed, inflated, or flattened cortex representations using fast automated 3D segmentation and cortex reconstruction tools (Goebel 2001).

2.2.4 Neurofeedback Based on Real-Time Pattern Analysis

As for offline analysis, multi-voxel pattern analysis (MVPA, see chapter fMRI analysis) is gaining increasing interest also for real-time fMRI neurofeedback studies (e.g., LaConte et al. 2007; Sorger et al. 2010). One reason of the popularity of MVPA is based on its potential to detect differences between neural correlates of mental states with higher sensitivity as conventional univariate statistical analysis. The higher sensitivity may also be beneficial for neurofeedback studies since distributed activity patterns may reflect activity modulations better than ROI-based approaches. In a recent application, volunteers achieved successful control of emotion-related activation patterns using a real-time support vector machine (SVM) classifier (Sitaram et al. 2011).

MVPA may also serve as a method to localize brain systems for subsequent neurofeedback as an alternative to using univariate GLM statistical analysis; as opposed to focal ROIs, MVPA may be especially useful to identify complex and interacting activity patterns over the whole brain as targets for neurofeedback (LaConte 2011; LaConte et al. 2007).

In the context of fMRI neurofeedback patient studies, MVPA may also be used in a novel way, namely, to let patients learn to "mimic desired brain states, or mimicking the brain states of others" (deCharms 2008). In this scenario, the feedback signal provided to a patient would not reflect activity levels in regions or networks, but it would reflect how much the currently evoked activity pattern resembles a desired target pattern. This approach requires more fundamental research in cognitive and affective neuroscience in order to explore whether it will be possible to define desirable brain states for specific patients. Since this would likely involve a transfer of desirable neural activation states to other participants, further research and application of advanced brain normalization schemes, such as cortex-based alignment (Fischl et al. 1999; Goebel et al. 2006), is important in order to relate corresponding brain regions across participants' brains as precisely as possible.

2.2.5 Real-Time Network Analysis and Feedback of Functional Connectivity

Besides univariate statistics and distributed pattern analyses, data-driven multivariate analysis tools may provide important complementary real-time information. As an example, windowed independent component analysis (ICA) has been introduced for real-time fMRI analysis (Esposito et al. 2003) allowing to detect and visualize dynamic activity changes in functional brain networks that occur at unpredictable moments during a real-time fMRI experiment.

It is also possible to use gradual values from multi-voxel pattern classification (or better regression) directly as a feedback signal with the rationale that optimally weighted voxel activity in distributed networks leads to integrated values that may differentiate fine-grained aspects of mental representations better than region-based methods. It has been, for example, shown that sparse regression-based online feedback leads to orientation-specific visual perceptual learning without presentation of an actual stimulus or subjects' awareness of what was to be learned (Shibata et al. 2011).

Another interesting avenue is to use functional or effective connectivity measures as a feedback signal instead of mean activation levels. Increasing or decreasing a visualized feedback signal would then reflect a stronger (upregulation) or weaker (downregulation) coupling between brain areas instead of (only) the mean activation level of the involved areas. This approach seems especially important since current research from clinical neuroscience on brain correlates of psychiatric and neurological disorders indicates that fMRI-based functional connectivity measures may belong to the most robust indicators of pathological brain processes. The online estimation of functional or effective connectivity between two or more brain areas requires, however, that a sufficiently long sliding window is used in order to calculate robust "instantaneous" coupling strength values. In a recent offline study, it has been shown that time windows of about 20 time points are sufficient to calculate robust (partial) correlation coefficients (Zilverstand et al. 2014). This study showed (in the context of various uni- and bimanual motor tasks) that instantaneous functional correlations may indeed provide relevant and unique information regarding ongoing brain processes, which is not captured equally well by standard activation level-based neurofeedback measures. Functional connectivity feedback will also benefit from recent technical advances such as newly developed "multiband" accelerated imaging sequences allowing a substantial increase in the number of collected data points per time unit (Feinberg and Yacoub 2012).

2.3 fMRI-NF Studies of Neural Circuits in the Healthy Brain

2.3.1 Motor Areas

Because a large body of functional imaging literature had shown activation of primary and higher motor cortices during motor imagery, motor areas were a natural target for some of the early studies of self-regulation training. In these early reports, feedback during both actual hand movements and imagined movements allowed participants to regulate activity in primary motor cortex to a desired level (Yoo and Jolesz 2002; deCharms et al. 2004; Yoo et al. 2008) and led to long-term plastic changes in motor networks that were specific to participants receiving neurofeedback (Yoo et al. 2008; Horovitz et al. 2010). However, a recent study, applying careful electromyography (EMG) monitoring for actual hand movements, has questioned whether upregulation of the primary motor cortex is possible by imagery alone (Berman et al. 2012). Higher motor areas, such as the ventral premotor cortex (Sitaram et al. 2012) and supplementary motor area (SMA) (Subramanian et al. 2011), may be more amenable to self-regulation in the absence of physical movement.

2.3.2 Sensory Areas

Visual cortex activity could be regulated by participants without visual stimulation in both primary (Shibata et al. 2011) and higher visual areas (Weiskopf et al. 2004a). For the auditory cortex, successful upregulation was demonstrated during auditory stimulation (Yoo et al. 2006), but downregulation also in the absence of external auditory stimuli (in a group of tinnitus patients) (Haller et al. 2010). An interesting aspect of the self-regulation training of primary visual cortex activity, which involved the modulation of classifier output related to gratings of a specific orientation, was that it actually improved visual perceptual learning for this orientation (Shibata et al. 2011). Another recent study found improved visual discrimination in those of the participants who successfully upregulated activity in specific parts of their visual cortex (Scharnowski et al. 2012).

2.3.3 "Emotion" Areas

The question whether limbic areas and other parts of the emotion-processing networks of the brain could be self-regulated through fMRI-NF has also attracted interest early on in the development of the field because of its potential use in psychiatric disorders and behavior modification. Several studies demonstrated upregulation of amygdala activity (Posse et al. 2003; Zotev et al. 2011) but also highlighted the problems of subcortical target areas that are vulnerable to physiological, movement, and susceptibility artifacts. Other target areas implicated in affective processing have included the insula (Caria et al. 2007; Johnston et al. 2010, 2011) and anterior cingulate cortex (Weiskopf et al. 2003; Hamilton et al. 2011). Target areas were identified on the basis of anatomical criteria or functional localizers, for example, using stimuli from the International Affective Picture System (Johnston et al. 2010, 2011; Lang et al. 1999). Although in most studies participants trained self-regulation in the absence of external stimuli, it is in principle also possible to pair it with specific, for example, affective, stimuli (Veit et al. 2012). Such pairings may be particularly promising for future clinical applications in disorders with altered cue reactivity, for example, anxiety disorders (Shin and Liberzon 2010) or substance misuse (Linden 2012; Ihssen et al. 2011). Generally, upregulation has been easier to achieve than downregulation, which may take longer training protocols (Veit et al. 2012). Some preliminary effects of these protocols have been demonstrated on mood states (Johnston et al. 2011), anxiety (Scheinost et al. 2013) and appraisal of affective stimuli (Caria et al. 2010), which might pave the way for clinical applications in mood disorders.

2.3.4 "Cognitive" Areas

The learned self-regulation of brain areas involved in core cognitive processes would also be of considerable interest for clinical applications, for example, in schizophrenia or dementia. Self-regulation of several prefrontal areas has been demonstrated, including the inferior frontal gyrus (Rota et al. 2009) and rostrolateral prefrontal cortex (McCaig et al. 2011). Successful self-regulation may have effects both on task performance, for example, on language tasks (identification of prosody) (Rota et al. 2009), and on the remodeling of cognitive networks (Rota et al. 2011).

2.4 Clinical Applications of fMRI-NF

Clinical studies of EEG neurofeedback (EEG-NF) training have been conducted in epilepsy, attention deficit/ hyperactivity disorder (ADHD), depression and anxiety as well as other neurological and psychiatric disorders (Birbaumer and Cohen 2007; Hammond 2005). Although some of the initial results were promising, the only field where EEG-NF has entered clinical practice is ADHD (using a variety of target parameters, e.g., theta/beta ratio or slow cortical potentials). For ADHD, several published trials found positive effects on symptoms as measured by parent or teacher questionnaires or clinical assessments (Lofthouse et al. 2012). One limitation of most trials has been the lack of blinding of participants to treatment condition, and the general difficulties of setting up of double-blind controlled trials of NF training (Lansbergen et al. 2011). Although these difficulties will apply similarly in fMRI-NF, there are good reasons to explore its use in clinical conditions, including and beyond those that respond to EEG-NF. Because of the strong cognitive component that can be implemented in fMRI-NF protocols, it may achieve its neurophysiological targets faster than EEG-NF and thus enhance patient motivation and compliance. Furthermore, fMRI-NF can directly target specific brain areas or networks that have been implicated in neuropsychiatric disorders, either because they show a dysfunction that might cause the disorder or because they may compensate for a primary dysfunction (Linden 2013). In the first case, the aim of neurofeedback would be to restore the function to normal levels and in the second to promote the recruitment of the compensatory network. Because the plastic effects of neurofeedback training are not confined to the area or network whose activation is explicitly used as the target for training, both processes can very well happen in tandem, making it a versatile tool for redressing imbalances within and between brain networks.

Up to now the clinical potential of fMRI-NF to improve symptoms or change behavior has only been explored in a small number of published studies, all with small patient numbers. FMRI-NF, targeting the anterior cingulate cortex, has shown effects on chronic pain in patients with fibromyalgia (deCharms et al. 2005). Two out of six patients with chronic tinnitus improved after training downregulation of auditory cortex activity (Haller et al. 2010). In our study of Parkinson's disease (PD), we trained patients in the early stages of the disease to upregulate their supplementary motor area (SMA) (Subramanian et al. 2011). The choice of target area was based on pathophysiological models implicating underactivity of the SMA in PD (Nachev et al. 2008) and based on our observation that reliable upregulation can be achieved through motor imagery. The five patients who received fMRI-NF, but not five control patients who engaged in motor imagery without feedback, achieved reliable SMA upregulation and improved motor fluency and clinical symptoms (Subramanian et al. 2011). We are currently conducting a randomized controlled trial comparing fMRI-NF and feedback-guided physical exercise with a recruitment target of 30 patients in the early stages of PD (registered on clinicaltrials.gov: NCT01867827).

In the first psychiatric application, we trained patients with depression to upregulate brain networks responsive to positive affective stimuli. This paradigm was modeled on our previous work with healthy participants, which had shown that the neurofeedback component is required for reliable control over emotion networks (Johnston et al. 2011). The eight patients who underwent this fMRI-NF protocol for four sessions improved significantly on the 17-item Hamilton Rating Scale for Depression, and this clinical improvement was not observed in eight control patients who engaged in a protocol of positive emotional imagery (matched to the fMRI-NF protocol for intervention and assessment times and affective stimuli) outside the scanner (Habes et al. 2010). Our ongoing randomized controlled trial of neurofeedback for depression (clinicaltrials.gov: NCT01544205) pits upregulation of emotion networks against upregulation of a higher visual areas, which is a rather conservative active control condition because it also involves mental imagery and the rewarding experience of brain self-control.

The feasibility of fMRI-NF has also been demonstrated for schizophrenia (upregulation of the anterior insula) (Ruiz et al. 2013) and stroke (upregulation of the ventral premotor cortex) (Sitaram et al. 2012), but data on clinical improvement are not available. The next steps for the development of fMRI-NF into a potential clinical tool will be the conversion of feasible paradigms into clinical protocols and, where these protocols are already available, as for depression (Habes et al. 2010), PD (Subramanian et al. 2011), fibromyalgia (deCharms et al. 2005), and tinnitus (Haller et al. 2010), their rigorous testing in blinded randomized controlled trials. It is encouraging that the analysis of data from several hundred fMRI-NF runs with over 100 chronic pain patients did not show an increase in adverse events due to fMRI-NF (Hawkinson et al. 2012). Suggestions for the further development of clinical fMRI-NF were incorporated in a recent consensus paper that arose from the first international real-time fMRI workshop, which took place at the ETH Zurich in 2012 (Sulzer et al. 2013). The design and reporting of clinical trials of fMRI-NF should generally follow the guidelines that have been established for complex interventions in medicine. In addition to any clinical and behavioral effects, trials and pilot studies of fMRI-NF should also document the degree of self-regulation success (in the target area or network) and any activation changes outside the directly targeted areas (whole-brain analysis).

After demonstrating robust clinical effects of fMRI-NF that are superior to those of appropriately designed control interventions, a next step would be to assess means of ensuring the sustainability of clinical effects and the transfer from the MRI laboratory to everyday practice. It is likely that any routine clinical use of fMRI-NF will be embedded in existing treatment and rehabilitation techniques that utilize pharmacological, psychological, and physical therapies and assistive technology, as appropriate.

References

Bagarinao E, Matsuo K, Nakai T, Sato S (2003) Estimation of general linear model coefficients for real-time application. Neuroimage 19:422–429

Bagarinao E, Nakai T, Tanaka Y (2006) Real-time functional MRI: development and emerging applications. Magn Reson Med Sci 5:157–165

Bandura A (1997) Self-efficacy: the exercise of control. W.H. Freeman, New York

Berman BD, Horovitz SG, Venkataraman G, Hallett M (2012) Self-modulation of primary motor cortex activity with motor and motor imagery tasks using real-time fMRI-based neurofeedback. Neuroimage 59(2): 917–925

Birbaumer N, Cohen L (2007) Brain-computer interfaces: communication and restoration of movement in paralysis. J Physiol 579(Pt 3):621–636

Birbaumer N, Ruiz S, Sitaram R (2013) Learned regulation of brain metabolism. Trends Cogn Sci 17(6): 295–302

Caria A, Veit R, Sitaram R et al (2007) Regulation of anterior insular cortex activity using real-time fMRI. Neuroimage 35(3):1238–1246

Caria A, Sitaram R, Veit R, Begliomini C, Birbaumer N (2010) Volitional control of anterior insula activity modulates the response to aversive stimuli. A real-time functional magnetic resonance imaging study. Biol Psychiatry 68(5):425–432

Caria A, Sitaram R, Birbaumer N (2012) Real-time fMRI: a tool for local brain regulation. Neuroscientist 18(5): 487–501

Cox RW, Jesmanowicz A, Hyde JS (1995) Realtime functional magnetic resonance imaging. Magn Reson Med 33(2):230–236

deCharms RC (2007) Reading and controlling human brain activation using real-time functional magnetic resonance imaging. Trends Cogn Sci 11:473–481

deCharms RC (2008) Applications of real-time fMRI. Nat Rev Neurosci 9:720–729

deCharms R, Christoff K, Glover G, Pauly J, Whitfield S, Gabrieli J (2004) Learned regulation of spatially localized brain activation using real-time fMRI. Neuroimage 21(1):436–443

deCharms R, Maeda F, Glover G et al (2005) Control over brain activation and pain learned by using real-time functional MRI. Proc Natl Acad Sci U S A 102(51): 18626–18631

Esposito F, Seifritz E, Formisano E, Morrone R, Scarabino T, Tedeschi G, Cirillo S, Goebel R, Di Salle F (2003) Real-time independent component analysis of fMRI time-series. Neuroimage 20:2209–2224

Feinberg DA, Yacoub E (2012) The rapid development of high speed, resolution and precision in fMRI. Neuroimage 62(2):720–725

Fischl B, Sereno MI, Tootell RBH, Dale AM (1999) High-resolution inter-subject averaging and a coordinate system for the cortical surface. Hum Brain Mapp 8:272–284

Goebel R (2001) Cortex-based real-time fMRI. Neuroimage 13:S129

Goebel R, Sorger B, Kaiser J, Birbaumer N, Weiskopf N (2004) BOLD brain pong: self regulation of local brain activity during synchronously scanned, interacting subjects. Presented at: 34th Annual Meeting of the Society for Neuroscience, CA

Goebel R, Esposito F, Formisano E (2006) Analysis of functional image analysis contest (FIAC) data with brainvoyager QX: from single-subject to cortically aligned group general linear model analysis and self-organizing group independent component analysis. Hum Brain Mapp 27:392–401

Goebel R, Zilverstand A, Sorger B (2010) Real-time fMRI-based brain computer interfacing for neurofeedback therapy and compensation of lost motor functions. Imaging Med 2:407–415

Habes I, Johnston SJ, Tatineni R et al (2010) Functional magnetic resonance (fMRI)-based neurofeedback as a potential treatment method for depression. FENS, Amsterdam

Haller S, Birbaumer N, Veit R (2010) Real-time fMRI feedback training may improve chronic tinnitus. Eur Radiol 20(3):696–703

Hamilton JP, Glover GH, Hsu JJ, Johnson RF, Gotlib IH (2011) Modulation of subgenual anterior cingulate cortex activity with real-time neurofeedback. Hum Brain Mapp 32(1):22–31

Hammond D (2005) Neurofeedback with anxiety and affective disorders. Child Adolesc Psychiatr Clin N Am 14(1):105–123, vii

Hawkinson JE, Ross AJ, Parthasarathy S et al (2012) Quantification of adverse events associated with functional MRI scanning and with real-time fMRI-based training. Int J Behav Med 19(3):372–81

Hinds O, Ghosh S, Thompson TW, Yoo JJ, Whitfield-Gabrieli S, Triantafyllou C, Gabrieli JDE (2011) Computing moment-to-moment BOLD activation for real-time neurofeedback. Neuroimage 54:361–368

Horovitz SG, Berman BD, Hallett M (2010) Real time BOLD functional MRI neuro-feedback affects functional connectivity. Conf Proc IEEE Eng Med Biol Soc 2010:4270–4273

Ihssen N, Cox WM, Wiggett A, Fadardi JS, Linden DE (2011) Differentiating heavy from light drinkers by neural responses to visual alcohol cues and other motivational stimuli. Cereb Cortex 21(6):1408–1415

Johnston SJ, Boehm SG, Healy D, Goebel R, Linden DE (2010) Neurofeedback: a promising tool for the self-regulation of emotion networks. Neuroimage 49(1):1066–1072

Johnston S, Linden DE, Healy D, Goebel R, Habes I, Boehm SG (2011) Upregulation of emotion areas through neurofeedback with a focus on positive mood. Cogn Affect Behav Neurosci 11(1):44–51

LaConte SM (2011) Decoding fMRI brain states in real-time. Neuroimage 56:440–454

LaConte SM, Peltier SJ, Hu XP (2007) Real-time fMRI using brain-state classification. Hum Brain Mapp 28:1033–1044

Lang PJ, Bradley MM, Cuthbert BN (1999) International Affective Picture System (IAPS): technical manual and affective ratings. University of Florida, Center for Research in Psychophysiology, Gainesville

Lansbergen MM, van Dongen-Boomsma M, Buitelaar JK, Slaats-Willemse D (2011) ADHD and EEG-neurofeedback: a double-blind randomized placebo-controlled feasibility study. J Neural Transm 118(2): 275–284

Linden DE (2012a) The challenges and promise of neuroimaging in psychiatry. Neuron 73(1):8–22

Linden D (2012b) The biology of psychological disorders. Palgrave Macmillan, Basingstoke

Linden D (2013) Biological psychiatry: time for new paradigms. Br J Psychiatry 202:166–167

Lofthouse N, Arnold LE, Hersch S, Hurt E, Debeus R (2012) A review of neurofeedback treatment for pediatric ADHD. J Atten Disord 16(5):351–72

McCaig RG, Dixon M, Keramatian K, Liu I, Christoff K (2011) Improved modulation of rostrolateral prefrontal cortex using real-time fMRI training and meta-cognitive awareness. Neuroimage 55(3):1298–1305

Nachev P, Kennard C, Husain M (2008) Functional role of the supplementary and pre-supplementary motor areas. Nat Rev Neurosci 9(11):856–869

Posse S, Fitzgerald D, Gao K et al (2003) Real-time fMRI of temporolimbic regions detects amygdala activation during single-trial self-induced sadness. Neuroimage 18(3):760–768

Rota G, Sitaram R, Veit R et al (2009) Self-regulation of regional cortical activity using real-time fMRI: the right inferior frontal gyrus and linguistic processing. Hum Brain Mapp 30(5):1605–1614

Rota G, Handjaras G, Sitaram R, Birbaumer N, Dogil G (2011) Reorganization of functional and effective connectivity during real-time fMRI-BCI modulation of prosody processing. Brain Lang 117(3):123–132

Ruiz S, Lee S, Soekadar SR et al (2013) Acquired self-control of insula cortex modulates emotion recognition and brain network connectivity in schizophrenia. Hum Brain Mapp 34(1):200–12

Scharnowski F, Hutton C, Josephs O, Weiskopf N, Rees G (2012) Improving visual perception through neurofeedback. J Neurosci 32(49):17830–17841

Scheinost D, Stoica T, Saksa J et al (2013) Orbitofrontal cortex neurofeedback produces lasting changes in contamination anxiety and resting-state connectivity. Transl Psychiatry 3:e250

Seitz RJ (2010) How imaging will guide rehabilitation. Curr Opin Neurol 23(1):79–86

Shibata K, Watanabe T, Sasaki Y, Kawato M (2011) Perceptual learning incepted by decoded fMRI neurofeedback without stimulus presentation. Science 334(6061):1413–1415

Shin L, Liberzon I (2010) The neurocircuitry of fear, stress, and anxiety disorders. Neuropsychopharmacology 35(1):169–191

Sitaram R, Caria A, Veit R, Gaber T, Kuebler A, Birbaumer N (2005) Real-time fMRI based brain–computer interface enhanced by interactive virtual worlds. In:

Proceedings of the 45th annual meeting Society for Psychophysiological Research, Lisbon

Sitaram R, Lee S, Ruiz S, Rana M, Veit R, Birbaumer N (2011) Real-time support vector classification and feedback of multiple emotional brain states. Neuroimage 56(2):753–765

Sitaram R, Veit R, Stevens B et al (2012) Acquired control of ventral premotor cortex activity by feedback training: an exploratory real-time FMRI and TMS study. Neurorehabil Neural Repair 26(3):256–265

Smyser C, Grabowski TJ, Frank RJ, Haller JW, Bolinger L (2001) Real-time multiple linear regression for fMRI supported by time-aware acquisition and processing. Magn Reson Med 45:289–298

Sorger B, Peters J, van den Boomen C, Zilverstand A, Reithler R, Goebel R (2010) Real-time decoding the locus of visuospatial attention using multi-voxel pattern classification. In: Human brain mapping conference, Barcelona, 16th edn

Subramanian L, Hindle JV, Johnston S et al (2011) Real-time functional magnetic resonance imaging neurofeedback for treatment of Parkinson's disease. J Neurosci 31(45):16309–16317

Sulzer J, Haller S, Scharnowski F, Weiskopf N, Birbaumer N, Blefari ML, Bruehl AB, Cohen LG, deCharms RC, Gassert R, Goebel R, Herwig U, LaConte S, Linden D, Luft A, Seifritz E, Sitaram R (2013) Real-time fMRI neurofeedback: progress and challenges. Neuroimage 76:386–399

Veit R, Singh V, Sitaram R, Caria A, Rauss K, Birbaumer N (2012) Using real-time fMRI to learn voluntary regulation of the anterior insula in the presence of threat-related stimuli. Soc Cogn Affect Neurosci 7(6): 623–34

Weiskopf N (2012) Real-time fMRI and its application to neurofeedback. Neuroimage 62:682–692

Weiskopf N, Veit R, Erb M et al (2003) Physiological self-regulation of regional brain activity using real-time functional magnetic resonance imaging (fMRI): methodology and exemplary data. Neuroimage 19(3): 577–586

Weiskopf N, Scharnowski F, Veit R, Goebel R, Birbaumer N, Mathiak K (2004a) Self-regulation of local brain activity using real-time functional magnetic resonance imaging (fMRI). J Physiol Paris 98: 357–373

Weiskopf N, Mathiak K, Bock S et al (2004b) Principles of a brain-computer interface (BCI) based on real-time functional magnetic resonance imaging (fMRI). IEEE Trans Biomed Eng 51(6):966–970

Yoo SS, Jolesz FA (2002) Functional MRI for neurofeedback: feasibility study on a hand motor task. Neuroreport 13(11):1377–1381

Yoo SS, O'Leary HM, Fairneny T et al (2006) Increasing cortical activity in auditory areas through neurofeedback functional magnetic resonance imaging. Neuroreport 17(12):1273–1278

Yoo SS, Lee JH, O'Leary H, Panych LP, Jolesz FA (2008) Neurofeedback fMRI-mediated learning and consolidation of regional brain activation during motor imagery. Int J Imaging Syst Technol 18(1):69–78

Zilverstand A, Sorger B, Zimmermann J, Kaas A, Goebel R (2014) Instantaneous correlation: a suitable tool for fMRI-based functional connectivity neurofeedback? PLoS One 9(1):e85929

Zotev V, Krueger F, Phillips R et al (2011) Self-regulation of amygdala activation using real-time FMRI neurofeedback. PLoS One 6(9):e24522

3 Modelling Effective Connectivity with Dynamic Causal Models

Yen Yu, William Penny, and Karl Friston

Abbreviations

ANOVA	Analysis of variance
BMA	Bayesian model averaging
BMS	Bayesian model selection
BOLD	Blood oxygenation level dependent
BPA	Bayesian parameter averaging
DCM	Dynamic causal modelling
FFX	Fixed effect analysis
GBF	Group Bayes factor
MAP	Maximum a posteriori
OMPFC	Orbitomedial prefrontal cortex
RFX	Random effect analysis
VL	Variational Laplace

3.1 Introduction

This chapter is about dynamic causal modelling (DCM) of the interactions between functionally elicited brain responses and its applications in the field of psychiatric neuroimaging. DCM was invented to test hypotheses about neural systems – as opposed to regionally specific correlates – and necessitates a predefined set of plausible structural models. In other words, each DCM embodies a specific hypothesis pertaining to how a neural system interacts and produces observed responses. To allow a hypothetical structure or network model to predict observed responses, it is crucial to understand and model how changes in neuronal states are manifested as observed haemodynamic responses.

In the first section of this chapter, we focus on the conceptual and operational constructs of DCM as a physiologically realistic forward model – exemplified by an up-to-date implementation of DCM. In the second section, we turn to Bayesian model selection of DCMs, where models can be considered as fixed effects (e.g. as low-level neurophysiological mechanisms that are concerned over subjects) or random effects (e.g. as high-level cognitive processes that are implemented with different strategies or networks) in the population. The third section considers inference about the parameters of a model, for example, how a connection from one region to another is changed by experimental context. We describe how such inferences can be made for the case of single models and for models derived from averaging over different models or subjects in a group. Finally, in the last section, we focus on recent DCM applications in psychiatric neuroimaging. Clinical symptoms of psychiatric disorders are thought to be the manifestation of disruptions to distributed neuronal processing. In order to accurately characterise psychiatric deficits in a refined way – rather than relying on conventional descriptive criteria – DCM is considered essential and complementary to classical analyses of regional responses. DCM allows inferences about group differences to be made at the level of interactions among neural networks.

Y. Yu (✉) • W. Penny • K. Friston, FRS
Wellcome Trust Centre for Neuroimaging, University College London, London, UK
e-mail: yen.yu.10@ucl.ac.uk; w.penny@ucl.ac.uk

C. Mulert, M.E. Shenton (eds.), *MRI in Psychiatry*,
DOI 10.1007/978-3-642-54542-9_3, © Springer Berlin Heidelberg 2014

A series of recent publications will be reviewed to discuss the insights that may be provided by future psychiatric neuroimaging studies.

3.2 Dynamic Causal Modelling for fMRI

This section presents the essential operational aspects of dynamic causal modelling (DCM). The theoretical basis of DCM rests on dynamic systems theory and Bayesian statistics. The primary objective of DCM appeals to a nonlinear system identification in which a set of differential equations are formulated to capture the (hidden) mechanistic structure of a neuronal system of interest. These equations specified how constituent nodes (or neuronal 'states') of a system exhibit time-varying and causal relations with one another. Specifically, this system is acted upon by exogenous *inputs* (e.g. visual stimuli) that engender regional neuronal activity that, in turn, generates outputs (e.g. BOLD signals). This necessarily requires DCM to be hierarchical – where a two-layered forward model translates neuronal states into haemodynamic states and the measured BOLD responses. The haemodynamic states are modelled in a regionally independent fashion. Neuronal dynamics emerge from designed experimental perturbations and directed interactions among regions. Specification of effective connectivity within the network of coupled nodes or regions depends on three sets of (neuronal) parameters: (a) parameters that mediate endogenous coupling among the states, (b) parameters that allow exogenous inputs to modulate the coupling, and (c) parameters that mediate the influences of exogenous inputs on the states. These parameters are embedded in a dynamic causal model that is motivated by a particular hypothesis about network structure, and can be estimated by fitting the ensuing forward model to observed data, using standard Bayesian procedures. This model inversion procedure provides posterior estimates of the parameters and an estimate of the model evidence. Critically, prior densities over parameters constrain parameter estimates to dynamic or physiologically realistic ranges. By default, 'shrinkage' priors (see section 3.2.5) are chosen for endogenous and modulatory parameters, while priors on haemodynamic parameters are derived from previous empirical studies.

In what follows, we first review the neuronal state equations, haemodynamic state equations and the priors over model parameters. We then briefly consider the standard Bayesian scheme used for model inversion. We will briefly review nonlinear DCMs, where one region can modulate the connectivity between another pair. Some of the more recent DCM developments are considered in the Conclusion.

3.2.1 Notation

Variables in bold face refer to matrices and vectors. States are functions of time, although the dependency on time t is not made explicit. The vector $\mathbf{z}=[z_1, z_2, \ldots]^{\mathrm{T}}$ denotes any number of neuronal states of interest. A neuronal state, say z_1, can be taken as the collective dynamics of neuronal activity in the first region. The remaining state variables are biophysical states $\{\mathbf{s}, \mathbf{f}, \mathbf{v}, \mathbf{q}\}$ that model haemodynamics. These haemodynamic states refer to (1) vasodilatory signal, (2) blood inflow, (3) blood volume and (4) deoxyhaemoglobin content, respectively. The vector $\mathbf{x}=\{\mathbf{z}, \mathbf{s}, \mathbf{f}, \mathbf{v}, \mathbf{q}\}$ denotes all the hidden (neuronal and haemodynamic) states collectively. The vector $\mathbf{u}=[u_1, u_2, \ldots]^{\mathrm{T}}$ denotes any number of exogenous inputs that are specified experimentally. Elements of $\mathbf{u}$ can be, for example, spike or boxcar functions of time that represent the onset/offset of task stimuli or contextual manipulations. Alternatively, exogenous inputs can also be motivated by a neurocomputational or model-based approach (O'Doherty et al. 2007). $\boldsymbol{\theta}$ denotes the collection of model parameters, including coupling parameters and haemodynamic parameters. Different model structures are indexed by m, that is, differences may exist in endogenous, modulatory or exogenous connections.

3.2.2 Neurodynamics

Assuming any number of neuronal states $\mathbf{z}$ and any number of exogenous inputs, one can posit a model of the general form

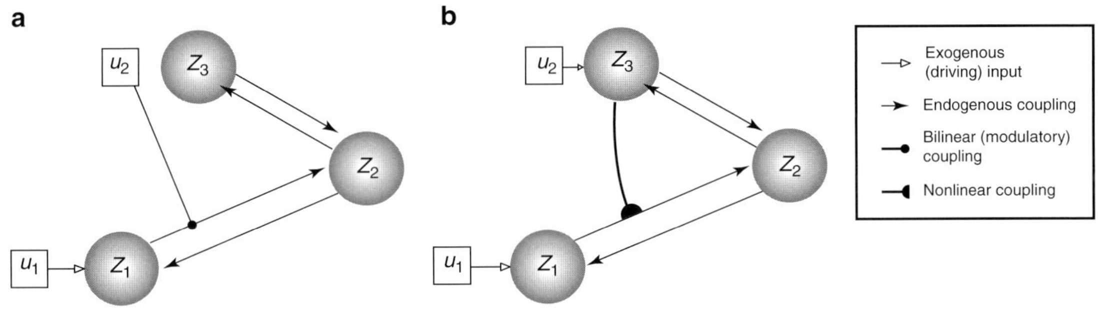

$$B^{(u_2)} = \begin{bmatrix} 0 & 0 & 0 \\ b_{21}^{(u_2)} & 0 & 0 \\ 0 & 0 & 0 \end{bmatrix} \qquad D^{(z_3)} \begin{bmatrix} 0 & 0 & 0 \\ d_{21}^{(z_3)} & 0 & 0 \\ 0 & 0 & 0 \end{bmatrix}$$

$$\frac{dz_2}{dt} = a_{21}z_1 + a_{22}z_2 + a_{23}z_3 + b_{21}^{(u_2)}u_2z_1 \qquad \frac{dz_2}{dt} = a_{21}z_1 + a_{22}z_2 + a_{23}z_3 + d_{21}^{(z_3)}z_1z_3$$

Fig. 3.1 This diagram illustrates two types of 'modulatory' effects – the bilinear (**a**) and the nonlinear (**b**) modulations. The target of modulation is made over the z_1-to-z_2 coupling. The difference between the neuronal state equations for z_2 is made explicit in the respective panels (see the last term). Specifically, the bilinear model allows an exogenous experimental manipulation (u_2) to induce connectivity change. The nonlinear model, on the other hand, manifests the neuronal origin (z_3, as now encodes u_2) of the modulatory effect

$$\frac{\mathrm{d}z}{\mathrm{d}t} = \mathrm{F}(\mathbf{z},\mathbf{u},\boldsymbol{\theta}) \tag{3.1}$$

where F is some nonlinear function describing the neurophysiological influences exerted by inputs **u** and the activity in all brain regions on the evolution of the neuronal states. A bilinear approximation provides a natural and useful re-parameterisation in terms of coupling parameters.

$$\begin{aligned} \frac{\mathrm{d}z}{\mathrm{d}t} &\approx \mathbf{Az} + \sum_j u_j \mathbf{B}^{(j)}\mathbf{z} + \mathbf{Cu} \\ &= \left(\mathbf{A} + \sum_j u_j \mathbf{B}^{(j)}\right)\mathbf{z} + \mathbf{Cu} \end{aligned} \tag{3.2}$$

The (effective) connectivity matrix **A** represents the first-order coupling among the regions in the absence of inputs. This can be thought of as the endogenous coupling in the absence of experimental perturbations. Note that the state, which is perturbed, depends on the experimental design (e.g. baseline or control state), and therefore, the endogenous coupling is specific to each experiment. The matrices $\mathbf{B}^{(j)}$ are the change in endogenous coupling induced by the jth input (Fig. 3.1 Finally, the matrix **C** encodes the exogenous (driving) influences of inputs on neuronal activity. The parameters $\boldsymbol{\theta}^c = \{\mathbf{A}, \mathbf{B}^{(j)}, \mathbf{C}\}$ are the coupling parameter matrices we wish to estimate and define the functional architecture and interactions among brain regions at a neuronal level. Note that the units of coupling are per unit time (Hz) and therefore correspond to rates. Because we are in a dynamic setting, a strong connection means an influence that is expressed quickly or with a small time constant. It is useful to appreciate this when interpreting estimates and thresholds quantitatively.

The neuronal activity in each region causes changes in volume and deoxyhaemoglobin which engender the observed BOLD response **y** as described below. The ensuing haemodynamic component of the model is specific to BOLD-fMRI and would be replaced by appropriate forward models for other imaging modalities, such as EEG or MEG.

The neuronal dynamics in Eq. (3.2) operate around a stable fixed point $z=0$ (strictly speaking, this will only be the case for certain ranges of parameter values – see Friston et al. 2003). This means that, in the absence of exogenous perturbations, the neuronal activity and consequently the fMRI activity will be zero. Briefly, a neuronal state in DCM predicts nothing but a flat line if it is not experimentally perturbed, directly or indirectly (but see Li et al. 2011a). This is because DCM for fMRI is based on a dynamic system with a fixed point attractor.

3.2.3 Nonlinear DCM

Nonlinear DCM (Stephan et al. 2008) introduces a parametric matrix **D** that allows neuronal activity in one region to change the connectivity between other regions (Fig. 3.1 This is in contrast to bilinear dynamics (Eq. 3.2) in which, perhaps unrealistically, effective connectivity can be changed via 'modulatory inputs'.

$$\frac{dz}{dt} = \left(\mathbf{A} + \sum_j u_j \mathbf{B}^{(j)} + \sum_i z_i \mathbf{D}^{(i)} \right) \mathbf{z} + \mathbf{Cu} \tag{3.3}$$

The motivation for this extension is to address 'neuronal gain control' between two neuronal states that are gated by other states (Stephan et al. 2008). The approach also models the neuronal origin of modulatory influences such as 'short-term synaptic plasticity' (Stephan et al. 2008). Applications based on nonlinear DCM can be found in den Ouden et al. (2010) and Table 3.1 (Desseilles et al. 2011; Neufang et al. 2011).

3.2.4 Haemodynamics

Neuronal activity is linked to fMRI signals via an extended Balloon model (Buxton et al. 1998, 2004) and BOLD signal model (Stephan et al. 2007b). The haemodynamic model specifies how changes in neuronal activity give rise to changes in blood oxygenation that is measured with fMRI. For each region i, neuronal activities are translated into BOLD signals via the interactions between the neuronal state z_i and haemodynamic state variables: the vasodilatory signal, the flow induced, changes in volume, and changes in deoxyhaemoglobin. The observed BOLD signals are produced by a nonlinear model that integrates over the states, $\mathbf{y} = g(\mathbf{v}, \mathbf{q})$, where the evolution of **v** and **q** over time depends on self-regulatory feedback as well as s and **f** (cf. to Fig. 3 in Friston et al. 2003). The equations for the haemodynamics are described in detail elsewhere (Grubb et al. 1974; Buxton et al. 1998; Mandeville et al. 1999; Friston et al. 2000).

3.2.5 Priors

Two classes of prior densities are used in DCM; they are placed over coupling and haemodynamic parameters $\boldsymbol{\theta} = \{\mathbf{A}, \mathbf{B}, \mathbf{C}, \boldsymbol{\theta}^{h}\}$. DCM uses 'shrinkage priors' over coupling parameters. These are zero-mean Gaussian priors with a variance that is chosen to reflect realistic ranges in effective connectivity seen in fMRI studies. These shrinkage priors move the posterior estimates towards zero, especially when the likelihood has a less precise distribution. For example, the posterior expectation will 'shrink' to its prior expectation given a likelihood with a very large variance. However, a likelihood that has high precision (inverse variance) enforces the posterior to deviate from zero. Prior variances can also be chosen to reflect anatomical knowledge, for example, probabilistic tractography (Stephan et al. 2009). Haemodynamic priors in DCM reflect empirical knowledge about blood flow and oxygenation dynamics in the brain (Buxton et al. 1998, 2004). The prior densities of the five haemodynamic parameters $\boldsymbol{\theta}^{h} = \{\kappa, \gamma, \tau, \alpha, \rho\}$ that mediate the interactions among these states are set based on empirical measures (see Eq. (3.3) and Table 3.1 in Friston et al. 2003). These priors have since been updated in light of more recent data (Penny 2012).

3.2.6 Model Fitting

DCMs are fitted to data using the Variational Laplace (VL) algorithm described in (Friston et al. 2007). This is an iterative algorithm which approximates the posterior distribution over parameters with a Gaussian distribution. The parameters of this distribution are updated so as to minimise the distance between the approximate and true posterior, in the sense of Kullback-Leibler divergence – a distance measure between probability densities (MacKay 2003). The VL

Table 3.1 List of DCM applications in psychiatric neuroimaging studies (shown in publication year-alphabetical order)

Study	Patient condition	Activation paradigm	Number of participants	N	M	Inference type	Findings
Almeida et al. (2009)	Bipolar disorder, major depressive disorder	Emotional regulation (labelling happy vs. sad faces)	15 BD, 16 MDD, 16 controls	2/BL	1	Parameter (classical)	Hemispheric and directional asymmetry in OMPFC-amygdala connectivity differentiates between BD and MDD abnormality
Schlösser et al. (2010)	Obsessive-compulsive disorder	Stroop task (colour-word congruency)	21 patients, 21 controls	6/BL	5	Model (RFX-BMS), parameter (classical)	OCD predominant fronto-cingulate connectivity enhancement subject to incongruent Stroop modulation
Almeida et al. (2011)	Major depressive disorder	Automatic attentional control of emotion	19 patients, 19 controls	3/BL	1	Parameter (classical)	Characteristic cortico-amygdala connectivity alterations to happy and fearful stimuli in female MDD
Bányai et al. (2011)	Schizophrenia	Associative learning (object-location)	11 patients, 11 controls	5/BL	24	Model (RFX-BMS), parameter (classical, BPA)	Information processing network in schizophrenia is different from that of controls. No specific impaired pathway can be identified
							Reduced fronto-hippocampal and hippocampo-IT connectivity in patients
Desseilles et al. (2011)	Major depressive disorder	Visual detection task	14 patients, 14 controls	3/NL	10 (2 families)	Family, model (RFX-BMS), parameter (BPA)	Control: high attentional load was associated with negative IPS-V4 modulation
							Patient: attention levels exerted modulations on IPS
van Leeuwen et al. (2011)	Synaesthesia	Colour-controlled grapheme viewing	19 synaesthetes ('projectors' vs. 'associators')	3/BL	2	Model (RFX-BMS), parameter (BPA)	Synaesthetic V4 cross-activation induced bottom-up and top-down pathways for projector and associator synaesthetes, respectively
Li et al. (2011b)	None included	TMS-triggered vs. visually cued thumb movement	0 patient, 25 (M1)/ 21 (PFC) controls	4(M1), 3(PFC)/BL	4 (M1), 3 (PFC)	Model (FFX-BMS), parameter (average)	LTG and VPA reduced motor circuitry connectivity, whereas VPA enhanced prefrontal circuitry connectivity
Neufang et al. (2011)	Prodromal Alzheimer's disease	Attentional network task	15 patients, 16 controls	4/NL	4	Model (RFX-BMS), parameter (classical)	Reduced frontoparietal connectivity corresponded to regional grey matter volume and attentional control

(continued)

Table 3.1 (continued)

Study	Patient condition	Activation paradigm	Number of participants	*N*	*M*	Inference type	Findings
Deserno et al. (2012)	Schizophrenia	Numeric n-back (2-back vs. 0-back)	41 patients, 42 controls	3/BL	48 (3 families)	Family, model (RFX-BMS), parameter (BMA, classical)	Identification of disrupted WM-dependent corticocortical dynamics contributing DLPFC inefficiency and cognitive deficits in schizophrenia
Diwadkar et al. (2012)	None included	Emotional processing	19 schizophrenic patient offspring, 24 controls	5/BL	136	Model (RFX-/FFX-BMS), parameter (classical)	High-risk schizophrenic group exhibited declined visual driving, declined frontolimbic intrinsic coupling and increased inhibitory frontolimbic modulation by negative valence
Passamonti et al. (2012)	None included	Emotional regulation with placebo-controlled ATD	0 patient, 19 controls	3/BL	49 (7 families, 3 'meta-families')	Family	ATD procedure reduced emotional processing within PFC-amygdala pathways

Keys: *N* neurodynamic model specification (nodes/linearity), *M* number of model in model space

Abbreviations: *ATD* acute tryptophan depletion, *BMA* Bayesian model averaging, *BPA* Bayesian parameter averaging, *BL* bilinear model, *DLPFC* dorsolateral prefrontal cortex, *FFX* fixed effects, *GBF* group Bayes factor, *IPS* intraparietal sulcus, *LTG* Lamotrigine, *M1* primary motor cortex, *NL* nonlinear model, *OCD* obsessive-compulsive disorder, *OMPFC* orbitomedial prefrontal cortex, *PFC* prefrontal cortex, *RFX* random effects, *TMS* transcranial magnetic stimulation, *V1* primary visual area, *V4* colour responsive visual area, *VPA* valproic acid, *WM* working memory algorithm provides estimates of two quantities. The first is the posterior density over model parameters $p(\boldsymbol{\theta}|m,\mathbf{y})$ that can be used to make inferences about model parameters $\boldsymbol{\theta}$. The second is the probability of the data given the model, otherwise known as the model evidence $p(\mathbf{y}|m)$.

3.2.7 Parameter Inference for Single Models

Inferences about DCM parameters – based on single models – require strong assumptions about network structure based on well-established neurophysiological, cognitive or anatomical evidence. For instance, Almeida and colleagues demonstrated that the uncinate fasciculus, which provides orbitofronto-amygdala circuitry, mediates emotional regulation and is abnormal in bipolar disorder patients, in terms of fractional anisotropy (Almeida et al. 2009). Both single-model studies reviewed here (Almeida et al. 2009, 2011) used a fully connected model to allow all possible coupling parameters to be estimated. They found that left-sided, top-down orbitomedial prefrontal cortex (OMPFC) to amygdala and right-sided bottom-up amygdala to OMPFC distinguished bipolar from major depression.

3.2.8 Model Evidence

In general, model evidence is not straightforward to compute, since this computation involves integrating out the dependence on model parameters:

$$p(\mathbf{y}\mid m)=\int p(\mathbf{y}\mid\boldsymbol{\theta},m)\,p(\boldsymbol{\theta}\mid m)\,\mathrm{d}\boldsymbol{\theta} \quad (3.4)$$

Therefore, an approximation to the model evidence is required. DCM uses the free-energy approximation to the model evidence provided by the VL algorithm. The model evidence, and the VL approximation to it, naturally embodies the accuracy-complexity trade-off that is the hallmark of a good model (Pitt and Myung 2002). The VL algorithm uses a 'free energy' approximation to the model evidence which has been shown to be superior to other information theoretical criteria (Penny 2012). By comparing the evidence of one model relative to another, a decision can be made as to which is the more veridical (Penny et al. 2003; Friston et al. 2008).

3.3 Model Inference

The model inference problem arises naturally in nearly every scientific discipline (Anderson 2008). Most importantly, it requires a well thought out specification of the model space – that is, the set of hypotheses that are to be considered. In the simplest case, one will have a null model and an alternative model and inference can proceed using Bayes factors. Once the evidence has been computed, a model (m_i) can be compared to another (m_j) by means of the Bayes factor (Raftery 1995):

$$\mathrm{BF}_{ij}=\frac{p(\mathbf{y}\mid m_i)}{p(\mathbf{y}\mid m_j)} \quad (3.5)$$

A Bayes factor of 20 (or log Bayes Factor of 3) corresponds to a posterior model probability of 0.95 and is used as the standard decision threshold (Penny et al. 2004).

More generally, one might be able to constrain the space of models to a small number. Model inference can then proceed using the posterior distribution over models, which can be obtained from Bayes rule for models:

$$p(m\mid\mathbf{y})=\frac{p(\mathbf{y}\mid m)\,p(m)}{p(\mathbf{y})} \quad (3.6)$$

The prior distribution over models, $p(m)$, is usually chosen to be a uniform distribution. In larger model spaces, it becomes increasingly unlikely that high posterior probability mass will be attributed to any single model. This is because there are likely to be many similar models in large model spaces – and they will share probability mass. This is known as *dilution* and can be dealt with by combining models into families (Penny et al. 2010). Models in the same family share the same characteristics, for example, nonlinearity, the same driving region or the same modulatory connection.

3.3.1 Group Inference

Next, we turn to the topic of model inference for data from a group of subjects. There are two approaches. Fixed effect analysis (FFX) (Stephan et al. 2007a) assumes that all subjects use the same model, whereas random effect (RFX) analysis assumes different subjects use different models (Stephan et al. 2009).

3.3.1.1 Fixed Effect Analysis

In FFX analysis, a group Bayes factor (GBF) is computed by multiplying together the Bayes factors from the group of subjects. As is considered in Stephan et al. (2009), the GBF approach implicitly assumes that every subject uses the same model (Fig. 3.2 This assumption is warranted when studying a basic physiological mechanism that is unlikely to vary across subjects, such as the role of forward and backward connections in visual processing (Chen et al. 2009). Li et al. (2011b), for example, studied the motor network by perturbing it with transcranial magnetic stimulation. With clearly defined timing in eliciting network responses and the homogeneity of motor circuitry over subjects, GBF was entirely suitable. In other words, inferences relying on GBF will – by default – neglect group heterogeneity, whereas functional tasks engaging higher cognitive processes may show group heterogeneity due to individual differences in cognitive strategies. Moreover, GBF is susceptible to outliers – towards which the inference may be heavily biased.

3.3.1.2 Random Effect Analysis

An alternative procedure for group level model inference allows for the possibility that different subjects use different models (Fig. 3.2 This is more realistic when investigating pathophysiological mechanisms in a spectrum disorder or when dealing with cognitive tasks that can be performed with different strategies. In random effect analyses, one makes inferences based on the posterior estimates of the model frequencies. For the kth model, r_k denotes the frequency with which it is used in the population. Inferences are therefore based on the posterior density

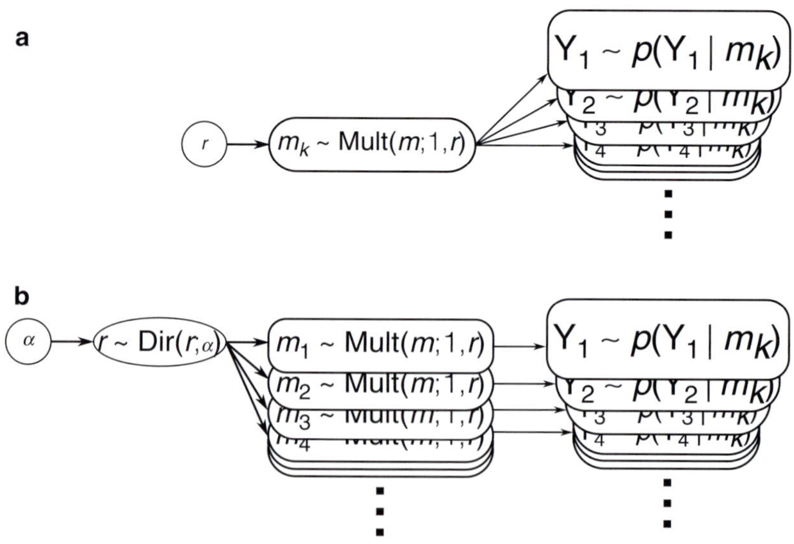

Fig. 3.2 Generative models of multi-subject data. The fixed effect model (**a**) assumes that subject-specific data are generated by a particular model. The random-effect model (**b**) suggests that the data-generating models (e.g. m_1–m_4) are treated as random variables. As a consequence, the random-effect model allows different structures of causal model across subjects (α = parameters of the Dirichlet distribution, or model 'occurrence' in the population level; r= parameters of the multinomial distribution, or the model)

$p(r_k|y)$. This can be computed by combining the table of model evidences with an uninformative prior, $p(r_k)$, using a Bayesian inversion scheme. Such an inversion can be implemented using a variational approach (Stephan et al. 2009) or Gibbs sampling (Penny et al. 2010). One should note that the variational approach is only valid for small numbers of models (small in relation to the number of subjects, e.g. 10 or so models for 20 or so subjects) and that Gibbs sampling is now the standard approach. Both algorithms produce approximations to the posterior density on which subsequent RFX model comparisons are based. One can report the result of RFX model comparison using (1) the posterior expected probability of observing the kth model or (2) the *exceedance probability* which reflects the belief that one model is more likely than any other in the model space.

Passamonti and colleagues provide an example of this approach: they investigated the neural mechanisms of emotion regulation and assumed the underlying cognitive processes would vary across the group (Passamonti et al. 2012). Thus, their adoption of random-effect BMS procedure was appropriate. Another example of model-level inference relates to different neural mechanisms giving rise to distinct synaesthetic experiences that can be explained in terms of alterations in the visual processing hierarchy (van Leeuwen et al. 2011). Generally, RFX is more conservative and is robust to outlying subjects.

3.3.1.3 Family Inference

Family inference can proceed using either an FFX or RFX approach. Passamonti et al. (2012) employed a tripartite model space ('meta-family') where numbers of driving exogenous inputs varied across each subspace. The variation of the location of driving inputs further constituted respective families within each meta-family. The authors performed BMS across all the meta-families, regardless of any other difference among the models considered – to establish how many inputs were needed. Models within the winning meta-family were further compared to determine the location of driving inputs (cf. Figure S1 in Passamonti et al. 2012). To summarise, model-level or family-level inference is appropriate when the hypothesis of interest can be answered in terms of overall model structure (i.e. the existence of multiple sets of parameters) rather than any specific model parameter.

3.4 Parameter Inference

Finally, we address inferences made on the basis of connectivity parameters in the context of a group analysis. Assessing the statistical significance of posterior estimates of individual model parameters is usually the last step in a DCM application (Almeida et al. 2009, 2011; Schlosser et al. 2010; Banyai et al. 2011; Li et al. 2011b; Neufang et al. 2011; van Leeuwen et al. 2011; Deserno et al. 2012; Passamonti et al. 2012).

If random effects on parameters are assumed in the population, a classical approach can be applied (e.g. t-test or ANOVA). Conceptually, this conforms to the classical (frequentist) *summary statistics* approach – using subject-specific MAP (maximum a posteriori) point estimates of the coupling parameters. This application is used widely (Schlosser et al. 2010; Banyai et al. 2011; Neufang et al. 2011; Deserno et al. 2012; Diwadkar et al. 2012) for identifying significant effects between patients and controls.

The summary statistical RFX approach is readily applied to the MAP parameter estimates for selected parameters from each subject. However, if one has multiple models per subject, then one also needs to average over models (accounting for the possibility that different subjects use different models). This can be implemented using Bayesian model averaging (see below) within subjects (over models). The resulting parameter estimates can then enter as summary statistics into a classical RFX analysis (e.g. t-test or ANOVA).

If fixed effects of parameters are assumed in the population, then one can compute a 'group' model by averaging over the models from subjects in that group. This is a FFX approach and can be implemented using Bayesian parameter averaging, as described in the next section and in Kasess et al. (2010).

3.4.1 Bayesian Parameter Averaging

Bayesian parameter averaging (BPA) has multiple uses. Generally, it is a procedure to combine parameter estimates from the same model of multiple datasets to produce parameter estimates from the entire dataset. The data could come from DCMs fitted to different sessions from the same subject. Or, most often, they could be the same model structure fitted to data from multiple subjects (Kasess et al. 2010). For example, van Leeuwen et al. (2011) summarised model parameters using BPA and found that V4 activation in synaesthetes was dependent on top-down – rather than bottom-up – inputs as a function of whether they were a 'projector' or an 'associator'. The common feature of all these applications is that variability over the model fitting is not taken into account. That is, the averaging procedure corresponds to a FFX analysis (because only one model is used). Mathematically, the posterior means from each model to be combined are weighted by their relative posterior precisions; this means that estimates with higher precision are given greater weight.

Low-level neurophysiological processing can be considered as fixed effects since they are unlikely to vary across populations, for example, Desseilles et al. (2011) and van Leeuwen et al. (2011) both interrogated selective colour vision processing mechanisms. If this is the case, Bayesian parameter averaging (BPA) can be used to summarise individual posterior densities of an identical optimal model across the entire group (Banyai et al. 2011; Desseilles et al. 2011; van Leeuwen et al. 2011).

3.4.2 Bayesian Model Averaging

Another approach to summarise parameters as random effects is through Bayesian model averaging (BMA). In this sort of averaging, there are multiple models of the same data (as opposed to a single model of multiple datasets). BMA is usually performed within a model family, where no model within the family clearly outperforms all others. It can also be applied to the whole model space. As such, parameter inference no longer depends on a particular model selection. For instance, Deserno et al. (2012) employed BMA in a DCM study of working memory in schizophrenia. They first performed a family-level inference and found that the family of models with modulation of backward connections from prefrontal to parietal cortex was the clear winner. They then performed BMA for each subject, entered the averaged parameters as summary statistics into a two-sample *t*-test and found reduced connectivity in the schizophrenic group (cf. Figures 3 and 4, Table 2 in Deserno et al. (2012)).

The relationships among the various model and parameter inference procedures perhaps seem complicated on a first reading but are clearly laid out in, for example, Fig. 3.1 of Stephan et al. (2010). Once one appreciates the simplicity of pooling evidence for different models and parameters, Bayesian model and parameter averaging can be a powerful approach to testing specific mechanistic hypotheses – particularly in psychiatric research.

3.5 DCM in Psychiatry

Modern neuropsychiatric studies aim to characterise clinical syndromes in terms of their 'biomarkers' (Gillihan and Parens 2011; Masdeu 2011; Linden 2012). This is because relying only on clinical descriptive criteria is sometimes inaccurate and may falsely categorise a patient (Linden and Thome 2011; Linden 2012). With the advent of functional neuroimaging techniques, regional responses elicited by neuropsychological tasks have provided a better characterisation of psychiatric disorders. However, regional activation only partially reflects cognitive processing. Recent developments in connectivity analysis are an important complement, which may provide biologically grounded descriptions of neuroimaging phenotypes (Rowe 2010; Linden 2012). In this section, we summarise some of the most recent psychiatric neuroimaging studies that use DCM.

Since the first DCM application in cognitive science by Mechelli et al. (2003), DCM-related studies have proliferated, with a PubMed search producing about 200 articles, of which, 30 specifically address findings related to psychiatric disorders or neuropsychiatric treatments. Most of these articles were published after 2010. We review some of these studies – focussing on whether differences between psychiatric and control groups were made based on parameter-, model- or family-level inference. We have examined four types of study, as characterised by their methodological approach:

(1) Parameter inference: single model, parameter difference between groups (patients vs. control or patients vs. patients) (Almeida et al. 2009, 2011)
(2) Parameter inference: a winning model from competing models, parameter difference between groups (Schlosser et al. 2010; Banyai et al. 2011; Desseilles et al. 2011; Li et al. 2011b; Neufang et al. 2011; Diwadkar et al. 2012)
(3) Parameter inference: a winning family from competing model families, parameter difference between groups in terms of within-family model averaging (Deserno et al. 2012)
(4) Model structure inference: distinctive winning models/model families across groups (van Leeuwen et al. 2011; Passamonti et al. 2012)

We have accordingly extracted 11 relevant studies as summarised in Table 3.1 (Almeida et al. 2009, 2011; Schlosser et al. 2010; Banyai et al. 2011; Desseilles et al. 2011; Li et al. 2011b; Neufang et al. 2011; van Leeuwen et al. 2011; Deserno et al. 2012; Diwadkar et al. 2012; Passamonti et al. 2012). All these studies were based on the (original) deterministic, single-state neuronal state equation, and two of the studies included a nonlinear model (Desseilles et al. 2011; Neufang et al. 2011). Patient-control or disorder-related comparisons were addressed in all but two studies (Li et al. 2011b; Passamonti et al. 2012). Li et al. (2011b) and Passamonti et al. (2012) recruited only healthy subjects but related results to psychiatric studies. Interestingly, the number of regions in each model varied from two to six nodes – according to the different neural systems being examined. Exogenous inputs were principally experimentally designed block/event stimulus functions and served to perturb domain or cognitively specific regions – except for Li et al. (2011b) that drove neural responses in non-primary sensory regions. Two of the studies (Almeida et al. 2009, 2011) drew their conclusions based on a single-model, group-wise parameter comparison. Others relied on the Bayesian model selection using multiple competing models, three of which (Desseilles et al. 2011; Deserno et al. 2012; Passamonti et al. 2012) used model space partitioning. This brief review of dynamic causal modelling in psychiatric research illustrates the diversity of hypotheses that has been addressed. It also highlights the potential usefulness of Bayesian model selection – and Bayesian parameter averaging – in testing mechanistic hypotheses about the pathophysiology of distributed processing that is manifested as changes in connectivity.

Conclusion

This chapter has described the basic principles of DCM – with a focus on how to implement parameter-, model- and family-level inferences in analyses of data from groups of subjects. We have not elaborated on some of the more recent developments in DCM that we expect will impact on the neuroimaging of psychiatric disorders. These include the use of two-state DCMs, in which neuronal activity is represented by separate populations of excitatory and inhibitory cells and stochastic DCMs in which neuronal activity is modelled via a combination of deterministic flow and stochastic innovations – thus, better describing the interaction between exogenous and endogenous brain activity. Moreover, there is a library of DCMs for the study of effective connectivity based on EEG, MEG and LFP data (Litvak et al. 2011) that may usefully complement the use of dynamic causal modelling in fMRI.

References

Almeida JR et al (2009) Abnormal amygdala-prefrontal effective connectivity to happy faces differentiates bipolar from major depression. Biol Psychiatry 66(5):451–459

Almeida JR et al (2011) Abnormal left-sided orbitomedial prefrontal cortical amygdala connectivity during happy and fear face processing: a potential neural mechanism of female MDD. Front Psychiatry 2:69

Anderson DR (2008) Model based inference in the life sciences: a primer on evidence. Springer, New York

Banyai M et al (2011) Model-based dynamical analysis of functional disconnection in schizophrenia. Neuroimage 58(3):870–877

Buxton RB et al (1998) Dynamics of blood flow and oxygenation changes during brain activation: the balloon model. Magn Reson Med 39(6):855–864

Buxton RB et al (2004) Modeling the hemodynamic response to brain activation. Neuroimage 23(Suppl 1):S220–S233

Chen CC et al (2009) Forward and backward connections in the brain: a DCM study of functional asymmetries. Neuroimage 45(2):453–462

den Ouden HE et al (2010) Striatal prediction error modulates cortical coupling. J Neurosci 30(9):3210–3219

Deserno L et al (2012) Reduced prefrontal-parietal effective connectivity and working memory deficits in schizophrenia. J Neurosci 32(1):12–20

Desseilles M et al (2011) Depression alters "top-down" visual attention: a dynamic causal modeling comparison between depressed and healthy subjects. Neuroimage 54(2):1662–1668

Diwadkar VA et al (2012) Disordered corticolimbic interactions during affective processing in children and adolescents at risk for schizophrenia revealed by functional magnetic resonance imaging and dynamic causal modeling. Arch Gen Psychiatry 69(3):231–242

Friston KJ et al (2000) Nonlinear responses in fMRI: the Balloon model, Volterra kernels, and other hemodynamics. Neuroimage 12(4):466–477

Friston KJ et al (2003) Dynamic causal modelling. Neuroimage 19(4):1273–1302

Friston K et al (2007) Variational free energy and the Laplace approximation. Neuroimage 34(1):220–234

Friston K et al (2008) Multiple sparse priors for the M/EEG inverse problem. Neuroimage 39(3):1104–1120

Gillihan SJ, Parens E (2011) Should we expect "neural signatures" for DSM diagnoses? J Clin Psychiatry 72(10):1383–1389

Grubb RLJ et al (1974) The effects of changes in PaCO2 on cerebral blood volume, blood flow, and vascular mean transit time. Stroke 5(5):630–639

Kasess CH et al (2010) Multi-subject analyses with dynamic causal modeling. Neuroimage 49(4):3065–3074

Li B et al (2011a) Generalised filtering and stochastic DCM for fMRI. Neuroimage 58(2):442–457

Li X et al (2011b) Using interleaved transcranial magnetic stimulation/functional magnetic resonance imaging (fMRI) and dynamic causal modeling to understand the discrete circuit specific changes of medications: lamotrigine and valproic acid changes in motor or prefrontal effective connectivity. Psychiatry Res 194(2):141–148

Linden DE (2012) The challenges and promise of neuroimaging in psychiatry. Neuron 73(1):8–22

Linden D, Thome J (2011) Modern neuroimaging in psychiatry: towards the integration of functional and molecular information. World J Biol Psychiatry 12(Suppl 1):6–10

Litvak V et al (2011) EEG and MEG data analysis in SPM8. Comput Intell Neurosci 2011:852961

MacKay DJC (2003) Information theory, inference and learning algorithms. Cambridge University Press, Cambridge, UK

Mandeville JB et al (1999) Evidence of a cerebrovascular postarteriole windkessel with delayed compliance. J Cereb Blood Flow Metab 19(6):679–689

Masdeu JC (2011) Neuroimaging in psychiatric disorders. Neurotherapeutics 8(1):93–102

Mechelli A et al (2003) A dynamic causal modeling study on category effects: bottom-up or top-down mediation? J Cogn Neurosci 15(7):925–934

Neufang S et al (2011) Disconnection of frontal and parietal areas contributes to impaired attention in very early Alzheimer's disease. J Alzheimers Dis 25(2):309–321

O'Doherty JP et al (2007) Model-based fMRI and its application to reward and decision making. Ann N Y Acad Sci 1104:35–53

Passamonti L et al (2012) Effects of acute tryptophan depletion on prefrontal-amygdala connectivity while viewing facial signals of aggression. Biol Psychiatry 71(1):36–43

Penny WD (2012) Comparing dynamic causal models using AIC, BIC and free energy. Neuroimage 59(1):319–330

Penny W et al (2003) Variational Bayesian inference for fMRI time series. Neuroimage 19(3):727–741

Penny WD et al (2004) Comparing dynamic causal models. Neuroimage 22(3):1157–1172

Penny WD et al (2010) Comparing families of dynamic causal models. PLoS Comput Biol 6(3):e1000709

Pitt MA, Myung IJ (2002) When a good fit can be bad. Trends Cogn Sci 6(10):421–425

Raftery AE (1995) Bayesian model selection in social research. Sociol Methodol 25:111–164

Rowe JB (2010) Connectivity analysis is essential to understand neurological disorders. Front Syst Neurosci 4:144

Schlosser RG et al (2010) Fronto-cingulate effective connectivity in obsessive compulsive disorder: a study with fMRI and dynamic causal modeling. Hum Brain Mapp 31(12):1834–1850

Stephan KE et al (2007a) Interhemispheric integration of visual processing during task-driven lateralization. J Neurosci 27(13):3512–3522

Stephan KE et al (2007b) Comparing hemodynamic models with DCM. Neuroimage 38(3):387–401

Stephan KE et al (2008) Nonlinear dynamic causal models for fMRI. Neuroimage 42(2):649–662

Stephan KE et al (2009) Bayesian model selection for group studies. Neuroimage 46(4):1004–1017

Stephan KE et al (2009) Tractography-based priors for dynamic causal models. Neuroimage 47(4):1628–1638

Stephan KE et al (2010) Ten simple rules for dynamic causal modeling. Neuroimage 49(4):3099–3109

van Leeuwen TM et al (2011) Effective connectivity determines the nature of subjective experience in grapheme-color synesthesia. J Neurosci 31(27):987 9–9884

EEG-fMRI

4

Gregor Leicht and Christoph Mulert

Abbreviations

AAS	Average artefact subtraction
ACC	Anterior cingulate cortex
ADHD	Attention deficit/hyperactivity disorder
BCG	Ballistocardiogram
BOLD	Blood oxygenation level dependent
DAN	Dorsal attention network
DMN	Default-mode network
EPSP	Excitatory postsynaptic potential
ERP	Event-related potential
GBR	Gamma-band response
GLM	General linear model
HRF	Haemodynamic response function
IPSP	Inhibitory postsynaptic potential
LFPs	Local field potentials
LORETA	Low-resolution brain electromagnetic tomography
PCA	Principal component analysis
RF	Radio frequency
SN	Salience network

4.1 Introduction

Multimodal brain imaging has already been shown to be a very important approach for gaining new insights into physiological brain functions and pathophysiological mechanisms in neuropsychiatric disorders over the last few years. Referring to this, the simultaneous measurement of electrical activity and the corresponding haemodynamic responses with combined recordings of electroencephalography (EEG) and functional magnetic resonance imaging (fMRI) represents one of the most powerful techniques. The complex dealing with the specific practical problems of this approach is easily outweighed by the opportunity to relate both modalities to actual brain events combining the high temporal resolution of EEG with the superior spatial information provided by fMRI. In this manner, it is possible to make progress in several important up-to-date research questions addressed in psychiatric neurosciences (Villringer et al. 2010).

Key principles in the organisation of brain function are functional segregation and integration (Friston 2009). Therefore, the characterisation of connectivity mechanisms in brain networks is the major goal in human neuroscience today (Fox and Friston 2012). Disturbances of connectivity are supposed to be a very relevant pathophysiological factor of neuropsychiatric disorders such as schizophrenia (Schmitt et al. 2011). fMRI is a powerful tool in characterising network structures (e.g. brain regions involved in a specific cognitive task) and interactions of brain regions. These interactions can be examined concerning the correlational relationship between brain regions (functional connectivity) or, even more sophisticated, with respect to the directed influence from one region to another

G. Leicht • C. Mulert (✉)
Psychiatry Neuroimaging Branch,
Department of Psychiatry and Psychotherapy,
University Medical Center Hamburg-Eppendorf,
Hamburg, Germany
e-mail: c.mulert@uke.de

C. Mulert, M.E. Shenton (eds.), *MRI in Psychiatry*,
DOI 10.1007/978-3-642-54542-9_4, © Springer Berlin Heidelberg 2014

(Friston et al. 2011). However, fMRI does not directly take the measurement of neuronal activity. In fact, functional coupling in the brain is realised by oscillation patterns in different frequency bands which can be directly assessed using EEG (Buzsaki and Watson 2012). For instance, low-frequency activity such as theta oscillations (4–8 Hz) is critically involved in the interaction between the hippocampus and the prefrontal cortex (Colgin 2011; Fujisawa and Buzsaki 2011), and high-frequency gamma oscillations (>30 Hz) are known to be critically relevant for the synchronisation of neural networks during perceptual processing (Fisch et al. 2009) or cognitive processes (Herrmann et al. 2010). Mechanisms of disturbed connectivity underlying the pathophysiology of neuropsychiatric disorders are related to new therapeutic strategies, e.g. in schizophrenia treatment: patients with schizophrenia show disturbances with respect to gamma oscillations (Leicht et al. 2010, 2011; Spencer et al. 2003). The generation of these oscillations in the brain is dependent of an intact function of the NMDA receptor (Carlen et al. 2012) providing the link to a promising new approach for treating schizophrenia which is based on an improvement of glutamatergic neurotransmission at the NMDA receptor (Javitt 2012). The simultaneous measurement of EEG and fMRI provides the possibility of bringing together both the spatial characterisation of the network and the description of the neurophysiological coupling mechanisms involved (Mulert and Lemieux 2010; Ullsperger and Debener 2010). Thus, EEG-fMRI represents an important tool for the investigation of disturbances of the interaction between brain regions in neuropsychiatric disorders.

4.2 Physiology

Being aware of the processes within the brain resulting in EEG or fMRI activity is crucial for the understanding of the physiological background of simultaneous EEG-fMRI measurements. EEG is a measure of voltage fluctuations resulting from ionic current flows within the neurons of the brain recorded along the scalp (Niedermeyer and Lopes Da Silva 2004). An EEG electrode placed on the scalp records the summed signal from millions of neurons reflecting the synchronous activity of a neuron population with similar spatial orientation. The electric potential generated by single neurons is far too small to be picked up (Murakami and Okada 2006). The orientation of apical dendrites of pyramidal neurons of the cortex is mainly perpendicular to the cortical surface which facilitates the cumulative summation of electric potentials generated by a simultaneously firing population of pyramidal cells. Therefore, most of the EEG signal is produced by cortical pyramidal cells (Davidson et al. 2000). The EEG signal of a certain generator within the brain decreases by the square root of the distance between source and EEG electrode. Thus, the EEG recording of activity generated in deep brain sources is more difficult than measuring currents generated near the skull.

EEG is well known to basically represent synaptic activity although there are some exceptions (e.g. high-frequency bursts representing action potentials). Next to the apical dendrite and the soma of a pyramidal cell, both excitatory and inhibitory postsynaptic potentials (EPSP and IPSP) produce current sinks and sources in the extracellular medium. This results in a dipolar configuration of sinks and sources. The in vivo measurement of this activity of a number of surrounding neurons reveals local field potentials (LFPs). LFPs represent a summation of synaptic events, including afferent inputs and synaptic inputs originating from local neurons. The electrocorticogram detects LFPs measured at the cortical surface, whereas the EEG records at longer distance even from the scalp (Mulert and Lemieux 2010; Niedermeyer and Lopes Da Silva 2004).

With respect to the question whether fMRI signals are related to similar or different aspects of neuronal activity, it turned out that in many cases the fMRI signal correlates equally well with LFPs and spiking activity. However, in a few studies, it has been successfully realised to differentiate between synaptic activity and spiking activity with regard to the related blood oxygenation level-dependent (BOLD) signal (Shmuel 2009) showing a much closer relationship between fMRI haemodynamic changes and presynaptic and postsynaptic processing of incoming afferents compared to output efferents of a brain region (Logothetis et al. 2001; Rauch et al. 2008; Thomsen et al. 2004; Viswanathan and Freeman 2007). In summary, though EEG and fMRI are obviously very different in terms of signal that is actually measured, they represent considerably overlapping neuronal activity, which is mainly synaptic activity.

Due to the sluggishness of the haemodynamics, the maximum of the BOLD response to a certain event does not appear until after a few seconds (Boynton et al. 1996). For this reason the temporal resolution of BOLD measurements is physically limited to a range of a few hundreds of milliseconds. The temporal resolution of EEG is in the range of 1 ms or less, but the spatial resolution of EEG source localisation approaches is physically limited by the distance of the neural sources from the electrodes and volume conductivity effects. Thus, the spatiotemporal patterns derived from simultaneously recorded EEG-fMRI are significantly enhanced by anatomically informed source spaces derived from fMRI data and temporal high-resolution neural activity representation from EEG. Using the high temporal resolution of simultaneously recorded EEG for the augmentation of the fMRI temporal model, it is possible to study neural events whose characterisation would otherwise be impossible at the fMRI temporal scale (Goebel and Esposito 2010).

4.3 The Inverse Problem of EEG

A basic problem in EEG research is how to estimate the neuronal sources underlying a certain distribution of electrical potentials recorded at the scalp. This so-called inverse problem of EEG has no unique solution because a given EEG scalp distribution of electric potentials can be due to an infinite number of possible source configurations within the brain (Von Helmholtz 2004). Lesion studies or intracranial recordings may offer general information about ERP generation, but are limited to specifically selected groups and not applicable to studies investigating healthy participants or the majority of neuropsychiatric patients. Therefore, several approaches of estimating the underlying generators of a given EEG signal have been suggested. Already available approximation procedures for EEG source localisation use specific assumptions concerning the properties of the generators, conductive media or recording electrodes like restricting the source space to the grey matter of realistic head models derived from subject-specific MRI information or the "smoothness assumption" implemented in the LORETA (low-resolution brain electromagnetic tomography) approach of Pascual-Marqui et al. (1994, 1999). The latter is based on neurophysiological findings of animal studies showing that neighbouring neurons are most likely to be active synchronously and simultaneously (Gray et al. 1989; Llinas 1988). However, compared to fMRI, these approaches lead to a worse spatial resolution of source localisation. This consideration, along with the limitations of EEG with respect to activity generated in deep brain sources, obviously suggests that the combination of the high spatial resolution and connectivity measures of fMRI with simultaneous EEG measurements providing high time resolution and direct assessment of functional coupling in the brain offers a powerful technique for the identification of electrical generators of the EEG signal (Debener et al. 2005; Eichele et al. 2005; Mulert et al. 2008, 2010). However, no general one-to-one relationship between EEG and fMRI signal can be premised: for instance, neural activity may be related to BOLD signal changes but not to scalp EEG changes due to spatial orientations of the electrical generators or origin in nonpyramidal neurons (Nunez and Silberstein 2000). On the other hand, haemodynamic changes related to highly synchronous activities of a small number of neurons detectable with EEG could be too small to survive statistical testing. Keeping these limitations in mind, it can be stated that EEG-fMRI investigations have the potential to significantly increase our knowledge of electrical generators of EEG signal and understanding of brain activity.

4.4 Technique: Safety and Quality Issues

The simultaneous measurement of EEG and fMRI requires careful consideration of participant's safety and precautions with respect to EEG and fMRI data quality. In the following, these requirements for a successful EEG-fMRI recording and an appropriate management of specific artefacts will be summarised. A comprehensive survey of these issues is provided by Allen (Allen 2010).

4.4.1 Recording

In addition to the specifications of a conventional EEG system, EEG recording devices for use in the MRI scanner may not include ferrous materials, need to limit radio frequency emissions to preserve fMRI image quality and have to deal with the presence of static and time-varying magnetic fields. Changes of the magnetic field induce EEG signal artefacts depending on the arrangement of EEG electrode leads. In order to reduce these artefacts, the area of loops formed by the EEG electrode leads has to be minimised. This can be achieved by bunching electrode leads together and twisting the wires together along their path from the subject's head to the EEG amplifiers (Vasios et al. 2006). Commonly used electrode caps further reduce loop areas inevitably present in multichannel EEG recordings (Baumann and Noll 1999; Laufs et al. 2003). EEG artefacts are also induced by variation in loop areas in the static field due to a movement of electrode leads. This might be caused by small head movements (Hill et al. 1995), scanner vibrations (Garreffa et al. 2004) or a ballistocardiogram (Debener et al. 2008). In order to significantly reduce these artefacts, techniques such as weighing down the electrode leads using sandbags and placing a tight bandage over the subject's head (Benar et al. 2003), using electrode caps (Kruggel et al. 2000) or immobilising the subject's head (Anami et al. 2003) have successfully been adopted.

The changing of magnetic fields applied for fMRI data acquisition causes another EEG artefact (imaging or gradient artefact) inducing electromotive forces in the electrode lead loops. The reduction and subsequent removal of these artefacts during postprocessing of EEG data requires low-pass filtering at the front end of the EEG instrumentation (Anami et al. 2003), a fast sampling rate (typically 5 kHz) (Allen et al. 2000) and a synchronisation of the EEG sampling to the MR scanner clock (Mandelkow et al. 2006). Moreover, saturation of the EEG instrumentation has to be avoided for a sufficient recording of EEG gradient artefacts which is mandatory for the removal of the artefact without losing the underlying physiological signal.

With respect to the quality of MRI data measured in a simultaneous EEG-fMRI set-up, any radio frequency (RF) emission from the EEG equipment will lead to artefacts in the MR images. Therefore, radio frequency signals have to be minimised by using low-power digital components and conductive enclosures of all active circuitries and eliminating the ingress of radio frequency from outside the scanner (e.g. by transmitting the EEG data to the receiver in the console room via fibre optic cables) (Allen et al. 2000; Garreffa et al. 2004).

4.4.2 Safety Issues

From the very beginning of EEG recordings in the MR scanner, safety issues have been an important aspect. First, the static magnetic field exerts a force on ferromagnetic materials. Fortunately, there are several commercially available MR-compatible EEG equipments using non-ferromagnetic materials. However, any new EEG instrumentation introduced into the MR scanner must be tested for displacement force and torque (Woods 2007). Secondly, changing gradient and RF fields applied during imaging induce currents in the EEG electrodes and the attached wires (Lemieux et al. 1997). The main hazard is the RF-related heating of electrodes or brain tissue which is dependent on several factors such as the field strength of the MRI scanner, the scanning sequence, the number and shape of EEG electrodes or the arrangement of electrode leads (Lemieux et al. 1997). In order to minimise this hazard, EEG instrumentation with electrode current limiting resistors and minimised electrode lead loops is commonly used. Allen provides a comprehensive overview of safety issues of simultaneous EEG-fMRI measurements (Allen 2010). Taking all aspects into account, simultaneous recordings of EEG and fMRI have been repeatedly and safely performed with scanners up to 3 Tesla and can be considered as a standard imaging technique. The specific safety instructions provided with the commercially available MR-compatible EEG equipment need to be regarded. However, EEG-fMRI measurements using higher-field scanners necessitate additional precautions and further investigation (Neuner et al. 2013).

4.4.3 EEG Artefacts

Two main artefacts have to be considered for the analysis of EEG data recorded simultaneously with fMRI: the cardiac pulse-related artefact and the image acquisition artefact. The cardiac pulse-related artefact which is often also referred to as ballistocardiogram or BCG artefact results from the interaction between the active cardiovascular system and the static field inside the scanner. Motions related to cardiac activity like the axial head rotation (Mullinger et al. 2013b; Nakamura et al. 2006), the pulsatile movements of scalp vessels on adjacent electrodes and the motion of the blood (i.e. changes in blood velocity) are supposed to contribute to the origin of the BCG artefact. The BCG artefact represents a rather complicated, dynamic contribution to the EEG signal adding a range of topographies and signatures (Debener et al. 2010). It consists of amplitudes and frequencies which are close to the range of the physiological EEG signal and shows synchrony with the cardiac rhythm apparent from a simultaneously recorded ECG (Debener et al. 2008).

The image acquisition artefact, which is often also referred to as gradient artefact, leads to a complete obscuration of the physiological EEG in simultaneously recorded EEG-fMRI (Allen et al. 2000; Ives et al. 1993). It results from the application of rapidly varying magnetic field gradients and RF pulses during fMRI echo planar imaging sequences which electromagnetically induce electromotive forces in electrodes, leads, subject and amplifier. The gradient artefact is characterised by amplitudes typically more than two orders of magnitude higher than the physiological EEG signal (Felblinger et al. 1999) and a power spectrum overlapping that of the physiological EEG (Ritter et al. 2008). However, caused by a repeated, constant and predefined RF and gradient switching sequence, gradient artefacts have a strong deterministic component (Ritter et al. 2010).

For both types of artefacts, today there are strategies available with sufficient efficacy trying to either avoid or reduce the artefacts during data acquisition or remove the artefacts during postprocessing.

In order to avoid the interference of BCG artefacts with the EEG signal of interest during simultaneous EEG-fMRI measurement, Ertl et al. suggested to present stimuli dependent on the presence of a cardiac pulse within pulse-free intervals (Ertl et al. 2010). Postprocessing strategies for the management of BCG artefacts are based either on waveform removal approaches such as the average artefact subtraction (AAS) algorithm (Allen et al. 1998) and variants of it (de Munck et al. 2013; Ellingson et al. 2004; Laufs et al. 2008) or on pattern removal approaches such as principal component analysis (PCA) or independent component analysis (ICA) (Benar et al. 2003; Liu et al. 2012) or on combinations of both approaches (Debener et al. 2007; Niazy et al. 2005). Furthermore, new strategies include the application of optical motion tracking systems (Levan et al. 2013).

In order to avoid the interference of gradient artefacts with the physiological EEG signal, interleaved protocols are a possible approach: MR acquisition is discontinued at regular intervals to gain intervals of gradient artefact-free EEG data (Ives et al. 1993; Mulert et al. 2008, 2010; Ritter et al. 2008). Interleaved protocols are generally less flexible. However, they are suitable for the investigation of certain forms of brain activity such as evoked responses or slowly varying rhythms. A reduction of the image acquisition artefacts at source during recording can be achieved by using proper EEG equipment, minimising the conductor loop area, reducing the conductor movement by immobilisation of the subject's head, weighing the EEG instrumentation and adjusting the subject's axial position (Mullinger et al. 2011, 2013a; Yan et al. 2009). Anami et al. introduced the so-called "stepping stone sampling" method which strongly attenuates the amplitude of the imaging artefact (Anami et al. 2003). By modifying a blip-type echo planar sequence, they were able to perform EEG sampling exclusively in the period in which the gradient artefact resided around the baseline level. The variability of the gradient artefact can be reduced by synchronising the internal clocks of both the MRI and the EEG acquisition system and setting the repetition time of the MRI sequence a multiple of the EEG sampling interval (Mandelkow et al. 2006, 2010). Thus, the artefact removal by common mean template subtraction methods during postprocessing can be facilitated which are very efficient and used today even as online artefact removal that allows immediate inspection of the ongoing EEG (Allen et al. 2000). This artefact template subtraction approach assumes that the shape of the artefact is quite constant over time and that the artefact is not correlated with the physiological EEG signal (Hill et al. 1995). For every EEG channel artefact templates are calculated as the mean of a certain number of single artefacts and subtracted from the EEG (Allen et al. 2000). Several enhancements of this approach (de Munck et al. 2013; Freyer et al. 2009; Negishi et al. 2004; Niazy et al. 2005; Sun and Hinrichs 2009) and its combination with other methods such as ICA (Mantini et al. 2007) are available.

Compared to the analysis of EEG components with slower frequencies, the investigation of high-frequency oscillations such as gamma oscillations causes additional difficulties. The amplitudes of these subtle EEG rhythms are relatively small because the skull tissue acts as a low-pass filter. Therefore, the signal-to-noise ratio is low, especially in the MRI environment with high artefact contamination. Concerning technical problems, a sufficient analysis of EEG gamma activity recorded in the MRI scanner may sometimes be difficult due to EEG artefacts caused by the MRI environment. One example is an EEG artefact in the frequency range between 30 and 60 Hz generated by the helium pump of the MRI scanner. Switching this pump off for some time during the simultaneous EEG-fMRI recording can solve this problem (Mulert et al. 2007). Another systematic artefact affecting physiologically relevant high-frequency brain rhythms in the EEG signal is induced by the internal ventilation system of certain MR scanners which can be eliminated by switching off the ventilator (Nierhaus et al. 2013). Furthermore, the availability of the high-frequency EEG range can be enhanced by the "stepping stone sampling" method (Anami et al. 2003) and the synchronisation of the MRI sequence with the sampling pattern of the EEG (Mandelkow et al. 2006).

4.4.4 FMRI Artefacts

In general, properly designed and tested EEG instrumentation should not substantially decrease MR image quality. However, the presence of an EEG system in the scanner room can interact with the magnetic fields used for MR data acquisition and, therefore, affect the fMRI signal-to-noise ratio. Even material with a weak magnetic susceptibility such as MRI-compatible EEG caps can introduce MR image artefacts due to an inhomogeneity of the magnetic field (Mullinger et al. 2008). However, carefully chosen and tested commercially available MR-compatible EEG equipments have only small deteriorating influence on image quality (Carmichael 2010).

4.5 Analysis Methods

The most effective way of combining EEG and fMRI data sets into one unique data model is still an active area of research trying to integrate knowledge derived from neuroscience, physics and computer science (Huster et al. 2012). In general, two different approaches have been suggested (Goebel and Esposito 2010): the integration of EEG and fMRI in the spatial domain and the integration of EEG and fMRI in the temporal domain.

4.5.1 fMRI-Informed EEG Source Localisation

In order to integrate information from simultaneously acquired EEG and fMRI data in the spatial domain, fMRI is used to map detailed spatial patterns of the brain activity of interest within the cortex source space. One approach for multimodal integration of EEG and fMRI has been to treat the simultaneously recorded EEG and fMRI data separately and compare the results in space for the purpose of cross-validation, e.g. by projection of the localisation results onto a common anatomical space (Mulert et al. 2004, 2005; Musso et al. 2011; Opitz et al. 1999).

The good spatial resolution of fMRI on the one hand and the inverse problem of the EEG on the other led to the development of approaches using spatial patterns derived from fMRI in combination with forward and inverse solutions for the source localisation of simultaneously recorded EEG phenomena. Based on the fMRI activation clusters, spatial constraints are introduced in EEG source localisation approaches for EEG spatial modelling (Bledowski et al. 2004). Thus, the information derived from fMRI selects a network of possible EEG regional sources that can be characterised with respect to specific time or time-frequency features. fMRI spatial information can either be incorporated into the estimation of the distributed inverse solution (Babiloni et al. 2003) or simply be used to identify the regions of activity in the cortex source space in order to analyse their electrophysiological properties. Approaches using Bayesian modelling have integrated fMRI-activated regions in the form of statistical priors and tested whether such fMRI constraints are compatible with the EEG data (Grova et al. 2008).

The potential discrepancy between EEG generator and BOLD localisation (e.g. regions that respond selectively to one of the modalities) could be a limitation of fMRI-informed EEG source localisation approaches, e.g. in the case of an actual EEG source that is not seen by fMRI. Moreover, for EEG dipolar source localisation, the selection of sources derived from fMRI to be used for EEG source localisation is a crucial point as the number of sources is a very sensitive parameter (Liu et al. 2006).

4.5.2 EEG-Informed fMRI

In order to integrate EEG and fMRI data in the temporal domain, a common approach is to use information derived from EEG to inform the fMRI analysis. This is done by introducing information derived from EEG into the prediction of fMRI signal time courses during the fitting of a general linear model (GLM) to the fMRI data from every brain voxel. By this means, voxels can be identified at which the BOLD signal is significantly correlated with a time course of a specific EEG phenomenon. This results in 3D maps representing brain regions specifically related to an EEG-derived temporal reference. Usually, EEG data are first analysed in the channel or source space and characterised in the time (e.g. amplitude peaks at a specific latency) or frequency (e.g. spectral power and phase distribution in certain frequency ranges) domain. The trial-by-trial variation or the spectrotemporal changes of one or more EEG parameters are extracted and used to create a temporal model for the fMRI analysis. Before introducing these time courses into a fMRI GLM, a correction for fMRI haemodynamics is required, e.g. the convolution with a haemodynamic response function (HRF) (Boynton et al. 1996). Moreover, the EEG-derived time courses have to be orthogonalised to the standard fMRI response (Eichele et al. 2005; Mulert et al. 2008)

One approach of EEG-informed fMRI has been to use the EEG in order to identify an onset of a certain event such as epileptic events. These event onsets are modelled as, e.g. a stick function of unitary amplitude, convolved with a HRF and then used in an event-related fMRI analysis to create regressors resulting in a model of the event-related haemodynamic change (Gotman et al. 2006; Lemieux 2004). Other strategies have been to include EEG parameters such as ERP amplitudes as a covariate into a fMRI GLM analysis (Regenbogen et al. 2012) or to examine correlations between parameters derived from simultaneously recorded but separately analysed EEG and fMRI data (Diukova et al. 2012; Huster et al. 2011; Mulert et al. 2005). The latter has been adopted in psychiatric studies investigating alcohol-dependent and ADHD patients (Karch et al. 2008, 2010).

The use of EEG-derived single-trial and spectral information for the integration of EEG and fMRI data has widely been applied in sensory and cognitive neuroscience. Intertrial fluctuations can be characterised by temporal structure, time-frequency structure (Wang et al. 2007) as well as spatial topography across sensors or combinations of those (Li et al. 2007). Compared to observation of certain processes in the channel space of the EEG, approaches such as ICA can help here to improve the separation of the activity from different sources and artefacts by decomposing channel time series into independent components (Lavric et al. 2011; Makeig et al. 2002). With respect to the fusion of EEG and fMRI data, introducing a parametric regressor derived from a single-trial analysis of the EEG data (e.g. in response to certain stimuli) offers the possibility of exploring the relationship between EEG and haemodynamic signals. For example, this strategy has been adopted to investigate BOLD correlates of ERP amplitudes (Debener et al. 2005; Eichele et al. 2005; Mulert et al. 2008; Scheibe et al. 2010) and latencies (Benar et al. 2007). Another approach is the examination of specific BOLD responses corresponding to the power amplitude of oscillations in a certain frequency range. The single-trial coupling of power amplitudes of the early auditory evoked EEG gamma-band response (GBR) and fMRI revealed GBR-specific activations in the auditory cortex, the ACC and the thalamus (see Fig. 4.1) (Mulert et al. 2010). These results show the possibility of the EEG-informed fMRI approach to detect subcortical structures related to the generation of EEG phenomena and validate GBR EEG source localisation results of a reduced activity of GBR generators in the auditory cortex and the ACC in patients with schizophrenia (Leicht et al. 2010).

Using EEG-fMRI, several studies also successfully tried to identify brain regions associated with changes in ongoing brain rhythms (Hanslmayr et al. 2011; Michels et al. 2012; Sadaghiani et al. 2010; Scheeringa et al. 2011). Therefore, in general, the spectral power in one or more specific frequency bands has been quantified for each of several short epochs of EEG. After convolving this power time series with a HRF, it is entered as a regressor in the fMRI GLM analysis.

The so-called joint ICA has been proposed as an alternative approach to the analysis techniques outlined above. This multivariate approach for the integration of EEG and fMRI measures tries to identify meaningful patterns in an EEG-fMRI joint data space considering EEG and fMRI signals together (Eichele et al. 2008; Kraut et al. 2003; Lei et al. 2010; Mangalathu-Arumana et al. 2012; Moosmann et al. 2008; Ostwald et al. 2012). EEG source localisation prior to the fusion of EEG and fMRI information may be another way of enriching the interpretation of EEG-fMRI patterns by separating the activity from different sources and additionally providing spatial information on the process of interest (Brookes et al. 2008). For example, Esposito et al. used the local source power trial-by-trial variation derived from EEG source localisation of the ERP of interest as an input for single-trial EEG-fMRI coupling (Esposito et al. 2009).

4.5.3 EEG-Informed fMRI and Functional Connectivity of Resting-State Networks

The functional connectivity between brain regions of intrinsic brain networks such as the default-mode network (DMN) is known to be increased. Patients with several neuropsychiatric diseases show alterations in resting-state networks, e.g. in Alzheimer's disease (Franciotti et al. 2013).

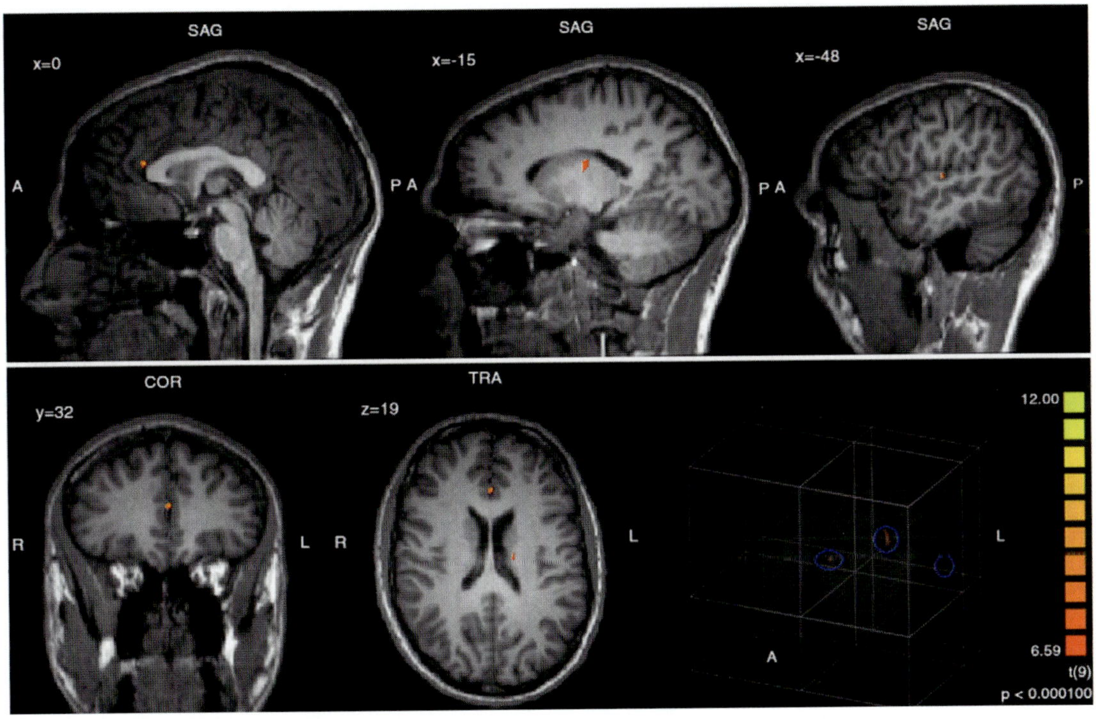

Fig. 4.1 Single-trial coupling of power amplitudes of the early auditory evoked EEG gamma-band response (*GBR*) and the corresponding BOLD activation (random effects analysis, uncorrected for multiple testing). GBR-specific BOLD activations can be seen in the left auditory cortex (gyrus temporalis superior, Brodmann area 41/22), the thalamus and the anterior cingulate cortex (Brodmann area 24). In the *lower right corner*, a glass brain view is provided demonstrating a 3D view of the three abovementioned clusters. *Upper row*: Talairach coordinates – *x*(*left*) = 0, *x*(*middle*) = −15, *x*(*right*) = −48. *Lower row*: *y*(*left*) = 32, *z*(*middle*) = 19. *SAG* sagittal, *COR* coronary, *TRA* transversal, *A* anterior, *P* posterior, *R* right, *L* left (Reprinted from Mulert et al. 2010. Copyright (2010), with permission from Elsevier)

Studies using simultaneous EEG-fMRI measurements investigated resting-state activity with respect to BOLD correlates of EEG oscillations of specific frequencies such as alpha-oscillations (Goldman et al. 2002; Laufs et al. 2003; Moosmann et al. 2003). Beyond that, the EEG signal has also been related to the functional connectivity within and between networks in order to characterise network dynamics. For example, delta and beta oscillations explain 70 % of the DMN variance of functional connectivity (Hlinka et al. 2010). An increase of alpha power leads to a decrease of fMRI connectivity between the primary visual cortex and occipital regions as well as a decrease of the negative coupling between visual areas and regions of the DMN (Scheeringa et al. 2012). The strength of connectivity between DMN and the dorsal attention network (DAN) has been found to be inversely correlated with alpha power, whereas alpha power correlated with the spatial extent of anticorrelation between DMN and DAN (Chang et al. 2013).

FMRI patterns have also been related with complex patterns of EEG organisation such as so-called microstates. These are quasi-stable and unique topographic representations of the EEG reflecting the summation of concurrent neuronal activity across brain regions rather than being specific for a certain frequency band (Lehmann et al. 1987). Recently, the relationship between EEG microstates and BOLD resting-state networks has been investigated using simultaneous EEG-fMRI (Britz et al. 2010; Huang et al. 1996; Musso et al. 2010). Microstates are known to be altered in different psychiatric disorders such as schizophrenia (Lehmann et al. 2005).

The relationship between EEG coherence and fMRI connectivity patterns has been investigated by Wang et al. using the combination of fMRI and invasive electrophysiological recordings. This study revealed that the coherence in low frequencies (<20 Hz) predominantly contributes to fMRI connectivity patterns in the resting state in monkeys (Wang et al. 2012). However, especially high-frequency oscillations (30–100 Hz) seem to be tightly linked to the BOLD signal (Mukamel et al. 2005; Niessing et al. 2005). In order to integrate findings from EEG coherence and fMRI connectivity in humans in the future, the combination of simultaneous EEG-fMRI and EEG source analysis represents a promising approach.

4.5.4 EEG-fMRI and High-Frequency Oscillations

Scalp EEG activity includes oscillations of a variety of distinct frequency ranges each of which is supposed to represent the synchronisation of varying neuronal networks. The rhythmic activities in the resting or "spontaneous" EEG are usually divided into several frequency bands (delta, <4 Hz; theta, 4–8 Hz; alpha, 8–12 Hz; beta, 12–30 Hz; and gamma, >30 Hz) which are associated with different behavioural states, ranging from sleep to relaxation, heightened alertness and mental concentration (Lindsley 1952; Niedermeyer and Lopes Da Silva 2004; Nunez 1995). A comprehensive analysis of the different types of brain rhythms can be found in Buzsáki's monography *Rhythms of the Brain* (Buzsaki 2006).

Several studies investigating the relationship between fMRI and special oscillatory EEG components reported especially high correlations between high- in comparison with low-frequency electric activity and BOLD signal. Recordings in cats revealed that correlations between LFPs and BOLD signal are especially high in the gamma-band frequency range (Niessing et al. 2005). Similar results were found in studies combining fMRI and intracranial EEG recordings in human epilepsy patients (Mukamel et al. 2005) showing also a predictive value of fMRI about the anatomical location of cross-condition gamma-band modulations (Brovelli et al. 2005; Lachaux et al. 2007). Foucher et al. reported a differential reactivity of event-related potentials (ERPs) and BOLD signal in response to the presentation of two kinds of rare events (i.e. target and novel stimulus in an oddball task) using combined measurement of EEG and fMRI. However, in accordance with fMRI results, target-related gamma oscillations were more intense than their novel-related counterparts (Foucher et al. 2003). The authors of this study propounded a physiological hypothesis underlying these findings: as opposed to ERPs reflecting synchronous activity of synapses within milliseconds (Baillet et al. 2001; Speckmann and Elger 1999) and similar to the BOLD signal, event-related gamma oscillations are less dependent on the small jitter of the neuronal response relative to a stimulus because they do not require to be time-locked to the stimulus that precisely (Tallon-Baudry and Bertrand 1999). Furthermore, ERPs reflect the synaptic input function of pyramidal cells only (Baillet et al. 2001; Speckmann and Elger 1999), whereas fMRI reflects the synaptic activity of all neural cells including inhibitory interneurons (Logothetis et al. 2001; Mathiesen et al. 2000; Matsuura and Kanno 2001) which are involved in the synchronisation of gamma oscillations (Traub et al. 1996), too. A third possible reason could be related to the assumption that ERPs might correspond to the simple phase resetting of ongoing cerebral activity (Makeig et al. 2002) which should not consume much energy (Foucher et al. 2003).

4.6 EEG-fMRI in Psychiatry

Using the EEG-informed fMRI approach, Mobascher et al. investigated the influence of nicotine on the EEG P300 component of the visual evoked potential elicited by an oddball-type task in patients with schizophrenia. The authors report a nicotine-associated increase in P300-informed fMRI activation in schizophrenia patients and healthy controls, mainly in the anterior cingulate

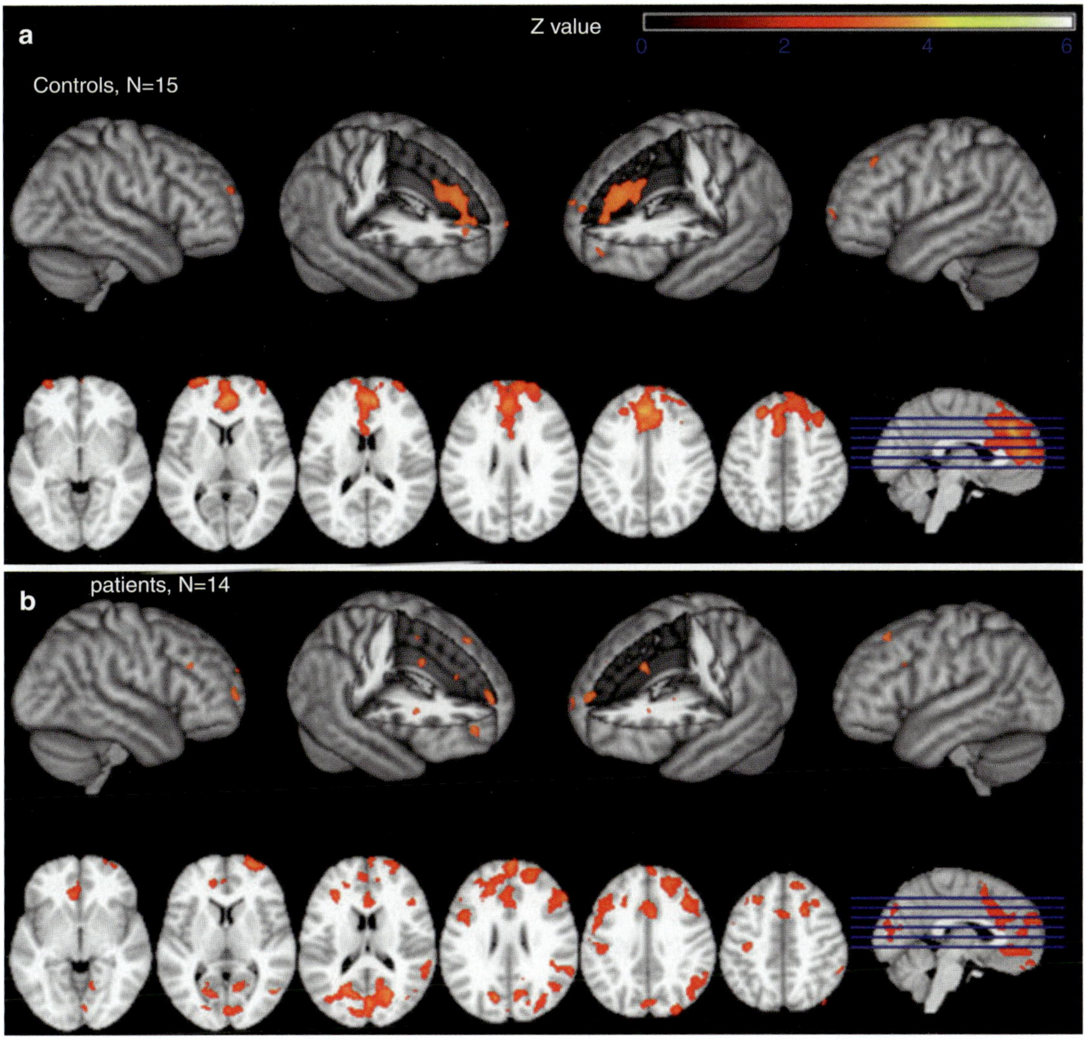

Fig. 4.2 Nicotine effects on EEG-informed fMRI BOLD activation specific for the EEG P300 component of the visual evoked potential elicited by a visual oddball task in habitual schizophrenia smokers (**b**) and healthy smoking controls (**a**). Nicotine (compared to placebo) induced an increase in P300-informed fMRI activation, mainly in the anterior cingulate cortex and adjacent medial frontal cortex (general linear model whole-brain analysis, cluster-corrected threshold $Z=2.3$, $p=0.05$). The results of separate analyses of EEG and fMRI measures were reported to be largely unaffected by nicotine (Reprinted from Mobascher et al. 2012. Copyright (2012), with permission from Elsevier)

cortex (ACC) and adjacent medial frontal cortex. The results of separate analyses of EEG and fMRI measures were reported to be largely unaffected by nicotine (see Fig. 4.2) (Mobascher et al. 2012). Musso et al. conducted an EEG-fMRI study on the human ketamine model of schizophrenia and report diminished P300 amplitudes (visual odd-ball task) and corresponding BOLD responses in the ketamine condition in cortical regions being involved in sensory processing and selective attention (Musso et al. 2011). The EEG-informed fMRI approach has recently also been used to describe the brain network involved in reward processing (Hauser et al. 2013; Plichta et al. 2013).

With respect to ongoing brain rhythms in the resting state, Balsters et al. identified neuroanatomical origins of EEG measures with high

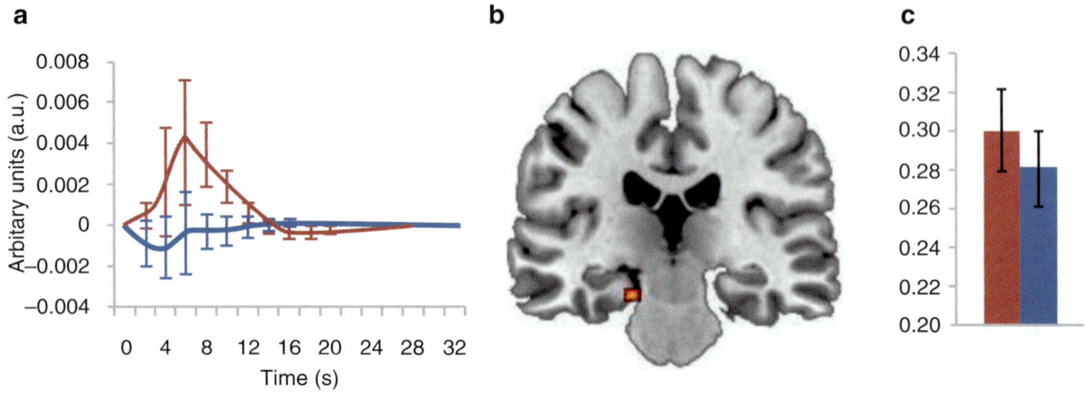

Fig. 4.3 Effects of acetylcholinesterase inhibitor donepezil on resting-state hippocampal delta (1.5–3 Hz) of healthy older participants assessed by means of EEG-informed fMRI: single-dose treatment with donepezil revealed an increase in EEG delta power and delta-specific activation of the hippocampus accompanied by a significantly worse performance on a working memory task. (**a**) Delta response function taken from the left hippocampus. (**b**) Hippocampal activation driven by differences in delta EEG power. (**c**) Relative delta band EEG power differences. In all graphs, *red* corresponds to donepezil and *blue* to placebo (Reprinted from Balsters et al. 2011)

sensitivity to the age-related cognitive decline and their manipulation by the acetylcholinesterase inhibitor donepezil. Investigating oscillatory patterns with an EEG-informed fMRI approach, the authors found drug-induced changes of delta, alpha and beta band power associated with activity differences within the hippocampus (delta, see Fig. 4.3), frontal-parietal network (alpha) and default-mode network (beta) (Balsters et al. 2011). In a study investigating resting-state networks with simultaneously recorded EEG-fMRI patients with a schizophrenia spectrum disorder showed lower EEG frequencies to be related to the default-mode and left-working-memory network compared to healthy controls (Razavi et al. 2013).

Based on recent advances in EEG-fMRI, a first successful approach of real-time integration of simultaneous fMRI and EEG has been made. Thus, it was possible to perform a simultaneous multimodal real-time fMRI and EEG neurofeedback study: participants were able to simultaneously regulate their BOLD fMRI activation in the left amygdale and frontal EEG power asymmetry in the high beta band. These results indicate possible applications of the EEG-fMRI approach in enhanced cognitive therapeutic strategies for major neuropsychiatric disorders (Zotev et al. 2014).

Conclusion

Multimodal brain imaging using simultaneous measurement of EEG and fMRI is a powerful technique combining the information from EEG on neurophysiological coupling mechanisms in high temporal resolution with the superior spatial information on the involved network provided by fMRI. Therefore, EEG-fMRI can strongly contribute to the investigation of disturbances of connectivity in neuropsychiatric disorders. Both safety and signal quality issues can be addressed sufficiently today.

References

Allen PJ (2010) EEG instrumentation and safety. In: Mulert C, Lemieux L (eds) EEG-fMRI – physiological basis, technique, and applications. Springer, Berlin, pp 115–133

Allen PJ, Polizzi G, Krakow K, Fish DR, Lemieux L (1998) Identification of EEG events in the MR scanner: the problem of pulse artifact and a method for its subtraction. Neuroimage 8:229–239

Allen PJ, Josephs O, Turner R (2000) A method for removing imaging artifact from continuous EEG recorded during functional MRI. Neuroimage 12:230–239

Anami K, Mori T, Tanaka F, Kawagoe Y, Okamoto J, Yarita M, Ohnishi T, Yumoto M, Matsuda H, Saitoh O (2003) Stepping stone sampling for retrieving artifact-free electroencephalogram during functional magnetic resonance imaging. Neuroimage 19:281–295

Babiloni F, Babiloni C, Carducci F, Romani GL, Rossini PM, Angelone LM, Cincotti F (2003) Multimodal integration of high-resolution EEG and functional magnetic resonance imaging data: a simulation study. Neuroimage 19:1–15

Baillet S, Mosher JC, Leahy RM (2001) Electromagnetic brain mapping. IEEE Signal Proc Mag 18:14–30

Balsters JH, O'Connell RG, Martin MP, Galli A, Cassidy SM, Kilcullen SM, Delmonte S, Brennan S, Meaney JF, Fagan AJ, Bokde AL, Upton N, Lai R, Laruelle M, Lawlor B, Robertson IH (2011) Donepezil impairs memory in healthy older subjects: behavioural, EEG and simultaneous EEG/fMRI biomarkers. PLoS One 6:e24126

Baumann SB, Noll DC (1999) A modified electrode cap for EEG recordings in MRI scanners. Clin Neurophysiol 110:2189–2193

Benar C, Aghakhani Y, Wang Y, Izenberg A, Al-Asmi A, Dubeau F, Gotman J (2003) Quality of EEG in simultaneous EEG-fMRI for epilepsy. Clin Neurophysiol 114:569–580

Benar CG, Schon D, Grimault S, Nazarian B, Burle B, Roth M, Badier JM, Marquis P, Liegeois-Chauvel C, Anton JL (2007) Single-trial analysis of oddball event-related potentials in simultaneous EEG-fMRI. Hum Brain Mapp 28:602–613

Bledowski C, Prvulovic D, Hoechstetter K, Scherg M, Wibral M, Goebel R, Linden DE (2004) Localizing P300 generators in visual target and distractor processing: a combined event-related potential and functional magnetic resonance imaging study. J Neurosci 24:9353–9360

Boynton GM, Engel SA, Glover GH, Heeger DJ (1996) Linear systems analysis of functional magnetic resonance imaging in human V1. J Neurosci 16: 4207–4221

Britz J, Van De Ville D, Michel CM (2010) BOLD correlates of EEG topography reveal rapid resting-state network dynamics. Neuroimage 52:1162–1170

Brookes MJ, Mullinger KJ, Stevenson CM, Morris PG, Bowtell R (2008) Simultaneous EEG source localisation and artifact rejection during concurrent fMRI by means of spatial filtering. Neuroimage 40:1090–1104

Brovelli A, Lachaux JP, Kahane P, Boussaoud D (2005) High gamma frequency oscillatory activity dissociates attention from intention in the human premotor cortex. Neuroimage 28:154–164

Buzsaki G (2006) Rhythms of the brain. Oxford University Press, New York

Buzsaki G, Watson BO (2012) Brain rhythms and neural syntax: implications for efficient coding of cognitive content and neuropsychiatric disease. Dialogues Clin Neurosci 14:345–367

Carlen M, Meletis K, Siegle JH, Cardin JA, Futai K, Vierling-Claassen D, Ruhlmann C, Jones SR, Deisseroth K, Sheng M, Moore CI, Tsai LH (2012) A critical role for NMDA receptors in parvalbumin interneurons for gamma rhythm induction and behavior. Mol Psychiatry 17:537–548

Carmichael D (2010) Image quality issues. In: Mulert C, Lemieux L (eds) EEG-fMRI – physiological basis, technique, and applications. Springer, Berlin, pp 173–199

Chang C, Liu Z, Chen MC, Liu X, Duyn JH (2013) EEG correlates of time-varying BOLD functional connectivity. Neuroimage 72:227–236

Colgin LL (2011) Oscillations and hippocampal-prefrontal synchrony. Curr Opin Neurobiol 21:467–474

Davidson RJ, Jackson DC, Larson CL (2000) Human electroencephalography. In: Cacioppo JT, Tassinary LG, Bernston GG (eds) Handbook of psychophysiology. Cambridge University Press, Cambridge, UK, pp 27–56

de Munck JC, van Houdt PJ, Goncalves SI, van Wegen E, Ossenblok PP (2013) Novel artefact removal algorithms for co-registered EEG/fMRI based on selective averaging and subtraction. Neuroimage 64:407–415

Debener S, Ullsperger M, Siegel M, Fiehler K, von Cramon DY, Engel AK (2005) Trial-by-trial coupling of concurrent electroencephalogram and functional magnetic resonance imaging identifies the dynamics of performance monitoring. J Neurosci 25:11730–11737

Debener S, Strobel A, Sorger B, Peters J, Kranczioch C, Engel AK, Goebel R (2007) Improved quality of auditory event-related potentials recorded simultaneously with 3-T fMRI: removal of the ballistocardiogram artefact. Neuroimage 34:587–597

Debener S, Mullinger KJ, Niazy RK, Bowtell RW (2008) Properties of the ballistocardiogram artefact as revealed by EEG recordings at 1.5, 3 and 7T static magnetic field strength. Int J Psychophysiol 67:189–199

Debener S, Kranczioch C, Gutberlet I (2010) EEG Quality: origin and reduction of the EEG cardiac-related artefact. In: Mulert C, Lemieux L (eds) EEG-fMRI – physiological basis, technique, and applications. Springer, Berlin, pp 135–151

Diukova A, Ware J, Smith JE, Evans CJ, Murphy K, Rogers PJ, Wise RG (2012) Separating neural and vascular effects of caffeine using simultaneous EEG-FMRI: differential effects of caffeine on cognitive and sensorimotor brain responses. Neuroimage 62:239–249

Eichele T, Specht K, Moosmann M, Jongsma ML, Quiroga RQ, Nordby H, Hugdahl K (2005) Assessing the spatiotemporal evolution of neuronal activation with single-trial event-related potentials and functional MRI. Proc Natl Acad Sci U S A 102:17798–17803

Eichele T, Calhoun VD, Moosmann M, Specht K, Jongsma ML, Quiroga RQ, Nordby H, Hugdahl K (2008) Unmixing concurrent EEG-fMRI with parallel independent component analysis. Int J Psychophysiol 67:222–234

Ellingson ML, Liebenthal E, Spanaki MV, Prieto TE, Binder JR, Ropella KM (2004) Ballistocardiogram artifact reduction in the simultaneous acquisition of auditory ERPS and fMRI. Neuroimage 22:1534–1542

Ertl M, Kirsch V, Leicht G, Karch S, Olbrich S, Reiser M, Hegerl U, Pogarell O, Mulert C (2010) Avoiding the ballistocardiogram (BCG) artifact of EEG data acquired simultaneously with fMRI by pulse-triggered presentation of stimuli. J Neurosci Methods 186: 231–241

Esposito F, Mulert C, Goebel R (2009) Combined distributed source and single-trial EEG-fMRI modeling: application to effortful decision making processes. Neuroimage 47:112–121

Felblinger J, Slotboom J, Kreis R, Jung B, Boesch C (1999) Restoration of electrophysiological signals distorted by inductive effects of magnetic field gradients during MR sequences. Magn Reson Med 41:715–721

Fisch L, Privman E, Ramot M, Harel M, Nir Y, Kipervasser S, Andelman F, Neufeld MY, Kramer U, Fried I, Malach R (2009) Neural "ignition": enhanced activation linked to perceptual awareness in human ventral stream visual cortex. Neuron 64:562–574

Foucher JR, Otzenberger H, Gounot D (2003) The BOLD response and the gamma oscillations respond differently than evoked potentials: an interleaved EEG-fMRI study. BMC Neurosci 4:22

Fox PT, Friston KJ (2012) Distributed processing; distributed functions? Neuroimage 61:407–426

Franciotti R, Falasca NW, Bonanni L, Anzellotti F, Maruotti V, Comani S, Thomas A, Tartaro A, Taylor JP, Onofrj M (2013) Default network is not hypoactive in dementia with fluctuating cognition: an Alzheimer disease/dementia with Lewy bodies comparison. Neurobiol Aging 34:1148–1158

Freyer F, Becker R, Anami K, Curio G, Villringer A, Ritter P (2009) Ultrahigh-frequency EEG during fMRI: pushing the limits of imaging-artifact correction. Neuroimage 48:94–108

Friston KJ (2009) Modalities, modes, and models in functional neuroimaging. Science 326:399–403

Friston KJ, Li B, Daunizeau J, Stephan KE (2011) Network discovery with DCM. Neuroimage 56:1202–1221

Fujisawa S, Buzsaki G (2011) A 4Hz oscillation adaptively synchronizes prefrontal, VTA, and hippocampal activities. Neuron 72:153–165

Garreffa G, Bianciardi M, Hagberg GE, Macaluso E, Marciani MG, Maraviglia B, Abbafati M, Carni M, Bruni I, Bianchi L (2004) Simultaneous EEG-fMRI acquisition: how far is it from being a standardized technique? Magn Reson Imaging 22:1445–1455

Goebel R, Esposito F (2010) The added value of EEG-fMRI in imaging neuroscience. In: Mulert C, Lemieux L (eds) EEG-fMRI – physiological basis, technique, and applications. Springer, Berlin, pp 97–112

Goldman RI, Stern JM, Engel J Jr, Cohen MS (2002) Simultaneous EEG and fMRI of the alpha rhythm. Neuroreport 13:2487–2492

Gotman J, Kobayashi E, Bagshaw AP, Benar CG, Dubeau F (2006) Combining EEG and fMRI: a multimodal tool for epilepsy research. J Magn Reson Imaging 23:906–920

Gray CM, Konig P, Engel AK, Singer W (1989) Oscillatory responses in cat visual cortex exhibit intercolumnar synchronization which reflects global stimulus properties. Nature 338:334–337

Grova C, Daunizeau J, Kobayashi E, Bagshaw AP, Lina JM, Dubeau F, Gotman J (2008) Concordance between distributed EEG source localization and simultaneous EEG-fMRI studies of epileptic spikes. Neuroimage 39:755–774

Hanslmayr S, Volberg G, Wimber M, Raabe M, Greenlee MW, Bauml KH (2011) The relationship between brain oscillations and BOLD signal during memory formation: a combined EEG-fMRI study. J Neurosci 31:15674–15680

Hauser TU, Iannaccone R, Stampfli P, Drechsler R, Brandeis D, Walitza S, Brem S (2013) The feedback-related negativity (FRN) revisited: new insights into the localization, meaning and network organization. Neuroimage 84C:159–168

Herrmann CS, Frund I, Lenz D (2010) Human gamma-band activity: a review on cognitive and behavioral correlates and network models. Neurosci Biobehav Rev 34:981–992

Hill RA, Chiappa KH, Huang-Hellinger F, Jenkins BG (1995) EEG during MR imaging: differentiation of movement artifact from paroxysmal cortical activity. Neurology 45:1942–1943

Hlinka J, Alexakis C, Diukova A, Liddle PF, Auer DP (2010) Slow EEG pattern predicts reduced intrinsic functional connectivity in the default mode network: an inter-subject analysis. Neuroimage 53:239–246

Huang Y, Koestner ML, Ackers GK (1996) Heterotropic effects of chloride on the ligation microstates of hemoglobin at constant water activity. Biophys J 71:2106–2116

Huster RJ, Eichele T, Enriquez-Geppert S, Wollbrink A, Kugel H, Konrad C, Pantev C (2011) Multimodal imaging of functional networks and event-related potentials in performance monitoring. Neuroimage 56:1588–1597

Huster RJ, Debener S, Eichele T, Herrmann CS (2012) Methods for simultaneous EEG-fMRI: an introductory review. J Neurosci 32:6053–6060

Ives JR, Warach S, Schmitt F, Edelman RR, Schomer DL (1993) Monitoring the patient's EEG during echo planar MRI. Electroencephalogr Clin Neurophysiol 87:417–420

Javitt DC (2012) Glycine transport inhibitors in the treatment of schizophrenia. Handb Exp Pharmacol 213:367–399

Karch S, Jager L, Karamatskos E, Graz C, Stammel A, Flatz W, Lutz J, Holtschmidt-Taschner B, Genius J, Leicht G, Pogarell O, Born C, Moller HJ, Hegerl U, Reiser M, Soyka M, Mulert C (2008) Influence of trait anxiety on inhibitory control in alcohol-dependent patients: simultaneous acquisition of ERPs and BOLD responses. J Psychiatr Res 42:734–745

Karch S, Thalmeier T, Lutz J, Cerovecki A, Opgen-Rhein M, Hock B, Leicht G, Hennig-Fast K, Meindl T, Riedel M, Mulert C, Pogarell O (2010) Neural correlates (ERP/fMRI) of voluntary selection in adult ADHD patients. Eur Arch Psychiatry Clin Neurosci 260:427–440

Kraut MA, Calhoun V, Pitcock JA, Cusick C, Hart J Jr (2003) Neural hybrid model of semantic object memory: implications from event-related timing using fMRI. J Int Neuropsychol Soc 9:1031–1040

Kruggel F, Wiggins CJ, Herrmann CS, von Cramon DY (2000) Recording of the event-related potentials during functional MRI at 3.0 Tesla field strength. Magn Reson Med 44:277–282

Lachaux JP, Fonlupt P, Kahane P, Minotti L, Hoffmann D, Bertrand O, Baciu M (2007) Relationship between task-related gamma oscillations and BOLD signal: new insights from combined fMRI and intracranial EEG. Hum Brain Mapp 28:1368–1375

Laufs H, Kleinschmidt A, Beyerle A, Eger E, Salek-Haddadi A, Preibisch C, Krakow K (2003) EEG-correlated fMRI of human alpha activity. Neuroimage 19:1463–1476

Laufs H, Daunizeau J, Carmichael DW, Kleinschmidt A (2008) Recent advances in recording electrophysiological data simultaneously with magnetic resonance imaging. Neuroimage 40:515–528

Lavric A, Bregadze N, Benattayallah A (2011) Detection of experimental ERP effects in combined EEG-fMRI: evaluating the benefits of interleaved acquisition and independent component analysis. Clin Neurophysiol 122:267–277

Lehmann D, Ozaki H, Pal I (1987) EEG alpha map series: brain micro-states by space-oriented adaptive segmentation. Electroencephalogr Clin Neurophysiol 67:271–288

Lehmann D, Faber PL, Galderisi S, Herrmann WM, Kinoshita T, Koukkou M, Mucci A, Pascual-Marqui RD, Saito N, Wackermann J, Winterer G, Koenig T (2005) EEG microstate duration and syntax in acute, medication-naive, first-episode schizophrenia: a multicenter study. Psychiatry Res 138:141–156

Lei X, Qiu C, Xu P, Yao D (2010) A parallel framework for simultaneous EEG/fMRI analysis: methodology and simulation. Neuroimage 52:1123–1134

Leicht G, Kirsch V, Giegling I, Karch S, Hantschk I, Moller HJ, Pogarell O, Hegerl U, Rujescu D, Mulert C (2010) Reduced early auditory evoked gamma-band response in patients with schizophrenia. Biol Psychiatry 67:224–231

Leicht G, Karch S, Karamatskos E, Giegling I, Moller HJ, Hegerl U, Pogarell O, Rujescu D, Mulert C (2011) Alterations of the early auditory evoked gamma-band response in first-degree relatives of patients with schizophrenia: hints to a new intermediate phenotype. J Psychiatr Res 45:699–705

Lemieux L (2004) Electroencephalography-correlated functional MR imaging studies of epileptic activity. Neuroimaging Clin N Am 14:487–506

Lemieux L, Allen PJ, Franconi F, Symms MR, Fish DR (1997) Recording of EEG during fMRI experiments: patient safety. Magn Reson Med 38:943–952

Levan P, Maclaren J, Herbst M, Sostheim R, Zaitsev M, Hennig J (2013) Ballistocardiographic artifact removal from simultaneous EEG-fMRI using an optical motion-tracking system. Neuroimage 75:1–11

Li R, Principe JC, Bradley M, Ferrari V (2007) Robust single-trial ERP estimation based on spatiotemporal filtering. Conf Proc IEEE Eng Med Biol Soc 2007:5206–5209

Lindsley DB (1952) Psychological phenomena and the electroencephalogram. Electroencephalogr Clin Neurophysiol 4:443–456

Liu Z, Kecman F, He B (2006) Effects of fMRI-EEG mismatches in cortical current density estimation integrating fMRI and EEG: a simulation study. Clin Neurophysiol 117:1610–1622

Liu Z, de Zwart JA, van Gelderen P, Kuo LW, Duyn JH (2012) Statistical feature extraction for artifact removal from concurrent fMRI-EEG recordings. Neuroimage 59:2073–2087

Llinas RR (1988) The intrinsic electrophysiological properties of mammalian neurons: insights into central nervous system function. Science 242:1654–1664

Logothetis NK, Pauls J, Augath M, Trinath T, Oeltermann A (2001) Neurophysiological investigation of the basis of the fMRI signal. Nature 412:150–157

Makeig S, Westerfield M, Jung TP, Enghoff S, Townsend J, Courchesne E, Sejnowski TJ (2002) Dynamic brain sources of visual evoked responses. Science 295:690–694

Mandelkow H, Halder P, Boesiger P, Brandeis D (2006) Synchronization facilitates removal of MRI artefacts from concurrent EEG recordings and increases usable bandwidth. Neuroimage 32:1120–1126

Mandelkow H, Brandeis D, Boesiger P (2010) Good practices in EEG-MRI: the utility of retrospective synchronization and PCA for the removal of MRI gradient artefacts. Neuroimage 49:2287–2303

Mangalathu-Arumana J, Beardsley SA, Liebenthal E (2012) Within-subject joint independent component analysis of simultaneous fMRI/ERP in an auditory oddball paradigm. Neuroimage 60:2247–2257

Mantini D, Perrucci MG, Cugini S, Ferretti A, Romani GL, Del Gratta C (2007) Complete artifact removal for EEG recorded during continuous fMRI using independent component analysis. Neuroimage 34:598–607

Mathiesen C, Caesar K, Lauritzen M (2000) Temporal coupling between neuronal activity and blood flow in rat cerebellar cortex as indicated by field potential analysis. J Physiol 523(Pt 1):235–246

Matsuura T, Kanno I (2001) Quantitative and temporal relationship between local cerebral blood flow and neuronal activation induced by somatosensory stimulation in rats. Neurosci Res 40:281–290

Michels L, Luchinger R, Koenig T, Martin E, Brandeis D (2012) Developmental changes of BOLD signal correlations with global human EEG power and synchronization during working memory. PLoS One 7:e39447

Mobascher A, Warbrick T, Brinkmeyer J, Musso F, Stoecker T, Jon Shah N, Winterer G (2012) Nicotine effects on anterior cingulate cortex in schizophrenia and healthy smokers as revealed by EEG-informed fMRI. Psychiatry Res 204:168–177

Moosmann M, Ritter P, Krastel I, Brink A, Thees S, Blankenburg F, Taskin B, Obrig H, Villringer A (2003) Correlates of alpha rhythm in functional magnetic resonance imaging and near infrared spectroscopy. Neuroimage 20:145–158

Moosmann M, Eichele T, Nordby H, Hugdahl K, Calhoun VD (2008) Joint independent component analysis for simultaneous EEG-fMRI: principle and simulation. Int J Psychophysiol 67:212–221

Mukamel R, Gelbard H, Arieli A, Hasson U, Fried I, Malach R (2005) Coupling between neuronal firing, field potentials, and FMRI in human auditory cortex. Science 309:951–954

Mulert C, Jager L, Schmitt R, Bussfeld P, Pogarell O, Moller HJ, Juckel G, Hegerl U (2004) Integration of fMRI and simultaneous EEG: towards a comprehensive understanding of localization and time-course of brain activity in target detection. Neuroimage 22:83–94

Mulert C, Jager L, Propp S, Karch S, Stormann S, Pogarell O, Moller HJ, Juckel G, Hegerl U (2005) Sound level dependence of the primary auditory cortex: simultaneous measurement with 61-channel EEG and fMRI. Neuroimage 28:49–58

Mulert C, Hepp P, Leicht G, Karch S, Lutz J, Moosmann M, Reiser M, Hegerl U, Pogarell O, Möller HJ, Jäger L (2007) High frequency oscillations in the gamma-band and the cor-responding BOLD signal: trial-by-trial coupling of EEG and fMRI reveals the involvement of the thalamic reticular nucleus (TRN). Neuroimage 36(Supplement 1):S1–S125

Mulert C, Seifert C, Leicht G, Kirsch V, Ertl M, Karch S, Moosmann M, Lutz J, Moller HJ, Hegerl U, Pogarell O, Jager L (2008) Single-trial coupling of EEG and fMRI reveals the involvement of early anterior cingulate cortex activation in effortful decision making. Neuroimage 42:158–168

Mulert C, Leicht G, Hepp P, Kirsch V, Karch S, Pogarell O, Reiser M, Hegerl U, Jager L, Moller HJ, McCarley RW (2010) Single-trial coupling of the gamma-band response and the corresponding BOLD signal. Neuroimage 49:2238–2247

Mulert C, Lemieux L (2010) EEG-fMRI - Physiological Basis, Technique, and Applications. Berlin Heidelberg, Springer

Mullinger K, Debener S, Coxon R, Bowtell R (2008) Effects of simultaneous EEG recording on MRI data quality at 1.5, 3 and 7 tesla. Int J Psychophysiol 67: 178–188

Mullinger KJ, Yan WX, Bowtell R (2011) Reducing the gradient artefact in simultaneous EEG-fMRI by adjusting the subject's axial position. Neuroimage 54:1942–1950

Mullinger KJ, Castellone P, Bowtell R (2013a) Best current practice for obtaining high quality EEG data during simultaneous FMRI. J Vis Exp, 76

Mullinger KJ, Havenhand J, Bowtell R (2013b) Identifying the sources of the pulse artefact in EEG recordings made inside an MR scanner. Neuroimage 71:75–83

Murakami S, Okada Y (2006) Contributions of principal neocortical neurons to magnetoencephalography and electroencephalography signals. J Physiol 575:925–936

Musso F, Brinkmeyer J, Mobascher A, Warbrick T, Winterer G (2010) Spontaneous brain activity and EEG microstates. A novel EEG/fMRI analysis approach to explore resting-state networks. Neuroimage 52:1149–1161

Musso F, Brinkmeyer J, Ecker D, London MK, Thieme G, Warbrick T, Wittsack HJ, Saleh A, Greb W, de Boer P, Winterer G (2011) Ketamine effects on brain function–simultaneous fMRI/EEG during a visual oddball task. Neuroimage 58:508–525

Nakamura W, Anami K, Mori T, Saitoh O, Cichocki A, Amari S (2006) Removal of ballistocardiogram artifacts from simultaneously recorded EEG and fMRI data using independent component analysis. IEEE Trans Biomed Eng 53:1294–1308

Negishi M, Abildgaard M, Nixon T, Constable RT (2004) Removal of time-varying gradient artifacts from EEG data acquired during continuous fMRI. Clin Neurophysiol 115:2181–2192

Neuner I, Arrubla J, Felder J, Shah NJ (2013) Simultaneous EEG-fMRI acquisition at low, high and ultra-high magnetic fields up to 9.4T: perspectives and challenges. Neuroimage [Epub ahead of print]

Niazy RK, Beckmann CF, Iannetti GD, Brady JM, Smith SM (2005) Removal of FMRI environment artifacts from EEG data using optimal basis sets. Neuroimage 28:720–737

Niedermeyer E, Lopes Da Silva F (2004) Electroencephalography: basic principles, clinical applications and related fields, 5th edn. Lippincott Williams & Wilkins, Philadelphia

Nierhaus T, Gundlach C, Goltz D, Thiel SD, Pleger B, Villringer A (2013) Internal ventilation system of MR scanners induces specific EEG artifact during simultaneous EEG-fMRI. Neuroimage 74:70–76

Niessing J, Ebisch B, Schmidt KE, Niessing M, Singer W, Galuske RA (2005) Hemodynamic signals correlate tightly with synchronized gamma oscillations. Science 309:948–951

Nunez PL (1995) Neocortical dynamics and human EEG rhythms. Oxford University Press, New York

Nunez PL, Silberstein RB (2000) On the relationship of synaptic activity to macroscopic measurements: does co-registration of EEG with fMRI make sense? Brain Topogr 13:79–96

Opitz B, Mecklinger A, Friederici AD, von Cramon DY (1999) The functional neuroanatomy of novelty processing: integrating ERP and fMRI results. Cereb Cortex 9:379–391

Ostwald D, Porcaro C, Mayhew SD, Bagshaw AP (2012) EEG-fMRI based information theoretic characterization of the human perceptual decision system. PLoS One 7:e33896

Pascual-Marqui RD, Michel CM, Lehmann D (1994) Low resolution electromagnetic tomography: a new method for localizing electrical activity in the brain. Int J Psychophysiol 18:49–65

Pascual-Marqui RD, Lehmann D, Koenig T, Kochi K, Merlo MC, Hell D, Koukkou M (1999) Low resolution brain electromagnetic tomography (LORETA) functional imaging in acute, neuroleptic-naive, first-episode, productive schizophrenia. Psychiatry Res 90:169–179

Plichta MM, Wolf I, Hohmann S, Baumeister S, Boecker R, Schwarz AJ, Zangl M, Mier D, Diener C, Meyer P, Holz N, Ruf M, Gerchen MF, Bernal-Casas D, Kolev V, Yordanova J, Flor H, Laucht M, Banaschewski T, Kirsch P, Meyer-Lindenberg A, Brandeis D (2013) Simultaneous EEG and fMRI reveals a causally connected subcortical-cortical network during reward anticipation. J Neurosci 33:14526–14533

Rauch A, Rainer G, Logothetis NK (2008) The effect of a serotonin-induced dissociation between spiking and perisynaptic activity on BOLD functional MRI. Proc Natl Acad Sci U S A 105:6759–6764

Razavi N, Jann K, Koenig T, Kottlow M, Hauf M, Strik W, Dierks T (2013) Shifted coupling of EEG driving frequencies and FMRI resting state networks in schizophrenia spectrum disorders. PLoS One 8:e76604

Regenbogen C, De Vos M, Debener S, Turetsky BI, Mossnang C, Finkelmeyer A, Habel U, Neuner I, Kellermann T (2012) Auditory processing under cross-modal visual load investigated with simultaneous EEG-fMRI. PLoS One 7:e52267

Ritter P, Freyer F, Curio G, Villringer A (2008) High-frequency (600Hz) population spikes in human EEG delineate thalamic and cortical fMRI activation sites. Neuroimage 42:483–490

Ritter P, Becker R, Freyer F, Villringer A (2010) EEG quality: the image acquisition artefact. In: Mulert C, Lemieux L (eds) EEG-fMRI - physiological basis, technique, and applications. Springer, Berlin, pp 153–171

Sadaghiani S, Scheeringa R, Lehongre K, Morillon B, Giraud AL, Kleinschmidt A (2010) Intrinsic connectivity networks, alpha oscillations, and tonic alertness: a simultaneous electroencephalography/functional magnetic resonance imaging study. J Neurosci 30:10243–10250

Scheeringa R, Mazaheri A, Bojak I, Norris DG, Kleinschmidt A (2011) Modulation of visually evoked cortical FMRI responses by phase of ongoing occipital alpha oscillations. J Neurosci 31:3813–3820

Scheeringa R, Petersson KM, Kleinschmidt A, Jensen O, Bastiaansen MC (2012) EEG alpha power modulation of FMRI resting-state connectivity. Brain Connect 2:254–264

Scheibe C, Ullsperger M, Sommer W, Heekeren HR (2010) Effects of parametrical and trial-to-trial variation in prior probability processing revealed by simultaneous electroencephalogram/functional magnetic resonance imaging. J Neurosci 30:16709–16717

Schmitt A, Hasan A, Gruber O, Falkai P (2011) Schizophrenia as a disorder of disconnectivity. Eur Arch Psychiatry Clin Neurosci 261(Suppl 2):S150–S154

Shmuel A (2009) Locally measured neuronal correlates of functional MRI signals. In: Mulert C, Lemieux L (eds) EEG-fMRI. Springer, Berlin, pp 63–82

Speckmann EJ, Elger CE (1999) Introduction to the neurophysiological basis of the EEG and DC potentials. In: Niedermeyer E, Lopes da Silva F (eds) Electroencephalography, basic principles, clinical applications and related fields, 5th edn. Lippincott. Williams & Wilkins, Philadelphia, pp 15–27

Spencer KM, Nestor PG, Niznikiewicz MA, Salisbury DF, Shenton ME, McCarley RW (2003) Abnormal neural synchrony in schizophrenia. J Neurosci 23:7407–7411

Sun L, Hinrichs H (2009) Simultaneously recorded EEG-fMRI: removal of gradient artifacts by subtraction of head movement related average artifact waveforms. Hum Brain Mapp 30:3361–3377

Tallon-Baudry C, Bertrand O (1999) Oscillatory gamma activity in humans and its role in object representation. Trends Cogn Sci 3:151–162

Thomsen K, Offenhauser N, Lauritzen M (2004) Principal neuron spiking: neither necessary nor sufficient for cerebral blood flow in rat cerebellum. J Physiol 560:181–189

Traub RD, Whittington MA, Colling SB, Buzsaki G, Jefferys JG (1996) Analysis of gamma rhythms in the rat hippocampus in vitro and in vivo. J Physiol 493(Pt 2):471–484

Ullsperger M, Debener S (2010) Simultaneous EEG and fMRI. Oxford University Press

Vasios CE, Angelone LM, Purdon PL, Ahveninen J, Belliveau JW, Bonmassar G (2006) EEG/(f)MRI measurements at 7 Tesla using a new EEG cap ("InkCap"). Neuroimage 33:1082–1092

Villringer A, Mulert C, Lemieux L (2010) Principles of multimodal functional imaging and data integration. In: Mulert C, Lemieux L (eds) EEG-fMRI - physiological basis, technique, and applications. Springer, Berlin, pp 3–17

Viswanathan A, Freeman RD (2007) Neurometabolic coupling in cerebral cortex reflects synaptic more than spiking activity. Nat Neurosci 10:1308–1312

Von Helmholtz HLF (2004) Some laws concerning the distribution of electric currents in volume conductors with applications to experiments on animal electricity. Proc IEEE 92:868–870

Wang Z, Maier A, Leopold DA, Logothetis NK, Liang H (2007) Single-trial evoked potential estimation using wavelets. Comput Biol Med 37:463–473

Wang L, Saalmann YB, Pinsk MA, Arcaro MJ, Kastner S (2012) Electrophysiological low-frequency coherence and cross-frequency coupling contribute to BOLD connectivity. Neuron 76:1010–1020

Woods TO (2007) Standards for medical devices in MRI: present and future. J Magn Reson Imaging 26: 1186–1189

Yan WX, Mullinger KJ, Brookes MJ, Bowtell R (2009) Understanding gradient artefacts in simultaneous EEG/fMRI. Neuroimage 46:459–471

Zotev V, Phillips R, Yuan H, Misaki M, Bodurka J (2014) Self-regulation of human brain activity using simultaneous real-time fMRI and EEG neurofeedback. Neuroimage 85 (Pt 3):985–995

Diffusion Tensor Imaging

5

Inga Katharina Koerte and Marc Muehlmann

Abbreviations

AD	Axial diffusivity
ADC	Apparent diffusion coefficient
ADHD	Attention deficit hyperactivity disorder
BET	Brain extraction tool
CSF	Cerebrospinal fluid
DKI	Diffusion kurtosis imaging
DTI	Diffusion tensor imaging
DWI	Diffusion-weighted imaging
FA	Fractional anisotropy
MD	Trace, mean diffusivity
MRI	Magnetic resonance imaging
RD	Radial diffusivity
ROI	Region of interest
TBSS	Tract-based spatial statistics

5.1 Introduction

Diffusion tensor imaging (DTI) is an advanced magnetic resonance imaging (MRI) technique that provides detailed information about tissue microstructure such as fiber orientation, axonal density, and degree of myelination. DTI is based on the measurement of the diffusion of water molecules. It was developed in the early 1990s and since then has been applied in a wide variety of scientific and clinical settings, especially, but not limited to, the investigation of brain pathology in schizophrenia (Shenton et al. 2001; Qiu et al. 2009, 2010), Alzheimer's disease (Damoiseaux et al. 2009; Mielke et al. 2009; Avants et al. 2010; Gold et al. 2010; Jahng et al. 2011), and autism (Pugliese et al. 2009; Cheng et al. 2010; Fletcher et al. 2010). This chapter describes the basics of the technique, outlines resources needed for acquisition, and focuses on post-processing techniques and statistical analyses with and without a priori hypotheses.

5.2 Basic Principles of Diffusion

5.2.1 Brownian Motion

Water molecules are constantly moving due to their thermal energy at temperatures above zero. In 1827, Robert Brown described the phenomena of particles suspended in fluid that move randomly (Brown 1827). Starting from the same position, the movement paths of single water molecules are unpredictable and end up in unforeseeable positions at time x. The process of diffusion depends upon a concentration gradient that was described by Fick's first law of diffusion. Fick's second law calculates the distance a particle diffuses in a certain time based on the diffusion coefficient (Fick 1855). Albert Einstein derived Fick's second law from thermodynamic laws, thereby building toward what has evolved

I.K. Koerte, MD (✉) • M. Muehlmann, MD
Division of Radiology, Ludwig-Maximilians-University, Munich, Germany
e-mail: inga.koerte@med.lmu.de; muehlmann@bwh.harvard.edu

C. Mulert, M.E. Shenton (eds.), *MRI in Psychiatry*,
DOI 10.1007/978-3-642-54542-9_5, © Springer Berlin Heidelberg 2014

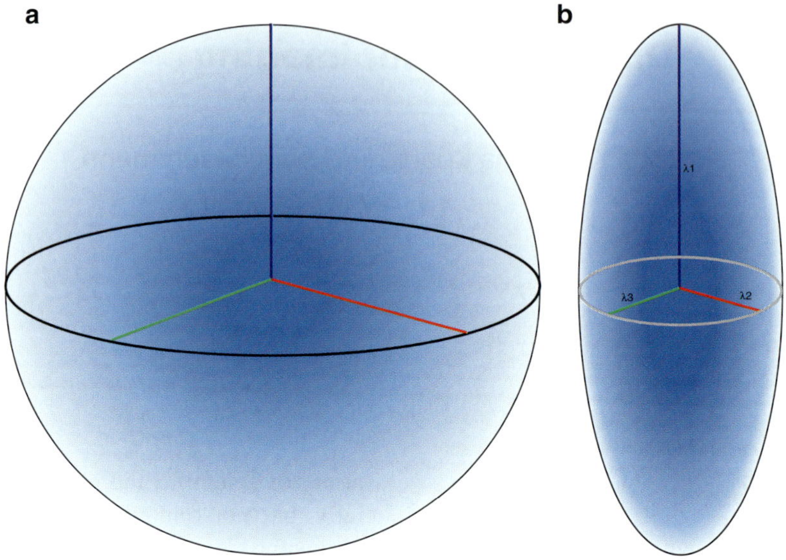

Fig. 5.1 Isotropic and anisotropic diffusion

to diffusion theory. Further, Einstein was able to mathematically explain Brownian motion by determining the mean squared displacement of a single particle (Einstein 1905).

5.2.2 Isotropic Diffusion

During the process of diffusion, the random nature of Brownian motion causes molecules to move passively from a region with high concentration to a region with low concentration. If a drop of ink is dropped in a glass filled with water, the drop seems to stay steady in place at first, before it enlarges spherically until the entire water in the glass becomes homogenously colored. While the mixing of the fluids ceases macroscopically as a result of the vanishing concentration gradient, the molecules keep randomly moving, which is also called self-diffusion. Looking at a large number of water molecules originating from the same position, we observe that the distribution of the end points after time x follows a 3D Gaussian distribution. This is also referred to as free, unrestricted, or isotropic diffusion (Fig. 5.1a).

Diffusion in biological tissues is affected by the microstructure of the respective tissue. In biological tissue, the water molecules still randomly diffuse in each 3D direction (isotropic) but are slowed down in their range due to collision with large-scaled biological molecules. Therefore, the diffusion coefficient for water in biological tissues, which is called apparent diffusion coefficient (ADC), is lower than in water. In an unorganized biological environment, ADC is independent from the direction of measurement, as a result of the spherical shape of diffusion. However, the radius of diffusion is reduced compared to unhindered diffusion when considering equal time points.

5.2.3 Anisotropic Diffusion

In highly structured biological tissues, such as white matter of the brain, the movement of water molecules is restricted by cell membranes, myelin sheaths, and microfilament (Jones et al. 2013). The diffusion probability of water molecules in structured tissue describes an ellipsoid based on the preferred diffusion direction parallel to the axon (Fig. 5.1b) (Basser et al. 1994a, b).

In organized tissues such as the white matter, the ADC depends on the direction from which it is measured. The ADC is higher if the diffusion gradients are aligned to the preferred diffusion direction and lower if measured perpendicular to it.

5.3 Magnetic Resonance Diffusion-Weighted Imaging

5.3.1 Magnetic Resonance Imaging

Magnetic resonance imaging (MRI) relies on the nuclear spin which is innate to all elements with an odd-numbered composition of nuclear particles in their nucleus (Rabi et al. 1938). One example is the hydrogen nucleus consisting of a single proton. The proton can be understood as a charged moving particle while the conoidal movement as described by its spin axis is called precession (Loeffler 1981). Tissues of the human body contain between 22 % (bones) and 95 % (blood plasma) water and are therefore rich in protons. In a static magnetic field, the spins are precessing parallel and antiparallel to the magnetic field axis. Both orientations can be understood as different states of energy, where the upward spins are on a lower energetic level compared to the downward spins. In order to keep the energetic balance, there are more upward-oriented spins. The surplus of spins oriented upward causes a constant magnetization along the axis of the static magnetic field (z-axis which is also called longitudinal magnetization). MRI uses radiofrequency pulses to disrupt the constant magnetization and tilt the spin axis into the xy-plane (transversal magnetization) (Loeffler 1981). The moving sum of magnetic vectors in the xy-plane induces an alternating voltage in the reception coil (MRI signal). The process in which the transversal magnetization returns to the longitudinal magnetization is accompanied with energy loss and is called spin-lattice relaxation or T1-relaxation. In contrast the T2-relaxation is caused by energetic shifts between the spins while dephasing, while no energy is getting lost to the surrounding (Loeffler 1981).

5.3.2 Diffusion-Weighted Imaging

In diffusion-weighted imaging (DWI), the strength of the MRI signal depends on the mean displacement of water molecules. Here a strong signal indicates a low diffusion in the direction of the magnetic gradient field. Accordingly, the signal loss is higher when the mean displacement of the water molecules is higher along the gradient. The mean displacement of water molecules is described as the apparent diffusion coefficient (ADC). In the clinical setting, DWI is often used in the assessment of stroke (Warach et al. 1995; Brunser et al. 2013) because extracellular diffusion restriction due to the swelling of neurons is a highly sensitive marker of cerebral ischemia. This was first described by Moseley in 1990 (1990).

5.3.2.1 Stejskal-Tanner Sequence

To obtain diffusion-weighted images, the Stejskal-Tanner sequence is applied (Stejskal and Tanner 1965). A basic scheme of this sequence is given in Fig. 5.2. The process starts by applying a 90° radiofrequency pulse after time (t) to shift the magnetization vector into the xy-plane. The equal precessing spins create a magnetic momentum that is rotating in the xy-plane thereby inducing a voltage that can be referred to as the MRI signal. Due to intermolecular interactions and also due to field inhomogeneity, the first equally precessing spins start to diphase, resulting in a decay of the T2 signal. Now a 180° radiofrequency pulse is applied to turn the whole spin magnetization around. Previously slower precessing spins are now put ahead of faster precessing spins. After time ($2t$ = TE), all spins precess in phase again, and the recovered magnetic momentum induces a voltage, the spin echo. One could also state that by using the 180° pulse, the effect of field inhomogeneities is reversed. For diffusion weighting, two magnetic gradients are applied in addition to the static magnetic field and the radiofrequency pulse (Stejskal and Tanner 1965). Here the first linear gradient weakens or strengthens the local field strength resulting in a certain position-dependent precession frequency of corresponding spins. For

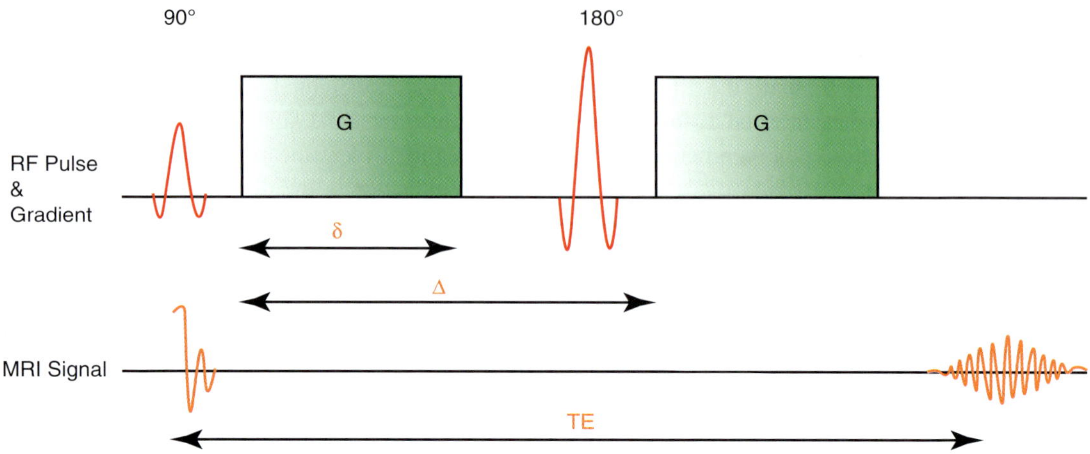

Fig. 5.2 Stejskal-Tanner Sequence

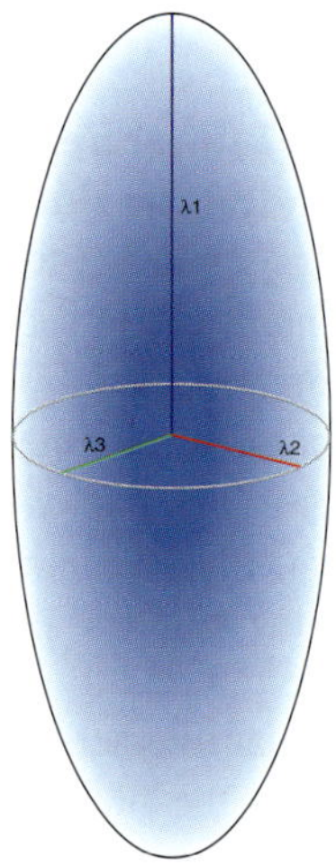

Fig. 5.3 Eigenvalues of the diffusion ellipsoid

spins in solid tissue, the effect of this gradient whether weakened or strengthened is reversed by the second gradient pulse (Stejskal and Tanner 1965). The signal loss is larger in tissues where the containing spins can move more rapidly between the applications of both gradient pulses.

5.3.3 Diffusion Tensor Imaging

Diffusion tensor imaging (DTI) emerged from DWI in 1994 (Basser et al. 1994a, b). DTI is also based on the diffusion of water molecules as acquired with the Stejskal-Tanner sequence. The main advantage over DWI is that it makes it possible to quantify not only the diffusion in total but also the directionality of diffusion, which is called the diffusion tensor. In order to quantify the diffusion tensor, six different diffusion measurements and directions are necessary to sufficiently satisfy the tensor equation (Pierpaoli et al. 1996). Information content and reliability of the calculation of the diffusion tensor can be further improved by increasing the number of diffusion directions. Moreover, for complex post-processing analyses such as tractography, at least 30 diffusion directions are recommended in order to obtain reliable results (Spiotta et al. 2012). Increasing the number of diffusion directions enhances the precision of the reconstruction while the statistical rotational variance is reduced (Jones et al. 2013). The 3D information obtained from various diffusion gradients is used to calculate a multilinear transformation, the diffusion tensor. The diffusion tensor can be described using eigenvalues and eigenvectors and visualized as the diffusion ellipsoid (Fig. 5.3). The axes of the three-dimensional coordinate system are called eigenvectors, while the length of their measure is called eigenvalues. The three eigenvalues are symbolized by the Greek letter lambda ($\lambda 1$, $\lambda 2$, $\lambda 3$). The eigenvalues are then used to calculate different parameters of the diffusion tensor.

5.3.4 Quantitative Parameters of the Diffusion Tensor

Commonly used quantitative parameters of the diffusion tensor are axial diffusivity (AD); radial diffusivity (RD); trace, mean diffusivity (MD); and fractional anisotropy (FA). These parameters are calculated for each voxel of the imaging data set and make it possible to characterize, noninvasively, tissue on a microscopic level.

5.3.4.1 Axial Diffusivity and Radial Diffusivity

The largest eigenvalue is named $\lambda 1$ and is oriented parallel to the axonal structures. It is therefore equal to the diffusion tensor parameter axial diffusivity:

$$\mathrm{AD} = \lambda 1$$

The eigenvalues $\lambda 2$ and $\lambda 3$ are the values along the two short axis of the coordinate system. They are therefore used to calculate the radial diffusivity which describes the diffusion perpendicular to the main diffusion direction. RD is calculated by dividing the sum of the short-axis eigenvalues by 2:

$$\mathrm{RD} = \frac{(\lambda 2 + \lambda 3)}{2}$$

5.3.4.2 Trace and Mean Diffusivity

The sum of all three eigenvalues is called trace. Mean diffusivity is obtained when trace is divided by three:

$$\mathrm{Trace} = \lambda 1 + \lambda 2 + \lambda 3$$

$$\mathrm{MD} = \frac{(\lambda 1 + \lambda 2 + \lambda 3)}{3}$$

5.3.4.3 Fractional Anisotropy

The calculation of fractional anisotropy weighs the preferred component of diffusivity to obtain information about the quantity of directionality. Therefore, the square root of the diffusion differences $\left(\sqrt{(\lambda 1-\lambda 2)^2+(\lambda 1-\lambda 3)^2+(\lambda 2-\lambda 3)^2}\right)$ is used and then averaged by the square root of sums of squares $\left(\sqrt{(\lambda 1)^2+(\lambda 2)^2+(\lambda 3)^2}\right)$. The first term $\left(\sqrt{\frac{1}{2}}\right)$ scales the FA to values between 0 and 1:

$$\mathrm{FA} = \sqrt{\frac{1}{2}}\frac{\sqrt{(\lambda 1-\lambda 2)^2+(\lambda 1-\lambda 3)^2+(\lambda 2-\lambda 3)^2}}{\sqrt{(\lambda 1)^2+(\lambda 2)^2+(\lambda 3)^2}}$$

Based on these DTI parameters, assumptions can be made regarding the microstructure of the brain such as the orientation and diameter of axons and their surrounding structures such as the myelin sheaths. For example, decreased RD and therefore higher FA can be based on increased axonal density, reduced axonal diameter, and thicker myelin sheaths (Song et al. 2002). A lower FA can be caused by a number of factors, for example, a larger axon diameter (Takahashi et al. 2002), a lower axon density (Takahashi et al. 2002), and/or increased membrane permeability. Increased mean diffusivity represents an increase in extracellular diffusion of water molecules and can be the result of, for example, vasogenic edema (Filippi et al. 2001).

5.4 Post-Processing of DTI Data

5.4.1 Quality of DTI Data

The first and one of the most important steps in post-processing diffusion MRI data is to check the quality of the acquired data. The importance of the quality check arises from the susceptibility of diffusion MR images to artifacts, for example, motion artifact and/or susceptibility artifacts. While there are no strict guidelines for the post-processing of DTI data or standardized quality assurance, it is recommended that manual inspection be performed on the data, slice by slice, in order to ensure that reliable results are obtained for all acquired diffusion directions. All regions of interest must also be free from any artifact to avoid possible confounds in the calculation of diffusion parameters.

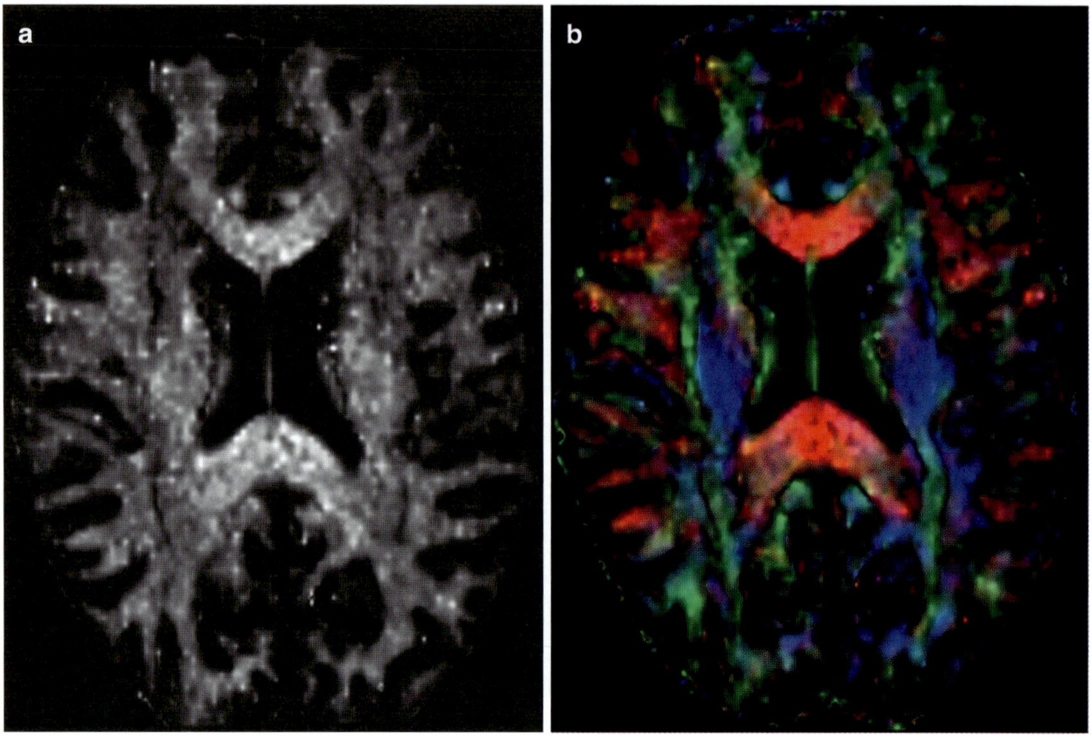

Fig. 5.4 Scalar (**a**) and color-coded (**b**) FA maps

5.4.2 Diffusion Tensor Masks

Most post-processing techniques require a mask for the individual data set that contains only the brain and the surrounding cerebrospinal fluid (CSF) space. These masks can be obtained automatically using a software such as the brain extraction tool (BET), which is part of the FSL software (FMRIB Software Library, The Oxford Centre of Functional MRI of the Brain – FMRIB). To guarantee a high quality of data and to take the individual's anatomy into account, the masks created automatically are then manually edited. The latter can be a labor-intensive process.

5.4.3 Visualization of DTI Parameters

The easiest way to visualize the DTI parameters is by displaying the value as the level of gray in each respective voxel (Fig. 5.4a). By adding the information of the main diffusion direction using a color scheme, color-coded maps can be obtained (Fig. 5.4b). Blue-colored voxels represent the main diffusion direction parallel to the *z*-axis, that is, inferior to superior in human individuals. In contrast, the green-colored voxels represent the main diffusion direction parallel to the *y*-axis, that is, anterior to posterior. Red-colored voxels represent a main diffusion direction parallel to the *x*-axis, that is, left to right. While the axis and the directionality of diffusion in a specific tissue can be calculated, it is not possible to obtain information about the direction of the flow of actual information, which can occur bidirectionally.

5.4.4 Tract-Based Spatial Statistics

To investigate differences in DTI parameters between groups, a statistical approach can be used which takes all voxels of each individual into account. The most commonly used method is called tract-based spatial statistics (TBSS) (Smith et al. 2006), which is part of the freely avail-

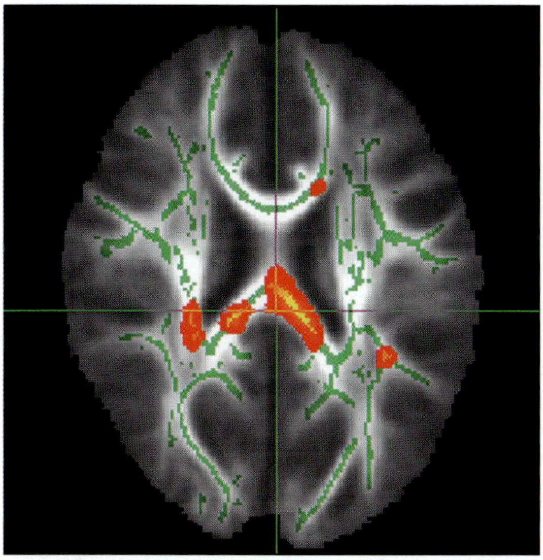

Fig. 5.5 Example of significant TBSS result

able software package FSL (FMRIB Software Library, The Oxford Centre of Functional MRI of the Brain – FMRIB).

The individual DTI images are scaled to a standard space and aligned either to each other or to a standard image. After aligning the individual images, a three-dimensional map is created which then contains the mean of all individuals. This map can be used to calculate the mean skeleton representing the center elements of the main white matter tracts of all included individuals. TBSS uses a nonparametric statistical test that analyzes all voxels corresponding to the mean skeleton. Additional co-variables such as age or gender can be included in the analysis allowing for the investigation of dependencies. Results are corrected for multiple testing and are displayed in a 3D image (Fig. 5.5).

Advantages of using TBSS analysis include the possibility of testing without an a priori hypothesis and the possibility of including co-variables. Drawbacks of this voxel-wise approach are the loss of individual information due to partial volume effects and interindividual differences. One should also keep in mind that only the skeleton representing the main white matter tracts is analyzed, whereas voxels containing peripheral white matter or gray matter are not included. Additional caution should be taken when investigating groups with a more heterogeneous anatomy such as children, especially when the age range is wide.

5.4.5 Region-of-Interest Analysis

Region-of-interest (ROI) analyses are commonly used to test a priori hypotheses. They are performed at the level of the individual's data set. The ROI can be placed either manually or semiautomatically. When placing ROIs, special care should be taken to only include the structure of interest. Therefore, it is recommended that group standardized thresholds be used during the selection process to minimize partial volume effects. Partial volume effects result from the fact that a three-dimensional voxel can include variable amounts of information from both the structure of interest and the surrounding tissue. This can be done by using freely available software packages such as 3DSlicer (SPL, BWH, Harvard, Boston) (Fedorov et al. 2012) or FreeSurfer (NMR, MGH, Harvard, Boston) (Dale et al. 1999). ROIs can be analyzed with respect to their size and the mean of the respective diffusion parameter on the level of the individual. Obtained values can then be used for offline statistical analysis.

5.4.6 Tractography

Tractography makes it possible to both visualize in three dimensions and quantify fiber tracts (Basser et al. 2000). Based on the preferred diffusion direction of neighboring voxels, the likelihood of (fiber) connectivity between the voxels can also be estimated. Fiber tracts can be identified by defining a single ROI as starting point or by using multiple ROIs that the fiber tract passes. The multiple ROI approach is especially useful to investigate long and complex tracts. Additional exclusion ROIs can be used to limit the obtained fibers, for example, when the corticospinal tract is of interest, fibers crossing to the other hemisphere may have to be excluded by an exclusion ROI in the sagittal midline. Tractography can be performed using the software 3DSlicer (Fedorov et al. 2012).

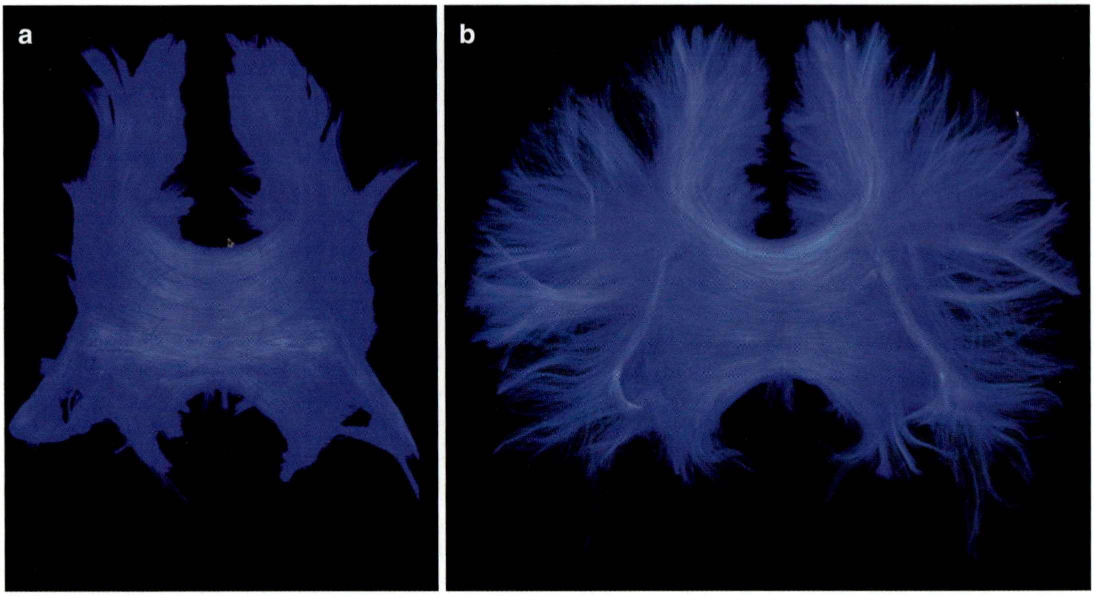

Fig. 5.6 Single and Two-Tensor Tractography

Identified tracts can then be visualized. Moreover, the tract can be analyzed regarding the number of reconstructed fibers as well as their DTI parameters RD, AD, trace, MD, and FA. This makes tractography a more specific approach compared to ROI-based analyses or TBSS because only the structure of interest is analyzed. However, tractography is limited when it comes to crossing of fibers, for example, transcallosal fibers crossing the downward projecting corticospinal tract. This is because the tensor includes all information on the diffusion in the respective voxel assuming that all fibers in this voxel travel in the same direction. However, in large parts of the brain's white matter, voxels contain crossing fiber tracts that result in decreased anisotropy in a given voxel. This false decrease in anisotropy leads to fiber tracts ending abruptly. Multi-tensor algorithms calculate more than one tensor per voxel thereby making it possible to follow and to analyze tracts that travel in different directions (Fig. 5.6) (Malcolm et al. 2010).

5.5 Limitations

Limitations of DTI include the vulnerability to artifacts such as motion artifact, susceptibility artifact, and distortion artifact (eddy current). To acquire DTI data covering the entire brain with high and reliable quality, a sequence takes between 5 and 10 min, which is relatively long compared to structural sequences such as T1-weighted and T2-weighted sequences. Especially in children and in the elderly, it can be difficult to obtain a complete DTI sequence without motion artifact.

Another limitation may be seen by the fact that the diffusion tensor is calculated for an entire voxel and therefore represents at least 1 mm^3, whereas the diameter of an axon is about 7 μm. This means that DTI parameters may represent a combination of different structures rather than a single axon or fiber. However, to date, DTI is the most sensitive noninvasive technique that makes it possible to characterize microscopic brain white matter (Jones et al. 2013).

5.6 Future Directions

Although diffusion tensor imaging already looks back at close to two decades of success, there are still promising further developments regarding the acquisition of DTI data as well as post-processing techniques. Advanced post-processing algorithms using compressed sensing have been developed by Rathi et al. to enhance the quality of images acquired with a lower number of diffusion directions in order to reduce the scanning time (Michailovich et al. 2011; Rathi et al. 2011). This advancement is of particular importance when investigating children, the elderly, and patients who have difficulties staying motionless in the scanner, for example, patients with Parkinson's disease, attention deficit hyperactivity disorder (ADHD), patients experiencing pain, or Alzheimer's disease.

Pasternak et al. have also introduced a new approach that makes it possible to differentiate two contributing compartments to the diffusion MRI signal (Pasternak et al. 2009). One compartment contains freely diffusing water molecules in the extracellular space, while the other compartment includes water molecules that are affected by structures in biological tissues. Therefore, this method enables the quantification of free water in the extracellular space and the assessment of tissue-specific, free-water-corrected FA and MD values (Pasternak et al. 2009). This method was recently applied to patients experiencing a first episode of schizophrenia where a widespread increase in free water was observed, suggesting a neuroinflammatory response, in addition to a reduction in free-water-corrected FA in the frontal regions of the brain, suggesting axonal degeneration (Pasternak et al. 2012).

Another promising approach is the diffusion kurtosis imaging (DKI). This approach is based on the multi-shell DWI, where multiple b-values are acquired in one session. In addition to common DTI parameters, the mean deviation from the Gaussian diffusion is calculated and taken into account. Cheung et al. were able to show higher sensitivity of kurtosis imaging compared with common DTI, in finding changes of white and gray matter during rat brain development (Cheung et al. 2009). Helpern used DKI to noninvasively compare the microstructural development of patients with ADHD compared to controls. Group comparison revealed that DKI-corrected RD and AD values stagnated in the ADHD group, while the control group showed an age-related increase in the complexity of white matter microstructure (Helpern et al. 2011). These preliminary results support the assumption that DKI may be a sensitive addition to DTI evaluation of white matter microstructure.

It is thus clear that there is much work to be done using diffusion measures in clinical populations and that this field is still in its infancy. A closer look at measures such as free water to delineate further possible neuroinflammatory versus neurodegenerative aspects of disease may provide important new insights as will the ability to more clearly specify the underlying changes by using DKI to specify further changes in the brain related to myelin disturbances. These advances, along with algorithms that greatly reduce the time in the magnet while providing the same information as acquisitions that by now take 60 min, will provide a level of sensitivity and specificity of pathology, in vivo, that is not possible today, although we are close.

References

Avants BB, Cook PA, Ungar L et al (2010) Dementia induces correlated reductions in white matter integrity and cortical thickness: a multivariate neuroimaging study with sparse canonical correlation analysis. Neuroimage 50(3):1004–1016

Basser PJ, Mattiello J, LeBihan D (1994a) Estimation of the effective self-diffusion tensor from the NMR spin echo. J Magn Reson B 103(3):247–254

Basser PJ, Mattiello J, LeBihan D (1994b) MR diffusion tensor spectroscopy and imaging. Biophys J 66(1): 259–267

Basser PJ, Pajevic S, Pierpaoli C et al (2000) In vivo fiber tractography using DT-MRI data. Magn Reson Med 44(4):625–632

Brown R (1827) A brief account of microscopical observations made in the month on June, July and August 1827 on the particles contained in the pollen of plants and on the general existence of active molecules in organic and inorganic bodies. Philos Mag 4(4): 161–173

Brunser AM, Hoppe A, Illanes S et al (2013) Accuracy of diffusion-weighted imaging in the diagnosis of stroke in patients with suspected cerebral infarct. Stroke 44(4):1169–1171

Cheng Y, Chou KH, Chen IY et al (2010) Atypical development of white matter microstructure in adolescents with autism spectrum disorders. Neuroimage 50(3):873–882

Cheung MM, Hui ES, Chan KC et al (2009) Does diffusion kurtosis imaging lead to better neural tissue characterization? A rodent brain maturation study. Neuroimage 45(2):386–392

Dale AM, Fischl B, Sereno MI (1999) Cortical surface-based analysis. I. Segmentation and surface reconstruction. Neuroimage 9(2):179–194

Damoiseaux JS, Smith SM, Witter MP et al (2009) White matter tract integrity in aging and Alzheimer's disease. Hum Brain Mapp 30(4):1051–1059

Einstein A (1905) Über die von der molekularkinetischen Theorie der Wärme geforderte Bewegung von in ruhenden Flüssigkeiten suspendierten Teilchen. Ann Phys 17:549–560

Fedorov A, Beichel R, Kalpathy-Cramer J et al (2012) 3D Slicer as an image computing platform for the quantitative imaging network. Magn Reson Imaging 30(9): 1323–1341

Fick A (1855) Ueber diffusion. Ann Phys 170(1):59–86

Filippi M, Cercignani M, Inglese M et al (2001) Diffusion tensor magnetic resonance imaging in multiple sclerosis. Neurology 56(3):304–311

Fletcher PT, Whitaker RT, Tao R et al (2010) Microstructural connectivity of the arcuate fasciculus in adolescents with high-functioning autism. Neuroimage 51(3):1117–1125

Gold BT, Powell DK, Andersen AH et al (2010) Alterations in multiple measures of white matter integrity in normal women at high risk for Alzheimer's disease. Neuroimage 52(4):1487–1494

Helpern JA, Adisetiyo V, Falangola MF et al (2011) Preliminary evidence of altered gray and white matter microstructural development in the frontal lobe of adolescents with attention-deficit hyperactivity disorder: a diffusional kurtosis imaging study. J Magn Reson Imaging 33(1):17–23

Jahng GH, Xu S, Weiner MW et al (2011) DTI studies in patients with Alzheimer's disease, mild cognitive impairment, or normal cognition with evaluation of the intrinsic background gradients. Neuroradiology 53(10):749–762

Jones DK, Knosche TR, Turner R (2013) White matter integrity, fiber count, and other fallacies: the do's and don'ts of diffusion MRI. Neuroimage 73: 239–254

Loeffler W, Oppelt A (1981) Physical principles of NMR tomography. Eur J Radiol 1(4):338–344

Malcolm JG, Shenton ME, Rathi Y (2010) Filtered multitensor tractography. IEEE Trans Med Imaging 29(9): 1664–1675

Michailovich O, Rathi Y, Dolui S (2011) Spatially regularized compressed sensing for high angular resolution diffusion imaging. IEEE Trans Med Imaging 30(5): 1100–1115

Mielke MM, Kozauer NA, Chan KC et al (2009) Regionally-specific diffusion tensor imaging in mild cognitive impairment and Alzheimer's disease. Neuroimage 46(1):47–55

Moseley ME, Cohen Y, Mintorovitch J et al (1990) Early detection of regional cerebral ischemia in cats: comparison of diffusion- and T2-weighted MRI and spectroscopy. Magn Reson Med 14(2): 330–346

Pasternak O, Sochen N, Gur Y et al (2009) Free water elimination and mapping from diffusion MRI. Magn Reson Med 62(3):717–730

Pasternak O, Westin CF, Bouix S et al (2012) Excessive extracellular volume reveals a neurodegenerative pattern in schizophrenia onset. J Neurosci 32(48): 17365–17372

Pierpaoli C, Jezzard P, Basser PJ et al (1996) Diffusion tensor MR imaging of the human brain. Radiology 201(3):637–648

Pugliese L, Catani M, Ameis S et al (2009) The anatomy of extended limbic pathways in Asperger syndrome: a preliminary diffusion tensor imaging tractography study. Neuroimage 47(2):427–434

Qiu A, Zhong J, Graham S et al (2009) Combined analyses of thalamic volume, shape and white matter integrity in first-episode schizophrenia. Neuroimage 47(4): 1163–1171

Qiu A, Tuan TA, Woon PS et al (2010) Hippocampal-cortical structural connectivity disruptions in schizophrenia: an integrated perspective from hippocampal shape, cortical thickness, and integrity of white matter bundles. Neuroimage 52(4):1181–1189

Rabi IIZ, Millman JR, Kusch SP (1938) A new method of measuring nuclear magnetic moment. Phys Rev 53(4):318

Rathi Y, Michailovich O, Setsompop K et al (2011) Sparse multi-shell diffusion imaging. Med Image Comput Comput Assist Interv 14(Pt 2):58–65

Shenton ME, Dickey CC, Frumin M et al (2001) A review of MRI findings in schizophrenia. Schizophr Res 49(1–2):1–52

Smith SM, Jenkinson M, Johansen-Berg H et al (2006) Tract-based spatial statistics: voxelwise analysis of multi-subject diffusion data. Neuroimage 31(4):1487–1505

Song SK, Sun SW, Ramsbottom MJ et al (2002) Dysmyelination revealed through MRI as increased radial (but unchanged axial) diffusion of water. Neuroimage 17(3):1429–1436

Spiotta AM, Bartsch AJ, Benzel EC (2012) Heading in soccer: dangerous play? Neurosurgery 70(1):1–11; discussion 11

Stejskal EO, Tanner J (1965) Spin diffusion measurements: spin echoes in the presence of time-dependent field gradient. J Chem Phys 42(1):288–292

Takahashi M, Hackney DB, Zhang G et al (2002) Magnetic resonance microimaging of intraaxonal water diffusion in live excised lamprey spinal cord. Proc Natl Acad Sci U S A 99(25):16192–16196

Warach S, Gaa J, Siewert B et al (1995) Acute human stroke studied by whole brain echo planar diffusion-weighted magnetic resonance imaging. Ann Neurol 37(2):231–241

Magnetic Resonance Spectroscopy

6

Sai Merugumala, Saadalah Ramadan, Walker Keenan, Huijun Liao, Luke Y-J. Wang, and Alexander Lin

Abbreviations

^{13}C	Carbon 13
2D COSY	Two-dimensional correlated spectroscopy
2D JPRESS	Two-dimensional J-resolved point-resolved spectroscopy
AD	Alzheimer's disease
BD	Bipolar disorder
CEST	Chemical exchange saturation transfer
Cho	Choline
Cr	Creatine
CRLB	Cramer–Rao lower bounds
CSI	Chemical shift imaging
FFT	Fast Fourier transform
FID	Free induction decay
fMRI	Functional magnetic resonance imaging
GABA	*Gamma*-aminobutyric acid
GABA-T	GABA transaminase
GAD	Glutamic acid decarboxylase
Gln	Glutamine
Glu	Glutamate
GluCEST	Glutamate chemical exchange saturation transfer
Glx	Both glutamate and glutamine
GPC	GlyceroPhosphoCholine
GPx	Glutathione peroxidase
GSH	Glutathione
GSSG	Glutathione disulfide
JPRESS	J-resolved point-resolved spectroscopy
LCModel	Linear compilation model
MCI	Mild cognitive impairment
MEGA-PRESS	Mescher–Garwood point-resolved spectroscopy
mI	Myoinositol
MRI	Magnetic resonance imaging
MRS	Magnetic resonance spectroscopy
NAA	*N*-Acetylaspartic acid
NAAG	*N*-Acetylaspartylglutamic acid
NMDA	*N*-Methyl-D-aspartic acid

S. Merugumala, MS • W. Keenan • H. Liao, BS
Department of Radiology, Brigham and Women's Hospital, Harvard Medical School, Boston, MA, USA

S. Ramadan, PhD
Faculty of Health, School of Health Sciences, University of Newcastle, Hunter Building, Callaghan, NSW 2308, Australia

L.Y.-J. Wang, MD
Department of Radiology, Brigham and Women's Hospital, Harvard Medical School, Boston, MA, USA

Department of Anesthesiology, Perioperative and Pain Medicine, Boston Children's Hospital, Harvard Medical School, Boston, MA, USA

A. Lin, PhD (✉)
Department of Radiology, Brigham and Women's Hospital, Harvard Medical School, Boston, MA, USA

Psychiatric Neuroimaging Laboratory, Brigham and Women's Hospital, Harvard Medical School, Boston, MA, USA
e-mail: aplin@partners.org

C. Mulert, M.E. Shenton (eds.), *MRI in Psychiatry*,
DOI 10.1007/978-3-642-54542-9_6, © Springer Berlin Heidelberg 2014

NMR	Nuclear magnetic resonance
PCP	Phencyclidine
PC	Phosphorylcholine
PCr	Phosphocreatine
PET	Positron emission tomography-PRESS, Point-resolved spectroscopy
ProFit	Prior knowledge fitting
RF	Radio frequency
ROI	Region of interest
ROS	Reactive oxygen species
STEAM	Stimulated echo acquisition mode
SVS	Single voxel spectroscopy
TCA	Tricarboxylic acid cycle
TE	Echo time
TI	Inversion time
TR	Relaxation time

6.1 Magnetic Resonance Spectroscopy: The Virtual Biopsy

6.1.1 Introduction

Research in neuroscience has slowly been trying to bridge the gap between understanding the difference between the brain and the mind. In this process, many of the concepts in cognitive psychology and psychiatry are being rewritten as development and application of new technologies further our understanding of the brain. The goal is to answer many of the questions that cognitive psychologists and psychiatrists have been seeking to understand. A particularly productive field of research has involved medical imaging that utilizes techniques such as functional magnetic resonance imaging (fMRI; as reviewed in this textbook (Posner and Raichle 1997)) or positron emission tomography (PET) (Friston 1997). Both techniques purport to measure neuronal activation; however, both only measure specific properties of the brain: blood flow in fMRI and glucose uptake with PET. The relationship between these measures and actual neuronal activity is a major assumption that is generally unquestioned. A growing number of scientists, psychiatrists, and philosophers are nonetheless beginning to question this assumption.

Until recently, there were no direct means to measure brain activity aside from animal studies in which neuronal activity was measured by electrodes implanted deep in the brain, clearly a very invasive technique (although electroencephalography as reviewed in this textbook can provide a related measure). A technique that could directly measure neuronal activity, could provide a noninvasive assay, and could explore brain systems as opposed to areas would be the ideal cognitive psychological tool: it exists as magnetic resonance spectroscopy (MRS), a noninvasive neurochemical assay that can directly measure neuronal activity, biochemical systems, and metabolic processes utilized by neurons.

6.1.2 Physics of MRS

A technique older than both fMRI and PET, nuclear magnetic resonance (NMR) spectroscopy, was first introduced in the early 1950s. Typically employed by the organic chemist to define chemical structures, NMR spectroscopy can in fact be applied to the human body, or more specifically to the human brain, providing quantitative and noninvasive studies of neurobiochemistry. The idea of using NMR to study the human body is by no means new. In fact, very soon after the first successful NMR experiments were conducted on test tube samples in 1946, Bloch, the pioneer of NMR, obtained a strong proton signal by placing his finger in the radio-frequency (RF) coil of his spectrometer (Andrew 1980). Although the potential of using NMR in biological systems was prevalent throughout the 1950s and early 1960s, it was not until the development of high-field superconducting magnets together with the emergence of fast Fourier transform (FFT) NMR did the potential of NMR in biological systems become realized (Ernst 1992). During this same time period, scientists began to realize that NMR machines could produce images of body structures that very quickly led to the advent of magnetic resonance imaging (MRI) as we know it today (Lauterbur 1989). This introduced whole body magnets with magnetic field strengths of 0.5–9.4 T, of which there are presently over 25,000 installed worldwide (Rinck 2012).

It was not until the late 1980s that industry standardization of spectroscopy occurred, heralding a new era of clinical applications of NMR, giving a new name (by dropping the word "nuclear"), magnetic resonance spectroscopy, or MRS, as well as a new lease to life to spectroscopy. The physics behind the technique is fundamentally the same as that utilized by early NMR: the sample (patient) slides into the bore of the magnetic where there is a uniform static magnetic field such that unbound nuclei are oriented parallel and antiparallel according to their spin state. A radio-frequency (RF) coil is then placed near the body part of the patient to maximize the amount of signal that can be obtained. The nuclei are then excited by electromagnetic radiation in the form of a pulse sequence via a transmitter. The nuclei absorb the energy, altering the nuclear spin. As the nuclei precess or "relax" back to their original state, energy is released and detected by the receiver in the RF coil as a free induction decay, which is promptly fast Fourier transformed into the resulting spectra (Fig. 6.1).

Each chemical resonates at established frequencies that upon Fourier transform results in peaks at specific locations, or chemical shifts, along the *x*-axis. The chemical shift of each chemical is governed by the structure of the chemical, in particular, the grouping of the hydrogen atoms as single or multiple peaks (singlets and multiplets), and proximity to other hydrogen-containing groups (*J*-coupling). These chemical shifts are often expressed as "parts per million," which can be confusing as it does not relate to the concentration of the chemical, but historically has been used to express the frequency. The use of this nomenclature is so that it is not dependent upon the field strength of the magnet used such that the resonance frequencies are the same between a 1.5 T MRI scanner and a 3.0 T scanner. The concentration of the metabolite is expressed along the *y*-axis as the height of the peak such that higher concentrations result in higher peaks and vice versa. Quantitation of these peak heights can provide an objective measure of brain biochemistry that is similar to a blood test

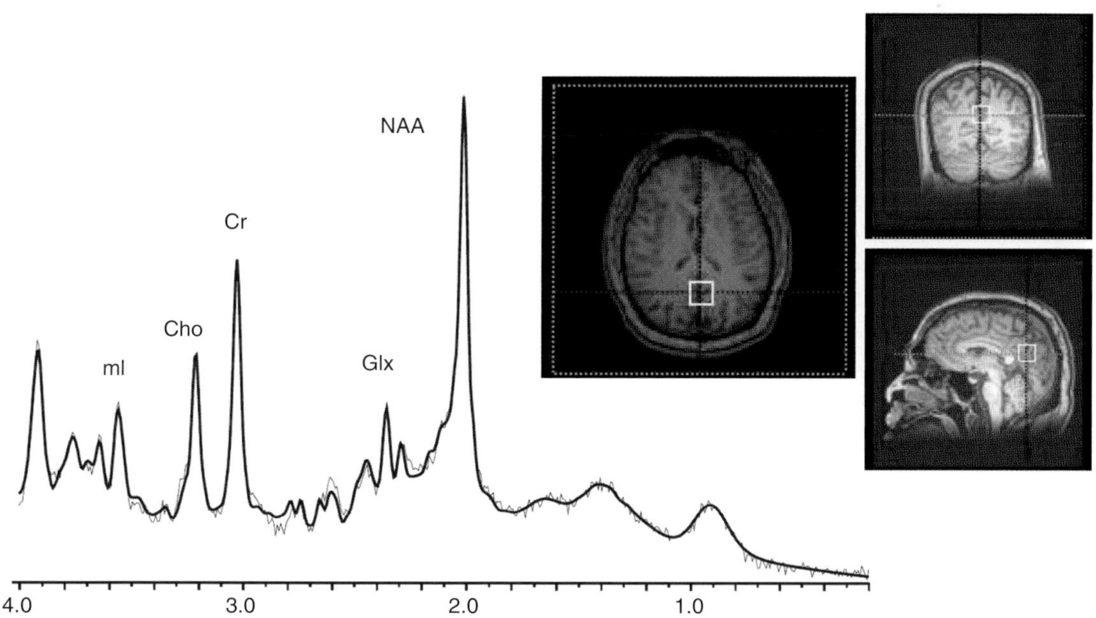

Fig. 6.1 Representative proton spectrum acquired from the posterior cingulate gyrus. Data were acquired using a 3 T clinical MRI scanner and point-resolved spectroscopy localization with an echo time of 30 ms and repetition time of 2 s. Voxel location is shown in the *inset* to the right in the axial, coronal, and sagittal planes. Major metabolites of *N*-acetylaspartate (*NAA*), glutamate and glutamine (*Glx*), creatine (*Cr*), choline (*Cho*), and myoinositol (*mI*) are labeled

or lab analysis: different concentrations are measured and reported, which can then be used for diagnosis or disease characterization. More importantly, however, the role of each of the chemicals is tied to metabolic and physiological processes within the brain that directly relate to cognitive or neuronal processes. The remainder of this chapter will be devoted to the major metabolites that are detected by MRS, as well as the different methods by which the data can be collected.

6.2 *N*-Acetylaspartate (NAA): Neuronal Marker

As indicated in the spectra in Fig. 6.1, one of the chemicals that can be measured by MRS is NAA. The primary resonance of NAA is at 2.02 ppm. It is an amino acid derivative synthesized in neurons and transported down axons. It is therefore a putative "marker" of viable neurons, axons, and dendrites. Studies where NAA was labeled with fluorescent tags demonstrated that NAA was distributed throughout the neuron and axon (Moffett et al. 1993). Studies have also correlated the concentration of NAA in the brain with the number of neurons measured (Urenjak et al. 1992). The ability to quantifiably measure neuron populations allows MRS to provide a diagnostic tool that no other radiological technique can match: an ability to literally "count" the number of active brain cells using a completely noninvasive and quantitative technique by simply measuring the peak height of the NAA chemical in the MRS spectra. This can be utilized in a number of practical means. For example, every metabolite has a "normal" concentration that generates a pattern of peaks that is the same from person to person unless there is an underlying pathology. Diagnosis with MRS can therefore be made either by comparing the numeric values of metabolite concentrations or by recognizing abnormal patterns of peaks in the spectra. With either method, it is the increase or decrease in the concentrations of the metabolites that is diagnostic for pathology as has been shown across a broad variety of diseases (Lin et al. 2005; Harris et al. 2006; Moffett et al. 2007). However, it should also be noted that NAA is a cerebral metabolite that participates in a number of metabolic processes, and therefore, the interpretation of NAA as solely a neuronal marker is somewhat of an oversimplification (Barker 2001). It is important to understand the context of NAA metabolism, as detailed below.

6.2.1 NAA Metabolism

NAA is formed by the transamination of glutamate (Glu)with oxaloacetate which leads to aspartate (Asp) (Cooper et al. 1970), and in the presence of acetyl coenzyme A and NAA synthase, (L-aspartate *N*-acetyltransferase), aspartate (Asp) is converted to NAA. This process occurs primarily, but not exclusively, in neurons, where it exhibits a very high intracellular–extracellular gradient. NAA is hydrolyzed back into aspartate and acetate by aspartoacylase (*N*-acetyl-L-ASPARTATE amidohydrolase) in mature oligodendrocytes only. In some neurons, a portion of NAA is turned into *N*-acetylaspartylglutamate (NAAG) in the presence of glutamate (Glu) by an NAAG synthase (*N*-acetylaspartate-L-glutamate ligase) (Baslow 2000). NAA and NAAG are major neuronal osmolytes, and the large amount of NAA and NAAG present in the brain can serve as cellular reservoirs for Asp and Glu. Since both are polar and ionizable hydrophilic molecules that undergo a regulated efflux into extracellular fluid, they can also play a role in water movement out of neurons (Baslow 2000). In addition, the fact that the NAA-metabolizing enzyme aspartoacylase is an integral component of the myelin sheath suggests that intraneuronal NAA may supply the acetyl groups for myelin lipid synthesis (Chakraborty et al. 2001). Finally, the cell type-specific metabolic enzymes in the NAA and NAAG cycle indicate a three-cell compartmentalization involving neurons, astrocytes, and oligodendrocytes. Those neurons that may synthesize NAAG from NAA and Glu target the release of NAAG to astrocytes,

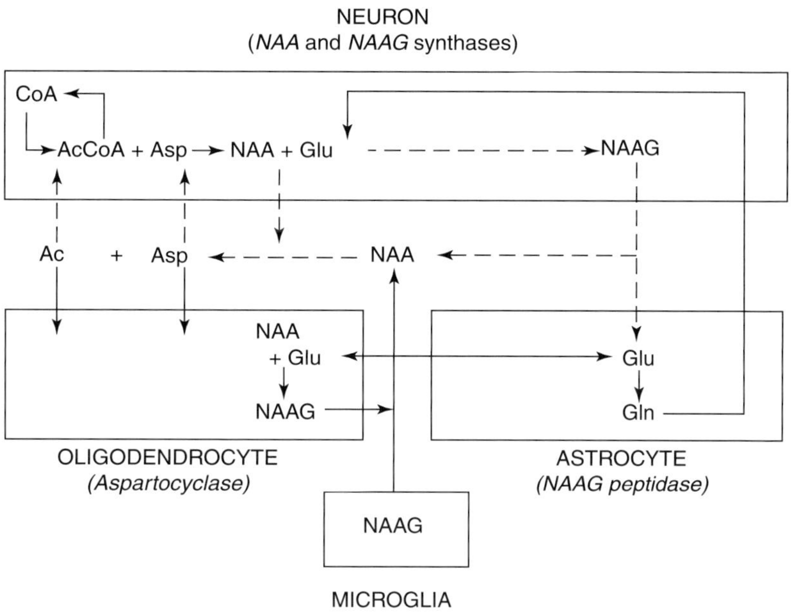

Fig. 6.2 Tricellular NAA–NAAG cycle operating between neurons, astrocytes, and oligodendrocytes (Modified from Baslow 2010). *CoA* coenzyme A, *Acetyl-CoA* acetyl coenzyme A, *Asp* aspartate, *Glu* glutamate, *Gln* glutamine, *NAA N*-Acetylaspartate, *NAAG N*-Acetylaspartylglutamate

where it is cleaved into NAA and Glu. The Glu is taken up by astrocytes, where it may be returned to neurons via the glutamate–glutamine cycle. The residual NAA extracellular products are removed and hydrolyzed by oligodendrocytes (Baslow 2000). This unique tricellular metabolic cycle, as shown in Fig. 6.2, involving NAA and NAAG, provides a potential glial cell-specific signaling pathway, which can also be reflected in changes in different diseases (Baslow 2010).

6.2.2 Measuring NAA Using Single Voxel Spectroscopy

As a "virtual biopsy," localization is required when examining specific regions of the brain involved in disease. Localization is defined by a region of interest, or voxel, from which the spectrum is acquired and, hence, described as single voxel spectroscopy (SVS). This region is a cube within the brain that can be visualized and placed in the sagittal, coronal, and axial planes as shown in Fig. 6.1. The localization is achieved by utilizing a pulse sequence that acquires the MRS signal. Though several sequences, and permutations of sequences, exist which are capable of yielding an in vivo MRS spectra, many of them are research techniques and not widely available (although discussed later in this chapter). The two most commonly used sequences are PRESS and STEAM as described below:

6.2.2.1 Point-Resolved Spectroscopy (PRESS)

The point-resolved spectroscopy (PRESS) sequence is a standard part of the software package that accompanies the two most popular MRI manufacturers, Siemens (Erlangen, Germany) and General Electric (Waukesha, Wisconsin). On Siemens scanners, it is known as "svs_se." On General Electric machines, the sequence is called "Probe-P." The PRESS sequence was originally developed in 1987 by Paul Bottomley at General Electric (Bottomley 1987) and utilizes three orthogonal magnetic field gradients along the x-, y-, and z-axes using slice-selective 90° pulse followed by two slice-selective 180° pulses to select a specific three-dimensional voxel, within which a proton MRS spectrum can be obtained.

6.2.2.2 Stimulated Echo Acquisition Mode (STEAM)

STEAM also uses a series of three consecutive slice-selective 90° pulses for localization. However, the sequence only yields half as much detectable magnetization; therefore, the signal-to-noise ratio is halved. Nonetheless, adequate water suppression—which is necessary for all MRS techniques, as the metabolites to be detected exist at much lower concentrations than water—is more easily accomplished via STEAM (Haase et al. 1986). Shorter echo times are also attainable with STEAM, which may provide an advantage when detecting short T2 metabolites such as γ-aminobutyric acid (GABA).

6.2.2.3 Post-processing Single Voxel Data

SVS data can be post-processed and analyzed in several different ways. All major MR platforms have their own methods of reconstructing the data where the details are somewhat different, but the end result is generally an automated or semiautomated fitting of the metabolite peaks and a quantitative measure of major metabolites (NAA, creatine, choline, myoinositol), usually as a ratio to creatine to provide a normalization factor to account for differences in the peak area or amplitude between subjects which will differ depending on parameters such as the voxel size, number of averages, transmit gain, etc. The MRS data can also be exported for further analysis using MRS post-processing packages such as LCModel (Provencher 1993), jMRUI/AMARES (Vanhamme et al. 1997), etc. The advantage of this more sophisticated post-processing is that it allows for "absolute quantitation" of the metabolites by incorporating prior knowledge or additional measures (Kreis et al. 1993) that provide a concentration in millimolar per kilogram wet weight that would be equivalent to in vitro measurements of metabolite concentrations. While the accuracy of the quantitation, in terms of "absolute" quantitation, is somewhat controversial, the advantage of this method is that the metabolites can be quantified without being a ratio to creatine. This is especially important in diseases that can cause a change in creatine, in which case it is unclear whether it is the major metabolite that has altered or creatine or both.

6.2.3 Measuring NAA Using Chemical Shift Imaging (CSI)

One of the drawbacks of SVS is the natural limits of a single area of acquisition. Obtaining spectra in many different areas of the lesion would be time consuming. Multivoxel spectroscopy, also called chemical shift imaging (CSI) or spectroscopic imaging, overcomes this issue by adding an additional phase-encoding step that allows for spatial encoding of a larger volume such that it is divided into smaller voxels that can be summed or selected chemical shifts can be color-coded into chemical shift maps. In experimental duration, some two to three times longer than that in which a single voxel method acquires the spectrum of a single ROI, the CSI technique (Maudsley et al. 1983) can collect an array of spectra from a single plane.

6.2.3.1 CSI Pulse Sequence

STEAM and PRESS are still utilized for localization; however, phase-encoding gradients are employed to encode the spatial dimensions, and the MR signal is collected in the absence of any gradient in order to maintain the spectroscopic information. Each acquired ROI contains an MR spectrum that allows for the assessment of the metabolic profile of a specific location or allows for visualization of the spatial distribution of specific metabolites of interest. CSI also allows for the acquisition of smaller volumes than in single voxel techniques (as small as 0.4 cm^3 at higher magnetic field strength). This has the advantage over single voxel MRS techniques, as multiple brain regions can be assessed using the CSI technique which can also be completed offline via post-processing routines that allow for positioning of different areas of interest, assuming that they are contained with the larger region of interest selected for CSI acquisitions. Furthermore, there are additional variants of CSI that utilize different phase-encoding methods to either reduce scan time or allow for additional spatial

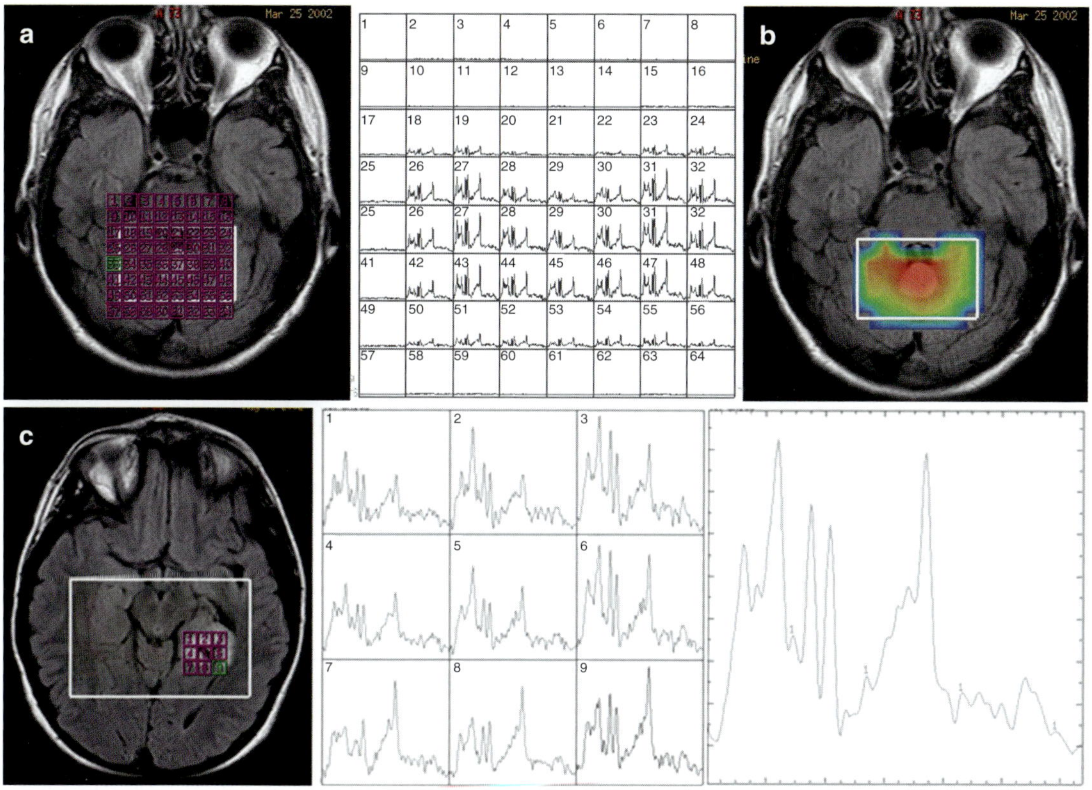

Fig. 6.3 Representative chemical shift imaging. (**a**) *Left*: MRI image with grid of voxels indicated in purple. *Right*: reconstruction of CSI data as individual spectra. (**b**) Metabolite map of NAA. (**c**) Simplified display of CSI data as a small grid of spectra as indicated on the image (*left*) and individual spectra (*middle*) and summed spectra (*right*)

resolution and information such as spiral phase encoding, echo planar sequences, multislice sequences, and parallel imaging methods (Posse et al. 2013).

6.2.3.2 Reconstruction of CSI Data

The main advantage of CSI is that the spatial information is retained such that during post-processing one can characterize the spectra in a number of different methods. The field of view can be partitioned into individual voxels determined by the resolution of the phase encoding. For example, if 16 frequency encodes and 16 phase encodes are used in a 16-cm^2 field of view and 1-cm slice thickness, each voxel will have a resolution of $1 \times 1 \times 1$ cm^3, and a spectrum can be reconstructed in each voxel as shown in Fig. 6.3a. Analyzing each voxel would be time consuming, and therefore, there are two different ways to display the data. First, one can create a "metabolite map" which selects a certain region of the spectrum, for example, 2.0–2.05 ppm for NAA, which creates a map based on the signal intensity of the spectrum from that spectral region, which is then displayed on top of the MRI image as shown in Fig. 6.3b. The spatial resolution of the metabolite maps are often interpolated which may result in over-interpretation of the true voxel resolution. There are also issues with signal correction where some areas may appear to have increased SNR but overall signal may be increased in that region, leading to possible misinterpretation of the data. It is therefore somewhat dangerous to rely solely on metabolite maps without evaluating the individual spectra. One can sum all of the spectra from a certain region of the spectrum, much like a single voxel acquisition as shown in Fig. 6.3c. This procedure is repeated at multiple regions in the CSI volume. The major advantage of CSI is that in post-processing, the ROI can be readily shifted to any location with the excitation volume.

However, CSI suffers from the disadvantage that the shape of individual ROIs is less well defined than in single voxel techniques. This can result in the adjacent ROIs being contaminated by large-amplitude signals from surrounding ROIs. Furthermore, CSI is not as reproducible as SVS (Rosen and Lenkinski 2007) and therefore is less sensitive to the more subtle changes. Finally, on most clinical scanners, the majority of CSI acquisitions are set up for long echo MRS which also eliminates important diagnostic metabolites such as glutamate Glu or myoinositol (mI), although for NAA, the focus of this section, it is sufficient.

6.2.4 Whole-Brain NAA

As described above, single voxel and CSI methods suffer from issues such as lipid contamination, voxel registration, as well as assumptions of T1 and T2 relaxation times. One method of addressing the aforementioned issues is to take advantage of the fact that NAA is the most prominent peak within the MRS spectrum (thus having a high signal-to-noise ratio) that acquires signal from the entire head and is thus named whole-brain NAA (WBNAA) spectroscopy (Rigotti et al. 2007). Unlike the single voxel and CSI methods, WBNAA does not utilize localization and instead acquires data immediately after excitation (TE = 0 ms) with a very long repetition time (TR = 10 s) and a short inversion time (TI = 940 ms). The inversion pulse is alternated between each acquisition where the TI is designed to null the NAA signal. The even and odd acquisitions are then subtracted from one another which nulls those metabolites, including lipid, that have a short T1 relaxation time. As NAA has a long T1 relaxation time (1.4 s), it remains visible. As a result of the long TR, there are no T1- or T2-weighting effects. The WBNAA signal is then normalized to the total brain volume as can be obtained from brain segmentation. This method operates under the assumption that all NAA arises from within the brain, and given the lack of localization, results of such studies would imply global effects of the disease.

6.2.5 NAA in Psychiatric Diseases

6.2.5.1 Schizophrenia

A recent meta-review of spectroscopy papers in schizophrenia revealed 103 papers alone that focused on MRS which included a total of 2,067 subjects and 2,115 controls of which the great majority of the studies focused on treated chronically ill patients (Kraguljac et al. 2012). All papers utilized either the SVS or the CSI methods, thereby allowing for regionalized differences to be measured. Decreased levels of NAA were found all across the brain including structures such as the hippocampus, thalamus, and frontal and temporal lobes. While the temporal lobes showed the greatest decreases in NAA with a standardized mean difference of −0.72, the variability of measurements in the temporal lobe were also greater than any other area of the brain. The reason for this variability is due to the susceptibility artifacts that occur due to the air and bone tissue interfaces within this brain region that result in distortions and signal loss (Olman et al. 2009). The basal ganglia and frontal lobe showed the most consistent decreases in NAA. While WBNAA measures have not been applied to schizophrenia, these global changes in NAA throughout the schizophrenic brain would indicate that this method would be sensitive to these changes.

6.2.5.2 Dementia

Given the relationship between NAA and neuronal viability, NAA is a key biomarker for dementia. Dozens of papers have shown decreases in NAA in both cortical and white matter regions of the brain including the posterior cingulate gyrus, the temporal lobe, and the occipital lobe (Graff-Radford and Kantarci 2013). In combination with myoinositol (mI), a putative glial marker, NAA provides up to 90 % sensitivity and 95 % specificity for distinguishing Alzheimer's disease (AD) from healthy subjects (Kantarci 2007). In a recent study, the ratio of NAA/mI predicted progression to mild cognitive impairment (MCI) or dementia in 214 subjects (Kantarci et al. 2013), which supports previous findings of reduced NAA in MCI patients. Furthermore, MRS is valuable in differentiating between different types of dementia such as the Lewy body, frontal lobe, and vascular dementia (Shonk et al. 1995; Kantarci et al. 2004). While the temporal lobe is often thought to be the ideal location for detecting Alzheimer's disease, it is the posterior cingulate gyrus that has been shown to be the most sensitive to AD. While there may be a neuropathological role for the posterior cingulate cortex in AD, as has been shown in PET studies (Minoshima et al. 1997), this region of the brain is also one of the most homogeneous parts of the brain that results in spectra of superior technical quality, which would also contribute to the diagnostic sensitivity of MRS to AD and other dementias.

6.2.5.3 Depression/Bipolar Disorder

Changes in NAA in bipolar disorder and major depression have also been studied extensively (Capizzano et al. 2007). Several studies have shown reductions in NAA/Cr in the temporal lobes as a result of bipolar disorder. Some studies have also shown reductions in NAA/Cr in the frontal lobe in patients suffering from major depression; however, there are also studies that were not able to replicate these results. This may be due to the fact that a majority of the studies used a ratio to Cr despite the fact that some studies have shown that Cr is altered as a result of major depression (Gruber et al. 2003).

6.3 Glutamate: Excitatory Neurotransmitter

Glutamate (Glu) is an amino acid with several important roles in the brain. First, it is the most abundant excitatory neurotransmitter in the human brain, where it plays a major role in neurotransmission and where it is released from presynaptic cells and then binds to postsynaptic receptors, thus inducing activation as shown in Fig. 6.4. As a result, many neurological and psychiatric diseases have an impact upon Glu. In particular, dysfunction reflected in excessive Glu release or reduced uptake can lead to an accumulation of Glu, which results in excitotoxicity. This second mechanism is key to not only understanding the underlying pathophysiology of different brain disorders but also providing a potential pathway for disease treatment. The existence of numerous Glu agonists and antagonists now allows for pharmaceutical interventions that can be used to modulate glutamate Glu levels and thus provide potential treatments. Finally, Glu is a key compound in brain metabolism via the citric acid cycle and therefore also tightly coupled to brain energetics.

6.3.1 Glutamate and Glutamine Metabolism

Metabolically, glutamate (Glu) is stored as glutamine (Gln) in the glia, and the balanced cycling between these two neurochemicals is essential for normal functioning of brain cells. Glu and Gln are compartmentalized in neurons and glia, respectively, and this chemical interconversion reflects an important aspect of metabolic interaction between these two types of cells. In vivo studies have revealed that the neuronal/glial Glu/Gln cycle is highly dynamic in the human brain and is the major pathway of both neuronal Glu repletion and astroglial Gln synthesis (Gruetter et al. 1994; Mason et al. 1995). After its release into the synaptic cleft, Glu is taken up by adjoining cells through excitatory amino acid transporters. Astrocytes are responsible for the uptake of most extracellular Glu via Glu transporters and additionally have a vital role in preserving the low extracellular concentration of Glu needed for proper receptor-mediated functions, as well as to prevent excitotoxicity (Schousboe 2003; Schousboe and Waagepetersen 2005). Once taken up into the astrocyte, Glu is rapidly converted to Gln by the enzyme Gln synthetase. Small quantities of Gln are also produced de novo or from GABA (Hertz and Zielke 2004; Bak et al. 2006). Gln is released from astrocytes, accrued by neurons, and converted to Glu by the neuron-specific

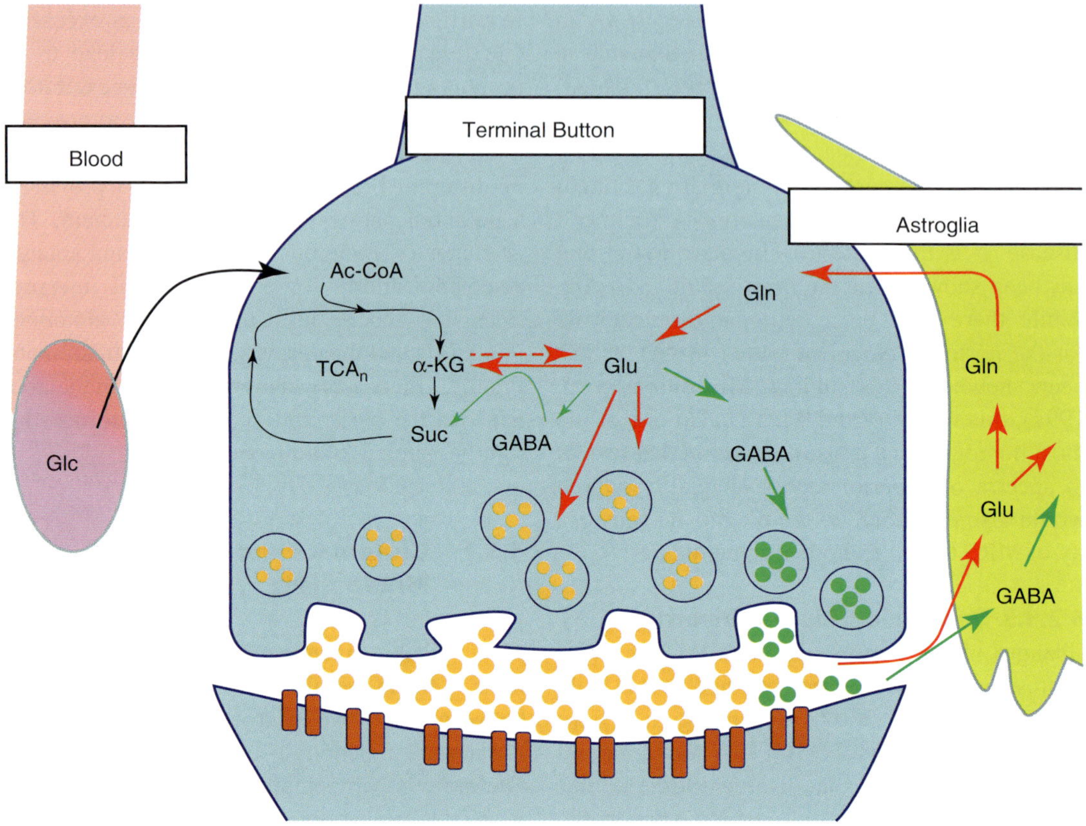

Fig. 6.4 Glutamate cycle in the neuron and glia. Glucose (*Glc*) from the capillaries is the primary source of fuel for the neurons which are then metabolized via the tricarboxylic acid cycle (*TCA*) to acetyl coenzyme A (*acetyl-CoA*), then to alpha-ketoglutarate (*α-KG*), and finally to succinate (*Suc*). α-KG is transaminated to glutamate (*Glu*), which can be further metabolized to gamma-aminobutyric acid (*GABA*). However, it is primarily released by the neuron and then taken up into the astroglia and metabolized to glutamine (*Gln*)

enzyme phosphate-activated glutaminase (Bak et al. 2006). Gln is the main precursor for neuronal Glu and GABA (Hertz and Zielke 2004), but Glu can also be synthesized de novo from tricarboxcylic acid cycle intermediates (De Graaf et al. 2011). The rate of Glu release into the synapse and subsequent processes are dynamically modulated by neuronal and metabolic activity via stimulation of extrasynaptic Glu receptors, and it has been estimated that the cycling between Gln and Glu accounts for more than 80 % of cerebral glucose consumption (Sibson et al. 1998). The tight coupling between the Glu/Gln cycle and brain energetics is largely tied to the nearly 1:1 stoichiometry between glucose oxidation and the rate of astrocytic Glu uptake. This relationship was first determined by Magistretti et al. in cultured astroglial cells where the addition of Glu resulted in increased glucose consumption (Pellerin and Magistretti 1994). These results provided the hypothesis that glycolysis in the astrocytes results in a production of two molecules of ATP, which are then consumed by the formation of Glu from Gln, which suggests a tight coupling between the two mechanisms.

6.3.2 Glutamate and Glutamine Structure and Spectroscopy

The molecular structures of Glu and Gln are very similar and, as a result, give rise to similar magnetic resonance spectra (Govindaraju et al. 2000). The four protons from the two methylene groups of glutamate Glu and Gln are located at 2.04–2.35 ppm and 2.12–2.46 ppm, respectively. Similarly, the methine group resonates at 3.74 ppm and 3.75 ppm for glutamate Glu and Gln, respectively. Thus, even though Glu has a relatively high concentration in the brain, its major resonances are usually contaminated by contributions from Gln, GABA, glutathione (GSH), and NAA. To avoid confusion in spectral assignment of Glu and Gln, the term "Glx" has traditionally been used to reflect the combined Glu and Gln concentrations. However, this approach does not allow for the evaluation of conditions where the concentrations of Gln and Glu are in opposing directions, nor does this approach allow for the evaluation of Gln and Glu separately. As our understanding of the importance of Glu/Gln system in the human brain has increased, much endeavor has been invested in being able to quantify Glu or Gln separately. The sections below detail methods for measuring Glu utilizing various pulse sequences to achieve the separation of Glu from co-resonating metabolites.

6.3.3 Single Voxel MRS and LCModel

LCModel (Provencher 1993) is a software package often used for post-processing of MRS data. It utilizes prior knowledge in the form of a basis set that is made up of a linear combination of in vitro, or simulated, spectra from individual metabolite solutions. The advantage of this method is that it is almost completely automated and utilizes algorithms for baseline correction and peak fitting without imposing restrictive parameterization and without subjective input. It is often used in the literature to calculate concentrations of individual metabolites including Glu and Gln. LCModel provides a quantitative measure of the accuracy of the measure by using the estimated standard deviations or Cramer–Rao lower bound (CRLB) that provides a measure of how reliable the measurement is. A standard deviation of <20 % has been used throughout the literature as the criteria for adequate reliability. While Glu standard deviation measures tend to fall below 20 %, Gln standard deviation measures do not. Given the overlap between the two molecules, it is unclear whether LCModel can accurately discern Glu from Gln.

In our lab we conducted a series of experiments where different solutions, or phantoms, were created that had different concentrations of Glu and Gln. In three of the phantoms, Glu was increased from 6 to 18 mM concentrations and Gln was maintained at 6 mM, which is equivalent to the physiological concentrations in the brain (3–5.8 mM (Govindaraju et al. 2000)). In the other three phantoms, Glu was maintained at 12 mM (equivalent to in vivo concentrations of 6–12.5 mM (Govindaraju et al. 2000)), but Gln was increased from 3 to 12 mM. Spectra were then acquired from these solutions using a clinical MRI scanner (Siemens 3 T TIM Trio) using conventional PRESS MRS (echo time of 30 ms, repetition time of 2 s). LCModel was then used to calculate the concentrations of each phantom as shown in Fig. 6.5. Our results showed that the variable concentration of Gln has a direct effect on the reported Glu concentrations and therefore demonstrates that this method is not sufficient for discerning Glu from Gln.

6.3.4 Optimal Echo Time

In order to better separate Glu and Gln, the echo time can be altered to minimize the contribution of the Gln signal to the Glx complex. Schubert et al. (Schubert et al. 2004) were able to selectively detect C4 proton resonances of Glu using a PRESS sequence at 3 T with a TE of 80 ms and analyzing the data using both the time and the frequency domains with prior knowledge obtained from phantom spectra. Using this approach, in vivo spectral features were similar to in vitro Glu spectral features collected from a phantom, and Glu signal was well resolved and separated from major interferences, such as Gln and NAA. Similarly, Jang et al. acquired PRESS spectra at 1.5 T in vitro and in vivo at four TE values: 30, 35, 40, and 144 ms (Jang et al. 2005), with the resulting spectra analyzed with LCModel (Provencher

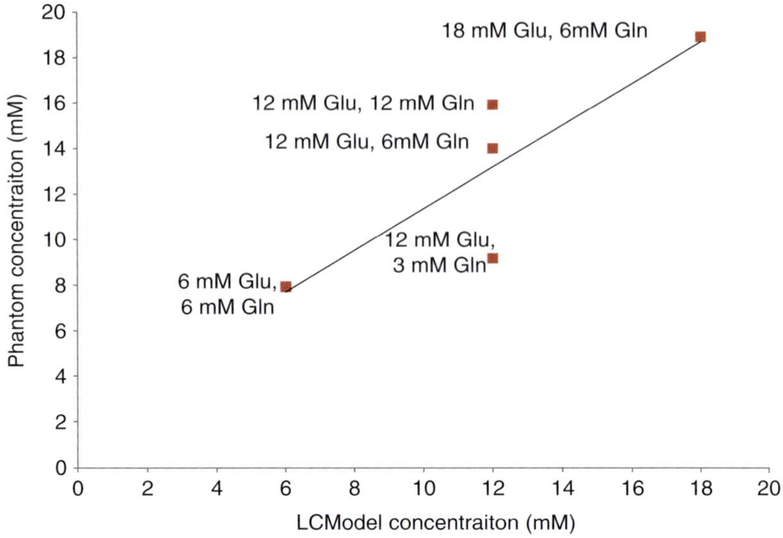

Fig. 6.5 Titration curve of solutions of glutamate and glutamine. Different concentrations of glutamate and glutamine were created as indicated by the data labels. The results show that the different concentrations of glutamine had a strong effect on the 12-mM Glu phantom solutions such that LCModel over- and underestimated the Glu concentration when Gln concentrations were high and low, respectively

2001). In vitro and in vivo spectra yielded the lowest Cramer–Rao lower bounds (CRLB) for Glu quantitation when TE was set to 40 ms. TE optimization for the purpose of Glu and Gln detection has also been demonstrated using stimulated echo acquisition mode (STEAM) spectroscopy by Yang et al. (2008), who optimized TE and mixing time (TM) to resolve the C4 proton resonances of Glu and Gln. Optimal TE methods are appealing acquisition strategies due to their ease of implementation for both acquisition and processing and the ease with which appropriate parameter timings can be selected at the scanner interface. Additionally, the resulting spectra can be analyzed using scanner software or commercially available software (e.g., LCModel, jMRUI) after simulating an appropriate basis set.

6.3.5 Ultrashort Echo Time

Due to the fact that Glu is a strongly coupled system, short echo time acquisitions are usually preferred over long echo time acquisitions for the purpose of reducing peak phase modulation. The ultrashort TE approach in such coupled systems gives rise to spectra where the majority of peaks are inphase and thus reduces signal cancellations due to J-modulations at longer echo times which in turn lead to more robust signal quantification. Wijtenburg and Knight-Scott (2011) quantified Glu/tCr at 3 T by combining short TE STEAM (6.5 ms) with phase rotation (Hennig 1992; Ramadan 2007) to detect Glu in human brain. This approach has been compared to 40 ms TE PRESS, 72 ms TE STEAM, and TE averaging and demonstrated that the short TE phase-rotation-STEAM approach yielded the greatest precision for Glu/tCr ratio quantification followed by TE averaging, PRESS 40 ms, and STEAM 72 ms (Wijtenburg and Knight-Scott 2011).

The implementation of very short echo time techniques requires substantial modification of standard vendor-supplied localization sequences. These spectra are usually complicated by a broad baseline extending over the whole spectral width, and care should be taken during analysis to eliminate the contribution of the baseline to the desired metabolite signal—which is usually done by utilizing T1 differences (Starcuk et al. 2001). The use of third-party spectral analysis software and, in some cases, the collection or simulation of metabolite basis set are essential to yield results of high accuracy (Wijtenburg and Knight-Scott 2011).

6.3.6 TE-Averaged PRESS

In this method, a number of 1D PRESS spectra are acquired at variable TE values and co-added

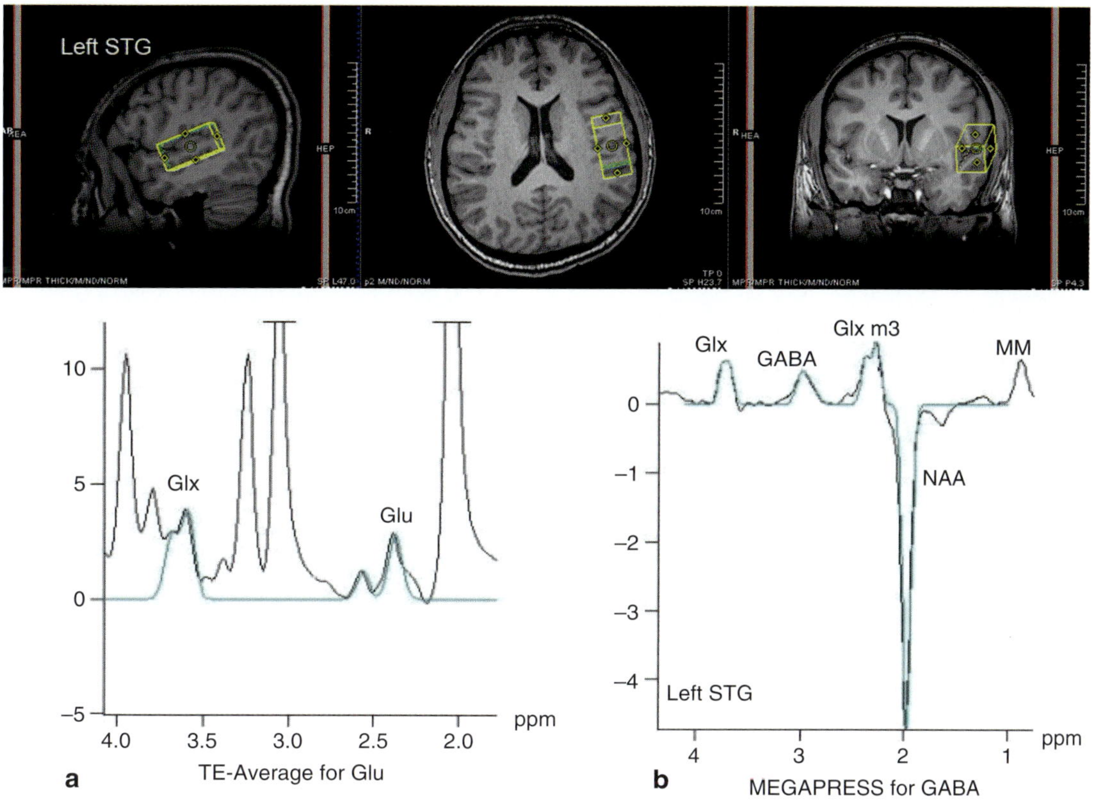

Fig. 6.6 In vivo spectra of schizophrenia. Representative spectrum of (**a**) glutamate-edited MRS (*left panel*) and (**b**) GABA-edited MRS (*right panel*) in the left superior temporal gyrus (*STG*) of a schizophrenia subject

(averaged) in real time to produce a single 1D, TE-averaged spectrum. Alternatively, the same TE-averaged spectrum can be obtained if these time-domain FIDs are ordered in a 2D matrix and a two-dimensional Fourier transform (2DFT) applied. The resultant 1D spectrum corresponding to the middle of the F1 axis (F1 = 0 Hz) is equivalent to the TE-averaged spectrum obtained above (Dreher and Leibfritz 1995; Bolan et al. 2002). This 1D spectrum offers relatively uncontaminated spectral peaks for C4 protons at 2.35 ppm for Glu and for C2 at 3.75 ppm for Glx (Hurd et al. 2004). Representative spectra are shown in Fig. 6.6a.

While TE averaging is a simple and reliable method for measuring Glu and Gln, it sacrifices spectral information arising from the J-evolution of other metabolites. The method also suffers from differential T2 weighting per individual TE value, which can be corrected for if a basis set is similarly simulated or experimentally acquired. TE-averaging sequences are not usually available from major MRI vendors but can be easily modified from standard PRESS sequence source code.

6.3.7 Glutamate Chemical Exchange Saturation Transfer (GLuCEST)

A large sensitivity enhancement for the detection of Glu can be achieved by careful design of MR exchange experiments. A recent example of this technique was recently implemented by Cai et al. where the chemical exchange saturation transfer (CEST) between bulk water and Glu was utilized for the detection of Glu (Cai et al. 2012). In such CEST experiments, advantage is taken from the protons on the Glu amine group, which are labile and in constant exchange with the protons of bulk water. When the Glu NH_2 protons are saturated

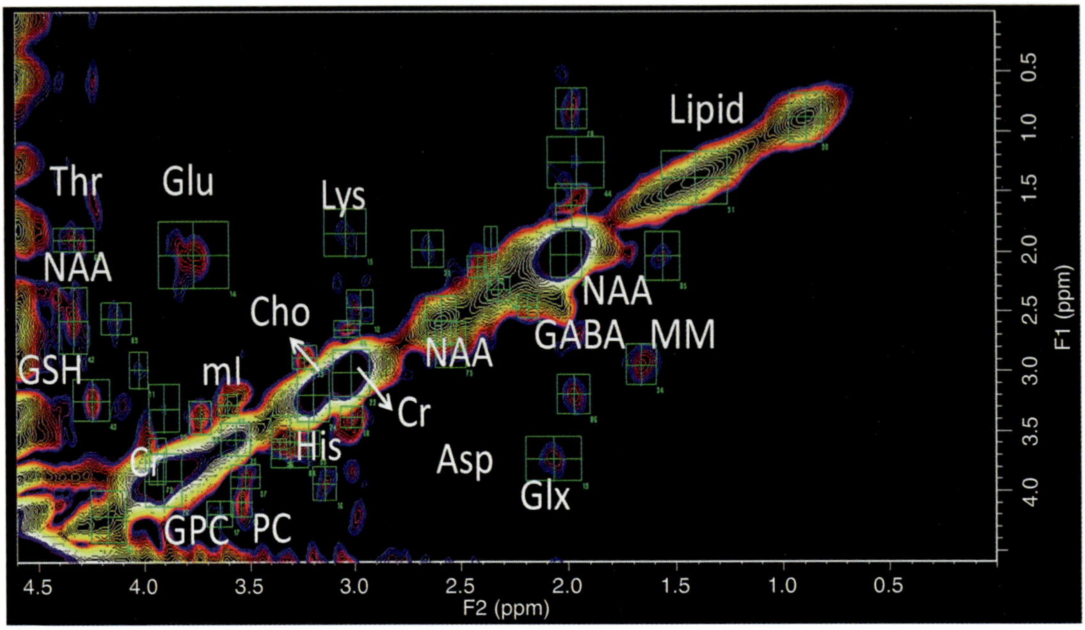

Fig. 6.7 2D correlated spectroscopy (COSY) of schizophrenia. Representative 2D COSY of left STG of SPD. *Green* ROIs show identification of multiple cross peaks such as threonine (*Thr*), *N*-acetylaspartate (*NAA*), glutathione (*GSH*), glutamate (*Glu*), myoinositol (*mI*), glycerophosphorylcholine (*GPC*), phosphorylcholine (*PC*), total choline (*Cho*), lysine (*Lys*), histidine (*His*), creatine (*Cr*), aspartate (*Asp*), gamma-amino butyric acid (*GABA*), glutamate–glutamine (*Glx*), macromolecules (*MM*), and lipids

by RF irradiation at 7.7 ppm (i.e., 3 ppm higher than water), the resultant saturation effect is transferred to water due to the ongoing chemical exchange between the protons of NH_2 and H_2O. This in turn leads to a reduction in the amplitude of water resonance due to the saturation of its protons and the saturation RF field applied to the Glu amine group. An excellent review of chemical exchange methodology can be found here (Zhou and van Zijl 2006).

6.3.8 Two-Dimensional MRS

An alternative to separating Glu from Gln is to use a second chemical shift dimension: 2D Fourier NMR spectroscopy utilizes a simple two-pulse sequence, 90x-t1-90x-Acq(t2), creating a dataset that is a series of one-dimensional (1D) spectra with traditional time readout (t2), but each with increments in delay (t1), inserted before the terminal readout 90° RF pulse (Jeener et al. 1979). Two-dimensional Fourier transformation (FT) of t1 and t2 produces a 2D spectrum where the x- and y-axes are the frequencies F2 and F1, respectively, known as correlated spectroscopy (COSY). In a 2D COSY spectrum, scalar coupling between protons in molecules results in cross peaks that allow for unambiguous identification of different metabolites (Thomas et al. 2001; Cocuzzo et al. 2011; Ramadan et al. 2011). This technology has been translated to clinical MR scanners and has been applied in vivo (Schulte et al. 2006), demonstrating detection of GABA, Glu, Gln, glutathione, as well as other metabolites (Fig. 6.7).

6.3.9 ^{13}C Spectroscopy

While the aforementioned methods have been utilized throughout the literature to study Glu, these measures provide static measures of Glu, measuring the total concentration of Glu in the brain without a sense of the dynamic changes that result from neurotransmission and brain energetics. Carbon 13 (^{13}C) MRS is presently

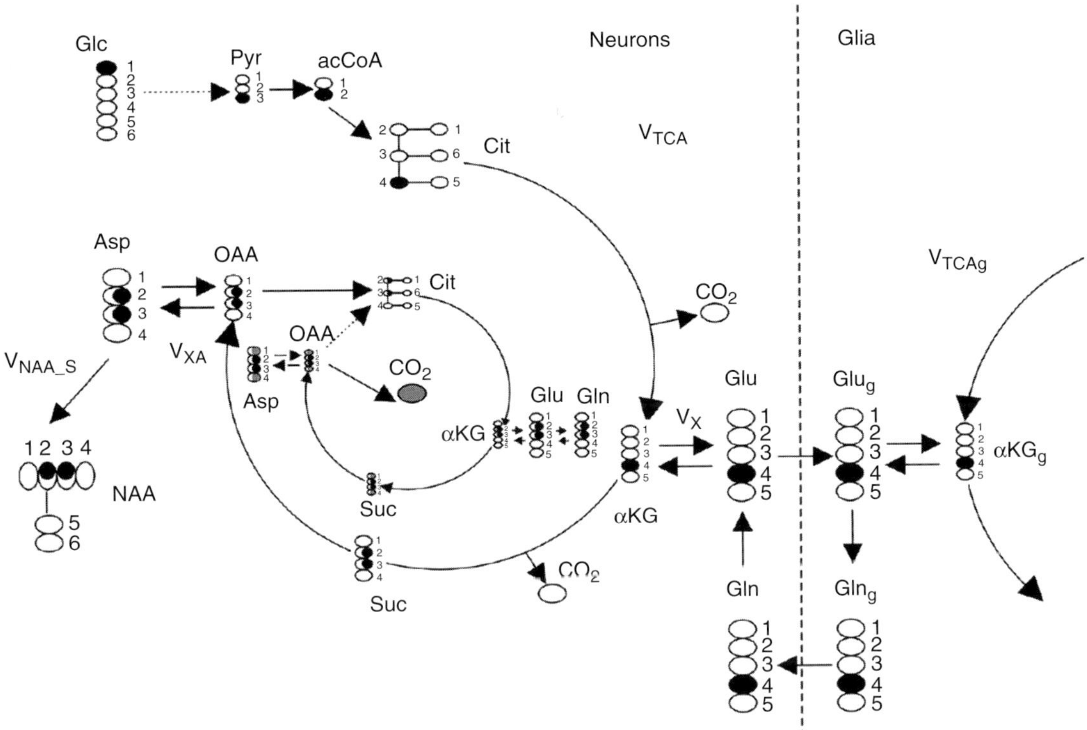

Fig. 6.8 Fate of 1-^{13}C atom of glucose through neuronal TCA and part of the neuronal–glial glutamate–glutamine cycle involved in glutamate neurotransmission. Two turns of the TCA cycle are represented. Carbon atoms enriched are shown in black. Carbons with 50 % chance of enrichment are shaded. For clarity the 3rd turn of neuronal TCA is represented only by bicarbonate ($^{13}CO_3$) release. In the paper, $^{13}CO_2$ is referred to as a bicarbonate ($H^{13}CO_3$). Other abbreviations: *Glc* glucose, *Pyr* pyruvate, *acetyl-CoA* acetyl coenzyme A, *Cit* citrate, V_{TCA} tricarboxylic acid cycle rate, *αKG* 2-oxoglutarate, $Vx_{,\alpha KG}$ Glu transaminase rate, V_{XA})transaminase or malate–aspartate shuttle rate, *Glu* glutamate, *Glug* glial glutamate, *Gln* glutamine, *Glng* glial glutamine, *αKGg* glial 2-oxoglutarate, *Suc* succinyl CoA, *OAA* oxaloacetate, *Asp* aspartate, *NAA* *N*-acetylaspartate, V_{NAA_S} NAA synthesis rate

the only method that provides noninvasive measurements of neuroenergetics and neurotransmitter cycling in the human brain. ^{13}C MRS, combined with the administration of ^{13}C-labeled substrates, allows the detection of ^{13}C incorporation from ^{13}C-labeled precursors into various carbon positions of metabolites, including the Glu cycles between neuronal and glial compartments (De Graaf et al. 2011). For example, if glucose, the primary fuel for the brain, is labeled with a nonradioactive isotope of ^{13}C on the first carbon, that label is retained as it is metabolized through the TCA cycle. At the point of alpha-ketoglutarate, the label is transaminated to Glu and Gln via the Glu/Gln cycle as shown in Fig. 6.8. As the label is incorporated separately into the fourth carbon group of Glu and Gln, real-time acquisition of the brain will result in the observation of the Glu C4and Gln C4 resonances at 34.4 ppm and 31.8 ppm, respectively. In the second turn of the TCA cycle, Glu C2 and C3 are labeled. Thus, ^{13}C MRS allows continuous, noninvasive monitoring of metabolic fluxes under different physiological or pathophysiological conditions.

The ^{13}C MRS acquisition consists of several different steps providing multiple means by which Glu can be measured. These include choice of substrate (glucose, acetate, label position), method of delivery (oral, IV, glucose clamped), data acquisition (direct detection, indirect detection), and method of analysis (dynamic difference spectroscopy, isotopomer analysis) (Ross et al. 2003). It is important to note that the natural abundance of ^{13}C is approximately 1.1 % which implies that there is very little background signal of indigenous ^{13}C Glu, therefore providing excellent contrast to noise ratio of signals that arise from the incorporation of the ^{13}C label via the TCA cycle and subsequent transamination from alpha-ketoglutarate to the Glu/Gln cycle.

6.3.10 Glutamate in Psychiatric Diseases

6.3.10.1 Schizophrenia

It is thought that schizophrenia is related to a dysfunction in *N*-methyl-D-aspartate (NMDA) receptors, the major subtype of Glu receptors. Evidence from studies of NMDA receptor agonists such as phencyclidine and ketamine has shown decreased Glu levels as well as the psychotic symptoms observed in schizophrenia (Javitt and Zukin 1991). A recent meta-review (Marsman et al. 2013) examined 28 publications for a total of 647 patients with schizophrenia compared with 608 healthy controls appears to support this hypothesis by showing evidence of reduced Glu. While several brain regions were explored including the medial frontal region, hippocampus, and thalamus, only significant differences were found in the medial frontal region with reduced Glu. Group-by-age associations showed that Glu decreased at a faster rate with age in patients with schizophrenia compared to controls. While there have been few longitudinal studies of Glu using MRS, the literature appears to reflect that first-episode schizophrenics show increased glutamate whereas in the chronic stage of disease, Glu is decreased (Port and Agarwal 2011). This provides the basis for the second hypothesis of excitotoxicity of Glu in schizophrenia. Proponents argue that initially Glu levels are increased leading to excitotoxicity, which in turn leads to neuronal death as reflected by decreases in NAA. In the chronic stages of the disease, neuronal loss would also be reflected in decreases in Glu.

The meta-review also showed several studies where Gln is increased in schizophrenia (Marsman et al. 2013), which may reflect deficiencies in glutaminase, the enzyme that converts Gln into Glu, thus leading to decreased Glu and increased Gln as reflected in studies in chronic schizophrenia. However, it is unclear whether the methods used to measure Gln are accurate as few, if any, of the aforementioned pulse sequences have been validated for their measure of Gln given the overlap between Glu and Gln. The only efficient method that clearly delineates Gln from Glu are ^{13}C MRS studies of which there is only one published study (Harris et al. 2006) that showed reductions in TCA cycle rate, but did not show differences in Gln metabolism rates between schizophrenics and controls.

6.3.10.2 Dementia

Several proton spectroscopy studies have demonstrated decreases in Glu in Alzheimer's dementia using traditional MRS methods (Ross et al. 1997; Jones and Waldman 2004; Fayed et al. 2011) as well as others such as GluCEST (Cai et al. 2012, 2013). Reduced Glu is also found in HIV-related dementia (Ernst et al. 2010). Perhaps of greatest interest is that some studies in dementia patients have shown how Glu levels can be reversed using pharmaceutical interventions such as galantamine, a cholinesterase inhibitor, which showed that Glu increased after treatment in AD patients (Penner et al. 2010).

The role of Glu in Alzheimer's disease has also been explored in detail using ^{13}C MRS. The initial study in AD utilized [1-^{13}C] glucose in patients clinically diagnosed with AD as well as age-matched controls (Lin et al. 2003). The results of the study showed reduced Glu neurotransmission as measured by examining the time course of enrichment of Glu via the TCA cycle as well as relative enrichment of Glu to Gln. The time course measurements reflect the neuroenergetics of the brain and effectively measure TCA cycles rates. In AD patients, these rates were found to be reduced. The measure of Glu and Gln enrichment is of particular interest as it is reflective of Glu neurotransmission in itself. For example, if the glial TCA cycle is operating faster, then an enrichment pattern with relatively increased Gln signal versus Glu signal would be expected. In this study, Glu/Gln ratios were found to be significantly decreased, reflective of decreased neurotransmission. More interestingly, when correlated with NAA measures as a surrogate marker for neuronal integrity, significant correlates were found with Glu/Gln ratios, further supporting the argument that this measure may be reflective of Glu neurotransmission in and of itself. As NAA, or the number of neurons decreased, Glu/Gln decreased, thus demonstrating that Glu neurotransmission may be decreased as a result in the reduction of the number of functioning brain cells. Sailasuta et al. also found that using [1-^{13}C] acetate, the third turn of the TCA cycle, which produces bicarbonate, is significantly slower in AD patients when compared to controls (Sailasuta et al. 2011).

6.3.10.3 Anxiety Disorders

Several studies have demonstrated that Glu is increased in the anterior cingulate cortex as a result of general (Strawn et al. 2013) and social anxiety disorder (Phan et al. 2005; Pollack et al. 2008) when compared with healthy controls. Excess Glu in anxiety is not only found in group differences and correlations with anxiety severity. One study showed that when anxiety is induced using cholecystokinin tetrapeptide, Glu levels in the anterior cingulate markedly increase within 2–10 min of the challenge (Zwanzger et al. 2013). Furthermore, when patients with anxiety are treated with medications, such as levetiracetam, Glu levels appear to decrease as a result of the treatment (Pollack et al. 2008).

6.3.10.4 Depression

A recent meta-review (Luykx et al. 2012) examined 16 publications of Glu in major depressive disorders (MDD) using MRS for a total of 281 patients and 301 controls. The anterior cingulate cortex and the prefrontal cortex were the two primary brain regions examined by most studies. The result of the meta-analysis showed that Glu and Glx were found to be significantly decreased in the anterior cingulate in MDD subjects. However, the literature remains mixed. For example, 2D COSY was used to study levels of metabolites in the dorsolateral prefrontal white matter regions to study MDD in the elderly. The study concluded that the depressed subjects had lower levels of NAA and higher levels of Glu/Gln, mI, and phosphoethanolamine (Binesh et al. 2004).

Perhaps the strongest evidence of the importance of using MRS to measure Glu in MDD comes from neuropharmacological literature where MRS has been used extensively to measure the effects of medication on brain metabolites. A recent review identified 15 articles where MRS was used to assess MDD treatments of which six specifically studied Glu and/or Glx. These findings showed several studies where Glx levels were initially decreased in MDD subjects, but after treatments such as electroconvulsive therapy and repetitive transcranial magnetic stimulation, Glx levels increased. These studies give rise to the hypothesis that Glu has a role in neuroplasticity (Sanacora et al. 2012), thus clearly demonstrate that Glx is an important component of MDD and that MRS is an effective means by which one can monitor therapeutic interventions.

6.3.10.5 Bipolar Disorder

Given the role of Glu in psychosis and depression, there is also a body of literature that has explored the use of MRS in bipolar disorder (BD). Similar hypotheses of Glu neurotoxicity and neuroplasticity arise in the literature for Glu MRS of BD as discussed in a recent review (Gigante et al. 2012), where 17 studies were identified. In most studies, the frontal lobe was the focus of MRS in regions such as the anterior cingulate cortex and the dorsal lateral prefrontal cortex. The meta-analysis showed elevated levels of Glx in BD. It is particularly interesting that this remained the case regardless of whether the patients were medicated. This is in contrast to MDD studies where Glx levels were decreased and modified by treatment.

6.4 γ-Amino Butyric Acid (GABA): Inhibitory Neurotransmitter

The counterpart to Glu is GABA, the major inhibitory neurotransmitter in the brain. Interestingly, in the developing brain, GABA is initially excitatory and later shifts to its inhibitory role as glutamatergic functions develop (Ben-Ari 2002). The balance between excitation and inhibition is reflected not only in the roles of these two neurotransmitters but also in their metabolism as Glu is a metabolic precursor of GABA. GABA acts as an inhibitor by binding to specific receptors that open ion channels that allow for the flow in of negatively charged ions or the flow out of positively charged ions, thus creating a negative membrane potential. This, in turn, hyperpolarizes the cells and prevents the formation of action potentials. GABA is of particular interest given the availability of drugs that modulate the binding of GABA receptors, such benzodiazepines, which have helped elucidate the role of GABA across a broad range of psychiatric conditions such as anxiety, memory loss, depression, and pain. Similar to Glu, GABA is difficult to detect using MRS due to its overlap with the resonances of Glu and Gln, therefore requiring specific sequences to measure. Further, the concentration of GABA in the brain ranges from 1.3 to 1.9 mM (Govindaraju et al. 2000) and therefore is difficult to detect accurately and consistently given its low concentration.

6.4.1 GABA Metabolism

GABA is synthesized via the GABA shunt (Olsen and DeLorey 1999): first, alpha-ketoglutarate from the TCA cycle is transaminated to glutamic acid by GABA transaminase (GABA-T). Glutamic acid decarboxylase (GAD) catalyzes the decarboxylation of glutamic acid to GABA. GABA can be metabolized to succinic semialdehyde by GABA-T but only in the presence of alpha-ketoglutarate in order to conserve the supply of GABA. Finally, succinic semialdehyde is oxidized into succinic acid which then reenters the TCA cycle, thus completing the shunt. This metabolism takes place in the GABAergic neuron where GABA is released as a neurotransmitter and then binds to receptors on the postsynaptic neuron. There are three different types of GABA receptors, $GABA_A$, $GABA_B$, and $GABA_C$ (also considered $GABA_A$-ρ), which have different properties where A and A-ρ are fast-acting and B is slow acting and hence different pharmacologies and diseases. Finally, GABA can also be taken up by the astrocytes, thus leading to a GABA–glutamine cycle following similar metabolic pathways as the GABA shunt.

6.4.2 GABA Structure

As with Glu, the GABA nuclear spins *J*-couple with other neighboring spins. This causes the molecule to appear as a number of small peaks spread out a broad frequency range on a standard PRESS spectrum. This results in multiplet resonances around 1.9, 2.28, and 3.0 ppm. The problem with all of those locations is that they overlap with other major metabolites within the spectrum such as NAA at 2.02 ppm, Glu at 2.34 ppm, and Cr at 3.02 ppm. Therefore, specialized methods are once again required to discern GABA and to reliably estimate its concentrations in vivo.

6.4.3 MEGA-PRESS

One technique to measure the GABA peaks involves utilizing a difference of spectra that selectively preserves the GABA signal while eliminating other metabolites. GABA has characteristic resonances at 3.0 and 1.9 ppm. A frequency-selective pulse that targets one of these regions would also enhance the signal in the other region due to the effect of *J*-coupling. Applying a frequency-selective pulse that is on resonance at 1.9 ppm would increase the signal in the 3.0 ppm region if GABA is indeed present. Likewise, a frequency-selective pulse that is off resonance would not have an effect on the 3.0 ppm peak. Subtracting the spectra from these two different pulse acquisitions should preserve the metabolite that only resonates at the characteristic frequencies of GABA as shown in Fig. 6.6b. MEGA-PRESS utilizes these frequency-selective pulses and combines them with PRESS localization. While this method has the selectivity to measure GABA, by utilizing difference spectra, there is a loss in SNR and a chance to introduce additional artifacts as a result of subtraction.

This method has been successfully applied to estimate the GABA concentration of several brain regions including the anterior cingulate cortex (ACC), occipital lobe, and parietal lobe (Terpstra et al. 2002; Öz et al. 2006; Bhattacharyya et al. 2007; Edden and Barker 2007; Kaiser et al. 2007; Waddell et al. 2007). This sequence combined with LCModel quantitation of the MEGA-PRESS spectra can provide an estimate of the absolute GABA concentration present in the brain in vivo.

6.4.4 Double Quantum Coherence (DQC)

A DQC filter technique can be used to eliminate uncoupled resonances from metabolites like choline, creatine, and NAA and leave strongly coupled resonances from GABA in a single acquisition without the need for subsequent spectral editing. This method exploits the difference in the phase-accumulation rate of coherences from single versus multiple quantum states. By selectively rephasing, only the magnetization from a multiple quantum coherence, the signal from the uncoupled metabolites, is removed. Because this method does not require the difference spectra from two scans, the SNR is improved and the acquisition time is reduced (Keltner et al. 1997). Quantifying GABA from DQC-filtered spectra has the potential to provide more reliable and reproducible estimates (McLean et al. 2002; Choi et al. 2004, 2005a, b). The technique can be applied with PRESS localization just as with MEGA-PRESS, and the same applications of study are possible. Comparison of subjects with occipital lobe epilepsy (OLE) versus healthy controls shows increased levels of GABA acquired from DQC spectra (Simister et al. 2003). An increased GABA concentration after treatment with the SSRI citalopram in the OCC with healthy subjects is also detected with DQC spectra (Bhagwagar et al. 2004).

6.4.5 2D JPRESS

Similar to 2D COSY, 2D JPRESS also acquires a series of TE increments to provide a 2D MR spectrum. 2D JPRESS differs from 2D COSY in that it utilizes the standard PRESS sequence for each increment with the addition of maximum echo sampling scheme that increases the sensitivity of the method (Ke et al. 2000; Schulte et al. 2006; Lymer et al. 2007). Additional consideration is also necessary to quantify the 2D JPRESS spectra. Unlike COSY where the cross-peak volumes can be measured as an analog of metabolite concentration, 2D JPRESS cross peaks are not as differentiated and therefore do not lend itself as readily for visual analysis. More sophisticated tools like ProFit (Schulte and Boesiger 2006) attempt to quantify the 2D spectra with prior knowledge basis sets of the metabolites similar to LCModel.

2D JPRESS can also be combined with the CSI method previously described in Sect. 2.3 where instead of using PRESS, 2D JPRESS is used instead (Jensen et al. 2005). 2D-JPRESS CSI can sample many regions of interest to determine the GABA concentration, but obtaining reliable spatial distribution maps of GABA in the brain requires overcoming technical challenges. The smaller voxel size greatly reduces the SNR of each spectrum making the GABA concentration estimates inherently less reliable relative to estimates from larger voxels. The 2D JPRESS can also be combined into a 1D TE-averaged JPRESS spectrum, as described earlier; however, while the TE-averaged JPRESS spectra can be used to reliably estimate other metabolites like Glu, this method is less reliable for quantifying GABA (Mullins et al. 2008).

6.4.6 GABA in Psychiatric Disease

6.4.6.1 Schizophrenia

Alterations in GABA neurotransmission underlying the pathophysiology of schizophrenia are evidenced by reduced GABA synthesis as reflected by decreased activity of GAD67 in parvalbumin-staining cortical neurons (Gonzalez-Burgos et al. 2010). Additionally, $GABA_A$ receptors may be upregulated, possibly reflecting a compensatory response to reduced GABA levels (Jarskog et al. 2007). Due to the relatively recent advances in MRS that has allowed for GABA measurements, MRS studies of GABA in schizophrenia are rapidly evolving. A study of chronic schizophrenics reported the MEGA-PRESS estimated GABA to be 10 % lower than in healthy controls and to correlate with the orientation-specific surround suppression – a quantitation of the level of visual inhibition that is thought to be regulated by GABAergic pathways (Yoon et al. 2010). A more recent study found similar reduction of GABA in chronic schizophrenia in the anterior cingulate which also correlated with poor attention performance (Rowland et al. 2012). In a study in early course schizophrenia, decreased GABA levels in the basal ganglia was also found (Goto et al. 2009). In contrast, however, another study found increases in GABA/Cr ratios in chronic SZ (Ongür et al. 2010). These discrepant results may have resulted from the use of Cr as studies have shown that Cr is reduced in schizophrenia thus possibly elevating the ratio if the reduction is greater than the reduction in GABA. It is also clear that there are significant medication effects on the GABA measures as another study in schizophrenia showed that the measured GABA concentration in the anterior cingulate was significantly negatively correlated with the dose of antipsychotics (Tayoshi et al. 2010). In this study, exclusion of patients on these compounds abolished the statistically significant effects.

6.4.6.2 Anxiety

There have only been a handful of studies that have used MRS for measuring GABA in anxiety. Comparing patients with panic disorder and healthy controls, MRS studies have found that GABA is reduced in the occipital cortex, anterior cingulate cortex, and medial prefrontal cortex (Goddard et al. 2001; Long et al. 2013). Interestingly, upon administration of benzodiazepine, GABA levels in the occipital cortex did not change (Goddard et al. 2004). Other studies however have not found differences in GABA levels in patients with panic disorder in the dorsal prefrontal or ventrolateral prefrontal regions (Hasler et al. 2005). Similarly, there were no significant differences in patients with social anxiety disorder (Pollack et al. 2008). It should be noted however that measuring GABA is challenging and can often result in large variability in the measurement which would result in the lack of significant findings, but should not necessarily be interpreted as a lack of change in GABA.

6.4.6.3 Depression

A recent review cited 14 papers that utilized MRS to measure GABA levels in major depressive disorder where there are general findings of decreased GABA in the occipital cortex, anterior cingulate, and dorsal prefrontal cortex but not in the frontal white matter (Maddock and Buonocore 2012). Furthermore, the review highlights four studies that examine the effect of treatment on GABA levels and show that treatments of selective serotonin reuptake inhibitors and electroconvulsive therapy both increased levels of GABA in the occipital cortex, although one study of cognitive–behavioral therapy did not show increased levels.

6.4.6.4 Bipolar Disorder

Two studies have shown that patients with bipolar disorder (BD) exhibit lower GABA in the anterior cingulate, parieto-occipital cortex, and occipital cortex (Bhagwagar et al. 2004; Brady et al. 2013). A CSI study, however, did not show any significant differences between patients with BD and healthy controls (Kaufman et al. 2009). It is likely that the low SNR of CSI sequences, as described earlier, resulted in the lack of statistical differences between the two cohorts.

6.5 Glutathione (GSH)

The primary role of GSH is as an antioxidant, providing the brain's defense system against damaging reactive oxygen species (ROS) that arise from oxidative stress such as super oxide, nitric oxide, hydrogen peroxide, and free radicals (Dringen et al. 2000). The damage caused by ROS include DNA modification, lipid peroxidation, and protein modification from which the brain is especially vulnerable due to the large quantity of ROS produced there, availability of lipids, and iron deposition that catalyzes ROS reactions. Oxidative stress is also strongly associated with neuroinflammatory processes (Mosley et al. 2006; Agostinho et al. 2010). As a result, compromises in GSH metabolism have been found to play an important role in the pathogenesis across a wide variety of diseases such as neurodegenerative disorders as well as psychiatric disorders.

6.5.1 Glutathione Metabolism

GSH is synthesized intracellularly by first combining Glu and cysteine to form γ-glutamylcysteine via synthetase enzyme followed by addition of glycine as catalyzed by GSH synthetase. Due to the limited supply of Glu, cysteine, and glycine, in different cell types, GSH is synthesized between different neurons, astrocytes, and glial cells, thus providing an interesting specificity to GSH measures to processes such as neuroinflammation. The thiol group on GSH is a reducing agent in that it can donate an electron to unstable molecules such as ROS. When ROS is present, GSH will reduce ROS by catalysis of glutathione peroxidase (GPx) to form glutathione disulfide (GSSG). GSSG can then regenerate GSH using glutathione reductase. GSSG resonates at a different frequency on the MRS spectrum (Satoh and Yoshioka 2006) and therefore reductions in GSH concentrations as measured by MRS would provide a putative measure of brain oxidative stress and neuroinflammation. However, it is important to note that defects in GSH metabolism can also reduce concentrations of GSH, and disruptions to GSH synthesis or GSH peroxidation can also reflect oxidative stress.

6.5.2 Measuring Glutathione

As a tripeptide of Glu, cysteine, and glycine, the GSH spectrum will have overlap with all three of those metabolites (Matsuzawa and Hashimoto 2011). From the Glu moiety, the methylene protons give rise to multiplets at 2.15–2.55 ppm and the methine at 3.77 ppm (which also has contributions from the glycine moiety). The cysteine moiety of GSH gives rise to resonances at 4.56, 2.93, and 2.98 ppm. Thus, there is significant overlap between Glu and GSH resonances. This is particularly of concern given that GSH concentrations in the brain are 1–3 mM, whereas Glu concentrations are 6–10 mM (Govindaraju et al. 2000). Furthermore, there is also overlap with NAA and creatine resonances as well. Fortunately, many of the same tools that are used for separating Glu and GABA from overlapping resonances can also be used for GSH. Minor modifications to those methods are required:

6.5.2.1 LCModel

Some studies have reported measuring GSH using conventional PRESS or STEAM MRS methods by using LCModel (Terpstra et al. 2005; Wood et al. 2009; Hermens et al. 2012; Lagopoulos et al. 2013; Duffy et al. 2013). Current versions of the LCModel basis set (versions 6.2 and above) contain glutathione within the basis set which can be used for quantifying levels of GSH in the brain. However, similar to the issue described for Glu and Gln, GSH measures are often reported to have <20 % CRLB in most publications; however; once again, our laboratory tests have shown that the Glu concentration can have a significant impact upon the estimation of the GSH concentration despite the fact that the CRLB are still reported to be less than 20 %. It is therefore unclear whether or not this is a reliable method by which one can measure GSH. Furthermore, it is important that those studies that utilize this method must report the CRLB and they should be below 20 %.

6.5.2.2 MEGA-PRESS

In this implementation of the pulse sequence, the selective editing pulses are targeted to the α-cysteinyl protons of GSH at 4.56 ppm, which after subtraction of the on and off resonance acquisitions results in detection of the β-cysteinyl protons of GSH (Choi et al. 2004; Terpstra et al. 2005; Matsuzawa et al. 2008; Mandal et al. 2012). LCModel is then used to fit this resonance. Unlike the PRESS or STEAM spectra, the spectral editing removes the contributions of metabolites such as Glu and Gln and thus provides a more accurate measure of GSH concentrations (Fig. 6.9).

6.5.2.3 Double Quantum Coherence

Similarly, DQC methods can also be modified to target the GSH molecule. In the one study reviewed here, the group also focused on the cysteinyl groups.

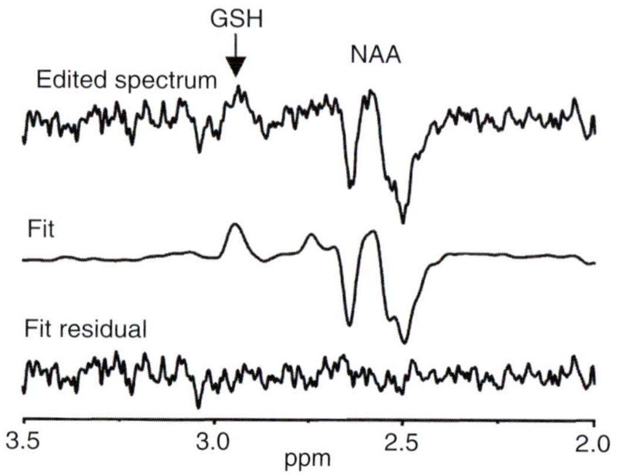

Fig. 6.9 MEGA-PRESS GSH spectrum. Representative spectrum of glutathione measured by MEGA-PRESS. The difference spectrum is shown at the *top*, with an *arrow* pointing to the GSH resonance. The *middle* spectrum shows the fit of the data and the bottom spectrum shows the residual. Note the NAA is co-edited with this implementation of the MEGA-PRESS sequences (Terpstra et al. 2005)

6.5.2.4 2D MRS

Although not utilized yet in psychiatric studies, 2D JPRESS and 2D COSY methods can accurately measure GSH in the brain (Thomas et al. 2001; Schulte et al. 2006). The advantage of the 2D methods is that unlike MEGA-PRESS or DQC which can only measure a single metabolite for one acquisition, 2D JPRESS and COSY can measure many metabolites. For example, in order to measure Glu, GABA, and GSH, using 1D methods would require three separate acquisitions of approximately 7 min each for a total of 21 min. 2D MRS could measure those metabolites and more in 12 min (Ramadan et al. 2011).

6.5.3 Glutathione in Psychiatry

6.5.3.1 Schizophrenia

A recent review identified four studies that measure GSH in patients with schizophrenia (Matsuzawa and Hashimoto 2011). The first study used DQC to measure GSH in the medial prefrontal cortex in 14 patients including schizophrenia and schizophreniform disorder and compared them to 14 controls (Do et al. 2000). GSH was found to be significantly lower which was further confirmed by a decrease in CSF GSH levels. The second study utilized MEGA-PRESS and STEAM in the medial prefrontal cortex in 11 patients with schizophrenia and 9 controls, and although they found decreased GSH, the difference was not statistically significant between the two groups (Terpstra et al. 2005). The third study also utilized MEGA-PRESS in the medial prefrontal cortex, and although they too did not find significant group differences between schizophrenic ($n=20$) and controls ($n=16$), they did find significant negative correlations between GSH measures and negative symptom scores (Matsuzawa et al. 2008). Finally, the fourth study utilized LCModel with conventional PRESS (TE = 30 ms) in the bilateral temporal cortex in first-episode psychosis ($n=30$), of which 13 were neuroleptic naïve and had various diagnoses of schizophrenia, schizopsychosis, schizoaffective disorder, and depression with psychotic symptoms (Wood et al. 2009). Their findings are surprisingly contrary to previous studies where they show increased GSH in the schizophrenia group when compared with controls. It is important to note however the authors used the criteria of <50 % CRLB which indicates that metabolite concentrations could range from zero to twice the estimated concentration. According to the LCModel guidelines, this would imply that the metabolite is undetectable. Coupled with findings of interference of Glu with GSH LCModel measures, it is likely that these results are not reliable.

6.5.3.2 Dementia

Two recent studies have utilized MRS to measure GSH in dementia. The first study utilized MEGA-PRESS and measured GSH in healthy young and old controls, 11 patients diagnosed with MCI, and 14 patients that suffered from Alzheimer's disease (Mandal et al. 2012). They explored several different brain regions including the frontal cortex, parietal cortex, hippocampus, and cerebellum. While the study found higher GSH in the young controls versus the elderly controls and patients, there did not seem be any significant differences between the elderly controls and MCI or AD patients. The second study

(Duffy et al. 2013), however, examined a much larger group of subjects (54 MCI and 41 healthy controls) using PRESS (TE=35 ms), MRS, and LCModel in the anterior and posterior cingulate gyri. They found elevated GSH levels in MCI patients when compared to age-matched healthy subjects in both brain regions. Furthermore, they also found the anterior and posterior cingulate GSH levels negatively correlated with executive and memory function, respectively. The authors hypothesize that the increased GSH may be a neuroprotective affect that arises from the upregulation of GSH in response to oxidative stress onset by MCI.

6.5.3.3 Bipolar Disorder

Fifty-four patients with bipolar disorder (13 bipolar I, 25 bipolar II, 15 bipolar spectrum) and 51 age- and gender-matched controls were studied with PRESS (TE-=30 ms) in the anterior cingulate cortex and GSH measured using LCModel (Lagopoulos et al. 2013). In this case, CRLB was reported to be less than 20 %. There were no significant differences found in the comparison of BD patients and controls nor were there any significant correlations with clinical scores of depression or mania. The authors conclude that oxidative stress does not play a role in bipolar disorder. It is interesting that an earlier paper from the same group utilizing the same methods but in a larger cohort ($n=80$; mixed psychiatric disease including depression, bipolar, and psychosis) found that GSH was accounted for by the 40 % of the clusters when cluster analysis was applied (Hermens et al. 2012). One of the clusters exhibited higher GSH; however, as a mix of different psychoses, it is difficult to determine the underlying pathophysiology, which might explain this finding.

6.6 Other Metabolites

The following metabolites can be measured using conventional MRS techniques. A description of their role in the brain and how they are affected by various psychiatric diseases will be described.

6.6.1 Creatine (Cr)

The major resonance at 3.02 ppm is comprised of creatine (Cr) and phosphocreatine (PCr), although in general, most publications refer to the combination as Cr. LCModel provides both a Cr measure and a PCr measure, although conventional proton MRS cannot readily distinguish Cr from PCr as they are in rapid molecular exchange with one another. Instead, ^{31}P MR spectroscopy must be used to discern the phosphorous resonances in PCr. The metabolism of Cr and PCr provided the basis of energy metabolism in the brain via the mitochondria where Cr is phosphorylated by the enzyme creatine kinase and adenosine triphosphate (ATP) to form PCr in reverse (Ross 2000). Used to maintain energy homeostasis in the brain, Cr is fairly uniform in different areas of the brain and between subjects and therefore is often used as a normalization factor between metabolites and subjects such that a ratio to Cr is often used. The danger of utilizing such a ratio is that it naturally assumes that the concentration of Cr in the brain (approximately 8 mM) is static and unaffected by disease (Kreis et al. 1993).

In many situations, this is not the case as energy utilization can be affected by disease and other Cr metabolic pathways. Mitochondrial dysfunction in the brain would likely cause changes in Cr and PCr and has been implicated across a range of neuropsychiatric disorders including schizophrenia and bipolar disorder (Rezin et al. 2009). ^{31}P MRS studies have shown reduced PCr in schizophrenia (Kegeles et al. 1998; Blüml et al. 1999) which supports findings from proton MRS studies where reduced Cr has been found using TE-averaged PRESS in schizophrenia (Ongür et al. 2009). Interestingly in the same study, bipolar disorder did not show changes in PCr; however, other studies using both proton MRS (Frey et al. 2007) and ^{31}P MRS (Kato et al. 1994) have shown reductions in Cr and PCr. Findings of changes in Cr in schizophrenia and bipolar disorder however are inconsistent. A recent study showed that meta-analysis failed to find significant abnormalities in Cr levels in both diseases (Kraguljac et al. 2012).

It is also important to note that in addition to mitochondrial dysfunction, Cr can be affected by other factors. For example, Cr is synthesized in the liver, and therefore, patients with liver failure will result in reduced Cr in the brain. Cr is also an osmolyte and therefore can be affected by osmotic equilibrium in the brain (Bluml et al. 1998). It is therefore important to screen for comorbid diseases such as liver disease and other conditions that may cause changes in brain metabolites secondary to the disease of interest.

6.6.2 Choline (Cho)

Similar to Cr, Cho is a term that includes several different metabolites that contain the choline moiety that give rise to the primary resonance at 3.2 ppm. These metabolites include choline, phosphorylcholine (PC), and glycerophosphorylcholine (GPC), which are the main constituents, as well as phosphorylethanolamine and glycerophosphorylethanolamine, to a lesser extent. All of these metabolites are constituents of sphingomyelin and lecithin which make up the myelin sheath critical for the propagation of signals throughout the brain. As a result, increased Cho is found in areas of greater myelin and membrane density such as in the white matter and in the pons. In addition, PC and GPC are involved in the metabolism of phosphatidylcholine and other membrane phospholipids. PC is a precursor of membrane phospholipid synthesis, and GPC is produced upon the breakdown of membrane phospholipids. Thus, the turnover of the membrane of myelin phospholipids as well as demyelination are closely tied to increases in Cho as observed by MRS. As a result of the involvement of Cho in both membrane turnover and degradation as well as density, it is a sensitive marker but not necessarily specific to membrane conditions.

Recent reviews in Alzheimer's dementia (Kantarci 2007), schizophrenia (Kraguljac et al. 2012), and depression (Caverzasi et al. 2012) have shown conflicting reports on Cho levels in these diseases. It is only in bipolar disorder where the literature has shown consistent elevated choline signal in the basal ganglia (Maddock and Buonocore 2012). The most likely reason for why Cho findings are discordant across different studies within the same disease is due to the sensitivity of Cho to the gray and white matter volumes in the regions of interest studied. For example, a meta-analysis has shown that duration of illness is associated with basal ganglia gray matter volume in bipolar patients (Bora et al. 2010), which may explain the increase in Cho as a function of membrane density. Within studies, Cho levels could be affected by placement of the voxel for MRS, which may include different volumes of gray and/or white matter when comparing controls with patients thus resulting in conflicting reports of Cho changes. The solution to this problem is to combine image segmentation within the voxel of interest for MRS to account for the relative contributions of gray and white matter to the Cho level.

6.6.3 Myoinositol (mI)

mI is a simple sugar alcohol; however, it is involved in a multitude of different metabolic processes, and as a result, its role cannot be defined to a single function (Ross and Bluml 2001). It is a cerebral osmolyte and also believed to be an astrocyte marker. Similar to Cho, mI has also been implicated in demyelination. While its role may not be well understood, the importance of mI however is the broad range of concentrations found across different conditions from 3× the normal adult values in newborns to nearly zero in hepatic encephalopathy. It has been found to be highly specific and sensitive within the context of disease diagnosis in patients when compared with controls and therefore provides great clinical value.

The importance of mI has long been recognized in studies of dementia. In early stages of dementia such as mild cognitive impairment, mI is elevated long before symptoms of dementia are obvious (Kantarci 2013). As described earlier in Sect. 2.5.2, mI in combination with NAA predicted outcome of MCI in patients and is also highly sensitive and specific in the diagnosis of Alzheimer's disease where much of the specificity arose from mI measures.

There does not appear to be strongly associated changes in mI in other psychiatric disorders including schizophrenia, bipolar disorder, or depression.

It should be noted that like Glu, mI can only be observed at short echo times (TE < 55 ms), and so studies that utilize long echo time MRS would not be capable of measuring mI. Also, given the changes in mI observed in hepatic encephalopathy, it is important to screen for liver disease to ensure that changes in mI are not masked by comorbid conditions.

Conclusion

MRS is a powerful neuroimaging tool that can provide noninvasive, quantitative, and objective measures of key neurometabolites *in vivo*. This insight can be utilized to reveal the underlying pathology of many neuropsychiatric disorders. Highly complementary to other imaging tools, MRS can be used in conjunction to better understand the different brain functions such as neurodegeneration (NAA), excitotoxicity (Glu), inhibition (GABA), neuroinflammation (GSH), mitochondrial function (Cr), membrane turnover and demyelination (Cho), and gliosis (mI). In schizophrenia, reduced NAA, increased Gln, and variable Glu (dependent on chronicity, reduced GABA, GSH, and Cr) have been the major findings providing insight into the neuronal function and the influence of the major neurotransmitters upon the disease. In dementia, the primary findings are reduced NAA and Glu and increased mI, which are not only diagnostic for Alzheimer's disease but also in predicting progression of dementia as well. Reduced NAA, Glu, and GABA are characteristic of depression. Those results provide an interesting contrast to bipolar disorder where NAA and GABA are also reduced but Cho and Glu are increased and GSH does not appear to change. Anxiety disorder is probably the least explored by MRS but have shown increased Glu but decreased GABA, thereby demonstrating an imbalance between the major excitatory and inhibitory neurotransmitters in the brain. As MRS methods evolve through the various pulse sequences and post-processing methods that are more and more rapidly becoming available to neuroscientists, psychiatrists, psychologists, and physicians, MRS may eventually cross over into the realm of clinical diagnosis where it would be a valuable tool to provide mechanistic and physiological insight into psychiatric diseases much as genetics and pathophysiology have provided for other diseases.

References

Agostinho P, Cunha RA, Oliveira C (2010) Neuroinflammation, axidative stress and the pathogenesis of Alzheimers disease. Curr Pharm Des 16:2766–2778. doi:10.2174/138161210793176572

Andrew ER (1980) N.m.r. imaging of intact biological systems. Philos Trans R Soc Lond B Biol Sci 289:471–481

Bak LK, Schousboe A, Waagepetersen HS (2006) The glutamate/GABA-glutamine cycle: aspects of transport, neurotransmitter homeostasis and ammonia transfer. J Neurochem 98:641–653. doi:10.1111/j.1471-4159.2006.03913.x

Barker PB (2001) N-acetyl aspartate – a neuronal marker? Ann Neurol 49:423–424

Baslow MH (2000) Functions of N-acetyl-L-aspartate and N-acetyl-L-aspartylglutamate in the vertebrate brain: role in glial cell-specific signaling. J Neurochem 75:453–459

Baslow MH (2010) Evidence that the tri-cellular metabolism of N-acetylaspartate functions as the brain's "operating system": how NAA metabolism supports meaningful intercellular frequency-encoded communications. Amino Acids 39:1139–1145. doi:10.1007/s00726-010-0656-6

Ben-Ari Y (2002) Excitatory actions of gaba during development: the nature of the nurture. Nat Rev Neurosci 3:728–739. doi:10.1038/nrn920

Bhagwagar Z, Wylezinska M, Taylor M et al (2004) Increased brain GABA concentrations following acute administration of a selective serotonin reuptake inhibitor. Am J Psychiatry 161:368–370

Bhattacharyya PK, Lowe MJ, Phillips MD (2007) Spectral quality control in motion-corrupted single-voxel J-difference editing scans: An interleaved navigator approach. Magn Reson Med 58:808–812

Binesh N, Kumar A, Hwang S et al (2004) Neurochemistry of late-life major depression: a pilot two-dimensional MR spectroscopic study. J Magn Reson Imaging 20:1039–1045. doi:10.1002/jmri.20214

Blüml S, Tan J, Harris K et al (1999) Quantitative proton-decoupled 31P MRS of the schizophrenic brain in vivo. J Comput Assist Tomogr 23:272–275

Bluml S, Zuckerman E, Tan J, Ross BD (1998) Proton-decoupled 31P magnetic resonance spectroscopy reveals osmotic and metabolic disturbances in human hepatic encephalopathy. J Neurochem 71:1564–1576

Bolan PJ, DelaBarre L, Baker EH et al (2002) Eliminating spurious lipid sidebands in 1H MRS of breast lesions. Magn Reson Med 48:215–222

Bora E, Fornito A, Yücel M, Pantelis C (2010) Voxelwise meta-analysis of gray matter abnormalities in bipolar disorder. Biol Psychiatry 67:1097–1105. doi:10.1016/j.biopsych.2010.01.020

Bottomley PA (1987) Spatial localization in NMR spectroscopy in vivo. Ann N Y Acad Sci 508:333–348

Brady RO, McCarthy JM, Prescot AP et al (2013) Brain gamma-aminobutyric acid (GABA) abnormalities in bipolar disorder. Bipolar Disord. doi:10.1111/bdi.12074

Cai K, Haris M, Singh A et al (2012) Magnetic resonance imaging of glutamate. Nat Med 18:302–306. doi:10.1038/nm.2615

Cai K, Singh A, Roalf DR et al (2013) Mapping glutamate in subcortical brain structures using high-resolution GluCEST MRI. NMR Biomed. doi:10.1002/nbm.2949

Capizzano AA, Jorge RE, Acion LC, Robinson RG (2007) In vivo proton magnetic resonance spectroscopy in patients with mood disorders: a technically oriented review. J Magn Reson Imaging: JMRI 26:1378–1389. doi:10.1002/jmri.21144

Caverzasi E, Pichiecchio A, Poloni GU et al (2012) Magnetic resonance spectroscopy in the evaluation of treatment efficacy in unipolar major depressive disorder: a review of the literature. Funct Neurol 27:13–22

Chakraborty G, Mekala P, Yahya D et al (2001) Intraneuronal N-acetylaspartate supplies acetyl groups for myelin lipid synthesis: evidence for myelin-associated aspartoacylase. J Neurochem 78:736–745

Choi I-Y, Lee S-P, Merkle H, Shen J (2004) Single-shot two-echo technique for simultaneous measurement of GABA and creatine in the human brain in vivo. Magn Reson Med 51:1115–1121. doi:10.1002/mrm.20082

Choi I-Y, Lee S-P, Shen J (2005a) In vivo single-shot three-dimensionally localized multiple quantum spectroscopy of GABA in the human brain with improved spectral selectivity. J Magn Reson (San Diego, Calif : 1997) 172:9–16. doi:10.1016/j.jmr.2004.09.021

Choi I-Y, Lee S-P, Shen J (2005b) Selective homonuclear Hartmann-Hahn transfer method for in vivo spectral editing in the human brain. Magn Reson Med 53:503–510. doi:10.1002/mrm.20381

Cocuzzo D, Lin A, Ramadan S et al (2011) Algorithms for characterizing brain metabolites in two-dimensional in vivo MR correlation spectroscopy. Conf Proc IEEE Eng Med Biol Soc 2011:4929–4934. doi:10.1109/IEMBS.2011.6091222

Cooper JR, Bloom FE, Roth RH (1970) The biochemical basis of neuropharmacology. Oxford University Press, New York

Do KQ, Trabesinger AH, Kirsten-Krüger M et al (2000) Schizophrenia: glutathione deficit in cerebrospinal fluid and prefrontal cortex in vivo. Eur J Neurosci 12:3721–3728

Dreher W, Leibfritz D (1995) On the use of 2-Dimensional-J NMR measurements for in-vivo proton MRS- measurement of homonuclear decoupled spectra without the need for short echo times. Magn Reson Med 34:331–337. doi:10.1002/mrm.1910340309

Dringen R, Gutterer JM, Hirrlinger J (2000) Glutathione metabolism in brain. Eur J Biochem 267:4912–4916. doi:10.1046/j.1432-1327.2000.01597.x

Duffy SL, Lagopoulos J, Hickie IB et al (2013) Glutathione relates to neuropsychological functioning in mild cognitive impairment. Alzheimers Dement. doi:10.1016/j.jalz.2013.01.005

Edden RAE, Barker PB (2007) Spatial effects in the detection of γ-aminobutyric acid: Improved sensitivity at high fields using inner volume saturation. Magn Reson Med 58:1276–1282

Ernst RR (1992) Nobel Lecture. Nuclear magnetic resonance Fourier transform spectroscopy. Biosci Rep 12:143–187

Ernst T, Jiang CS, Nakama H et al (2010) Lower brain glutamate is associated with cognitive deficits in HIV patients: a new mechanism for HIV-associated neurocognitive disorder. J Magn Reson Imaging: JMRI 32:1045–1053. doi:10.1002/jmri.22366

Fayed N, Modrego PJ, Rojas-Salinas G, Aguilar K (2011) Brain glutamate levels are decreased in Alzheimer's disease: a magnetic resonance spectroscopy study. Am J Alzheimers Dis Other Demen 26:450–456. doi:10.1177/1533317511421780

Frey BN, Stanley JA, Nery FG et al (2007) Abnormal cellular energy and phospholipid metabolism in the left dorsolateral prefrontal cortex of medication-free individuals with bipolar disorder: an in vivo 1H MRS study. Bipolar Disord 9(Suppl 1):119–127. doi:10.1111/j.1399-5618.2007.00454.x

Friston KJ (1997) Analyzing brain images: principles and overview. In: Frackowiak RSJ, Friston Frith CD, Dolan RJMS (eds) Human brain function. Academic, San Diego, pp 25–41

Gigante AD, Bond DJ, Lafer B et al (2012) Brain glutamate levels measured by magnetic resonance spectroscopy in patients with bipolar disorder: a meta-analysis. Bipolar Disord 14:478–487. doi:10.1111/j.1399-5618.2012.01033.x

Goddard AW, Mason GF, Almai A et al (2001) Reductions in occipital cortex GABA levels in panic disorder detected with 1 h-magnetic resonance spectroscopy. Arch Gen Psychiatry 58:556

Goddard AW, Mason GF, Appel M et al (2004) Impaired GABA neuronal response to acute benzodiazepine administration in panic disorder. Am J Psychiatry 161:2186–2193

Gonzalez-Burgos G, Hashimoto T, Lewis DA (2010) Alterations of cortical GABA neurons and network oscillations in schizophrenia. Curr Psychiatry Rep 12:335–344. doi:10.1007/s11920-010-0124-8

Goto N, Yoshimura R, Moriya J et al (2009) Reduction of brain gamma-aminobutyric acid (GABA) concentrations in early-stage schizophrenia patients: 3T Proton MRS study. Schizophr Res 112:192–193. doi:10.1016/j.schres.2009.04.026

Govindaraju V, Young K, Maudsley AA (2000) Proton NMR chemical shifts and coupling constants for brain metabolites. NMR Biomed 13:129–153

De Graaf RA, Rothman DL, Behar KL (2011) State of the art direct 13C and indirect 1H-[13C] NMR spectroscopy in vivo. A practical guide. NMR Biomed 24:958–972. doi:10.1002/nbm.1761

Graff-Radford J, Kantarci K (2013) Magnetic resonance spectroscopy in Alzheimer's disease. Neuropsychiatr Dis Treat 9:687–696. doi:10.2147/NDT.S35440

Gruber S, Frey R, Mlynárik V et al (2003) Quantification of metabolic differences in the frontal brain of depressive patients and controls obtained by 1H-MRS at 3 Tesla. Invest Radiol 38:403–408. doi:10.1097/01.rli.0000073446.43445.20

Gruetter R, Novotny EJ, Boulware SD et al (1994) Localized 13C NMR spectroscopy in the human brain of amino acid labeling from d-[1-^{13}C]glucose. J Neurochem 63:1377–1385. doi:10.1046/j.1471-4159.1994.63041377.x

Haase A, Frahm J, Matthaei D et al (1986) MR imaging using stimulated echoes (STEAM). Radiology 160:787–790

Harris K, Lin A, Bhattacharya P et al (2006) Regulation of NAA-synthesis in the human brain in vivo: Canavan's disease, Alzheimer's disease and schizophrenia. Adv Exp Med Biol 576:263–273. doi:10.1007/0-387-30172-0_18; discussion 361–3

Hasler G, Neumeister A, Van der Veen JW et al (2005) Normal prefrontal gamma-aminobutyric acid levels in remitted depressed subjects determined by proton magnetic resonance spectroscopy. Biol Psychiatry 58:969–973

Hennig J (1992) The application of phase rotation for localized in vivo proton spectroscopy with short echo times. J Magn Reson 40–49

Hermens DF, Lagopoulos J, Naismith SL et al (2012) Distinct neurometabolic profiles are evident in the anterior cingulate of young people with major psychiatric disorders. Transl Psychiatry 2:e110. doi:10.1038/tp.2012.35

Hertz L, Zielke HR (2004) Astrocytic control of glutamatergic activity: astrocytes as stars of the show. Trends Neurosci 27:735–743

Hurd R, Sailasuta N, Srinivasan R et al (2004) Measurement of brain glutamate using TE-averaged PRESS at 3T. Magn Reson Med 51:435–440

Jang DP, Lee JM, Lee E et al (2005) Interindividual reproducibility of glutamate quantification using 1.5-T proton magnetic resonance spectroscopy. Magn Reson Med 53:708–712. doi:10.1002/mrm.20387

Jarskog LF, Miyamoto S, Lieberman JA (2007) Schizophrenia: new pathological insights and therapies. Annu Rev Med 58:49–61. doi:10.1146/annurev.med.58.060904.084114

Javitt DC, Zukin SR (1991) Recent advances in the phencyclidine model of schizophrenia. Am J Psychiatry 148:1301–1308

Jeener J, Meier BH, Bachmann P, Ernst RR (1979) Investigation of exchange processes by two-dimensional NMR spectroscopy. J Chem Phys 71:4546. doi:10.1063/1.438208

Jensen JE, Frederick BD, Wang L et al (2005) Two-dimensional, J-resolved spectroscopic imaging of GABA at 4 Tesla in the human brain. Magn Reson Med 54:783–788. doi:10.1002/mrm.20644

Jones RS, Waldman AD (2004) 1H-MRS evaluation of metabolism in Alzheimer's disease and vascular dementia. Neurol Res 26:488–495. doi:10.1179/016164104225017640

Kaiser LG, Young K, Matson GB (2007) Elimination of spatial interference in PRESS-localized editing spectroscopy. Magn Reson Med 58:813–818

Kantarci K (2007) 1H magnetic resonance spectroscopy in dementia. Br J Radiol 80(Spec No 2):S146–S152. doi:10.1259/bjr/60346217

Kantarci K (2013) Proton MRS in mild cognitive impairment. J Magn Reson Imaging: JMRI 37:770–777. doi:10.1002/jmri.23800

Kantarci K, Petersen RC, Boeve BF et al (2004) 1H MR spectroscopy in common dementias. Neurology 63:1393–1398

Kantarci K, Weigand SD, Przybelski SA et al (2013) MRI and MRS predictors of mild cognitive impairment in a population-based sample. Neurology 81:126–133. doi:10.1212/WNL.0b013e31829a3329

Kato T, Takahashi S, Shioiri T et al (1994) Reduction of brain phosphocreatine in bipolar II disorder detected by phosphorus-31 magnetic resonance spectroscopy. J Affect Disord 31:125–133

Kaufman RE, Ostacher MJ, Marks EH et al (2009) Brain GABA levels in patients with bipolar disorder. Prog Neuropsychopharmacol Biol Psychiatry 33:427–434

Ke Y, Cohen BM, Bang JY et al (2000) Assessment of GABA concentration in human brain using two-dimensional proton magnetic resonance spectroscopy. Psychiatry Res 100:169–178

Kegeles LS, Humaran TJ, Mann JJ (1998) In vivo neurochemistry of the brain in schizophrenia as revealed by magnetic resonance spectroscopy. Biol Psychiatry 44:382–398

Keltner JR, Wald LL, Frederick BD, Renshaw PF (1997) In vivo detection of GABA in human brain using a localized double-quantum filter technique. Magn Reson Med 37:366–371

Kraguljac NV, Reid M, White D et al (2012) Neurometabolites in schizophrenia and bipolar disorder - a systematic review and meta-analysis. Psychiatry Res 203:111–125. doi:10.1016/j.pscychresns.2012.02.003

Kreis R, Ernst T, Ross BD (1993) Absolute quantitation of water and metabolites in the human brain. II. Metabolite concentrations. J Magn Reson - Ser B 102:9–19. doi:10.1006/jmrb.1993.1056

Lagopoulos J, Hermens DF, Tobias-Webb J et al (2013) In vivo glutathione levels in young persons with bipolar disorder: A magnetic resonance spectroscopy study. J Psychiatr Res 47:412–417. doi:10.1016/j.jpsychires.2012.12.006

Lauterbur PC (1989) Image formation by induced local interactions. Examples employing nuclear magnetic resonance. 1973. Clin Orthop Relat Res (244): 3–6

Lin A, Ross BD, Harris K, Wong W (2005) Efficacy of proton magnetic resonance spectroscopy in neurological diagnosis and neurotherapeutic decision making. NeuroRx 2:197–214. doi:10.1602/neurorx.2.2.197

Lin AP, Shic F, Enriquez C, Ross BD (2003) Reduced glutamate neurotransmission in patients with Alzheimer's disease–an in vivo 13C magnetic resonance spectroscopy study. Magn Reson Mater Phys, Biol Med 16:29–42

Long Z, Medlock C, Dzemidzic M et al (2013) Decreased GABA levels in anterior cingulate cortex/medial prefrontal cortex in panic disorder. Prog Neuropsychopharmacol Biol Psychiatry 44:131–135. doi:10.1016/j.pnpbp.2013.01.020

Luykx JJ, Laban KG, Van den Heuvel MP et al (2012) Region and state specific glutamate downregulation in major depressive disorder: a meta-analysis of (1) H-MRS findings. Neurosci Biobehav Rev 36:198–205. doi:10.1016/j.neubiorev.2011.05.014

Lymer K, Haga K, Marshall I et al (2007) Reproducibility of GABA measurements using 2D J-resolved magnetic resonance spectroscopy. Magn Reson Imaging 25:634–640. doi:10.1016/j.mri.2006.10.010

Maddock RJ, Buonocore MH (2012) MR spectroscopic studies of the brain in psychiatric disorders. Curr Top Behav Neurosci. doi:10.1007/7854_2011_197

Mandal PK, Tripathi M, Sugunan S (2012) Brain oxidative stress: detection and mapping of anti-oxidant marker "Glutathione" in different brain regions of healthy male/female, MCI and Alzheimer patients using non-invasive magnetic resonance spectroscopy. Biochem Biophys Res Commun 417:43–48. doi:10.1016/j.bbrc.2011.11.047

Marsman A, Van den Heuvel MP, Klomp DWJ et al (2013) Glutamate in schizophrenia: a focused review and meta-analysis of ^{1}H-MRS studies. Schizophr Bull 39:120–129. doi:10.1093/schbul/sbr069

Mason GF, Gruetter R, Rothman DL et al (1995) Simultaneous determination of the rates of the TCA cycle, glucose utilization, μ-ketoglutarate/Glutamate exchange, and glutamine synthesis in human brain by NMR. J Cereb Blood Flow Metab 15:12–25

Matsuzawa D, Hashimoto K (2011) Magnetic resonance spectroscopy study of the antioxidant defense system in schizophrenia. Antioxid Redox Signal 15:2057–2065. doi:10.1089/ars.2010.3453

Matsuzawa D, Obata T, Shirayama Y et al (2008) Negative correlation between brain glutathione level and negative symptoms in schizophrenia: a 3T 1H-MRS study. PloS one 3:e1944. doi:10.1371/journal.pone.0001944

Maudsley AA, Hilal SK, Perman WH, Simon HE (1983) Spatially resolved high resolution spectroscopy by "Four-Dimensional" NMR. J Magn Reson 147–152

McLean MA, Busza AL, Wald LL et al (2002) In vivo GABA+ measurement at 1.5T using a PRESS-localized double quantum filter. Magn Reson Med 48:233–241

Minoshima S, Giordani B, Berent S et al (1997) Metabolic reduction in the posterior cingulate cortex in very early Alzheimer's disease. Ann Neurol 42:85–94. doi:10.1002/ana.410420114

Moffett JR, Namboodiri MA, Neale JH (1993) Enhanced carbodiimide fixation for immunohistochemistry: application to the comparative distributions of N-acetylaspartylglutamate and N-acetylaspartate immunoreactivities in rat brain. J Histochem Cytochem 41:559–570

Moffett JR, Ross B, Arun P et al (2007) N-Acetylaspartate in the CNS: from neurodiagnostics to neurobiology. Prog Neurobiol 81:89–131. doi:10.1016/j.pneurobio.2006.12.003

Mosley RL, Benner EJ, Kadiu I et al (2006) Neuroinflammation, oxidative stress and the pathogenesis of Parkinson's disease. Clin Neurosci Res 6:261–281. doi:10.1016/j.cnr.2006.09.006

Mullins PG, Chen H, Xu J et al (2008) Comparative reliability of proton spectroscopy techniques designed to improve detection of J-coupled metabolites. Magn Reson Med 60:964–969. doi:10.1002/mrm.21696

Olman CA, Davachi L, Inati S (2009) Distortion and signal loss in medial temporal lobe. PloS one 4:e8160. doi:10.1371/journal.pone.0008160

Olsen RW, DeLorey TM (1999) GABA synthesis, uptake and release. In: Siegel GJ, Agranoff BW, Albers RW, Fisher SK, Uhler MD (eds) Basic neurochemistry: molecular, cellular and medical aspects. Lippincott-Raven, Philadelphia

Ongür D, Prescot AP, Jensen JE et al (2009) Creatine abnormalities in schizophrenia and bipolar disorder. Psychiatry Res 172:44–48. doi:10.1016/j.pscychresns.2008.06.002

Ongür D, Prescot AP, McCarthy J et al (2010) Elevated gamma-aminobutyric acid levels in chronic schizophrenia. Biol Psychiatry 68:667–670. doi:10.1016/j.biopsych.2010.05.016

Öz G, Terpstra M, Tkáč I et al (2006) Proton MRS of the unilateral substantia nigra in the human brain at 4 tesla: detection of high GABA concentrations. Magn Reson Med 55:296–301

Pellerin L, Magistretti P (1994) Glutamate uptake into astrocytes stimulates aerobic glycolysis: A mechanism coupling neuronal activity to glucose utilization. Proc Natl Acad Sci U S A 91:10625–10629

Penner J, Rupsingh R, Smith M et al (2010) Increased glutamate in the hippocampus after galantamine treatment for Alzheimer disease. Prog Neuropsychopharmacol Biol Psychiatry 34:104–110. doi:10.1016/j.pnpbp.2009.10.007

Phan KL, Fitzgerald DA, Cortese BM et al (2005) Anterior cingulate neurochemistry in social anxiety disorder: 1H-MRS at 4 Tesla. Neuroreport 16:183–186

Pollack MH, Jensen JE, Simon NM et al (2008) High-field MRS study of GABA, glutamate and glutamine in social anxiety disorder: response to treatment with levetiracetam. Prog Neuropsychopharmacol Biol Psychiatry 32:739–743

Port JD, Agarwal N (2011) MR spectroscopy in schizophrenia. J Magn Reson Imaging: JMRI 34:1251–1261. doi:10.1002/jmri.22787

Posner MJ, Raichle ME (1997) Images of mind. WH Freeman, New York

Posse S, Otazo R, Dager SR, Alger J (2013) MR spectroscopic imaging: principles and recent advances. J Magn Reson Imaging 37:1301–1325. doi:10.1002/jmri.23945

Provencher SW (2001) Automatic quantitation of localized in vivo 1H spectra with LCModel. NMR Biomed 14:260–264
Provencher SW (1993) Estimation of metabolite concentrations from localized in vivo proton NMR spectra. Magn Reson Med 30:672–679
Ramadan S (2007) Phase-rotation in in-vivo localised spectroscopy. Concepts Magn Reson 30:147–153
Ramadan S, Andronesi OC, Stanwell P et al (2011) Use of in vivo two-dimensional MR spectroscopy to compare the biochemistry of the human brain to that of glioblastoma. Radiology 259:540–549. doi:10.1148/radiol.11101123
Rezin GT, Amboni G, Zugno AI et al (2009) Mitochondrial dysfunction and psychiatric disorders. Neurochem Res 34:1021–1029. doi:10.1007/s11064-008-9865-8
Rigotti DJ, Inglese M, Gonen O (2007) Whole-brain N-acetylaspartate as a surrogate marker of neuronal damage in diffuse neurologic disorders. AJNR Am J Neuroradiol 28:1843–1849. doi:10.3174/ajnr.A0774
Rinck P (2012) Magnetic resonance in medicine. The Basic Textbook of the European Magnetic Resonance Forum. 6th edn. Electronic version 6.1. Available at www.magnetic-resonance.org.
Rosen Y, Lenkinski RE (2007) Recent advances in magnetic resonance neurospectroscopy. Neurotherapeutics 4:330–345. doi:10.1016/j.nurt.2007.04.009
Ross B, Bluml S (2001) Magnetic resonance spectroscopy of the human brain. Anat Rec 265:54–84
Ross B, Lin A, Harris K et al (2003) Clinical experience with 13C MRS in vivo. NMR Biomed 16:358–369
Ross BD (2000) Real or imaginary? Human metabolism through nuclear magnetism. IUBMB Life 50:177–187. doi:10.1080/152165400300001499
Ross BD, Bluml S, Cowan R et al (1997) In vivo magnetic resonance spectroscopy of human brain: the biophysical basis of dementia. Biophys Chem 68:161–172
Rowland LM, Kontson K, West J et al (2012) In vivo measurements of glutamate, GABA, and NAAG in schizophrenia. Schizophr Bull. doi:10.1093/schbul/sbs092
Sailasuta N, Harris K, Tran T, Ross B (2011) Minimally invasive biomarker confirms glial activation present in Alzheimer's disease: a preliminary study. Neuropsychiatr Dis Treat 7:495–499
Sanacora G, Treccani G, Popoli M (2012) Towards a glutamate hypothesis of depression: an emerging frontier of neuropsychopharmacology for mood disorders. Neuropharmacology 62:63–77. doi:10.1016/j.neuropharm.2011.07.036
Satoh T, Yoshioka Y (2006) Contribution of reduced and oxidized glutathione to signals detected by magnetic resonance spectroscopy as indicators of local brain redox state. Neurosci Res 55:34–39. doi:10.1016/j.neures.2006.01.002
Schousboe A (2003) Role of astrocytes in the maintenance and modulation of glutamatergic and GABAergic neurotransmission. Neurochem Res 28:347–352. doi:10.1023/a:1022397704922
Schousboe A, Waagepetersen H (2005) Role of astrocytes in glutamate homeostasis: Implications for excitotoxicity. Neurotox Res 8:221–225. doi:10.1007/bf03033975
Schubert F, Gallinat J, Seifert F, Rinneberg H (2004) Glutamate concentrations in human brain using single voxel proton magnetic resonance spectroscopy at 3 Tesla. Neuroimage 21:1762–1771
Schulte RF, Boesiger P (2006) ProFit: two-dimensional prior-knowledge fitting of J-resolved spectra. NMR Biomed 19:255–263. doi:10.1002/nbm.1026
Schulte RF, Lange T, Beck J et al (2006) Improved two-dimensional J-resolved spectroscopy. NMR Biomed 19:264–270. doi:10.1002/nbm.1027
Shonk TK, Moats RA, Gifford P et al (1995) Probable Alzheimer disease: diagnosis with proton MR spectroscopy. Radiology 195:65–72
Sibson NR, Dhankhar A, Mason GF et al (1998) Stoichiometric coupling of brain glucose metabolism and glutamatergic neuronal activity. Proc Natl Acad Sci U S A 95:316–321
Simister RJ, McLean MA, Barker GJ, Duncan JS (2003) A proton magnetic resonance spectroscopy study of metabolites in the occipital lobes in epilepsy. Epilepsia 44:550–558
Starcuk Z, Starcuk Z Jr, Horky J (2001) "Baseline" problems in very short echo-time proton MR spectroscopy of low molecular weight metabolites in the brain. Meas Sci Rev 1:17–20
Strawn JR, Chu W-J, Whitsel RM et al (2013) A pilot study of anterior cingulate cortex neurochemistry in adolescents with generalized anxiety disorder. Neuropsychobiology 67:224–229. doi:10.1159/000347090
Tayoshi S, Nakataki M, Sumitani S et al (2010) GABA concentration in schizophrenia patients and the effects of antipsychotic medication: a proton magnetic resonance spectroscopy study. Schizophr Res 117:83–91. doi:10.1016/j.schres.2009.11.011
Terpstra M, Ugurbil K, Gruetter R (2002) Direct in vivo measurement of human cerebral GABA concentration using MEGA-editing at 7 Tesla. Magn Reson Med 47:1009–1012
Terpstra M, Vaughan TJ, Ugurbil K et al (2005) Validation of glutathione quantitation from STEAM spectra against edited 1H NMR spectroscopy at 4 T: application to schizophrenia. Magma (New York, NY) 18:276–282. doi:10.1007/s10334-005-0012-0
Thomas MA, Yue K, Binesh N et al (2001) Localized two-dimensional shift correlated MR spectroscopy of human brain. Magn Reson Med 46:58–67
Urenjak J, Williams SR, Gadian DG, Noble M (1992) Specific expression of N-acetylaspartate in neurons, oligodendrocyte-type-2 astrocyte progenitors, and immature oligodendrocytes in vitro. J Neurochem 59:55–61
Vanhamme L, Van den Boogaart A, Van Huffel S (1997) Improved method for accurate and efficient quantification of MRS data with use of prior knowledge. J Magn Reson 129:35–43. doi:10.1006/jmre.1997.1244
Waddell KW, Avison MJ, Joers JM, Gore JC (2007) A practical guide to robust detection of GABA in human

brain by J-difference spectroscopy at 3 T using a standard volume coil. Magn Reson Imaging 25:1032–1038. doi:10.1016/j.mri.2006.11.026

Wijtenburg SA, Knight-Scott J (2011) Very short echo time improves the precision of glutamate detection at 3T in 1H magnetic resonance spectroscopy. J Magn Reson Imaging 34:645–652

Wood SJ, Berger GE, Wellard RM et al (2009) Medial temporal lobe glutathione concentration in first episode psychosis: a 1H-MRS investigation. Neurobiol Dis 33:354–357. doi:10.1016/j.nbd.2008.11.018

Yang S, Hu J, Kou Z, Yang Y (2008) Spectral simplification for resolved glutamate and glutamine measurement using a standard STEAM sequence with optimized timing parameters at 3, 4, 4.7, 7, and 9.4 T. Magn Reson Med// 59:236–244. doi:10.1002/mrm.21463

Yoon JH, Maddock RJ, Rokem A et al (2010) GABA concentration is reduced in visual cortex in schizophrenia and correlates with orientation-specific surround suppression. J Neurosci 30:3777–3781

Zhou J, van Zijl PCM (2006) Chemical exchange saturation transfer imaging and spectroscopy. Prog Nucl Magn Reson Spectrosc 48:109–136. doi:10.1016/j.pnmrs.2006.01.001

Zwanzger P, Zavorotnyy M, Gencheva E et al (2013) Acute shift in glutamate concentrations following experimentally induced panic with cholecystokinin tetrapeptide-A 3T-MRS study in healthy subjects. Neuropsychopharmacology 38:1648–1654. doi:10.1038/npp.2013.61

7 Imaging Genetics: Unraveling the Neurogenetic Risk Architecture of Mental Illness

Heike Tost, Andreas Böhringer, and Andreas Meyer-Lindenberg

Abbreviations

ACC	Anterior cingulate cortex
ADHD	Attention-deficit/hyperactivity disorder
ADNI	Alzheimer's disease neuroimaging initiative
CNV	Copy number variations
COMT	Catechol-*O*-methyltransferase
D2	Dopamine D2 receptor
DAOA	d-amino acid oxidase activator
DLPFC	Dorsolateral prefrontal cortex
ENIGMA	Enhancing neuro imaging genetics through meta-analysis
FDR	False discovery rate
GSEA	Gene-set enrichment analysis
GWA	Genome-wide analysis
GWAS	Genome-wide association studies
ISC	International schizophrenia Consortium
MAOA	Monoamine oxidase A
PFC	Prefrontal cortex
SNP	Single nucleotide polymorphism
WHO	World Health Organization

H. Tost, Dipl-Psych (✉) • A. Böhringer, Dipl-Psych
A. Meyer-Lindenberg
Department of Psychiatry and Psychotherapy, Medical Faculty Mannheim, Central Institute of Mental Health, University of Heidelberg, Square J5, 68159 Mannheim, Germany
e-mail: heike.tost@zi-mannheim.de; a.meyer-lindenberg@zi-mannheim.de

7.1 Introduction

Apart from a keen interest in neurobiology, the basic motivation of neuroimaging researchers in psychiatry is to provide insights that help clinicians to effectively prevent and treat mental disorders to avert the immense associated personal and societal burdens. A particularly important topic is that of schizophrenia, a brain disorder with a lifetime prevalence of 1 % that manifests in early adulthood and disturbs a wide range of brain functions, many of which are insufficiently addressed by current guideline therapies (e.g., negative symptoms, cognitive and social deficits). As a consequence, a sizeable proportion of patients do not fully recover, never enter or fall out of the labor market, fall victim to stigma and social marginalization, live under poor financial conditions, die prematurely, and contribute to a public health burden that exceeds that of any other disorder of similar prevalence by far (Rössler et al. 2005; WHO 2001). It is generally hoped that many of these aftereffects can be averted if the pathophysiology of mental illness is better understood (Insel and Scolnick 2006). In particular, the brain mechanisms of risk and resilience are of interest since they operate at a point in time when the disease could, in principle, be prevented or at least treated earlier than current clinical diagnostic criteria allow (Meyer-Lindenberg 2010a, b).

Many psychiatric disorders are highly heritable (Sullivan et al. 2012, 2003) and disturb the structural and functional organization of

C. Mulert, M.E. Shenton (eds.), *MRI in Psychiatry*,
DOI 10.1007/978-3-642-54542-9_7, © Springer Berlin Heidelberg 2014

the human brain, which in turn is under strong genetic control (Glahn et al. 2010. Thus, not surprisingly, the impact of genetic risk factors for mental disorders on the brain has gained much interest in the recent history of psychiatry research (Meyer-Lindenberg and Weinberger 2006; Pezawas and Meyer-Lindenberg 2010). A particular fruitful approach has been "imaging genetics," the combination of neuroimaging phenotypes and molecular genetics made possible by the unprecedented technological progress in both these fields in the previous decade. In this chapter, we review the basic principles, applications, caveats, and future prospects of this approach. Naturally, due to the large number of disorders, risk genes, phenotypes, and genetic models, we cannot provide a full representation of the imaging genetics literature in psychiatry here. The reader is referred to a recent special issue of NeuroImage for a comprehensive survey of the field (Pezawas and Meyer-Lindenberg 2010). Instead, we focus on highlighting some of the main tenets, genetic analysis strategies, and landmark findings, as well as the renewed interest of the field in the role of the environment. While we mostly draw from examples from schizophrenia research, the majority of concepts equally apply to other highly heritable psychiatric conditions with a complex genetic architecture, such as autism, bipolar disorder, or ADHD.

7.2 Imaging Genetics

The etiology of psychiatric disorders such as schizophrenia is complex. While genetic factors play a major role, as demonstrated by the estimated heritability of schizophrenia in the range of 0.65–0.855 (Sullivan et al. 2003), the environment and gene-environment interactions must also be considered. The genetics of schizophrenia is complex and not consistent with a Mendelian, mono- or oligogenic architecture. Instead, current evidence suggests a role for single nucleotide polymorphisms (SNPs) that are frequent in the normal population and confer only small increments on risk (odds ratio, OR, <1.2) as well as rare structural genomic alterations conferring high risk in the few cases who are carriers (e.g., copy number variations, CNVs, OR 3-30) (Malhotra and Sebat 2012). Since schizophrenia is a devastating disorder that markedly reduces fecundity, CNVs are under negative selection pressure and rare in the population (Stefansson et al. 2008). In the majority of individuals, therefore, increased risk results from the complex interplay of a large number of common genetic variants with small effect size that interact with one another and with the environment (Gottesmann and Shields 1967). For complex behavioral and clinical course constructs such as those captured by currently defined psychiatric diagnoses, the identification of causal genes and their operating mode has proven to be extraordinarily difficult (Sullivan et al. 2012). Among others, this relates to the complexity of the observable clinical phenotypes, the structural and functional complexity of the organ brain, the limited biological validity of current diagnostic categories, the unknown biological function of most statistically associated variants, and the seemingly long and convoluted way (or large number of poorly defined biological processes) from altered DNA to psychopathology (Meyer-Lindenberg and Weinberger 2006).

7.2.1 Intermediate Phenotypes

Given these challenges psychiatry faces, alternative approaches in the study of the genetic underpinnings of psychiatric disorders have attracted much attention, in particular the analysis of "intermediate phenotypes" (Meyer-Lindenberg and Weinberger 2006; Preston and Weinberger 2005) or "endophenotypes" (Gottesman and Gould 2003). Inherent to this concept is the assumption that complex disease phenotypes (or "exophenotypes") can be deconstructed in genetically simpler neural subprocesses (e.g., neurophysiological, neuroanatomical, or cognitive traits) that are quantifiable and are more directly and strongly influenced by the genetic mechanism that cause mental illness. If this notion is true (c.f. Flint and Munafo 2007; Stein et al. 2012), then it follows that the penetrance of genetic effects should be higher at the level of these biological traits than at the level of behavior, which increases the odds for successful identification of disease-related loci and causal biological processes by genetic linkage and association analyses.

Importantly, not all quantifiable biological markers are necessarily heritable and thus equally useful for the dissection of the genetic risk architecture of mental illness. To date, different but largely overlapping requirements for biologically meaningful intermediate phenotypes have been defined (Preston and Weinberger 2005; Gottesman and Gould 2003; Cannon and Keller 2006; Kendler and Neale 2010), including those explored by imaging genetics. Among others, the definitions require the examined phenotypes to be:

(a) Quantitative in nature
(b) Heritable
(c) Reliably measurable
(d) Continuously distributed in the general population
(e) Associated with the illness in the population
(f) State independent (i.e., traceable in carriers of genetic risk variants whether or not the illness is manifest)
(g) Linked to the genetic risk for the illness (i.e., found in unaffected relatives of patients at higher rates than in the general population)
(h) Reflective of a mediator in the causal pathway from risk genes to psychopathology

The last point is rarely explicitly stated but of particular interest. While it is challenging to establish that a given phenotype functions in the role of a biological mediator (Green et al. 2013), this requirement conceptually differentiates "true" intermediate phenotypes from other correlated phenomena such as liability indices or "biomarkers." While the latter may be otherwise informative (e.g., for early diagnosis or prediction of treatment response), and fulfill several other of the above defined criteria, they are of limited value for the dissection of the causal biological mechanisms that interrelate genetic variation and disease risk (see Gottesman and Gould 2003 for a detailed discussion on this topic).

7.2.2 Methodological Challenges

In recent years, neuroimaging methods have been extensively used as means to explore how risk-associated variants modulate illness-related structural and functional processes in the human brain. The spectrum of applied techniques and examined outcome measures is large and includes markers of brain structure, brain function, and biochemistry derived from various imaging modalities, tasks, and analysis strategies, some of which are highlighted in the upcoming sections. While meta-analyses suggest that genetic effects are indeed more penetrant at the neural systems level than on the clinical level (Mier et al. 2010; Munafo et al. 2008), recent evidence also points to a substantial between-study heterogeneity in findings (Murphy et al. 2013), an observation that may reflect some of the remaining challenges in the imaging genetics field.

Firstly, as outlined above, the mere establishment of a statistical link between a genotype and some imaging phenotype is not sufficient to inform the field about the biological mechanism of a disorder (Meyer-Lindenberg and Weinberger 2006; Rasetti and Weinberger 2011). At the same time, the majority of the criteria for intermediate phenotypes listed in the preceding section have not been established yet for many structural and most functional phenotypes in use, in particular when it comes to their quality criteria, heritability, and their link to the genetic risk for the illness (Rasetti and Weinberger 2011; Bigos and Weinberger 2010). In contrast, for other imaging outcome measures, this groundwork has been laid (Callicott et al. 2003; Plichta et al. 2012). Specifically, as detailed later, some prefrontal cortex (PFC) activation and connectivity phenotypes have established quality criteria (Plichta et al. 2012; Bilek et al. 2013) and have repeatedly been shown to be altered in unaffected relatives of schizophrenia patients (Callicott et al. 2003; Rasetti et al. 2011) and healthy carriers of genome-wide supported genetic risk variants (Rasetti et al. 2011; Esslinger et al. 2009). Given the heterogeneity of studies mentioned above, it would be desirable that the field moves to the application of such established paradigms wherever possible.

Secondly, just as behavioral phenotypes, imaging phenotypes are also sensitive to a variety of nongenetic confounds such as those that relate to demographic differences, the ethnic composition of samples (e.g., genetic stratification), the presence of brain-active substances and conditions (e.g., medication, nicotine, recreational drugs, neurological conditions, illness chronicity), preexisting differences in task performance, and a multitude of method-inherent noise sources. Many of these confounds can be addressed better in imaging genetics studies of healthy individuals than in studies examining patient populations. To address the remaining issues, many but not all studies control as many of these factors as possible. It is also useful to establish familiarity (as required for a heritable phenotype) in matched healthy first-grade relatives of affected patients (which share per definition 50 % of the genetic makeup of their affected relatives, including half of that conferring risk), and then examine the neural effects of genetic risk variants in carefully matched groups of healthy volunteers that are stratified by genotype (Meyer-Lindenberg and Weinberger 2006; Rasetti and Weinberger 2011; Bigos and Weinberger 2010; Meyer-Lindenberg 2010a, b).

Thirdly, although the effect size of risk genes is larger in brain, imaging genetics still calls for the examination of large sample sizes (Waldman 2005), a standard that is still not reached by many studies in the field. The appropriate number depends hereby on several factors, such as the adopted research strategy (e.g., single-variant candidate approach vs. genome-wide association analysis, GWAS), the quality criteria of the examined imaging phenotype, or the minor allele frequency of a given variant of interest, and may range from about 70 up to several thousand subjects (Stein et al. 2010, 2012; Mier et al. 2010). Here, the rate-limiting factor has been the high cost of neuroimaging experiments. Analogous to the molecular genetics field, this has lately led to a trend toward large-scale multisite studies (e.g., ENIGMA consortium). First evidence suggests that the implementation of this strategy can be successful (Stein et al. 2012), although it also bears its own methodological challenges such as the execution of comparable data acquisition protocols and quality control procedures over sites.

7.3 Single-Variant Approaches

7.3.1 Candidate Gene Variants

Historically, psychiatric imaging genetics has focused on the hypothesis-driven analysis of single genetic variants that were chosen based on candidacy, i.e., prior evidence for a potential role for disorder-related neural alterations coming from smaller genetic association studies or a plausible mechanism derived from basic neuroscience research (Pezawas and Meyer-Lindenberg 2010). Here, the empirical link of the genetic variant to the disease entity is typically not been established at currently accepted levels of statistical significance for hypothesis-free approaches such as GWAS. While candidate gene approaches are still informative and topical, this hypothesis-driven process implies that the studies need to be carefully weighed by considering the quality of the prior knowledge in molecular, cellular, and neural systems biology in order to deliver meaningful results for pathophysiology research (Meyer-Lindenberg and Weinberger 2006; Bigos and Weinberger 2010; Meyer-Lindenberg 2010a, b).

Typical examples for candidate gene approaches are functional variants in dopamine-related genes and their examination in the context of schizophrenia risk. The dopamine hypothesis is the oldest neurochemical theory of the disorder and traces back to the observation that the antipsychotic efficacy of neuroleptics relates to their affinity for the dopamine D2 receptor (Bigos and Weinberger 2010; Callicott et al. 2003). Contemporary extensions of the model suggest that the formation of positive symptoms (and in particular delusions) is promoted by decreased efficiency of the PFC, which leads to excessive and chaotic subcortical dopamine release (Meyer-Lindenberg et al. 2002), which in turn promotes attribution of motivational significance to inappropriate stimuli ("aberrant

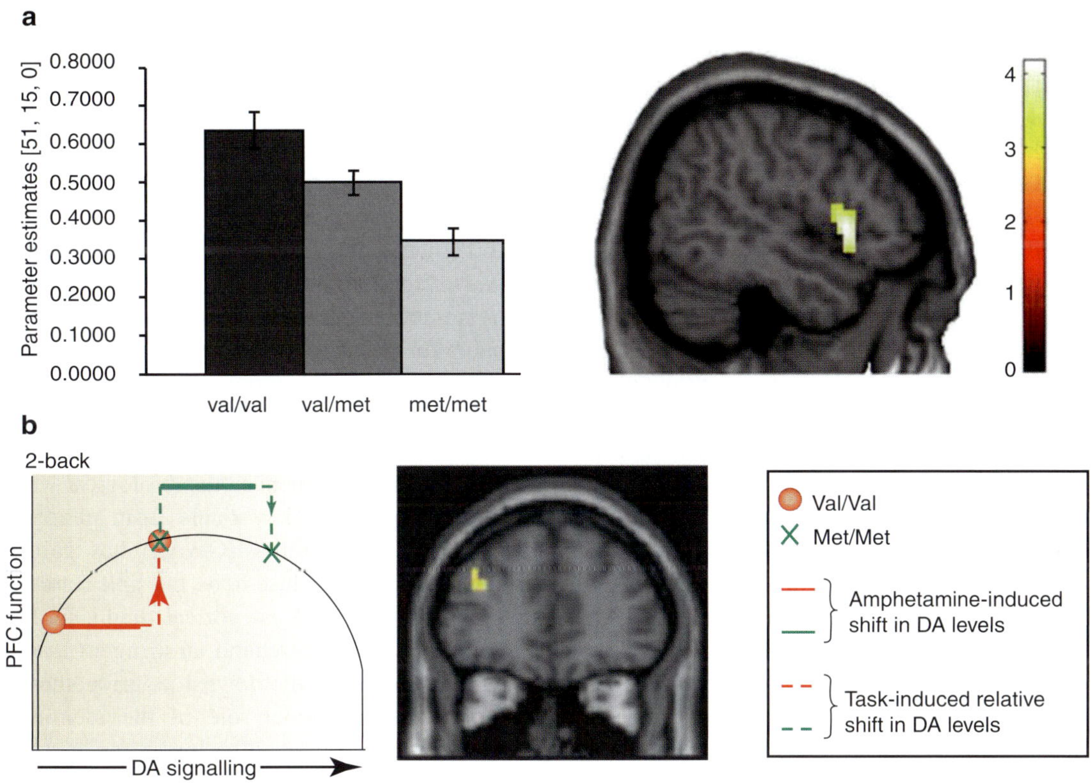

Fig. 7.1 Effects of the COMT Val158Met polymorphism on PFC activation. (**a**) Significant linear effect of the COMT variant on prefrontal cortex activation during a working memory task in 308 healthy controls with increased activation (or decreased efficiency) in Val-allele carriers. (**b**) At medium working memory load level (2-back task), increasing synaptic dopamine by the administration of amphetamine increases prefrontal efficiency in Val-allele carriers but not Met-allele carriers. *Abbreviations*: *PFC* prefrontal cortex (Reprinted with permission from Meyer-Lindenberg and Weinberger (2006) and Tost et al. (2010))

salience") (Kapur 2003; Kapur et al. 2005). The most-studied dopaminergic gene is the catechol-O-methyltransferase (COMT) gene on chromosome 22q11.22–23, a region implicated in schizophrenia by linkage studies (Owen et al. 2004) and 22q11 syndrome, a hemideletion syndrome with strongly increased risk for psychosis (Zinkstok and van Amelsvoort 2005). A common functional variant in COMT (rs4680) results in an amino acid substitution (Val158Met) of the dopamine-catabolizing enzyme (Chen et al. 2004). In Met-allele carriers, the resulting conformation change in the encoded protein prompts a decrease in the activity of the enzyme (Chen et al. 2004) and consequently higher levels of extracellular dopamine (Bilder et al. 2004), particularly in the PFC where COMT plays a dominant role in the degradation of dopamine.

To date, a large body of imaging genetics evidence demonstrates that COMT Val158Met modulates cognitive and affective neural functions in humans differentially (see Mier et al. 2010 for an in-depth review). Specifically, beneficial effects of the Val allele on PFC function have been repeatedly observed in paradigms challenging emotion processing (Drabant et al. 2006; Smolka et al. 2005). Conversely, detrimental effects are seen in Val-allele carriers in cognition, as demonstrated by increased PFC activation (i.e., decreased efficiency) during *n*-back working memory (Diaz-Asper et al. 2008; Egan et al. 2001; Goldberg et al. 2003) (Fig. 7.1a) and episodic memory (Bertolino et al. 2006, 2008). At intermediate task loads, these functional deficits are compensated by the administration of pro-dopaminergic drugs such as amphetamines (Mattay et al.

2003), which is in line with prior evidence for an inverted U-shaped relationship between dopaminergic tone and prefrontal signal-to-noise ratio ratios during working memory performance (Williams and Goldman-Rakic 1995; Sawaguchi and Goldman-Rakic 1991) (Fig. 7.1b). To date, only one study (Green et al. 2013) explicitly tested one of the key assumptions of the intermediate phenotype concept, namely, that prefrontal cortex function is a causal mediator in the relationship between COMT Val158Met genotype and cognitive behavior. Using a multiple mediator model, the authors demonstrate that PFC activity during a cognitive paradigm mediates the effects of COMT genetic variation on intelligence, thereby providing direct support for a gene-brain-cognition mediation. Taken together, these data are consistent with the idea that the Val allele contributes to prefrontal cognitive deficits in schizophrenia and provide evidence for a neural substrate for the pleiotropic behavioral effects of COMT genetic variation on cognitive functions and emotional stability ("warrior/worrier" hypothesis by Stein et al. 2006).

Up to date, an impressive number of single-variant candidate gene approaches have been published in psychiatric imaging genetics, and a full representation of this literature exceeds the scope and possibilities of this chapter. For example, other candidate approaches in the dopaminergic signaling pathway examined risk variants in the dopamine D2 receptor gene (DRD2) and AKT1 (Tan et al. 2008), a gene that plays a crucial role in the D2-dependent noncanonical signaling cascade (AKT1/GSK-3). Risk variants in both genes have been associated with increased risk for schizophrenia (Tan et al. 2008; Cho et al. 2012) and modulation of antipsychotic drug response (Tan et al. 2012; Blasi et al. 2011). Moreover, imaging genetics provided evidence that these variants alter the functional (Tan et al. 2008; Blasi et al. 2011; Bertolino et al. 2010, 2012) and structural (Tan et al. 2008; Markett et al. 2013) properties of frontostriatal circuits in healthy humans. Taken together, these data provide a biological validation of the effects of dopaminergic gene variants in prefronto-neostriatal neural circuits and support the idea for a crucial role of these genetic factors in the pathophysiology and pharmacology of psychosis.

7.3.2 Genome-Wide Supported Risk Variants

Although candidate gene studies offer a principled method to explore promising loci of the genome, these approaches provides only limited support for the pathophysiological relevance of the identified neurogenetic effects if the impact of the genetic variant on the disease phenotype itself is minor (Pezawas and Meyer-Lindenberg 2010). As a consequence, the biological validation of genetic risk variants from genome-wide association studies (GWAS) has gained much popularity because here, the link between genetic variation and categorical diagnosis has been established beforehand at stringent statistical thresholds. On the downside, prior knowledge on the functional role of the identified GWAS variants is typically sparse, an empirical void that can be filled in part by imaging genetics. To date, several studies have identified effects (or the lack thereof (Tost et al. 2010; Wei et al. 2013; Cousijn et al. 2012) of GWAS risk variants for schizophrenia and/or bipolar disorders on structural and functional neuroimaging phenotypes, in particularly for risk variants in the genes ZNF804A (Rasetti et al. 2011; Esslinger et al. 2009, 2011; Thurin et al. 2013; Linden et al. 2013), CACN1AC (Erk et al. 2010; Bigos et al. 2010; Wessa et al. 2010), RELN (Tost et al. 2010), and ANK3 (Linke et al. 2012).

The first such study concerned PFC functional coupling and its modulation by rs1344706A, a genetic variant in ZNF804A conferring risk for schizophrenia and possibly also bipolar disorder (O'Donovan et al. 2008, 2009). The gene maps to chromosome 2q32.1 and encodes for the zinc finger protein 804A, which is abundantly expressed in the developing neocortex and hippocampus (Johnson et al. 2009). Here, connectivity phenotypes derived from functional magnetic resonance imaging (fMRI) are of high interest, as modern neurodevelopmental theories (O'Donovan et al. 2008; Johnson et al. 2009) of schizophrenia

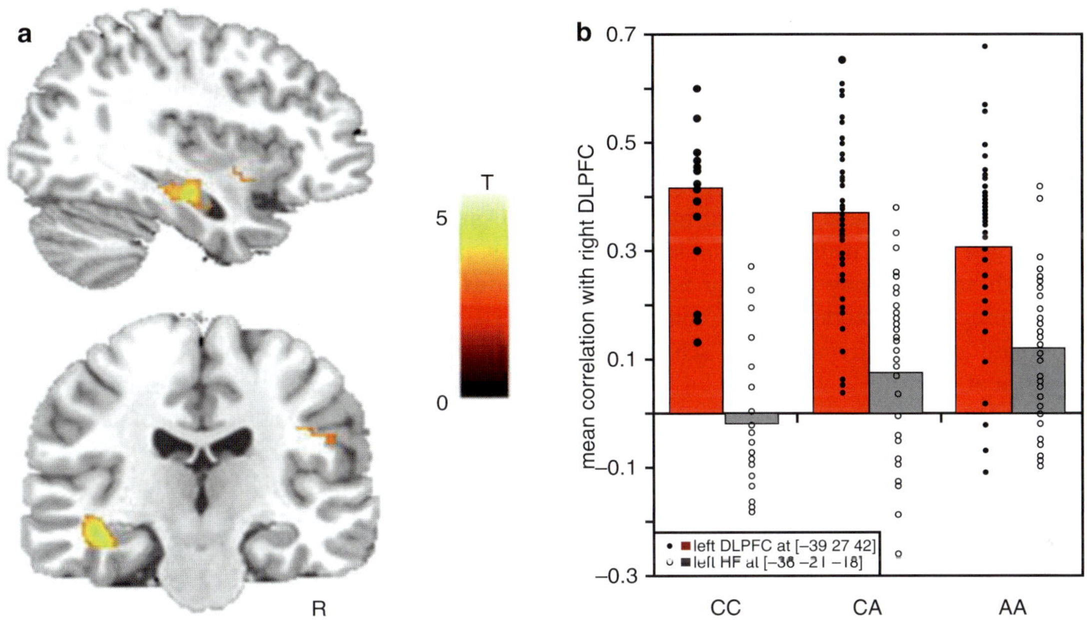

Fig. 7.2 A genome-wide supported variant in ZNF804A modulates PFC functional coupling. (**a**) Brain regions in temporal lobe where genotype predicts increased correlation with right PFC. (**b**) Correlation coefficients reflecting connectivity with right DLPFC by genotype: connectivity with left DLPFC shown as *red columns* and *solid diamonds*, connectivity with left HF as *gray columns* and *open diamonds*. *Abbreviations*: *PFC* prefrontal cortex, *HF* hippocampus formation (Reprinted with permission from Esslinger et al. (2009))

conceptualize the disorder as a neural "disconnection syndrome" (Friston and Frith 1995; Stephan et al. 2006; Meyer-Lindenberg 2009), in which genetic and environmental risk factors disturb the formation (and subsequently also the functional interaction and experience-dependent plasticity) of neural regulatory circuits connecting the PFC and hippocampus. Among the available imaging phenotypes, PFC-hippocampus functional coupling during fMRI working memory is the best established intermediate connectivity risk phenotype to date, owing to its established quality criteria (Bilek et al. 2013) and the robust alterations observed in schizophrenia patients (Rasetti et al. 2011; Meyer-Lindenberg et al. 2005), their unaffected first-grade relatives (Rasetti et al. 2011), and genetic animal models of the disorder (Sigurdsson et al. 2010). Using this phenotype, subjects at genetic risk displayed gene dose-dependent alterations in the functional coupling of the PFC and hippocampus (Esslinger et al. 2009) (Fig. 7.2a, b) that mirror the coupling deficits seen in schizophrenia patients (Meyer-Lindenberg et al. 2005). This finding has since been replicated in a large independent sample (Rasetti et al. 2011) and extended (Esslinger et al. 2011) by demonstrating that the ZNF804A coupling effect is cognitive state dependent (i.e., present during working memory function but not during emotion processing or at rest).

Interestingly, we (Esslinger et al. 2009) and others (Rasetti et al. 2011) have recently shown that healthy carriers of the rs1344706 A-allele do not exhibit differences in PFC regional activation. Moreover, altered PFC-hippocampus connectivity (likely reflecting disturbed plasticity) has been shown to be statistically independent from the well-known fMRI activation or "PFC inefficiency" phenotype (likely reflecting disturbed dopamine release) (Rasetti et al. 2011). Consistent with this, an own recent study (Bilek et al. 2013) has demonstrated that PFC-hippocampus functional connectivity during working memory performance is modifiable by repetitive transcranial magnetic stimulation (rTMS), a plasticity-enhancing intervention that did not impact other neuroimaging phenotypes (i.e., PFC efficiency during working memory, PFC-hippocampus resting-state connectivity). Taken together, these data provide first direct evidence for the "nonredundancy" of different intermediate phenotypes in imaging genetics and point to dissociable biological mechanisms of risk genes at the neural systems level, some of which are promising targets for novel therapies aiming at the modulation of compromised prefrontal network dynamics in schizophrenia (Bilek et al. 2013).

7.3.3 Rare High-Risk Variants

One important lesson from GWAS is the increased occurrence of structural genetic variations (i.e., microdeletions or microduplications) in schizophrenia. Of these, only a copy number variant at 22q11, causing velocardiofacial or 22q11 syndrome, was previously known. These newly discovered microdeletions are associated with a range of brain phenotypes such as schizophrenia but also mental disability, epilepsy, attention-deficit/hyperactivity disorder, and autism (Ben-Shachar et al. 2009). Despite their relative rarity and the complexity of the associated phenotypes, identifying and characterizing structural variations holds considerable potential because they are well-defined genetic alterations with a very high associated risk, which often exceeds that from common genetic variants by more than an order of magnitude (Karayiorgou et al. 2010). While none of the newly identified CNVs have been characterized on the neural systems level, previous work on the 22q11 microdeletion (Karayiorgou et al. 2010) shows that a multimodal imaging approach is feasible and holds the promise to identify neural abnormalities associated with high genetic risk. Further, since it appears likely that copy number variant risk is unlikely to be explained by deletion or duplication of single genes, but rather interactions of genes jointly affected in their expression (Meechan et al. 2007), imaging genetics can ask whether variants in such genes converge on neural systems implicated in schizophrenia. An example of this is the 22q11 microdeletion, which includes, besides COMT and several other candidate genes for schizophrenia, the gene PRODH encoding proline oxidase. A recent study showed that schizophrenia-associated functional polymorphisms in PRODH impact the structure, function, and connectivity of striatum and prefrontal cortex (Kempf et al. 2008). As more and more of these copy number variants are discovered and larger imaging genetics samples permit the study of reasonably sized groups of carriers even for these rare variants, the application of neuroimaging should be very informative in defining neural mechanisms for these high-risk states.

7.4 Accounting for Genetic Diversity

7.4.1 Epistasis

The genetic risk architecture of psychiatric illness is shaped by a large number of risk alleles that are distributed over 3.2 billion base pairs and about 25,000 genes and interact with each other in a largely unknown but surely complex fashion. One aspect of this complexity is captured by the term epistasis, which means that the biological expression of a gene is altered in the presence of one or several other modifying genes. Epistasis is of high relevance to pathophysiology, as seemingly inconsistent association findings for single variants may in fact result from the interaction with other risk factors, including other genes. To date, even larger research consortiums have focused on the examination of pre-hypothesized gene-gene interactions in candidate genes (e.g., "Epistasis Project" on Alzheimer's disease (Lehmann et al. 2012; Combarros et al. 2009). This relates to the enormous number of potential combinations of genetic risk variants and the resulting high statistical threshold of forward-genetics approaches that exceed the statistical power of even the most comprehensive GWAS to date by far.

Analogous to molecular genetics, research strategies for epistatic interactions of candidate genes have been implemented in imaging genetics. Here, the resulting increase in the number of genotype cells and the multitude of potential confounding factors call for exceptionally careful study designs and the availability of very large neuroimaging databases, which may explain the relative paucity of published imaging epistasis work to date (Tost and Meyer-Lindenberg 2011). A particularly good example for the success of this strategy is a prior study (Nixon et al. 2011) on the effects of the interaction of COMT Val158Met polymorphism and a risk variant (M24) in the d-amino-acid oxidase activator (DAOA) gene. Similar to COMT, DAOA modulates prefrontal functional efficacy during working memory performance (Goldberg et al. 2006) and has been implicated in schizophrenia risk (Chumakov et al. 2002) through epistatic interaction with COMT (Nicodemus et al. 2007), possibly by its detrimental effects on *N*-methyl-d-aspartate receptor function (Chumakov et al. 2002) and downstream alterations in the glutamate-dopamine balance of frontal-striatal networks (Tost and Meyer-Lindenberg 2011).

The carefully designed study (Nixon et al. 2011) demonstrated significant effects of DAOA by COMT epistasis on prefrontal efficacy during working memory, as indicated by a disproportionately higher PFC activation (or decreased efficiency) in carriers of both deleterious alleles (i.e., DAOA T/T and COMT Val/Val) at a fixed level of performance. In the context of COMT Met, however, efficient PFC signaling was seen in DAOA T/T carriers. The study is compelling, as it provides a functional validation of a previously reported statistical epistasis of DAOA and COMT on a well-established intermediate phenotype linked to the genetic risk for schizophrenia. Other epistasis studies on dopamine- and glutamate-regulating genes in PFC demonstrated complex functional effects COMT Val158Met in interaction with risk variants in AKT1 (Tan et al. 2008), the dopamine transporter gene (DAT, 3′ VNTR) (Bertolino et al. 2008), and the type II metabotropic glutamate receptor 3 gene (GRM3) (Tan et al. 2007). Taken together, these findings highlight the nonadditive nature of genetic architecture of the brain and draw attention to the fact that imaging genetics, meaningful signals may be easily overlooked if the research scope is restricted to the examination of the main effects of variants in single genes or their linear combination.

7.4.2 Haplotypes

Haplotype analyses go one step further in characterizing the genetic complexity that can be addressed with imaging genetics (Consortium 2005). The term refers to a set of statistically associated SNPs at adjacent loci of the chromosome that reflects the linkage disequilibrium structure of genetic variants in a given population (Hinds et al. 2005). Compared to single alleles, haplotypes may show stronger associations with genetically complex disorders as they contain more information about the underlying causal variants. The study of haplotypes is particularly important when several functional variants are implicated within a single gene or haplotype block, as here, the net effect on brain biology will be critically determined by the specific constellation of coexisting alleles.

The COMT gene is again a very good example. Here, prior evidence for a disease association of the Val158Met coding variant (rs4680) alone has been conflicting (Fan et al. 2005). Also, multiple functional variants have been described in COMT that impact prefrontal dopamine signaling and show association with schizophrenia risk (Bray et al. 2003; Shifman et al. 2002) (e.g., rs2097603 in the P2 promoter (Chen et al. 2004), rs165599 in the 3′ region (Bray et al. 2003)). Two prior imaging genetics studies built upon this prior knowledge and tested whether complex genetic variation in COMT modulates brain function (Meyer-Lindenberg et al. 2006) and structure (Honea et al. 2009) in a pattern that is consistent with interacting functional variants and developed a general methodological solution to the problem that haplotypes are usually only known probabilistically. Among others, these studies demonstrated that a 2-SNP haplotype composed of rs4680 (G/A) and rs2097603 (A/G) showed associations with hippocampus volume (Honea et al. 2009) and prefrontal functional efficiency

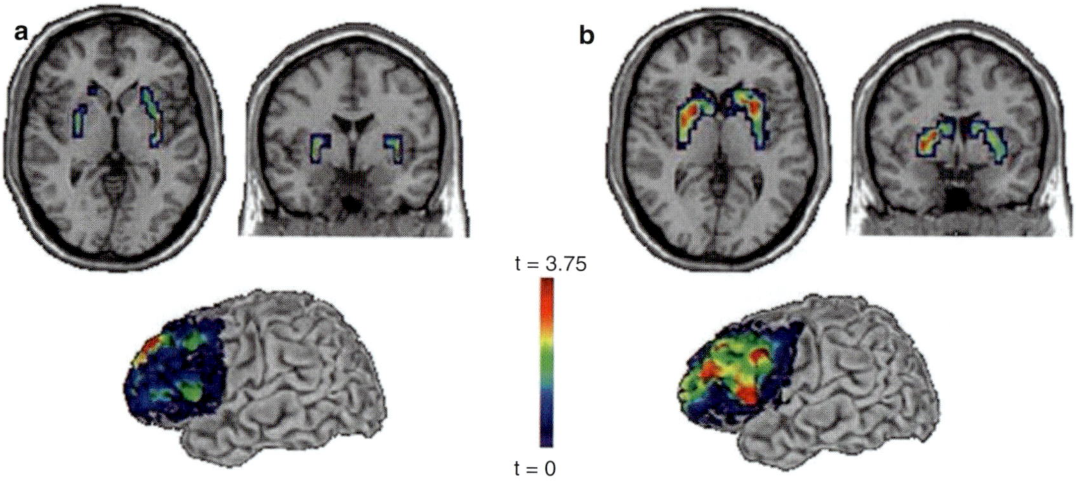

Fig. 7.3 Effects of a PPP1R1B haplotype on frontostriatal activation and connectivity. (**a**) Working memory: significantly reduced activity in putamen (*top*) and greater functional connectivity between prefrontal cortex and striatum (*bottom*) for homozygotes for the frequent (CGCACTC) haplotype. (**b**) Emotional face matching: significantly reduced reactivity in striatum (*top*) and greater functional connectivity between prefrontal cortex and striatum (*bottom*) for homozygotes for the frequent (CGCACTC) haplotype (Reprinted with permission from Meyer-Lindenberg et al. (2007))

during working memory (Meyer-Lindenberg et al. 2006) in a pattern that is consistent with a nonlinear (inverted U-shaped) effect of extracellular dopamine on these schizophrenia-related phenotypes.

Another multimodal imaging genetics haplotype study focused on PPP1R1B (Meyer-Lindenberg et al. 2007). The gene maps to chromosome 17q12 and encodes for the dopamine-regulated and cyclic adenosine 3′,5′-monophosphate-regulated phosphoprotein of molecular weight 32,000 (DARPP-32), which is expressed in the striatum and integrates dopamine neurotransmission with the signals from other transmitter pathways (e.g., glutamate, serotonin, neuropeptides) (Svenningsson et al. 2003, 2004). The multimodal neuroimaging study (Meyer-Lindenberg et al. 2007) identified a frequent 7-SNP PPP1R1B haplotype (CGCACTC) that was associated with schizophrenia risk in a family-based association analysis and the expression of PPP1R1B isoforms in postmortem human brain (Meyer-Lindenberg et al. 2007). Moreover, the risk haplotype modulated cognitive performance, striatal gray matter volume, striatal activation during working memory and emotion processing, as well as striatal functional connectivity to the PFC in a large sample of healthy volunteers (Fig. 7.3) (Meyer-Lindenberg et al. 2007). These data demonstrate the value of haplotype analysis in imaging genetics to identify complex and nonadditive genetic interactions in candidate risk genes in frontostriatal regulatory systems and highlight their potential role in the pathogenesis of schizophrenia.

7.4.3 Genome-Wide Strategies on Imaging Phenotypes

An original hope of the endophenotype concept was to use these quantitative traits to identify new genes that are relevant for mental illness (Gottesman and Gould 2003). Lately, the broad availability of high-throughput genotyping in combination with ever-larger imaging datasets has made this possible in imaging genetics. For example, the imaging GWAS approach has been advanced by large multinational consortia such as the Enhancing Neuro Imaging Genetics Through Meta-Analysis (ENIGMA) (The ENIGMA Network 2010) or Alzheimer's

Disease Neuroimaging Initiative (ADNI) (Mueller et al. 2005) work groups, which seek to share databases and analysis strategies to attain sufficiently powered cohorts and aid the replication of neurogenetic association findings. Common outcome measures are well-established structural intermediate phenotypes (e.g., hippocampus volume (Peper et al. 2007; Kremen et al. 2010)) derived from the spatial reduction of high-resolution 3D MRI datasets, for example, by calculating summary measures over automated anatomical parcellations. A high-profiled imaging GWAS study (Stein et al. 2012) included over 21,000 datasets and identified a very strong ($P=6.7\times10^{-16}$) and genome-wide significant ($P<1.25\times10^{-8}$) association of an intergenic SNP (rs7294919) at 12q24.22 with bilateral hippocampus volume. Using three independent samples of human brain tissue, the study also provided evidence for a functional role of the variant by demonstrating converging effects of rs7294919 (or SNPs in allelic identity) on the expression of TESC (Stein et al. 2012), a positional candidate gene involved in cell proliferation and differentiation during neurodevelopment (Bao et al. 2009).

A conceptual extension of imaging GWAS, voxel-wise GWAS (vGWAS), identifies associations between genetic variants and neuroimaging phenotypes in a whole-brain whole-genome approach, i.e., without a priori limitation of the genetic survey to spatially circumscribed neural regions of interest. An advantage of vGWAS over is that it offers greater anatomical detail of the detected genomic associations, thereby potentially increasing the statistical power of the approach. The first published vGWAS study (Stein et al. 2010) in the field quantified individual differences in brain structure with tensor-based morphometry and explored 448,293 SNPs for associations with 31,622 brain voxels in a sample of 740 elderly subjects with high-resolution structural MRI data from the ADNI consortium. Here, only the P value of the most highly associated ("winning") variant was retained for each voxel, and multiple comparison correction across SNPs and voxels was achieved by modeling the distribution of minimum P values under the null hypothesis and subsequent false discovery rate (FDR) correction of the corrected P-maps. Although no variant was detected that survived the strict significance criterion, several SNPs were identified that merit further investigation such as a variant in CADPS2, a gene involved in neuronal monoamine uptake that was associated with brain structure in the lateral temporal lobe (rs2429582, 7q31.32, uncorrected $P_{min}=4.23\times10^{-10}$) (Stein et al. 2010).

Several other genome-wide strategies are currently in translation from molecular genetics to imaging genetics. Among others, this includes the association of polygenic risk scores with established neuroimaging intermediate phenotypes, a method initially applied by the International Schizophrenia Consortium (ISC) (Purcell et al. 2009) to test the polygenic theory of psychiatric disorders (Gottesman and Shields 1967), which postulates that numerous common genetic risk variants with small effect sizes have an important biological role in combination (Wray et al. 2007). Here, GWAS "training datasets" are used to identify a very large number of putative risk alleles distributed across the genome that are associated with an established schizophrenia intermediate phenotype at very relaxed thresholds (e.g., $P<0.5$). In subsequent independent "test datasets," individual polygenic risk scores are assigned, based upon the cumulative number of putative risk alleles and their weighted effect sizes, and subsequently tested for associations with the imaging phenotype. Another method relevant to imaging genetics is gene-set enrichment analysis (GSEA) (Potkin et al. 2010). The approach takes advantage of the fact that common genetic risk variants for psychiatric disorders lie among sets of genes with overlapping functions. By combining whole-genome genotype information, neuroimaging, and a priori knowledge of empirical databases on the genetic contributions to molecular functions or biological processes, GSEA tests whether variants, which are more strongly associated with a given imaging intermediate phenotype of interest, tend to significantly aggregate in "gene sets" and biological pathways that are of relevance to the illness (Mootha et al. 2003; Subramanian et al. 2005).

7.5 The Role of the Environment

Consistent with earlier (Gottesman and Shields 1972; McEwen and Stellar 1993) and contemporary (McEwen 2012; Booij et al. 2013) accounts on the diathesis-stress model, adverse environmental factors clearly contribute to psychiatric disorders such as schizophrenia, notwithstanding the fact that they are highly heritable (van Os et al. 2010). Many of these factors (e.g., childhood adversity) tend to increase the risk for mental illness and other unfavorable developmental outcomes in a rather unspecific fashion. However, two stressors linked to the social environment have been primarily associated with schizophrenia, urban upbringing and migration. Prior epidemiological research has repeatedly demonstrated that there is at least a twofold increase in schizophrenia risk in individuals that have been raised in urban environments (van Os et al. 2005; Krabbendam and van Os 2005) or belong to ethnic minority populations (van Os et al. 2010; Cantor-Graae and Selten 2005; Bourque et al. 2011). The adverse effects of urbanicity and migration have been established in various countries and ethnicities worldwide, are particularly detrimental when the exposure occurs in childhood and early adolescence (Krabbendam and van Os 2005; Veling et al. 2011), and interact with genetic risk factors for the illness (Krabbendam and van Os 2005; van Os et al. 2008). These observations, especially the likely early impact of these risk factors, are consistent with the proposed neurodevelopmental origin of the disorder (Weinberger 1987) and suggest that the pathway to psychosis is, at least in parts, socioneurodevelopmental in nature (Morgan et al. 2010). A commonly proposed intermediate of the association of social risk factors and psychosis is chronic social stress induced by, for example, repeated exposure to social fragmentation (urbanicity) or racial discrimination (migration) in early life. This may impact the functional organization of neural regulatory circuits during development (Niwa et al. 2013; Burghy et al. 2012; McEwen 2012) and dispose the brain to abnormal stress responses and psychotic experiences in adulthood (see Fig. 7.4) (van Os et al. 2010; Cantor-Graae and Selten 2005; Bourque et al. 2011; Selten and Cantor-Graae 2005).

Only recently, neuroimaging studies have begun to identify neural mechanisms that mediate the effects of social environmental risk factors such as urban upbringing (Lederbogen et al. 2011) or social status (Zink et al. 2008; Gianaros et al. 2007) on the integrity of neural regulatory circuits (Meyer-Lindenberg and Tost 2012; Tost and Meyer-Lindenberg 2012). These initial data point to a vulnerability of the perigenual anterior cingulate cortex (pACC), a key region for regulation of limbic activity and negative emotion, to adverse experiences in the social environment. Interestingly, neural abnormalities in ACC have also been established in schizophrenia patients (Radua et al. 2012; Sambataro et al. 2013), individuals at high genetic risk for the disorder (Sambataro et al. 2013), and healthy carriers of genome-wide supported psychosis risk variants in CACNA1C (Erk et al. 2010) or ZNF804A (Thurin et al. 2013). The same area was also prominently implicated in a recent meta-analysis of functional and structural alterations in early schizophrenia (Radua et al. 2012). These observations support prior epidemiological evidence for interacting genetic and environmental susceptibility, suggest a specific neural system where the effects of genetic and social environmental risk factors may converge, and underscore the potential of neuroimaging to interrogate the neural basis of both genetic and environmental risk constellations. The same set of data also motivates an interest in the adverse interaction of both sources of risk. For example, it is well-known that the individual susceptibility to social adversity is moderated by functional polymorphisms in the promoter regions of the serotonin transporter (SLC6A4) (Caspi et al. 2003) and monoamine oxidase A (MAOA) (Kim-Cohen et al. 2006) genes. Thus, future successful attempts of unraveling the sociogenetic risk architecture of mental illness with neuroimaging will require a substantial extension of established imaging genetics approaches to include nongenetic factors and state-of-the-art methods of epigenetics, environics, epidemiology, and social psychology

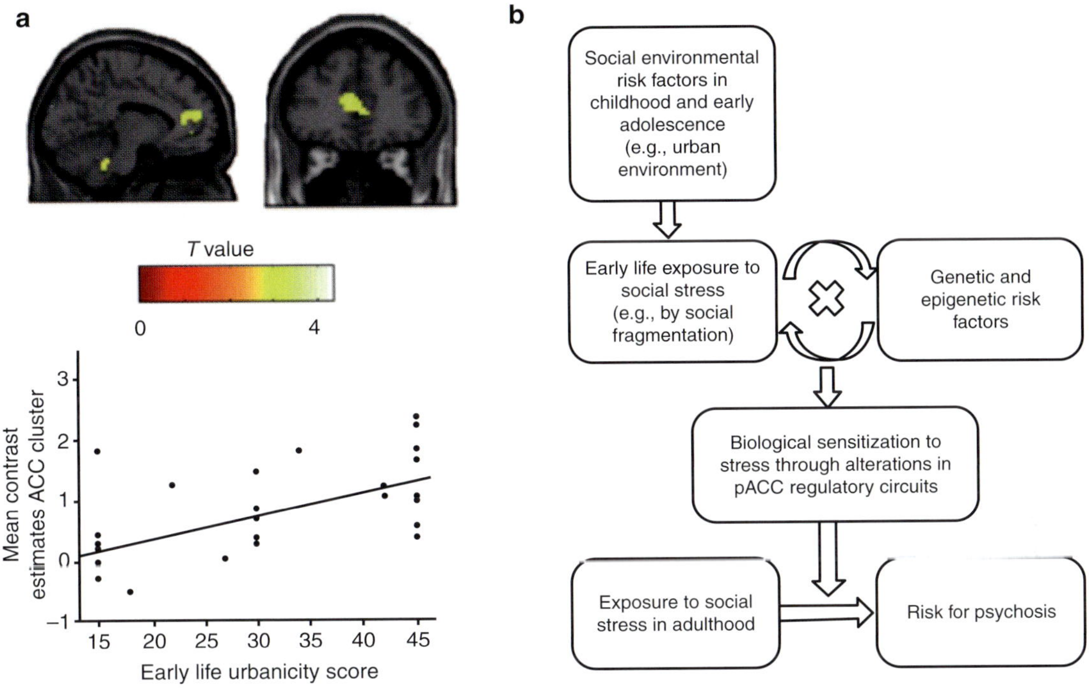

Fig. 7.4 Neural mechanisms of social environmental risk factors. (**a**) Relationship between urban upbringing and pACC function in humans. Individuals raised in urban environments show increased brain response to social evaluative stress in pACC. (**b**) Specification of the diathesis-stress model to the risk factor urban upbringing: the model proposes that early exposure to social stress and the adverse interaction with genetic and epigenetic risk factors result in a sensitization of disease-related neural regulatory circuits. This may lead to abnormal neural stress responses in adulthood and promote the manifestation of psychotic phenomena. *Abbreviations*: *pACC* perigenual cingulate cortex (Reprinted with permission from Lederbogen et al. (2011) and Haddad and Meyer-Lindenberg (2012))

(Meyer-Lindenberg and Tost 2012; Tost and Meyer-Lindenberg 2012).

7.6 Outlook

Since its first application (Heinz et al. 2000) more than a decade ago, the combination of modern neuroimaging methods and genetic mapping techniques has evolved to a successful neuroscience field that utilizes a variety of newly developed research methods to uncover the biological pathways that lead from risk genes over neural circuit abnormalities to psychopathology (Pezawas and Meyer-Lindenberg 2010; Meyer-Lindenberg 2010). To date, imaging genetics has identified several intermediate phenotypes that are reliably altered in healthy individuals at genetic risk and patient populations, thereby identifying relevant pathophysiological processes and providing a biological confirmation of prior molecular genetic associations. As exemplified by the altered activation and functional connectivity of the DLPFC in the context of schizophrenia, this work has already prompted subsequent successful bench work aiming to validate the effects of established and novel treatment strategies for mental illness (Bilek et al. 2013; Blasi et al. 2011; Rasetti et al. 2010; Apud et al. 2007). While imaging genetics has thus proven to be very fruitful, the true potential of the field has not been fully exploited yet, and several remaining challenges need to addressed in order to prompt true clinical innovations in the years to come (Pezawas and Meyer-Lindenberg 2010; Meyer-Lindenberg 2010a, b).

Current limitations include, among others, the predominance of single-variant approaches that fail to account for the true genetic complexity of psychiatric disorders, the high proportion of single-site studies not powered to detect small effect sizes, the relative lack of coordinated attempts for independent replication of association findings, the small amount of translational approaches designed to exploit the benefits of both human neuroimaging and animal research, and the remaining gaps in knowledge on many of the proposed neuroimaging "intermediate phenotypes" in use (e.g., unspecified task quality criteria, inconsistent directionality of effects, unknown link to the genetic risk for disorders, biological redundancy vs. independence of different phenotypes, etc.) (Rasetti and Weinberger 2011). Moreover, the renewed interest of biological psychiatry in the environment underscores the need for a conceptual expansion of current imaging genetics approaches to include nongenetic risk factors. To account for the true complexity of biology, this conceptual redesign will require large multisite studies with sizeable samples, extensive coverage of the "risk matrix" over multiple observational levels, and the implementation of sophisticated multivariate methods for cross-level biomarker identification. As sample sizes expand, the important study of rare high-risk (e.g., copy number) variants will also become feasible. Translational neuroimaging approaches combined with genetically engineered rodent models will allow a more direct translation of systems-level imaging genetic findings into animal models that may be useful for drug discovery (Meyer-Lindenberg 2010a, b). Finally, embracing a lifetime perspective will be crucial, as the causal interplay of molecular, individual, and societal risk factors is expected to predispose the brain for mental illness in some vulnerable periods of life but not others. While challenging, neuroimaging will certainly play an important role in all these efforts, and there is good reason to be optimistic that a conceptually extended approach will continue to pioneer new discoveries in pathophysiology and advance the development of new and effective measures for the diagnosis, treatment, and prevention of mental illness (Meyer-Lindenberg 2010a, b; Insel 2010).

References

Apud JA, Mattay V, Chen J, Kolachana BS, Callicott JH, Rasetti R, Alce G, Iudicello JE, Akbar N, Egan MF, Goldberg TE, Weinberger DR (2007) Tolcapone improves cognition and cortical information processing in normal human subjects. Neuropsychopharmacology 32(5):1011–1020

Bao Y, Hudson QJ, Perera EM, Akan L, Tobet SA, Smith CA, Sinclair AH, Berkovitz GD (2009) Expression and evolutionary conservation of the tescalcin gene during development. Gene Expr Patterns 9(5):273–281

Ben-Shachar S, Lanpher B, German JR, Qasaymeh M, Potocki L, Nagamani SC, Franco LM, Malphrus A, Bottenfield GW, Spence JE, Amato S, Rousseau JA et al (2009) Microdeletion 15q13.3: a locus with incomplete penetrance for autism, mental retardation, and psychiatric disorders. J Med Genet 46(6):382–388

Bertolino A, Rubino V, Sambataro F, Blasi G, Latorre V, Fazio L, Caforio G, Petruzzella V, Kolachana B, Hariri A, Meyer-Lindenberg A, Nardini M et al (2006) Prefrontal-hippocampal coupling during memory processing is modulated by COMT val158met genotype. Biol Psychiatry 60(11):1250–1258

Bertolino A, Di Giorgio A, Blasi G, Sambataro F, Caforio G, Sinibaldi L, Latorre V, Rampino A, Taurisano P, Fazio L, Romano R, Douzgou S et al (2008) Epistasis between dopamine regulating genes identifies a nonlinear response of the human hippocampus during memory tasks. Biol Psychiatry 64(3):226–234

Bertolino A, Fazio L, Caforio G, Blasi G, Rampino A, Romano R, Di Giorgio A, Taurisano P, Papp A, Pinsonneault J, Wang D, Nardini M et al (2010) Functional variants of the dopamine receptor D2 gene modulate prefronto-striatal phenotypes in schizophrenia. Brain 132(Pt 2):417–425

Bertolino A, Taurisano P, Pisciotta NM, Blasi G, Fazio L, Romano R, Gelao B, Lo Bianco L, Lozupone M, Di Giorgio A, Caforio G, Sambataro F et al (2012) Genetically determined measures of striatal D2 signaling predict prefrontal activity during working memory performance. PLoS One 5(2):e9348, 2010

Bigos KL, Weinberger DR (2010) Imaging genetics-days of future past. Neuroimage 53(3):804–809

Bigos KL, Mattay VS, Callicott JH, Straub RE, Vakkalanka R, Kolachana B, Hyde TM, Lipska BK, Kleinman JE, Weinberger DR (2010) Genetic variation in CACNA1C affects brain circuitries related to mental illness. Arch Gen Psychiatry 67(9):939–945

Bilder RM, Volavka J, Lachman HM, Grace AA (2004) The catechol-O-methyltransferase polymorphism: relations to the tonic-phasic dopamine hypothesis and neuropsychiatric phenotypes. Neuropsychopharmacology 29(11):1943–1961

Bilek E, Schäfer A, Ochs E, Esslinger C, Zangl M, Plichta MM, Braun U, Kirsch P, Schulze TG, Rietschel M, Meyer-Lindenberg A, Tost H (2013) Application of high-frequency rTMS to the DLPFC alters human prefrontal-hippocampal functional interaction. J Neurosci 33:(16)7050–6

Blasi G, Napolitano F, Ursini G, Taurisano P, Romano R, Caforio G, Fazio L, Gelao B, Di Giorgio A, Iacovelli L, Sinibaldi L, Popolizio T et al (2011) DRD2/AKT1 interaction on D2 c-AMP independent signaling, attentional processing, and response to olanzapine treatment in schizophrenia. Proc Natl Acad Sci U S A 108(3):1158–1163

Booij L, Wang D, Levesque ML, Tremblay RE, Szyf M (2013) Looking beyond the DNA sequence: the relevance of DNA methylation processes for the stress-diathesis model of depression. Philos Trans R Soc Lond B Biol Sci 368(1615):20120251

Bourque F, van der Ven E, Malla A (2011) A meta-analysis of the risk for psychotic disorders among first- and second-generation immigrants. Psychol Med 41(5):897–910

Bray NJ, Buckland PR, Williams NM, Williams HJ, Norton N, Owen MJ, O'Donovan MC (2003) A haplotype implicated in schizophrenia susceptibility is associated with reduced COMT expression in human brain. Am J Hum Genet 73(1):152–161

Burghy CA, Stodola DE, Ruttle PL, Molloy EK, Armstrong JM, Oler JA, Fox ME, Hayes AS, Kalin NH, Essex MJ, Davidson RJ, Birn RM (2012) Developmental pathways to amygdala-prefrontal function and internalizing symptoms in adolescence. Nat Neurosci 15(12):1736–1741

Callicott JH, Egan MF, Mattay VS, Bertolino A, Bone AD, Verchinksi B, Weinberger DR (2003) Abnormal fMRI response of the dorsolateral prefrontal cortex in cognitively intact siblings of patients with schizophrenia. Am J Psychiatry 160(4):709–719

Cannon TD, Keller MC (2006) Endophenotypes in the genetic analyses of mental disorders. Annu Rev Clin Psychol 2:267–290

Cantor-Graae E, Selten JP (2005) Schizophrenia and migration: a meta-analysis and review. Am J Psychiatry 162(1):12–24

Caspi A, Sugden K, Moffitt TE, Taylor A, Craig IW, Harrington H, McClay J, Mill J, Martin J, Braithwaite A, Poulton R (2003) Influence of life stress on depression: moderation by a polymorphism in the 5-HTT gene. Science 301(5631):386–389

Chen J, Lipska BK, Halim N, Ma QD, Matsumoto M, Melhem S, Kolachana BS, Hyde TM, Herman MM, Apud J, Egan MF, Kleinman JE et al (2004) Functional analysis of genetic variation in catechol-O-methyltransferase (COMT): effects on mRNA, protein, and enzyme activity in postmortem human brain. Am J Hum Genet 75(5):807–821

Cho AR, Lee SM, Kang WS, Kim SK, Chung JH (2012) Assessment between dopamine receptor D2 (DRD2) polymorphisms and schizophrenia in Korean population. Clin Psychopharmacol Neurosci 10(2):88–93

Chumakov I, Blumenfeld M, Guerassimenko O, Cavarec L, Palicio M, Abderrahim H, Bougueleret L, Barry C, Tanaka H, La Rosa P, Puech A, Tahri N et al (2002) Genetic and physiological data implicating the new human gene G72 and the gene for D-amino acid oxidase in schizophrenia. Proc Natl Acad Sci U S A 99(21):13675–13680

Combarros O, van Duijn CM, Hammond N, Belbin O, Arias-Vasquez A, Cortina-Borja M, Lehmann MG, Aulchenko YS, Schuur M, Kolsch H, Heun R, Wilcock GK et al (2009) Replication by the epistasis project of the interaction between the genes for IL-6 and IL-10 in the risk of Alzheimer's disease. J Neuroinflammation 6:22

Consortium IH (2005) A haplotype map of the human genome. Nature 437(7063):1299–1320

Cousijn H, Rijpkema M, Harteveld A, Harrison PJ, Fernandez G, Franke B, Arias-Vasquez A (2012) Schizophrenia risk gene ZNF804A does not influence macroscopic brain structure: an MRI study in 892 volunteers. Mol Psychiatry 17(12):1155–1157

Diaz-Asper CM, Goldberg TE, Kolachana BS, Straub RE, Egan MF, Weinberger DR (2008) Genetic variation in catechol-O-methyltransferase: effects on working memory in schizophrenic patients, their siblings, and healthy controls. Biol Psychiatry 63(1):72–79

Drabant EM, Hariri AR, Meyer-Lindenberg A, Munoz KE, Mattay VS, Kolachana BS, Egan MF, Weinberger DR (2006) Catechol O-methyltransferase val158met genotype and neural mechanisms related to affective arousal and regulation. Arch Gen Psychiatry 63(12):1396–1406

Egan MF, Goldberg TE, Kolachana BS, Callicott JH, Mazzanti CM, Straub RE, Goldman D, Weinberger DR (2001) Effect of COMT Val108/158 Met genotype on frontal lobe function and risk for schizophrenia. Proc Natl Acad Sci U S A 98(12):6917–6922

Erk S, Meyer-Lindenberg A, Schnell K, Opitz von Boberfeld C, Esslinger C, Kirsch P, Grimm O, Arnold C, Haddad L, Witt SH, Cichon S, Nothen MM et al (2010) Brain function in carriers of a genome-wide supported bipolar disorder variant. Arch Gen Psychiatry 67(8):803–811

Esslinger C, Walter H, Kirsch P, Erk S, Schnell K, Arnold C, Haddad L, Mier D, Opitz von Boberfeld C, Raab K, Witt SH, Rietschel M et al (2009) Neural mechanisms of a genome-wide supported psychosis variant. Science 324(5927):605

Esslinger C, Kirsch P, Haddad L, Mier D, Sauer C, Erk S, Schnell K, Arnold C, Witt SH, Rietschel M, Cichon S, Walter H et al (2011) Cognitive state and connectivity effects of the genome-wide significant psychosis variant in ZNF804A. Neuroimage 54(3):2514–2523

Fan JB, Zhang CS, Gu NF, Li XW, Sun WW, Wang HY, Feng GY, St Clair D, He L (2005) Catechol-O-methyltransferase gene Val/Met functional polymorphism and risk of schizophrenia: a large-scale association study plus meta-analysis. Biol Psychiatry 57(2):139–144

Flint J, Munafo MR (2007) The endophenotype concept in psychiatric genetics. Psychol Med 37(2):163–180

Friston KJ, Frith CD (1995) Schizophrenia: a disconnection syndrome? Clin Neurosci 3(2):89–97

Gianaros PJ, Horenstein JA, Cohen S, Matthews KA, Brown SM, Flory JD, Critchley HD, Manuck SB,

Hariri AR (2007) Perigenual anterior cingulate morphology covaries with perceived social standing. Soc Cogn Affect Neurosci 2(3):161–173

Glahn DC, Winkler AM, Kochunov P, Almasy L, Duggirala R, Carless MA, Curran JC, Olvera RL, Laird AR, Smith SM, Beckmann CF, Fox PT et al (2010) Genetic control over the resting brain. Proc Natl Acad Sci U S A 107(3):1223–1228

Goldberg TE, Egan MF, Gscheidle T, Coppola R, Weickert T, Kolachana BS, Goldman D, Weinberger DR (2003) Executive subprocesses in working memory: relationship to catechol-O-methyltransferase Val158Met genotype and schizophrenia. Arch Gen Psychiatry 60(9):889–896

Goldberg TE, Straub RE, Callicott JH, Hariri A, Mattay VS, Bigelow L, Coppola R, Egan MF, Weinberger DR (2006) The G72/G30 gene complex and cognitive abnormalities in schizophrenia. Neuropsychopharmacology 31(9):2022–2032

Gottesman II, Gould TD (2003) The endophenotype concept in psychiatry: etymology and strategic intentions. Am J Psychiatry 160(4):636–645

Gottesman II, Shields J (1967) A polygenic theory of schizophrenia. Proc Natl Acad Sci U S A 58(1):199–205

Gottesman II, Shields J (1972) Schizophrenia and genetics: a twin study vantage point. Academic, New York

Green AE, Kraemer DJ, Deyoung CG, Fossella JA, Gray JR (2013) A gene-brain-cognition pathway: prefrontal activity mediates the effect of COMT on cognitive control and IQ. Cereb Cortex 23(3):552–9

Haddad L, Meyer-Lindenberg A (2012) Social environmental risk factors and mental disorders: insights into underlying neural mechanisms drawing on the example of urbanicity. Nervenarzt 83(11):1403–1409

Heinz A, Goldman D, Jones DW, Palmour R, Hommer D, Gorey JG, Lee KS, Linnoila M, Weinberger DR (2000) Genotype influences in vivo dopamine transporter availability in human striatum. Neuropsychopharmacology 22(2):133–139

Hinds DA, Stuve LL, Nilsen GB, Halperin E, Eskin E, Ballinger DG, Frazer KA, Cox DR (2005) Whole-genome patterns of common DNA variation in three human populations. Science 307(5712):1072–1079

Honea R, Verchinski BA, Pezawas L, Kolachana BS, Callicott JH, Mattay VS, Weinberger DR, Meyer-Lindenberg A (2009) Impact of interacting functional variants in COMT on regional gray matter volume in human brain. Neuroimage 45(1):44–51

Insel TR (2010) Rethinking schizophrenia. Nature 468(7321):187–193

Insel TR, Scolnick EM (2006) Cure therapeutics and strategic prevention: raising the bar for mental health research. Mol Psychiatry 11(1):11–17

Johnson MB, Kawasawa YI, Mason CE, Krsnik Z, Coppola G, Bogdanovic D, Geschwind DH, Mane SM, State MW, Sestan N (2009) Functional and evolutionary insights into human brain development through global transcriptome analysis. Neuron 62(4):494–509

Kapur S (2003) Psychosis as a state of aberrant salience: a framework linking biology, phenomenology, and pharmacology in schizophrenia. Am J Psychiatry 160(1):13–23

Kapur S, Mizrahi R, Li M (2005) From dopamine to salience to psychosis–linking biology, pharmacology and phenomenology of psychosis. Schizophr Res 79(1):59–68

Karayiorgou M, Simon TJ, Gogos JA (2010) 22q11.2 microdeletions: linking DNA structural variation to brain dysfunction and schizophrenia. Nat Rev Neurosci 11(6):402–416

Kempf L, Nicodemus KK, Kolachana B, Vakkalanka R, Verchinski BA, Egan MF, Straub RE, Mattay VA, Callicott JH, Weinberger DR, Meyer-Lindenberg A (2008) Functional polymorphisms in PRODH are associated with risk and protection for schizophrenia and fronto-striatal structure and function. PLoS Genet 4(11):e1000252

Kendler KS, Neale MC (2010) Endophenotype: a conceptual analysis. Mol Psychiatry 15(8):789–797

Kim-Cohen J, Caspi A, Taylor A, Williams B, Newcombe R, Craig IW, Moffitt TE (2006) MAOA, maltreatment, and gene-environment interaction predicting children's mental health: new evidence and a meta-analysis. Mol Psychiatry 11(10):903–913

Krabbendam L, van Os J (2005) Schizophrenia and urbanicity: a major environmental influence–conditional on genetic risk. Schizophr Bull 31(4):795–799

Kremen WS, Prom-Wormley E, Panizzon MS, Eyler LT, Fischl B, Neale MC, Franz CE, Lyons MJ, Pacheco J, Perry ME, Stevens A, Schmitt JE et al (2010) Genetic and environmental influences on the size of specific brain regions in midlife: the VETSA MRI study. Neuroimage 49(2):1213–1223

Lederbogen F, Kirsch P, Haddad L, Streit F, Tost H, Schuch P, Wust S, Pruessner JC, Rietschel M, Deuschle M, Meyer-Lindenberg A (2011) City living and urban upbringing affect neural social stress processing in humans. Nature 474(7352):498–501

Lehmann DJ, Schuur M, Warden DR, Hammond N, Belbin O, Kolsch H, Lehmann MG, Wilcock GK, Brown K, Kehoe PG, Morris CM, Barker R et al (2012) Transferrin and HFE genes interact in Alzheimer's disease risk: the epistasis project. Neurobiol Aging 33(1):202.e201–213

Linden DE, Lancaster TM, Wolf C, Baird A, Jackson MC, Johnston SJ, Donev R, Thome J (2013) ZNF804A genotype modulates neural activity during working memory for faces. Neuropsychobiology 67(2):84–92

Linke J, Witt SH, King AV, Nieratschker V, Poupon C, Gass A, Hennerici MG, Rietschel M, Wessa M (2012) Genome-wide supported risk variant for bipolar disorder alters anatomical connectivity in the human brain. Neuroimage 59(4):3288–3296

Malhotra D, Sebat J (2012) CNVs: harbingers of a rare variant revolution in psychiatric genetics. Cell 148(6):1223–1241

Markett S, Reuter M, Montag C, Weber B (2013) The dopamine D2 receptor gene DRD2 and the nicotinic acetylcholine receptor gene CHRNA4 interact on striatal gray matter volume: evidence from a genetic imaging study. Neuroimage 64:167–172

Mattay VS, Goldberg TE, Fera F, Hariri AR, Tessitore A, Egan MF, Kolachana B, Callicott JH, Weinberger DR (2003) Catechol O-methyltransferase val158-met genotype and individual variation in the brain response to amphetamine. Proc Natl Acad Sci U S A 100(10):6186–6191

McEwen BS (2012) Brain on stress: how the social environment gets under the skin. Proc Natl Acad Sci U S A 109(Suppl 2):17180–17185

McEwen BS, Stellar E (1993) Stress and the individual. Mechanisms leading to disease. Arch Intern Med 153(18):2093–2101

Meechan DW, Maynard TM, Gopalakrishna D, Wu Y, LaMantia AS (2007) When half is not enough: gene expression and dosage in the 22q11 deletion syndrome. Gene Expr 13(6):299–310

Meyer-Lindenberg A (2009) Neural connectivity as an intermediate phenotype: brain networks under genetic control. Hum Brain Mapp 30(7):1938–1946

Meyer-Lindenberg A (2010a) From maps to mechanisms through neuroimaging of schizophrenia. Nature 468(7321):194–202

Meyer-Lindenberg A (2010b) Imaging genetics of schizophrenia. Dialogues Clin Neurosci 12(4):449–456

Meyer-Lindenberg A, Tost H (2012) Neural mechanisms of social risk for psychiatric disorders. Nat Neurosci 15(5):663–668

Meyer-Lindenberg A, Weinberger DR (2006) Intermediate phenotypes and genetic mechanisms of psychiatric disorders. Nat Rev Neurosci 7(10):818–827

Meyer-Lindenberg A, Miletich RS, Kohn PD, Esposito G, Carson RE, Quarantelli M, Weinberger DR, Berman KF (2002) Reduced prefrontal activity predicts exaggerated striatal dopaminergic function in schizophrenia. Nat Neurosci 5(3):267–271

Meyer-Lindenberg AS, Olsen RK, Kohn PD, Brown T, Egan MF, Weinberger DR, Berman KF (2005) Regionally specific disturbance of dorsolateral prefrontal-hippocampal functional connectivity in schizophrenia. Arch Gen Psychiatry 62(4):379–386

Meyer-Lindenberg A, Nichols T, Callicott JH, Ding J, Kolachana B, Buckholtz J, Mattay VS, Egan M, Weinberger DR (2006) Impact of complex genetic variation in COMT on human brain function. Mol Psychiatry 11(9):867–877, 797

Meyer-Lindenberg A, Straub RE, Lipska BK, Verchinski BA, Goldberg T, Callicott JH, Egan MF, Huffaker SS, Mattay VS, Kolachana B, Kleinman JE, Weinberger DR (2007) Genetic evidence implicating DARPP-32 in human frontostriatal structure, function, and cognition. J Clin Invest 117(3):672–682

Mier D, Kirsch P, Meyer-Lindenberg A (2010) Neural substrates of pleiotropic action of genetic variation in COMT: a meta-analysis. Mol Psychiatry 15(9):918–927

Mootha VK, Lindgren CM, Eriksson KF, Subramanian A, Sihag S, Lehar J, Puigserver P, Carlsson E, Ridderstrale M, Laurila E, Houstis N, Daly MJ et al (2003) PGC-1alpha-responsive genes involved in oxidative phosphorylation are coordinately downregulated in human diabetes. Nat Genet 34(3):267–273

Morgan C, Charalambides M, Hutchinson G, Murray RM (2010) Migration, ethnicity, and psychosis: toward a sociodevelopmental model. Schizophr Bull 36(4):655–664

Mueller SG, Weiner MW, Thal LJ, Petersen RC, Jack CR, Jagust W, Trojanowski JQ, Toga AW, Beckett L (2005) Ways toward an early diagnosis in Alzheimer's disease: the Alzheimer's Disease Neuroimaging Initiative (ADNI). Alzheimers Dement 1(1):55–66

Munafo MR, Brown SM, Hariri AR (2008) Serotonin transporter (5-HTTLPR) genotype and amygdala activation: a meta-analysis. Biol Psychiatry 63(9):852–857

Murphy SE, Norbury R, Godlewska BR, Cowen PJ, Mannie ZM, Harmer CJ, Munafo MR (2013) The effect of the serotonin transporter polymorphism (5-HTTLPR) on amygdala function: a meta-analysis. Mol Psychiatry 18(4):512–520

Nicodemus KK, Kolachana BS, Vakkalanka R, Straub RE, Giegling I, Egan MF, Rujescu D, Weinberger DR (2007) Evidence for statistical epistasis between catechol-O-methyltransferase (COMT) and polymorphisms in RGS4, G72 (DAOA), GRM3, and DISC1: influence on risk of schizophrenia. Hum Genet 120(6):889–906

Niwa M, Jaaro-Peled H, Tankou S, Seshadri S, Hikida T, Matsumoto Y, Cascella NG, Kano S, Ozaki N, Nabeshima T, Sawa A (2013) Adolescent stress-induced epigenetic control of dopaminergic neurons via glucocorticoids. Science 339(6117):335–339

Nixon DC, Prust MJ, Sambataro F, Tan HY, Mattay VS, Weinberger DR, Callicott JH (2011) Interactive effects of DAOA (G72) and catechol-O-methyltransferase on neurophysiology in prefrontal cortex. Biol Psychiatry 69(10):1006–1008

O'Donovan MC, Craddock N, Norton N, Williams H, Peirce T, Moskvina V, Nikolov I, Hamshere M, Carroll L, Georgieva L, Dwyer S, Holmans P et al (2008) Identification of loci associated with schizophrenia by genome-wide association and follow-up. Nat Genet 40(9):1053–1055

O'Donovan MC, Craddock NJ, Owen MJ (2009) Genetics of psychosis; insights from views across the genome. Hum Genet 126(1):3–12

Owen MJ, Williams NM, O'Donovan MC (2004) The molecular genetics of schizophrenia: new findings promise new insights. Mol Psychiatry 9(1):14–27

Peper JS, Brouwer RM, Boomsma DI, Kahn RS, Hulshoff Pol HE (2007) Genetic influences on human brain structure: a review of brain imaging studies in twins. Hum Brain Mapp 28(6):464–473

Pezawas L, Meyer-Lindenberg A (2010) Imaging genetics: progressing by leaps and bounds. Neuroimage 53(3):801–803

Plichta MM, Schwarz AJ, Grimm O, Morgen K, Mier D, Haddad L, Gerdes AB, Sauer C, Tost H, Esslinger C, Colman P, Wilson F et al (2012) Test-retest reliability of evoked BOLD signals from a cognitive-emotive fMRI test battery. Neuroimage 60(3):1746–1758

Potkin SG, Macciardi F, Guffanti G, Fallon JH, Wang Q, Turner JA, Lakatos A, Miles MF, Lander A, Vawter

MP, Xie X (2010) Identifying gene regulatory networks in schizophrenia. Neuroimage 53(3):839–847
Preston GA, Weinberger DR (2005) Intermediate phenotypes in schizophrenia: a selective review. Dialogues Clin Neurosci 7(2):165–179
Purcell SM, Wray NR, Stone JL, Visscher PM, O'Donovan MC, Sullivan PF, Sklar P (2009) Common polygenic variation contributes to risk of schizophrenia and bipolar disorder. Nature 460(7256):748–752
Radua J, Borgwardt S, Crescini A, Mataix-Cols D, Meyer-Lindenberg A, McGuire PK, Fusar-Poli P (2012) Multimodal meta-analysis of structural and functional brain changes in first episode psychosis and the effects of antipsychotic medication. Neurosci Biobehav Rev 36(10):2325–2333
Rasetti R, Weinberger DR (2011) Intermediate phenotypes in psychiatric disorders. Curr Opin Genet Dev 21(3):340–348
Rasetti R, Mattay VS, Stankevich B, Skjei K, Blasi G, Sambataro F, Arrillaga-Romany IC, Goldberg TE, Callicott JH, Apud JA, Weinberger DR (2010) Modulatory effects of modafinil on neural circuits regulating emotion and cognition. Neuropsychopharmacology 35(10):2101–2109
Rasetti R, Sambataro F, Chen Q, Callicott JH, Mattay VS, Weinberger DR (2011) Altered cortical network dynamics: a potential intermediate phenotype for schizophrenia and association with ZNF804A. Arch Gen Psychiatry 68(12):1207–1217
Rössler W, Salize HJ, van Os J, Riecher-Rössler A (2005) Size of burden of schizophrenia and psychotic disorders. Eur Neuropsychopharmacol 15(4):399–409
Sambataro F, Mattay VS, Thurin K, Safrin M, Rasetti R, Blasi G, Callicott JH, Weinberger DR (2013) Altered cerebral response during cognitive control: a potential indicator of genetic liability for schizophrenia. Neuropsychopharmacology 38(5):846–853
Sawaguchi T, Goldman-Rakic PS (1991) D1 dopamine receptors in prefrontal cortex: involvement in working memory. Science 251(4996):947–950
Selten JP, Cantor-Graae E (2005) Social defeat: risk factor for schizophrenia? Br J Psychiatry 187:101–102
Shifman S, Bronstein M, Sternfeld M, Pisante-Shalom A, Lev-Lehman E, Weizman A, Reznik I, Spivak B, Grisaru N, Karp L, Schiffer R, Kotler M et al (2002) A highly significant association between a COMT haplotype and schizophrenia. Am J Hum Genet 71(6):1296–1302
Sigurdsson T, Stark KL, Karayiorgou M, Gogos JA, Gordon JA (2010) Impaired hippocampal-prefrontal synchrony in a genetic mouse model of schizophrenia. Nature 464(7289):763–767
Smolka MN, Schumann G, Wrase J, Grusser SM, Flor H, Mann K, Braus DF, Goldman D, Buchel C, Heinz A (2005) Catechol-O-methyltransferase val158met genotype affects processing of emotional stimuli in the amygdala and prefrontal cortex. J Neurosci 25(4):836–842
Stefansson H, Rujescu D, Cichon S, Pietilainen OP, Ingason A, Steinberg S, Fossdal R, Sigurdsson E, Sigmundsson T, Buizer-Voskamp JE, Hansen T, Jakobsen KD et al (2008) Large recurrent microdeletions associated with schizophrenia. Nature 455(7210):232–236
Stein DJ, Newman TK, Savitz J, Ramesar R (2006) Warriors versus worriers: the role of COMT gene variants. CNS Spectr 11(10):745–748
Stein JL, Hua X, Lee S, Ho AJ, Leow AD, Toga AW, Saykin AJ, Shen L, Foroud T, Pankratz N, Huentelman MJ, Craig DW et al (2010) Voxelwise genome-wide association study (vGWAS). Neuroimage 53(3):1160–1174
Stein JL, Medland SE, Vasquez AA, Hibar DP, Senstad RE, Winkler AM, Toro R, Appel K, Bartecek R, Bergmann O, Bernard M, Brown AA et al (2012) Identification of common variants associated with human hippocampal and intracranial volumes. Nat Genet 44(5):552–561
Stephan KE, Baldeweg T, Friston KJ (2006) Synaptic plasticity and dysconnection in schizophrenia. Biol Psychiatry 59(10):929–939
Subramanian A, Tamayo P, Mootha VK, Mukherjee S, Ebert BL, Gillette MA, Paulovich A, Pomeroy SL, Golub TR, Lander ES, Mesirov JP (2005) Gene set enrichment analysis: a knowledge-based approach for interpreting genome-wide expression profiles. Proc Natl Acad Sci U S A 102(43):15545–15550
Sullivan PF, Kendler KS, Neale MC (2003) Schizophrenia as a complex trait: evidence from a meta-analysis of twin studies. Arch Gen Psychiatry 60(12):1187–1192
Sullivan PF, Daly MJ, O'Donovan M (2012) Genetic architectures of psychiatric disorders: the emerging picture and its implications. Nat Rev Genet 13(8):537–551
Svenningsson P, Tzavara ET, Carruthers R, Rachleff I, Wattler S, Nehls M, McKinzie DL, Fienberg AA, Nomikos GG, Greengard P (2003) Diverse psychotomimetics act through a common signaling pathway. Science 302(5649):1412–1415
Svenningsson P, Nishi A, Fisone G, Girault JA, Nairn AC, Greengard P (2004) DARPP-32: an integrator of neurotransmission. Ann Rev Pharmacol Toxicol 44:269–296
Tan HY, Chen Q, Sust S, Buckholtz JW, Meyers JD, Egan MF, Mattay VS, Meyer-Lindenberg A, Weinberger DR, Callicott JH (2007) Epistasis between catechol-O-methyltransferase and type II metabotropic glutamate receptor 3 genes on working memory brain function. Proc Natl Acad Sci U S A 104(30):12536–12541
Tan HY, Nicodemus KK, Chen Q, Li Z, Brooke JK, Honea R, Kolachana BS, Straub RE, Meyer-Lindenberg A, Sei Y, Mattay VS, Callicott JH et al (2008) Genetic variation in AKT1 is linked to dopamine-associated prefrontal cortical structure and function in humans. J Clin Invest 118(6):2200–2208
Tan HY, Chen AG, Kolachana B, Apud JA, Mattay VS, Callicott JH, Chen Q, Weinberger DR (2012) Effective connectivity of AKT1-mediated dopaminergic working memory networks and pharmacogenetics of anti-dopaminergic treatment. Brain 135(Pt 5): 1436–1445

The ENIGMA Network (2010) http://enigma.loni.ucla.edu

Thurin K, Rasetti R, Sambataro F, Safrin M, Chen Q, Callicott JH, Mattay VS, Weinberger DR (2013) Effects of ZNF804A on neurophysiologic measures of cognitive control. Mol Psychiatry 18(8):852–4

Toga AW, Thompson PM (2005) Genetics of brain structure and intelligence. Annu Rev Neurosci 28:1–23

Tost H, Meyer-Lindenberg A (2011) Dopamine-glutamate interactions: a neural convergence mechanism of common schizophrenia risk variants. Biol Psychiatry 69(10):912–913

Tost H, Meyer-Lindenberg A (2012) Puzzling over schizophrenia: schizophrenia, social environment and the brain. Nat Med 18(2):211–213

Tost H, Lipska BK, Vakkalanka R, Lemaitre H, Callicott JH, Mattay VS, Kleinman JE, Marenco S, Weinberger DR (2010) No effect of a common allelic variant in the reelin gene on intermediate phenotype measures of brain structure, brain function, and gene expression. Biol Psychiatry 68(1):105–107

van Os J, Krabbendam L, Myin-Germeys I, Delespaul P (2005) The schizophrenia envirome. Curr Opin Psychiatry 18(2):141–145

van Os J, Rutten BP, Poulton R (2008) Gene-environment interactions in schizophrenia: review of epidemiological findings and future directions. Schizophr Bull 34(6):1066–1082

van Os J, Kenis G, Rutten BP (2010) The environment and schizophrenia. Nature 468(7321):203–212

Veling W, Hoek HW, Selten JP, Susser E (2011) Age at migration and future risk of psychotic disorders among immigrants in the Netherlands: a 7-year incidence study. Am J Psychiatry 168(12):1278–1285

Waldman ID (2005) Statistical approaches to complex phenotypes: evaluating neuropsychological endophenotypes for attention-deficit/hyperactivity disorder. Biol Psychiatry 57(11):1347–1356

Wei Q, Kang Z, Diao F, Guidon A, Wu X, Zheng L, Li L, Guo X, Hu M, Zhang J, Liu C, Zhao J (2013) No association of ZNF804A rs1344706 with white matter integrity in schizophrenia: a tract-based spatial statistics study. Neurosci Lett 532:64–69

Weinberger DR (1987) Implications of normal brain development for the pathogenesis of schizophrenia. Arch Gen Psychiatry 44(7):660–669

Wessa M, Linke J, Witt SH, Nieratschker V, Esslinger C, Kirsch P, Grimm O, Hennerici MG, Gass A, King AV, Rietschel M (2010) The CACNA1C risk variant for bipolar disorder influences limbic activity. Mol Psychiatry 15(12):1126–1127

WHO (2001) The WHO world health report: new understanding, new hope. World Health Organization, Geneva

Williams GV, Goldman-Rakic PS (1995) Modulation of memory fields by dopamine D1 receptors in prefrontal cortex. Nature 376(6541):572–575

Wray NR, Goddard ME, Visscher PM (2007) Prediction of individual genetic risk to disease from genome-wide association studies. Genome Res 17(10):1520–1528

Zink CF, Tong Y, Chen Q, Bassett DS, Stein JL, Meyer-Lindenberg A (2008) Know your place: neural processing of social hierarchy in humans. Neuron 58(2):273–283

Zinkstok J, van Amelsvoort T (2005) Neuropsychological profile and neuroimaging in patients with 22Q11.2 deletion syndrome: a review. Child Neuropsychol 11(1):21–37

Brain Stimulation and Imaging 8

Alexander T. Sack and Teresa Schuhmann

Abbreviations

BOLD	Blood oxygenation level dependent
DLPFC	Dorsolateral prefrontal cortex
DWI	Diffusion-weighted imaging
EPI	Echo-planar imaging
FEF	Frontal eye field
IPS	Intraparietal sulcus
M1/S1	Primary sensorimotor cortex
MDD	Major depressive disorder
MFG	Middle frontal gyrus
PFC	Prefrontal cortex
PMd	Dorsal premotor cortex
RF	Radio frequency
SPL	Superior parietal lobule
TBS	Theta burst stimulation
TDCS	Transcranial direct current stimulation
TMS	Transcranial magnetic stimulation

8.1 Brain Stimulation and Imaging

8.1.1 Brain Imaging: Possibilities and Limitations

Functional magnetic resonance imaging (fMRI) is a noninvasive imaging method, capable of visualizing brain areas that are active during different behavioral or cognitive functions. Yet, although functional brain imaging provides evidence for task-dependent changes in brain activity, it is limited in revealing direct causal relationships between these brain activity changes and their respective behavioral or cognitive consequences. Thus, the question remains: is the change in brain activity observed actually functionally relevant for successful task performance? To answer this question, the experimental design must somehow be inverted. Where in functional neuroimaging the cognition or behavior is the independent variable, and the brain activity the dependent variable, we wish to turn this around. Ideally, we should manipulate brain activity, making this the experimental factor, and observe the effects of this manipulation on cognition or behavior. If the experimentally induced brain activity change has effects on task performance, only then can one conclude that the brain activity involved is functionally relevant. The direction of behavioral effects moreover provides information on the specific role of the targeted brain region in the task at hand. To achieve this sort of controlled experimental setup, a method of transient and local brain activity manipulation is required. Such methods exist and are collectively referred to as functional brain interference, or brain stimulation techniques.

8.1.2 Brain Stimulation Techniques

Brain stimulation techniques (also referred to as brain perturbation or brain interference

A.T. Sack (✉) • T. Schuhmann
Faculty of Psychology and Neuroscience,
Maastricht University, Maastricht, The Netherlands
e-mail: a.sack@maastrichtuniversity.nl

C. Mulert, M.E. Shenton (eds.), *MRI in Psychiatry*,
DOI 10.1007/978-3-642-54542-9_8, © Springer Berlin Heidelberg 2014

techniques) can be divided into invasive and noninvasive approaches. Invasive methods, such as cooling and microstimulation, are mainly limited to animal studies, whereas transcranial direct current stimulation (tDCS) and transcranial magnetic stimulation (TMS) are noninvasive brain stimulation techniques which can be safely used in human volunteers and patients. TMS allows for controlled manipulation of brain activity in several ways: (1) inducing transient disruptions of neural activity ("virtual lesions"), (2) enhancing or decreasing cortical excitability, (3) stimulating neural populations, or (4) inducing local oscillations. By transiently changing activity in the stimulated brain area and revealing a subsequent change in a particular behavior, TMS can be regarded as a unique research tool for the investigation of causal structure-function relationships (see, e.g., Sack and Linden 2003).

8.1.2.1 Transcranial Magnetic Stimulation (TMS): Basic Mechanisms of Action

Any TMS device consists of a bank of capacitors capable of producing high discharge currents and an electromagnetic stimulating coil to apply magnetic pulses of up to several Tesla. The high and rapidly changing currents are discharged into the coil, thereby creating a strong and time-varying magnetic field (pulse). This pulse can reach its peak in a few hundred microseconds and induce an electric field in the neuronal tissue underneath the coil; the strength of which depends mainly on the rate of change of the magnetic field. Due to the electrical conductivity of the living tissue, the induced electric field results in an electrical (eddy) current in the cortex, in a parallel but opposite direction to the current in the coil (Lenz's law), and subsequently in a depolarization of the underlying neurons. The magnetic stimulation indirectly creates a transmembrane potential by moving a charge across the cellular membrane which can lead to membrane depolarization and to an action potential of the respective axon.

Physical parameters of the magnetic field (e.g., rise time and spatial field distribution) determine the temporal-spatial characteristics of the magnetic pulse sent into the brain, but the induced electric field characteristics in neural tissue depend on some additional factors. The shape of the skull, the distance from TMS coil to the gyrating cortical layers, the shape of coil and intensity of stimulation, and whether pulses are monophasic or biphasic all influence the final effective strength and extent of stimulation at the cortical level. Moreover, the magnetic field strength decreases exponentially with distance and the cortical surface is convoluted. Magnetic coils have different possible geometric shapes, affecting focality and induced current direction. All these characteristics, of stimulation coils and underlying neuronal tissue, interact to determine the actualized neuronal depolarization of mostly superficial levels of the brain (within a few cm of the coil). And that is considering the effects of one magnetic pulse only (Sack and Linden 2003).

8.1.2.2 Transcranial Magnetic Stimulation (TMS): Basic Protocols

TMS pulses can be applied one at a time (single-pulse TMS), in pairs separated by a variable interval (paired-pulse TMS), or in multiples, ranging from triple-pulse up to quintuple-pulse TMS. Importantly, for these application methods, the pulses are usually locked to an external event (e.g., task onset), therefore potentially revealing information about the chronometry of a cognitive process. We can, therefore, refer to these approaches as chronometric, or event-related, TMS. By applying chronometric TMS at variable times during task execution, it is possible to investigate not only whether a given brain region is necessary for the tested behavior but also at what time point (with a temporal resolution of 5–10 ms) the neural activity at the stimulation site is critical for successful task performance (chronometry of functional relevance (see also Walsh and Pascual-Leone 2003)).

In contrast, TMS pulses can also be applied in a repetitive manner (repetitive TMS; rTMS) using either "conventional" or "patterned" protocols of repetitive stimulation (Rossi et al. 2009). The important feature of both conventional and patterned rTMS is that it is capable of modulating the excitability of the stimulated area for some time after the TMS application itself. The nature of these aftereffects, whether they are inhibitory or excitatory, mainly depends on the frequency of stimulation. In conventional rTMS protocols, single TMS pulses are applied in a regular rhythm, with a distinction between low-frequency rTMS (stimulation frequency of 1 Hz or less) and high-frequency rTMS (stimulation frequency >1 Hz). Patterned rTMS refers to repetitive application of short high-frequency bursts of rTMS, interleaved by short pauses of no stimulation. In theta burst stimulation (TBS), short bursts of 50 Hz rTMS are repeated with a rate in the theta range (5 Hz) as a continuous (cTBS) or intermittent (iTBS) train (Di Lazzaro 2008; Huang et al. 2005). Both 1 Hz rTMS and cTBS are consistently found to produce lasting inhibitory aftereffects, whereas high-frequency rTMS and iTBS induce lasting facilitatory aftereffects on motor corticospinal output in healthy participants.

The ability of rTMS to induce longer-lasting excitability changes has opened the door for the clinical applications of TMS in treating various neuropsychiatric disorders, for example, by "down- or upregulating" pathologically hyper- or hypoactive brain areas (Brighina et al. 2003; Haraldsson et al. 2004; Hoffman 2003; Hoffman and Becker 2005; Martin et al. 2003; Paus and Barrett 2004).

8.1.2.3 Clinical TMS

Over the past 15 years, increasing numbers of studies of the potential therapeutic effects of TMS have been published. Disorders including addiction (Camprodon et al. 2007; Eichhammer et al. 2003), obsessive compulsive disorder (Martin et al. 2003; Sachdev et al. 2001), pain (Khedr et al. 2005; Lefaucheur et al. 2001), schizophrenia (Chibbaro et al. 2005; Lee et al. 2005), and depression (George et al. 1995; Pascual-Leone et al. 1996) have been studied; however, of all the psychiatric disorders, TMS in major depressive disorder (MDD) has been studied most thoroughly.

Repetitive TMS above the dorsolateral prefrontal cortex (DLPFC) has been proposed as a potential new treatment option for depression. Numerous studies have been carried out—stimulating either left DLPFC with high-frequency TMS or right DLPFC with low-frequency TMS—however, with diverse results (for review see, e.g., (Schonfeldt-Lecuona et al. 2010). O'Reardon and colleagues (2007) published a large multicenter trial of daily left pre-frontal TMS in medication-free patients with MDD, reporting encouraging results. Herwig and colleagues, on the other hand, found no difference in responder rates or depression rating scales between real TMS and sham treatment groups in 209 their multicenter trial (Herwig et al. 2007). Meta-analyses of the antidepressant effect of rTMS (Burt et al. 2002; Gross et al. 2007; Holtzheimer et al. 2001; Kozel and George 2002; Martin et al. 2003; McNamara et al. 2001) have also revealed mixed results, with differences between findings perhaps relating to small sample sizes as well as their heterogeneous designs. Thus, at this point in time, the validity of TMS for the treatment of depression in clinical practice still needs further investigation. While TMS certainly seems to have beneficial effects with therapeutic potential, the inconsistency of results needs explanation, so that consensus can be reached on which TMS protocols are effective for which types of depression patients. Currently, our understanding of TMS effects in general, and in depression in particular, appears to be too limited to afford any strong predictions about the chance of success in therapeutic application (see also Ridding and Rothwell 2007). Nevertheless, in 2008 the first rTMS device (NeuroStar TMS Therapy System) received FDA approval for the treatment of resistant major refractory depression in adults.

Recent TMS studies have also harnessed this technique with the aim of alleviating behavioral or cognitive deficits in patients suffering from brain injury, lesions, and stroke (see, e.g., Brighina et al. 2003; Koch et al. 2008; Oliveri et al. 2001, 1999). By suppressing the intact hemisphere of stroke patients, the damaged hemisphere is (to an extent) released from the strong interhemispheric inhibition. This allows the damaged hemisphere to express its remaining functionality. TMS studies based on this logic have delivered some encouraging results, demonstrating that the counterintuitive strategy of decreasing neural excitability of the healthy hemisphere actually improves deficits following unilateral brain damage to the other hemisphere (Brighina et al. 2003; Cazzoli et al. 2010; Koch et al. 2008; Nyffeler et al. 2009; Oliveri et al. 2001, 2000a, b; Shindo et al. 2006; Song et al. 2009).

8.1.3 Combining Brain Stimulation and Brain Imaging

Brain imaging and brain stimulation offer highly complementary methods for studying the healthy and diseased human brain. It is, therefore, sensible to combine these approaches in human fundamental and clinical neuroscience. TMS and functional imaging can be combined either during simultaneous measurements or by using the same paradigm and participant sample during separate TMS and imaging sessions. Both simultaneous combination and experimental combination methods are useful for the investigation of functional brain-behavior relationships, but they have different applications, advantages, and limitations.

8.1.3.1 Brain Imaging Before Brain Stimulation

When applying TMS in cognitive studies, the brain areas of interest do not always have a behavioral signature output, as is the case for TMS over motor cortex or visual cortex. For these brain regions, and associate cognitive research questions, it is not straightforward to determine the precise scalp location where TMS pulses should be administered. Functional imaging before TMS can be used to address this problem by precisely localizing a task-related area of cortical activation for subsequent use with a frameless stereotaxic TMS neuronavigation system, thus optimizing the exact coil positioning for TMS. In this way, the combination of brain imaging and subsequent brain stimulation permits the assessment of whether, in a given participant, this task-related functional activity (shown using brain imaging) is actually functionally relevant to that individual's successful task performance (Andoh et al. 2006; Sack et al. 2006; Thiel et al. 2005). There are now several commercially available stereotaxic systems for TMS neuronavigation. Most of them allow for fMRI-TMS co-registration procedures so that events occurring around the head of the participant in real space are registered online and visualized in real time at correct positions relative to the participant's anatomical reconstruction of the brain. By superimposing the functional data on the anatomical reconstruction of the brain, the TMS coil can be neuronavigated to a specific functional activation area of every participant (see Sack et al. 2009) (Fig. 8.1).

Using such neuronavigation systems, TMS coil positioning can become highly accurate, targeting anatomical or functional "hotspots" in individual participants with millimeter precision. This is relevant since, despite the limited spatial resolution of the applied magnetic field, spatial TMS coil shifts in the order of millimeters have been shown to sometimes result in a complete loss of behavioral or cognitive impairment effects (Beckers and Homberg 1992; d'Alfonso et al. 2002). Comparing different localization strategies for TMS-based primary motor cortex mappings in terms of accuracy and efficiency, Sparing and colleagues (2008) found that fMRI-guided stimulation was most precise (accuracy was concluded to be in the millimeter range). Feredoes and colleagues (2007) used fMRI to localize TMS sites for disruption of short-term verbal information retention. Sack et al. (2009) investigated the behavioral impact of right parietal TMS on a number comparison task, when TMS localization was based on (1) individual fMRI-guided TMS neuronavigation, (2) individual MRI-guided TMS neuronavigation, (3) group functional Talairach coordinates, or (4) the 10–20 EEG position P4. They quantified the behavioral effect of each TMS localization approach, calculated the standardized experimental effect sizes, and conducted a statistical power analysis, which revealed that the individual fMRI-guided TMS neuronavigation yielded the strongest behavioral effect size (Sack et al. 2009). This increased effect size of TMS when using (f)MRI-guided coil positioning has also been shown in the context of clinical TMS applications for various

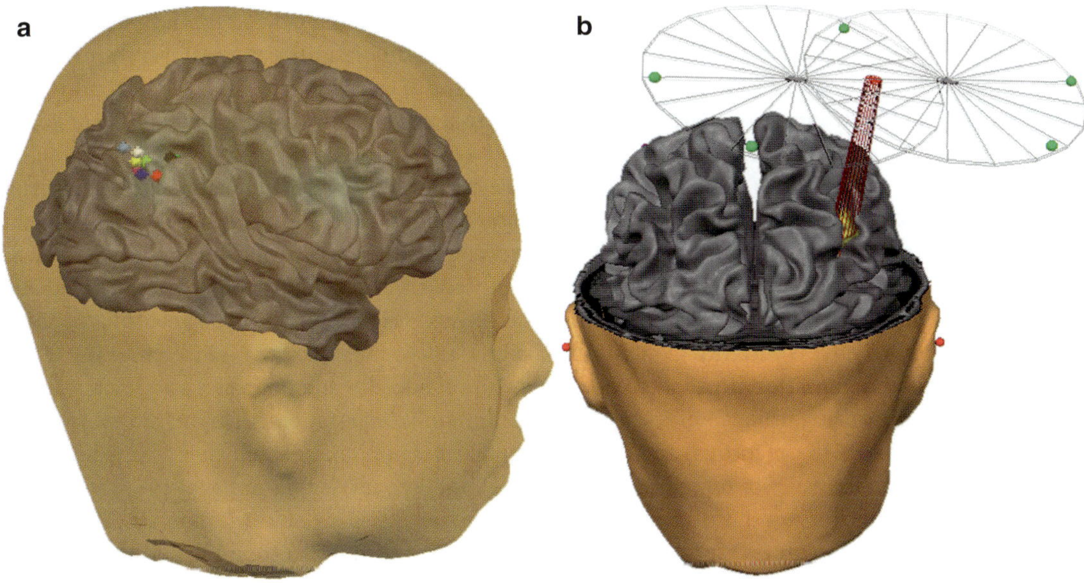

Fig. 8.1 fMRI-guided TMS neuronavigation. Panel (**a**) shows several color-coded fMRI activity clusters superimposed on a reconstruction of the cortical surface, projected within a transparent mesh of a reconstructed head in Talairach space. Each of these clusters represents an individual fMRI "hotspot," i.e., strongest task-related activity, of a given individual participant obtained in a separate fMRI measurement. The spatial distribution between these individual fMRI activity clusters accounts for the interindividual variability in the exact structure-function correspondence. Panel (**b**) shows a snapshot of the Brain Voyager TMS neuronavigation system used to guide TMS coil positioning based on one of these activity clusters of a given participant. The *red beam* indicates where the magnetic field of TMS is strongest and is navigated in real time to the *orange color-coded* individual fMRI hotspot of this particular participant. The exact positioning of the TMS coil and thus the target area for the magnetic brain stimulation are therefore individually defined based on the fMRI data obtained in a separate session prior to TMS

psychiatric disorders (Ahdab et al. 2010; De Ridder et al. 2011; Herbsman et al. 2009).

8.1.3.2 Brain Imaging After Brain Stimulation

Certain brain stimulation protocols, such as rTMS or TBS, are capable of modulating neural excitability of a region beyond the TMS stimulation itself. Functional imaging can then be used to investigate these prolonged TMS aftereffects. Imaging the immediate and longer-lasting aftereffects of TMS is paramount for revealing the underlying neurobiological mechanisms that lead to the observed behavioral changes and clinical treatment effects of TMS stimulation.

An elegant example of this approach comes from Hubl and colleagues (2008). Here, the right frontal eye field (FEF) was stimulated outside the MR scanner using continuous theta burst rTMS (TBS). Then fMRI was used to map the TBS-induced effects and assess their temporal persistence across the brain during a saccade task. The results showed a TBS-induced suppression of local BOLD activity that appeared 20–35 min (but not immediately) after stimulation (Hubl et al. 2008). Suppression, albeit weaker, was also evident in more remote regions, including the (pre)supplementary and parietal eye fields. Similarly, Cárdenas-Morales and colleagues (2011) used fMRI for exploring the aftereffects of iTBS over primary motor cortex.

Several studies have used functional imaging to visualize TMS aftereffects in prefrontal cortex (PFC), in order to explore the underlying mechanisms of potential therapeutic applications for depression (Fitzgerald et al. 2007). The implication is that prefrontal rTMS in normal and depressed participants has profound effects on both local and remote brain regions implicated in depression, including bilateral frontal, limbic, and paralimbic areas (Fitzgerald et al. 2007; Kimbrell et al. 1999, 2002; Pogarell et al.

2006, 2007; Speer et al. 2000, 2009; Teneback et al. 1999). Importantly, these rTMS-induced effects appear to be frequency dependent, with low-frequency rTMS leading to bilateral reduction in frontal activation (Fitzgerald et al. 2007).

8.1.3.3 Brain Stimulation During Brain Imaging

While useful, functional imaging after TMS application remains fundamentally limited in elucidating the neuronal effects of TMS. Concurrent TMS and neuroimaging offer a broader range of in vivo information regarding the actual and immediate effects of TMS on cortical activation, both local and remotely. Simultaneous TMS and imaging can thus be used to online track the TMS effects in the brain or probe intracerebral connectivity (Bestmann et al. 2003b, 2004, 2005; Bohning et al. 1999, 2000b; Ruff et al. 2006; Sack et al. 2007). Therefore, even in the absence of overt behavior, TMS during fMRI facilitates the imaging of pathways of activity spreading within and between brain networks. Furthermore, in simultaneous TMS/fMRI, brain stimulation can be applied while concurrently recording changes in brain activity and behavior. This simultaneous approach allows the investigation of the local and remote brain responses at a neurophysiological level. Thus, it can be determined, in vivo, which brain areas—either directly or transsynaptically affected by TMS—underlie the observed TMS-induced behavioral changes during active task execution. However, the simultaneous combination of TMS and functional imaging poses great technical challenges. Therefore, it is routinely used by only few research groups, and the number of simultaneous TMS/fMRI publications is still considerably small (Reithler et al. 2011).

Besides the need for specific hardware (e.g., an MR-compatible TMS system), simultaneous TMS and BOLD fMRI requires appropriate temporal synchronization between MRI acquisition and TMS pulse application. Furthermore, the discharge, and even mere presence, of MR-compatible TMS coils in the bore of the magnet produces artifacts in the echo-planar imaging (EPI) images that need to be resolved before the simultaneous combination of functional imaging and brain stimulation becomes feasible.

Setup, Experimental Procedures, and Artifacts

The use of TMS inside the MR scanner during simultaneous TMS/fMRI studies requires several modifications to TMS hardware, specific TMS/fMRI interleaved experimental designs, and the consideration or removal of several artifacts. Most importantly, the standard TMS coils routinely used outside the MR scanner are not appropriate for simultaneous TMS/fMRI studies. Instead, specific MR-compatible non-ferromagnetic TMS coils are required. MR-compatible TMS coils are characterized by several main modifications: (1) removal of ferromagnetic materials and electronic elements from the coil, (2) strengthened casing to withstand the large forces of the MR scanner without cracking, (3) a connection cable long enough to feed through a wave guide leaving the radiofrequency (RF)-shielded cabin, and (4) removed or modified TMS coil handle to ease positioning within the spatially restricted MR environment. Finally, since frameless stereotaxy is not applicable inside the scanner, TMS coils are often fitted with specific MR markers in order to post hoc identify the position of the coil relative to the simultaneously acquired structural and functional data. The MR signal of these markers on the TMS coil can be used to estimate and reconstruct, by triangulation, the exact position and orientation of the coil inside the scanner.

Although necessary, these TMS coil modifications are by no means sufficient to avoid further technical problems and measurement artifacts during simultaneous TMS/fMRI. One principle problem of combined TMS/fMRI studies is a direct consequence of the standard TMS/fMRI setup described above. In this setup, the MR-compatible TMS coil is connected to the stimulator outside the RF-shielded cabin via a cable running through a wave guide. Therefore, the RF shield of the MR scanner is pierced by the TMS cable, which acts as an antenna transmitting RF noise into the scanner. Special RF noise filters then need to be installed for simultaneous TMS/fMRI studies as an additional hardware component (Fig. 8.2).

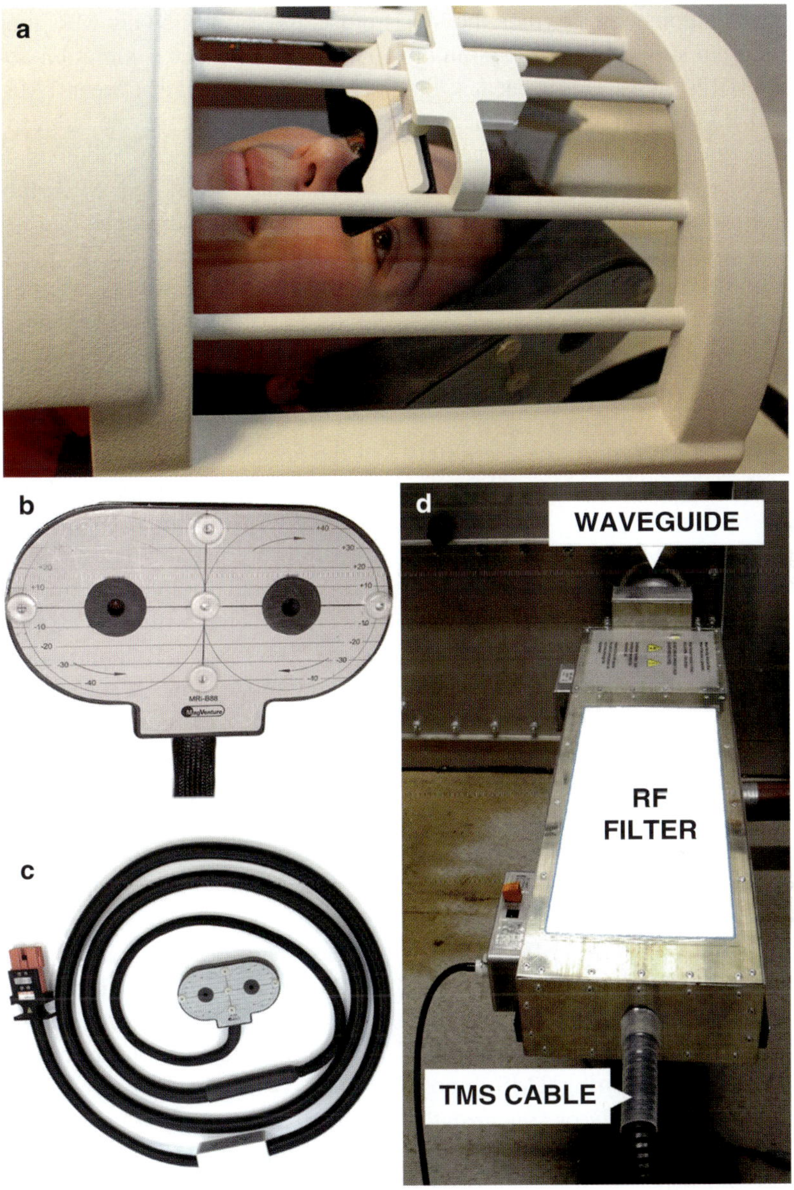

Fig. 8.2 TMS during fMRI. Panel (**a**) shows a participant inside the MR scanner during simultaneous TMS and fMRI measurements. The participant's head is fixated within the MR head coil, while an MR-compatible non-ferromagnetic TMS coil is positioned on the scalp and fixated in order to apply noninvasive brain stimulation during functional brain imaging. Panel (**b**) shows a top view of the MR-compatible non-ferromagnetic TMS coil, which is fitted with five specific MR markers. The MR signal of these markers can be used to estimate and reconstruct, by triangulation, the exact position and orientation of the coil inside the scanner and to thus post hoc identify the position of the coil relative to the simultaneously acquired structural and functional data. Panel (**c**) depicts the long connection cable of the MR-compatible TMS coil. This cable is used to connect the MR-compatible TMS coil to the stimulator outside the RF-shielded cabin via a wave guide. Panel (**d**) shows the special RF noise filters that need to be installed for simultaneous TMS/fMRI studies as an additional hardware component. This is necessary because the connecting TMS cable running through the wave guide pierces the RF shield and acts as an antenna transmitting RF noise into the scanner. Therefore, the TMS cable outside the RF cabin is connected to this specific RF noise filter device which connects via the wave guide to the inner RF cabin wall, to then connect to the MR-compatible TMS coil

Despite the installation of an RF filter, the MR image quality is often still decreased in simultaneous TMS and fMRI studies. This is because the mere presence of a TMS coil in the scanner can result in static magnetic field inhomogeneities, which particularly affect EPI scans (commonly used for fMRI). Baudewig and colleagues (2000) systematically investigated the type and extent of the artifacts induced by the TMS coil during MR measurements. The authors revealed that although the anatomical images were unaffected, there were pronounced signal losses and geometric distortions in EPI acquisitions perpendicular to the plane of the coil. However, these artifacts could be markedly reduced by using an EPI orientation parallel to the coil plane. Furthermore, these signal losses and geometric distortions attenuate with increasing distance from the coil and so are restricted to the area very close to the coil. Therefore it is unlikely that functional images of the human cortex are largely affected, given the scalp-cortex distance of >1 cm.

After having addressed the technical challenges discussed above, one can progress to the most important step of applying TMS pulses during actual MR EPI data acquisition. Although, it must be noted that simultaneous or concurrent TMS/fMRI is not possible in the strictest sense. In reality, TMS pulses and MRI acquisitions must be appropriately interleaved in order to avoid the inevitable artifacts produced by the TMS-induced currents, which would otherwise make artifact-free scanning during TMS impossible. Therefore, simultaneously or concurrently combined TMS/fMRI studies, generally refers to interleaved TMS and fMRI measurements, during which the MR sequence must send a trigger signal to the TMS apparatus with every RF pulse excitation. TMS pulses are thus temporally separated from MR imaging. Still, distortions can even occur when pulses are applied up to 100 ms before slice acquisition onset (Bestmann et al. 2003a; Shastri et al. 1999). These lasting artifacts are purportedly related to residual currents in the TMS coil and to currents induced by the vibrations in the TMS coil following a pulse (Shastri et al. 1999). The current standard is, therefore, to leave at least 100 ms between each TMS pulse and any following MR image acquisition. However, with better vibration absorption in the TMS coil, the delay between TMS pulse and MR image acquisition may be reduced considerably.

There are various methods for temporally interleaving TMS and MRI for simultaneous experiments. For example, TMS pulses and MR images can be interleaved by insertion of temporal gaps after each volume (Ruff et al. 2006; Sack et al. 2007). Sack and colleagues (2007) applied bursts of rTMS at ~13.3 Hz over 560 ms at the end of each MR volume. In this study, a delay of 200 ms from the last TMS pulse to the beginning of the next MR volume acquisition protected the subsequent MR acquisition from pulse-related artifacts. Alternatively, TMS pulses can be separated, not by placing them at the end (or beginning) of each volume, but by interleaving them after each slice within one volume (Bestmann et al. 2004, 2005; Bohning et al. 2000a). This method still requires a sufficient delay between TMS pulses and slice acquisition so that subsequent slices are not perturbed. Finally, single slices might also be deliberately perturbed by the TMS pulse and then be identified and replaced, either by interpolation between pre- and post-pulse acquisition of the same slice or by including affected slices as covariates in a general linear model analysis. When employing any of these methods with modified EPI sequences to optimize interleaved TMS/fMRI measurements, it is also recommended to introduce oversampling of the phase-encoding direction of EPI images in order to shift the so-called "ghosting" artifact outside the volume of interest.

One additional problem for simultaneous, or interleaved, TMS/fMRI studies was shown by Weiskopf and colleagues (2009). The authors reported that leakage currents may be generated when switching stimulation intensities. In a phantom measurement, these leakage currents in the TMS coil varied parametrically with the TMS output intensity (its capacitor charge) and induced magnetic field inhomogeneities which led to false-positive fMRI findings. In other words, BOLD signal increased parametrically with TMS intensity in their phantom measurement (Weiskopf et al. 2009). Following this report, a technical solution has been pioneered which introduces a relay in parallel (and diodes in series) with the TMS coil. When the relay is closed, leakage current primarily flows through this relay, rather than the TMS coil. A trigger signal then briefly opens the relay so that a TMS pulse can be applied. However, although these (or similar) solutions are now standard in MR-compatible TMS systems, appropriate test measurements should be run in order to identify any remaining artifacts or false positives due to leakage current.

TMS Has Local and Remote Effects

Generally, all reported studies using concurrent TMS-fMRI show that TMS has task-specific effects on the BOLD signal in the targeted site. This is encouraging, given the widespread assumption that TMS affects excitability/activity in the region directly underneath the coil and that this activity change reflects behavioral effects of TMS (see Reithler et al. 2011, for an exhaustive overview). However, one of the most important additional conclusions from combined TMS and functional imaging studies is that locally applied focal TMS does not exclusively affect neural activity at the stimulation site, but can also be shown to affect remote and interconnected brain regions (Bestmann et al. 2003b; Blankenburg et al. 2008; Bohning et al. 2000a; Denslow et al. 2005; Ruff et al. 2006; Rushworth et al. 2002; Sack et al. 2007). This includes cortical as well as subcortical brain areas, as revealed by early application to the human motor system (Baudewig et al. 2001; Bestmann et al. 2004; Bohning et al. 1999, 2000a). It seems that application of TMS in essence involves inserting energy into a system and that TMS to an isolated neuron population will excite not only that population, but a connected brain area will propagate the inserted energy throughout its anatomical (Boorman et al. 2007) and functional (Sack 2006) network. It is precisely the value of TMS-fMRI that this spread of TMS excitation can be tracked throughout the brain. Bohning et al. (1999) showed that the BOLD signal resulting from TMS correlated to the TMS intensity both in local (targeted) and remote brain areas. Moreover, Bohning and colleagues (2000a) could show that TMS-induced finger movements resulted in BOLD signals throughout the brain that were similar to BOLD signals resulting from voluntary finger tapping. This constituted an early demonstration of the validity of using TMS-fMRI to probe functional/anatomical networks in the brain. Bestmann and colleagues (2004) confirmed this notion, stimulating with high-frequency rTMS the left primary sensorimotor cortex (M1/S1) at supra- and subthreshold intensities (no finger movements induced in the latter condition) and measuring the BOLD signals throughout the brain. A network of distinct cortical and subcortical motor system structures was activated in response to the TMS, again involving the same regions activated by voluntary finger movements. Interestingly, this was the case even for subthreshold stimulation, showing that TMS can probe an anatomical network even in the absence of overt behavioral response, although subthreshold stimulation in the absence of induced muscle contractions mainly led to enhanced BOLD responses in supplementary and premotor cortices and not in the local M1/S1 region that was actually stimulated (see Hanakawa et al. 2009 for similar intensity-dependent remote activation changes based on spTMS). This suggests that the local BOLD effects, directly underneath the coil, may constitute a special case: they depend on actually induced muscle contractions, while remote-connected motor network regions also involved in voluntary movements are activated by M1/S1 TMS even subthreshold (Bestmann et al. 2004; Denslow et al. 2005). Based on modeling work, Esser and colleagues (2005) suggest that TMS locally stimulates both excitatory and inhibitory neural populations (ergo the net activation and thus BOLD is weaker here), but remotely results mainly in excitatory responses which are easier to detect. However, the matter is not settled, given the still ill-defined intricacies of TMS effects on local neural circuits and moreover the connection between such effects and the BOLD signal (Logothetis 2008; Logothetis et al. 2010). Still, the anatomical and functional specificity of the observed remote network effects argues against a nonspecific (water ripple-like) spread of TMS-induced activity. Moreover, the observed networks closely resemble the brain systems involved in natural tasks involving the same regions. For a more elaborate review of these issues, see Reithler and colleagues (2011).

Local and Remote TMS Effects Are State/Task Dependent

Focal TMS can therefore lead to both local and remote neural effects, within anatomically or functionally connected brain regions. However, several combined TMS/fMRI studies have also found that these local and remote network effects are state or task dependent. In other words, the state of the brain at the moment of TMS, as induced by task demands or external sensory stimulation, or even by naturally occurring fluctuations, can influence the local and remote network response to TMS. An excellent example of these state-dependent effects comes from Bestmann and colleagues (2008), who applied TMS over left dorsal premotor cortex (PMd) at two intensities (low vs. high) and two motor states (grip vs. no grip). Participants were stimulated over left PMd either when performing a handgrip task with their left hand or during rest (nogrip). The authors revealed a significant crossover interaction between motor state and TMS intensity over left PMd, arising in right M1 and right PMd. TMS over left PMd during rest led to an activation decrease in right PMd and M1 of the contralateral hemisphere. This contralateral decrease following TMS has been observed in most (Bestmann et al. 2004; Kemna and Gembris 2003), but not all, simultaneous TMS/fMRI studies over the motor cortex (Bohning et al. 2000a; Hanakawa et al. 2009). However, more importantly, Bestmann et al. (2008) also showed that this contralateral decrease after TMS over left PMd during rest then becomes a contralateral increase in activation during a left-handed grip task, with stronger functional coupling following TMS (when comparing high vs. low intensity). Thus, the direction of remote effects (activation increases/decreases) was reversed depending on the state of the system. This reversal is likely caused by differences in the initial brain states, in relation to interregional mutual inhibition/facilitation mechanisms. Another demonstration of the task dependency of TMS-induced activation changes comes from O'Shea and colleagues (2007). These authors applied 15 min of offline 1 Hz rTMS over left PMd and reported compensatory activation increases in the contralateral (right) PMd. However, this effect was specific to an action selection motor task that otherwise significantly engaged (the now disrupted) left PMd. The compensation effect was not observed during a simpler motor execution task (repetitive finger movements). Importantly, when dpTMS was applied to the right PMd after 1Hz rTMS over left PMd, behavioral performance on the action selection task suffered. In other words, the compensatory right PMd activation increases after left PMd disruption were causally relevant for the task.

In a more cognitive application, Sack and colleagues (2007) revealed that TMS over the right intraparietal sulcus (IPS) only results in right hemispheric frontoparietal network effects of TMS (i.e., neural effects in local and remote regions within a functionally connected frontoparietal network) when the participant is engaged in a task that requires the proper functioning of the targeted brain region. Conversely, the authors showed that this same parietal TMS protocol did not lead to such frontoparietal network effects when the task did not rely on parietal cortex (Sack et al. 2007). Thus, parietal TMS led to significantly different local and remote brain effects depending whether, or not, the stimulated region was engaged in task-relevant processes at the time of the experimentally induced brain perturbation. These findings indicate that TMS-induced neural activity is particularly likely to spread to nodes of a (currently active) functional network and that activity does not necessarily spread to regions that are only anatomically connected to the target site. These state- and task-dependent modulations of TMS effects should not be underestimated and could also partially explain differences in remote effects between target sites when the same TMS protocol is used.

Local and Remote TMS Effects Are Functionally Relevant

The demonstration of remote neural effects of TMS raises the question of whether (and to what extent) these indirect remote effects are also relevant and functionally related to the TMS-induced behavioral changes, in other words, whether reported behavioral effects of TMS that are seemingly specific to a particular target site do actually relate to TMS-induced neural activity changes at that target site or whether these behavioral effects might relate to a widely distributed network effect. Ruff and colleagues (2006, 2008, 2009) applied TMS over right FEF inside the MR scanner and revealed remote BOLD effects in two bilateral sets of occipital brain regions within retinotopic visual areas V1–V4. Right FEF-TMS led to BOLD increases for peripheral visual field representations, but BOLD decreases for the central visual field. Assuming that higher BOLD signal equals higher-contrast sensitivity, the authors concluded that FEF-TMS may enhance peripheral, relative to central, vision. Interestingly, these behavioral predictions following the remote neural effects of FEF-TMS within early visual cortex were later confirmed by the authors in a psychophysical study outside the MR scanner. Sack and colleagues (2007) showed that simultaneous TMS/fMRI during active task execution potentially allows in vivo imaging of the neural network effects underlying TMS-induced behavioral changes. The authors applied TMS over right and left parietal cortex during whole-brain BOLD fMRI of spatial cognition performances. The authors found that right, but not left, parietal TMS (i) behaviorally impairs spatial cognition, (ii) induces neural activity changes across a right hemispheric network of frontoparietal regions, and (iii) results in significant correlations between TMS-induced behavioral impairments and neural activity changes in the directly stimulated parietal region as well as ipsilateral frontal brain regions. Thus, it appears that neural activity, not just in the stimulated right superior parietal lobule (SPL), but also in the remote ipsilateral middle frontal gyrus (MFG), was influenced by right parietal TMS (during a spatial cognition task) and contributed to a reduction in task performance (Sack et al. 2007). Importantly, these task-specific TMS-induced BOLD reductions correlated with behavioral impairment: the stronger the reduction, the slower participants responded. Again, this raises the question of whether these remote effects of right parietal TMS (e.g., at right MFG) are functionally relevant or causally related to the observed behavioral deficit. The TMS/fMRI study by Sack and colleagues (2007) does strongly suggest that the right parietal TMS-induced behavioral deficits are not exclusively caused by neural activity changes at the site of stimulation, but rather caused by neural network effects within a right hemispheric frontoparietal network consisting of right MFG and SPL. However, in this study, the functional relevance of these remote regions has to be assumed based on a correlation between the remote activation change in MFG and the behavioral impairments in spatial task performance. Therefore, in a follow-up study, the authors directly tested the functional relevance of MFG by now targeting this region directly with TMS. Causal evidence was thus provided for the functional relevance of the remote TMS activation change identified earlier (de Graaf et al. 2009). Only such an iterative approach can directly verify the functional role of revealed response profiles in distant network nodes (Fig. 8.3).

8.2 Conclusion and Outlook

The combination of brain stimulation with brain imaging offers unique experimental possibilities for understanding the functional architecture of the healthy and diseased human brain. Brain imaging before brain stimulation is useful for the identification (in individual participants or patients) of an exact TMS stimulation site. Here, the fMRI data of an individual is used to place the TMS coil above the exact brain area that has shown activation changes during the task performance of this particular participant. Brain imaging after brain stimulation is useful for identifying the spatial pattern and persistency of rTMS-induced neural activity changes that last beyond the stimulation itself (TMS aftereffects). Finally, brain imaging during brain stimulation enables to stimulate a particular brain region while simultaneously monitoring whole-brain changes in brain activity and behavior, thereby

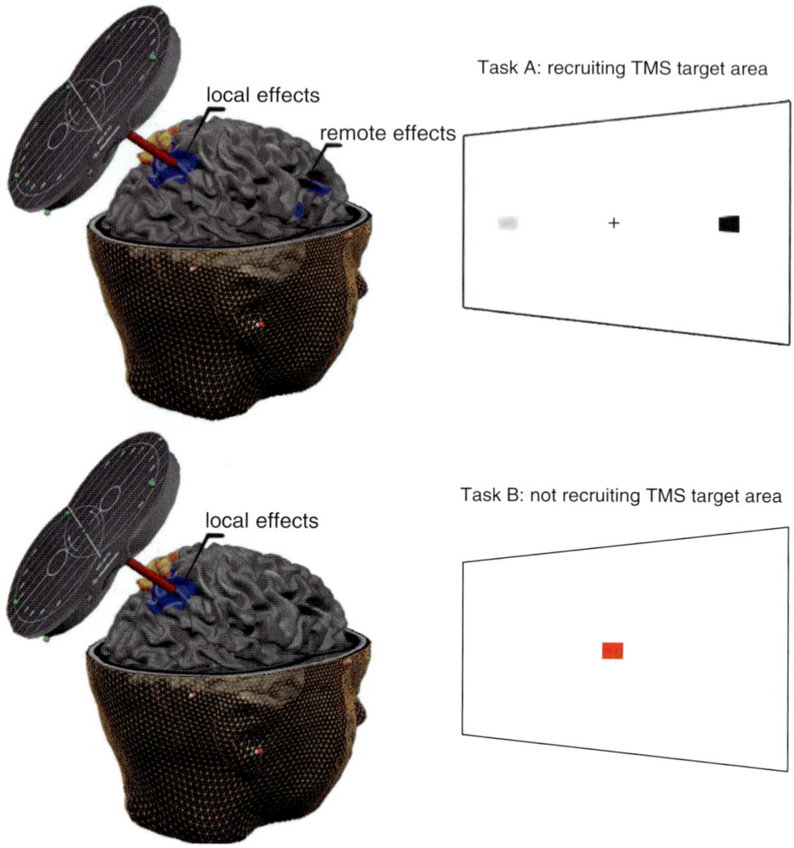

Fig. 8.3 Simultaneous fMRI and TMS during active behavior. This figure conceptualizes and generalizes the main findings of simultaneous TMS and fMRI during behaviorally controlled task execution. During execution of Task A (*upper panel*), a spatial visual detection task, fMRI reveals task-related bilateral neural activity within posterior parietal cortex. Yet, only right (but not left) parietal TMS induces functional deficits in Task A (reduced detection of left visual stimulus during bilateral stimulus presentation). These right parietal TMS-induced functional deficits in Task A are mirrored by task-specific neural activity changes in the brain (color coded in *blue*). These neural activity changes occur in the directly stimulated posterior parietal cortex and within functionally connected ipsilateral remote frontal brain areas. These remote frontal brain areas are also functionally relevant for successful execution of Task A. In contrast, the same brain stimulation protocol applied to the same cortical target site during execution of Task B (*lower panel*), a color discrimination task not recruiting the stimulated parietal brain area, does not result in functional deficits in Task B and also does not induce the specific right hemispheric frontoparietal network effects of TMS. This illustration thus depicts that (i) focal TMS applied to a particular brain region has both local and remote neural effects in the brain, (ii) these local and remote neural effects of TMS are state and task dependent, and (iii) the state-/task-dependent remote neural effects of TMS are functionally relevant for behavior. In this sense, simultaneous fMRI and TMS during active behavior may be a means of identifying effective brain connectivity networks of functional relevance or network accounts of behavior and cognition

potentially allowing causal brain-behavior inferences across the entire brain. These simultaneous, or more precisely interleaved, TMS/fMRI studies appear to converge on the following conclusions: (1) focal TMS applied to a particular brain region has both local and remote neural effects, (2) these local and remote neural effects of TMS are state and task dependent, and (3) these state- and task-dependent remote neural effects of TMS are functionally relevant for behavior.

These results seem to have troubling implications for the interpretation of purely behavioral TMS (without concurrent imaging) studies. After all, if TMS has been shown to have remote effects and these effects have been shown to be functionally relevant, what is left of the starting assumption that TMS has local effects and that these local effects underlie observed behavioral effects? Several alternative mechanisms underlying TMS-induced cognitive/behavioral impairments can now be suggested. For example, perhaps the remote effects of TMS are effectively responsible for the behavioral effects, rather than the local effects. Or, the network changes as a whole (i.e., local + remote effects) may be responsible for the behavioral effects. Alternatively, maybe the disruption of the connectivity itself between the local and remote regions caused the behavioral effects. All in all, these conclusions prompt us to move away from modular views of brain function and TMS disruption thereof, forcing us to consider a new conceptualization that involves functional interactions between remote, connected brain regions. Of course, a very positive consequence of this body of work is that TMS imaging can be used to investigate exactly these mechanisms to show how interactions within remote brain network nodes may support perception and cognition. But, does this mean that behavioral TMS studies without concurrent imaging are still useful as tools to reveal functional relevance of particular, stimulated, brain regions? Sack (2010) concludes, in brief: "Yes." While strictly speaking it is possible that the remote rather than local TMS-induced activity changes are responsible for behavioral effects, there is currently no conclusive evidence for this interpretation. Several alternative interpretations concerning the remote TMS-induced effects can be entertained, such as remote activity changes being the consequence of altered behavior, rather than the cause, or remote activity changes reflecting incidental covariations driven by different physiological processes. Basically, we are left with the question that we started out with, which is that we must somehow disentangle the neural activity changes that causally relate to the observed behavioral effects and those that do not. We have seen some examples of this above; it involves separate follow-up measurements to simultaneous TMS imaging, in which the remote regions affected by TMS during the simultaneous measurement are targeted to see if behavioral effects persist during stimulation of these regions also. Considering the necessity of such a follow-up, isolated behavioral, TMS study and simultaneous TMS imaging work should be regarded as truly complementary. It helps to refine the causal topography of structure-function relationships across the brain. New target areas for follow-up TMS studies can be identified and investigated. By systematically exploring in this manner the various network nodes of brain systems underlying perception, cognition, and behavior, revealed by simultaneous TMS-fMRI, these systems and interactions within and between them can be better understood. Simultaneous TMS imaging in this way substantially adds information and insight to purely behavioral TMS experiments, without taking away any of the original relevance of such work. In fact, this enterprise can only be enriched by work employing further complementary techniques in combination with brain stimulation, for instance, MR spectroscopy (Stagg et al. 2009), functional near-infrared spectroscopy (Hada et al. 2006; Kozel et al. 2009; Mochizuki et al. 2006), and diffusion-weighted imaging (DWI) of white matter bundles (Boorman et al. 2007; Kloppel et al. 2008).

To complete this viewpoint and support the system-level investigations outlined above, investigations at a more fine-grained level will likely be required. This is achieved most informatively through invasive animal research (e.g., see Funke and Benali 2010), helping us understand the neurophysiological mechanisms underlying the local and remote effects observed in human research. Work with cats (Allen et al. 2007; Aydin-Abidin et al. 2006; de Labra et al. 2007; Moliadze et al. 2005, 2003; Pasley et al. 2009; Valero-Cabre et al. 2007, 2005), rodents (Aydin-Abidin et al. 2008; Trippe et al. 2009), and monkeys (Ohnishi et al. 2004; Hayashi et al. 2004) has already delivered important contributions in this regard, although not yet into the remote effects of TMS. Also, considering the intrinsic intricacies of neural circuits, a multimodal approach with complementary methods (Logothetis 2008) will likely be required to achieve a cross-level understanding of TMS effects in the brain.

The role and potential of TMS in research and therapeutic settings has, thanks in part to the advances described here, not only been validated but actually increased over the years. With the multimodal research facilities now in place in several labs all over the world, the analysis on several levels from animal work to human whole-brain analysis to computational modeling, we are starting to improve our understanding of TMS-induced changes in brain and behavior. As such, TMS has begun to provide unique insights into the causal relations and interactions within and between system-level networks in the human brain, all in vivo and noninvasively. Ultimately, we remain confident that better understanding of the neural effects of TMS will lead to more informed clinical applications. Effective and well-controlled therapeutic interventions may thus become possible in the near future.

References

Ahdab R, Ayache SS, Brugieres P et al (2010) Comparison of "standard" and "navigated" procedures of tms coil positioning over motor, premotor and prefrontal targets in patients with chronic pain and depression. Neurophysiol Clin 40:27–36

Allen EA, Pasley BN, Duong T et al (2007) Transcranial magnetic stimulation elicits coupled neural and hemodynamic consequences. Science 317:1918–1921

Andoh J, Artiges E, Pallier C et al (2006) Modulation of language areas with functional mr image-guided magnetic stimulation. Neuroimage 29:619–627

Aydin-Abidin S, Moliadze V, Eysel UT et al (2006) Effects of repetitive tms on visually evoked potentials and EEG in the anaesthetized cat: dependence on stimulus frequency and train duration. J Physiol 574:443–455

Aydin-Abidin S, Trippe J, Funke K et al (2008) High- and low-frequency repetitive transcranial magnetic stimulation differentially activates c-fos and zif268 protein expression in the rat brain. Exp Brain Res 188:249–261

Baudewig J, Paulus W, Frahm J (2000) Artifacts caused by transcranial magnetic stimulation coils and EEG electrodes in t(2)*-weighted echo-planar imaging. Magn Reson Imaging 18:479–484

Baudewig J, Siebner HR, Bestmann S et al (2001) Functional mri of cortical activations induced by transcranial magnetic stimulation (tms). Neuroreport 12:3543–3548

Beckers G, Homberg V (1992) Cerebral visual motion blindness: transitory akinetopsia induced by transcranial magnetic stimulation of human area v5. Proc Biol Sci 249:173–178

Bestmann S, Baudewig J, Frahm J (2003a) On the synchronization of transcranial magnetic stimulation and functional echo-planar imaging. J Magn Reson Imaging 17:309–316

Bestmann S, Baudewig J, Siebner HR et al (2003b) Subthreshold high-frequency tms of human primary motor cortex modulates interconnected frontal motor areas as detected by interleaved fMRI-TMS. Neuroimage 20:1685–1696

Bestmann S, Baudewig J, Siebner HR et al (2004) Functional mri of the immediate impact of transcranial magnetic stimulation on cortical and subcortical motor circuits. Eur J Neurosci 19:1950–1962

Bestmann S, Baudewig J, Siebner HR et al (2005) Bold mri responses to repetitive tms over human dorsal premotor cortex. Neuroimage 28:22–29

Bestmann S, Swayne O, Blankenburg F et al (2008) Dorsal premotor cortex exerts state-dependent causal influences on activity in contralateral primary motor and dorsal premotor cortex. Cereb Cortex 18:1281–1291

Blankenburg F, Ruff CC, Bestmann S et al (2008) Interhemispheric effect of parietal tms on somatosensory response confirmed directly with concurrent tms-fMRI. J Neurosci 28:13202–13208

Bohning DE, Shastri A, McConnell KA et al (1999) A combined tms/fMRI study of intensity-dependent tms over motor cortex. Biol Psychiatry 45:385–394

Bohning DE, Shastri A, McGavin L et al (2000a) Motor cortex brain activity induced by 1-hz transcranial magnetic stimulation is similar in location and level to that for volitional movement. Invest Radiol 35:676–683

Bohning DE, Shastri A, Wassermann EM et al (2000b) Bold-fMRI response to single-pulse transcranial magnetic stimulation (tms). J Magn Reson Imaging 11:569–574

Boorman ED, O'Shea J, Sebastian C et al (2007) Individual differences in white-matter microstructure reflect variation in functional connectivity during choice. Curr Biol 17:1426–1431

Brighina F, Bisiach E, Oliveri M et al (2003) 1 hz repetitive transcranial magnetic stimulation of the unaffected hemisphere ameliorates contralesional visuospatial neglect in humans. Neurosci Lett 336:131–133

Burt T, Lisanby SH, Sackeim HA (2002) Neuropsychiatric applications of transcranial magnetic stimulation: a meta analysis. Int J Neuropsychopharmacol 5:73–103

Camprodon JA, Martinez-Raga J, Alonso-Alonso M et al (2007) One session of high frequency repetitive transcranial magnetic stimulation (RTMS) to the right prefrontal cortex transiently reduces cocaine craving. Drug Alcohol Depend 86:91–94

Cardenas-Morales L, Gron G, Kammer T (2011) Exploring the after-effects of theta burst magnetic stimulation on the human motor cortex: a functional imaging study. Hum Brain Mapp 32:1948–1960

Cazzoli D, Muri RM, Hess CW et al (2010) Treatment of hemispatial neglect by means of RTMS – a review. Restor Neurol Neurosci 28:499–510

Chibbaro G, Daniele M, Alagona G et al (2005) Repetitive transcranial magnetic stimulation in schizophrenic patients reporting auditory hallucinations. Neurosci Lett 383:54–57

d'Alfonso AA, van Honk J, Schutter DJ et al (2002) Spatial and temporal characteristics of visual motion perception involving v5 visual cortex. Neurol Res 24:266–270

de Graaf TA, Jacobs C, Roebroeck A et al (2009) FMRI effective connectivity and tms chronometry: complementary accounts of causality in the visuospatial judgment network. PLoS One 4:e8307

de Labra C, Rivadulla C, Grieve K et al (2007) Changes in visual responses in the feline dLGN: selective thalamic suppression induced by transcranial magnetic stimulation of v1. Cereb Cortex 17:1376–1385

De Ridder D, Vanneste S, Kovacs S et al (2011) Transcranial magnetic stimulation and extradural electrodes implanted on secondary auditory cortex for tinnitus suppression. J Neurosurg 114:903–911

Denslow S, Lomarev M, George MS et al (2005) Cortical and subcortical brain effects of transcranial magnetic stimulation (tms)-induced movement: an interleaved tms/functional magnetic resonance imaging study. Biol Psychiatry 57:752–760

Di Lazzaro V (2008) The physiological basis of the effects of intermittent theta burst stimulation of the human motor cortex. J Physiol 586:3871–3871

Eichhammer P, Johann M, Kharraz A et al (2003) High-frequency repetitive transcranial magnetic stimulation decreases cigarette smoking. J Clin Psychiatry 64:951–953

Esser SK, Hill SL, Tononi G (2005) Modeling the effects of transcranial magnetic stimulation on cortical circuits. J Neurophysiol 94:622–639

Feredoes E, Tononi G, Postle BR (2007) The neural bases of the short-term storage of verbal information are anatomically variable across individuals. J Neurosci 27:11003–11008

Fitzgerald PB, Sritharan A, Daskalakis ZJ et al (2007) A functional magnetic resonance imaging study of the effects of low frequency right prefrontal transcranial magnetic stimulation in depression. J Clin Psychopharmacol 27:488–492

Funke K, Benali A (2010) Cortical cellular actions of transcranial magnetic stimulation. Restor Neurol Neurosci 28:399–417

George MS, Wassermann EM, Williams WA et al (1995) Daily repetitive transcranial magnetic stimulation (RTMS) improves mood in depression. Neuroreport 6:1853–1856

Gross M, Nakamura L, Pascual-Leone A et al (2007) Has repetitive transcranial magnetic stimulation (RTMS) treatment for depression improved? A systematic review and meta-analysis comparing the recent vs. the earlier RTMS studies. Acta Psychiatr Scand 116:165–173

Hada Y, Abo M, Kaminaga T et al (2006) Detection of cerebral blood flow changes during repetitive transcranial magnetic stimulation by recording hemoglobin in the brain cortex, just beneath the stimulation coil, with near-infrared spectroscopy. Neuroimage 32:1226–1230

Hanakawa T, Mima T, Matsumoto R et al (2009) Stimulus-response profile during single-pulse transcranial magnetic stimulation to the primary motor cortex. Cereb Cortex 19:2605–2615

Haraldsson HM, Ferrarelli F, Kalin NH et al (2004) Transcranial magnetic stimulation in the investigation and treatment of schizophrenia: a review. Schizophr Res 71:1–16

Hayashi T, Ohnishi T, Okabe S et al (2004) Long-term effect of motor cortical repetitive transcranial magnetic stimulation [correction]. Ann Neurol 56:77–85

Herbsman T, Avery D, Ramsey D et al (2009) More lateral and anterior prefrontal coil location is associated with better repetitive transcranial magnetic stimulation antidepressant response. Biol Psychiatry 66:509–515

Herwig U, Fallgatter AJ, Hoppner J et al (2007) Antidepressant effects of augmentative transcranial magnetic stimulation: randomised multicentre trial. Br J Psychiatry 191:441–448

Hoffman RE (2003) Variations on the chemical shift of tms. J Magn Reson 163:325–331

Hoffman RE, Becker ED (2005) Temperature dependence of the 1h chemical shift of tetramethylsilane in chloroform, methanol, and dimethylsulfoxide. J Magn Reson 176:87–98

Holtzheimer PE 3rd, Russo J, Avery DH (2001) A meta-analysis of repetitive transcranial magnetic stimulation in the treatment of depression. Psychopharmacol Bull 35:149–169

Huang Y-Z, Edwards MJ, Rounis E et al (2005) Theta burst stimulation of the human motor cortex. Neuron 45:201–206

Hubl D, Nyffeler T, Wurtz P et al (2008) Time course of blood oxygenation level-dependent signal response after theta burst transcranial magnetic stimulation of the frontal eye field. Neuroscience 151:921–928

Kemna LJ, Gembris D (2003) Repetitive transcranial magnetic stimulation induces different responses in different cortical areas: a functional magnetic resonance study in humans. Neurosci Lett 336:85–88

Khedr EM, Kotb H, Kamel NF et al (2005) Longlasting antalgic effects of daily sessions of repetitive transcranial magnetic stimulation in central and peripheral neuropathic pain. J Neurol Neurosurg Psychiatry 76:833–838

Kimbrell TA, Little JT, Dunn RT et al (1999) Frequency dependence of antidepressant response to left prefrontal repetitive transcranial magnetic stimulation (RTMS) as a function of baseline cerebral glucose metabolism. Biol Psychiatry 46:1603–1613

Kimbrell TA, Ketter TA, George MS et al (2002) Regional cerebral glucose utilization in patients with a range of severities of unipolar depression. Biol Psychiatry 51:237–252

Kloppel S, Baumer T, Kroeger J et al (2008) The cortical motor threshold reflects microstructural properties of cerebral white matter. Neuroimage 40:1782–1791

Koch G, Oliveri M, Cheeran B et al (2008) Hyperexcitability of parietal-motor functional connections in the intact left-hemisphere of patients with neglect. Brain 131:3147–3155

Kozel FA, George MS (2002) Meta-analysis of left prefrontal repetitive transcranial magnetic stimulation (RTMS) to treat depression. J Psychiatr Pract 8:270–275

Kozel FA, Tian F, Dhamne S et al (2009) Using simultaneous repetitive transcranial magnetic stimulation/functional near infrared spectroscopy (rTMS/fNIRS) to measure brain activation and connectivity. Neuroimage 47:1177–1184

Lee SH, Kim W, Chung YC et al (2005) A double blind study showing that two weeks of daily repetitive tms over the left or right temporoparietal cortex reduces symptoms in patients with schizophrenia who are having treatment-refractory auditory hallucinations. Neurosci Lett 376:177–181

Lefaucheur JP, Drouot X, Nguyen JP (2001) Interventional neurophysiology for pain control: duration of pain relief following repetitive transcranial magnetic stimulation of the motor cortex. Neurophysiol Clin 31:247–252

Logothetis NK (2008) What we can do and what we cannot do with fMRI. Nature 453:869–878

Logothetis NK, Augath M, Murayama Y et al (2010) The effects of electrical microstimulation on cortical signal propagation. Nat Neurosci 13:1283–1291

Martin JL, Barbanoj MJ, Perez V et al (2003) Transcranial magnetic stimulation for the treatment of obsessive-compulsive disorder. Cochrane Database Syst Rev 3:CD003387

McNamara B, Ray JL, Arthurs OJ et al (2001) Transcranial magnetic stimulation for depression and other psychiatric disorders. Psychol Med 31:1141–1146

Mochizuki H, Ugawa Y, Terao Y et al (2006) Cortical hemoglobin-concentration changes under the coil induced by single-pulse tms in humans: a simultaneous recording with near-infrared spectroscopy. Exp Brain Res 169:302–310

Moliadze V, Zhao Y, Eysel U et al (2003) Effect of transcranial magnetic stimulation on single-unit activity in the cat primary visual cortex. J Physiol 553:665–679

Moliadze V, Giannikopoulos D, Eysel UT et al (2005) Paired-pulse transcranial magnetic stimulation protocol applied to visual cortex of anaesthetized cat: effects on visually evoked single-unit activity. J Physiol 566:955–965

Nyffeler T, Cazzoli D, Hess CW et al (2009) One session of repeated parietal theta burst stimulation trains induces long-lasting improvement of visual neglect. Stroke 40:2791–2796

Ohnishi T, Hayashi T, Okabe S et al (2004) Endogenous dopamine release induced by repetitive transcranial magnetic stimulation over the primary motor cortex: an [11c]raclopride positron emission tomography study in anesthetized macaque monkeys. Biol Psychiatry 55:484–489

Oliveri M, Rossini PM, Traversa R et al (1999) Left frontal transcranial magnetic stimulation reduces contralesional extinction in patients with unilateral right brain damage. Brain 122(Pt 9):1731–1739

Oliveri M, Caltagirone C, Filippi MM et al (2000a) Paired transcranial magnetic stimulation protocols reveal a pattern of inhibition and facilitation in the human parietal cortex. J Physiol 529(Pt 2):461–468

Oliveri M, Rossini PM, Filippi MM et al (2000b) Time-dependent activation of parieto-frontal networks for directing attention to tactile space. A study with paired transcranial magnetic stimulation pulses in right-brain-damaged patients with extinction. Brain 123(Pt 9):1939–1947

Oliveri M, Bisiach E, Brighina F et al (2001) RTMS of the unaffected hemisphere transiently reduces contralesional visuospatial hemineglect. Neurology 57:1338–1340

O'Reardon JP, Solvason HB, Janicak PG et al (2007) Efficacy and safety of transcranial magnetic stimulation in the acute treatment of major depression: a multisite randomized controlled trial. Biol Psychiatry 62:1208–1216

O'Shea J, Sebastian C, Boorman ED et al (2007) Functional specificity of human premotor-motor cortical interactions during action selection. Eur J Neurosci 26:2085–2095

Pascual-Leone A, Rubio B, Pallardo F et al (1996) Rapid-rate transcranial magnetic stimulation of left dorsolateral prefrontal cortex in drug-resistant depression. Lancet 348:233–237

Pasley BN, Allen EA, Freeman RD (2009) State-dependent variability of neuronal responses to transcranial magnetic stimulation of the visual cortex. Neuron 62:291–303

Paus T, Barrett J (2004) Transcranial magnetic stimulation (tms) of the human frontal cortex: implications for repetitive tms treatment of depression. J Psychiatry Neurosci 29:268–279

Pogarell O, Koch W, Popperl G et al (2006) Striatal dopamine release after prefrontal repetitive transcranial magnetic stimulation in major depression: preliminary results of a dynamic [123i] ibzm spect study. J Psychiatr Res 40:307–314

Pogarell O, Koch W, Popperl G et al (2007) Acute prefrontal rTMS increases striatal dopamine to a similar degree as d-amphetamine. Psychiatry Res 156:251–255

Reithler J, Peters JC, Sack AT (2011) Multimodal transcranial magnetic stimulation: using concurrent neuroimaging to reveal the neural network dynamics of noninvasive brain stimulation. Prog Neurobiol 94:149–165

Ridding MC, Rothwell JC (2007) Is there a future for therapeutic use of transcranial magnetic stimulation? Nat Rev Neurosci 8:559–567

Rossi S, Hallett M, Rossini PM et al (2009) Safety, ethical considerations, and application guidelines for the use of transcranial magnetic stimulation in clinical practice and research. Clin Neurophysiol 120:2008–2039

Ruff CC, Blankenburg F, Bjoertomt O et al (2006) Concurrent tms-fMRI and psychophysics reveal frontal influences on human retinotopic visual cortex. Curr Biol 16:1479–1488

Ruff CC, Bestmann S, Blankenburg F et al (2008) Distinct causal influences of parietal versus frontal areas on human visual cortex: evidence from concurrent tms-fMRI. Cereb Cortex 18:817–827

Ruff CC, Driver J, Bestmann S (2009) Combining tms and fMRI: from 'virtual lesions' to functional-network accounts of cognition. Cortex 45:1043–1049

Rushworth MF, Hadland KA, Paus T et al (2002) Role of the human medial frontal cortex in task switching: a combined fMRI and TMS study. J Neurophysiol 87:2577–2592

Sachdev PS, McBride R, Loo CK et al (2001) Right versus left prefrontal transcranial magnetic stimulation for obsessive-compulsive disorder: a preliminary investigation. J Clin Psychiatry 62:981–984

Sack AT (2006) Transcranial magnetic stimulation, causal structure-function mapping and networks of functional relevance. Curr Opin Neurobiol 16:593–599

Sack AT (2010) Does tms need functional imaging? Cortex 46:131–133

Sack AT, Linden DE (2003) Combining transcranial magnetic stimulation and functional imaging in cognitive brain research: possibilities and limitations. Brain Res Cogn Brain Res 43:41–56

Sack AT, Kohler A, Linden DE et al (2006) The temporal characteristics of motion processing in hmt/v5+: combining fMRI and neuronavigated tms. Neuroimage 29:1326–1335

Sack AT, Kohler A, Bestmann S et al (2007) Imaging the brain activity changes underlying impaired visuospatial judgments: simultaneous fMRI, tms, and behavioral studies. Cereb Cortex 17:2841–2852

Sack AT, Cohen Kadosh R, Schuhmann T et al (2009) Optimizing functional accuracy of tms in cognitive studies: a comparison of methods. J Cogn Neurosci 21:207–221

Schonfeldt-Lecuona C, Cardenas-Morales L, Freudenmann RW et al (2010) Transcranial magnetic stimulation in depression–lessons from the multicentre trials. Restor Neurol Neurosci 28:569–576

Shastri A, George MS, Bohning DE (1999) Performance of a system for interleaving transcranial magnetic stimulation with steady-state magnetic resonance imaging. Electroencephalogr Clin Neurophysiol Suppl 51:55–64

Shindo K, Sugiyama K, Huabao L et al (2006) Long-term effect of low-frequency repetitive transcranial magnetic stimulation over the unaffected posterior parietal cortex in patients with unilateral spatial neglect. J Rehabil Med 38:65–67

Song W, Du B, Xu Q et al (2009) Low-frequency transcranial magnetic stimulation for visual spatial neglect: a pilot study. J Rehabil Med 41:162–165

Sparing R, Buelte D, Meister IG et al (2008) Transcranial magnetic stimulation and the challenge of coil placement: a comparison of conventional and stereotaxic neuronavigational strategies. Hum Brain Mapp 29:82–96

Speer AM, Kimbrell TA, Wassermann EM et al (2000) Opposite effects of high and low frequency rTMS on regional brain activity in depressed patients. Biol Psychiatry 48:1133–1141

Speer AM, Benson BE, Kimbrell TK et al (2009) Opposite effects of high and low frequency rTMS on mood in depressed patients: relationship to baseline cerebral activity on pet. J Affect Disord 115:386–394

Stagg CJ, Wylezinska M, Matthews PM et al (2009) Neurochemical effects of theta burst stimulation as assessed by magnetic resonance spectroscopy. J Neurophysiol 101:2872–2877

Teneback CC, Nahas Z, Speer AM et al (1999) Changes in prefrontal cortex and paralimbic activity in depression following two weeks of daily left prefrontal tms. J Neuropsychiatry Clin Neurosci 11:426–435

Thiel A, Haupt WF, Habedank B et al (2005) Neuroimaging-guided rTMS of the left inferior frontal gyrus interferes with repetition priming. Neuroimage 25:815–823

Trippe J, Mix A, Aydin-Abidin S et al (2009) Theta burst and conventional low-frequency rTMS differentially affect gabaergic neurotransmission in the rat cortex. Exp Brain Res 199:411–421

Valero-Cabre A, Payne BR, Rushmore J et al (2005) Impact of repetitive transcranial magnetic stimulation of the parietal cortex on metabolic brain activity: a 14c-2dg tracing study in the cat. Exp Brain Res 163:1–12

Valero-Cabre A, Payne BR, Pascual-Leone A (2007) Opposite impact on 14c-2-deoxyglucose brain metabolism following patterns of high and low frequency repetitive transcranial magnetic stimulation in the posterior parietal cortex. Exp Brain Res 176:603–615

Walsh V, Pascual-Leone A (2003) Transcranial magnetic stimulation: a neurochronometrics of mind. MIT Press, Cambridge, MA

Weiskopf N, Josephs O, Ruff CC et al (2009) Image artifacts in concurrent transcranial magnetic stimulation (tms) and fMRI caused by leakage currents: modeling and compensation. J Magn Reson Imaging 29:1211–1217

Part II

Systems

9 Imaging Perception

Assaf Harel and Chris I. Baker

Abbreviations

BA	Brodmann area
BOLD	Blood oxygenation dependent
EPI	Echoplanar imaging
FFA	Fusiform face area
LGN	Lateral geniculate nucleus
MVPA	Multi-voxel pattern analysis
OTC	Occipitotemporal cortex
PPA	Parahippocampal place area
SVM	Support vector machines
V1	Primary visual cortex
V2	Prestriate cortex
V3	Extrastriate visual area V3
V4	Extrastriate visual area V4
V5/MT	Extrastriate visual area V5/MT (middle temporal)

9.1 Introduction

The question of how we translate physical signals from the environment (e.g., light, sound waves) into an internal description of the external world that facilitates adaptive behavior has occupied the minds of many great thinkers (Boring 1942). A major approach to the study of perception focuses on understanding its underlying neurophysiology. Specifically, the goal is to elucidate how physical energy is transduced into an electrical signal at the peripheral receptors and transmitted to the central nervous system where it is processed and integrated with other concurrent signals to form a subjectively real percept. The past two decades have seen a great advance in the neurosciences, particularly with the advent of advanced noninvasive brain imaging techniques (Bandettini 2009), key among which is magnetic resonance imaging (MRI). In this chapter, we will describe how MRI and in particular functional MRI (fMRI) can be used to map and elucidate sensory and perceptual processes in the different modalities and shed light on the rules governing perception and brain organization.

This chapter is not meant to be a comprehensive review of the entirety of fMRI studies of perception. Rather, we will describe and discuss several themes and issues underlying the neural substrates of perception as revealed by fMRI. While we will predominately focus on vision, being the dominant sensory modality in human and nonhuman primates (Felleman and Van Essen 1991), we will also provide key examples from other sensory modalities (audition, somatosensation, olfaction, gustation). Indeed, a recurrent theme throughout this chapter will be whether we can apply general principles learned from one sensory modality to the others. Further, while focusing on human perception, we will also consider nonhuman primate studies. Given that some

A. Harel • C.I. Baker, PhD (✉)
Laboratory of Brain and Cognition,
National Institute of Mental Health,
National Institutes of Health,
10 Center Drive, Room 3 N228,
Bethesda, MD 20892, USA
e-mail: assaf.harel@nih.gov;
bakerchris@mail.nih.gov

C. Mulert, M.E. Shenton (eds.), *MRI in Psychiatry*,
DOI 10.1007/978-3-642-54542-9_9, © Springer Berlin Heidelberg 2014

fundamental perceptual mechanisms are likely shared across species, combining knowledge from physiological and anatomical nonhuman primate studies with neuroimaging (of both human and nonhuman primates) provides an important tool for understanding perceptual processing. For example, the study of a seemingly unique human ability such as speech perception has benefited extensively from research in nonhuman primates (for review, see Rauschecker and Scott 2009).

We start by describing the major pathways for perceptual processing from the peripheral sensory surfaces to high-level cortex, focusing on how neural representations develop along these pathways and the general principles observed (e.g., parallel and hierarchical processing). We then highlight the insights fMRI has provided into the functional organization of the cortex, from the topographic maps observed in primary sensory regions to the selectivity for complex stimuli (e.g., faces) found in higher-level areas. Perception is an "active," constructive process, and accordingly we will discuss the importance of both bottom-up (stimulus-driven) and top-down (internally generated) signals highlighting that the line between "perception" and "cognition" is often not clear. Finally, we will elaborate on the role of learning and plasticity in shaping perceptual representations both throughout development and during adulthood.

9.2 Pathways of Perceptual Processing

9.2.1 Sensory Surfaces

To understand perceptual processing in the cortex, it is important to understand the form in which information from the environment first enters the nervous system. The initial stage of processing occurs at the sensory surface, an array of specialized receptors that transduce the external stimuli into an electrical signal propagated to the central nervous system (Adrian 1950). There are five major sensory surfaces: the retina (vision), cochlea (audition), dermis (somatosensation), epithelium (olfaction), and tongue (gustation). One of the challenges of using MRI to study perception is delivering carefully controlled sensory stimuli to these surfaces inside the confined space of the scanner (see Box 9.1).

A key feature of many sensory surfaces that impacts all subsequent stages of processing is the systematic organization of the receptors, which preserves some structure of the stimulus. For example, in vision, light from adjacent locations stimulates adjacent photoreceptors on the retina forming a retinotopic map of visual space. Similarly, in the cochlea, neighboring frequencies stimulate adjacent hair cells forming a frequency map of auditory stimuli. Thus, the spatial organization of the peripheral receptor arrays often captures specific stimulus dimensions, and these dimensions are carried forward and influence subsequent processing stages even at high levels of processing in the cortex, forming topographic maps (Kaas 1997).

Box 9.1. Studying Perception in the Scanner

There are a number of challenges to studying the neural substrates of perception with fMRI. The initial challenge is simply stimulus presentation inside the scanner. In vision, this may seem trivial – projecting the stimuli onto a screen that the subject views through a mirror on top of the head coil. However, if one considers other modalities – for example, audition – the challenge becomes clearer. The echoplanar imaging (EPI) sequence that is typically used in fMRI produces loud (~100 dB) bursts of noise that not only interfere with the fidelity of acoustic stimuli but also excite auditory cortex (Bandettini et al. 1998; Seifritz et al. 2006). To avoid this problem, auditory stimuli are often presented at high intensity. This, however, creates a new problem since at high intensity levels neurons respond to a broad range of frequencies (Recanzone et al. 2000), making it difficult to assess frequency selectivity (Tanji et al. 2010). Several solutions have been proposed over the years to tackle these complexities, such as active noise cancellation headphones, and specialized acquisition sequences, including "silent" acquisition protocols that include silent periods during which the stimulus can be presented (Gonzalez-Castillo et al. 2011).

In the somatosensory domain, early studies involved the experimenter manually stimulating body parts (Polonara et al. 1999). The difficulty here is of standardization; it is hard to control the parameters of stimulation, such as force and speed, enabling reproducibility of stimulation. Further, access to many body parts is limited by the restricted space inside the bore of the scanner. Finally, stimulating body parts may spontaneously trigger motor movements, confounding the imaging data. To overcome these difficulties, current somatosensory research commonly uses MRI-compatible machines that allow the stimulation of multiple body parts while delivering standardized computer-controlled stimuli such as air puffs (Huang and Sereno 2007), vibrations (Briggs et al. 2004), electrical stimulation (Nihashi et al. 2005), and punctate stimulation using monofilaments (Dresel et al. 2008). Systems for presenting complex high-level tactile stimuli involving form discrimination, such as texture surfaces (Ingeholm et al. 2006) and 3D objects (Culham et al. 2003), have also been developed.

For gustation, one major obstacle is that consumption of food involves jaw movements and swallowing, which can produce head motion artifacts. To minimize these effects and to control stimulus timing, gustatory stimuli are often delivered as fluids through tubes that are fixed in place in the participants' mouth and connected to a computer-controlled pump system that administers different chemical solutions at different rates and intensities, with inter-stimulus rinsing solutions (Frank et al. 2003; Haase et al. 2007; Kami et al. 2008). Similar systems are used to deliver olfactory stimuli using odorants rather than solutions, although sometimes solutions are also used to assess retronasal olfaction (Lorig et al. 1999; Marciani et al. 2006; Johnson and Sobel 2007).

While fMRI can provide critical insights into perception, it is important to consider the constraints of the scanner environment, which limit the types of processes that can be studied easily.

9.2.2 Subcortical Structures

The sensory surfaces project directly to a number of subcortical structures, including the thalamus, superior colliculus, and inferior colliculus. For all senses except olfaction, the major pathway includes specific thalamic nuclei (e.g., vision, lateral geniculate nucleus and inferior pulvinar; audition, medial geniculate nucleus; somatosensation, ventral posterior nucleus), each separately projecting to distinct cortical sensory regions. The olfactory epithelium instead projects to the olfactory bulb, which in turn projects to cortical regions in the frontal and medial temporal lobes (Gottfried 2010).

Until recently, fMRI studies of the neural substrates of perception were almost entirely focused on cortical structures. The reasons for this are twofold. First, it is often assumed that the cortex is the seat of high-level processing, whereas subcortical structures, which are evolutionarily older, serve primarily a relay or regulatory function (Collins et al. 2013). Second, the deep location of subcortical structures, their relatively small size, and susceptibility to cardiac pulsatile motion artifacts pose severe technical difficulties that prevented successful functional imaging of these regions. However, with the advent of high field strength MRI and sophisticated denoising techniques (e.g., Wall et al. 2009), the potential for investigating the role of subcortical structures in perception is increasing. In vision, studies of both the lateral geniculate nucleus (LGN) (Chen et al. 1999; Schneider et al. 2004) and the superior colliculus (DuBois and Cohen 2000; Schneider and Kastner 2005) have revealed underlying retinotopic maps, consistent with studies in nonhuman primates (Cynader and Berman 1972; Sherman and Guillery 2001), which reflect ordered projections from the retina. Further, eye-specific biases can

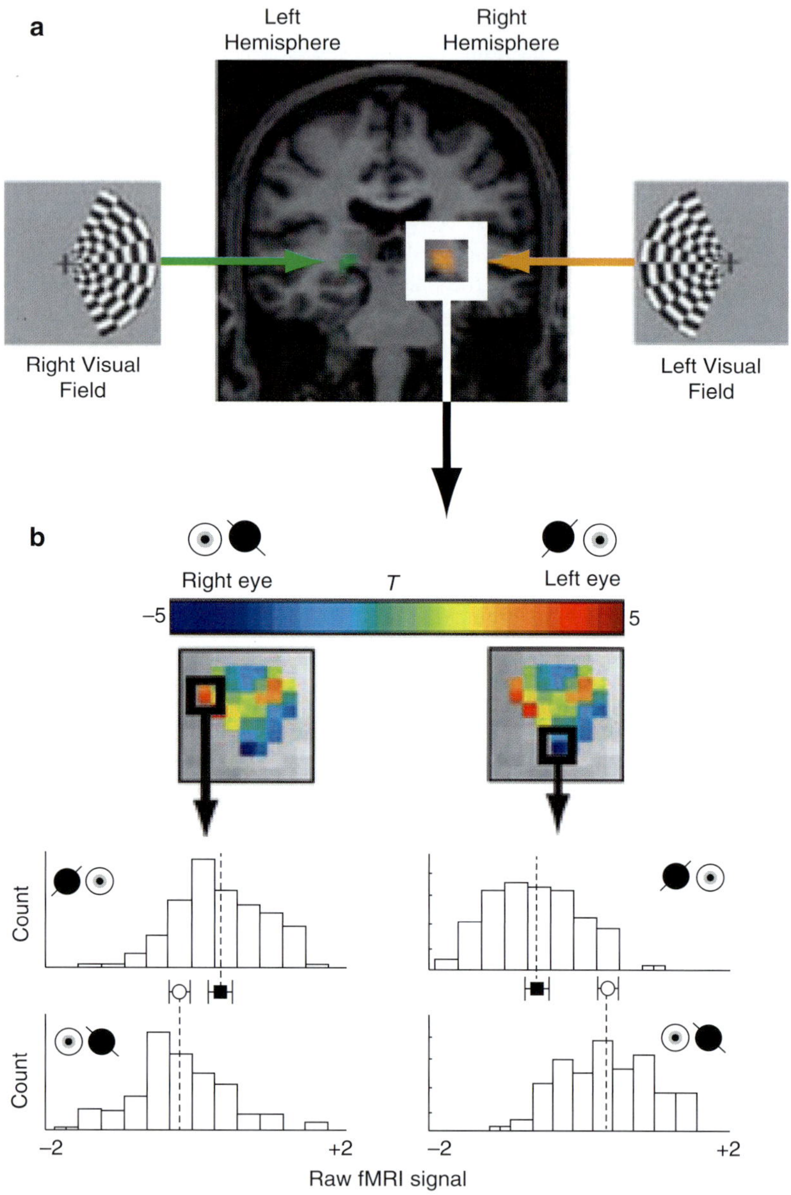

Fig. 9.1 fMRI of the thalamic lateral geniculate nucleus. (**a**) Localization of LGN. The left and right LGN respond to stimuli in the contralateral visual field and were localized by comparing the response to flickering checkerboards presented in either the left or right field. (**b**) Eye-biased signals in voxels of a single participant's right LGN (*white box* in (**a**)). The LGN is a six-layered structure with each layer responding to stimuli from either the left or right eye only. Although the individual layers are not easily resolved with fMRI, individual voxels showed a stronger response to stimulation from the right or left eye (*color scale* shows *t*-values for the direct comparison). The *histograms* show the raw fMRI signal distribution for two sample voxels with stimulation of each eye separately, one voxel with a left eye bias and one with a right eye bias (Adapted with permission from Haynes et al. (2005)

be measured in the LGN reflecting the segregated input into separate layers (Fig. 9.1, Haynes et al. 2005). Finally, it is worth noting that fMRI studies of subcortical structures are challenging the notion that they are simply passive relays (Saalmann and Kastner 2011, see Sect. 9.4.1 below).

9.2.3 Primary Sensory Cortices

The cortical areas receiving the major input from the sensory surfaces, predominantly through the thalamus, are often referred to as the primary sensory cortices (Fig. 9.2). We first describe the location and major characteristics of these cortices before discussing in detail their output pathways (Sect. 9.2.4) and functional properties (Sect. 9.3).

The primary sensory cortices can be distinguished based on cytoarchitectonic (e.g., dense cellular structure and myelination in layer IV reflecting strong input), connectivity (e.g., direct projections from thalamic nuclei), and functional criteria (see Sect. 9.2.4 below). The primary visual cortex (V1), often referred to as the

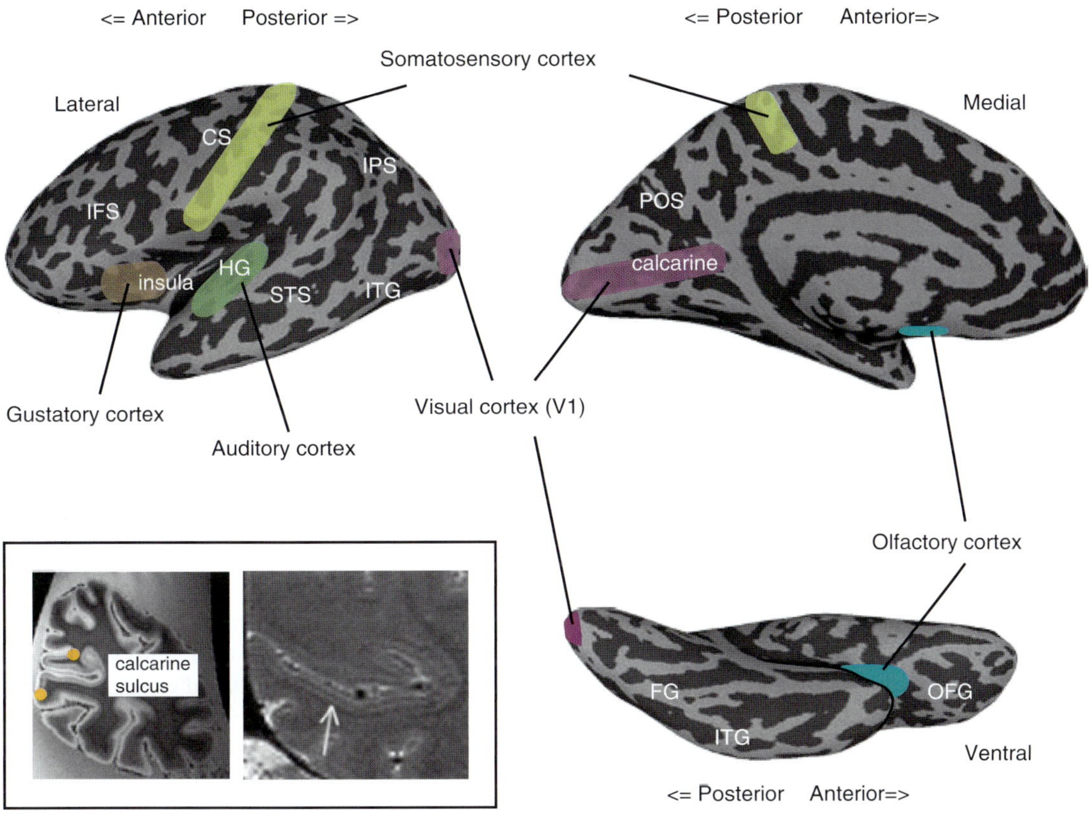

Fig. 9.2 Primary sensory cortices. Inflated views of the cortex showing the location of the primary sensory cortices. The *inset* shows post-mortem (*left*) and in vivo (*right*) images of the calcarine sulcus showing the stria of Gennari – the band of dense myelination corresponding to layer IV in primary visual cortex. The *orange circles* denote the V1/V2 boundary estimated from the location of the stria (Adapted with permission from Hinds et al. (2008) and Trampel et al. (2011)), respectively. *CS* central sulcus, *FG* fusiform gyrus, *HG* Heschl's gyrus, *IFS* inferior frontal sulcus, *IPS* intraparietal sulcus, *ITG* inferior temporal gyrus, *OFG* orbitofrontal gyrus, *POS* parieto-occipital sulcus, *STS* superior temporal sulcus

Brodmann area (BA) 17, is located within the banks of the calcarine sulcus on the medial wall of each hemisphere. Using MRI, V1 can be identified in vivo both by detecting the strong layer IV myelination (stria of Gennari, Turner et al. 2008; Sanchez-Panchuelo et al. 2012) and by using the cortical folding patterns (Hinds et al. 2008). The primary auditory cortex is located on Heschl's gyrus, close to the lateral sulcus (BA 41), and appears to follow very closely the shape of Heschl's gyrus, which has high morphological variability (Da Costa et al. 2011). The somatosensory cortices are in the anterior parietal lobe comprising four cytoarchitectonic areas (BA 1, 2, 3a, 3b) parallel to the central sulcus. Collectively, these areas are often referred to as the primary somatosensory cortex or S1, with BA3a and 3b contained within the posterior bank of the central sulcus, immediately posterior to the primary motor cortex, and BA 1 and 2 extending onto the postcentral gyrus (Keysers et al. 2010). While these areas are often considered collectively, somatosensory activation can be separately localized to them using fMRI (Moore et al. 2000). The determination of the primary gustatory cortex is complicated by the presence of two gustatory-related projections from the thalamus in nonhuman primates, but in humans it is typically localized to the middorsal insula and overlying operculum (for a competing view, see Ogawa et al. 2005; Small 2010; Veldhuizen et al. 2011b). Similarly, in the absence of direct thalamic projections, the precise location of the primary olfactory cortex is also debated. However, the area receiving most projections from the epithelium is the posterior piriform cortex, and the term primary olfactory cortex is often applied collectively to the piriform cortex and associated structures including the olfactory nucleus and rostral entorhinal cortex (Gottfried 2010).

Together, these primary sensory cortices provide the major source of sensory input for the rest of the cortex – they are the starting point of all modality-specific cortical circuits, each with their own unique pathways (Mesulam 1998).

9.2.4 Parallel and Hierarchical Processing

One of the prominent features of primate sensory systems is parallel processing in which independent circuits or pathways extract distinct types of information simultaneously. The separate sensory systems are themselves an example of parallel processing, but parallel processing is also common within individual sensory systems (somatosensation, Johnson and Hsiao 1992; audition, Kaas and Hackett 2000; olfaction, Savic et al. 2000; vision, Nassi and Callaway 2009). This sometimes even is evident early in the processing stream (e.g., multiple populations of retinal ganglion cells each processing input from the same part of the visual field) and extends into the cortex.

In vision, cortical processing beyond V1 thought to proceed along two major pathways (Ungerleider and Mishkin 1982; Milner and Goodale 2006; Kravitz et al. 2011b): a dorsal pathway (often termed "where" or "how" pathway) projecting into the parietal cortex, primarily involved in processing spatial information, and a ventral pathway (often termed "what" pathway) projecting into the inferior temporal cortex, primarily involved in processing stimulus quality or object information (Fig. 9.3a). These two pathways are readily observed with fMRI (Fig. 9.3b). A similar division of is also well established for auditory processing with a ventral auditory pathway projecting into the anterior temporal cortex involved in processing spectrally complex sounds (e.g., speech, Rauschecker and Scott 2009) and a dorsal auditory pathway projecting into the posterior parietal cortex involved in processing spatial and motion information as well as sensorimotor integration (Rauschecker 2011) (Fig. 9.3a). Finally, in the somatosensory system, there is also some evidence for a separation of spatial processing from stimulus quality processing in both human fMRI (Reed et al. 2005; Van Boven et al. 2005) and nonhuman primate studies (e.g., Murray and Mishkin 1984), but the pathways have not been well characterized (for review, see Dijkerman and de Haan 2007).

A second prominent feature of cortical sensory systems is their hierarchical nature. Responses in the primary sensory cortices represent simple modality-specific features, while regions further removed from direct sensory input represent complex stimulus features. For example, in audition, human fMRI studies show that the primary auditory cortex responds strongly to simple tones (Wessinger et al. 2001). Selective activation for speech sounds and words, however, is found further along the ventral auditory pathway (Binder et al. 2000; Chevillet et al. 2011; for review and meta-analysis, see DeWitt and Rauschecker 2012). Similarly in vision, V1 can be driven by simple visual features such as oriented lines or contrast, but more complex stimuli (such as faces and objects) are required to drive higher-level visual regions along the ventral pathway (Grill-Spector and Malach 2004). Further, the impact of image scrambling (cutting an image into smaller components) on the magnitude of the observed response increases along the ventral pathway suggesting a growing reliance on global rather than local form (Lerner et al. 2001).

One of the challenges of perception is to enable identification of the same entity despite vast changes in the physical information that may be impinging on the sensory surfaces.

For example, in audition, the spectral frequency of a spoken word may vary substantially depending on the gender, accent, and size of the speaker, yet can still be recognized as the same word even though the early levels of auditory processing are extremely sensitive to changes in physical properties such as frequency (DeWitt and Rauschecker 2012). Similarly, in vision, an object appearing at different locations in the

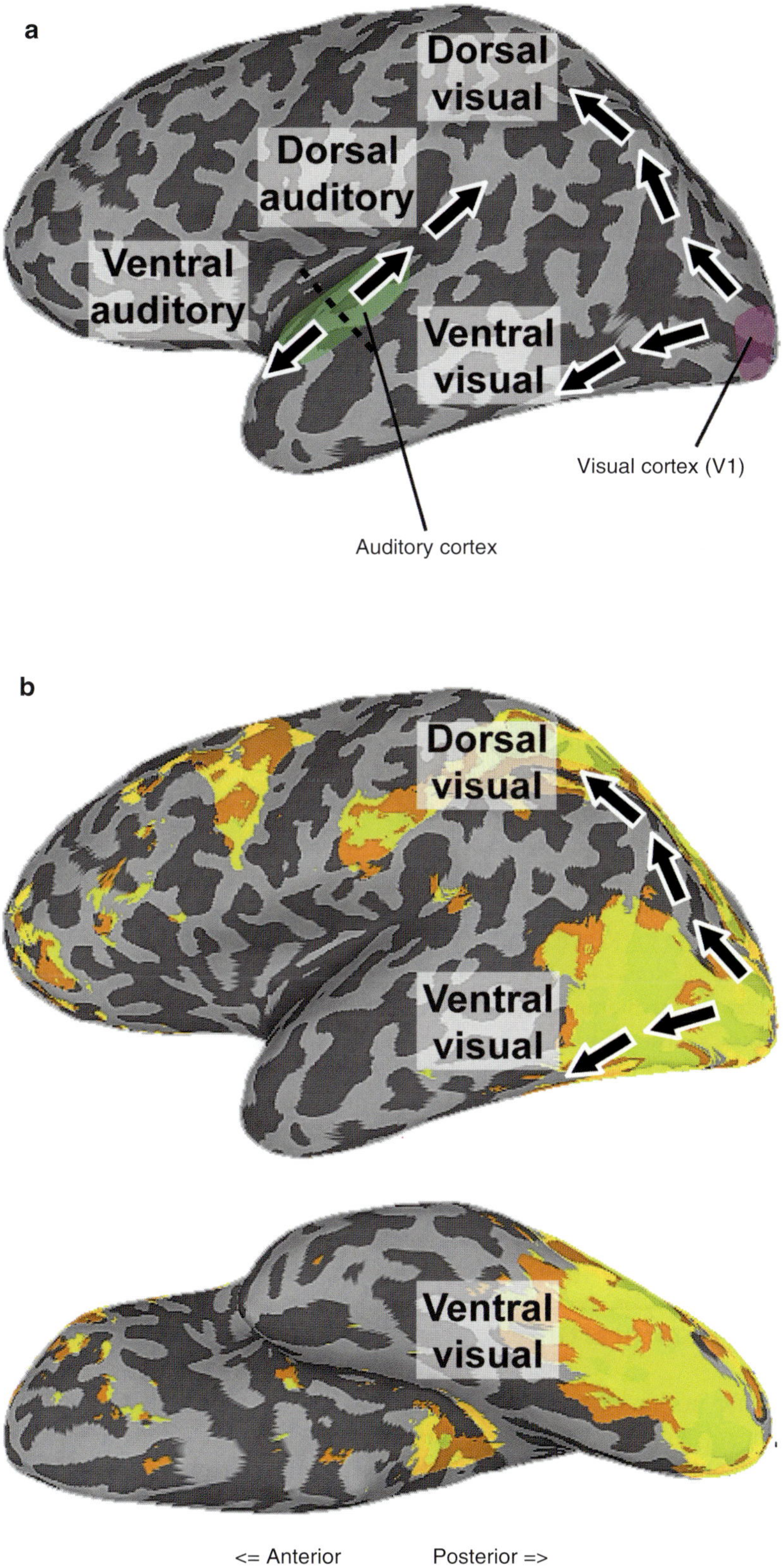

Fig. 9.3 Dorsal and ventral visual and auditory cortical pathways. (**a**) Illustration of the dorsal and ventral visual and auditory cortical pathways. *Black dotted line* shows the distinction between anterior and posterior parts of the auditory cortex. (**b**) Positive activation elicited by viewing movies of body parts relative to a fixation baseline (unthresholded) with the schematic pathways overlaid. Two streams of activation are clearly visible, one extending into the superior parietal cortex and other onto the lateral and ventral surfaces of occipitotemporal cortex (With thanks to A. Chan)

visual field or from different viewing angles can be perceived as the same object even though these variations produce very different patterns of activity on the retina (DiCarlo et al. 2012). It is often assumed that these challenges are overcome through hierarchical processing by progressively abstracting away from the specifics of the input at the sensory surface to produce "invariant" representations (DeWitt and Rauschecker 2012; DiCarlo et al. 2012). For example, the increase in the size of receptive fields (area of space a neuron is responsive to) along the ventral visual pathway in macaque was taken as evidence that the problem of position invariance was somehow solved by the time the signals reached the highest levels of processing (Kravitz et al. 2008). However, even in the anterior temporal cortex, the receptive fields of the neurons only cover a small portion of the visual field (Op De Beeck and Vogels 2000), and position information is present throughout the visual processing hierarchy (Schwarzlose et al. 2008). Further, at least in the case of stimulus position, the behavioral invariance may have been overestimated (Kravitz et al. 2008), and fMRI studies comparing the response elicited by the same object in different locations show a strong effect of position even in anterior regions of the ventral pathway (e.g., Kravitz et al. 2010). Thus, while invariance does increase along the ventral visual pathway, the hierarchy may never produce fully abstract visual object representations. Instead, our ability to identify objects despite changes in the precise nature of the input may reflect the learning of associations between the neural activations elicited by separate occurrences of the same object over time (Cox et al. 2005; Wallis et al. 2009).

9.3 Cortical Organization

So far in this chapter, we have focused on the general structure, principles, and pathways in perceptual systems. In the next section, we will focus in more detail on the functional properties of different regions (see Box 9.2) within these pathways and highlight two aspects of cortical organization: topographic maps and higher-order organization, such as by category.

9.3.1 Maps

The organization of the sensory surfaces is reflected even at the level of the cerebral cortex (Fig. 9.4). For example, in vision, separate representations of the two eyes are maintained through the LGN and into V1, forming ocular dominance columns that can be visualized in humans using fMRI (Cheng et al. 2001; Yacoub et al. 2007). Further, many regions in the visual cortex exhibit retinotopic organization with adjacent neurons responding to stimuli presented at adjacent locations in the visual field and thus stimulating adjacent points on the retina. Finally, the spatial sampling at the retina is much greater at the fovea than in the periphery, and this is reflected in a disproportionately large area of the cortex devoted to foveal vision compared to peripheral vision (Sereno et al. 1995).

Prior to fMRI, most of the knowledge of retinotopic maps in the brain was based on studies of human patients with lesions to the visual cortex (Holmes 1931; Horton and Hoyt 1991) and invasive studies in nonhuman primates (Tootell et al. 1982). However, fMRI has enabled the identification of a multitude of such retinotopic maps both in subcortical structures (LGN, Schneider et al. 2004; e.g., superior colliculus, Katyal et al. 2010) and throughout visual cortex from V1 onwards (e.g., Sereno et al. 1995). The presence of a complete map of visual space is one of the criteria used to define a distinct cortical area (for review, see Wandell et al. 2007).

Topographic maps can be identified by systematically varying specific stimulus dimensions and measuring the corresponding changes in the BOLD (blood oxygenation level dependent) response (Engel 2012). Typically, the dimensions manipulated reflect the dimensions mapped by the sensory surface. For example, visual field maps can be identified by systematically varying either the eccentricity (distance from the center of gaze) or polar angle (angular distance from the horizontal or vertical meridian) of a flickering

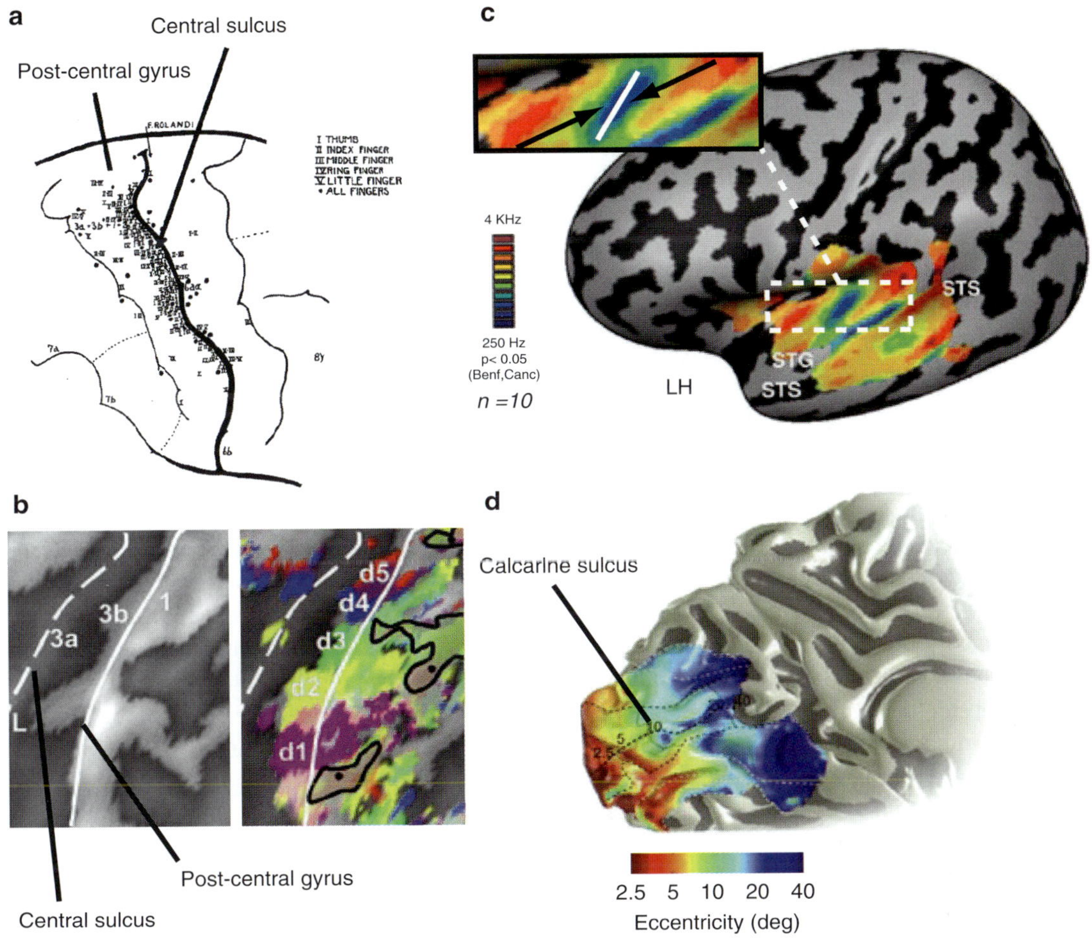

Fig. 9.4 Topographic maps in primary sensory cortices. (**a**) Stimulation sites eliciting sensations in the fingers. Sites were largely confined to the posterior bank of the central sulcus and postcentral gyrus (Adapted with permission from (Penfield and Boldrey 1937)). (**b**) Inflated cortical representation of the central sulcus and postcentral gyrus showing fingertip activations elicited by each of the digits. There is an orderly progression from the thumb (*d1*) to the little finger (*d5*) (Adapted with permission from (Schweisfurth et al. 2011)). (**c**) Tonotopic maps in the auditory cortex. Average relative frequency preference map on the left hemisphere revealing multiple frequency bands within and outside Heschl's gyrus. The enlarged inset shows the mirror-symmetric frequency gradients from high to low to high running across Heschl's gyrus (Adapted from (Striem-Amit et al. 2011)). (**d**) Retinotopic eccentricity maps in the visual cortex. Medial view of the left hemisphere showing the increase in eccentricity preference from the fovea to the periphery, moving more anteriorly along the calcarine sulcus (Adapted with permission from Wandell and Winawer (2011)

checkerboard and estimating each voxel's preferred visual field position along both dimensions (Engel et al. 1994; for review, see Wandell and Winawer 2011). An alternative model-based approach estimates the population receptive field for each voxel (Dumoulin and Wandell 2008). Using these methods, field maps have been identified not only in the occipital lobe (corresponding to V1, V2, and V3) (e.g., Sereno et al. 1995) but also more anteriorly along both the dorsal (e.g., Swisher et al. 2007) and ventral pathways (e.g., Larsson and Heeger 2006; Arcaro et al. 2009).

In audition, mapping studies have systematically varied the frequency of tones revealing tonotopic (or "cochleotopic") maps, reflecting the organization of hair cells in the cochlea

(Vonbekesy 1949). Such mappings reveal a mirror-symmetric tonotopic organization (high-low-low-high) in the human primary auditory cortex (Woods et al. 2009; Da Costa et al. 2011; Langers and van Dijk 2012) that runs across or perpendicular to Heschl's gyrus (but see Formisano et al. 2003; Da Costa et al. 2011). Additional frequency gradients have been found in multiple sites in the auditory cortex (Talavage et al. 2000, 2004; Humphries et al. 2010), extending to higher-order auditory regions in the temporal lobe (Striem-Amit et al. 2011).

Perhaps the most renowned cortical maps are the somatotopic maps forming a somatosensory homunculus in the cortex. Consistent with Penfield's groundbreaking electrical stimulation studies (Penfield and Boldrey 1937), human fMRI has revealed a highly detailed systematic representation of body surface along the postcentral gyrus and posterior bank of the central sulcus (Medina and Coslett 2010). The homunculus shows an upside-down medial-to-lateral organization in the primary somatosensory cortex: the feet, legs, and genitals are represented medially, whereas the face and hands are represented along the lateral aspects of the postcentral gyrus (Huang and Sereno 2007). The resolution of current fMRI methods is sufficient to distinguish between representations of the individual digits on the hand (Nelson and Chen 2008; Schweizer et al. 2008; Sanchez-Panchuelo et al. 2010; Stringer et al. 2011), with an ordered mapping from the thumb (lateral) to the little finger (medial) and identify distinct intradigit representations (Overduin and Servos 2004; Schweisfurth et al. 2011).

The overall medial-lateral topographic organization extends to the secondary somatosensory cortex (SII) in the parietal operculum, although the representation is less fine grained than in S1 (Disbrow et al. 2000; Ruben et al. 2001; Eickhoff et al. 2007).

The properties of these topographic maps are directly reflected in behavior. For example, in both somatosensory and visual processing, the size of the cortical representation in V1 or S1 for that location in visual space or for a given finger correlates with behavioral acuity (Duncan and Boynton 2003, 2007) and with the change in psychophysical thresholds following learning (Pleger et al. 2003). In turn, these cortical representations likely reflect both the density of receptors on the sensory surfaces and the density of connections between the thalamus and cortex.

In contrast to vision, audition, and somatosensation, the evidence for cortical topographic maps in the chemical senses (olfaction and gustation) is less clear. In olfaction, studies in rodents suggest a rough spatial organization at the level of the olfactory epithelium, with different olfactory receptors representing perceptually similar odorants clustered together that is also reflected in the organization of the olfactory bulb (for review, see Murthy 2011). This organization is not, however, retained in the cortex with olfactory bulb neurons projecting to very broad regions of the piriform cortex (Illig and Haberly 2003). Correspondingly, chemotopic organization has not been reported in human fMRI studies of the piriform cortex (Arzi and Sobel 2011). Nevertheless, fMRI has been used to uncover other forms of functional organization in the piriform cortex. For example, different parts of the piriform cortex are sensitive to different odorant attributes: the anterior piriform cortex is sensitive to the chemical structure of odorants, while the posterior piriform cortex is sensitive to the perceptual quality of odorants (Gottfried et al. 2006).

Finally, the human gustatory cortex has also not been found to contain systematic sensory maps. One major factor contributing to the lack of such an observation may be the large extent to which the gustatory cortex is multimodal and integrative (Small and Prescott 2005; Rolls 2006). fMRI studies suggest that besides encoding the basic tastes (bitter, sweet, salt, sour, umami: de Araujo et al. 2003), the gustatory cortex also responds preferentially to odors (Small et al. 2005), to the combination of odor and taste (Small et al. 2004), to somatosensory oral textures (e.g., viscosity and lubricity), and to their combination with particular tastes as well as other chemical properties, such as fat concentration (Eldeghaidy et al. 2011).

9.3.2 Higher-Order Representations

Beyond maps, which may reflect the organization of the sensory surfaces, neural representations in the cortex may cluster according to the properties or features of the stimuli themselves. This has been most extensively investigated in vision with specific visual areas or patches of cortex that are primarily responsive to features such as motion (Huk et al. 2002) or color (Conway et al. 2007; Wade et al. 2008).

Further, by contrasting the magnitude of response elicited during presentation of different categories of objects, fMRI has revealed a number of visual category-selective regions in the human occipitotemporal cortex (OTC) (Fig. 9.5, for review, see Kanwisher and Dilks 2013) that respond more strongly to one category (e.g., faces) than others (e.g., scenes). These include regions selective for faces (Puce et al. 1996; Kanwisher et al. 1997), objects (Malach et al. 1995; Kourtzi and Kanwisher 2000), letter strings or words (Puce et al. 1996; Cohen and Dehaene 2004; Baker et al. 2007), body parts (Downing et al. 2001), scenes (Aguirre et al. 1998; Epstein and Kanwisher 1998), animals (Chao et al. 1999), and tools (Chao et al. 1999). In many cases, there are at least two regions selective for a given category, one on lateral OTC and one on ventral OTC (Hasson et al. 2003; Taylor and Downing 2011). Support for a fundamental role of these regions in behavior comes from neuropsychological studies of patients with focal lesions in OTC producing specific impairments in the recognition of particular categories (faces, Barton and Cherkasova 2003; bodies, Moro et al. 2008; objects, Behrmann et al. 1992; scenes, Aguirre and D'Esposito 1999; words, Mycroft et al. 2009). Thus, while defining an object is a complex philosophical question, the striking clustering and functional specialization of neurons in OTC by object categories highlight the centrality of objects to our everyday experience (Martin 2009).

Strikingly, these different category-selective regions show a remarkable consistency in their topographical organization across individuals suggesting they may reflect some underlying principle of brain organization. For example, supported by the finding of category selective in the occipitotemporal cortex in the blind (Mahon et al. 2009), their locations may be indicative of cortical connectivity patterns between functionally related regions (Martin 2006; Mahon and Caramazza 2011). Similarly, their location may reflect retinotopic biases in high-level visual cortex. Specifically, the ventral surface of the temporal lobe shows an eccentricity bias with foveal stimuli biased to lateral regions and peripheral stimuli biased to more medial regions (Levy et al. 2001; Hasson et al. 2002). Since different categories and different tasks are associated with different goals requiring different types of visual information (Harel and Bentin 2009), such underlying principles of organization may constrain the location of category-selective regions. For example, given the necessity of foveating faces and words compared with scenes, the observed eccentricity biases may account for the lateral and medial locations of face-selective and scene-selective regions, respectively (Hasson et al. 2002). Consistent with this eccentricity bias, in which larger objects would stimulate more peripheral vision, a large-scale organization of visual representations according to real-world object size across the lateral and ventral surfaces has been reported (Konkle and Oliva 2012). Importantly, these principles are not mutually exclusive, and there may be multiple organizational principles underlying the observed pattern of responses (Op de Beeck et al. 2008).

While category appears to be a dominant organizing feature of high-level visual processing, it does not appear to be as dominant in the other modalities. In addition, many studies have emphasized distributed rather than localized representations of auditory categories (Formisano et al. 2008; Staeren et al. 2009).

There is some evidence for stronger responses to vocalizations, especially those of conspecifics, than other complex natural sounds in the auditory ventral pathway (Belin and Zatorre 2000;

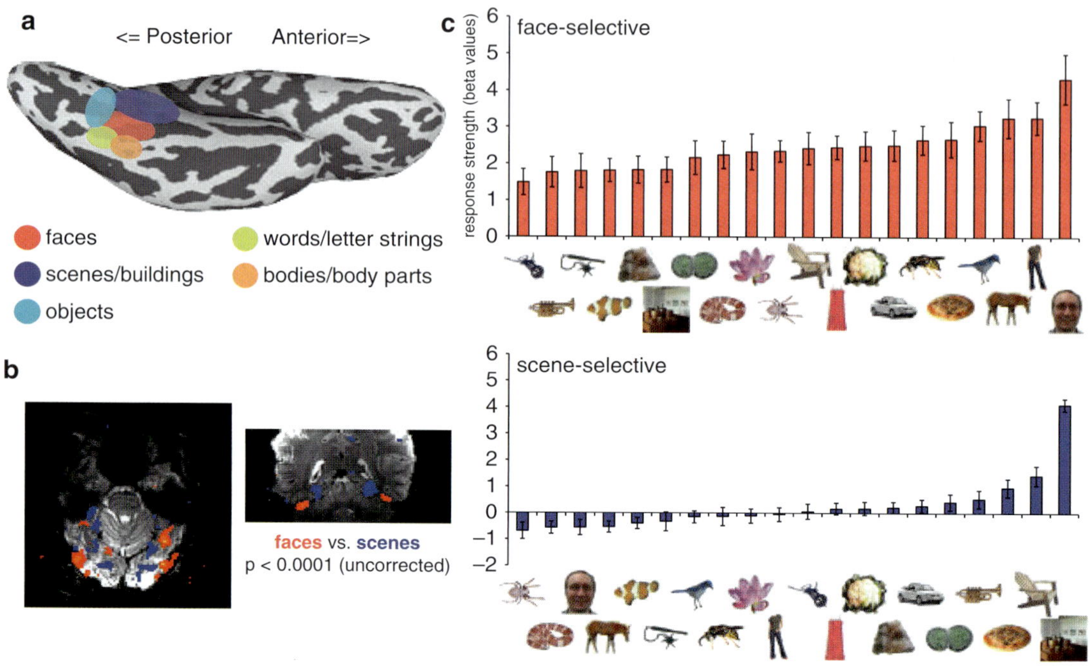

Fig. 9.5 Category selectivity in the visual cortex. (**a**) Ventral view of the inflated left hemisphere showing the relative locations of category-selective regions. (**b**) Axial (*left*) and coronal (*right*) slices from a single subject at 7 T showing the contrast between viewing faces and houses overlaid on the original functional data. Regions showing greater activation to scenes relative to faces are more medial than those showing the opposite preference. (**c**) Average response profile of face-selective and scene-selective regions on the ventral surface of the right hemisphere across 20 different categories (Adapted with permission from Downing et al. 2006 with thanks to A. Chan)

Lewis et al. 2005; Doehrmann et al. 2008; Talkington et al. 2012) with subregions selective for musical instrument sounds, human speech, and acoustic phonetic content (Leaver and Rauschecker 2010). However, the extent to which such distinctions simply reflect the spectrotemporal properties of the stimuli is unclear (Samson et al. 2011).

Most of the studies on higher-order representations discussed so far have focused on regional differences in the magnitude of response. While such an approach can readily provide evidence for gross distinctions in the cortex (e.g., face selectivity), it does not necessarily provide insight into the functional role of regions or the specific information being processed and may miss distinctions at a finer spatial scale.

The nature of representations within regions can be investigated by focusing on aspects of the fMRI response other than simple magnitude (see Boxes 9.2 and 9.3). For example, the scene-selective parahippocampal place area (PPA) on the ventral OTC responds more strongly to scenes than to other categories such as objects and faces (Fig. 9.5). Studies focusing on the specific pattern of response across this region using either hypothesis-driven or data-driven approaches have demonstrated that PPA contains information about both spatial layout (Fig. 9.6, Kravitz et al. 2011a; Park et al. 2011) and object identity (Harel et al. 2013), consistent with the inputs it receives from both the ventral and dorsal pathways (Kravitz et al. 2011b) and suggesting a key role in navigation.

In olfaction, while there is no evidence for distinct patches coding odor categories, the pattern of response in the posterior piriform cortex does reflect odor category, correlating with perceptual judgments of odor quality and similarity (Howard et al. 2009). The more similar odorants were perceived, the more similar the patterns of response elicited (Fig. 9.7).

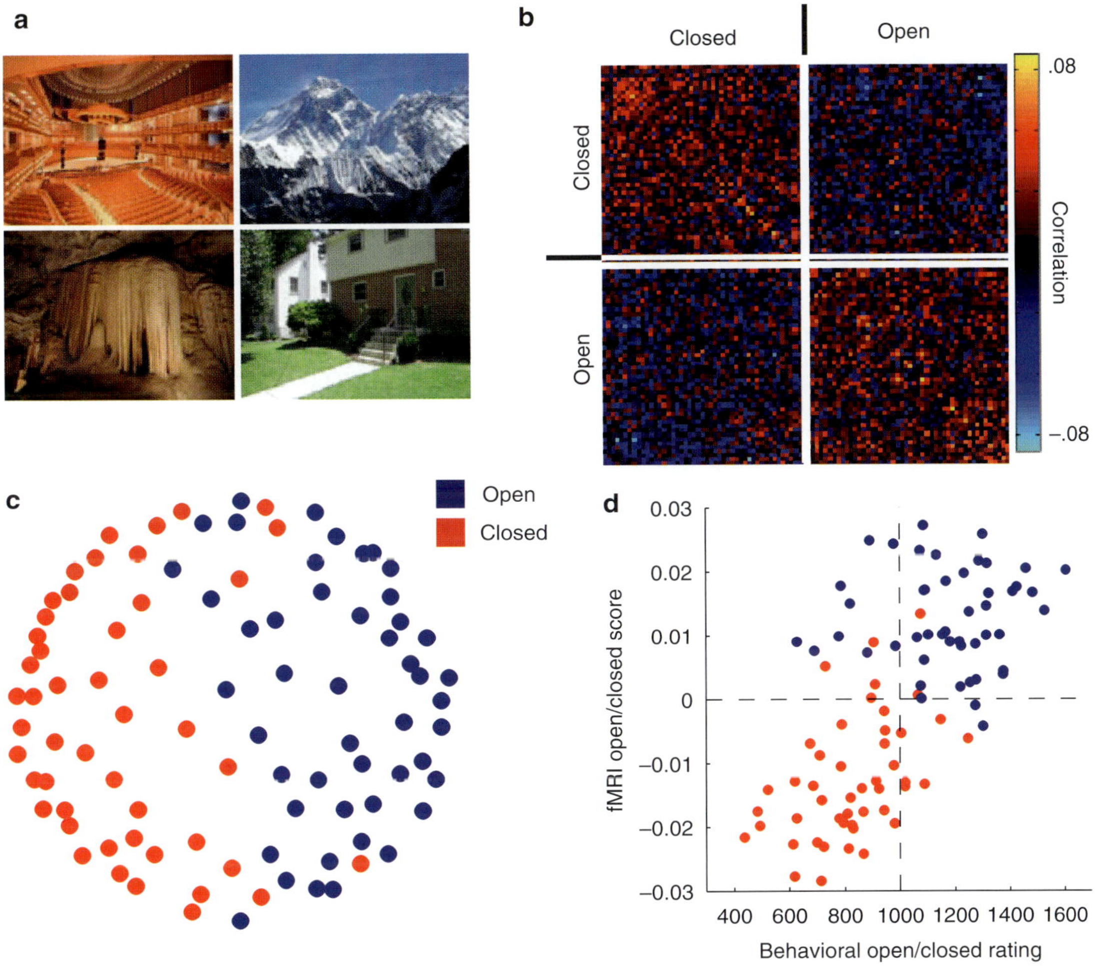

Fig. 9.6 Determining the structure of perceptual representations. To investigate the nature of representations in the scene-selective region on the ventral surface, participants were presented with 96 individual scenes, varying widely in semantic category (e.g., church, beach) and properties (e.g., open, closed, natural man-made). The multi-voxel response patterns within this region were used to determine which properties of the scenes are encoded (see Box 9.2). (**a**) Four sample scene images. (**b**) Similarity matrix showing the correlation in the multi-voxel pattern of response for all 96 scenes. Each individual point in the matrix shows the correlation in the pattern of response for two scenes. Correlations between any two open scenes and between any two closed scenes were on average higher than between open and closed scenes, suggesting that the distinction between open and closed scenes is critical in this region. (**c**) Multidimensional scaling of the correlation data. Each point corresponds to a single scene, and the distance between points reflects the similarity between their patterns of response. The grouping of the points clearly reflects whether they were open or closed. (**d**) There was a significant correlation between behavioral ratings of openness for individual scenes and the fMRI measures (With thanks to D. Kravitz)

Insights into cortical organization can also be revealed by searching the entire brain for the information contained within patterns of response, rather than just simply differences in response magnitude (Kriegeskorte et al. 2006). For example, studies searching for regions distinguishing between the faces of different individuals have highlighted the importance of anterior temporal lobe regions (Kriegeskorte et al. 2007; Nestor et al. 2011), even though face selectivity tends to be weak in this region. Similarly, a study searching for areas containing head-view-invariant representation of gaze direction identified regions in both the anterior and posterior superior temporal

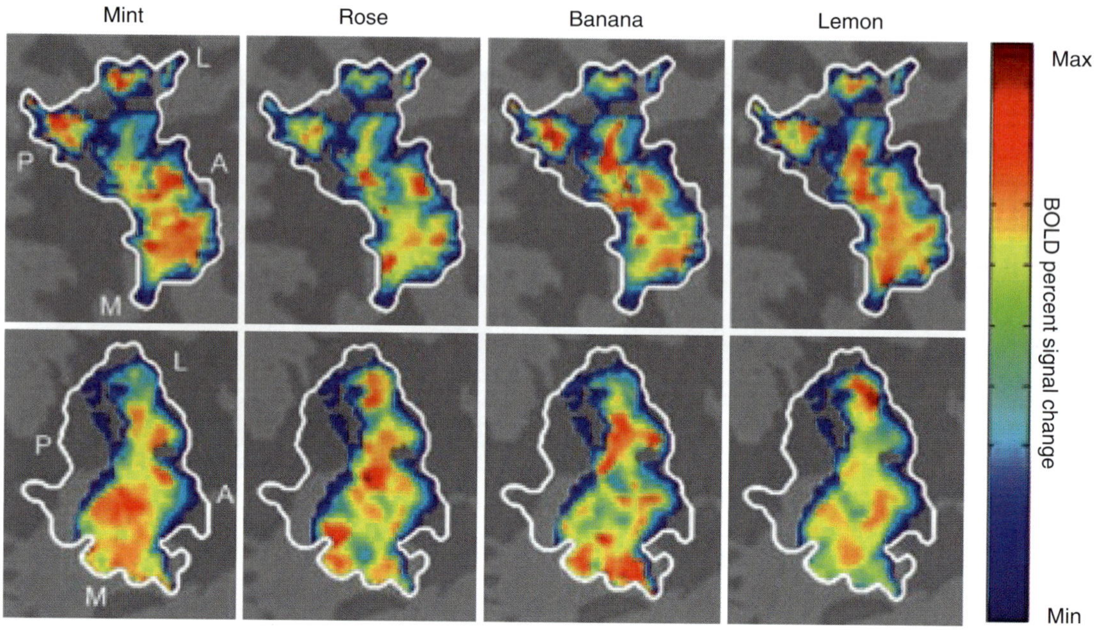

Fig. 9.7 Distributed patterns of response to different odorants in the piriform cortex. Distributed response patterns of activation in the left posterior piriform cortex (outlined in *white*) projected onto flattened cortical maps of two representative participants (*top* and *bottom rows*) in response to four odorants (mint, rose, banana, lemon). Note the distinct, yet spatially overlapping activation patterns evoked by the different odorants as well as the lack of consistency in patterns of response across the two subjects. These distinctive patterns can be detected using multi-voxel pattern analysis (see Box 9.2). Color scale represents the range of BOLD percent signal change from minimum (*blue*) to maximum (*red*). *A* anterior, *L* lateral, *M* medial, *P* posterior (Adapted with permission from Howard et al. 2009)

sulcus, although the posterior region was more dependent on the physical properties of the stimuli (Carlin et al. 2011, see Fig. 9.8).

Overall, there is a pronounced organization in high visual cortex (i.e., categories) that is not clearly evident in the other modalities. However, there are clearly high-order representations in other modalities, and any such organizational principles may occur at a spatial scale below the resolution of current fMRI.

Box 9.2. Probing Perceptual Representations

The responses observed with fMRI provide several different means of evaluating the neural representations of sensory stimuli. First, traditional fMRI studies focus on the magnitude of the BOLD response. Differences in response magnitude across different stimuli, different categories of stimuli, or along particular stimulus dimensions are taken to reflect differences in the neural representation and thus coding of those stimulus differences. However, given the large population of neurons contributing to the response in each individual voxel, the sensitivity of this measure may be limited, particularly when a voxel contains heterogeneous populations of neurons.

Second, fMRI adaptation has been developed as a method to evaluate sub-voxel neural populations (Grill-Spector and Malach 2001). Based on the observation that the BOLD response reduces upon repetition of a given stimulus (Grill-Spector et al. 2006), consistent with the reduction in firing rate observed in electrophysiological studies (e.g., Li et al. 1993; Sobotka and Ringo 1994), the

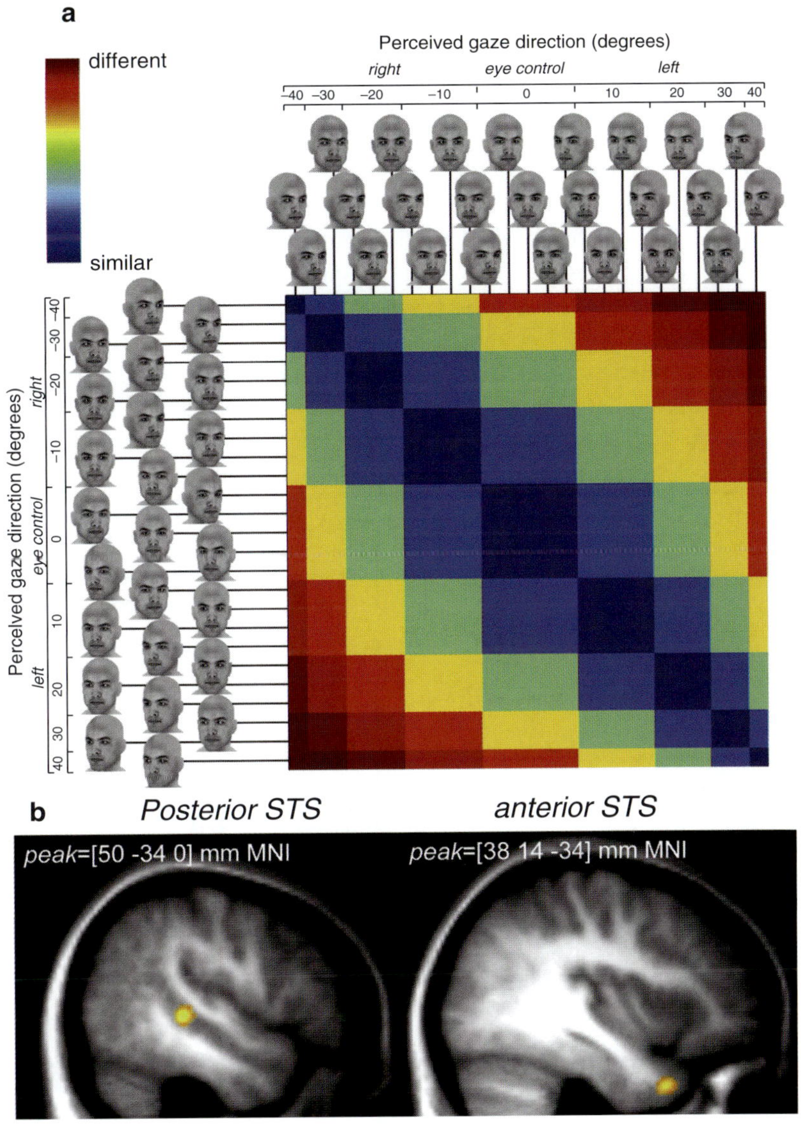

Fig. 9.8 View-invariant gaze direction coding in the anterior superior temporal sulcus (*STS*). To identify regions encoding gaze direction independent of head direction, a whole-brain search was conducted based on predicted similarities in the pattern of response across a range of head stimuli. (**a**) Predicted similarity matrix. For 25 computer-generated face images varying in head and gaze direction, the predicted similarity of response was estimated for a region encoding gaze direction. Color scale represents the predicted similarity from minimum ("different," *red*) to maximum ("similar," *blue*). (**b**) A whole-brain search identified two distinct regions that show the predicted view-invariant gaze direction response patterns: posterior STS and anterior STS (Adapted with permission from Carlin et al. 2011)

extent of adaptation can be used as a measure of neural similarity (see also Sect. 9.5.2). Adaptation is often measured during the presentation of blocks (e.g., Grill-Spector et al. 1999; Belin and Zatorre 2003) or pairs of stimuli (e.g., Kourtzi and Kanwisher 2001; Gottfried et al. 2006; Doehrmann et al. 2008) that are either the same or vary along a particular dimension with the extent of reduction for different stimuli relative to the presentation of the same stimulus indicating the degree to which the neural representations are similar.

Finally, analyses can focus on the distributed pattern of response across voxels (multi-voxel pattern analysis or MVPA) (Norman et al. 2006). In this approach, the similarity in the pattern of response is used as a measure of neural similarity and can be evaluated using a variety of measures (Cox and Savoy 2003; Pereira et al. 2009) including correlation (e.g., Haxby et al. 2001; Howard et al. 2009), linear discriminant analysis (Carlson et al. 2003), or support vector machines (SVM) (Kamitani and Tong 2005; De Martino et al. 2008; Beauchamp et al. 2009), which are all forms of linear classifiers. In MVPA, the similarity of neural representations is commonly expressed in terms of the ability to predict novel stimuli based on activity patterns and is often referred to as decoding (Haynes and Rees 2006, see also Box 9.3). Importantly, measures of neural similarity across large sets of stimuli can be directly contrasted with other measures of similarity (e.g., behavioral, physical properties) allowing quantitative comparisons across modalities, methodologies, and even species (Kriegeskorte et al. 2008a, b). While it has been suggested that MVPA assesses sub-voxel information by relying on subtle biases in individual voxels (Kamitani and Tong 2005), it seems more likely that MVPA relies on larger-scale information (Op de Beeck 2010; Freeman et al. 2011).

These three different approaches (magnitude, adaptation, and MVPA) are often applied separately, but they provide complementary information and can be measured simultaneously (Aguirre 2007).

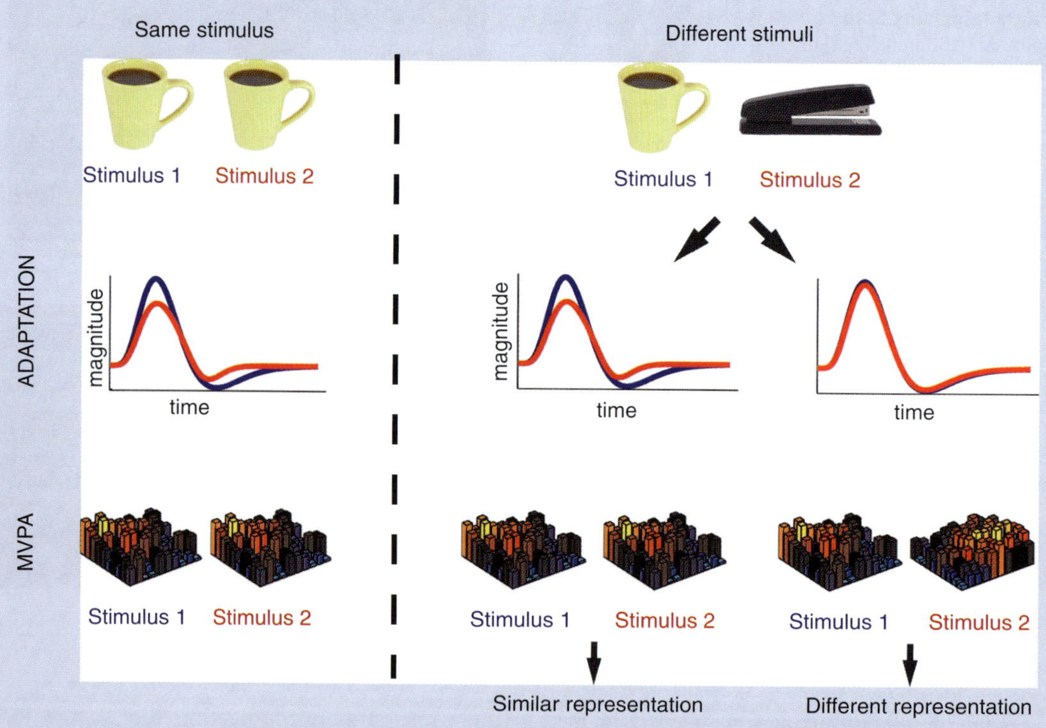

Using adaptation and MVPA to probe perceptual representations. Both adaptation and MVPA have been used as tools to assess the degree of similarity between the neural representations of pairs of stimuli. A repeated presentation of the same stimulus (left column) can produce: (i) a decrease in the BOLD response, known as adaptation (middle row) and (ii) a high correlation between the patterns of response elicited (bottom row). These relations are used as a benchmark to assess the representation of two different stimuli (middle and right columns). If the two stimuli have similar representations (middle column), adaptation or similar patterns of response would be observed. In contrast, if the two stimuli have different representations (right column), no adaptation or distinct patterns of response would be observed

Box 9.3. Decoding, Encoding, and Reconstructing Perception

The use of linear classifiers to probe perceptual representations (see Box 9.2) can be thought of as building a decoding model: using voxel patterns to predict perceptual stimuli. If voxel patterns differ between particular types of stimuli, then the pattern of response for a novel stimulus can be used to predict or decode its identity.

An alternative approach is to build an encoding or forward model: predict activity that will be evoked in voxels by different perceptual stimuli (Naselaris et al. 2011). For example, by presenting a wide range of natural images, Kay and colleagues built a quantitative model for each individual voxel in early visual areas (V1, V2, and V3) that characterized the stimulus information eliciting a response in terms of spatial position, orientation, and spatial frequency (Kay et al. 2008). Using these voxel models, predicted activity for a candidate set of novel images was compared with actual activity elicited by one of those images to identify which specific image was presented. Similarly, Schönwiesner and Zatorre presented auditory stimuli varying in spectral and temporal modulation rates and adapted a quantitative model developed from electrophysiology of auditory cortex to characterize the stimulus information eliciting a response in individual voxels in the primary and secondary auditory cortex (Schonwiesner and Zatorre 2009). One advantage of encoding models over decoding models is that they allow prediction of voxel activity patterns from stimuli outside the set used to build the model, even stimuli from a completely different stimulus space (Naselaris et al. 2011).

Such decoding and encoding models can be used to try and reconstruct presented stimuli from voxel activity patterns, to produce a literal representation of the specific stimulus presented. For example, using simple black-and-white geometric patterns to develop encoding and decoding models, it has been possible to reconstruct novel patterns (Thirion et al. 2006) or letters (Miyawaki et al. 2008). Similarly, an encoding of color space was used to reconstruct novel colors using voxels in multiple visual areas (Brouwer and Heeger 2009). Finally, the encoding model of Kay and colleagues has been extended, incorporating semantic features, to enable reconstruction of natural images (Naselaris et al. 2009) and a similar approach with dynamic stimuli allows reconstruction of movie clips (Nishimoto et al. 2011).

9.4 Top-Down Processing in Perception

Most of this chapter so far has treated perception as primarily a bottom-up stimulus-driven process, whereby neural representations of increasing complexity are hierarchically assembled, resulting in a coherent percept. One extreme version of this hierarchical view is that the underlying perceptual representations are independent from or "impenetrable" to top-down factors such as attention, knowledge, memory, or expectations (Pylyshyn 1999). However, most connections between regions are bidirectional, and even the primary sensory cortices receive large amounts of input from higher-level areas. These feedback projections allow the operation of recurrent modulatory mechanisms on sensory representations (Lamme and Roelfsema 2000) throughout the sensory processing hierarchies. This interaction between top-down and bottom-up factors makes the distinction between "perception" and "cognition" less obvious than it may at first seem. For example, retinotopic activity in V1 – which is almost by definition bound to the stimulus physical properties – has been shown to vary based on the perceived size of the stimulus presented for physically identical stimuli (Murray et al. 2006; Sperandio et al. 2012). In the following section, we will focus on two different forms of top-down influences on perceptual representations: attention and imagery.

9.4.1 Attention

Top-down attention involves enhancement (or prioritization) of particular aspects of incoming sensory information based on the specific goals of the observer. Attention can be directed to spatial locations (space-based attention) or features of the sensory input (e.g., object-based attention in vision). At a neural level, attention manifests in several different forms including lowering of firing thresholds, scaling of response magnitude, and enhanc synchronization (Reynolds and Chelazzi 2004; Lee and Maunsell 2009; Reynolds and Heeger 2009).

Effects of top-down attention have been reported with fMRI even as early as the subcortical structures. For example, the visual response in LGN is modulated by spatial attention with enhanced responses to attended visual stimuli and suppressed responses to ignored stimuli (O'Connor et al. 2002). Attention has also been shown to modulate visual responses in the pulvinar (Kastner et al. 2004) and superior colliculus (Schneider and Kastner 2005; Katyal et al. 2010). These effects of top-down input highlight that the subcortical structures are more than just passive relay structures, as has often been assumed (for review and further examples, see Saalmann and Kastner 2011).

Similar effects of attention have been reported in the early visual cortex including V1. Thus, directing attention to a particular spatial location results in superior task performance for stimuli appearing in that location, concomitant with an enhancement of the BOLD response in the corresponding retinotopic regions of the cortex (Tootell et al. 1998; Brefczynski and DeYoe 1999). Further, the attentional effects on the BOLD response are constrained by the receptive size of the regions along ventral visual pathway (Kastner et al. 1999; Muller et al. 2003).

How attention is allocated can be influenced by features of the visual stimuli and not just by space. For example, following a spatial cue on an object, retinotopic activity in early visual areas is enhanced for other locations on that same object, but not equally distant locations outside the object (Muller and Kleinschmidt 2003), a phenomenon known as object-based attention (see also Martinez et al. 2006; Ciaramitaro et al. 2011). Effects of attention to objects and features have also been demonstrated in higher-order category-selective regions. For example, attending to a particular object category (faces or houses) enhances activity in the respective face- or scene-selective regions (O'Craven et al. 1999). The top-down modulatory effect on category-selective regions has been suggested to originate in the prefrontal cortex and can be seen even before stimulus presentation in the form of increased activity with the expectation of incoming stimuli (for review, see Gazzaley and Nobre 2012). Such object-based attentional effects are thought to reflect increased neural firing rate for the attended stimulus, as well as increased selectivity for features of the attended stimulus (Murray and Wojciulik 2004).

Top-down modulations of the sensory cortex by attention, expectations, and task have also been observed with fMRI in modalities other than vision. For example, activity in the olfactory cortex (Fig. 9.9) was found to increase when participants sniffed non-odorant air but were expecting an odor ("attended sniff" condition) relative to when they sniffed the same non-odorant air but were not expecting an odor ("unattended sniff" condition) (Zelano et al. 2005). Further, activity in part of the piriform cortex in the "attended sniff" condition was enhanced in response to auditory task instructions, before the stimulus was even presented, indicating a strong effect of expectation. Higher-order olfactory regions, such as the orbitofrontal cortex, are also modulated by odor content expectations (de Araujo et al. 2005), and attention to odors appears to modulate functional connectivity between the piriform and orbitofrontal cortex through the mediodorsal nucleus of the thalamus (Plailly et al. 2008).

Similar to the effects observed for olfaction, trying to detect taste in a tasteless solution results in enhanced activation in the middorsal insula and frontal operculum, corresponding to the primary gustatory cortex (Veldhuizen et al. 2007). Further, manipulation of expectation also modulates fMRI responses in the gustatory cortex (Nitschke et al. 2006; Small et al. 2008;

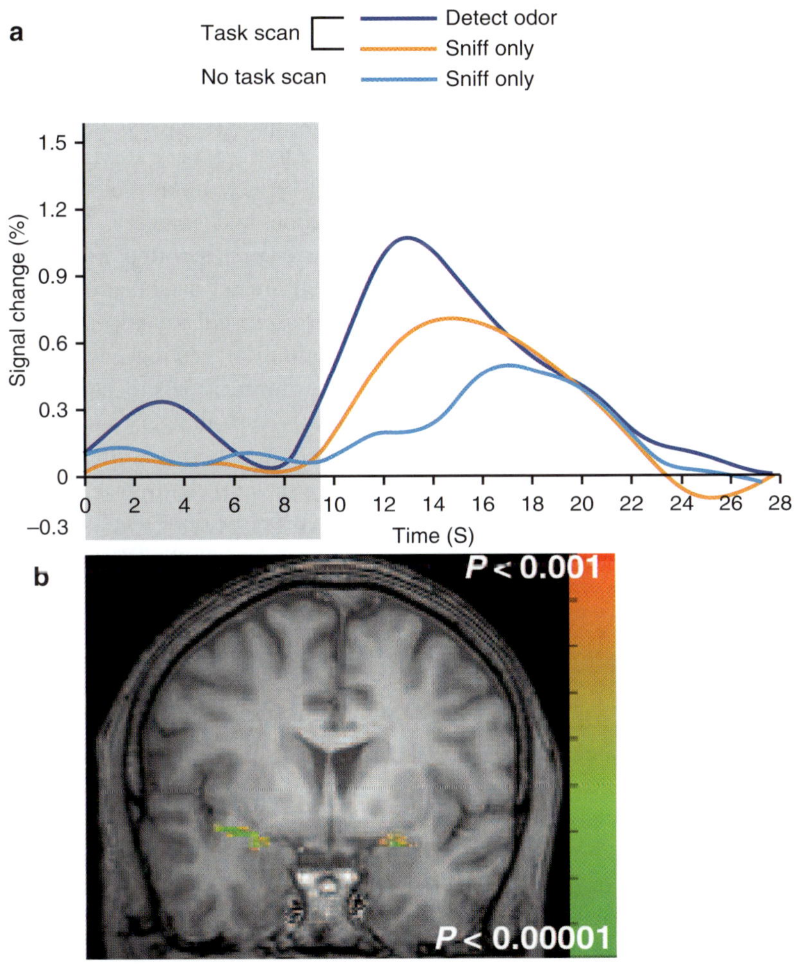

Fig. 9.9 *Top-down* attentional effects in olfaction. (**a**) Average response in the anatomically defined piriform cortex as a function of three conditions varying in their attentional demands. In the task scan, participants were required either to detect the presence of an odor ("detect odor" condition) or to simply inhale ("sniff only"). In a separate scan (no task scan), participants were only required to inhale. Critically, only non-odorant air was presented in all conditions, and so differences in response can only be attributed to the *top-down* attentional factors. Note the stronger responses during detection than during simply sniffing. Further, an anticipatory effect is also apparent in the period before the stimulus was delivered (*gray area*) in the "detect odor" relative to the other two conditions. (**b**) Group analysis from the same study showing functional activation in the piriform cortex during the odorant condition (Images adapted with permission from Zelano et al. 2005)

Veldhuizen et al. 2011a). Importantly, these attentional effects in olfaction and gustation are modality specific (Veldhuizen and Small 2011).

In sum, top-down attentional effects can be seen across modalities and are one mechanism underlying the active, constructive process of perception, which utilizes our prior knowledge and experience of environmental regularities to form hypotheses and expectations about the external world.

9.4.2 Imagery

Even in the absence of external sensory stimulation, it is still possible to generate internal representations using top-down signals, a process that is often referred to as mental imagery (Kosslyn et al. 2006). Such mental imagery can provide important insights into the nature of the top-down signals and their potential effect on perceptual processing along the sensory hierarchies. A general finding across modalities is that mental imagery elicits activity in similar cortical regions to those involved in perception (e.g., olfaction, Levy et al. 1999; somatosensory, Yoo et al. 2003; vision, Ganis et al. 2004; gustation, Kobayashi et al. 2004; audition, Bunzeck et al. 2005), and it is often suggested that perception and imagery share the same neural substrate. However, simply showing activation in the same regions is not sufficient to show that imagery and perception share similar mechanisms. Such activation could simply reflect a nonspecific engagement of a region. For example, in the case of olfaction, odor imagery is associated with sniffing, which is known to elicit activation on its own (Sobel et al. 1998). To demonstrate that activation of the olfactory cortex during imagery is odor specific rather than a general modulation, activations elicited by pleasant and unpleasant odors during both perception and imagery were compared, revealing similar hedonic specificity in both cases (Bensafi et al. 2007). In vision, studies have revealed category specificity in high-level visual areas during imagery (O'Craven and Kanwisher 2000; Reddy et al. 2010) and even decoding of individual objects (Lee et al. 2012). Furthermore, the specific representations elicited are similar during perception and imagery suggesting extensive overlap in the neural substrate of each (Stokes et al. 2009; Lee et al. 2012). Thus, imagery not only does engage the same regions as perception but can evoke highly specific representations (Fig. 9.10).

One area of contention is whether imagery-elicited activation extends all the way back to the primary sensory cortices. This debate has been entwined with the discussion of whether imagery is symbolic or depictive (Pylyshyn 2002; Kosslyn et al. 2006). In addition, it has been reported that only the secondary auditory cortices and not the primary auditory cortex is involved in auditory imagery (Bunzeck et al. 2005). However, in vision, retinotopically specific activity has been reported in the primary visual cortex (Slotnick et al. 2005; Thirion et al. 2006) and the activity in early visual cortex correlates with the vividness of imagery (Lee et al. 2012). In olfaction, the extent to which the primary olfactory cortex is active correlates with experience, with professional perfumers showing stronger activation than novices (Plailly et al. 2012).

So far we have highlighted the similarities between perception and imagery, but introspectively, they are very different phenomena. The strength of activation is clearly different between perception and imagery, but other differences have also been reported. For example, one study reported greater deactivation in the auditory cortex during visual imagery compared with visual perception, suggesting enhanced unimodal processing during imagery (Amedi et al. 2005). In another visual imagery study, the similarity of representations across different regions along the ventral visual pathway was greater during imagery than perception, suggesting different neural dynamics (Lee et al. 2012).

In sum, the study of mental imagery shows that top-down processes are able to elicit highly specific responses in sensory areas. The relationship between imagery and perception is complex, with overlap in the neural substrates involved but differences in the interregional relationships.

9.5 Experience-Dependent Perceptual Representations

Sensory information processing in mammalian brains is highly malleable both during development and adulthood, and over both short and long time scales (Buonomano and Merzenich 1998). In this section, we will focus on several ways in which prior experience impacts the nature of perceptual representations.

9.5.1 Development

While the large-scale circuitry of the brain is present at birth, it is clear that there is the potential for extensive plasticity early in life and the nature of perceptual representations is dependent on experience throughout development. Even in primary sensory cortices, many of the properties of neurons and maps depend on experience early in life. For example, in V1, neural properties such as orientation selectivity

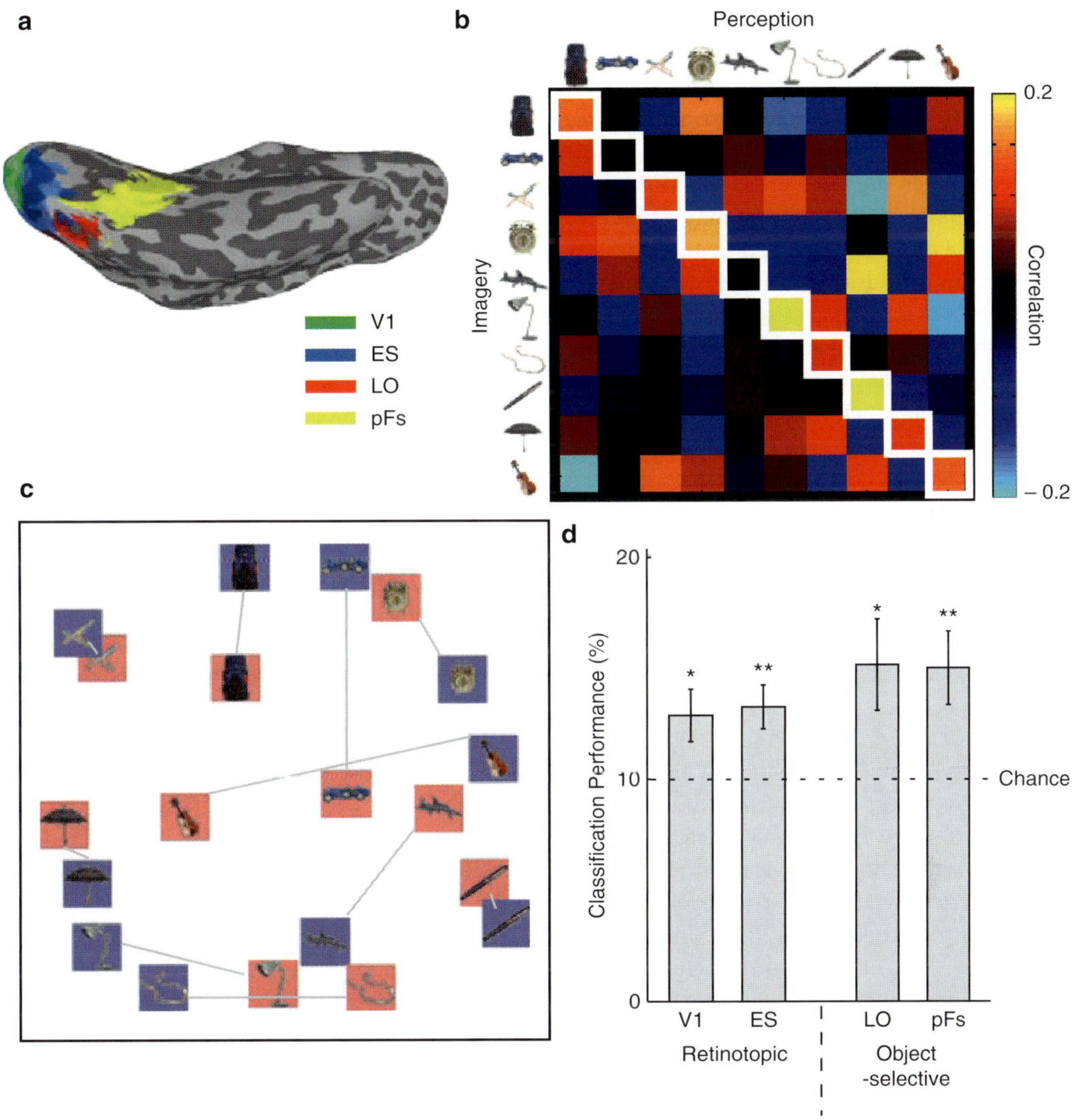

Fig. 9.10 Visual imagery and perception of real-world objects. Participants were familiarized with photographs of ten objects and during the fMRI scan were either presented with those same images or asked to imagine them. The results demonstrate the similarity between representations elicited during perception and imagery. (**a**) fMRI responses were analyzed in V1, extrastriate retinotopic cortex (*ES*), and two object-selective regions (LO in lateral OTC and pFS on ventral OTC). (**b**) Example data from one participant showing the correlation in the pattern of response across pFS between imagery and perception. In most cases, the strongest correlation is for the same object in the two conditions (cells along the main diagonal, highlighted in *white squares*). (**c**) Multidimensional scaling of the average data across participants. The perception (*blue*) and imagery (*red*) conditions for the same object are close to each other. (**d**) Classification of object identity between imagery and perception using a linear classifier (support vector machine – see Box 9.2) (With thanks to SH Lee)

and ocular dominance are dependent on early postnatal experience (Espinosa and Stryker 2012). Such changes are difficult to observe with fMRI due to the substantial practical challenges posed by scanning young children (but see, e.g., Arichi et al. 2010; Perani et al. 2010;

Raschle et al. 2012). fMRI studies of visual development in infant monkeys have suggested that while there is strong activation in V1 early in life, there is only weak activation in extrastriate areas such as MT and V4 (Kourtzi et al. 2006). Studies in older children suggest a prolonged time course for developmental changes that may extend into adulthood. For example, in audition, a cross-sectional study of music perception in 5–33-year-olds revealed selective activation for melody and rhythm processing that increases over time even in the absence of any musical training (Ellis et al. 2012). In vision, much work has focused on high-level category-selective regions and in particular the face-selective FFA (fusiform face area). At a behavioral level, many of the markers of specialized face processing are present by 3–4 years of age (Sangrigoli and de Schonen 2004; Cassia et al. 2009), although performance improves through adolescence (e.g., Mondloch et al. 2003; for review, see McKone et al. 2012). Many fMRI studies report increases in size, response magnitude, or selectivity of FFA (e.g., Golarai et al. 2007; Scherf et al. 2007, 2011) with a smaller FFA in 12–16-year-olds than in adults (Golarai et al. 2010). However, it has been questioned whether methodological issues might account for such effects, rather than changes in neural processing (McKone et al. 2012). In particular, a major concern is the reduced signal strength observed when putting a child's head inside an adult-sized head coil – it may be critical to match the size of head coil to the size of the child's head (Keil et al. 2011).

9.5.2 Repetition Suppression or Adaptation

One of the simplest forms of experience is prior exposure to a given perceptual stimulus. Such stimulus repetition leads both to decreased BOLD responses (variously termed repetition suppression or adaptation) throughout brain regions active in a given task, not just sensory areas, and also improvement in behavioral performance (termed repetition priming) (Grill-Spector et al. 2006). Such repetition suppression has been reported in multiple modalities and has been used as a way of probing neural representations with fMRI (see Box 9.2). The reduction in the BOLD response is immediate (Kourtzi and Kanwisher 2001), increases with multiple presentations (Sayres and Grill-Spector 2006), survives intervening stimuli (Vuilleumier et al. 2002; Weiner et al. 2010), and occurs even with delays of 3 days (van Turennout et al. 2000). However, the relationship between the decreased BOLD response, behavioral improvement, and the underlying neural mechanisms is unclear (Wiggs and Martin 1998; Henson 2003; Grill-Spector et al. 2006). How is it that reduced activity can be associated with enhanced performance? Proposals have included a more rapid neural response (James and Gauthier 2006), a sharper or more-selective response (Desimone 1996), or some suppression of bottom-up sensory responses by top-down predictions (Friston 2005). One intriguing suggestion is that the reduced neural activity is associated with enhanced neuronal synchronization producing more effective long-range communication between brain regions (Gotts et al. 2012).

9.5.3 Learning and Perceptual Expertise

Beyond repetition suppression, perceptual representations can change as a result of active training or learning. One way to investigate the effect of learning is to use a cross-sectional approach and compare subjects with varying amounts of perceptual expertise. Such studies have focused on differences in both functional and anatomical properties. For example, studies of musicians compared with nonmusicians have reported increased gray matter volume in the auditory cortex that correlates with musical aptitude (Schneider et al. 2002; Gaser and Schlaug 2003). Further, musical training correlates with stronger activation in the superior temporal gyrus (Koelsch et al. 2005) and in the planum temporale, just posterior to the primary auditory areas (Ohnishi et al. 2001). In gustation/olfaction, wine sommeliers show enhanced activation in the insula and orbitofrontal cortex (Castriota-Scanderbeg et al.

2005), and in somatosensation, Mah-Jong experts, who have learned tactile discrimination of the 2D shapes on Mah-Jong tiles, show enhanced activation of V1 (Saito et al. 2006) suggesting changes in cross-modal associations.

In the visual domain, cross-sectional studies have focused on expertise in the recognition of object categories, such as cars and birds. Faces can be thought of as a perceptual category we are all experts in processing (Diamond and Carey 1986), and it has been proposed that processing objects of expertise should engage the same neural structures that are engaged by faces (Tarr and Gauthier 2000).

While some studies have reported enhanced activation in FFA when car or bird experts view objects from their domain of expertise relative to control objects (Gauthier et al. 2000; Xu 2005), other studies failed to replicate these findings (Grill-Spector et al. 2004; Rhodes et al. 2004). Further, the enhanced activity for car experts viewing cars is both widespread, extending well beyond the FFA and even into parietal and prefrontal regions, and diminished when the experts are not engaged with their objects of expertise, suggesting that it may reflect enhanced attention (Harel et al. 2010). This same group of car experts also showed increased gray matter volume in the prefrontal cortex and not in the visual cortex (Gilaie-Dotan et al. 2012) suggesting that the neural changes involved in expert recognition are not necessarily perceptual and may involve top-down factors, such as knowledge and attention.

Since studies of real-world expertise are correlational in nature, it is impossible to avoid the "chicken and egg problem": perhaps the specialized neural correlates exhibited by experts do not reflect the acquisition of expertise, but are in fact a precondition for the successful acquisition of expertise. Thus, an alternative approach is to provide laboratory training and measure the changes in perceptual representations directly. The advantage of such longitudinal studies is that they provide causal evidence for the relationship between training and neural changes. Across modalities, these studies have revealed changes in cortical responses following training, in the form of both increases and decreases. For example, in audition, frequency discrimination training over a week leads to reduced BOLD responses in the auditory cortex (Jancke et al. 2001). In contrast, odorant learning produces increased responses in the piriform and orbitofrontal cortex (Li et al. 2006). In vision, training on a simple visual texture discrimination task produces enhanced activity in the early visual cortex after only a single session of practice (Schwartz et al. 2002). However, both increases and decreases in visual responses across distributed regions have been reported for trained versus untrained shapes (Kourtzi et al. 2005), novel objects ("smoothies," "spikies," "cubies" Op de Beeck et al. 2006), and bird types (van der Linden et al. 2008). In the context of comparing object expertise to face expertise, some studies have reported changes in FFA following training on novel objects ("Greebles" Gauthier et al. 1999), while others have not (Op de Beeck et al. 2006; Wong et al. 2009), even with the same training paradigm (Brants et al. 2011). Although the presence of both increases and decreases in response following perceptual learning both within and across studies may seem contradictory, there are two major factors that need to be considered. First, the nature of the stimuli may determine which preexisting representations training modifies, by either strengthening or weakening. For example, whether responses in FFA change during training may depend on how face-like the stimuli are (Brants et al. 2011). Second, the nature of the task and the informativeness of any given neural population for that task will determine their likely modulation in learning (Op de Beeck and Baker 2010). Neural populations that are engaged by and used in performing a task will be more likely to show increased responsiveness.

9.5.4 Loss of Sensory Input

Changes in functional response properties have also been reported following loss of sensory input. Complete loss of input in a given modality seems to produce some form of cross-modal plasticity (Merabet and Pascual-Leone 2010). For example, in the blind, a number of studies have demonstrated enhanced responsiveness to auditory (Gougoux et al. 2005; Poirier et al. 2006) and tactile (Burton et al. 2002) stimuli in visual areas (V1, MT), and others have reported activity even during verbal memory tasks (Amedi et al. 2003). Similarly, visual and somatosensory activation of the auditory cortex have been reported in the early deaf (Finney et al. 2001; Karns et al. 2012). However, these cross-modal effects are stronger with loss of input early in life and may reflect developmental plasticity mechanisms (Merabet and Pascual-Leone 2010). Further, such plasticity may be driven by enhancement of preexisting cross-modal responsiveness, which is normally masked by the dominant modality. Support for this "unmasking" comes from studies showing tactile responses in V1 of sighted subjects following blindfolding (Merabet et al. 2007, 2008). However, cross-modal responses in the early blind cannot be predicted from cross-modal responses in sighted subjects (Lewis et al. 2010), and it may be that unmasking of preexisting connections is accompanied by more permanent structural changes with long-term loss of sensory input (Merabet and Pascual-Leone 2010).

In contrast to the cross-modal plasticity described above, partial loss of sensory input within a modality leads to changes in the modality-specific functional maps. For example, in arm amputees, lip movements appear to elicit activation not only in lip somatosensory regions but also in regions corresponding to the missing limb (Lotze et al. 2001). Similarly, in people with macular degeneration and subsequent loss of central vision, peripheral visual stimuli have been reported to elicit activation in early visual areas of the cortex that would normally represent central stimuli only (Baker et al. 2005, 2008; Masuda et al. 2008; but see Baseler et al. 2011). The perceptual consequences of such changes in activation are not clear. However, a patient with disrupted input to V1 from a stroke was found to have distorted perception that appeared to directly reflect functional reorganization in V1 (Dilks et al. 2007). It is possible that the changes following partial loss of sensory input are related to disorders such as phantom limb pain (Flor et al. 2006) and Charles Bonnet syndrome, visual hallucinations experienced by those with vision loss (Yacoub and Ferrucci 2011). It has also been suggested that tinnitus may also reflect lesion-induced plasticity of the auditory pathways (Rauschecker et al. 2010).

9.6 Summary/Discussion

In this chapter, we have provided an overview of perceptual processing from the sensory surfaces to high-level areas of the cortex, highlighting the insights provided by MRI. The sensory pathways are characterized by parallel, hierarchical processing, with a progressively decreasing influence of the spatial distribution of receptors on the sensory surfaces on the representations observed. Topographic maps are a common feature at lower levels of the hierarchies, whereas higher-order representations appear to capture specific properties of the stimuli that likely reflect the specific information required for particular functions. Finally, we have emphasized the integration of both bottom-up and top-down signals in producing a percept and the dynamic nature of perceptual representations throughout the lifetime (Box 9.4).

While we primarily focused on unimodal perception, it is important to realize that our perception of the external world is multimodal and there are major interactions between the different sensory pathways. For example, viewing appetizing foods elicits activation in the gustatory cortex (Simmons et al. 2005). Similarly, haptic object recognition appears to engage not only secondary somatosensory cortex but also areas of OTC that are primarily considered visual (Amedi et al. 2001; Pietrini et al. 2004; Reed et al. 2004; for further discussion see James et al. 2007). However, the precise role of such cross-modal interactions remains unclear.

Finally, a complete account of the neural substrates of perception will require a detailed understanding of the large-scale connectivity and circuits across the brain. Here, we focused on fMRI, but techniques such as diffusion-weighted MRI and functional connectivity promise to provide important insights into these aspects of brain function.

Box 9.4. Summary

- fMRI provides insights into how sensory stimuli are processed across the brain to form coherent percepts.
- Physical stimuli in the environment (e.g., light, sound waves) are processed by specialized sensory surfaces (e.g., retina, cochlea) and project to distinct primary sensory cortices.
- Sensory systems are both parallel and hierarchical, with distinct large-scale pathways extending from subcortical structures to high-level areas of cortex.
- Subcortical and early cortical areas often contain topographic maps reflecting the organization of receptors on the sensory surfaces (e.g., retinotopic maps in primary visual cortex).
- High-level areas show complex forms of organization, such as by category or other perceptual qualities.
- Perceptual processing within cortical regions reflects both bottom-up (properties of the stimuli) and top-down (internally generated) task requirements.
- Sensory representations are highly dependent on experience from development through adulthood and can be modified by prior exposure, training, or deprivation of sensory input (e.g., blindness).

References

Adrian E (1950) Sensory discrimination: with some recent evidence from the olfactory organ. Br Med Bull 6:330–332

Aguirre GK (2007) Continuous carry-over designs for fMRI. Neuroimage 35:1480–1494

Aguirre GK, Zarahn E, D'Esposito M (1998) An area within human ventral cortex sensitive to "building" stimuli: evidence and implications. Neuron 21: 373–383

Aguirre GK, D'Esposito M (1999) Topographical disorientation: a synthesis and taxonomy. Brain 122(Pt 9):1613–1628

Amedi A, Malach R, Hendler T, Peled S, Zohary E (2001) Visuo-haptic object-related activation in the ventral visual pathway. Nat Neurosci 4(3):324–330

Amedi A, Raz N, Pianka P, Malach R, Zohary E (2003) Early 'visual' cortex activation correlates with superior verbal memory performance in the blind. Nat Neurosci 6:758–766

Amedi A, Malach R, Pascual-Leone A (2005) Negative BOLD differentiates visual imagery and perception. Neuron 48:859–872

Arcaro MJ, McMains SA, Singer BD, Kastner S (2009) Retinotopic organization of human ventral visual cortex. J Neurosci 29:10638–10652

Arichi T, Moraux A, Melendez A, Doria V, Groppo M, Merchant N, Combs S, Burdet E, Larkman DJ, Counsell SJ, Beckmann CF, Edwards AD (2010) Somatosensory cortical activation identified by functional MRI in preterm and term infants. Neuroimage 49:2063–2071

Arzi A, Sobel N (2011) Olfactory perception as a compass for olfactory neural maps. Trends Cogn Sci 15:537–545

Baker CI, Peli E, Knouf N, Kanwisher NG (2005) Reorganization of visual processing in macular degeneration. J Neurosci 25:614–618

Baker CI, Liu J, Wald LL, Kwong KK, Benner T, Kanwisher N (2007) Visual word processing and experiential origins of functional selectivity in human extrastriate cortex. Proc Natl Acad Sci U S A 104: 9087–9092

Baker CI, Dilks DD, Peli E, Kanwisher N (2008) Reorganization of visual processing in macular degeneration: replication and clues about the role of foveal loss. Vision Res 48:1910–1919

Bandettini PA (2009) What's new in neuroimaging methods? Ann N Y Acad Sci 1156:260–293

Bandettini PA, Jesmanowicz A, Van Kylen J, Birn RM, Hyde JS (1998) Functional MRI of brain activation induced by scanner acoustic noise. Magn Reson Med 39:410–416

Barton JJ, Cherkasova M (2003) Face imagery and its relation to perception and covert recognition in prosopagnosia. Neurology 61(2):220–225

Baseler HA, Gouws A, Haak KV, Racey C, Crossland MD, Tufail A, Rubin GS, Cornelissen FW, Morland AB (2011) Large-scale remapping of visual cortex is absent in adult humans with macular degeneration. Nat Neurosci 14:649–655

Beauchamp MS, Laconte S, Yasar N (2009) Distributed representation of single touches in somatosensory and visual cortex. Hum Brain Mapp 30:3163–3171

Behrmann M, Winocur G, Moscovitch M (1992) Dissociation between mental imagery and object recognition in a brain-damaged patient. Nature 359(6396):636–637

Belin P, Zatorre RJ (2000) 'What', 'where' and 'how' in auditory cortex. Nat Neurosci 3:965–966

Belin P, Zatorre RJ (2003) Adaptation to speaker's voice in right anterior temporal lobe. Neuroreport 14: 2105–2109

Bensafi M, Sobel N, Khan RM (2007) Hedonic-specific activity in piriform cortex during odor imagery mimics that during odor perception. J Neurophysiol 98: 3254–3262

Binder JR, Frost JA, Hammeke TA, Bellgowan PS, Springer JA, Kaufman JN, Possing ET (2000) Human temporal lobe activation by speech and nonspeech sounds. Cereb Cortex 10:512–528

Boring EG (1942) Sensation and perception in the history of experimental psychology. D. Appleton-Century company, incorporated, New York

Brants M, Wagemans J, Op de Beeck HP (2011) Activation of fusiform face area by Greebles is related to face similarity but not expertise. J Cogn Neurosci 23: 3949–3958

Brefczynski JA, DeYoe EA (1999) A physiological correlate of the 'spotlight' of visual attention. Nat Neurosci 2:370–374

Briggs RW, Dy-Liacco I, Malcolm MP, Lee H, Peck KK, Gopinath KS, Himes NC, Soltysik DA, Browne P, Tran-Son-Tay R (2004) A pneumatic vibrotactile stimulation device for fMRI. Magn Reson Med 51: 640–643

Brouwer GJ, Heeger DJ (2009) Decoding and reconstructing color from responses in human visual cortex. J Neurosci 29:13992–14003

Bunzeck N, Wuestenberg T, Lutz K, Heinze HJ, Jancke L (2005) Scanning silence: mental imagery of complex sounds. Neuroimage 26:1119–1127

Buonomano DV, Merzenich MM (1998) Cortical plasticity: from synapses to maps. Annu Rev Neurosci 21:149–186

Burton H, Snyder AZ, Conturo TE, Akbudak E, Ollinger JM, Raichle ME (2002) Adaptive changes in early and late blind: a fMRI study of braille reading. J Neurophysiol 87:589–607

Carlin JD, Calder AJ, Kriegeskorte N, Nili H, Rowe JB (2011) A head view-invariant representation of gaze direction in anterior superior temporal sulcus. Curr Biol 21:1817–1821

Carlson TA, Schrater P, He S (2003) Patterns of activity in the categorical representations of objects. J Cogn Neurosci 15:704–717

Cassia VM, Picozzi M, Kuefner D, Bricolo E, Turati C (2009) Holistic processing for faces and cars in preschool-aged children and adults: evidence from the composite effect. Dev Sci 12:236–248

Castriota-Scanderbeg A, Hagberg GE, Cerasa A, Committeri G, Galati G, Patria F, Pitzalis S, Caltagirone C, Frackowiak R (2005) The appreciation of wine by sommeliers: a functional magnetic resonance study of sensory integration. Neuroimage 25: 570–578

Chao LL, Haxby JV, Martin A (1999) Attribute-based neural substrates in temporal cortex for perceiving and knowing about objects. Nat Neurosci 2:913–919

Chen W, Zhu XH, Thulborn KR, Ugurbil K (1999) Retinotopic mapping of lateral geniculate nucleus in humans using functional magnetic resonance imaging. Proc Natl Acad Sci U S A 96:2430–2434

Cheng K, Waggoner RA, Tanaka K (2001) Human ocular dominance columns as revealed by high-field functional magnetic resonance imaging. Neuron 32:359–374

Chevillet M, Riesenhuber M, Rauschecker JP (2011) Functional correlates of the anterolateral processing hierarchy in human auditory cortex. J Neurosci 31:9345–9352

Ciaramitaro VM, Mitchell JF, Stoner GR, Reynolds JH, Boynton GM (2011) Object-based attention to one of two superimposed surfaces alters responses in human early visual cortex. J Neurophysiol 105:1258–1265

Cohen L, Dehaene S (2004) Specialization within the ventral stream: the case for the visual word form area. Neuroimage 22:466–476

Collins CE, Leitch DB, Wong P, Kaas JH, Herculano-Houzel S (2013) Faster scaling of visual neurons in cortical areas relative to subcortical structures in non-human primate brains. Brain Struct Funct 218(3): 805–816

Conway BR, Moeller S, Tsao DY (2007) Specialized color modules in macaque extrastriate cortex. Neuron 56:560–573

Cox DD, Savoy RL (2003) Functional magnetic resonance imaging (fMRI) "brain reading": detecting and classifying distributed patterns of fMRI activity in human visual cortex. Neuroimage 19:261–270

Cox DD, Meier P, Oertelt N, DiCarlo JJ (2005) 'Breaking' position-invariant object recognition. Nat Neurosci 8:1145–1147

Culham JC, Danckert SL, DeSouza JF, Gati JS, Menon RS, Goodale MA (2003) Visually guided grasping produces fMRI activation in dorsal but not ventral stream brain areas. Exp Brain Res 153:180–189

Cynader M, Berman N (1972) Receptive-field organization of monkey superior colliculus. J Neurophysiol 35:187–201

Da Costa S, van der Zwaag W, Marques JP, Frackowiak RS, Clarke S, Saenz M (2011) Human primary auditory cortex follows the shape of Heschl's gyrus. J Neurosci 31:14067–14075

Diamond R, Carey S (1986) Why faces are and are not special: an effect of expertise. J Exp Psychol Gen 115(2):107

De Araujo IET, Kringelbach ML, Rolls ET, Hobden P (2003) Representation of umami taste in the human brain. J Neurophysiol 90:313–319

De Araujo IE, Rolls ET, Velazco MI, Margot C, Cayeux I (2005) Cognitive modulation of olfactory processing. Neuron 46:671–679

De Martino F, Valente G, Staeren N, Ashburner J, Goebel R, Formisano E (2008) Combining multivariate voxel selection and support vector machines for mapping and classification of fMRI spatial patterns. Neuroimage 43:44–58

Desimone R (1996) Neural mechanisms for visual memory and their role in attention. Proc Natl Acad Sci U S A 93:13494–13499

DeWitt I, Rauschecker JP (2012) Phoneme and word recognition in the auditory ventral stream. Proc Natl Acad Sci U S A 109:E505–E514

DiCarlo JJ, Zoccolan D, Rust NC (2012) How does the brain solve visual object recognition? Neuron 73:415–434

Dijkerman HC, de Haan EH (2007) Somatosensory processes subserving perception and action. Behav Brain Sci 30:189–201, Discussion 201-139

Dilks DD, Serences JT, Rosenau BJ, Yantis S, McCloskey M (2007) Human adult cortical reorganization and consequent visual distortion. J Neurosci 27:9585–9594

Disbrow E, Roberts T, Krubitzer L (2000) Somatotopic organization of cortical fields in the lateral sulcus of Homo sapiens: evidence for SII and PV. J Comp Neurol 418:1–21

Doehrmann O, Naumer MJ, Volz S, Kaiser J, Altmann CF (2008) Probing category selectivity for environmental sounds in the human auditory brain. Neuropsychologia 46:2776–2786

Downing PE, Jiang Y, Shuman M, Kanwisher N (2001) A cortical area selective for visual processing of the human body. Science 293:2470–2473

Downing PE, Chan AW, Peelen MV, Dodds CM, Kanwisher N (2006) Domain specificity in visual cortex. Cereb Cortex 16:1453–1461

Dresel C, Parzinger A, Rimpau C, Zimmer C, Ceballos-Baumann AO, Haslinger B (2008) A new device for tactile stimulation during fMRI. Neuroimage 39: 1094–1103

DuBois RM, Cohen MS (2000) Spatiotopic organization in human superior colliculus observed with fMRI. Neuroimage 12:63–70

Dumoulin SO, Wandell BA (2008) Population receptive field estimates in human visual cortex. Neuroimage 39:647–660

Duncan RO, Boynton GM (2003) Cortical magnification within human primary visual cortex correlates with acuity thresholds. Neuron 38:659–671

Duncan RO, Boynton GM (2007) Tactile hyperacuity thresholds correlate with finger maps in primary somatosensory cortex (S1). Cereb Cortex 17: 2878–2891

Eickhoff SB, Grefkes C, Zilles K, Fink GR (2007) The somatotopic organization of cytoarchitectonic areas on the human parietal operculum. Cereb Cortex 17: 1800–1811

Eldeghaidy S, Marciani L, McGlone F, Hollowood T, Hort J, Head K, Taylor AJ, Busch J, Spiller RC, Gowland PA, Francis ST (2011) The cortical response to the oral perception of fat emulsions and the effect of taster status. J Neurophysiol 105:2572–2581

Ellis RJ, Norton AC, Overy K, Winner E, Alsop DC, Schlaug G (2012) Differentiating maturational and training influences on fMRI activation during music processing. Neuroimage 60:1902–1912

Engel SA (2012) The development and use of phase-encoded functional MRI designs. Neuroimage 62: 1195–1200

Engel SA, Rumelhart DE, Wandell BA, Lee AT, Glover GH, Chichilnisky EJ, Shadlen MN (1994) fMRI of human visual cortex. Nature 369:525

Epstein R, Kanwisher N (1998) A cortical representation of the local visual environment. Nature 392:598–601

Espinosa JS, Stryker MP (2012) Development and plasticity of the primary visual cortex. Neuron 75: 230–249

Felleman DJ, Van Essen DC (1991) Distributed hierarchical processing in the primate cerebral cortex. Cereb Cortex 1:1–47

Finney EM, Fine I, Dobkins KR (2001) Visual stimuli activate auditory cortex in the deaf. Nat Neurosci 4:1171–1173

Flor H, Nikolajsen L, Staehelin Jensen T (2006) Phantom limb pain: a case of maladaptive CNS plasticity? Nat Rev Neurosci 7:873–881

Formisano E, Kim D-S, Di Salle F, van de Moortele P-F, Ugurbil K, Goebel R (2003) Mirror-symmetric tonotopic maps in human primary auditory cortex. Neuron 40:859–869

Formisano E, De Martino F, Bonte M, Goebel R (2008) “Who” is saying “what”? Brain-based decoding of human voice and speech. Science 322:970–973

Frank GK, Kaye WH, Carter CS, Brooks S, May C, Fissell K, Stenger VA (2003) The evaluation of brain activity in response to taste stimuli–a pilot study and method for central taste activation as assessed by event-related fMRI. J Neurosci Methods 131:99–105

Freeman J, Brouwer GJ, Heeger DJ, Merriam EP (2011) Orientation decoding depends on maps, not columns. J Neurosci 31:4792–4804

Friston K (2005) A theory of cortical responses. Philos Trans R Soc Lond B Biol Sci 360:815–836

Ganis G, Thompson WL, Kosslyn SM (2004) Brain areas underlying visual mental imagery and visual perception: an fMRI study. Brain Res Cogn Brain Res 20:226–241

Gaser C, Schlaug G (2003) Brain structures differ between musicians and non-musicians. J Neurosci 23: 9240–9245

Gauthier I, Tarr MJ, Anderson AW, Skudlarski P, Gore JC (1999) Activation of the middle fusiform ‘face area’ increases with expertise in recognizing novel objects. Nat Neurosci 2:568–573

Gauthier I, Skudlarski P, Gore JC, Anderson AW (2000) Expertise for cars and birds recruits brain areas involved in face recognition. Nat Neurosci 3: 191–197

Gazzaley A, Nobre AC (2012) Top-down modulation: bridging selective attention and working memory. Trends Cogn Sci 16:129–135

Gilaie-Dotan S, Harel A, Bentin S, Kanai R, Rees G (2012) Neuroanatomical correlates of visual car expertise. Neuroimage 62:147–153

Golarai G, Ghahremani DG, Whitfield-Gabrieli S, Reiss A, Eberhardt JL, Gabrieli JD, Grill-Spector K (2007) Differential development of high-level visual cortex correlates with category-specific recognition memory. Nat Neurosci 10:512–522

Golarai G, Liberman A, Yoon JM, Grill-Spector K (2010) Differential development of the ventral visual cortex extends through adolescence. Front Hum Neurosci 3:80

Gonzalez-Castillo J, Olulade OA, Talavage TM (2011) Using functional MRI to study auditory comprehension. Imaging Med 4:137–143

Gottfried JA (2010) Central mechanisms of odour object perception. Nat Rev Neurosci 11:628–641

Gottfried JA, Winston JS, Dolan RJ (2006) Dissociable codes of odor quality and odorant structure in human piriform cortex. Neuron 49:467–479

Gotts S, Chow C, Martin A (2012) Repetition priming and repetition suppression: multiple mechanisms in need of testing. Cogn Neurosci 3:250–259

Gougoux F, Zatorre RJ, Lassonde M, Voss P, Lepore F (2005) A functional neuroimaging study of sound localization: visual cortex activity predicts performance in early-blind individuals. PLoS Biol 3:e27

Grill-Spector K, Malach R (2001) fMR-adaptation: a tool for studying the functional properties of human cortical neurons. Acta Psychol (Amst) 107: 293–321

Grill-Spector K, Malach R (2004) The human visual cortex. Annu Rev Neurosci 27:649–677

Grill-Spector K, Kushnir T, Edelman S, Avidan G, Itzchak Y, Malach R (1999) Differential processing of objects under various viewing conditions in the human lateral occipital complex. Neuron 24:187–203

Grill-Spector K, Knouf N, Kanwisher N (2004) The fusiform face area subserves face perception, not generic within-category identification. Nat Neurosci 7:555–562

Grill-Spector K, Henson R, Martin A (2006) Repetition and the brain: neural models of stimulus-specific effects. Trends Cogn Sci 10:14–23

Haase L, Cerf-Ducastel B, Buracas G, Murphy C (2007) On-line psychophysical data acquisition and event-related fMRI protocol optimized for the investigation of brain activation in response to gustatory stimuli. J Neurosci Methods 159:98–107

Harel A, Bentin S (2009) Stimulus type, level of categorization, and spatial-frequencies utilization: implications for perceptual categorization hierarchies. J Exp Psychol Hum Percept Perform 35:1264–1273

Harel A, Gilaie-Dotan S, Malach R, Bentin S (2010) Top-down engagement modulates the neural expressions of visual expertise. Cereb Cortex 20:2304–2318

Harel A, Kravitz DJ, Baker CI (2013) Deconstructing visual scenes in cortex: gradients of object and spatial layout information. Cereb Cortex 23:947–957

Hasson U, Levy I, Behrmann M, Hendler T, Malach R (2002) Eccentricity bias as an organizing principle for human high-order object areas. Neuron 34: 479–490

Hasson U, Harel M, Levy I, Malach R (2003) Large-scale mirror-symmetry organization of human occipito-temporal object areas. Neuron 37:1027–1041

Haxby JV, Gobbini MI, Furey ML, Ishai A, Schouten JL, Pietrini P (2001) Distributed and overlapping representations of faces and objects in ventral temporal cortex. Science 293:2425–2430

Haynes JD, Rees G (2006) Decoding mental states from brain activity in humans. Nat Rev Neurosci 7: 523–534

Haynes J-D, Deichmann R, Rees G (2005) Eye-specific effects of binocular rivalry in the human lateral geniculate nucleus. Nature 438:496–499

Henson RN (2003) Neuroimaging studies of priming. Prog Neurobiol 70:53–81

Hinds OP, Rajendran N, Polimeni JR, Augustinack JC, Wiggins G, Wald LL, Diana Rosas H, Potthast A, Schwartz EL, Fischl B (2008) Accurate prediction of V1 location from cortical folds in a surface coordinate system. Neuroimage 39:1585–1599

Holmes G (1931) A contribution to the cortical representation of vision. Brain 54:470–479

Horton JC, Hoyt WF (1991) The representation of the visual field in human striate cortex. A revision of the classic Holmes map. Arch Ophthalmol 109:816–824

Howard JD, Plailly J, Grueschow M, Haynes J-D, Gottfried JA (2009) Odor quality coding and categorization in human posterior piriform cortex. Nat Neurosci 12:932–938

Huang RS, Sereno MI (2007) Dodecapus: an MR-compatible system for somatosensory stimulation. Neuroimage 34:1060–1073

Huk AC, Dougherty RF, Heeger DJ (2002) Retinotopy and functional subdivision of human areas MT and MST. J Neurosci 22:7195–7205

Humphries C, Liebenthal E, Binder JR (2010) Tonotopic organization of human auditory cortex. Neuroimage 50:1202–1211

Illig KR, Haberly LB (2003) Odor-evoked activity is spatially distributed in piriform cortex. J Comp Neurol 457:361–373

Ingeholm JE, Dold GR, Pfeffer LE, Ide D, Goldstein SR, Johnson KO, Van Boven RW (2006) The Helix: a multi-modal tactile stimulator for human functional neuroimaging. J Neurosci Methods 155:217–223

James TW, Gauthier I (2006) Repetition-induced changes in BOLD response reflect accumulation of neural activity. Hum Brain Mapp 27:37–46

James TW, Kim S, Fisher JS (2007) The neural basis of haptic object processing. Can J Exp Psychol (Revue canadienne de psychologie expérimentale) 61(3):219

Jancke L, Gaab N, Wustenberg T, Scheich H, Heinze HJ (2001) Short-term functional plasticity in the human auditory cortex: an fMRI study. Brain Res Cogn Brain Res 12:479–485

Johnson KO, Hsiao SS (1992) Neural mechanisms of tactual form and texture perception. Annu Rev Neurosci 15:227–250

Johnson BN, Sobel N (2007) Methods for building an olfactometer with known concentration outcomes. J Neurosci Methods 160:231–245

Kaas JH (1997) Topographic maps are fundamental to sensory processing. Brain Res Bull 44:107–112

Kaas JH, Hackett TA (2000) Subdivisions of auditory cortex and processing streams in primates. Proc Natl Acad Sci U S A 97:11793–11799

Kami YN, Goto TK, Tokumori K, Yoshiura T, Kobayashi K, Nakamura Y, Honda H, Ninomiya Y, Yoshiura K (2008) The development of a novel automated taste stimulus delivery system for fMRI studies on the human cortical segregation of taste. J Neurosci Methods 172:48–53

Kamitani Y, Tong F (2005) Decoding the visual and subjective contents of the human brain. Nat Neurosci 8:679–685

Kanwisher N, Dilks DD (2013) The functional organization of the ventral visual pathway in humans. In: Chalupa LM, Werner JS (eds) The new visual neurosciences. MIT Press, Cambridge, MA

Kanwisher N, McDermott J, Chun MM (1997) The fusiform face area: a module in human extrastriate cortex specialized for face perception. J Neurosci 17:4302–4311

Karns CM, Dow MW, Neville HJ (2012) Altered cross-modal processing in the primary auditory cortex of congenitally deaf adults: a visual-somatosensory FMRI study with a double-flash illusion. J Neurosci 32:9626–9638

Kastner S, Pinsk MA, De Weerd P, Desimone R, Ungerleider LG (1999) Increased activity in human visual cortex during directed attention in the absence of visual stimulation. Neuron 22:751–761

Kastner S, O'Connor DH, Fukui MM, Fehd HM, Herwig U, Pinsk MA (2004) Functional imaging of the human lateral geniculate nucleus and pulvinar. J Neurophysiol 91:438–448

Katyal S, Zughni S, Greene C, Ress D (2010) Topography of covert visual attention in human superior colliculus. J Neurophysiol 104:3074–3083

Kay KN, Naselaris T, Prenger RJ, Gallant JL (2008) Identifying natural images from human brain activity. Nature 452:352–355

Keil B, Alagappan V, Mareyam A, McNab JA, Fujimoto K, Tountcheva V, Triantafyllou C, Dilks DD, Kanwisher N, Lin W, Grant PE, Wald LL (2011) Size-optimized 32-channel brain arrays for 3 T pediatric imaging. Magn Reson Med 66:1777–1787

Keysers C, Kaas JH, Gazzola V (2010) Somatosensation in social perception. Nat Rev Neurosci 11:417–428

Kobayashi M, Takeda M, Hattori N, Fukunaga M, Sasabe T, Inoue N, Nagai Y, Sawada T, Sadato N, Watanabe Y (2004) Functional imaging of gustatory perception and imagery: "top-down" processing of gustatory signals. Neuroimage 23:1271–1282

Koelsch S, Fritz T, Schulze K, Alsop D, Schlaug G (2005) Adults and children processing music: an fMRI study. Neuroimage 25:1068–1076

Konkle T, Oliva A (2012) A real-world size organization of object responses in occipitotemporal cortex. Neuron 74:1114–1124

Kosslyn SM, Thompson WL, Ganis G (2006) The case for mental imagery. Oxford University Press, Oxford

Kourtzi Z, Kanwisher N (2000) Cortical regions involved in perceiving object shape. J Neurosci 20:3310–3318

Kourtzi Z, Kanwisher N (2001) Representation of perceived object shape by the human lateral occipital complex. Science 293:1506–1509

Kourtzi Z, Betts LR, Sarkheil P, Welchman AE (2005) Distributed neural plasticity for shape learning in the human visual cortex. PLoS Biol 3:e204

Kourtzi Z, Augath M, Logothetis NK, Movshon JA, Kiorpes L (2006) Development of visually evoked cortical activity in infant macaque monkeys studied longitudinally with fMRI. Magn Reson Imaging 24:359–366

Kravitz DJ, Vinson LD, Baker CI (2008) How position dependent is visual object recognition? Trends Cogn Sci 12:114–122

Kravitz DJ, Kriegeskorte N, Baker CI (2010) High-level visual object representations are constrained by position. Cereb Cortex 20:2916–2925

Kravitz DJ, Peng CS, Baker CI (2011a) Real-world scene representations in high-level visual cortex: it's the spaces more than the places. J Neurosci 31: 7322–7333

Kravitz DJ, Saleem KS, Baker CI, Mishkin M (2011b) A new neural framework for visuospatial processing. Nat Rev Neurosci 12:217–230

Kriegeskorte N, Goebel R, Bandettini P (2006) Information-based functional brain mapping. Proc Natl Acad Sci U S A 103:3863–3868

Kriegeskorte N, Formisano E, Sorger B, Goebel R (2007) Individual faces elicit distinct response patterns in human anterior temporal cortex. Proc Natl Acad Sci U S A 104:20600–20605

Kriegeskorte N, Mur M, Bandettini P (2008a) Representational similarity analysis - connecting the branches of systems neuroscience. Front Syst Neurosci 2:4

Kriegeskorte N, Mur M, Ruff DA, Kiani R, Bodurka J, Esteky H, Tanaka K, Bandettini PA (2008b) Matching categorical object representations in inferior temporal cortex of man and monkey. Neuron 60:1126–1141

Lamme VA, Roelfsema PR (2000) The distinct modes of vision offered by feedforward and recurrent processing. Trends Neurosci 23:571–579

Langers DRM, van Dijk P (2012) Mapping the tonotopic organization in human auditory cortex with minimally salient acoustic stimulation. Cereb Cortex 22:2024–2038

Larsson J, Heeger DJ (2006) Two retinotopic visual areas in human lateral occipital cortex. J Neurosci 26:13128–13142

Leaver AM, Rauschecker JP (2010) Cortical representation of natural complex sounds: effects of acoustic features and auditory object category. J Neurosci 30:7604–7612

Lee J, Maunsell JH (2009) A normalization model of attentional modulation of single unit responses. PLoS One 4:e4651

Lee SH, Kravitz DJ, Baker CI (2012) Disentangling visual imagery and perception of real-world objects. Neuroimage 59:4064–4073

Lerner Y, Hendler T, Ben-Bashat D, Harel M, Malach R (2001) A hierarchical axis of object processing stages in the human visual cortex. Cereb Cortex 11:287–297

Levy LM, Henkin RI, Lin CS, Hutter A, Schellinger D (1999) Odor memory induces brain activation as measured by functional MRI. J Comput Assist Tomogr 23:487–498

Levy I, Hasson U, Avidan G, Hendler T, Malach R (2001) Center-periphery organization of human object areas. Nat Neurosci 4:533–539

Lewis JW, Brefczynski JA, Phinney RE, Janik JJ, DeYoe EA (2005) Distinct cortical pathways for processing tool versus animal sounds. J Neurosci 25:5148–5158

Lewis LB, Saenz M, Fine I (2010) Mechanisms of cross-modal plasticity in early-blind subjects. J Neurophysiol 104:2995–3008

Li L, Miller EK, Desimone R (1993) The representation of stimulus familiarity in anterior inferior temporal cortex. J Neurophysiol 69:1918–1929

Li W, Luxenberg E, Parrish T, Gottfried JA (2006) Learning to smell the roses: experience-dependent neural plasticity in human piriform and orbitofrontal cortices. Neuron 52:1097–1108

Lorig TS, Elmes DG, Zald DH, Pardo JV (1999) A computer-controlled olfactometer for fMRI and electrophysiological studies of olfaction. Behav Res Methods Instrum Comput 31:370–375

Lotze M, Flor H, Grodd W, Larbig W, Birbaumer N (2001) Phantom movements and pain. An fMRI study in upper limb amputees. Brain 124:2268–2277

Mahon BZ, Caramazza A (2011) What drives the organization of object knowledge in the brain? Trends Cogn Sci 15:97–103

Mahon BZ, Anzellotti S, Schwarzbach J, Zampini M, Caramazza A (2009) Category-specific organization in the human brain does not require visual experience. Neuron 63:397–405

Malach R, Reppas JB, Benson RR, Kwong KK, Jiang H, Kennedy WA, Ledden PJ, Brady TJ, Rosen BR, Tootell RB (1995) Object-related activity revealed by functional magnetic resonance imaging in human occipital cortex. Proc Natl Acad Sci U S A 92: 8135–8139

Marciani L, Pfeiffer JC, Hort J, Head K, Bush D, Taylor AJ, Spiller RC, Francis S, Gowland PA (2006) Improved methods for fMRI studies of combined taste and aroma stimuli. J Neurosci Methods 158:186–194

Martin A (2006) Shades of Dejerine – forging a causal link between the visual word form area and reading. Neuron 50:173–175

Martin A (2009) Circuits in mind: the neural foundations for object concepts. In: Gazzaniga MS (ed) The cognitive neurosciences, 4th edn. MIT Press, Cambridge, pp 1031–1046

Martinez A, Teder-Salejarvi W, Vazquez M, Molholm S, Foxe JJ, Javitt DC, Di Russo F, Worden MS, Hillyard SA (2006) Objects are highlighted by spatial attention. J Cogn Neurosci 18:298–310

Masuda Y, Dumoulin SO, Nakadomari S, Wandell BA (2008) V1 projection zone signals in human macular degeneration depend on task, not stimulus. Cereb Cortex 18:2483–2493

McKone E, Crookes K, Jeffery L, Dilks DD (2012) A critical review of the development of face recognition: experience is less important than previously believed. Cogn Neuropsychol 29:174–212

Medina J, Coslett HB (2010) From maps to form to space: touch and the body schema. Neuropsychologia 48: 645–654

Merabet LB, Pascual-Leone A (2010) Neural reorganization following sensory loss: the opportunity of change. Nat Rev Neurosci 11:44–52

Merabet LB, Swisher JD, McMains SA, Halko MA, Amedi A, Pascual-Leone A, Somers DC (2007) Combined activation and deactivation of visual cortex during tactile sensory processing. J Neurophysiol 97: 1633–1641

Merabet LB, Hamilton R, Schlaug G, Swisher JD, Kiriakopoulos ET, Pitskel NB, Kauffman T, Pascual-Leone A (2008) Rapid and reversible recruitment of early visual cortex for touch. PLoS One 3:e3046

Mesulam MM (1998) From sensation to cognition. Brain 121(Pt 6):1013–1052

Milner AD, Goodale MA (2006) The visual brain in action. Oxford University Press, Oxford

Miyawaki Y, Uchida H, Yamashita O, Sato MA, Morito Y, Tanabe HC, Sadato N, Kamitani Y (2008) Visual image reconstruction from human brain activity using a combination of multiscale local image decoders. Neuron 60:915–929

Mondloch CJ, Geldart S, Maurer D, Le Grand R (2003) Developmental changes in face processing skills. J Exp Child Psychol 86:67–84

Moore CI, Stern CE, Corkin S, Fischl B, Gray AC, Rosen BR, Dale AM (2000) Segregation of somatosensory activation in the human rolandic cortex using fMRI. J Neurophysiol 84:558–569

Moro V, Urgesi C, Pernigo S, Lanteri P, Pazzaglia M, Aglioti SM (2008) The neural basis of body form and body action agnosia. Neuron 60(2):235–246

Muller NG, Kleinschmidt A (2003) Dynamic interaction of object- and space-based attention in retinotopic visual areas. J Neurosci 23:9812–9816

Muller NG, Bartelt OA, Donner TH, Villringer A, Brandt SA (2003) A physiological correlate of the "Zoom Lens" of visual attention. J Neurosci 23:3561–3565

Murray EA, Mishkin M (1984) Relative contributions of SII and area 5 to tactile discrimination in monkeys. Behav Brain Res 11:67–83

Murray SO, Wojciulik E (2004) Attention increases neural selectivity in the human lateral occipital complex. Nat Neurosci 7:70–74

Murray SO, Boyaci H, Kersten D (2006) The representation of perceived angular size in human primary visual cortex. Nat Neurosci 9:429–434

Murthy VN (2011) Olfactory maps in the brain. Annu Rev Neurosci 34:233–258

Mycroft RH, Behrmann M, Kay J (2009) Visuoperceptual deficits in letter-by-letter reading? Neuropsychologia 47(7):1733–1744

Naselaris T, Prenger RJ, Kay KN, Oliver M, Gallant JL (2009) Bayesian reconstruction of natural images from human brain activity. Neuron 63:902–915

Naselaris T, Kay KN, Nishimoto S, Gallant JL (2011) Encoding and decoding in fMRI. Neuroimage 56:400–410

Nassi JJ, Callaway EM (2009) Parallel processing strategies of the primate visual system. Nat Rev Neurosci 10:360–372

Nelson AJ, Chen R (2008) Digit somatotopy within cortical areas of the postcentral gyrus in humans. Cereb Cortex 18:2341–2351

Nestor A, Plaut DC, Behrmann M (2011) Unraveling the distributed neural code of facial identity through spatiotemporal pattern analysis. Proc Natl Acad Sci U S A 108:9998–10003

Nihashi T, Naganawa S, Sato C, Kawai H, Nakamura T, Fukatsu H, Ishigaki T, Aoki I (2005) Contralateral and ipsilateral responses in primary somatosensory cortex

following electrical median nerve stimulation–an fMRI study. Clin Neurophysiol 116:842–848
Nishimoto S, Vu AT, Naselaris T, Benjamini Y, Yu B, Gallant JL (2011) Reconstructing visual experiences from brain activity evoked by natural movies. Curr Biol 21:1641–1646
Nitschke JB, Dixon GE, Sarinopoulos I, Short SJ, Cohen JD, Smith EE, Kosslyn SM, Rose RM, Davidson RJ (2006) Altering expectancy dampens neural response to aversive taste in primary taste cortex. Nat Neurosci 9:435–442
Norman KA, Polyn SM, Detre GJ, Haxby JV (2006) Beyond mind-reading: multi-voxel pattern analysis of fMRI data. Trends Cogn Sci 10:424–430
O'Connor DH, Fukui MM, Pinsk MA, Kastner S (2002) Attention modulates responses in the human lateral geniculate nucleus. Nat Neurosci 5:1203–1209
O'Craven KM, Kanwisher N (2000) Mental imagery of faces and places activates corresponding stimulus-specific brain regions. J Cogn Neurosci 12: 1013–1023
O'Craven KM, Downing PE, Kanwisher N (1999) fMRI evidence for objects as the units of attentional selection. Nature 401:584–587
Ogawa H, Wakita M, Hasegawa K, Kobayakawa T, Sakai N, Hirai T, Yamashita Y, Saito S (2005) Functional MRI detection of activation in the primary gustatory cortices in humans. Chem Senses 30:583–592
Ohnishi T, Matsuda H, Asada T, Aruga M, Hirakata M, Nishikawa M, Katoh A, Imabayashi E (2001) Functional anatomy of musical perception in musicians. Cereb Cortex 11:754–760
Op de Beeck HP (2010) Against hyperacuity in brain reading: spatial smoothing does not hurt multivariate fMRI analyses? Neuroimage 49:1943–1948
Op de Beeck HP, Baker CI (2010) The neural basis of visual object learning. Trends Cogn Sci 14:22–30
Op De Beeck H, Vogels R (2000) Spatial sensitivity of macaque inferior temporal neurons. J Comp Neurol 426:505–518
Op de Beeck HP, Baker CI, DiCarlo JJ, Kanwisher NG (2006) Discrimination training alters object representations in human extrastriate cortex. J Neurosci 26:13025–13036
Op de Beeck HP, Haushofer J, Kanwisher NG (2008) Interpreting fMRI data: maps, modules and dimensions. Nat Rev Neurosci 9:123–135
Overduin SA, Servos P (2004) Distributed digit somatotopy in primary somatosensory cortex. Neuroimage 23:462–472
Park S, Brady TF, Greene MR, Oliva A (2011) Disentangling scene content from spatial boundary: complementary roles for the parahippocampal place area and lateral occipital complex in representing real-world scenes. J Neurosci 31:1333–1340
Penfield W, Boldrey E (1937) Somatic motor and sensory representation in the cerebral cortex of man as studied by electrical stimulation. Brain 60:389–443
Perani D, Saccuman MC, Scifo P, Spada D, Andreolli G, Rovelli R, Baldoli C, Koelsch S (2010) Functional specializations for music processing in the human newborn brain. Proc Natl Acad Sci U S A 107: 4758–4763
Pereira F, Mitchell T, Botvinick M (2009) Machine learning classifiers and fMRI: a tutorial overview. Neuroimage 45:S199–S209
Pietrini P, Furey ML, Ricciardi E, Gobbini MI, Wu WHC, Cohen L, Haxby JV (2004) Beyond sensory images: object-based representation in the human ventral pathway. Proc Natl Acad Sci U S A 101(15):5658–5663
Plailly J, Howard JD, Gitelman DR, Gottfried JA (2008) Attention to odor modulates thalamocortical connectivity in the human brain. J Neurosci 28:5257–5267
Plailly J, Delon-Martin C, Royet JP (2012) Experience induces functional reorganization in brain regions involved in odor imagery in perfumers. Hum Brain Mapp 33:224–234
Pleger B, Foerster AF, Ragert P, Dinse HR, Schwenkreis P, Malin JP, Nicolas V, Tegenthoff M (2003) Functional imaging of perceptual learning in human primary and secondary somatosensory cortex. Neuron 40:643–653
Poirier C, Collignon O, Scheiber C, Renier L, Vanlierde A, Tranduy D, Veraart C, De Volder AG (2006) Auditory motion perception activates visual motion areas in early blind subjects. Neuroimage 31:279–285
Polonara G, Fabri M, Manzoni T, Salvolini U (1999) Localization of the first and second somatosensory areas in the human cerebral cortex with functional MR imaging. AJNR Am J Neuroradiol 20:199–205
Puce A, Allison T, Asgari M, Gore JC, McCarthy G (1996) Differential sensitivity of human visual cortex to faces, letterstrings, and textures: a functional magnetic resonance imaging study. J Neurosci 16:5205–5215
Pylyshyn Z (1999) Is vision continuous with cognition? The case for cognitive impenetrability of visual perception. Behav Brain Sci 22:341–365, discussion 366–423
Pylyshyn ZW (2002) Mental imagery: in search of a theory. Behav Brain Sci 25:157–182, discussion 182–237
Raschle N, Zuk J, Ortiz-Mantilla S, Sliva DD, Franceschi A, Grant PE, Benasich AA, Gaab N (2012) Pediatric neuroimaging in early childhood and infancy: challenges and practical guidelines. Ann N Y Acad Sci 1252:43–50
Rauschecker JP (2011) An expanded role for the dorsal auditory pathway in sensorimotor control and integration. Hear Res 271:16–25
Rauschecker JP, Scott SK (2009) Maps and streams in the auditory cortex: nonhuman primates illuminate human speech processing. Nat Neurosci 12:718–724
Rauschecker JP, Leaver AM, Muhlau M (2010) Tuning out the noise: limbic-auditory interactions in tinnitus. Neuron 66:819–826
Recanzone GH, Guard DC, Phan ML (2000) Frequency and intensity response properties of single neurons in the auditory cortex of the behaving macaque monkey. J Neurophysiol 83:2315–2331
Reddy L, Tsuchiya N, Serre T (2010) Reading the mind's eye: decoding category information during mental imagery. Neuroimage 50:818–825

Reed CL, Klatzky RL, Halgren E (2005) What vs. where in touch: an fMRI study. Neuroimage 25:718–726
Reed CL, Shoham S, Halgren E (2004) Neural substrates of tactile object recognition: an fMRI study. Hum Brain Mapp 21(4):236–246
Reynolds JH, Chelazzi L (2004) Attentional modulation of visual processing. Annu Rev Neurosci 27:611–647
Reynolds JH, Heeger DJ (2009) The normalization model of attention. Neuron 61:168–185
Rhodes G, Byatt G, Michie PT, Puce A (2004) Is the fusiform face area specialized for faces, individuation, or expert individuation? J Cogn Neurosci 16:189–203
Rolls ET (2006) Brain mechanisms underlying flavour and appetite. Phil Trans R Soc B: Biol Sci 361:1123–1136
Ruben J, Schwiemann J, Deuchert M, Meyer R, Krause T, Curio G, Villringer K, Kurth R, Villringer A (2001) Somatotopic organization of human secondary somatosensory cortex. Cereb Cortex 11:463–473
Saalmann YB, Kastner S (2011) Cognitive and perceptual functions of the visual thalamus. Neuron 71:209–223
Saito DN, Okada T, Honda M, Yonekura Y, Sadato N (2006) Practice makes perfect: the neural substrates of tactile discrimination by Mah-Jong experts include the primary visual cortex. BMC Neurosci 7:79
Samson F, Zeffiro TA, Toussaint A, Belin P (2011) Stimulus complexity and categorical effects in human auditory cortex: an activation likelihood estimation meta-analysis. Front Psychol 1:241
Sanchez-Panchuelo RM, Francis S, Bowtell R, Schluppeck D (2010) Mapping human somatosensory cortex in individual subjects with 7T functional MRI. J Neurophysiol 103:2544–2556
Sanchez-Panchuelo RM, Francis ST, Schluppeck D, Bowtell RW (2012) Correspondence of human visual areas identified using functional and anatomical MRI in vivo at 7 T. J Magn Reson Imaging 35:287–299
Sangrigoli S, de Schonen S (2004) Effect of visual experience on face processing: a developmental study of inversion and non-native effects. Dev Sci 7:74–87
Savic I, Gulyas B, Larsson M, Roland P (2000) Olfactory functions are mediated by parallel and hierarchical processing. Neuron 26:735–745
Sayres R, Grill-Spector K (2006) Object-selective cortex exhibits performance-independent repetition suppression. J Neurophysiol 95:995–1007
Scherf KS, Behrmann M, Humphreys K, Luna B (2007) Visual category-selectivity for faces, places and objects emerges along different developmental trajectories. Dev Sci 10:F15–F30
Scherf KS, Luna B, Avidan G, Behrmann M (2011) "What" precedes "which": developmental neural tuning in face- and place-related cortex. Cereb Cortex 21:1963–1980
Schneider KA, Kastner S (2005) Visual responses of the human superior colliculus: a high-resolution functional magnetic resonance imaging study. J Neurophysiol 94:2491–2503
Schneider P, Scherg M, Dosch HG, Specht HJ, Gutschalk A, Rupp A (2002) Morphology of Heschl's gyrus reflects enhanced activation in the auditory cortex of musicians. Nat Neurosci 5:688–694
Schneider KA, Richter MC, Kastner S (2004) Retinotopic organization and functional subdivisions of the human lateral geniculate nucleus: a high-resolution functional magnetic resonance imaging study. J Neurosci 24:8975–8985
Schonwiesner M, Zatorre RJ (2009) Spectro-temporal modulation transfer function of single voxels in the human auditory cortex measured with high-resolution fMRI. Proc Natl Acad Sci U S A 106:14611–14616
Schwartz S, Maquet P, Frith C (2002) Neural correlates of perceptual learning: a functional MRI study of visual texture discrimination. Proc Natl Acad Sci U S A 99:17137–17142
Schwarzlose RF, Swisher JD, Dang S, Kanwisher N (2008) The distribution of category and location information across object-selective regions in human visual cortex. Proc Natl Acad Sci U S A 105:4447–4452
Schweisfurth MA, Schweizer R, Frahm J (2011) Functional MRI indicates consistent intra-digit topographic maps in the little but not the index finger within the human primary somatosensory cortex. Neuroimage 56:2138–2143
Schweizer R, Voit D, Frahm J (2008) Finger representations in human primary somatosensory cortex as revealed by high-resolution functional MRI of tactile stimulation. Neuroimage 42:28–35
Seifritz E, Di Salle F, Esposito F, Herdener M, Neuhoff JG, Scheffler K (2006) Enhancing BOLD response in the auditory system by neurophysiologically tuned fMRI sequence. Neuroimage 29:1013–1022
Sereno MI, Dale AM, Reppas JB, Kwong KK, Belliveau JW, Brady TJ, Rosen BR, Tootell RB (1995) Borders of multiple visual areas in humans revealed by functional magnetic resonance imaging. Science 268:889–893
Sherman SM, Guillery RW (2001) Exploring the thalamus. Academic, San Diego
Simmons WK, Martin A, Barsalou LW (2005) Pictures of appetizing foods activate gustatory cortices for taste and reward. Cereb Cortex 15:1602–1608
Slotnick SD, Thompson WL, Kosslyn SM (2005) Visual mental imagery induces retinotopically organized activation of early visual areas. Cereb Cortex 15:1570–1583
Small DM (2010) Taste representation in the human insula. Brain Struct Funct 214:551–561
Small D, Prescott J (2005) Odor/taste integration and the perception of flavor. Exp Brain Res 166:345–357
Small DM, Voss J, Mak YE, Simmons KB, Parrish T, Gitelman D (2004) Experience-dependent neural integration of taste and smell in the human brain. J Neurophysiol 92:1892–1903
Small DM, Gerber JC, Mak YE, Hummel T (2005) Differential neural responses evoked by orthonasal versus retronasal odorant perception in humans. Neuron 47:593–605
Small DM, Veldhuizen MG, Felsted J, Mak YE, McGlone F (2008) Separable substrates for anticipatory and consummatory food chemosensation. Neuron 57:786–797

Sobel N, Prabhakaran V, Desmond JE, Glover GH, Goode RL, Sullivan EV, Gabrieli JD (1998) Sniffing and smelling: separate subsystems in the human olfactory cortex. Nature 392:282–286

Sobotka S, Ringo JL (1994) Stimulus specific adaptation in excited but not in inhibited cells in inferotemporal cortex of macaque. Brain Res 646:95–99

Sperandio I, Chouinard PA, Goodale MA (2012) Retinotopic activity in V1 reflects the perceived and not the retinal size of an afterimage. Nat Neurosci 15:540–542

Staeren N, Renvall H, De Martino F, Goebel R, Formisano E (2009) Sound categories are represented as distributed patterns in the human auditory cortex. Curr Biol 19:498–502

Stokes M, Thompson R, Cusack R, Duncan J (2009) Top-down activation of shape-specific population codes in visual cortex during mental imagery. J Neurosci 29:1565–1572

Striem-Amit E, Hertz U, Amedi A (2011) Extensive cochleotopic mapping of human auditory cortical fields obtained with phase-encoding fMRI. PLoS One 6:e17832

Stringer EA, Chen LM, Friedman RM, Gatenby C, Gore JC (2011) Differentiation of somatosensory cortices by high-resolution fMRI at 7 T. Neuroimage 54:1012–1020

Swisher JD, Halko MA, Merabet LB, McMains SA, Somers DC (2007) Visual topography of human intraparietal sulcus. J Neurosci 27:5326–5337

Talavage TM, Ledden PJ, Benson RR, Rosen BR, Melcher JR (2000) Frequency-dependent responses exhibited by multiple regions in human auditory cortex. Hear Res 150:225–244

Talavage TM, Sereno MI, Melcher JR, Ledden PJ, Rosen BR, Dale AM (2004) Tonotopic organization in human auditory cortex revealed by progressions of frequency sensitivity. J Neurophysiol 91:1282–1296

Talkington WJ, Rapuano KM, Hitt LA, Frum CA, Lewis JW (2012) Humans mimicking animals: a cortical hierarchy for human vocal communication sounds. J Neurosci 32:8084–8093

Tanji K, Leopold DA, Ye FQ, Zhu C, Malloy M, Saunders RC, Mishkin M (2010) Effect of sound intensity on tonotopic fMRI maps in the unanesthetized monkey. Neuroimage 49:150–157

Tarr MJ, Gauthier I (2000) FFA: a flexible fusiform area for subordinate-level visual processing automatized by expertise. Nat Neurosci 3:764–769

Taylor JC, Downing PE (2011) Division of labor between lateral and ventral extrastriate representations of faces, bodies, and objects. J Cogn Neurosci 23:4122–4137

Thirion B, Duchesnay E, Hubbard E, Dubois J, Poline JB, Lebihan D, Dehaene S (2006) Inverse retinotopy: inferring the visual content of images from brain activation patterns. Neuroimage 33:1104–1116

Tootell RB, Silverman MS, Switkes E, De Valois RL (1982) Deoxyglucose analysis of retinotopic organization in primate striate cortex. Science 218:902–904

Tootell RB, Hadjikhani N, Hall EK, Marrett S, Vanduffel W, Vaughan JT, Dale AM (1998) The retinotopy of visual spatial attention. Neuron 21:1409–1422

Trampel R, Ott DV, Turner R (2011) Do the congenitally blind have a stria of Gennari? First intracortical insights in vivo. Cereb Cortex 21:2075–2081

Turner R, Oros-Peusquens AM, Romanzetti S, Zilles K, Shah NJ (2008) Optimised in vivo visualisation of cortical structures in the human brain at 3 T using IR-TSE. Magn Reson Imaging 26:935–942

Ungerleider LG, Mishkin M (1982) Two cortical visual systems. In: Ingle DJ, Mansfield RJW, Goodale MS (eds) The analysis of visual behavior. MIT Press, Cambridge, pp 549–586

Van Boven RW, Ingeholm JE, Beauchamp MS, Bikle PC, Ungerleider LG (2005) Tactile form and location processing in the human brain. Proc Natl Acad Sci USA 102:12601–12605

van der Linden M, Murre JM, van Turennout M (2008) Birds of a feather flock together: experience-driven formation of visual object categories in human ventral temporal cortex. PLoS One 3:e3995

van Turennout M, Ellmore T, Martin A (2000) Long-lasting cortical plasticity in the object naming system. Nat Neurosci 3:1329–1334

Veldhuizen MG, Small DM (2011) Modality-specific neural effects of selective attention to taste and odor. Chem Senses 36:747–760

Veldhuizen MG, Bender G, Constable RT, Small DM (2007) Trying to detect taste in a tasteless solution: modulation of early gustatory cortex by attention to taste. Chem Senses 32:569–581

Veldhuizen MG, Douglas D, Aschenbrenner K, Gitelman DR, Small DM (2011a) The anterior insular cortex represents breaches of taste identity expectation. J Neurosci 31:14735–14744

Veldhuizen MG, Albrecht J, Zelano C, Boesveldt S, Breslin P, Lundstrom JN (2011b) Identification of human gustatory cortex by activation likelihood estimation. Hum Brain Mapp 32:2256–2266

Vonbekesy G (1949) The vibration of the cochlear partition in anatomical preparations and in models of the inner ear. J Acoust Soc Am 21:233–245

Vuilleumier P, Henson RN, Driver J, Dolan RJ (2002) Multiple levels of visual object constancy revealed by event-related fMRI of repetition priming. Nat Neurosci 5:491–499

Wade A, Augath M, Logothetis N, Wandell B (2008) fMRI measurements of color in macaque and human. J Vis 8(6):1–19

Wall MB, Walker R, Smith AT (2009) Functional imaging of the human superior colliculus: an optimised approach. Neuroimage 47:1620–1627

Wallis G, Backus BT, Langer M, Huebner G, Bulthoff H (2009) Learning illumination- and orientation-invariant representations of objects through temporal association. J Vis 9:6

Wandell BA, Winawer J (2011) Imaging retinotopic maps in the human brain. Vision Res 51:718–737

Wandell BA, Dumoulin SO, Brewer AA (2007) Visual field maps in human cortex. Neuron 56:366–383

Weiner KS, Sayres R, Vinberg J, Grill-Spector K (2010) fMRI-adaptation and category selectivity in human ventral temporal cortex: regional differences across time scales. J Neurophysiol 103:3349–3365

Wessinger CM, VanMeter J, Tian B, Van Lare J, Pekar J, Rauschecker JP (2001) Hierarchical organization of the human auditory cortex revealed by functional magnetic resonance imaging. J Cogn Neurosci 13:1–7

Wiggs CL, Martin A (1998) Properties and mechanisms of perceptual priming. Curr Opin Neurobiol 8:227–233

Wong AC, Palmeri TJ, Rogers BP, Gore JC, Gauthier I (2009) Beyond shape: how you learn about objects affects how they are represented in visual cortex. PLoS One 4:e8405

Woods DL, Stecker GC, Rinne T, Herron TJ, Cate AD, Yund EW, Liao I, Kang X (2009) Functional maps of human auditory cortex: effects of acoustic features and attention. PLoS One 4:e5183

Xu Y (2005) Revisiting the role of the fusiform face area in visual expertise. Cereb Cortex 15:1234–1242

Yacoub R, Ferrucci S (2011) Charles Bonnet syndrome. Optometry 82:421–427

Yacoub E, Shmuel A, Logothetis N, Ugurbil K (2007) Robust detection of ocular dominance columns in humans using Hahn spin echo BOLD functional MRI at 7 Tesla. Neuroimage 37:1161–1177

Yoo SS, Freeman DK, McCarthy JJ 3rd, Jolesz FA (2003) Neural substrates of tactile imagery: a functional MRI study. Neuroreport 14:581–585

Zelano C, Bensafi M, Porter J, Mainland J, Johnson B, Bremner E, Telles C, Khan R, Sobel N (2005) Attentional modulation in human primary olfactory cortex. Nat Neurosci 8:114–120

Emotions

10

Birgit Derntl, Frank Schneider, and Ute Habel

Abbreviations

aMCC	Anterior midcingulate cortex
CS	Conditioned stimulus
DCM	Dynamic causal modeling
MTL	Medial temporal lobe
PET	Positron emission tomography
rCBF	Regional cerebral blood flow
US/UCS	Unconditioned stimulus
V5	Extrastriatal visual area V5/MT (middle temporal)
VPFC	Ventral prefrontal cortex

10.1 Emotional Experience

Emotions usually cannot be measured under realistic conditions in neuroimaging settings. A fundamental problem of research in this area is, therefore, the application of effective methods in experimental mood induction. To be specific, the induced emotional state to be analyzed has to be genuine. So far, this has led to the development of very different experimental approaches. The disadvantage of such a wide variety of methods is the diversity of published results and the problem of comparability between findings and studies. For that reason, a major problem is still the lack of standardization of materials and approaches.

10.1.1 Mood Induction Methods

The emotional experience is investigated mostly with the help of experimental mood induction methods. Various forms of experimental mood induction can be distinguished, such as the following:

- The ability to put oneself in a certain emotional state based on the presented emotional material (text, movies, music, odors) according to predetermined instructions
- The free recall of subjects' own experiences
- The presentation of emotional material without explicit instructions to empathize with emotions
- The feedback of success or failure to induce satisfaction or frustration
- Experimental physiological changes (e.g., administration of medication)

The first imaging studies to investigate emotional experience were performed with positron emission tomography (PET): Examining self-induced sadness and inferior and orbitofrontal activity has been demonstrated (Pardo et al. 1993). In another study, sadness and happiness were triggered by watching affective facial expressions (George et al. 1995). Here, sadness

Adapted from Derntl B, Schneider F, Habel U (2012) Emotionen. In: Schneider & Fink (Eds), Funktionelle MRT in Psychiatrie und Neurologie. Springer, pp. 483–504

B. Derntl (✉) • F. Schneider • U. Habel
Department of Psychiatry, Psychotherapy and Psychosomatics, RWTH Aachen University, Pauwelsstrasse 30, Aachen D-52074, Germany
e-mail: bderntl@ukaachen.de; fschneider@ukaachen.de; uhabel@ukaachen.de

C. Mulert, M.E. Shenton (eds.), *MRI in Psychiatry*,
DOI 10.1007/978-3-642-54542-9_10, © Springer Berlin Heidelberg 2014

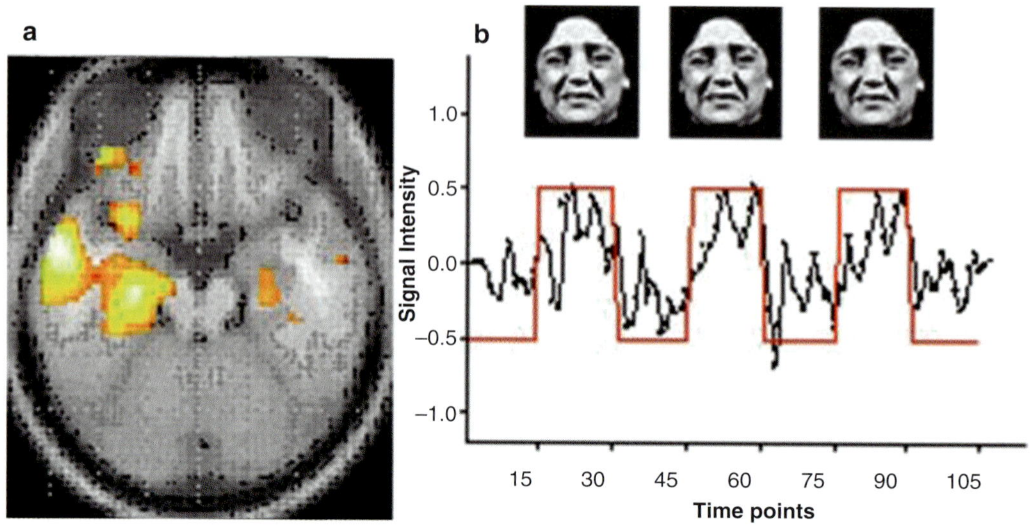

Fig. 10.1 Amygdala activity during sad mood induction in 26 healthy volunteers (**a**). Besides whole brain analysis, analysis of amygdala activation was also performed individually. Activation could be localized in the amygdala in 19 subjects. The averaged signal during sadness induction (**b**) of these 19 subjects substantially corresponds to the reference function, which represents the experimental design. The activity was also significantly correlated with subjective reports of the experienced mood state (see Habel et al. 2005)

was associated with diffuse blood flow changes in limbic and paralimbic areas, whereas happiness was associated with bilateral rCBF decreases in the temporoparietal and right frontal cortex.

As mentioned above, the different induction methods were followed by very different results. In an effort to avoid such inconsistency, a standardized method for mood induction (Weiss et al. 1999) was developed (see also Fig. 10.1), which has been validated upfront through studies on healthy volunteers (Schneider et al. 1994) and psychiatric patients (Habel et al. 2000).

10.1.2 Amygdala

The central role of this structure, in terms of emotional abilities, could be certainly confirmed for the first time by the impressive descriptions of behavioral changes in monkeys by Heinrich Klüver and Paul Bucy. Based on animal studies, Joseph LeDoux (2000) postulated his well-known model of the two input routes of the amygdala: On the one hand, the amygdala has a fast subcortical route that provides coarse information, thereby allowing the first assessment of the situation or stimuli. On the other hand, a second slower route based on a subcortical-cortical circuit contains detailed information that allows an accurate classification and discrimination of the stimulus.

The abovementioned mood induction method (Schneider et al. 1994) has been widely used in PET and fMRI studies in the meantime. Activation of the amygdala has been frequently observed during sad mood induction, which was detected with PET and fMRI (Schneider et al. 1998, 2000).

Due to the high correlation with the subjective emotional experience, these studies additionally confirmed the critical role of the amygdala for emotional experience (Schneider et al. 1998, 2000; Habel et al. 2004, 2005; see also Fig. 10.1).

The importance of the amygdala as one neural correlate of sad mood was also demonstrated in one of the first real-time fMRI studies showing that activation of the amygdala can be modulated via feedback (Posse et al. 2003). Six healthy subjects participated in a mood induction study applying the abovementioned paradigm. During the fMRI measurements, a randomized neutral mood induction with simultaneous presentation of neutral faces or a sad mood induction with the presentation of sad faces was used. After each condition, subjects rated their subjective experience and also received verbal feedback about their level of activation in the amygdala to enhance the mood induction effect. This feedback was based on an online real-time analysis of the data carried out outside the scanner room. The data showed a strong concordance between self-assessed sadness and the extracted amygdala activity. The involvement of the amygdala was also demonstrated in a regional analysis conducted after the experiment. A total of 120 subjects were scanned and in which 78 % showed a match between mainly left-sided amygdala activity and self-rated sadness as compared to only 14 % of the neutral mood conditions. Moreover, findings from Zotev et al. (2011) point out the training ability of the amygdala activation during pleasure induction using real-time fMRI neurofeedback (see Fig. 10.2). In this study, participants were confronted with visual feedback

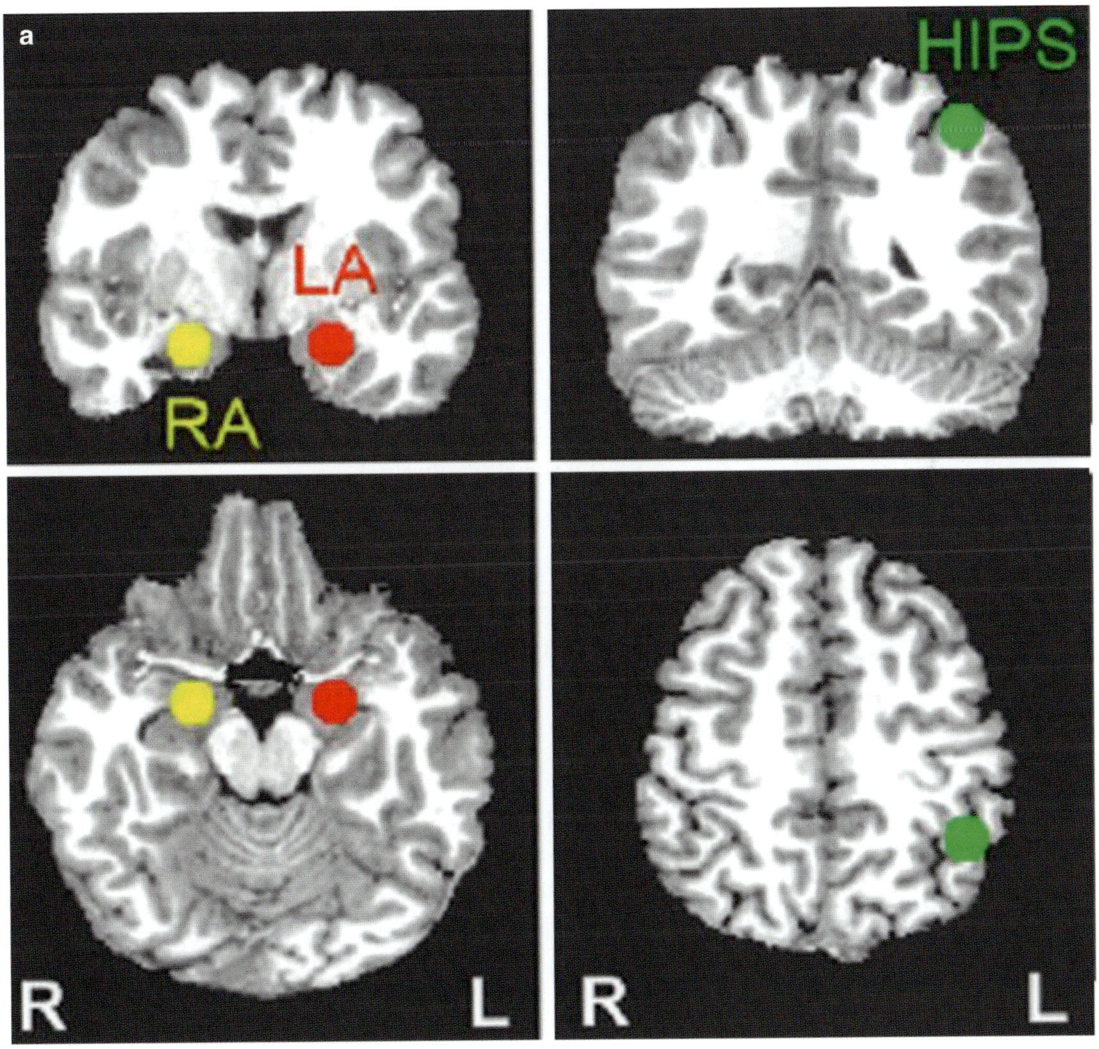

Fig. 10.2 (**a**) The location of the left and right amygdala (*red* = left, right = *yellow*) and the horizontal intraparietal sulcus (*HIPS*, *green*). (**b**) The percentile of BOLD signal change at the beginning (*RE*) for each feedback passage (R1 to R3) and the temporal resolution (*TR*). These results show a significant effect of neurofeedback during pleasure induction via retrieval of autobiographical memories especially for the left amygdala (see Zotev et al. 2011)

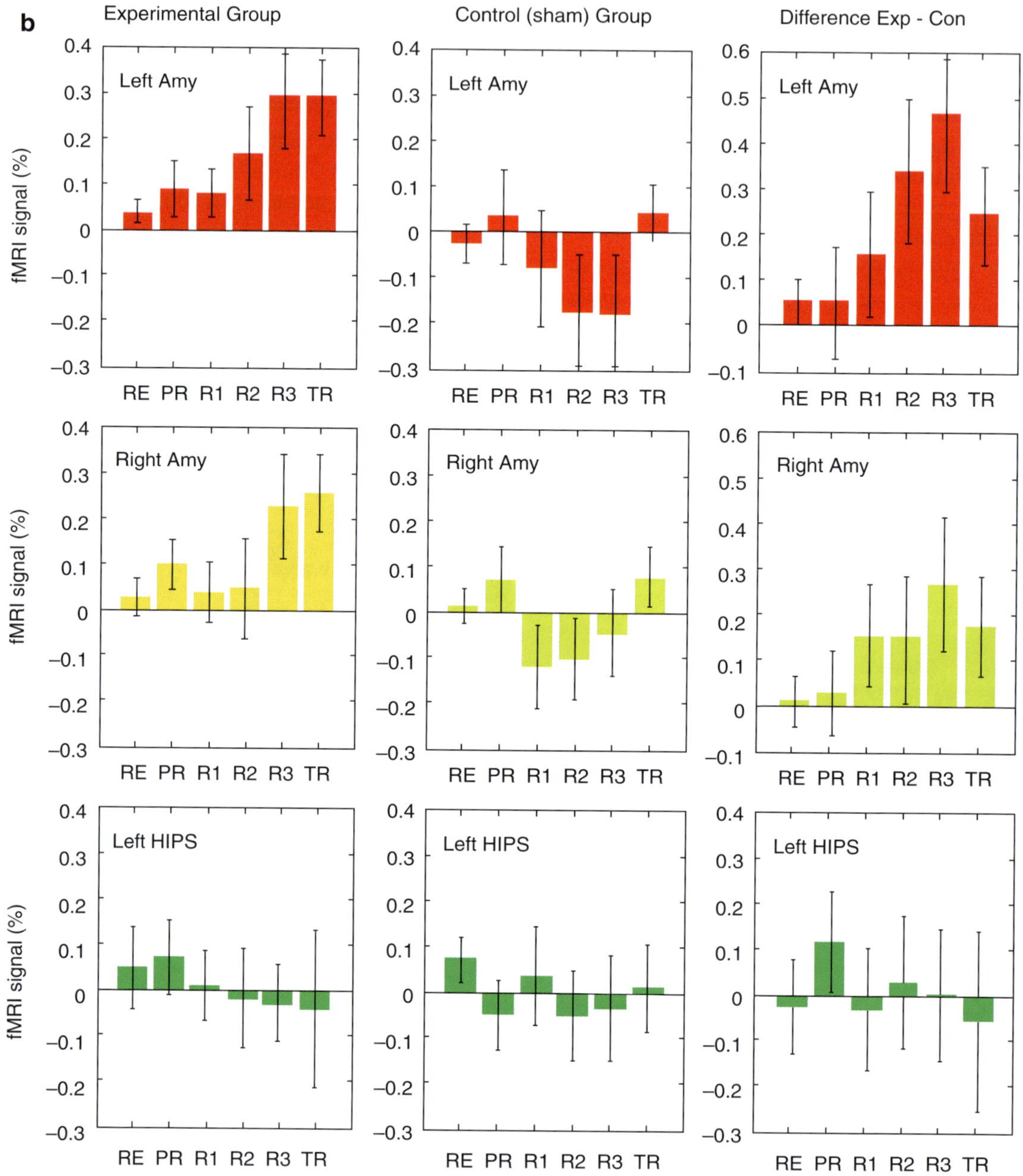

Fig. 10.2 (continued)

about their amygdala activation and they were asked to enhance the activation via positive, autobiographical memories, which successfully served the purpose.

A direct comparison of different mood induction methods in neuroimaging has rarely been carried out (Baumgartner et al. 2006; Falkenberg et al. 2012). Recently, Dyck et al. (2011) compared mood induction via presentation of faces with a combination of face presentation and music. They showed that the left amygdala was strongly activated in both methods, but the right amygdala showed significantly increased activity during the combination of both modalities.

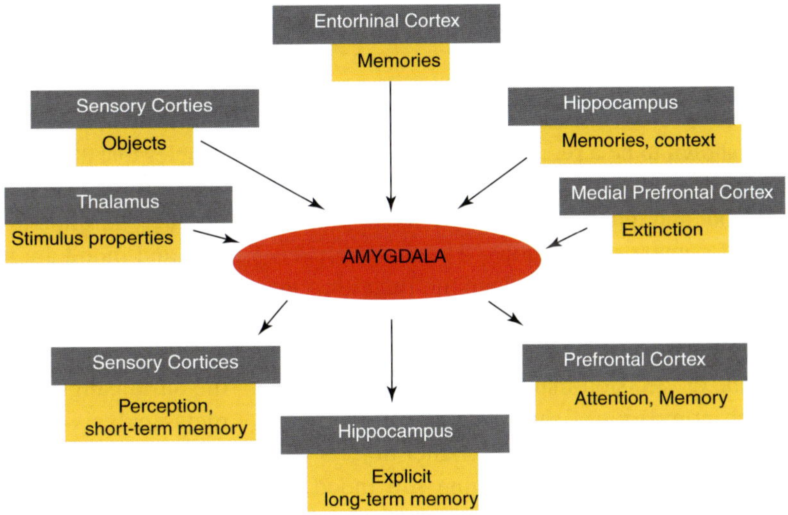

Fig. 10.3 Excerpt from an exemplary recorded connectivity of the amygdala in the context of emotional-cognitive functionality

This task specific lateralization suggests that the left amygdala contributes to more conscious and controlled mood, while the right amygdala seems to be more involved in the automatic induction of emotions. These results support the assumption that the amygdala also plays an essential role in the regulation of emotions.

As can be seen in Fig. 10.3, it is the placement amidst all the areas of the brain involved in emotional life as well as the interconnection with all of them, which makes the amygdala appear as an integrated switch point. The importance of the amygdala for emotional processes becomes obvious in terms such as "headganglion of the emotional-motivational system" (Doty 1989) or "sensory gateway to the emotions" (Aggleton and Mishkin 1986). Due to its connection to the hypothalamus, the amygdala seems to control the rudimentary, rather hypothalamically, controlled emotional reactions and allows the emergence of a hypothalamic-engaged drive satisfaction due to its evolutionary function. Thus, its functionality forms the basis of an emotional reaction as a result of the attribution of valence to the stimulus or the situation, respectively. This valence attribution is in turn the starting point for a differentiated emotional response.

However, not all studies were able to successfully demonstrate amygdala activation during emotional experience. Instead, involvement of the anterior cingulate cortex, the prefrontal cortex, the precuneus, and the hippocampus has been found repeatedly (Jeong et al. 2011; Mitterschiffthaler et al. 2007; Koepp et al. 2009; Teasdale et al. 1999) as well as activity in cerebellar regions (Brattico et al. 2011; Hofer et al. 2006). Methodological differences are likely responsible for these divergent results. Different methods of mood induction have been used, predominantly visual material and music (Baumgartner et al. 2006; Dyck et al. 2011; Hofer et al. 2006; Kohn et al. 2011; Teasdale et al. 1999; Mitterschiffthaler et al. 2007), but also memories of personal emotion-loaded real-life events (Zotev et al. 2011) or emotional words/phrases (Colibazzi et al. 2010; Hofer et al. 2007) have been applied. However, it could be demonstrated that internally and externally generated emotions activate different cortical and subcortical networks (Reiman et al. 1997). Also, the task instruction during the presentation of emotional material significantly modulates limbic activation (Gur et al. 2002; Habel et al. 2007).

Whether emotions are evoked as a result of cognitive processes or cognitive processes are involved, one often finds involvement of the anterior cingulate cortex and prefrontal areas. Given its attention-modulating function and its influence on executive processes as well as its connection to subcortical-limbic areas (especially the amygdala), the involvement of the anterior cingulate cortex is clearly comprehensible in this context.

10.1.3 Medial Prefrontal Cortex

In several meta-analyses (Etkin et al. 2010; Wager et al. 2003), the medial prefrontal cortex was shown to be a region that is involved regardless of the specific emotion and mood induction method and thus demonstrated its importance in the context of emotions. There are conceivably some emotion-overlapping functions that can be modulated by the frontal cortex such as attention focusing, assessment/evaluation, regulation, or decision-making (Etkin et al. 2010). When comparing a sad mood induction (using films) and a condition in which the subjects were asked to suppress an emerging sad mood, the temporal pole, midbrain, insula, amygdala, and ventrolateral prefrontal cortex were involved in sadness, while suppression of this mood led to dorsolateral prefrontal and orbitofrontal activity (Levesque et al. 2003).

10.1.4 Individual Factors

In addition to methodological aspects, individual factors influencing previous findings may be of importance in this context (Eugene et al. 2003): Two identical fMRI studies of sad mood induction were performed in two different experimental groups. In one case, the subjective experience of sadness was correlated with activity in the anterior temporal cortex and the insular, and in the other case, with the activity in the orbitofrontal and medial prefrontal cortex. In addition, a significant amount of interindividual variability was observed. Gender (Hofer et al. 2006, 2007; Kohn et al. 2011) or personality factors, such as extraversion or neuroticism can act as moderator variables (Hamann and Canli 2004) as well.

10.1.5 Emotion, Mood, and Feeling

Another important distinction in this context concerns the one between emotion and mood (see Dolan 2002; Lammers 2007): Emotions can be defined as mental representations of physiological changes due to processing of emotion-inducing conditions or stimuli. Emotions are more specific and conscious, short term, and intense and rely on rather automatic reaction patterns.

Mood refers to longer-lasting response tendencies that may favor the occurrence of a certain emotion.

There is some evidence to suggest that emotion and mood are mediated by different neural systems (Dolan 2002). Specific correlates for certain emotions were identified in the subgenual anterior cingulate cortex for sadness and in the basal ganglia for happiness according to the meta-analysis by Phan et al. (2002), in which 55 PET and fMRI studies were included. Fear, however, is strongly associated with an activation of the amygdala. Regarding the role of the subgenual cingulate cortex in negative emotion, a clinical study reported decreased activation in this area for depressed patients (Drevets et al. 2008). Notably, reduced activation of this region could be normalized during the course of a successful antidepressant therapy (Mayberg et al. 2000).

Recent studies of Keedwell and colleagues (2009, 2010) were also able to show that the activation of this particular region while viewing sad faces represents the best predictor for clinical improvement (see also Pizzagalli 2011 for review).

Due to the high concentration of dopaminergic neurons and their role in the reward system, the basal ganglia are extremely relevant for positive emotions. However, this functional attribution cannot be considered comprehensive and exclusive. It rather represents a special role of this region in the context of specific emotions and is part of a differential and emotion-specific activation pattern. Nevertheless, the involvement of the basal ganglia has been described, for example, in the context of disgust, and many studies demonstrated activity of the amygdala in positive emotions, such as happiness (Habel et al. 2005; Kohn et al. 2011; Zotev et al. 2011), although one can assume that the involvement in negative emotions predominates.

Taken together, previous studies indicated a wide-ranging network of participating regions in the experience of sadness and happiness, which involves prefrontal, orbitofrontal, anterior

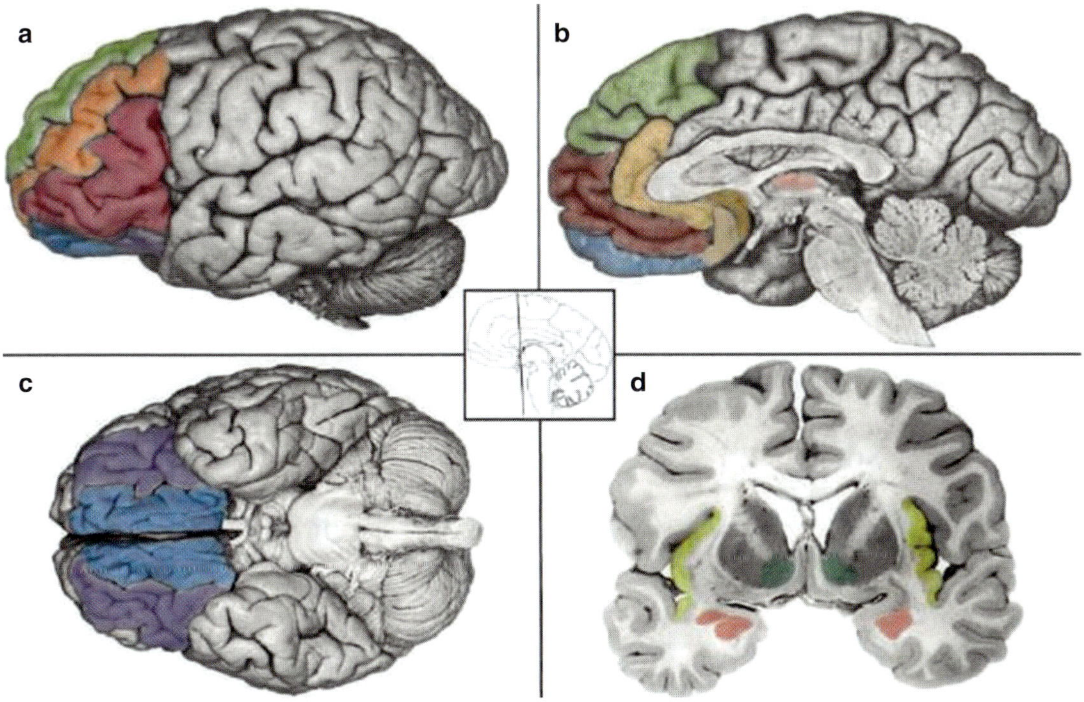

Fig. 10.4 Regions that play a role in emotions viewed from sagittal (**a**, **b**), axial (**c**), and coronal (**d**) by Barrett et al. (2007): orbitofrontal cortex (*purple*), ventromedial prefrontal cortex (*blue*), insular cortex (*yellow*), (subgenual) anterior cingulate cortex (*beige* and *brown*), dorsolateral prefrontal cortex (*orange* and *pink*), dorsomedial prefrontal cortex (*light green*), ventral striatum (*dark green*), thalamus (*pink*), and the amygdala (*red*)

cingulate, and temporal cortical areas and the amygdala in comparison to cognitive control tasks suggested by individual findings and meta-analytical results (Barrett et al. 2007; Phan et al. 2002, see Fig. 10.4).

The direct comparison of happiness and sadness, however, points to different valence-specific activation focuses within this network: While sadness rather evokes activation of the anterior cingulate area and the ventrolateral prefrontal and temporal cortex, happiness involves the dorsolateral prefrontal area, the inferior temporal cortex, the dorsal posterior cingulate cortex, and the cerebellum. This suggests that specific cerebral components within a common emotion network are characteristic for each emotion and thus may be responsible for the unique emotional quality and tinge. By presenting emotional sentences, Colibazzi et al. (2010) were able to reveal that all emotions can be divided into two dimensions of valence and arousal, also recruiting different neural networks: While medial brain regions, particularly medial temporal areas such as the amygdala, are rather associated with arousal, dorsally located cortical areas and mesolimbic activation differentiate between the valences.

To bring the diverse results in line is a challenge due to the variety of methods and experimental designs. In terms of gender, some studies only examined female (George et al. 1995; Pardo et al. 1993; Reske et al. 2010) or male subjects (Habel et al. 2005). However, recent studies indicate gender differences regarding the neural correlates of positive and negative mood (Hofer et al. 2006, 2007; Kohn et al. 2011; Schneider et al. 2000), mainly pointing to stronger neural activation in females compared to males. It will be up to future studies to further investigate whether these activation differences indicate different strategies for the generation or regulation of emotions. On the other hand, the mood induction methods vary considerably: While in some studies, the subjects were asked to change their mood on their own using the presented stimuli (Habel et al. 2005; Hofer et al. 2006; Reske et al. 2007, 2010), this occurred rather passively in other studies (Mitterschiffthaler et al. 2007) and yet unstructured (Koepp et al. 2009) or unconscious methods (Teasdale et al. 1999) were also applied. Taken together, previous studies consistently showed involvement of cortical and subcortical regions during emotional experience.

However, the observed differences between them also point towards a complex interplay of the regions involved in emotional experience depending on the emotion and the task demands of the experiments.

Emotional experience relies on a widely stretched network of cortical and subcortical areas wherein the amygdala plays a special role. The findings further suggest that emotion-specific correlates within this network may be the basis for qualitatively different emotions.

10.2 Emotion Recognition

Another frequently studied emotional ability in the scanner is identification or discrimination of emotional facial expressions. Many studies have examined implicit recognition performance and processing of emotional facial expressions by presenting stimuli without further instruction (passive viewing) or using a distracting task (e.g., gender discrimination) (Fitzgerald et al. 2006). This method is in contrast to an explicit emotion recognition task where the subject needs to decide if an emotional expression is shown and if so, which expression of the face is presented (Derntl et al. 2009; Fitzgerald et al. 2006; Gur et al. 2002).

Emotion discrimination is generally understood as the ability to recognize emotions in facial expressions. In imaging studies, there is a distinction between implicit and explicit processing. In the explicit emotion recognition task, subjects have to name or characterize the emotion or to choose from various alternatives the correct facial emotion.

During the implicit processing, emotional faces are presented along with distracting tasks (e.g., gender or age discrimination), which do not draw substantial attention specifically to the emotional content and do not require any decision on the facial content.

10.2.1 Explicit vs. Implicit Processing

Directly comparing explicit vs. implicit emotional processing, previous studies reported some overlapping but also distinct neural activation. In the study by Critchley and colleagues (2000), explicit processing elicited greater temporal involvement, while the implicit task rather prompted amygdala activation. The findings of other research groups, however, exactly showed the opposite. Here, a stronger amygdala and hippocampus activity was observed when the emotional facial expressions were task-relevant (emotion discrimination) as opposed to a condition of age discrimination in which they were task-irrelevant (Gur et al. 2002; Habel et al. 2007; see Fig. 10.5).

10.2.2 Emotion-Specific Amygdala Activation

In addition to the task setup (explicit vs. implicit), the presented emotions as well as character properties or the nature of the stimulus material were shown to modulate neural activation during emotion recognition. Even different task requirements of explicit emotion discrimination tasks resulted in different activation patterns: While visual matching resulted in amygdala activation, verbal labeling of facial affect led to reduced activation of the amygdala but increased frontal activity (Hariri et al. 2000). Moreover, the response of the amygdala is stronger when confronted with emotional facial expressions compared to other affective visual stimuli (e.g., IAPS, Hariri et al. 2002), even though both stimuli recruit similar networks (Britton et al. 2006).

In one of our own studies, we were interested in knowing which basic emotions (happiness, sadness, anger, disgust, and neutral facial expressions) trigger an activation of the amygdala during an explicit emotion recognition task

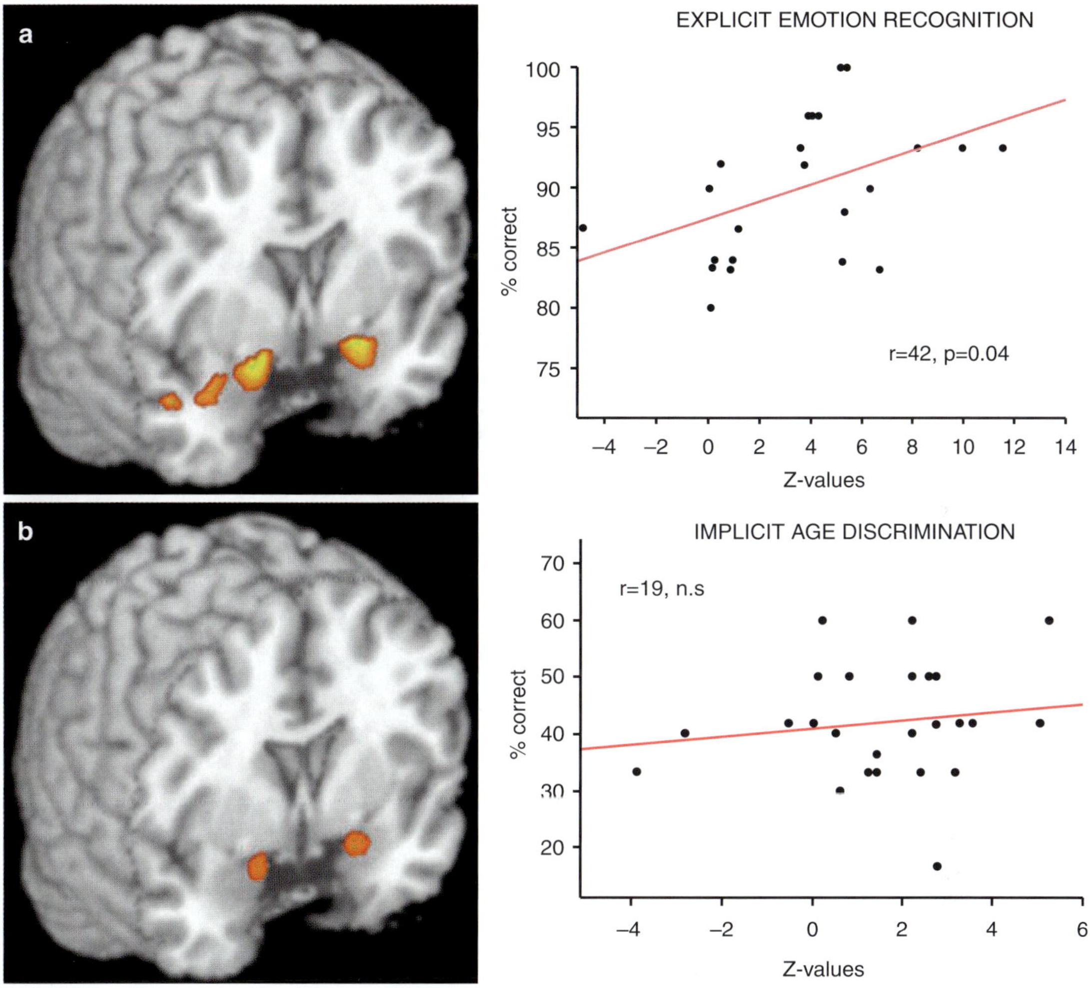

Fig. 10.5 Bilateral amygdala activation in explicit emotion recognition (**a**) and implicit age discrimination (**b**) in 29 healthy volunteers. Only explicit emotion recognition reveals a significant positive correlation between amygdale activation and recognition performance (Habel et al. 2007)

(Derntl et al. 2009). We observed bilateral amygdala activation to facial expressions of all emotions and neutral expressions. Thus, our results extended the former assumption that the amygdala is involved only in the processing of threatening stimuli (Morris et al. 1996) and supported models, which postulate an evaluation function and the role of the amygdala in detection of relevance (cf. Sander et al. 2003).

However, the amygdala is only a node in the neural network of emotion processing and the question of which other structures are relevant in processing emotions has long remained unanswered. A meta-analysis of Fusar-Poli and colleagues (2009) tackled this very question, and for each emotion, the respective networks could be identified (see Fig. 10.6). While amygdala and cerebellar activations are evident during recognition of negative and positive emotions, there are some other structures that show a more emotion-specific pattern. Sad, happy, and disgusted faces activate among other regions especially the insula, while threatening emotions, that is, anger and fear, amplified activity in the inferior frontal gyrus.

Thus, it is conceivable that the frontal cortex has cognitive as well as decision-making and evaluative functions in this context. The fact that

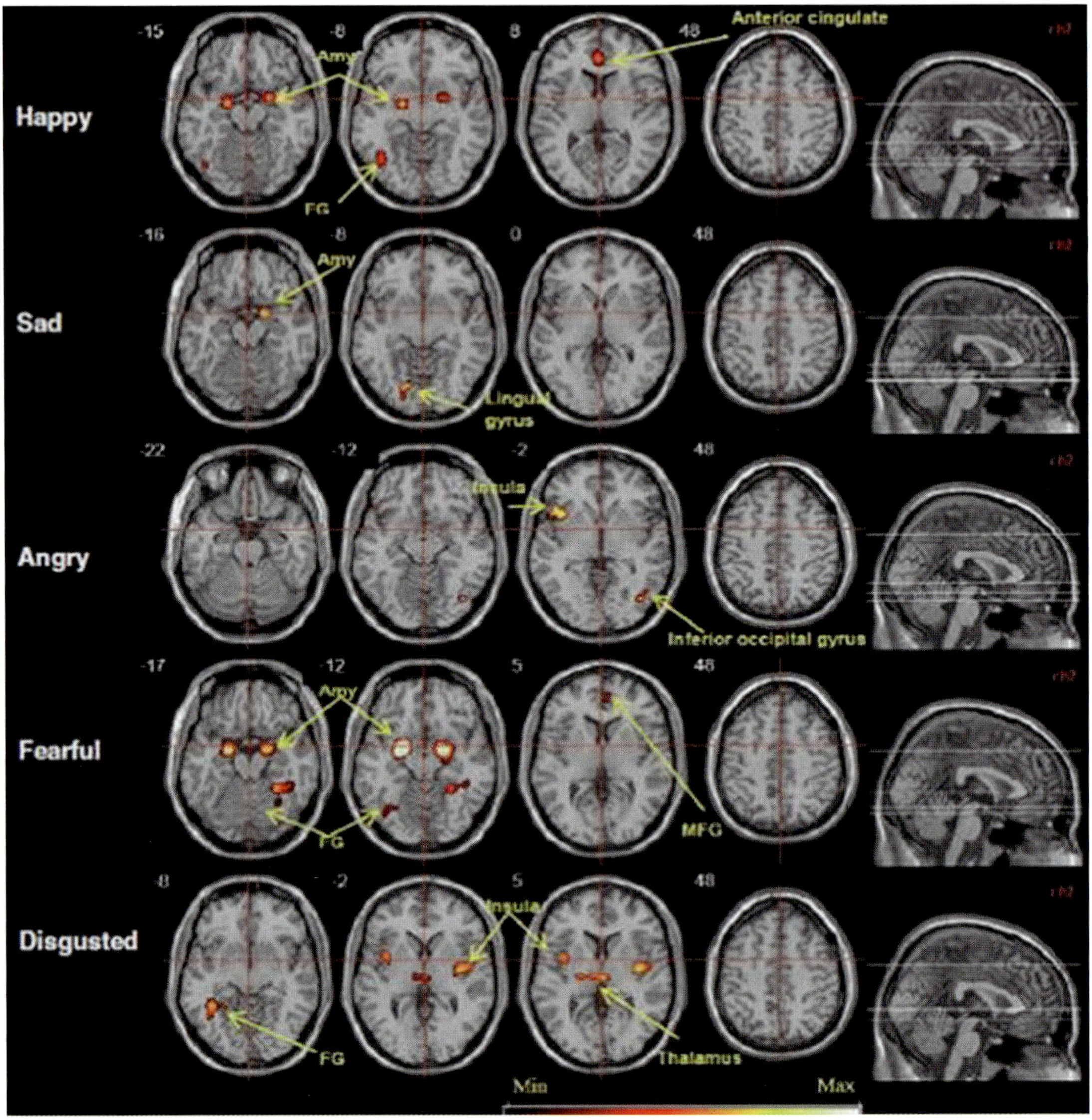

Fig. 10.6 Activation patterns for the individual basic emotions like happiness (happy), sadness (sad), anger (angry), anxiety (fearful), and disgust (disgusted) in direct comparison with neutral faces. *AMY* amygdala, *FG* fusiform gyrus, *MFG* medial prefrontal gyrus (Fusar-Poli et al. 2009)

emotional compared to neutral faces elicit stronger activity also supports this notion (Fusar-Poli et al. 2009; see Fig. 10.6).

In addition to a general emotion-processing network, which responds to positive and negative emotions and includes the amygdala and the cerebellum, other structures show rather emotion-specific patterns such as the insula.

It seems that attention also has a modulating function with respect to the amygdala: Especially a reduced amygdala activation was observed during attention-intensive tasks (Morawetz et al. 2011; Williams et al. 2005) and thus supports the assumption of the "attentional load theory" (Lavie 1995; Lavie and De Fockert 2005) which postulates that more neural capacity is assigned to the system which contributes to the processing of the task, that is, the more difficult the task, the higher the attentional load and the less task-irrelevant aspects can be processed.

In addition to the recognizability (Pessoa et al. 2005), the location of the stimuli (Morawetz et al. 2010) affects the amygdala. A point worth mentioning here is that the activation of the amygdala has a survival function, in the sense that emotional faces are instantly being evaluated in terms of their menace and valence (positive vs. negative) which earned it the denotation as "relevance detector" (see also Chap. 2.4 on subconscious stimulus processing).

When emotional stimuli are presented under cognitively demanding conditions (e.g., a cognitive task has to be solved), decreased activation in the amygdala can be observed, thereby possibly acting as a filter for negative and potentially threatening stimuli.

10.2.3 Habituation Processes

Due to the rapid habituation of the amygdala during repeated stimulus presentation (Britton et al. 2008; Williams et al. 2004), it has been difficult to prove its involvement in the context of emotional processing. However, this can be prevented by relying on an "event-related" paradigm in which the stimuli are presented in a random fashion. Regarding "boxcar" or block paradigms, it should be noted that recent studies have also shown that shorter blocks (up to 20 s in duration) could prevent habituation of the amygdala (Haas et al. 2009; Morawetz et al. 2011).

Nevertheless, the time course of activation should be observed: Previous studies showed that the temporal pattern of response was different between the right and left amygdala (Phillips et al. 2001) which also holds true for habituation processes (Wright et al. 2001). Moreover, besides habituation processes, sensitization processes must be taken into account as well. Feinstein and colleagues (2002) found habituation processes in attention-related regions of the right hemisphere, as in the posterior parietal cortex and the frontal eye field. In contrary, sensitization was observed on the left side of the angular gyrus, the posterior superior temporal cortex, and the insula while participants performed a gender discrimination task with emotional faces.

10.2.4 Backward Masking

Since even the unconscious perception ("backward masking") of emotional stimuli is able to elicit the subcortical-limbic emotion network, the amygdala can be referred to as a warning and protection system that automatically responds adaptively to changing environmental requirements.

"Backward masking" refers to a short stimulus presentation below the threshold of conscious perception (subliminal, 17–30 ms), followed by a longer presentation of another stimulus, so that the perception of the first stimulus is masked by the presentation of the second stimulus.

In "backward-masking" paradigms, the brief presentation of emotional faces masked with neutral faces was sufficient to demonstrate amygdala activity in healthy participants (e.g., Duan et al. 2010; Juruena et al. 2010; Kim et al. 2010; Whalen et al. 1998).

Whalen and colleagues (2004) have also shown that only the masked presentation of the fear expression of the eye region is sufficient to elicit reliable amygdala activation. A recent study by Kim and colleagues (2010) points not only to the influence of the emotion of the masked face, but also shows an effect of the mask itself. While masking with neutral faces, as applied in most studies, leads to an increased activation of the amygdala, a graphic pattern as mask leads to a reduction in the amygdala response, especially for masked fearful faces (see Fig. 10.7).

In initial studies, it was also questioned whether the valence of the emotion has an effect on the extent of amygdala activation as Whalen et al. (1998) found a stronger stimulus response in the amygdala for masked fearful faces than for happy ones, while Killgore and Yurgelun-Todd (2004) were able to record a stronger amygdala activation to masked happy faces than masked sad faces. These results are also supported by more recent findings (Juruena et al. 2010) according to which masked happy and sad faces elicited bilateral amygdala activation, but happy faces to a significantly greater extent. Masked surprised faces vs. happy or neutral masked stimuli triggered not only a response in the amygdala but also increased activation in the parahippocampal gyrus and the

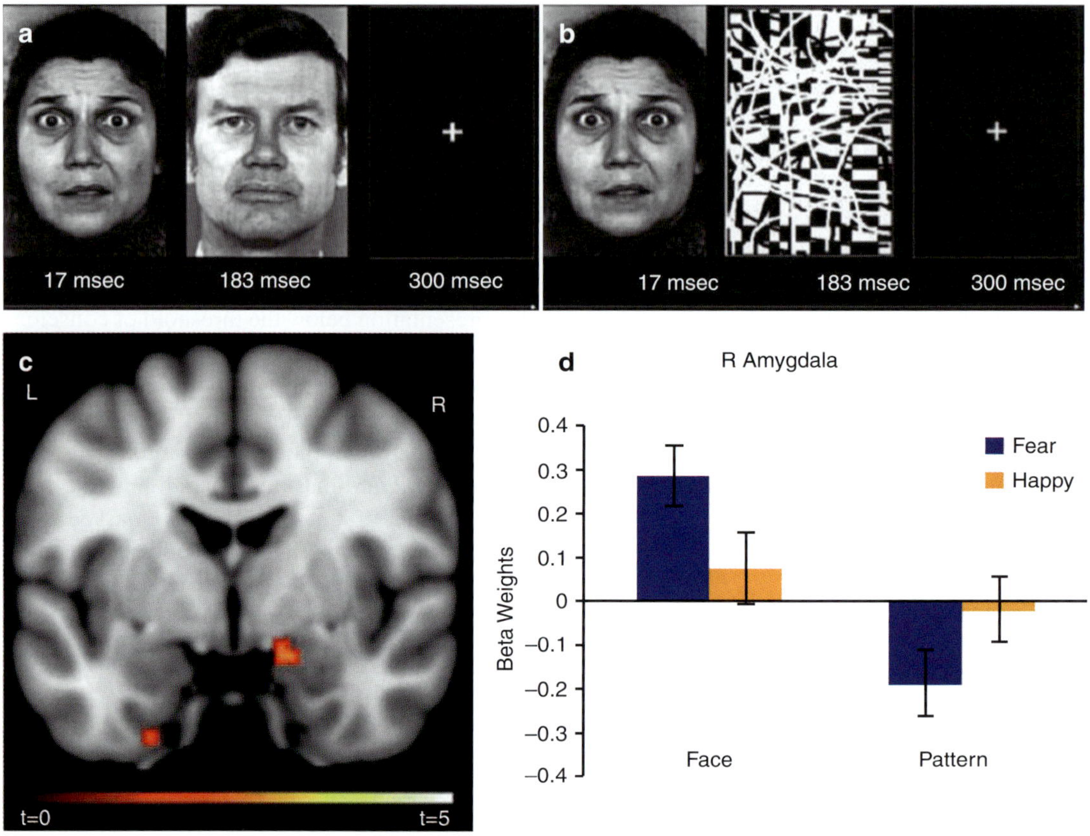

Fig. 10.7 A classical example of a backward-masking paradigm with a short presentation of the emotional stimulus followed by a neutral face (**a**). In comparison to a backward masking with a graphic pattern as a mask (**b**). While face masks lead to a significant increase in amygdala activation in fearful as compared to happy faces (**c**), the pattern mask leads to a significantly reduced activation of the amygdala, especially for fearful faces, which is also apparent in the parameter estimates (**d**)

fusiform gyrus (Duan et al. 2010). These results extend the central role of the amygdala to unconscious processing of emotional stimuli.

10.2.5 Modulation of Amygdala Activity

Overall, the existing findings could identify and characterize a wide range of relevant factors that modulate the activity of the amygdala. For example, the eye gaze and face direction of the stimulus play a role, because directly facing angry faces are associated with stronger amygdala reaction than averted expressions (Sato et al. 2004a), but contradictory findings have also been shown (Adams et al. 2003). This divergence could possibly be due to the different stimulus material: While Adams et al. used static faces, Sato and colleagues have presented dynamic representations. These dynamic and thus biologically more relevant stimuli are not only associated with a higher arousal (Sato and Yoshikawa 2007) but also prompt increased amygdala involvement as compared to static facial expressions (Pelphrey et al. 2007; Sato et al. 2004b).

The same applies to some other areas, mainly the middle temporal cortex (V5), the superior temporal sulcus, and the frontal regions (Kessler et al. 2011). Also, schematic emotional compared to neutral facial expressions can act as suitable stimuli to trigger responses in the amygdala, hippocampus, and prefrontal cortex (Wright et al. 2002). A direct comparison of artificial faces, so-called avatars, with human emotional facial expressions showed no significant difference in the extent of amygdala activation, only differences in the fusiform gyrus were apparent, where human faces elicited a stronger response (Moser et al. 2007). The fact that the amygdala reacts to schematic and avatar facial expressions has an experimental advantage because "artificial" stimuli can be produced more easily and are much more controllable than human facial expressions. Emotion recognition is thus not only much tied to the involvement of the amygdala but also closely linked to other cortical and subcortical structures that are relevant for intact emotion recognition.

Application of newer analytical methods such as effective connectivity analysis ("dynamic causal modeling," DCM) mostly supported assumptions about neural network models for the recognition of emotion (e.g., Pessoa and Adolphs 2010; Phillips et al. 2003). Above all, the amygdala has been shown to be responsible for the rapid detection and evaluation of affective stimuli, whereas the ventral prefrontal cortex takes the more detailed encoding and differentiation of stimuli (Dima et al. 2011). Recently, Tettamanti and colleagues (2012) were the first to be able to show that the networks for the emotions fear, disgust, and happiness indeed all contain the angular gyrus and the amygdala but also display emotion-specific neural connections: Whereas the fear network involved frontoparietal regions, disgust activated somatosensory cortices, pleasure involved a network of medial prefrontal cortex and temporoparietal regions (see Fig. 10.8).

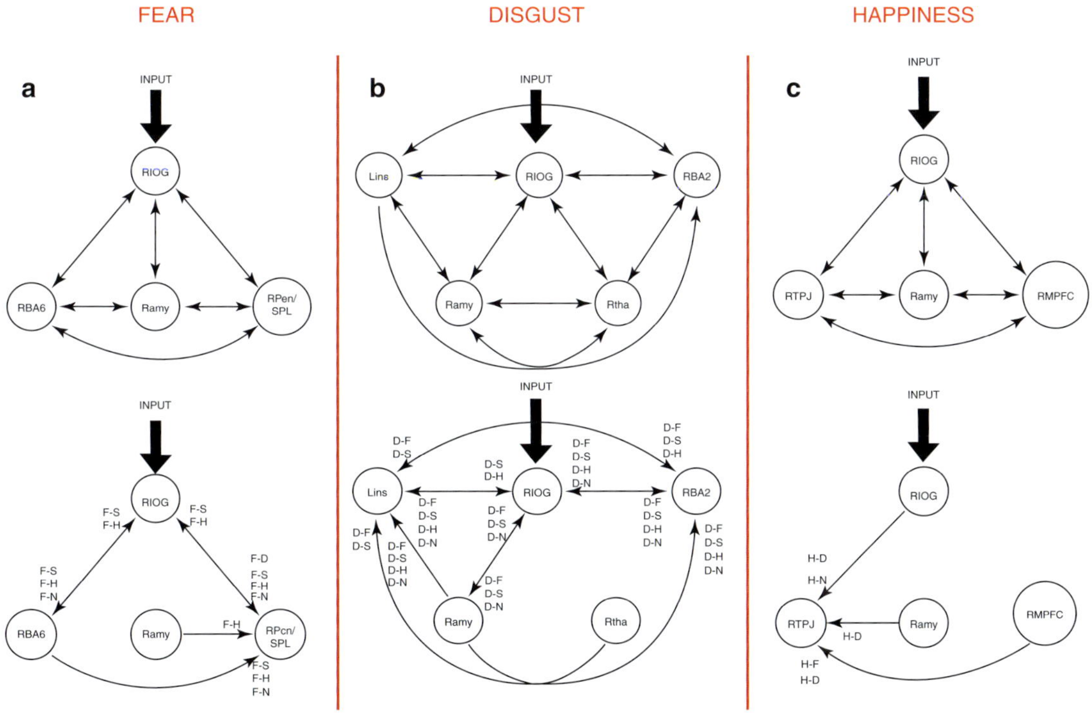

Fig. 10.8 Functional integration between amygdala and emotion-specific sensorimotor, somatosensory, and cognitive systems. The *top row* depicts connectivity patterns for the three specified models/emotions. The *bottom row* illustrates the significant condition-specific connection strength modulations ($p<.05$ FDR corr.). (**a**) The DCM-model for fear, testing the hypothesis of stronger positive modulatory effects in a brain network subserving action representation (**b**) the disgust-DCM model, and testing the hypothesis of stronger positive modulatory effects in a brain network subserving somatosensory representations. (**c**) The DCM model for happiness, testing the hypothesis of stronger positive modulatory effects in a brain network subserving mentalizing and representation of others' mental states. *RPcn/SPL* right precuneus/superior parietal lobule complex, *RBA6* right premotor cortex (BA 6), *Rtha* right thalamus, *Lins* left insula, *RBA2*, right somatosensory cortex (Brodmann's area 2), *RIOG* right inferior occipital gyrus, *Ramy* right amygdala, *RMPFC* right medial prefrontal cortex, and *RTPJ* right temporoparietal junction

10.3 Empathy

Relying on accurate emotion recognition, the ability to show empathic behavior plays a critical role within social communities particularly regarding the complexity of structures and networks in our society today. Empathy and empathic behavior have various definitions probably due to the complexity of the construct (see Preston and de Waal 2002; de Vignemont and Singer 2006); however, according to most models, one can derive at least three core components (cf. Decety and Jackson 2004):

- Emotion recognition as a differentiation between self-experienced emotions and those expressed by others in that emotions can be recognized via facial, verbal, or behavioral expressions
- Perspective taking (cognitive component) describing the competency to take over the perspective of another person, though the distinction between self and others remains intact
- Affective responsiveness (affective component), meaning sharing of emotional states with others or the ability to experience similar emotions as others

Correctly inferring emotional states and intentions via the observation of others' behavior are prerequisites for successful social interaction. Decety and Lamm (2006) proposed a model in which bottom-up and top-down information processes are intertwined in the generation and modulation of empathy. Bottom-up processes are mainly responsible for affect generation and automatic responses and are rather associated with limbic and temporal activation, while top-down regulation is related to prefrontal and cingulate cortices.

Application of neuroimaging tools enables investigation of the different empathic components and analysis of their neural underpinnings. Most studies rely on the so-called perception-action model by Preston and de Waal (2002) who proclaim that observation as well as imagination of another person in a particular emotional state automatically activates a representation of that state in the observer, along with its associated automatic and somatic responses. In other words, when we try to understand how someone is feeling in a certain situation, we simulate the feelings by activating our own affective program prompting shared neural representations (cf. Singer and Lamm 2009).

Previous neuroimaging studies on empathy have covered a wide range of emotions (from pain to disgust and happiness) and tasks (from passive viewing to imagination and evaluation). Due to this diversity in study design, various regions were found to be activated during empathic behavior leaving open the question which brain areas form the empathy network and whether such a core network really exists. Evidence showed that both the prefrontal and the temporal cortices are implicated in empathic behavior, but patterns of associations are different depending upon various factors, such as whether cognitive or affective empathy was investigated (Lee et al. 2004). According to a recent meta-analysis on 40 fMRI studies on empathy, Fan and colleagues (2011) assign a key role in cognitive empathy to the left dorsal anterior midcingulate cortex (aMCC) and to the anterior insula bilaterally in affective empathy irrespective of emotional category. Focusing on empathy for pain, Lamm et al. (2011) reported activation of the anterior insula bilaterally, the anterior medial cingulate cortex, and the posterior cingulate cortex, partly supporting findings from Fan et al. (2011, see also Fig. 10.9). More recently, Kennedy and Adolphs (2012) described several core social cognition networks, differentiating between a so-called mentalizing network and an empathy network.

Concerning empathy and empathic behavior, several studies suggested that females might be more empathic than males, but mostly relied on self-report data (Rueckert and Naybar 2008). Besides self-report, gender differences in empathy have also been shown using functional neuroimaging, for example, Singer and colleagues (2006) studied brain activity while female and male subjects underwent mild electric shocks or witnessed a confederate receiving a similar shock. While females showed a response in pain-related areas even when an unfairly behaving confederate received a shock, males exhibited activation in brain regions associated with reward, that is,

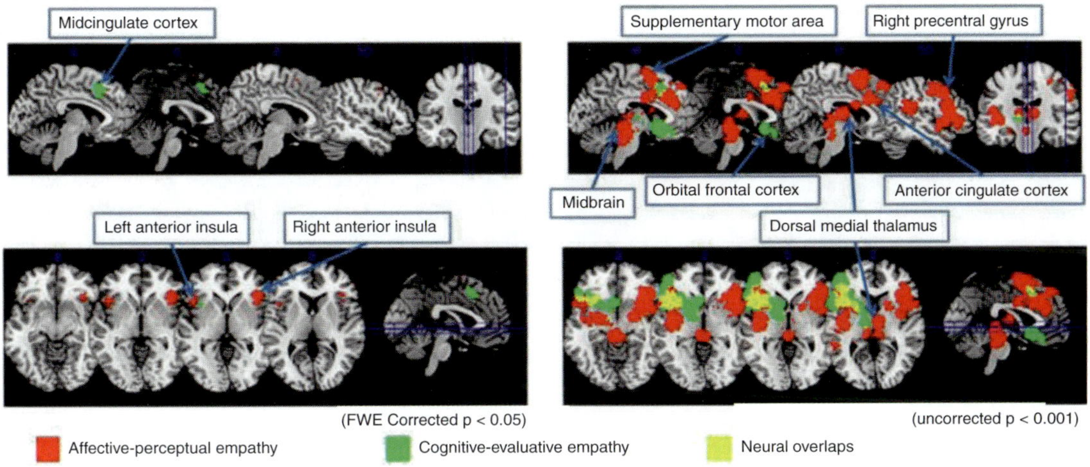

Fig. 10.9 Results of a meta-analysis by Fan et al. (2011), relying on 40 fMRI studies investigating empathy. Regions that are predominately involved in processing affective aspects are colored in *red*; areas that are mainly recruited during cognitive processes are colored in *green*. Regions that are involved in both empathic processes are colored in *yellow*

nucleus accumbens and orbitofrontal cortex, under this condition. Recently, gender differences have also been reported for the neural correlates of emotional perspective taking (Schulte-Rüther et al. 2008) using an emotional attribution task. The results suggest that the better performance of females is related to an enhanced recruitment of inferior frontal and superior temporal regions, while males activate the left temporoparietal junction instead when assessing emotional states of others and themselves.

This finding was further extended by a study from our lab (Derntl et al. 2010), where we investigated all three empathy components separately in females and males. Despite similar behavioral performance, females rated themselves as more empathic than males, and functional data analyses revealed that females and males not only showed common but also distinct activation patterns: Females exhibited stronger activation of emotion-associated regions such as the amygdala, whereas males rather recruited cognition-related areas, for example, temporoparietal junction (see Fig. 10.10 for details). Thus, our findings support and extend previous results on divergent neural processing strategies as well as sensitivity to emotional stimuli (e.g., Schulte-Rüther et al. 2008).

10.4 Emotional Learning and Memory

A biologically relevant and adaptive mechanism to facilitate and enhance the efficiency of human responses is classical conditioning. This learning phenomenon is particularly important when it comes to emotions and again the amygdala seems to be critically involved.

To optimally study classical conditioning, an event-related paradigm should be applied. In a typical classical conditioning paradigm, Büchel and colleagues (1999) presented faces as conditioned stimuli (CS) and aversive sounds as unconditioned stimuli (US, see also Fig. 10.11). Fifty percent of the CS+ (duration of 3 s) were paired with an US. This partial enhancement allowed to separate the response to the CS+ and the US, and only those were further analyzed that were not associated with the US. Differential effects of conditioning (e.g., CS+ vs. CS−) demonstrated activation in the amygdala and the anterior cingulate cortex.

All experiments focusing on fear conditioning demonstrated a decrease in the conditioned response. This supports data from animal studies that clearly indicated that the amygdala is critically involved in the process of forming

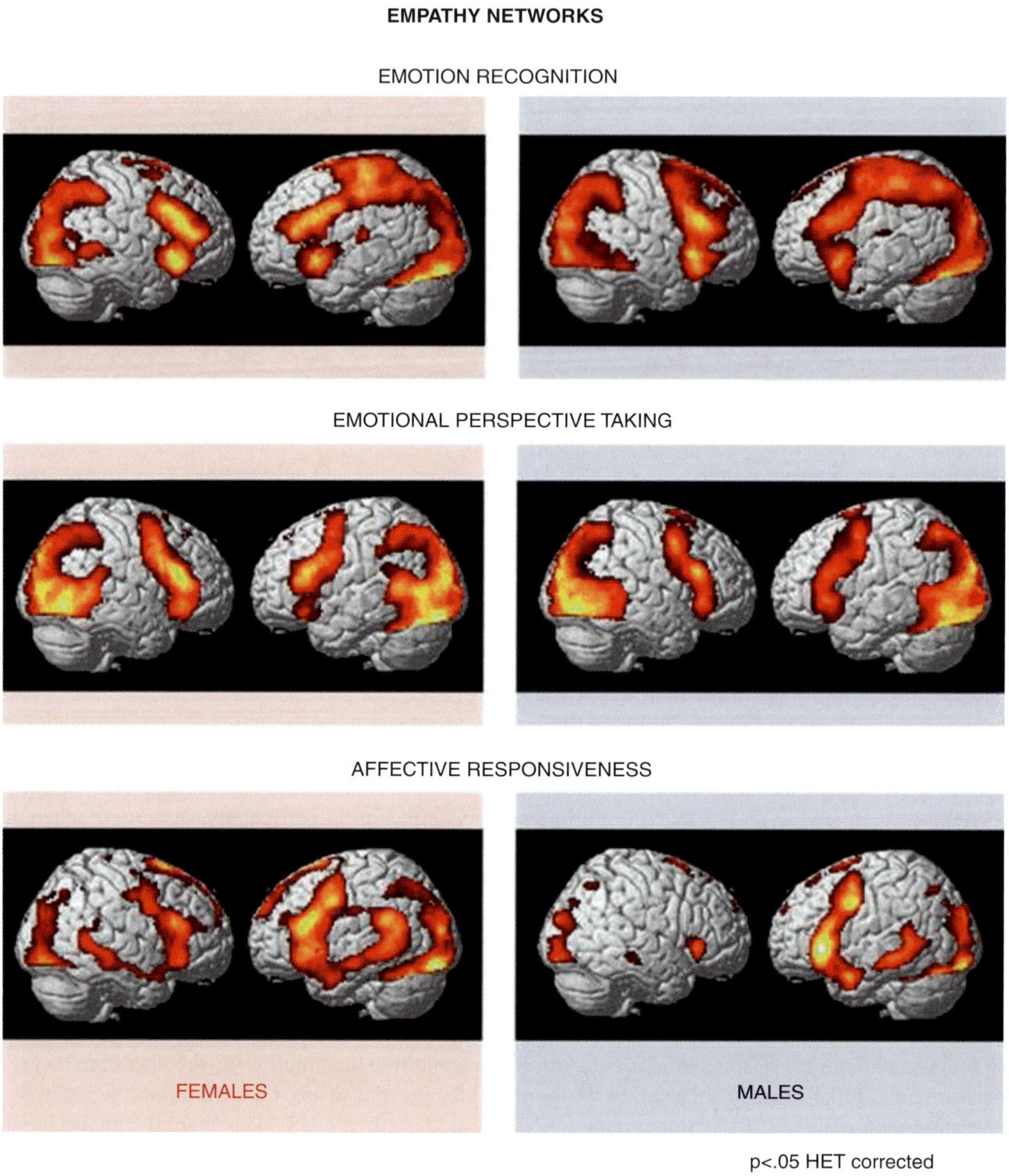

Fig. 10.10 Illustration of the neural networks underlying the empathic behavior in females and males, showing similar activation in the frontotemporal and occipital regions and brain stem, with pronounced activation in the inferior and superior frontal region bilaterally (Derntl et al. 2010)

associations and then shows a habituation of its response (Phelps et al. 2004; Quirk and Beer 2006). The role of the amygdala for extinction is rather elusive. Some studies report heightened activation whenever contingencies change, including extinction (Knight et al. 2004), and others showed that conditioning effects are less reversible for fear stimuli in the amygdala. While the orbitofrontal cortex responded very fast to changes in contingencies (CS+ and CS− were

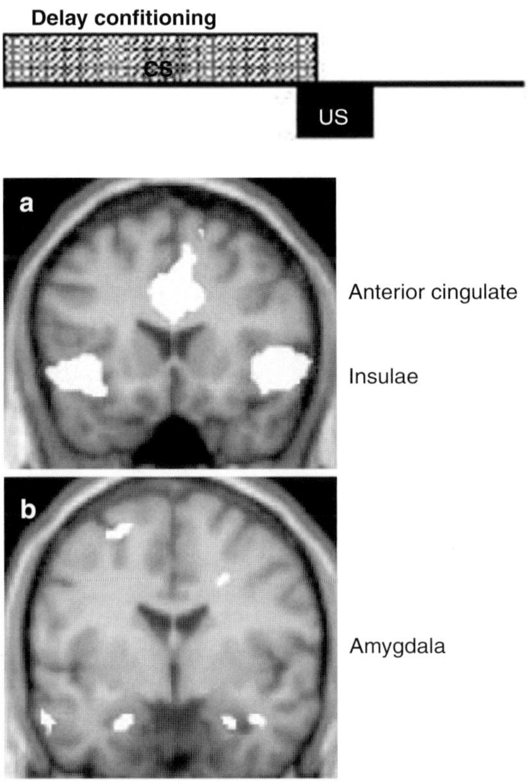

Fig. 10.11 Schematic illustration of the timely pairing of US and CS for classical conditioning (**a**). Activation within the anterior cingulated cortex, insula, and amygdala as response to CS+ (*top*) and compared to CS− (*bottom*) (**b**, Büchel and Dolan 2000)

exchanged), the amygdala reacted most strongly to the new CS−, which used to be CS+ before (Morris and Dolan 2004). Regarding extinction, the hippocampus seems to be a very critical region (Ji and Maren 2007).

Indovina et al. (2011) investigated the effect of trait anxiety on the neural network underlying fear conditioning. For this purpose, high and low anxious participants were visually confronted with three computerized environments or "rooms." The contingency between the conditioned stimulus (CS) and presentation of the unconditioned stimulus (UCS) differed between rooms. In the "predictable" (cued fear) room, the CS was predictive of the UCS; in the "unpredictable" (background contextual fear) room, the CS was nonpredictive of UCS occurrence; and in the "safe" (control) room, CS presentation occurred in the absence of the UCS. Each room was presented for approximately 40 s. The CS was a virtual actor (male or female) putting hands to ears as if to protect him- or herself from a loud sound.

This gesture terminated after 4–6 s. In the predictable room, CS offset was always accompanied by UCS presentation. The UCS was a 103 dB scream lasting 750 ms. In the unpredictable room, UCS presentation was randomized with regard to CS presentation. Each predictable and unpredictable room presentation contained three CS and three UCS occurrences. Each safe room presentation contained three CS occurrences (see Fig. 10.12a). Authors found evidence for two independent dimensions of neurocognitive function associated with trait vulnerability to anxiety. The first entailed increased amygdala responsivity to phasic fear cues (see Fig. 10.12b). The second involved impoverished VPFC recruitment to downregulate both cued and contextual fear prior to omission (extinction) of the aversive unconditioned stimulus (see Fig. 10.12c). These results have several implications and may contribute to symptomatology differences across anxiety disorders with the amygdala mechanism affecting the development of phobic fear and the frontal mechanism influencing the maintenance of both specific fears and generalized anxiety by deficits in negative affect regulation.

10.4.1 Cerebral Correlates of Encoding and Retrieval

The effect of emotions on memory processes has been demonstrated in several studies showing that emotional contents were not only easily encoded but also better remembered (e.g., Dolcos et al. 2004, 2005; Sharot and Yonelinas 2008; Sharot et al. 2004). Since all the studies relied on different paradigms, it was unclear whether these effects were due to the valence or arousal or both of the emotional stimuli, that is, whether the specific emotions or the increase of arousal led to the observed effects. Some authors speculate that the intensity of emotions is essential for these memory effects, which has been shown with text material and visual stimuli.

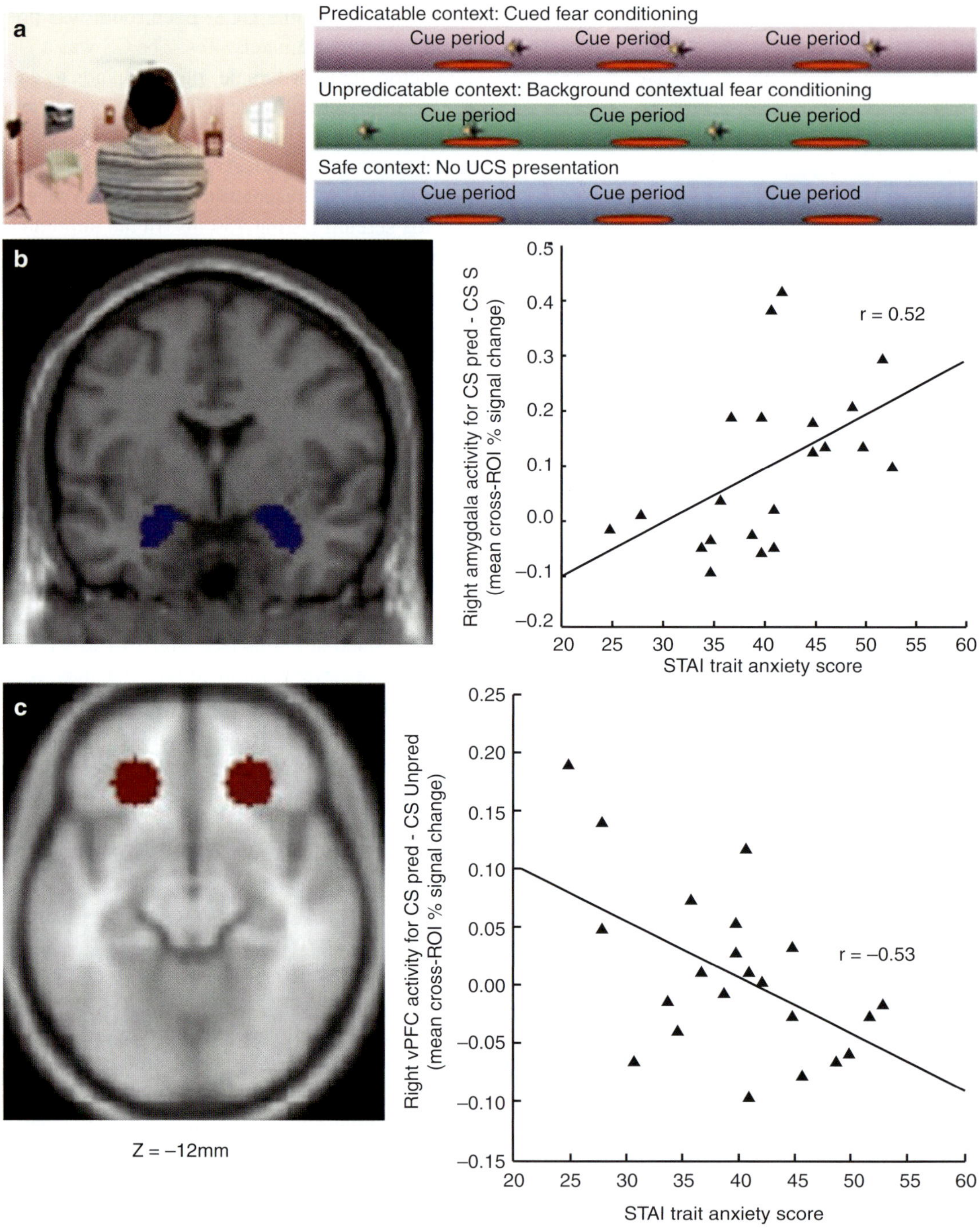

Fig. 10.12 Schematic illustration of the experimental setup to study fear conditioning (**a**). Significant positive correlation between amygdala activation in the CS condition and anxiety of participants, indicating that the more anxious the person describes himself/herself, the higher the amygdala response (**b**). The effect of trait anxiety on activation of the ventral prefrontal cortex (**c**). Here, a significant negative correlation was observed, indicating that the more anxious a person, the weaker the response of this region during viewing of fear-conditioned stimuli

Most neuroimaging studies investigated neural activation during encoding and analyzed its relation to behavioral performance (Dolcos et al. 2004; Sergerie et al. 2005). Rarely have fMRI studies examined neural activation during retrieval (Smith et al. 2004) or analyzed the neural underpinnings of encoding and retrieval in the same participants (Tabert et al. 2001). Based on their findings, it seems that the amygdala plays an essential role in emotional learning and memory (LaBar and Cabeza 2006; Murty et al. 2010); this further strengthens the assumption of the amygdala being a highly critical brain region in the limbic network. Regarding emotions and memory, most researchers refer to the medial temporal lobe (MTL) memory system, which comprises the following regions: the hippocampus, the entorhinal cortex, the perirhinal cortex, and the amygdala (see Fig. 10.13 for an illustration).

Some studies even report a significant correlation between activation intensity and memory performance (Canli et al. 2000; Tabert et al. 2001). The correlation with the subsequent behavioral performance seems to be affected by the subjectively rated emotional arousal (Canli et al. 2000), further supporting the assumption that arousal critically modulates memory performance.

During encoding, the amygdala seems to function as a booster for memory performance for emotional material. This has been shown in studies with amygdala lesion patients, where the size of the lesion was associated with the memory deficit for emotional but nor for neutral stimuli (Richardson et al. 2004). It has further been assumed that the response of the amygdala to emotional material modulates the early processing of the stimuli, intensifying information processing and evaluation of stimuli. This assumption has been strengthened by correlations between activation of the amygdala and occipital regions (Tabert et al. 2001). More recent findings using the beta-adrenergic antagonist propranolol indicate a noradrenergic-modulated amygdala involvement during encoding (van Stegeren et al. 2010).

Most studies investigated amygdala activation during encoding or consolidation of emotional stimuli. Recent studies also examined its function during retrieval and particularly focus on the long-term memory where the effect of the amygdala is proclaimed to be strongest. One option to

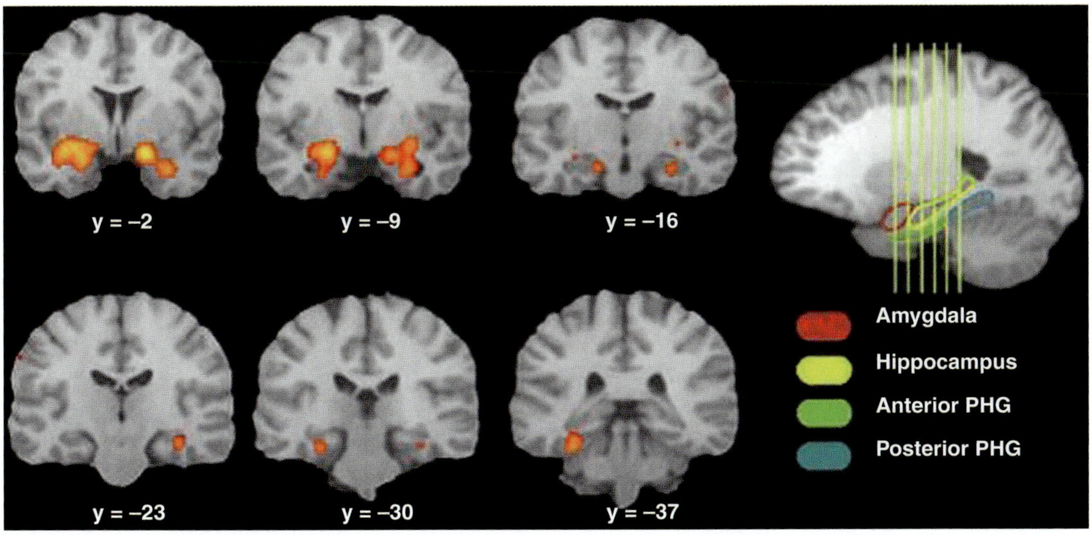

Fig. 10.13 Regions that are assigned to the MTL system. According to a recent meta-analysis, these regions show consistent activation during emotional memory studies ($p < .05$ FDR corr.)

investigate memory retrieval during fMRI is to directly compare the successfully retrieved stimuli ("hits") with the incorrectly retrieved items ("misses"). The influence of emotions can then be analyzed by directly comparing the amount of hits for emotional vs. neutral items. However, it is essential to differentiate between "real" recollection, familiarity, and recognition, since each of these methods relies on different neural networks (Yonelinas et al. 2005) and particularly recognition seems to be strongly affected by emotions (Jackson et al. 2008; Ochsner 2000). Dolcos et al. (2005) investigated the influence of emotions on two of these retrieval mechanisms. All participants were scanned twice, with a 1-year follow-up (retention interval), and authors reported that successful retrieval of emotional pictures elicited greater activity than successful retrieval of neutral pictures in the amygdala, entorhinal cortex, and hippocampus. Moreover, in the amygdala and hippocampus, the emotion effect was greater for recollection than for familiarity, whereas in the entorhinal cortex, it was similar for both forms of retrieval. Jackson et al. (2008) investigated the visual short-term memory for emotional faces and showed that the basal ganglia seem to be critically involved in the interaction of emotion and cognition. Additionally, authors reported a strong positive correlation between amygdala response and other regions of the MTL system during retrieval of emotional vs. neutral stimuli, thereby supporting previous findings. These results further underscore the notion that the MTL system is not only essential for encoding and consolidation but also plays a critical role for retrieval, especially of emotional material, and thus seems to be particularly relevant for long-term memory.

A recent meta-analysis including 18 fMRI studies (Murty et al. 2010) used activation likelihood estimates to assess the anatomical specificity and reliability of event-related fMRI activations related to successful memory encoding for emotional vs. neutral information. The meta-analysis revealed consistent clusters within the amygdala bilaterally, the anterior hippocampus, the anterior and posterior parahippocampal gyrus, the ventral visual stream, the left lateral prefrontal cortex, and the right ventral parietal cortex. The results within the amygdala and the MTL system support a wealth of findings from the animal literature linking these regions to arousal-mediated memory effects. The consistency of findings in cortical targets, particularly prefrontal and parietal cortices, underscores the interaction of cortical and subcortical-limbic regions during emotional memory formation.

In particular, Murty and colleagues (2010) propose that the amygdala interacts with these cortical structures to promote enhancements in perceptual processing, semantic elaboration, and attention, which serve to benefit subsequent memory for emotional material.

Evidence has accumulated that the amygdala is not only involved in encoding and consolidation but also plays a role in retrieval of emotional material. Here, the strongest activation has been observed during a successful recollection of emotional items. Moreover, the response of the amygdala correlated positively with activation of other regions of the medial temporal lobe (MTL) memory system (hippocampus, entorhinal and perirhinal cortices) during retrieval of emotional vs. neutral stimuli.

Concerning emotional memory, a strong interaction between subcortical-limbic and cortical regions has been reported, particularly regarding arousal-mediated memory effects which are linked to the amygdala but also to other regions of the MTL memory system and the prefrontal cortex (LaBar and Cabeza 2006).

10.5 Clinical Relevance

Many patients suffering from mental disorders are characterized by deficits in emotional competencies. Due to the high prevalence and the immense personal burden, these deficits have high clinical relevance.

On the neural level, most of these disorders show dysfunctions of regions in the limbic system and particularly the amygdala. Pathological changes in amygdala volume have been reported for patients suffering from Alzheimer's disease, depression, bipolar disorder, or schizophrenia. Each year, more and more fMRI studies are conducted that investigate neural dysfunctions during emotional behavior and social cognition in psychiatric patients. Emotional abnormalities have been demonstrated for patients with depression; bipolar disorder; schizophrenia; autism; anxiety disorder; addiction; personality disorder, for example, borderline or antisocial personality disorder; and dementia.

Mainly, studies focused on disorder-specific dysfunctions, such as emotional experience in schizophrenia and depression, emotion recognition in schizophrenia, affective disorder or personality disorder or neural responses to disorder-specific stimuli in patients with anxiety disorder, personality disorder or depression, as well as investigation of craving in addiction patients. Based on these findings, general as well as specific neural dysfunctions characterizing various patient groups can be examined. For example, schizophrenia patients are characterized by amygdala dysfunctions during sad mood induction (adult patients, Habel et al. 2004; juvenile patients, Habel et al. 2006), emotion discrimination (Habel et al. 2010a), as well as empathy (Derntl et al. 2012), thus supporting the assumption that the amygdala is particularly involved in affective disturbances in schizophrenia.

The next step would be to draw implications for diagnostics and treatment. For instance, dysfunctional neural activation during emotion recognition in schizophrenia patients can be trained using a standardized emotion discrimination training (Habel et al. 2010b).

Recently, Keedwell and colleagues (2009, 2010) showed that activation of the subgenual cingulate cortex during viewing of sad faces is a valid predictor for therapy response and thus reduction of clinical symptoms in depressed patients.

A whole new era for neuroimaging tools is their application for intervention, that is, neurofeedback (real-time fMRI or EEG neurofeedback) which both have showed valuable results recently (for reviews, see Birbaumer et al. 2009; Weisskopf 2012).

10.6 Summary

Due to their high degree of subjectivity and the lack of objective tests, emotions have long been neglected by scientists. However, with the advent of neuroimaging tools, the scientific investigation of emotional behavior received more attention and became more prominent.

The paradigms applied measured emotional experience, emotion recognition, empathy, and social cognition or examined the influence of emotions on learning and memory formation. Hence, this strikingly underscores that emotions seem to affect almost all cognitive processes and emotion and cognition are constructs that can be hardly separated. Here, neuroimaging research has to further characterize this interaction and the underlying neural correlates in healthy participants as well as psychiatric patients.

Concerning emotions, previous findings indicate a widespread neural network including subcortical-limbic as well as cortical structures, with the amygdala as a key node. Though many studies indicate the amygdala as the central and integrative node for emotion processing, our knowledge about its function is far from being conclusive. Due to several functional but also methodological factors, it is still challenging to reliably measure robust amygdala activation with fMRI.

Other core regions in the emotional network are the prefrontal cortex and the anterior cingulate cortex as well as temporal regions.

Studying the neural correlates of emotional behavior in patients is of particular importance, because deficits in these domains may explain the major dysfunctions in psychiatric disorders that prevent effective (re)integration into work and social life. It has also become clearer that emotional dysfunctions and their neural underpinnings may represent trait markers and endophenotypes of the diseases.

References

Adams RB Jr, Gordon HL, Baird AA, Ambady N, Kleck RE (2003) Effects of gaze on amygdala sensitivity to anger and fear faces. Science 300:1536

Aggleton JP, Mishkin M (1986) The amygdala: sensory gateway to the emotions. In: Plutchik R, Kellermann H (eds) Emotion: theory, research, and experience, vol 3. Academic, New York, pp 281–299

Barrett LF, Mesquita B, Ochsner KN, Gross JJ (2007) The experience of emotion. Ann Rev Psychol 58:373–403

Baumgartner T, Lutz K, Schmidt CF, Jäncke L (2006) The emotional power of music: how music enhances the feeling of affective pictures. Brain Res 1075:151–164

Birbaumer N. Ramos MA, Weber C, Montoya P (2009) Neurofeedback and brain-computer interface clinical applications. International Reviews in Neurobiology 86:107–117

Brattico E, Alluri V, Bogert B, Jacobsen T, Vartiainen N, Nieminen S, Tervaniemi M (2011) A functional MRI study of happy and sad emotions in music with and without lyrics. Front Psychol 2:308

Britton JC, Taylor SF, Sudheimer KD, Liberzon I (2006) Facial expressions and complex IAPS pictures: common and differential networks. Neuroimage 31: 906–919

Britton JC, Shin LM, Barrett LF, Rauch SL, Wright CI (2008) Amygdala and fusiform gyrus temporal dynamics: responses to negative facial expressions. BMC Neurosci 9:44

Büchel C, Dolan RJ (2000) Classical fear conditioning in functional neuroimaging. Curr Opin Neurobiol 10: 219–223

Büchel C, Dolan RJ, Armony JL, Friston KJ (1999) Amygdala-hippocampal involvement in human aversive trace conditioning revealed through event-related functional magnetic resonance imaging. J Neurosci 19:10869–10876

Canli T, Zhao Z, Brewer J, Gabrieli JD, Cahill L (2000) Event-related activation in the human amygdala associates with later memory for individual emotional experience. J Neurosci 20:RC99

Colibazzi T, Posner J, Wang Z, Gorman D, Gerber A, Yu S, Zhu H, Kangarlu A, Duan Y, Russell JA, Peterson BS (2010) Neural systems subserving valence and arousal during the experience of induced emotions. Emotion 10:377–389

Critchley HD, Daly E, Phillips M, Brammer M, Bullmore E, Williams S, van Amelsvoort T, Rob-ertson D, David A, Murphy D (2000) Explicit and implicit neural mechanisms for processing of social information from facial expressions: a functional magnetic imaging study. Hum Brain Mapp 9:93–105

Decety J, Jackson PL (2004) The functional architecture of human empathy. Behav Cogn Neurosci Rev 3:71–100

Decety J, Lamm C (2006) Human empathy through the lens of social neuroscience. Sci World J 6:1146–1163

Derntl B, Habel U, Windischberger C, Robinson S, Kryspin-Exner I, Gur RC, Moser E (2009) General and specific responsiveness of the amygdala during explicit emotion recognition in females and males. BMC Neurosci 10:91

Derntl B, Finkelmeyer A, Eickhoff SB, Kellermann T, Falkenberg DI, Schneider F, Habel U (2010) Multidimensional assessment of empathic abilities: neural correlates and gender differences. Psychoneuroendocrinology 35:67–82

Derntl B, Finkelmeyer A, Voss B, Eickhoff SB, Kellermann T, Schneider F, Habel U (2012) Neural correlates of the core facets of empathy in schizophrenia. Schizophr Res 136:70–81

de Vignemont F, Singer T (2006) The empathic brain: how, when and why? Trends in Cognitive Sciences 10:435–441

Dima D, Klaas ES, Roiser JP, Friston KL, Frangou S (2011) Effective connectivity during processing of facial affect: evidence for multiple parallel pathways. J Neurosci 31:14378–14385

Dolan RJ (2002) Emotion, cognition, and behavior. Science 298:1191–1194

Dolcos F, LaBar KS, Cabeza R (2004) Dissociable effects of arousal and valence on prefrontal activity indexing emotional evaluation and subsequent memory: an event-related fMRI study. NeuroImage 23:64–74

Dolcos F, LaBar KS, Cabeza R (2005) Remembering one year later: role of the amygdala and the medial temporal lobe memory system in retrieving emotional memories. Proc Natl Acad Sci U S A 102:2626–2631

Doty RW (1989) Some anatomical substrates of emotion, and their bihemispheric coordination. In: Gainotti G, Caltagirone C (eds) Emotion and the dual brain. Springer, Berlin, pp 55–81

Drevets WC, Price JL, Furey ML (2008) Brain structural and functional abnormalities in mood disorders: implications for neurocircuitry models of depression. Brain Struct Funct 213:93–118

Duan X, Dai Q, Gong Q, Chen H (2010) Neural mechanism of unconscious perception of surprised facial expression. Neuroimage 52:401–407

Dyck M, Loughead J, Kellermann T, Boers F, Gur RC, Mathiak K (2011) Cognitive versus automatic mechanisms of mood induction differentially activate left and right amygdala. Neuroimage 54:2503–2513

Etkin A, Egner T, Kalisch R (2010) Emotional processing in anterior cingulated and medial prefrontal cortex. Trends Cogn Sci 15:85–93

Eugene F, Levesque J, Mensour B, Leroux JM, Beaudoin G, Bourgouin P, Beauregard M (2003) The impact of individual differences on the neural circuitry underlying sadness. NeuroImage 19:354–364

Falkenberg I, Kohn N, Schoepker R, Habel U (2012) Mood induction in depressive patients: a comparative multidimensional approach. PLoS One 7:e30016

Fan Y, Duncan NW, de Greck M, Northoff G (2011) Is there a core neural network in empathy? An fMRI based quantitative meta-analysis. Neurosci Biobehav Rev 35:903–911

Feinstein JS, Goldin PR, Stein MB, Brown GG, Paulus MP (2002) Habituation of attentional networks during emotion processing. Neuroreport 13:1255–1258

Fitzgerald DA, Angstadt M, Jelsone LM, Nathan PJ, Phan KL (2006) Beyond threat: amygdala reactivity across multiple expressions of facial affect. Neuroimage 30: 1441–1448

Fusar-Poli P, Placentino A, Carletti F, Landi P, Allen P, Surguladze S, Benedetti F, Abbamonte M, Gasparotti R, Barale F, Perez J, McGuire P, Politi P (2009) Functional atlas of emotional faces processing: a voxel-based meta-analysis of 105 functional magnetic resonance imaging studies. J Psychiatry Neurosci 34:418–432

George MS, Ketter TA, Parekh PI, Horwitz B, Herscovitch P, Post RM (1995) Brain activity during transient sadness and happiness in healthy women. Am J Psychiatry 152:341–351

Gur RC, Schroeder L, Turner T, McGrath C, Chan RM, Turetsky BI, Alsop D, Maldjian J, Gur RE (2002) Brain activation during facial emotion processing. NeuroImage 16:651–662

Haas BW, Constable RT, Canli T (2009) Functional magnetic resonance imaging of temporally distinct responses to emotional facial expressions. Soc Neurosci 4:121–134

Habel U, Gur RC, Mandal MK, Salloum JB, Gur RE, Schneider F (2000) Emotional processing in schizophrenia across cultures: standardized measures of discrimination and experience. Schizophr Res 42: 57–66

Habel U, Klein M, Shah NJ, Toni I, Zilles K, Falkai P, Schneider F (2004) Genetic load on amygdala hypofunction during sadness in non-affected brothers of schizophrenia patients. Am J Psychiatry 161:1806–1813

Habel U, Klein M, Kellermann T, Shah NJ, Schneider F (2005) Same or different? Neural correlates of happy and sad mood in healthy males. NeuroImage 26: 206–214

Habel U, Krasenbrink I, Bowi U, Ott G, Schneider F (2006) A special role of negative emotion in children and adolescents with schizophrenia and other psychoses. Psychiatry Res 145:9–19

Habel U, Windischberger C, Derntl B, Robinson S, Kryspin-Exner I, Gur RC, Moser E (2007) Amygdala activation and facial expressions: explicit emotion discrimination versus implicit emotion processing. Neuropsychologia 45:2369–2377

Habel U, Chechko N, Pauly K, Koch K, Backes V, Seiferth N, Shah NJ, Stöcker T, Schneider F, Kellermann T (2010a) Neural correlates of emotion recognition in schizophrenia. Schizophr Res 112:113–123

Habel U, Koch K, Kellermann T, Reske M, Frommann N, Wölwer W, Zilles K, Shah NJ, Schneider F (2010b) Training of affect recognition in schizophrenia: neurobiological correlates. Soc Neurosci 5:92–104

Hamann S, Canli T (2004) Individual differences in emotion processing. Curr Opin Neurobiol 14:233–238

Hariri AR, Bookheimer SY, Mazziotta JC (2000) Modulating emotional responses: effects of a neocortical network on the limbic system. Neuroreport 11:43–48

Hariri AR, Tessitore A, Mattay VS, Fera F, Weinberger DR (2002) The amygdala response to emotional stimuli: a comparison of faces and scenes. NeuroImage 17:317–323

Hofer A, Siedentopf CM, Ischebeck A, Rettenbacher MA, Verius M, Felber S, Fleischhacker WW (2006) Gender differences in regional cerebral activity during the perception of emotion: a functional MRI study. Neuroimage 32:854–862

Hofer A, Siedentopf CM, Ischebeck A, Rettenbacher MA, Verius M, Felber S, Wolfgang Fleischhacker W (2007) Sex differences in brain activation patterns during processing of positively and negatively valenced emotional words. Psychol Med 37:109–119

Indovina I, Robbins TW, Núñez-Elizalde AO, Dunn BD, Bishop SJ (2011) Fear-conditioning mechanisms associated with trait vulnerability to anxiety in humans. Neuron 69:563–571

Jackson MC, Wolf C, Johnston SJ, Raymond JE, Linden DE (2008) Neural correlates of enhanced visual short-term memory for angry faces: an FMRI study. PLoS One 3:e3536

Jeong JW, Diwadkar VA, Chugani CD, Sinsoongsud P, Muzik O, Behen ME, Chugani HT, Chugani DC (2011) Congruence of happy and sad emotion in music and faces modifies cortical audiovisual activation. Neuroimage 54:2973–2982

Ji J, Maren S (2007) Hippocampal involvement in contextual modulation of fear extinction. Hippocampus 17:749–758

Juruena MF, Giampietro VP, Smith SD, Surguladze SA, Dalton JA, Benson PJ, Cleare AJ, Fu CH (2010) Amygdala activation to masked happy facial expressions. J Int Neuropsychol Soc 16:383–387

Keedwell P, Drapier D, Surguladze S, Giampietro V, Brammer M, Phillips M (2009) Neural markers of symptomatic improvement during antidepressant therapy in severe depression: subgenual cingulate and visual cortical responses to sad, but not happy, facial stimuli are correlated with changes in symptom score. J Psychopharmacol 23:775–788

Keedwell PA, Drapier D, Surguladze S, Giampietro V, Brammer M, Phillips M (2010) Subgenual cingulate and visual cortex responses to sad faces predict clinical outcome during antidepressant treatment for depression. J Affect Disord 120:120–125

Kennedy DP, Adolphs R (2012) The social brain in psychiatric and neurological disorders. Trends Cogn Sci 16:559–572

Kessler H, Taubner S, Buchheim A, Münte TF, Stasch M, Kächele H, Roth G, Heinecke A, Erhard P, Cierpka M, Wiswede D (2011) Individualized and clinically derived stimuli activate limbic structures in depression: an fMRI study. PLoS ONE 6:e15712

Killgore WD, Yurgelun-Todd DA (2004) Activation of the amygdala and anterior cingulate during nonconscious processing of sad versus happy faces. NeuroImage 21:1215–1223

Kim MJ, Loucks RA, Neta M, Davis FC, Oler JA, Mazzulla EC, Whalen PJ (2010) Behind the mask: the influence of mask-type on amygdala response to fearful faces. Soc Cogn Affect Neurosci 5:363–368

Knight DC, Smith CN, Cheng DT, Stein EA, Helmstetter FJ (2004) Amygdala and hippocampal activity during acquisition and extinction of human fear conditioning. Cogn Affect Behav Neurosci 4:317–325

Koepp MJ, Hammers A, Lawrence AD, Asselin MC, Grasby PM, Bench CJ (2009) Evidence for endogenous opioid release in the amygdala during positive emotion. Neuroimage 44:252–256

Kohn N, Kellermann T, Gur RC, Schneider F, Habel U (2011) Gender differences in the neural correlates of humor processing: implications for different processing modes. Neuropsychologia 49:888–897

LaBar KS, Cabeza R (2006) Cognitive neuroscience of emotional memory. Nat Rev Neurosci 7:54–64

Lamm C, Decety J, Singer T (2011) Meta-analytic evidence for common and distinct neural networks associated with directly experienced pain and empathy for pain. NeuroImage 54:2492–2502

Lammers CH (2007) Emotionsbezogene Psychotherapie. Schattauer, Stuttgart

Lavie N (1995) Perceptual load as a necessary condition for selective attention. J Exp Psychol Hum Percept Perform 21:451–468

Lavie N, De Fockert J (2005) The role of working memory in attentional capture. Psychol Bull Rev 12:669–674

Lee KH, Farrow TF, Spence SA, Woodruff PW (2004) Social cognition, brain networks and schizophrenia. Psychological Medicine 34:391–400

LeDoux JE (2000) Emotion circuits in the brain. Annu Rev Neurosci 23:155–184

Levesque J, Eugene F, Joanette Y, Paquette V, Mensour B, Beaudoin G, Leroux JM, Bourgouin P, Beauregard M (2003) Neural circuitry underlying voluntary suppression of sadness. Biol Psychiatry 53:502–510

Mayberg HS, Brannan SK, Tekell JL, Silva JA, Mahurin RK, McGinnis S, Jerabek PA (2000) Regional metabolic effects of fluoxetine in major depression: serial changes and relationship to clinical response. Biol Psychiatry 48:830–843

Mitterschiffthaler MT, Fu CH, Dalton JA, Andrew CM, Williams SC (2007) A functional MRI study of happy and sad affective states induced by classical music. Hum Brain Mapp 28:1150–1162

Morawetz C, Baudewig J, Treue S, Dechent P (2010) Diverting attention suppresses human amygdala responses to faces. Front Hum Neurosci 3:226

Morawetz C, Baudewig J, Treue S, Dechent P (2011) Effects of spatial frequency and location of fearful faces on human amygdala activity. Brain Res 31:87–99

Morris JS, Dolan RJ (2004) Dissociable amygdala and orbitofrontal responses during reversal fear conditioning. NeuroImage 22:372–380

Morris JS, Frith CD, Perrett DI, Rowland D, Young AW, Calder AJ, Dolan RJ (1996) A differential neural response in the human amygdala to fearful and happy facial expressions. Nature 383:812–815

Moser E, Derntl B, Robinson S, Fink B, Gur RC, Grammer K (2007) Amygdala activation at 3T in response to human and avatar facial expressions of emotions. J Neurosci Methods 161:126–133

Murty VP, Ritchey M, Adcock RA, LaBar KS (2010) fMRI studies of successful emotional memory encoding: a quantitative meta-analysis. Neuropsychologia 48:3459–3469

Ochsner KN (2000) Are affective events richly recollected or simply familiar? The experience and process of recognizing feelings past. J Exp Psychol Gen 129: 242–261

Pardo JV, Pardo PJ, Raichle ME (1993) Neural correlates of self-induced dysphoria. Am J Psychiatry 150:713–719

Pelphrey KA, Morris JP, McCarthy G, LaBar K (2007) Perception of dynamic changes in facial affect and identity in autism. Soc Cogn Affect Neurosci 2:140–149

Pessoa L, Adolphs R (2010) Emotion processing and the amygdala: from a 'low road' to 'many roads' of evaluating biological significance. Nat Rev Neurosci 11:773–783

Pessoa L, Japee S, Sturman D, Ungerleider LG (2005) Target visibility and visual awareness modulate amygdala responses to fearful faces. Cereb Cortex 16:366–375

Phan KL, Wager T, Taylor SF, Liberzon I (2002) Functional neuroanatomy of emotion: a meta-analysis of emotion activation studies in PET and fMRI. NeuroImage 16:331–348

Phelps EA, Delgado MR, Nearing KI, LeDoux JE (2004) Extinction learning in humans: role of the amygdala and vmPFC. Neuron 16:897–905

Phillips ML, Medford N, Young AW, Williams L, Williams SC, Bullmore ET, Gray JA, Brammer MJ (2001) Time courses of left and right amygdalar responses to fearful facial expressions. Hum Brain Mapp 12:193–202

Phillips ML, Drevets WC, Rauch SL, Lane R (2003) Neurobiology of emotion perception I: the neural basis of normal emotion perception. Biol Psychiatry 54:504–514

Pizzagalli F (2011) Frontocingulate dysfunction in depression: toward biomarkers of treatment response. Neuropsychopharmacology 36:183–206

Posse S, Fitzgerald D, Gao K, Habel U, Rosenberg D, Moore GJ, Schneider F (2003) Real-time fMRI of temporolimbic regions detects amygdala activation during single-trial self-induced sadness. NeuroImage 18:760–768

Preston SD, de Waal FBM (2002) Empathy: its ultimate and proximate bases. Behav Brain Sci 25:1–72

Quirk GJ, Beer JS (2006) Prefrontal involvement in the regulation of emotion: convergence of rat and human studies. Curr Opin Neurobiol 16:723–727

Reiman EM, Lane RD, Ahern GL, Schwartz GE, Davidson RJ, Friston KJ, Yun LS, Chen K (1997) Neuroanatomical correlates of externally and internally generated human emotion. Am J Psychiatry 154:918–925

Reske M, Kellermann T, Habel U, Jon Shah N, Backes V, von Wilmsdorff M, Stöcker T, Gaebel W, Schneider F (2007) Stability of emotional dysfunctions? A long-term fMRI study in first-episode schizophrenia. J Psychiatr Res 41:918–927

Reske M, Kellermann T, Shah NJ, Schneider F, Habel U (2010) Impact of valence and age on olfactory induced brain activation in healthy women. Behav Neurosci 124:414–422

Richardson MP, Strange BA, Dolan RJ (2004) Encoding of emotional memories depends on amygdala and hippocampus and their interactions. Nat Neurosci 7:278–285

Rueckert L, Naybar N (2008) Gender differences in empathy: the role of the right hemisphere. Brain Cogn 67:162–167

Sander D, Grafman J, Zalla T (2003) The human amygdala: an evolved system for relevance detection. Rev Neurosci 14:303–316

Sato W, Yoshikawa S (2007) Enhanced experience of emotional arousal in response to dynamic facial expressions. J Nonverbal Behav 31:119–135

Sato W, Yoshikawa S, Kochiyama T, Matsumura M (2004a) The amygdala processes the emotional significance of facial expressions: an fMRI investigation using the interaction between expression and face direction. NeuroImage 22:1006–1013

Sato W, Kochiyama T, Yoshikawa S, Naito E, Matsumura M (2004b) Enhanced neural activity in response to dynamic facial expressions of emotion: an fMRI study. Brain Res Cogn Brain Res 20:81–91

Schneider F, Gur RC, Gur RE, Muenz LR (1994) Standardized mood induction with happy and sad facial expressions. Psychiatry Res 51:19–31

Schneider F, Weiss U, Kessler C, Salloum JB, Posse S, Grodd W, Müller-Gärtner H-W (1998) Differential amygdala activation in schizophrenia during sadness. Schizophr Res 34:133–142

Schneider F, Habel U, Kessler C, Salloum JB, Posse S (2000) Gender differences in regional cerebral activity during sadness. Hum Brain Mapp 9:226–238 [Erratum: 2001, 13:124]

Schulte-Rüther M, Markowitsch HJ, Shah NJ, Fink GR, Piefke M (2008) Gender differences in brain networks supporting empathy. Neuroimage 42:393–403

Sergerie K, Lepage M, Armony JL (2005) A face to remember: emotional expression modulates prefrontal activity during memory formation. NeuroImage 24: 580–585

Sharot T, Yonelinas AP (2008) Differential time-dependent effects of emotion on recollective experience and memory for contextual information. Cognition 106:538–547

Sharot T, Delgado MR, Phelps EA (2004) How emotion enhances the feeling of remembering. Nat Neurosci 7:1376–1380

Singer T, Lamm C (2009) The social neuroscience of empathy. Ann N Y Acad Sci 1156:81–96

Singer T, Seymour B, O'Doherty JP, Stephan KE, Dolan RJ, Frith CD (2006) Empathic neural responses are modulated by the perceived fairness of others. Nature 439:466–469

Smith AP, Henson RN, Dolan RJ, Rugg MD (2004) FMRI correlates of the episodic retrieval of emotional contexts. NeuroImage 22:868–878

Tabert MH, Borod JC, Tang CY, Lange G, Wei TC, Johnson R, Nusbaum AO, Buchsbaum MS (2001) Differential amygdala activation during emotional decision and recognition memory tasks using unpleasant words: an fMRI study. Neuropsychologia 39:556–573

Teasdale JD, Howard RJ, Cox SG, Ha Y, Brammer MJ, Williams SC, Checkley SA (1999) Functional MRI study of the cognitive generation of affect. Am J Psychiatry 156:209–215

Tettamanti M, Rognoni E, Cafiero R, Costa T, Galati D, Perani D (2012) Distinct pathways of neural coupling for different basic emotions. Neuroimage 59: 1804–1817

van Stegeren AH, Roozendaal B, Kindt M, Wolf OT, Joëls M (2010) Interacting noradrenergic and corticosteroid systems shift human brain activation patterns during encoding. Neurobiol Learn Mem 93:56–65

Wager TD, Phan KL, Liberzon I, Taylor SF (2003) Valence, gender, and lateralization of functional brain anatomy in emotion: a meta-analysis of findings from neuroimaging. Neu-roImage 19:513–531

Weiskopf N (2012) Real-time fMRI and its application to neurofeedback. Neuroimage 62:682–692

Weiss U, Salloum JB, Schneider F (1999) Correspondence of emotional self-rating with facial expression. Psychiatry Res 86:175–184

Whalen PJ, Rauch SL, Etcoff NL, McInerney SC, Lee MB, Jenike MA (1998) Masked presentations of emotional facial expressions modulate amygdala activity without explicit knowledge. J Neurosci 18:411–418

Whalen PJ, Kagan J, Cook RG, Davis FC, Kim H, Polis S, McLaren DG, Somerville LH, McLean AA, Maxwell JS, Johnstone T (2004) Human amygdala responsivity to masked fearful eye whites. Science 306:2061

Williams MA, Morris AP, McGlone F, Abbott DF, Mattingley JB (2004) Amygdala responses to fearful and happy facial expressions under conditions of binocular suppression. J Neurosci 24:2898–2904

Williams MA, McGlone F, Abbott DF, Mattingley JB (2005) Differential amygdale responses to happy and fearful facial expressions depend on selective attention. Neuroimage 24:417–425

Wright CI, Fischer H, Whalen PJ, McInerney SC, Shin LM, Rauch SL (2001) Differential prefrontal cortex and amygdala habituation to repeatedly presented emotional stimuli. Neuroreport 12:379–383

Wright CI, Martis B, Shin LM, Fischer H, Rauch SL (2002) Enhanced amygdala responses to emotional versus neutral schematic facial expressions. Neuroreport 13:785–790

Yonelinas AP, Otten LJ, Shaw KN, Rugg MD (2005) Separating the brain regions involved in recollection and familiarity in recognition memory. J Neurosci 25:3002–3008

Zotev V, Krueger F, Phillips R, Alvarez RP, Simmons WK, Bellgowan P, Drevets WC, Bodurka J (2011) Self-regulation of amygdala activation using real-time FMRI neurofeedback. PLoS One 6:e24522

Imaging Cognition

11

Roberto Goya-Maldonado and Oliver Gruber

Abbreviations

ACC	Anterior cingulate cortex
ALE	Activation likelihood estimation
BA	Brodmann area
DN	Default network
FEF	Frontal eye field
IFG	Inferior frontal gyrus
IFJ	Inferior frontal junction
IPC	Intraparietal cortex
OFC	Orbitofrontal cortex
PET	Positron emission tomography
PFC	Prefrontal cortex
ROI	Region of interest
SMA	Supplementary motor area

11.1 Introduction

Understanding the correlates of cognitive functions in the prefrontal cortex is a topic that remains highly relevant. This proves true both from the side of a puzzling complexity achieved by this area of the brain (what makes us such a distinct group in comparison to other species and so individual in contrast to one another) and from the potential of new therapeutic approaches in disorders presenting dysfunctional cognitive mechanisms. And nowadays, this not a matter of science fiction or very distant future plans. It is reality, for instance, to directly apply neuroimaging data to subserve the anatomical choice of prefrontal electrode implantation to treat pharmacoresistant depression (Mayberg et al. 2005). Differently from decades ago, our generation has been gifted to have hands on a myriad of noninvasive technological tools to access the human brain in vivo. Currently, imaging research protocols can empirically test existing theories about cognitive functions in neurophysiological and psychopathological settings, enabling possible reformulations quicker than ever before. Cutting-edge protocols and novel methods of analysis are available individually or in combination to extend the frontiers of knowledge about the working brain. In this chapter, we review the data collected and present the state of the art about imaging cognition. We start with a brief overview of the anatomical distribution of the prefrontal cortex, passing through the functional boundaries revealed by lesion studies, laboratory paradigms applied in fMRI investigation of cognition and regional activations, and ending with more recent assumptions about functional connectivity, network organization, and future perspectives. Above all, we understand that patients with neuropsychiatric disorders, usually featuring several cognitive limitations, will enormously benefit from mechanism-guided therapeutics still to be developed.

R. Goya-Maldonado (✉) • O. Gruber
Department of Psychiatry and Psychotherapy, Center for Translational Research in Systems Neuroscience and Psychiatry, Georg August University, Von-Siebold-Str. 5, Goettingen, D-37075, Germany
e-mail: roberto.goya@med.uni-goettingen.de; oliver.gruber@medizin.uni-goettingen.de

C. Mulert, M.E. Shenton (eds.), *MRI in Psychiatry*,
DOI 10.1007/978-3-642-54542-9_11, © Springer Berlin Heidelberg 2014

11.2 The Architecture of the Prefrontal Cortex

The organization of the human brain has been addressed for long date with the aim to comprehend the complexity of this organ. Postmortem neuroanatomical and histological seminal works of many researchers, among them Ramon y Cajal and Brodmann, have made great achievements about the histological organization and characteristics of different regions. Here, we will briefly address the most relevant anatomical aspects within the scope of this chapter.

11.2.1 Brief Neuroanatomical Aspects

The prefrontal cortex (PFC) has been the main region of interest when considering cognitive functions. A classification of areas presented by Brodmann (1908) has been used since then and until today to support the communication of findings in modern imaging literature. As a general principle to comprehend complex constructs, the PFC was initially divided into smaller parts with the hope to reach simpler cortical equivalences of unitary, basic functions behind such intricate high-order cognitive abilities. Also in direction of simplification, relative anatomical comparisons across other species could give important insights about the evolutionary modifications incorporated into more sophisticated systems. Therefore, Petrides and Pandya (1994) reported the anatomical homology between the human and monkey PFC. Although they may share some similarities, there are consistent modifications in the localization of the Brodmann areas (BA), considering expansion, rotation, and gyrification. As seen in Fig. 11.1, the PFC, located in front of the motor/premotor cortex (BA 4 and 6, not shown), is usually divided into dorsolateral PFC (DLPFC), BA 9 and 46, Broca and ventrolateral PFC (VLPFC), BA 44 and 45, orbitofrontal cortex (OFC), BA 12 and 47, and anterior PFC, BA 10. The anterior cingulate cortex (ACC), BA 24 and 32, frequently referred in the literature of cognitive processes, cannot be seen in the lateral, but in the medial surface of the PFC.

11.2.2 Learning from Lesion Studies

The functional impairments derived from lesions have been one of the first forms to access the

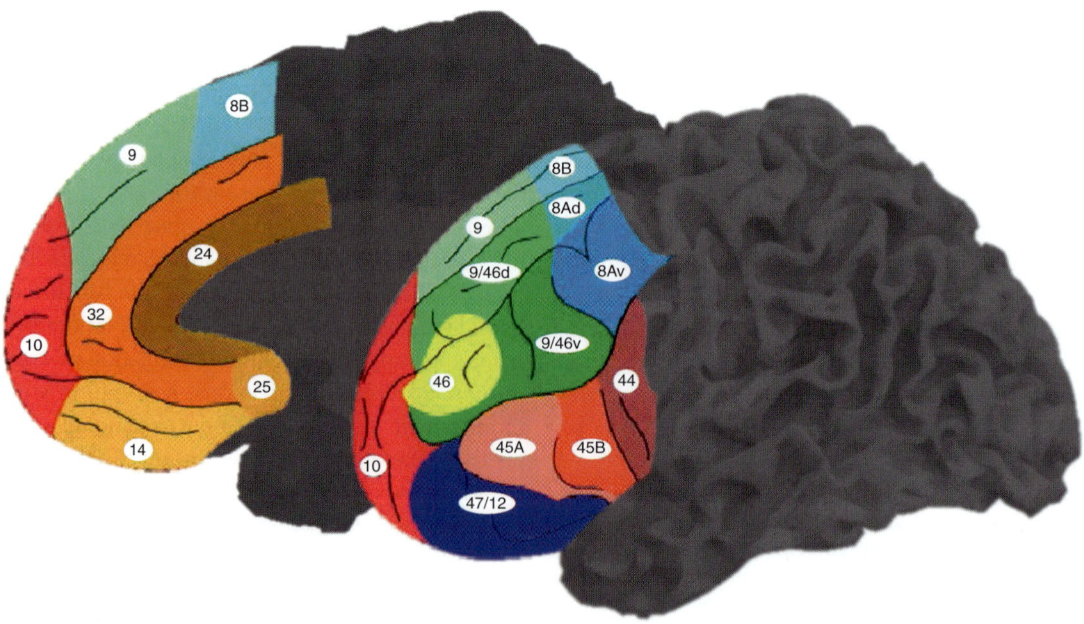

Fig. 11.1 The prefrontal cortex division based on Brodmann areas (Modified from Petrites and Pandaya (1994))

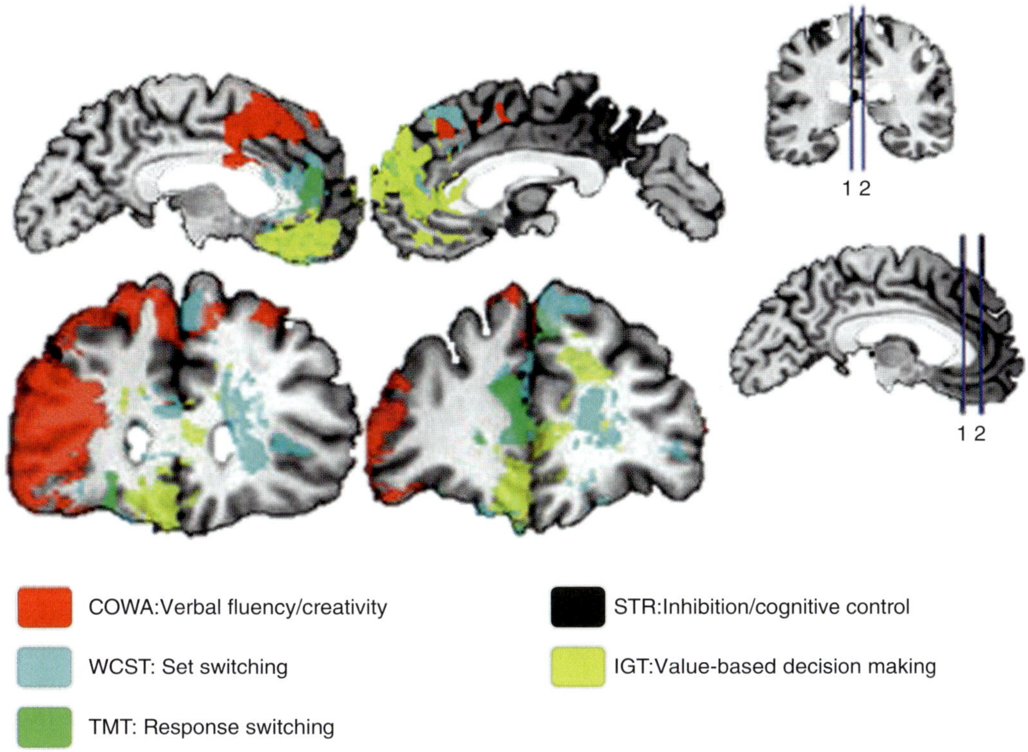

Fig. 11.2 Spatial lesion mapping associated with cognitive impairments. *1*, *2* spatially represent the sections presented in the left side of the figure (Modified from Gläscher et al. (2012))

functional neuroanatomy of the PFC. Multiple studies evaluated the consequences of surgical lesions in animals. Obviously the readout restrictions of an animal behavior compose the main limitation and interpretation pitfall. Even when studying nonhuman primates, presenting more similar cognitive skills to humans, this limitation is reduced but still present. Therefore, studies of human PFC spontaneous lesions became so valuable (Milner 1982; Leimkuhler and Mesulam 1985; Owen et al. 1990; Bechara et al. 2000; Stuss et al. 2000, 2001; Fellows and Farah 2005a). A more recent publication with a vast sample of patients systematically addressing cognitive impairments (Gläscher et al. 2012) re-incited the discussion about the specificity of cognitive correlates (Fig. 11.2). The color overlays represent affected regions, each color associated with reduced performance in a specific task. Here, the PFC correlates related to different cognitive functions such as cognitive control, conflict monitoring, response inhibition, flexibility, planning, working memory (functions detailed over the next section), and decision-making supported the segregation of two distinct groups. The first involved the DLPFC and VLPFC in association with cognitive control, namely, conflict monitoring, response inhibition, and switching/flexibility. The second comprised the medial/anterior PFC and the OFC and was associated to value-based decision-making in a gambling task, which requires assessment of internal information such as individual preferences. Lesions in the rostral ACC could not be classified in one or another of these groups. Also, the dorsal ACC has not been specifically related to Stroop impairment corroborating to previous findings (Fellows and Farah 2005b). This reinforces a certain degree of participation across tasks in these ACC regions, instead of showing clear-cut task divisions. After all, integration should be advantageous for the optimization of resources regarding the multiplicity of stimuli repeatedly perceived by our sensory system. Nevertheless, one should always consider that lesion-deficit studies can give valuable insights about the necessity of certain region(s) during behavioral performance, but will rarely allow more specific inferences about the functional significance of this region(s) in the network. Therefore, functional studies are imperative to further elucidate the neural correlates of cognition.

11.3 Imaging Cognitive Functions

We are daily challenged to respond and adapt to the environment around us. The capacity to evaluate and decide over multiple alternatives is essential for optimizing resources and surviving. Hence, it is demanded from the brain to be functionally a dynamic organ, in which components interact and solutions are achieved by a product of perception and experience. Laboratory functional imaging studies have been used toward further understanding of the complex construct of cognition.

11.3.1 Executive Functions

There has been so far little consensus on the taxonomy of attention and executive control functions. Uncertainties about the classification of such functions derive at least partially from the overlapping correlates behind the psychological constructs of selective attention, response inhibition, cognitive control, and working memory. For instance, numerous neurophysiological and functional imaging studies have collected evidence for very similar prefrontoparietal and prefrontotemporal neural correlates involved in both short-term maintenance of task-relevant information in working memory and selective attention processes (Gruber and Goschke 2004; Ikkai and Curtis 2011; Gazzaley and Nobre 2012; Niendam et al. 2012). Selective attention, the ability of focusing on specific relevant information over the background, is gained through sensory upregulation guided by the PFC (Hopfinger et al. 2000; Kastner and Ungerleider 2000; Petersen and Posner 2012). The sensory areas processing irrelevant and possibly interfering information will be downregulated. At this point, a close relationship between the psychological constructs of attention, executive, and cognitive control can be observed. Such modulatory processes can interfere on sensory perception and processing of a certain object or aspect from the environment. These are top-down-oriented processes that control bottom-up-oriented attention processes (Fig. 11.3). The first can allow us to succeed daily in different cognitive tasks; the second can alert us from any threat unexpectedly arriving that may endanger our integrity. A frequent example is the one about a student preparing for an exam (top-down), while the fire alarm goes off (bottom-up). These two forces compete for the attention resources as shown (Corbetta and Shulman 2002; Gruber and Goschke 2004; Posner 2012; Petersen and Posner 2012). In the laboratory, the bottom-up component of attention can be investigated with a task called oddball paradigm (Bledowski et al. 2004; Melcher and Gruber 2006; Gruber et al. 2009). Here, an infrequent stimulus is presented among other frequent stimuli. The fact that it is rarer triggers attention, a process called stimulus-driven attention (Corbetta and Shulman 2002).

A paradigm called Go-Nogo demands response inhibition of infrequent stimuli. Subjects are instructed to respond to frequent stimuli in such a regular and fast fashion that a tendency to respond is developed, making it difficult to avoid responding even when a Nogo infrequent stimulus is presented. The inferior frontal gyrus has been implied in the capacity to suspend this initiated response as part of a supramodal network (Chikazoe et al. 2007; Walther et al. 2010). Also, the activation in this region during inhibition has been correlated with the variability in impulsivity traits measured from a behavioral scale (Goya-Maldonado et al. 2010). These findings suggest that the IFG, as an area involved in cognitive control, regulates the expression of inherent impulsive tendencies in healthy subjects. Most interestingly, the same area has been reported to be involved in both response inhibition and stimulus-driven attention processes (Fig 11.4).

Another laboratory approach of inhibition called Stroop paradigm deals with conflict detection and cognitive control. While responding to the color of letters displayed, one reads from the letters, another color name leading to

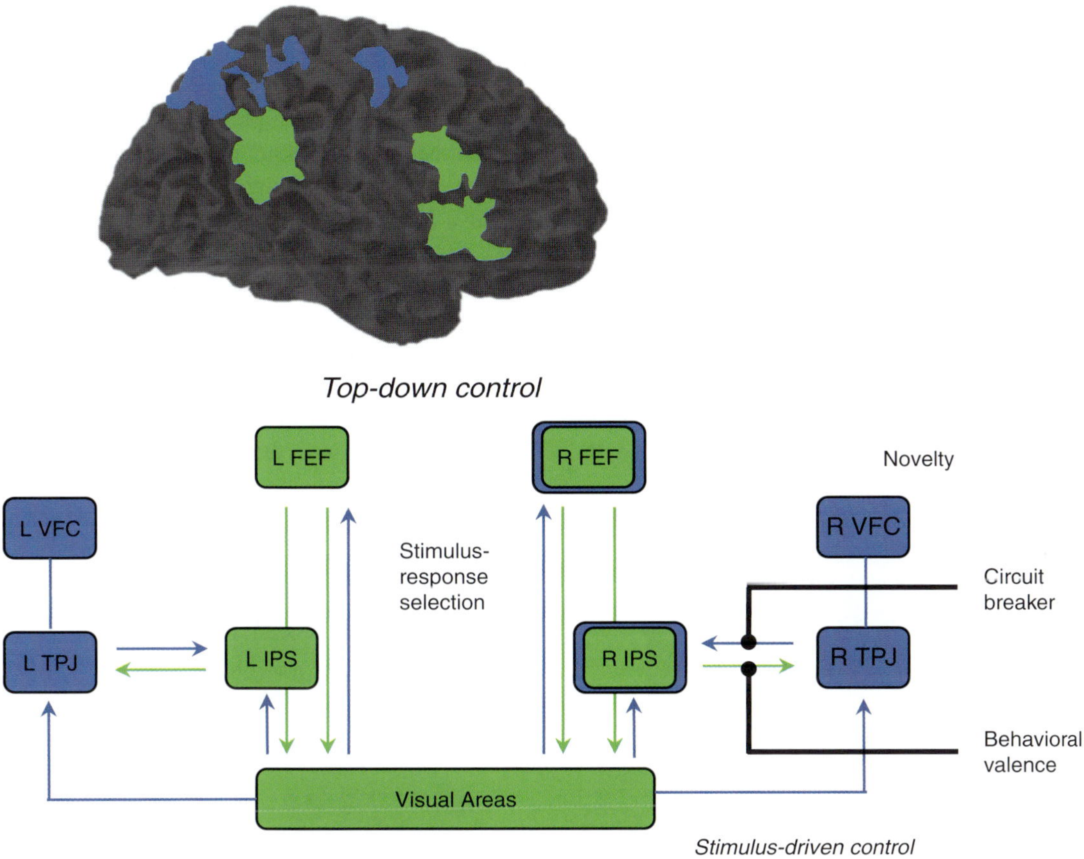

Fig. 11.3 Dorsal and ventral frontoparietal areas involved in *top-down* and *bottom-up* attention processes (Modified from Corbetta and Shulman (2002)). *FEF* frontal eye field, *IPS* intraparietal sulcus, *TPJ* temporoparietal junction, *VFC* ventral frontal cortex

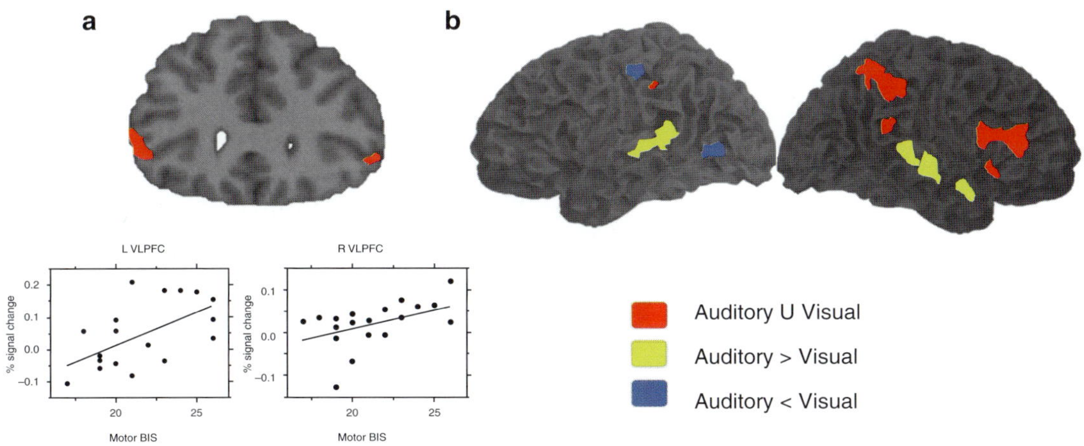

Fig. 11.4 Neural correlates of response inhibition elicited by Go-Nogo paradigm (**a**) which correlated to impulsivity scores, (**b**) presented supramodal (*red areas*) characteristics (Modified from Goya-Maldonado et al. (2010), Walther et al. (2010)). *L VLPFC* left ventrolateral prefrontal cortex, *R VLPFC* right ventrolateral prefrontal cortex, *BIS* Barratt Impulsiveness Scale

conflict. Response conflict and semantic conflict were identified in Stroop paradigm (Melcher and Gruber 2009), and there are now many variations of the paradigm (Stroop 1935, 1992; MacLeod 1991; Kerns et al. 2004; Matsumoto and Tanaka 2004; Melcher and Gruber 2006), but the principle remains that contradictory information is displayed and conflict has to be identified and solved to achieve the correct answer. With the final purpose of succeeding, adjustments based on additional information from the environment are demanded. This recruits the DLPFC for increased cognitive control, which is again deactivated once the conflict disappears (Fig. 11.5).

To address the dynamic mechanisms involved in bottom-up and top-down components of behavioral control, a more recent paradigm called the desire-reason-dilemma was used (Diekhof and Gruber 2010; Diekhof et al. 2011). It consists of two settings, one that allows the acceptance (desire) of previously conditioned rewards while the subject performs a simple task and another that restricts the acceptance (reason) of rewards. The nucleus accumbens and ventral tegmental area, regions involved in the dopaminergic bottom-up of the reward system, have shown high activation during desire, which was drastically top-down suppressed during reason by specific prefrontal regions.

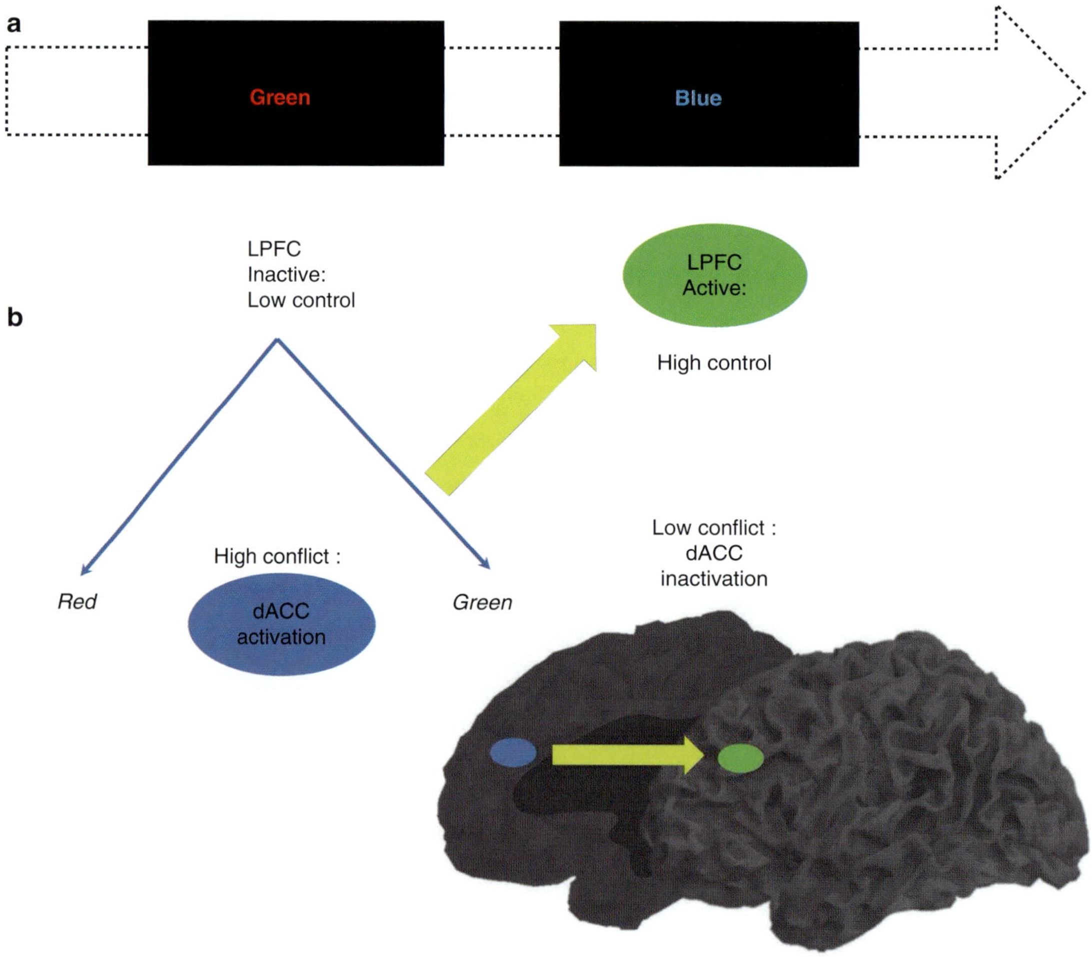

Fig. 11.5 (**a**) Color and names of colors presented as conflict stimuli in the Stroop paradigm and (**b**) the medial and lateral areas involved in conflict detection and cognitive control, respectively (Modified from Matsumoto and Tanaka (2004), Kerns et al. (2004)). *LPFC* lateral prefrontal cortex, *dACC* dorsal anterior cingulate cortex

11.3.2 Working Memory

Working memory or short-term memory is a cognitive function that allows holding an amount of information for a limited period of time so a certain goal can be achieved. The information can be captured from the environment and may be discarded right after it has served its purpose. This enables a faster and more effective performance of daily tasks. Analogous to the input of external information, parts of consolidated memory can be retrieved and put online during the performance of language, problem solving, and task planning.

Multiple cognitive models have been proposed, among which the model of Baddeley and Hitch (1974) was highly influential. Here and in a more actual version (Baddeley 2000), short-term memory processes could be partitioned in different components. Studies addressing the functional anatomy of working memory were then conducted in the early 1990s, based on this cognitive model (Baddeley and Hitch 1974). These components described as part of the working memory were suggested to be separable in PET and fMRI studies (Jonides et al.1993; Paulesu et al.1993; D'Esposito et al. 1995), and the verbal and visual components could be later compared (Fig. 11.6b). However, this segregation was not confirmed by following fMRI studies using the n-back paradigm (D'Esposito et al. 1998; Nystrom et al. 2000; Zurowski et al. 2002; Owen et al. 2005). This task involves responding with a button press every time a stimulus is repeated in the sequential presentation. The n represents the position in which this repetition should be tracked and is instructed right before the task block begins; for instance, 1-back means that a button press is demanded every time a stimulus is immediately repeated, 2-back every time a stimulus is repeated in an interval of another stimulus, 3-back every time it is repeated in an interval of two other stimuli, and so on.

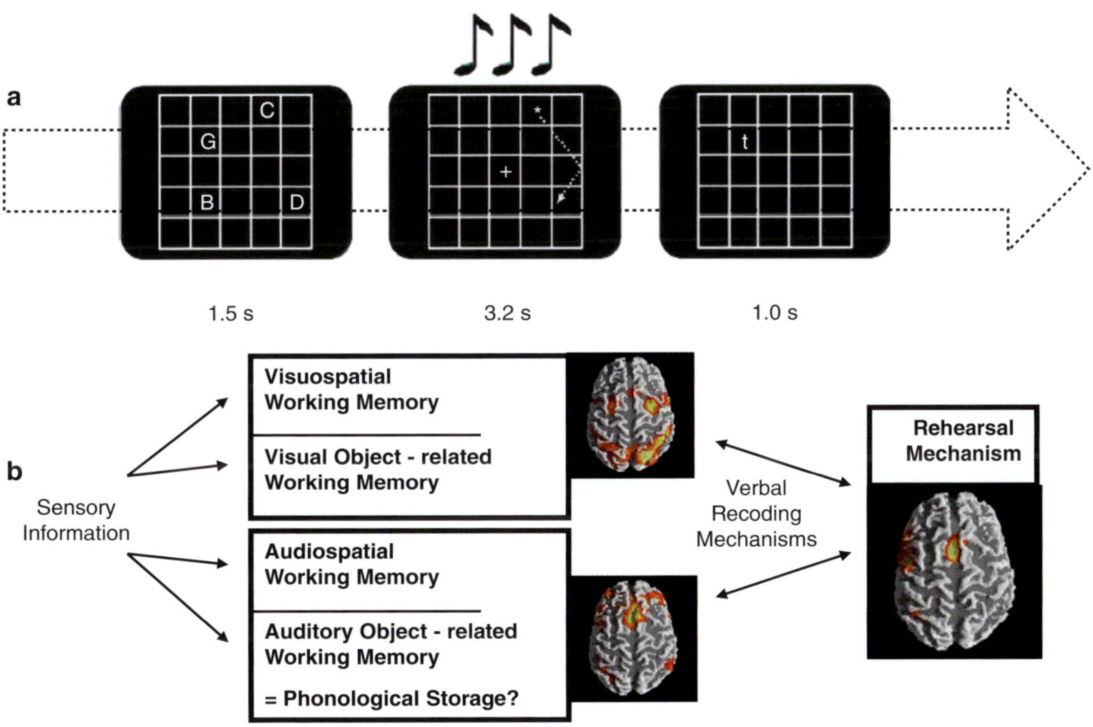

Fig. 11.6 (**a**) Visuospatial item recognition or verbal item recognition with or without visual or articulatory suppression; visual suppression was performed by following a moving star or articulatory suppression by subvocalizing numbers. (**b**) Scheme depicting prefrontoparietal and prefrontotemporal circuits according to information domains and the phonological loop (Modified from Gruber and von Cramon (2003), Gruber and Goschke (2004))

Clearly the level of difficulty increases with the n, so that the information has to be hold for longer and across more in-between stimuli. The use of this task in a psychiatric sample withholds the disadvantage of not allowing the investigation of process-specific working memory functions due to its complexity in regard to the traditional instruments for neuropsychological testing. In this paradigm, sustaining the information online is mixed with processes involved in sequencing, updating, and manipulating the information. Obviously, the simultaneous activation of brain areas comprising a variety of related processes to working memory will confound the differentiation of even well-established correlates such as the human visuospatial working memory (see Fig. 11.6b). Differently, the use of process- or network-specific tasks like the Sternberg paradigm, a delayed match-to-sample task, allows the functional dissociation of working memory components (Gruber and von Cramon 2003). In the first moment a list of stimuli is presented, followed by the presentation of a single stimulus in a second moment. The subject has to respond whether this stimulus has been shown before or not. So, one mechanism frequently used here is the articulatory rehearsal, related to the verbal working memory, eliciting the activation of left-dominant regions, which include Broca, ventrolateral prefrontal cortex (PFC), pre-supplementary motor area (SMA), intraparietal cortex (IPC), and part of the cerebellum (Gruber 2001; Gruber and von Cramon 2003).

The fMRI investigation of non-articulatory mechanisms for sustaining phonological information confirmed the cognitive model (Baddeley and Hitch 1974), describing a dual architecture of the linguistic working memory in terms of a phonological loop. Unlike suggested by lesions that the non-articulatory phonological memory is localized in a single brain area, namely, the inferior parietal lobe, functional imaging studies reinforced the implementation of this mechanism in a bilateral network of several brain regions (Gruber 2001; Gruber and von Cramon 2001, 2003). These mainly include anterior parts of middle frontal gyri and the inferior parietal lobules. Additionally, the functional significance of the anterior middle frontal gyrus for the non-articulatory maintenance of phonological information could be confirmed with the experimental and neuropsychological assessment of patients with lesions in this region (Gruber et al. 2005; Trost and Gruber 2012).

Out of these two partially dissociable and functionally interacting verbal working memory components (Gruber et al. 2007), a visuospatial component could be delineated. fMRI investigation of the latter could consistently show that the frontal eye fields (FEF) and the IPC play a crucial role (Gruber and von Cramon 2003). Working memory neural correlates for object form seem to be differently organized from the visual working memory for spatial information (Gruber and von Cramon 2001, 2003), mirroring the dorsal and ventral pathways of visual processing (see Chap. 9). Overall, grounded in systematic evaluation of the functional anatomy of working memory in humans and converging evidence from research with primates, working memory would be essentially composed of two neurofunctional systems originated from different evolutionary contexts (Goschke and Gruber 2004). From one side is the working memory of different sensory modalities, e.g., visual, auditory, and possibly others, being specifically implemented in this domain, prefrontal-parietal and prefrontal-temporal networks. Considering that these networks are also identifiable in nonhuman primates (Goldman-Rakic 1988, 1996; Romanski et al. 1999), it is plausible that this is an older phylogenetic system (Gruber and Goschke 2004). However, in humans the substantial development of a particular left-dominant language-based system permeated the use of articulatory rehearsal mechanism (Fig. 11.6b) as a predominant working memory tool involved in many higher cognitive functions (Gruber et al. 2000).

11.3.3 Meta-analysis of Executive Functions Including Working Memory

In the previous two sections, many functional studies have been presented and strikingly the neural correlates subserving different executive functions frequently overlapped. Also, lesion studies have supported the existence of a unique correlate behind initiation, inhibition, flexibility, and working memory. However, according to classic cognitive theory, each of these executive functions has been conceptualized as a different and specific domain. As an attempt to present more robust results, the activation likelihood estimation (ALE), a valuable tool to collect and compile imaging information from multiple studies, has been performed (Niendam et al. 2012). Different tasks addressing vigilance, planning, initiation, flexibility, inhibition, and working memory were evaluated as a whole and separately for initiation, flexibility, inhibition, and working memory domains using conjunction analysis in a very large sample (Fig. 11.7a). A common pattern of activation involving the medial and lateral PFC and superior and inferior parietal regions is represented with green and red overlays. Additionally, when results are presented in domain-specific patterns represented by the colors in Fig. 11.7b, the cortical overlapping is even more representative. This reinforces the idea of a general and superordinate cognitive network across different executive domains.

11.3.4 Episodic Memory

Episodic memory is a part of the broad concept of long-term memory. Long-term memory can be generally divided into non-declarative or implicit memory, in which automated motor, cognitive performance, and unconscious associations can be learned, and declarative or explicit memory,

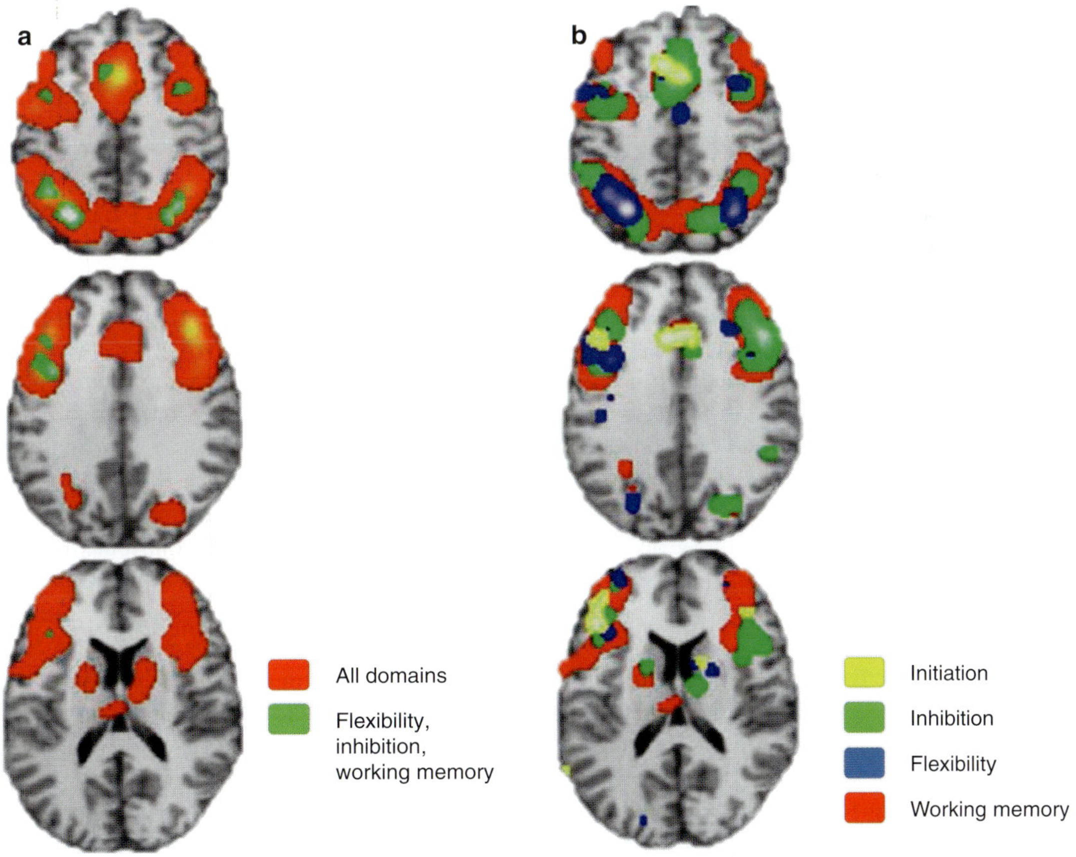

Fig. 11.7 Meta-analytic results supporting a superordinate characteristic of cognitive control network with overlapping results of (**a**) all studies (*red*) and the intersection of some domains (*green*) and (**b**) separation by domains (Modified from Niendam et al. (2012))

in which the content is consciously learned and can be rehearsed and taught. Declarative memory can be further divided into semantic and episodic memory. Semantic memory is the pure factual knowledge about the world, whereas episodic memory is responsible for storing the personal experiences accumulated.

The famous case of the patient H.M. (Scoville and Milner 1957) was a tragic milestone for the comprehension of neural correlates of memory. Due to severe drug-resistant epilepsy, the patient had both hippocampi surgically removed. Since then, immediate rehearsal of information was preserved, but the consolidation into long-term memory was persistently impaired, resulting in permanent anterograde amnesia. This means that it was impossible for H.M. to store new information, only holding it for shorter than a couple of minutes. The hippocampus and extended hippocampal formation are essential for memory consolidation, which was confirmed by behavioral experiments in nonhuman primates. Also, functional imaging studies posed the medial temporal lobe as crucial in storing new content in episodic memory, involved in long-term consolidation and retrieval of information (Brewer et al. 1998; Wagner et al. 1998). Frequently, however, additional activations displayed in subregions of the prefrontal cortex were associated with successful memory consolidation. It has been considered plausible that these regions played an important role in information selection and organization, as well as strategy selection (Fletcher and Henson 2001). Moreover, it should be clearly stated that the neural prefrontal correlates involved in encoding and retrieving episodic memory or general long-term memory are not exclusive. This correlates are also involved in working memory and recognition processes.

In principle, the retrieval of information from long-term memory can be seen as an update of working memory contents and therefore an activation of working memory itself. Consequently it is not surprising for instance to observe the activation of the anterior middle frontal gyrus during information retrieval in the context of both working memory and long-term memory (Gruber 2001; Fletcher and Henson 2001). Although recent imaging literature is full of speculation about the specialization of subregions of the PFC activated during episodic memory tasks with eventual lateralization, it must be emphasized that in the overall view of the empirical evidence, the structures of the medial temporal lobe likely play a more important role in the establishment and maintenance of episodic memory. The hippocampus per se receives information from sensory association areas over the other mediotemporal structures such as the parahippocampal, perirhinal, and entorhinal cortices (Lavenex and Amaral 2000). While the activation of the entorhinal cortex may be sufficient to ascertain familiarity to specific stimuli (Henson et al. 2003), the hippocampus retains an essential role in making associations with existing memory contents, which means processing contextual information (Weis et al. 2004). This feature allows a genuine recognition of someone's identity beyond the recognition of a familiar face by retrieving additional information such as the name, situation where we met, and what else is known about this person (Fig. 11.8).

A recent ALE meta-analysis (Viard et al. 2012) addressed the variability in the correlates of episodic memory driven by different tasks applied in the laboratory setting. Multiple paradigms exist, generally demanding the subjects to recall past experiences or imagine future events cued by words, sounds, or pictures with personal or generic content (Crovitz and Schiffma 1974). It has been shown that the type of cue, type of task, nature of the information retrieved, and age of the participants contribute for the elicited imaging correlates to vary. This supports the assumption of some pattern of specialization in parts of the hippocampus and medial temporal structures in dealing with stored information. Future studies addressing networking processes behind this specialization, e.g., with high-resolution functional connectivity and/or structural tractography, could extend the comprehension of whether spatial specificities have to do with perceptual specialization or encoding/recall or both.

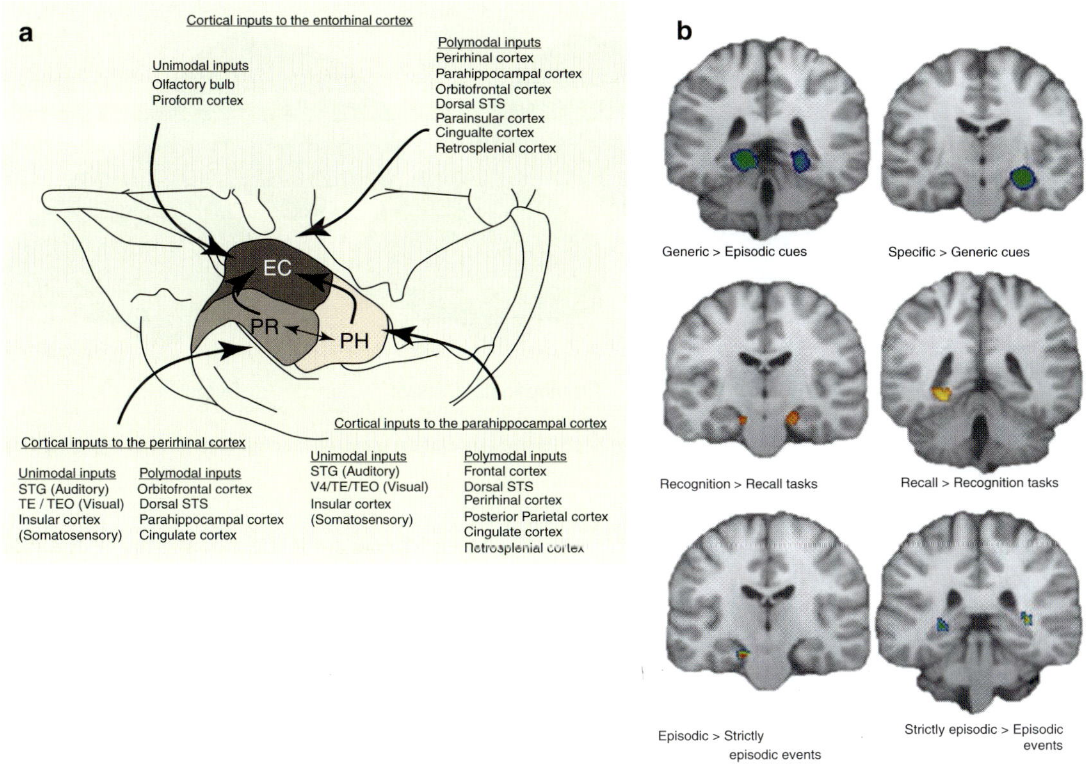

Fig. 11.8 (a) Scheme of cortical afferents and local interchange of information and (b) meta-analysis of multiple studies addressing episodic memory correlates (Modified from Lavenex and Amaral (2000), Viard et al. (2012))

11.4 The Prefrontal Cortex in Networks

Different methods of analysis of fMRI data allow the investigation of connectivity patterns. Functional and effective connectivity are based, differently from the activity analysis already seen, on the activation patterns along the time or timecourses of certain regions of interest (ROIs). Other regions presenting similar or opposite timecourse patterns, namely, positive or negative correlations, to these ROIs can be identified and are assumed to play a role in the network. Functional connectivity is a very important complementary approach that allows focusing on the investigation of regions that not necessarily present wide and robust activation, but even though might encompass significant functional importance, e.g., parts of the hippocampus. It is valid to reinforce that this field is newer and therefore more speculative, although with persistently increasing interest and importance.

11.4.1 Task-Based and Resting-State Insights

The brain is organized by highly coordinated networks. Positive activations are seen in the dorsal attention and frontoparietal networks (Fig. 11.9) during the performance of external goal-oriented tasks (task-positive), whereas the so-called default network (DN) is generally deactivated (task-negative). Conversely, during rest or introspective cognitive processes such as internal monitoring, the DN is activated, while the other two networks deactivate. The interplay between task-positive and task-negative networks has been shown in many studies (Biswal et al. 1997; Biswal 2012; Fox et al. 2005; Greicius et al.

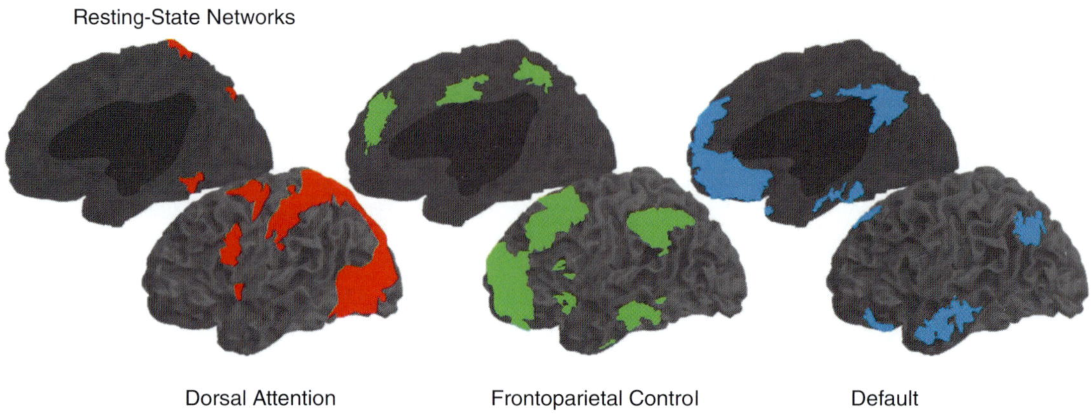

Fig. 11.9 Identification of resting-state networks by seed-based correlation maps of functional connectivity (Modified from Spreng et al. (2010))

2003; Margulies et al. 2007; Raichle et al. 2001) and happens continually in the healthy brain. The physiological meaning of this interplay and the importance of such a robust network, the DN, have been questioned since then in the academic community. Descriptions of hub properties within a small-world organization (for review, Bassett and Bullmore 2006) of functional networks have made promising advancement in the comprehension of cost/effectiveness balance in the central neurosystem. Although offering a clear insight about local-distant interconnection of regions, it could not entirely explain the physiological significance of the DN and its interplay with task-positive networks. An interesting study addressed working memory functional connectivity of task-relevant information distracted by task-irrelevant information (Chadick and Gazzaley 2011). They identified that to succeed the task the visual regions elicited by irrelevant information were coupled with the DN, whereas task-relevant information areas were coupled to the frontoparietal network. This implies that a possible functional role of the DN during external goal-oriented performance is to enable the suppression of distracting information, analogous to the suppression of internally generated distracting information, e.g., hunger or tiredness during an interview. Another argument in favor of the cognitive support given by the DN is the emergence of this network around 2 years after birth (Gao et al. 2009). If it would simply reflect the basal of energy balance of the brain, the DN would be expected already in intrauterine life. Two years coincide with the improvement in more complex cognitive abilities such as symbolization and language, followed by a highly growing interaction with the world. And the occurrence of inhibitory control in children has been shown to generally happen only after the age of 3 years old (Dowsett and Livesey 2000). Moreover, the prefrontal component of the DN, the rostral ACC, has been associated with errors in the antisaccade inhibition task, whenever it failed to properly deactivate (Polli et al. 2005).

Another study (Spreng et al. 2010) investigated the engagement of the frontoparietal control network during the performance of autobiographical planning (internal-oriented) task in contrast to a visuospatial planning (external-oriented) task. Connectivity analyses indicated the coupling of this network with the DN in the first and with the dorsal attention network in the second task. These findings support the dynamic interaction of the frontoparietal network with the other networks in order to achieve the goal successfully. More substantial findings driven by internally oriented tasks such as remembering the past as well as envisioning the future have displayed the DN activation (Szpunar et al. 2007; Cabeza and St Jacques 2007; Maguire 2001; Wagner et al. 2005) including medial temporal regions. These studies support the idea that imagining the future involves many of the

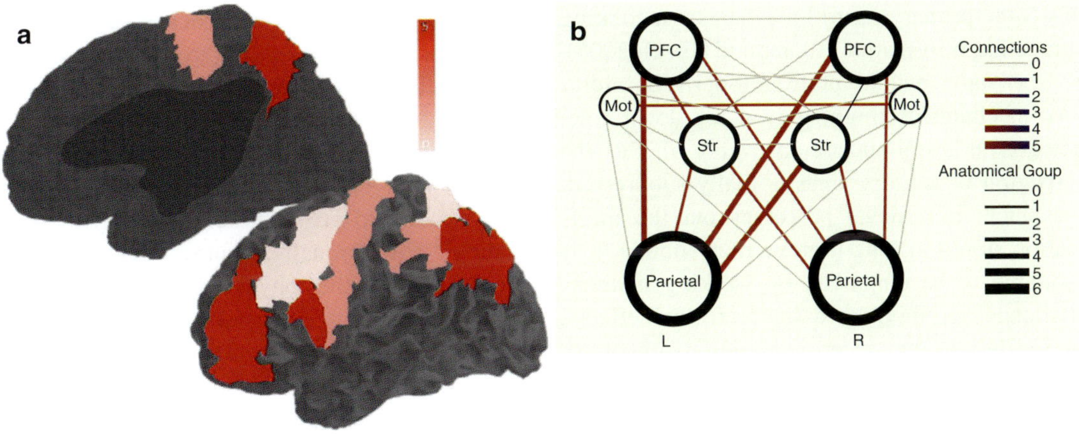

Fig. 11.10 Longitudinal evaluation of functional connectivity changes between regions after intensive reasoning training in real-world setting showing (**a**) regions with increased number of interconnections and (**b**) intensity of changes within and across regions in the diagram (Modified from Mackey et al. (2013))

same neural correlates present in the network of episodic rehearsal. The ability of remembering the past would then carry adaptive value in building the future with information stored from previous experiences.

11.5 Future Perspectives

Functional connectivity analysis has opened another perspective for the comprehension of the PFC than pure topographic descriptions of functional activity. This is especially relevant here, considering the predominance of superordinate cognitive characteristics. In the close future, one should expect not only a better description of developmental plasticity of such networks but also a day-to-day comprehension of the plastic boundaries involving cognitive demands. An interesting example of this plastic potential is offered by Mackey and colleagues (2013), in which changes in the basal connectivity patterns are identified and measured after intensive reasoning training. The identified regions are predominantly those taking part in the frontoparietal network (Fig. 11.10). This opens new dimensions for cognition investigation in order to extend the knowledge of mechanisms behind such a flexible and adaptable organ as the brain.

The potential application of shedding light on such mechanisms remains very promising toward psychopathology. The pathophysiology of multiple disorders should comprise the malfunctioning of cognitive mechanisms generally spared in the healthy brain. For instance, research in psychosis-related disorders such as schizophrenia and mood disorders has shown associated impairment in cognitive domains such as executive functions, attention, language, verbal, and visual memory (Green 1996; Hoff et al. 1999; Reichenberg et al. 2009). It is expected that several cognitive aspects that pertain to multiple neuropsychiatric disorders could be more effectively approached and improved with mechanism-guided therapies.

11.6 Summary

The classical psychological construct of cognition comprises many distinguishable cognitive functions as parts of a complex construct. Starting with anatomical descriptions, lesion imaging studies, and more recently based on fMRI studies, specific neural correlates to each of these functions do not seem to represent the reality.

Selective attention elicits the activation of prefrontoparietal and prefrontotemporal regions. Stimuli can elicit the prefrontal modulation to enhance activity in supplementary and primary areas perceiving and processing relevant sensory information. According to recent imaging data, this is the primary control mechanism mediating competing information (Egner and Hirsch 2005). To the same extent, prefrontal-derived inhibitory mechanisms might play a top-down role; however, it is still under controversy whether the activation of sensory areas is really suppressed. Inasmuch as retrieval processes put the information stored in long-term memory online, it is available to be employed by working memory challenges or modulated by prefrontal effects. In parallel to the selective attention processes, there is a continuous background monitoring of the environment of unexpected and potentially behavior-relevant stimuli (Gruber and Goschke 2004) outside of the focus of attention. Bottom-up mechanisms are essential to interrupt the current behavior, e.g., in case of emergency, and adapt it to surprising or threatening events. This works through stimulus-elicited activation of the VLPFC (Corbetta and Schulman 2002; Ridderinkhof et al. 2004; Brown and Braver 2005; Gruber et al. 2009) among other regions involved in the network. In other words, sensory events can as well modulate the individual's behavior. Whenever the current expectation and behavioral tendencies are in disagreement with the situation perceived from the environment, the dorsal ACC identifies conflict and signalizes the necessity of increasing cognitive control performed by the DLPFC. Other inferior frontal and intraparietal regions are involved in cognitive control processes, especially the inferior frontal junction (IFJ) (Derrfuss et al. 2005; Brass et al. 2005; Gruber et al. 2009; Goya-Maldonado et al. 2010). This region is adjacent to the Broca area, suggesting a certain functional relation between basal cognitive control and language processes according to the proximity. Also, IFG neural processes are recruited during preparation for an upcoming task or during announcement of task change, under which verbalization mechanisms play an important functional role (Gruber et al. 2006).

All these cognitive abilities have been shown to elicit overlapping regions that do not represent specific functions. Instead, a superordinate collection of regions seem to be responsible for the orchestration of executive functions. Newer functional connectivity approaches suggest that the interactions between the frontoparietal, dorsal attention, and default networks support the dynamic development/adjustment achieved by the cognitive functions. But this field is relatively new and both replication studies and future meta-analyses should prove the consistency of these findings. Additionally, cutting-edge studies should clarify the physiological mechanisms behind interplaying networks in order to foster cognitive processes. The uttermost importance of these neurocognitive mechanisms is to accomplish flexible adaptations of behavior in a constantly changing environment.

References

Baddeley AD, Hitch GJ (1974) Working memory. In: Bower G (Hrsg.). Recent advances in learning and motivation, vol. VIII. Academic, New York: 47–90

Baddeley A (2000) The episodic buffer: a new component of working memory? Trends Cogn Sci 4(11):417–423

Bassett DS, Bullmore E (2006) Small-world brain networks. Neuroscientist 12(6):512–523

Bechara A, Tranel D, Damasio H (2000) Characterization of the decision-making deficit of patients with ventromedial prefrontal cortex lesions. Brain 123:2189–2202

Biswal BB (2012) Resting state fMRI: a personal history. Neuroimage 62(2):938–944

Biswal BB, VanKylen J, Hyde JS (1997) Simultaneous assessment of flow and BOLD signals in resting-state functional connectivity maps. NMR Biomed 10(4–5):165–170

Bledowski C, Prvulovic D, Goebel R, Zanella FE, Linden DEJ (2004) Attentional systems in target and distractor processing: a combined ERP and fMRI study. Neuroimage 22(2):530–540

Brass M, Derrfuss J, Forstmann B, von Cramon DY (2005) The role of the inferior frontal junction area in cognitive control. Trends Cogn Sci 9(7):314–316

Brewer JB, Zhao Z, Desmond JE, Glover GH, Gabrieli JDE (1998) Making memories: brain activity that predicts how well visual experience will be remembered. Science 281(5380):1185–1187

Brodmann K (1908) Beitraege zur histologischen Lokalisation der Grosshirnrinde. VI Mitteilung. Die Cortexgliederung des Menschen. J Psychol Neurol 10:231–246

Brown JW, Braver TS (2005) Learned predictions of error likelihood in the anterior cingulate cortex. Science 307(5712):1118–1121

Cabeza R, St Jacques P (2007) Functional neuroimaging of autobiographical memory. Trends Cogn Sci 11(5):219–227

Chadick JZ, Gazzaley A (2011) Differential coupling of visual cortex with default or frontal-parietal network based on goals. Nat Neurosci 14(7):830–832

Chikazoe J, Konishi S, Asari T, Jimura K, Miyashita Y (2007) Activation of right inferior frontal gyrus during response inhibition across response modalities. J Cogn Neurosci 19(1):69–80

Corbetta M, Shulman GL (2002) Control of goal-directed and stimulus-driven attention in the brain. Nat Rev Neurosci 3(3):201–215

Crovitz HF, Schiffma H (1974) Frequency of episodic memories as a function of their age. Bull Psychon Soc 4(NB5):517–518

D'Esposito M, Detre JA, Alsop DC, Shin RK, Atlas S, Grossman M (1995) The neural basis of the central executive system of working-memory. Nature 378(6554):279–281

D'Esposito M, Aguirre GK, Zarahn E, Ballard D, Shin RK, Lease J (1998) Functional MRI studies of spatial and nonspatial working memory. Cogn Brain Res 7(1):1–13

Derrfuss J, Brass M, Neumann J, von Cramon DY (2005) Involvement of the inferior frontal junction in cognitive control: meta-analyses of switching and stroop studies. Hum Brain Mapp 25(1):22–34

Diekhof EK, Gruber O (2010) When desire collides with reason: functional interactions between anteroventral prefrontal cortex and nucleus accumbens underlie the human ability to resist impulsive desires. J Neurosci 30(4):1488–1493

Diekhof EK, Nerenberg L, Falkai P, Dechent P, Baudewig J, Gruber O (2011) Impulsive personality and the ability to resist immediate reward: an fMRI study examining interindividual differences in the neural mechanisms underlying self-control. Hum Brain Mapp 33(12):2768–2784

Dowsett SM, Livesey DJ (2000) The development of inhibitory control in preschool children: effects of "executive skills" training. Dev Psychobiol 36(2):161–174

Egner T, Hirsch J (2005) The neural correlates and functional integration of cognitive control in a Stroop task. Neuroimage 24(2):539–547

Fellows LK, Farah MJ (2005a) Different underlying impairments in decision-making following ventromedial and dorsolateral frontal lobe damage in humans. Cereb Cortex 15(1):58–63

Fellows LK, Farah MJ (2005b) Is anterior cingulate cortex necessary for cognitive control? Brain 128:788–796

Fletcher PC, Henson RNA (2001) Frontal lobes and human memory – insights from functional neuroimaging. Brain 124:849–881

Fox MD, Snyder AZ, Vincent JL, Corbetta M, Van Essen DC, Raichle ME (2005) The human brain is intrinsically organized into dynamic, anticorrelated functional networks. Proc Natl Acad Sci U S A 102(27): 9673–9678

Gao W, Zhu HT, Giovanello KS, Smith JK, Shen DG, Gilmore JH et al (2009) Evidence on the emergence of the brain's default network from 2-week-old to 2-year-old healthy pediatric subjects. Proc Natl Acad Sci U S A 106(16):6790–6795

Gazzaley A, Nobre AC (2012) Top-down modulation: bridging selective attention and working memory. Trends Cogn Sci 16(2):129–135

Glascher J, Adolphs R, Damasio H, Bechara A, Rudrauf D, Calamia M et al (2012) Lesion mapping of cognitive control and value-based decision making in the prefrontal cortex. Proc Natl Acad Sci U S A 109(36):14681–14686

Goldman-Rakic PS (1988) Topography of cognition – parallel distributed networks in primate association cortex. Annu Rev Neurosci 11:137–156

Goldman-Rakic PS (1996) The prefrontal landscape: implications of functional architecture for understanding human mentation and the central executive. Phil Trans Royal Soc Lond B Biol Sci 351(1346):1445–1453

Goya-Maldonado R, Walther S, Simon J, Stippich C, Weisbrod M, Kaiser S (2010) Motor impulsivity and the ventrolateral prefrontal cortex. Psychiatry Res Neuroimaging 183(1):89–91

Green MF (1996) What are the functional consequences of neurocognitive deficits in schizophrenia? Am J Psychiatr 153(3):321–330

Greicius MD, Krasnow B, Reiss AL, Menon V (2003) Functional connectivity in the resting brain: a network analysis of the default mode hypothesis. Proc Natl Acad Sci U S A 100(1):253–258

Gruber O (2001) Effects of domain-specific interference on brain activation associated with verbal working memory task performance. Cereb Cortex 11(11):1047–1055

Gruber O, Goschke T (2004) Executive control emerging from dynamic interactions between brain systems mediating language, working memory and attentional processes. Acta Psychol 115:105–121

Gruber O, von Cramon DY (2001) Domain-specific distribution of working memory processes along human prefrontal and parietal cortices: a functional magnetic resonance imaging study. Neurosci Lett 297(1):29–32

Gruber O, von Cramon DY (2003) The functional neuroanatomy of human working memory revisited – evidence from 3-T fMRI studies using classical domain-specific interference tasks. Neuroimage 19(3):797–809

Gruber O, Kleinschmidt A, Binkofski F, Steinmetz H, von Cramon DY (2000) Cerebral correlates of working memory for temporal information. Neuroreport 11(8):1689–1693

Gruber O, Gruber E, Falkai R (2005) Neural correlates of working memory deficits in schizophrenic patients. Ways to establish neurocognitive endophenotypes of psychiatric disorders. Radiologe 45(2):153–160

Gruber O, Karch S, Schlueter EK, Falkai P, Goschke T (2006) Neural mechanisms of advance preparation in task switching. Neuroimage 31(2):887–895

Gruber O, Mueller T, Falkai P (2007) Dynamic interactions between neural systems underlying different components of verbal working memory. J Neural Transm 114(8):1047–1050

Gruber O, Melcher T, Diekhof EK, Karch S, Falkai P, Goschke T (2009) Brain mechanisms associated with background monitoring of the environment for potentially significant sensory events. Brain Cogn 69(3):559–564

Henson RN, Cansino S, Herron JE, Robb WG, Rugg MD (2003) A familiarity signal in human anterior medial temporal cortex? Hippocampus 13:301–304

Hoff AL, Sakuma M, Wieneke M, Horon R, Kushner M, DeLisi LE (1999) Longitudinal neuropsychological follow-up study of patients with first-episode schizophrenia. Am J Psychiatr 156(9):1336–1341

Hopfinger JB, Buonocore MH, Mangun GR (2000) The neural mechanisms of top-down attentional control. Nat Neurosci 3(3):284–291

Ikkai A, Curtis CE (2011) Common neural mechanisms supporting spatial working memory, attention and motor intention. Neuropsychologia 49(6):1428–1434

Jonides J, Smith EE, Koeppe RA, Awh E, Minoshima S, Mintun MA (1993) Spatial working-memory in humans as revealed by PET. Nature 363(6430):623–625

Kastner S, Ungerleider LG (2000) Mechanisms of visual attention in the human cortex. Annu Rev Neurosci 23:315–341

Kerns JG, Cohen JD, MacDonald AW, Cho RY, Stenger VA, Carter CS (2004) Anterior cingulate conflict monitoring and adjustments in control. Science 303(5660):1023–1026

Lavenex P, Amaral DG (2000) Hippocampal-neocortical interaction: a hierarchy of associativity. Hippocampus 10(4):420–430

Leimkuhler ME, Mesulam MM (1985) Reversible go-no go deficits in a case of frontal lobe tumor. Ann Neurol 18(5):617–619

Mackey AP, Singley ATM, Bunge SA (2013) Intensive reasoning training alters patterns of brain connectivity at rest. J Neurosci 33(11):4796–4803

Macleod CM (1991) Half a century of research on the Stroop effect – and integrative review. Psychol Bull 109(2):163–203

Maguire EA (2001) Neuroimaging studies of autobiographical event memory. Philos Trans Royal Soc B Biol Sci 356(1413):1441–1451

Margulies DS, Kelly AMC, Uddin LQ, Biswal BB, Castellanos FX, Milham MP (2007) Mapping the functional connectivity of anterior cingulate cortex. Neuroimage 37(2):579–588

Matsumoto K, Tanaka K (2004) Neuroscience. Conflict and cognitive control. Science 303(5660):969–970

Mayberg HS, Lozano AM, Voon V, McNeely HE, Seminowicz D, Hamani C et al (2005) Deep brain stimulation for treatment-resistant depression. Neuron 45(5):651–660

Melcher T, Gruber O (2006) Oddball and incongruity effects during Stroop task performance: a comparative fMRI study on selective attention. Brain Res 1121:136–149

Melcher T, Gruber O (2009) Decomposing interference during Stroop performance into different conflict factors: an event-related fMRI study. Cortex 45(2):189–200

Milner B (1982) Some cognitive effects of frontal-lobe lesions in man. Philos Trans R Soc Lond B Biol Sci 298(1089):211–226

Niendam TA, Laird AR, Ray KL, Dean YM, Glahn DC, Carter CS (2012) Meta-analytic evidence for a superordinate cognitive control network subserving diverse executive functions. Cogn Affect Behav Neurosci 12(2):241–268

Nystrom LE, Braver TS, Sabb FW, Delgado MR, Noll DC, Cohen JD (2000) Working memory for letters, shapes, and locations: fMRI evidence against stimulus-based regional organization in human prefrontal cortex. Neuroimage 11(5):424–446

Owen AM, Downes JJ, Sahakian BJ, Polkey CE, Robbins TW (1990) Planning and spatial working memory following frontal lobe lesions in man. Neuropsychologia 28(10):1021–1034

Owen AM, McMillan KM, Laird AR, Bullmore E (2005) N-back working memory paradigm: a meta-analysis of normative functional neuroimaging. Hum Brain Mapp 25(1):46–59

Paulesu E, Frith CD, Frackowiak RSJ (1993) The neural correlates of the verbal component of working memory. Nature 362(6418):342–345

Petersen SE, Posner MI (2012) The attention system of the human brain: 20 years after. Annu Rev Neurosci 35:73–89

Petrides M, Pandya DN (1994) Comparative architechtonic analysis of the human and macaque frontal cortex. In: Boller F, Grafman J (eds) Handbook of neuropsychology, vol 9. Elsevier, Amsterdam, pp 17–58

Polli FE, Barton JJ, Cain MS, Thakkar KN, Rauch SL, Manoach DS (2005) Rostral and dorsal anterior cingulate cortex make dissociable contributions during antisaccade error commission. Proc Natl Acad Sci U S A 102(43):15700–15705

Posner MI (2012) Imaging attention networks. Neuroimage 61(2):450–456

Raichle ME, MacLeod AM, Snyder AZ, Powers WJ, Gusnard DA, Shulman GL (2001) A default mode of brain function. Proc Natl Acad Sci U S A 98(2):676–682

Reichenberg A, Harvey PD, Bowie CR, Mojtabai R, Rabinowitz J, Heaton RK et al (2009) Neuropsychological function and dysfunction in schizophrenia and psychotic affective disorders. Schizophr Bull 35(5):1022–1029

Ridderinkhof KR, Ullsperger M, Crone EA, Nieuwenhuiss S (2004) The role of the medial frontal cortex in cognitive control. Science 306(5695):443–447

Romanski LM, Tian B, Fritz J, Mishkin M, Goldman-Rakic PS, Rauschecker JP (1999) Dual streams of auditory afferents target multiple domains in the primate prefrontal cortex. Nat Neurosci 2(12):1131–1136

Scoville WB, Milner B (1957) Loss of recent memory after bilateral hippocampal lesions. J Neurol Neurosurg Psychiatry 20(1):11–21

Spreng RN, Stevens WD, Chamberlain JP, Gilmore AW, Schacter DL (2010) Default network activity, coupled with the frontoparietal control network, supports goal-directed cognition. Neuroimage 53(1):303–317

Stroop JR (1992) Studies of interference in serial verbal reactions. J Exp Psychol Gen 121(1):15–23 (reprinted from Journal Experimental-Psychology, vol 18, pp 643–662, 1935)

Stroop JR (1935) Studies of interference in serial verbal reactions. J Exp Psychol 121:15–23

Stuss DT, Levine B, Alexander MP, Hong J, Palumbo C, Hamer L et al (2000) Wisconsin Card Sorting Test performance in patients with focal frontal and posterior brain damage: effects of lesion location and test structure on separable cognitive processes. Neuropsychologia 38(4):388–402

Stuss DT, Floden D, Alexander MP, Levine B, Katz D (2001) Stroop performance in focal lesion patients: dissociation of processes and frontal lobe lesion location. Neuropsychologia 39(8):771–786

Szpunar KK, Watson JM, McDermott KB (2007) Neural substrates of envisioning the future. Proc Natl Acad Sci U S A 104(2):642–647

Trost S, Gruber O (2012) Evidence for a double dissociation of articulatory rehearsal and non-articulatory maintenance of phonological information in human verbal working memory. Neuropsychobiology 65(3):133–140

Viard A, Desgranges B, Eustache F, Piolino P (2012) Factors affecting medial temporal lobe engagement for past and future episodic events: an ALE meta-analysis of neuroimaging studies. Brain Cogn 80(1):111–125

Wagner AD, Schacter DL, Rotte M, Koutstaal W, Maril A, Dale AM et al (1998) Building memories: remembering and forgetting of verbal experiences as predicted by brain activity. Science 281(5380):1188–1191

Wagner AD, Shannon BJ, Kahn I, Buckner RL (2005) Parietal lobe contributions to episodic memory retrieval. Trends Cogn Sci 9(9):445–453

Walther S, Goya-Maldonado R, Stippich C, Weisbrod M, Kaiser S (2010) A supramodal network for response inhibition. Neuroreport 21(3):191–195

Weis S, Specht K, Klaver P, Tendolkar I, Willmes K, Ruhlmann J, Elger CE, Fernández G (2004) Process dissociation between contextual retrieval and item recognition. Neuroreport 15:2729–2733

Zurowski B, Gostomzyk J, Gron G, Weller R, Schirrmeister H, Neumeier B et al (2002) Dissociating a common working memory network from different neural substrates of phonological and spatial stimulus processing. Neuroimage 15(1):45–57

fMRI Investigations of the Mesolimbic Dopaminergic Reward System in Schizophrenia

12

Georg Juckel

Abbreviations

5-HT2	Serotonin-2 receptor
BOLD	Blood oxygen level dependent
COMT	Catechol-O-methyltransferase
D2	Dopamine D2 receptor
DA	Dopamine
DOPA	Dopamine
EPI	Echo planar imaging
FDR	False discovery rate
FWHM	Full width at half maximum
IBZM	Iodobenzamide (-SPECT)
MET	Methionine
MID	Monetary incentive delay
mPFC	Medial prefrontal cortex
PANSS	Positive and negative syndrome scale
PET	Positron emission tomography
ROI	Region of interest
SPECT	Single photon emission computed tomography
TE	Echo time
TR	Volume time to repeat
VAL	Valine
VTA	Ventral tegmental area

G. Juckel
Department of Psychiatry,
Ruhr University Bochum, LWL University Hospital,
Alexandrinenstr. 1, 44791 Bochum, Germany
e-mail: georg.juckel@wkp-lwl.org

12.1 Mesolimbic Dopaminergic Reward System in Schizophrenia

A dysfunction of the dopaminergic reward or motivation system in patients with schizophrenia was nowdays often postulated (Chau et al. 2004; Goldstein and Volkow 2002; Green 2005; Heinz et al. 1998, 2003, 2004; Juckel et al. 2003; Wise 1982). A long time ago, schizophrenia was conceptualized as a dysfunction of evolutionary younger and "higher developed" brain centers, which is assumed to result in a disinhibition of evolutionary "higher developed" centers, which assumably results in a disinhibition of evolutionary older and "more primitive" centers (Heinz et al. 2004). The deficiency of the function of "higher developed centers" was related to the development of negative symptoms, e.g., apathy, affective flattening, and anhedonia, whereas disinhibition of "more primitive centers" was related to the occurrence of positive symptoms, e.g., delusion and hallucinations. A more modern version of this conception, the neurobiological hypothesis for the development of schizophrenia of Weinberger (1987), is based on the assumption that an early ontogenetic mediotemporal lesion results in a disinhibition of the striatal release of dopamine caused by a dysfunction of the fronto-cortical control (Heinz and Weinberger 2000; Marenco and Weinberger 2000). Brain imaging studies furnish evidence of a dopaminergic dysfunction in schizophrenic patients. At the same time, replicated diagnostic findings show an

C. Mulert, M.E. Shenton (eds.), *MRI in Psychiatry*,
DOI 10.1007/978-3-642-54542-9_12, © Springer Berlin Heidelberg 2014

increased DOPA metabolism in the striatum (Meyer-Lindenberg et al. 2002; Reith et al. 1994) as well as an increased amphetamine-induced striatal DA release (Abi-Dargham et al. 2000; Laruelle et al. 1996). Evidence for the hypothesis of elevated dopamine in schizophrenia primarily caused by the effect of neuroleptics, dysfunction of the striatal dopaminergic system by PET and SPECT studies in regard to increase of the presynaptic synthesis by dopamine in patients with schizophrenia (Gründer et al. 1999; Hietala et al. 1999), increased synaptic concentration of dopamine in competition with IBZM at D2 receptor (Abi-dargham et al. 2000), as well as the study of Meyer-Lindenberg et al. (2002) in combination with dopamine increase (PET) and low function of the prefrontal cortex measured by fMRI during a working memory task.

The mesolimbic dopaminergic reward system is a network of different cortical and subcortical structures. A core region is the ventral striatum including the nucleus accumbens, which receives ascending dopaminergic projections from the midbrain (VTA). The reward system is an ancient system concerning its historical development, which is stimulated by "primary intensifiers," e.g., eating, drinking, or sexual activity (Robbins and Everitt 1996). Its task is to draw attention to reward indicative stimuli and to forecast potential reward ("salience"). Schultz (2002) and Schultz et al. (1993) examined dopaminergic neurons in the VTA of monkeys by solitary cell derivation in different reward situations: On receipt of an unexpected reward, the activity rate of the dopaminergic neurons was increased. After the monkeys were exposed to a conditioning experiment, whereas the unconditioned stimulus (reward) was combined with a conditioned stimulus (photic stimulation), the dopaminergic signal was transposed. It did not emerge on receipt of the reward, but on occurrence of the conditioned stimulus (photic stimulation). Schultz summarized these results by describing the dopaminergic signal as a "reward prediction error." This error signal encodes the difference between anticipated reward and actual reward. Thus, what does actually signify an increased dopaminergic activity in the mesolimbic dopaminergic reward system (ventral striatum, nucleus accumbens)? Internal or external stimuli, as, e.g., alcohol, cocaine, sexuality causing happiness, and sense of well-being (good salience) for the organism, are increased. These stimuli should be new and unpredictable. The central task of the reward system is to depict the salience of a stimulus from the environment and its internal representation for the organism. Dopamine is related to motivation, but not to happiness (happiness is more related to serotonin and the opioidergic system): "wanting not liking." This is related to the motivation in order to receive a reward; it is not related to the effect causing happiness by the reward itself, if you finally get it. This can be triggered by money, reward prediction error (see also O'Doherty et al. 2003), nice faces, sports cars, pleasant music, humor, even placebos, romantic love, as well as suffering from punishment for somebody else.

How can we deal with the traced assumption of a hyperdopaminergic state of schizophrenic patients? In 1991, Gray and Hamsley postulated that one of the first prodromal symptoms in schizophrenia is to try to find out how to create essential connections of new irritating experiences and perceptions. This complies with the spirit of delusion, that is, apophenia according to Conrad: "Something is changing, something is happening, something mysterious, suspicious, there is something behind all the things." Each organism is trying to find an explanation for it. At the time when more and more stimuli are gaining in importance due to the incipient filter dysfunction at the outset of schizophrenia, the organism is trying to find any kind of interpretation and consistent theory resulting in an abundance and breakdown of the information processing system. What does a hypodopaminergic state in patients suffering from schizophrenia signify? It signifies amotivational syndrome, apathy, anhedonia, that is, negative symptomatology, and social withdrawal. There are only a few or no stimuli being important for the organism. However, caution is recommended concerning the donation of antipsychotics: Antipsychotics may abate the motivational "salience" of usual occurrences. The main problem in this connection is the blockade of the D2 receptor especially by classical neuroleptics. This consequently results in a depression in 20–40 % of the patients, when taking classical neuroleptics blocking D2 receptors, or in a secondary negative symptomatology, loss of drive, energy, and motivation; apathy; and anhedonia, and finally in noncompliance of the patients. This can be explained by the fact that classical neuroleptics are completely blocking the endogenous dopaminergic system in the ventral striatum, thus the mesolimbic dopaminergic reward system, which means that it is completely paralyzed; consequently, an affective drive stimulation is no longer existing resulting in a severe secondary affective negative symptomatology. The advantage of new atypical neuroleptics, i.e. antipsychotics of the second generation, is that they are only partially blocking the D2 receptors in the ventral striatum (especially concerning clozapine; refer to the work of Meltzer 2013). Particularly with regard to the presynaptic 5 HT2 antagonist as heteroreceptor mechanism, this results in a sufficient permanent flow in the ventral striatum, which consequently results in maintenance of affectivity and drive.

Therefore, the rationale of our studies was if the dopaminergic reward system is dysregulated in acute ill patients with schizophrenia and there is a close meshed correlation to motivational salience, this dysfunction should be successfully examined by fMRI and a motivation reward paradigm, which is activating the ventral striatum, the Knutson paradigm. Patients taking typical neuroleptics will not show any activation of the ventral striatum during the fMRI motivation reward paradigm, whereas such an activity can be detected after changing the medication to an atypical neuroleptic agent.

12.2 "Monetary Incentive Delay Task" in Functional Magnetic Resonance Imaging

By means of functional magnetic resonance tomography (fMRI), the function of the reward system can be visualized. For this purpose, it was used as a paradigm developed by Brian Knutson, the so-called monetary incentive delay (MID) task (Knutson et al. 2001). This is a simple monetary gain game, which consists of a sequence of geometric figures, which were projected onto the screen of the MRT scanner when the fMRI measurement was carried out. This "game" (Fig. 12.1)

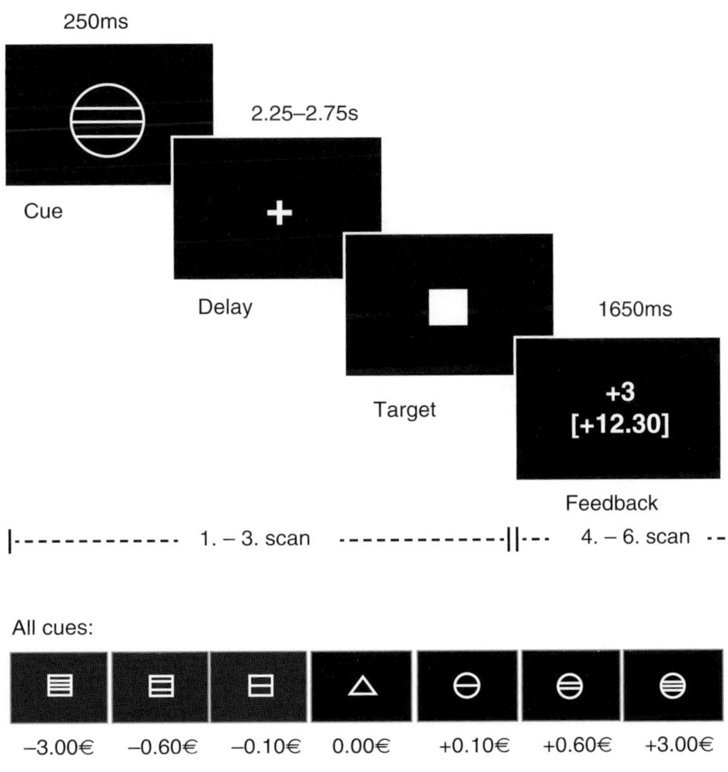

Fig. 12.1 "Monetary incentive delay task" (According to Knutson et al. (2001))

consists of a frequently repeated number of runs, at the beginning of which a geometric figure respectively informs the volunteers during the actual runs, if money can be gained (circle), loss of money can be hold off (square), or it is a neutral run (triangle) without any monetary consequences. After these cues have been given, it follows a delay phase, during which the volunteers are waiting for the appearance of a target. The task is to press a button with the thumb, as long as the brief presentation of the target is visible. Immediately after that, the success or failure is indicated and the actual cumulative total is shown. The chance of winning was set to 66 % by means of an adaption mechanism, so that the volunteers anticipated the receipt of monetary gain on appearance of a reward-indicating cue. After the examination the money was actually paid out. According to the studies of Schultz, the occurrence of the dopaminergic signal is to be expected during the anticipation phase. The examinations were performed on a 1.5 tesla scanner (Magnetom Vision) with an EPI sequence (TR = 1.9 s, TE = 40 ms, $a = 90°$, voxel size = $4 \times 4 \times 3.3$ mm^3). Data were analyzed with SPM2 (http://www.fil.ion.ucl.ac.uk/spm). According to the normalized preprocessed data (slice timing, motion correction, spatial normalization, spatial smoothing with 8 mm FWHM), single models were developed for each volunteer according to the general linear model by defining the different gain, loss, and neutral cues as single conditions. On the basis of the random effect analysis, the contrast of "anticipation of gain > neutral condition" was examined (for detailed descriptions, Juckel et al. 2006a, b).

12.3 Unmedicated Schizophrenic Patients

In a first study (Juckel et al. 2006a), ten unmedicated schizophrenic patients (ages, 26.8 ± 7.8 years) and ten healthy controls (ages, 31.7 ± 8.4 years) were examined. Both groups were matched according to gender, age, left or right handed, as well as task performance concerning the monetary game. Of the schizophrenic patients, seven have never been treated with neuroleptics and three had taken neuroleptics 2 years ago. There was no difference between both groups concerning total monetary gain or the reaction times. In accordance with the findings of Knutson et al. (2001), healthy controls showed a BOLD response at both sides of the ventral striatum during the anticipation of monetary gain versus the neutral condition (contrast "anticipation of monetary gain > neutral condition" in healthy control subjects at the left side (($x\ y\ z$) = (−21 6 −3), $t = 5.63$) and at the right side (($x\ y\ z$) = (9 6 −5), $t = 4.26$)). In contrast, untreated schizophrenic patients displayed no significant activation in the ventral striatum, though the achievements did not show any differences for both groups (Fig. 12.2). In comparing both groups a significant higher activation was observed in the left ventral striatum in healthy controls compared to the schizophrenic patients concerning the anticipation of monetary gain (($x\ y\ z$) = (−15 9 −3), $t = 3.29$) and loss (($x\ y\ z$) = (−18 6 −5), $t = 3.24$) compared with the neutral conditions. Within the group of schizophrenic patients, the BOLD response correlated in the left ventral striatum inversely with the severity of the negative symptomatology (measured with the PANSS negative score; Spearman's $R = -0.66$, $P < 0.05$), i.e., a reduced activity of the ventral striatum concerning reward-indicating stimuli was correlated with a higher negative symptomatology.

In a recently published study (Juckel et al 2012), it could be shown that in ultra high-risk persons (prodromal schizophrenia) (mean age, 25.5 ± 4.6 years), matched with healthy controls in regard of age, gender, and test performance, the activity of the ventral striatum in prodromal patients during the Knutson paradigm is approximately positioned in the center compared to completely manifested schizophrenic patients on the one side and healthy probands on the other side. A remaining activation concerning monetary anticipation, but also concerning avoidance of loss, was significantly different concerning the left and right striatum compared with healthy persons (Juckel et al. 2012).

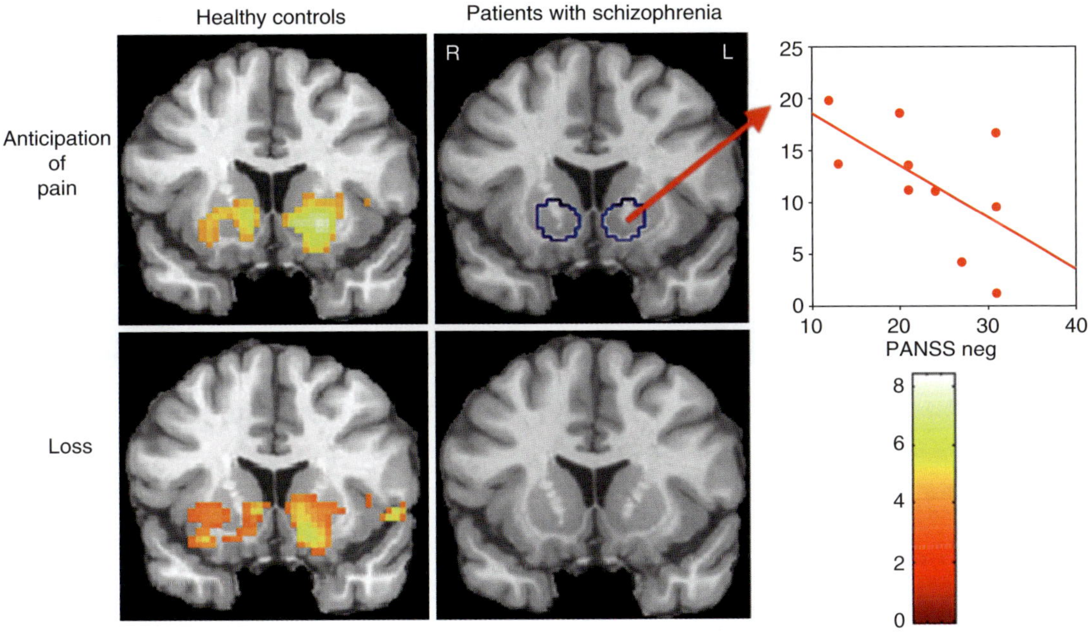

Fig. 12.2 Unmedicated patients with schizophrenia and healthy controls as well as the relationship to negative symptoms

12.4 Typical Versus Atypical Neuroleptics: A Cross-Sectional Study

In a further study (Juckel et al. 2006b) the influence of typical and atypical neuroleptics concerning the reward system in schizophrenic patients was examined. Three groups were tested: one group of ten schizophrenic patients, which received typical neuroleptics (four flupenthixol 12±4 mg, four haloperidol 10±5 mg, and two fluphenazine 12±4 mg); one group of schizophrenic patients, which were treated with atypical neuroleptics (four with risperidone 5±1 mg, four with olanzapine 19±6 mg, one with aripiprazole 30 mg, and one with amisulpride 300 mg); and a healthy control group. There were no significant differences found between the three groups concerning age, gender, and total gain. There were no significant differences concerning the psychopathology of both patient groups (PANSS total in patients under typical neuroleptics, 70.11±20.37, and those under atypical neuroleptics, 64.44±22.59). The healthy control subjects showed again an activation concerning the contrast between gain anticipation versus neutral condition at both sides of the ventral striatum (left, $(x\ y\ z)=(-21\ 5\ -3)$, $t=9.53$, and right, $(x\ y\ z)=(12\ 2\ -10)$, $t=4.25$). In schizophrenic patients under typical neuroleptics, no activation of reward-indicating stimuli was observed, whereas in schizophrenic patients with atypical neuroleptic medication, an activation of the right ventral striatum was observed ($(x\ y\ z)=(12\ 12\ -1)$, $t=3.58$) (Fig. 12.3). In comparing the groups, a significant difference between the healthy subjects and the schizophrenic patients under typical neuroleptics ($(x\ y\ z)=(-21\ 9\ -3)$, $t=3.39$) was observed, but there was no difference between healthy controls and patients under atypical neuroleptics. Within the group of persons being treated with typical neuroleptics – as well as unmedicated schizophrenic patients – an inverse correlation between reduced activation concerning reward-indicating stimuli and the severity of the negative symptomatology was observed (Spearman's $R=-0.67$, $P<0.05$).

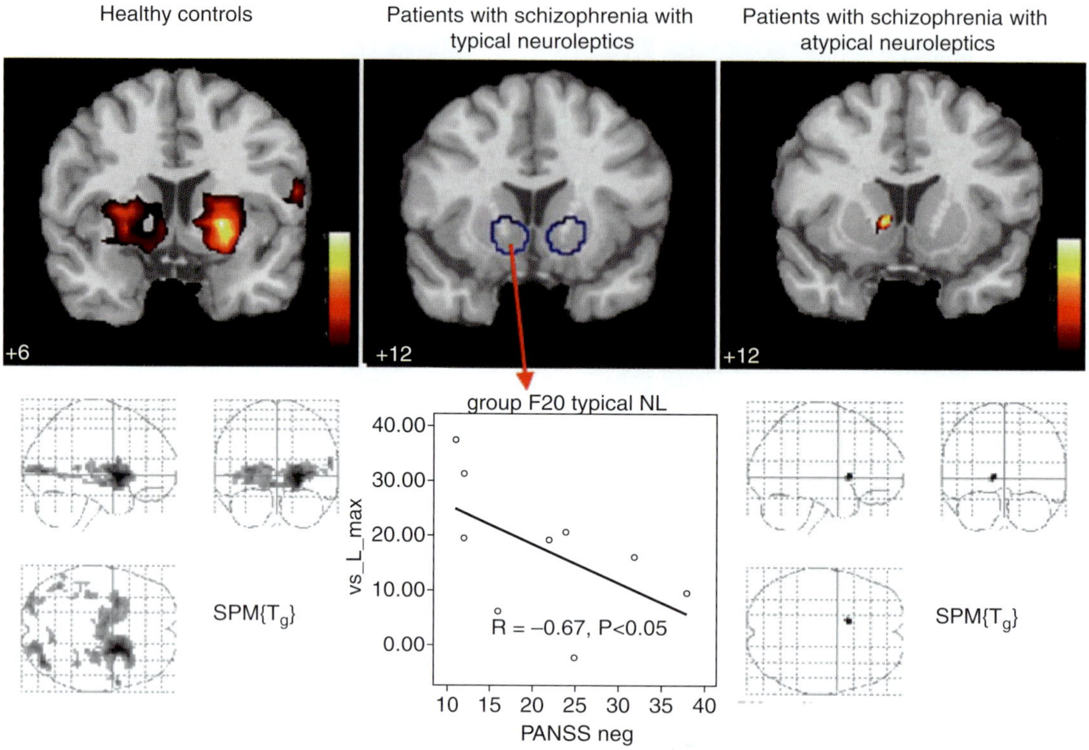

Fig. 12.3 Effects of typical versus atypical neuroleptics on the ventral striatum in patients with schizophrenia (cross-sectional study)

12.5 Switch From Typical to Atypical Neuroleptics: Reactivation of the Ventral Striatum in Schizophrenic Patients

As cross-sectional studies are not sufficiently valid in order to evaluate effects of antipsychotic medication on ventral striatum activation within the MID task, further studies have been carried out by treating schizophrenic patients with a typical neuroleptic agent (haloperidol or flupenthixol) for at least 2 weeks and afterwards with an atypical neuroleptic medication (risperidone, olanzapine, or aripiprazole) for another 2 weeks. After the respective treatment for 2 weeks, the patients were examined by fMRI and the "monetary incentive delay task" and compared with healthy controls, who were examined twice by fMRI, too, after similar time intervals. Concerning changes to olanzapine (Schlagenhauf et al. 2008), ten schizophrenic patients were first examined by fMRI under typical neuroleptics (haloperidol, 10.8±4.3 mg/day; flupenthixol, 7.0±5.1 mg/day) and were afterwards treated with olanzapine (18.3±7.5 mg/day). The patients were treated with typical neuroleptics for 17.8±15.0 days and with olanzapine for 18.5±7.5 days. The repeated examination by fMRI under olanzapine was carried out after 31.7±17.3 days after the first examination under typical neuroleptics. Ten healthy subjects were also examined twice after a period of 32.7±15.5 days in order to control a potential time effect. In anticipation of a monetary gain, healthy volunteers showed a significant activation of the ventral striatum (left, $(x\ y\ z)=(-19\ 6\ -9)$, $t=3.87$, and right, $(x\ y\ z)=(18\ 6\ -3)$, $t=4.7$), whereas the schizophrenic patients did not show any activity under the typical neuroleptics (Fig. 12.4). However, under the treatment of olanzapine, an activation of the ventral striatum could be observed on the right side ($(x\ y\ z)=(15\ 6\ -12)$, $t=4.36$) in schizophrenic patients. In the healthy

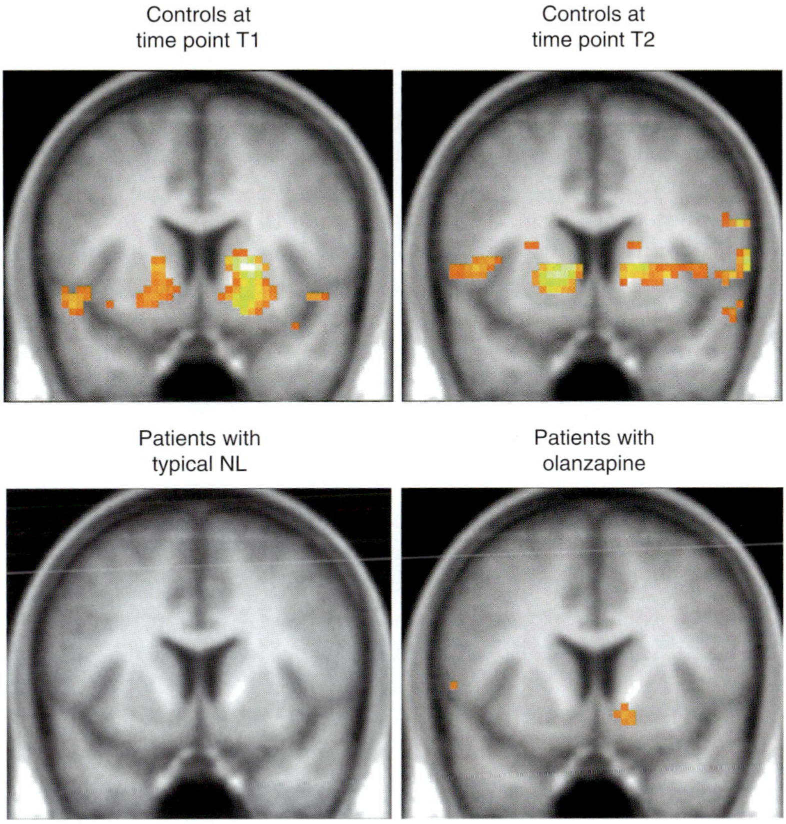

Fig. 12.4 Activation of the ventral striatum under several weeks of treatment with an atypical neuroleptic agent (olanzapine) after switching from a classical typical neuroleptic (mostly haloperidol, flupenthixol) in patients with schizophrenia in comparison to healthy subjects with two recordings in similar time interval distance

control subjects, activation of the ventral striatum due to anticipation of reward remains quite stable over time. Presumably due to the secondary negative symptomatology, in the patients, there was an observed significant correlation between the activation of the ventral striatum and the PANSS negative score (r=–0.721, P=0.019) only under treatment with typical neuroleptics in patients.

In a further study (Juckel, Schlagenhauf et al.; not published yet) in eight patients suffering from schizophrenia, who received a typical medication with haloperidol or flupenthixol, the medication was changed to risperidone. The age average of these male patients was 35 ± 13 years versus the examinations of 33.6 ± 12.5 years. The test performance did not show any differences for both groups. In the course of this study, the patients were tested again after a period of nearly 2 months in order to minimize effects of classical neuroleptics, e.g., haloperidol and flupenthixol, which are acting a long time. With T1 under a classical neuroleptic, schizophrenic patients showed a significantly lower activation compared to healthy controls (t-values 2.5–3.0, t=0.03–0.04, corrected for ventral striatum ROI FDR). After changing the medication to risperidone, the MID paradigm showed a visible increase of the activity in the ventral striatum concerning the gain anticipation compared with healthy subjects at the left side (t=3.6, P=0.04 corrected for ROI) as well as at the right side (t=3.6, P=0.02 corrected for ROI), which substantiates a regeneration of the dopaminergic reward system in the ventral striatum after change over to an atypical neuroleptic. The difference between the administration of risperidone compared to a typical neuroleptic in schizophrenic persons was not significant due to the small number of cases; in the ventral striatum a t-value of 2.1 and a p-value of 0.186 were ascertained. A change from typical neuroleptics to aripiprazole (Schlagenhauf et al. 2010) also resulted in an increased activation under aripiprazole compared to typical medication in the ventral striatum t=2.17, P<0.05.

12.6 Discussion

For the first time, we could substantiate a dysfunction of the reward system in patients with schizophrenia. They showed a reduced activation of the ventral striatum, a core region of the reward system, compared to healthy controls, and the severity of the negative symptomatology was associated with the reduced activation in the ventral striatum. The reduced ventral striatal activation in schizophrenic patients could be potentially an explanation for an increased dopaminergic "noise" in this region interfering with the processing of cue-elicited signals. In unmedicated schizophrenic patients, an increased dopaminergic transmission in the ventral striatum was observed again (Abi-Dargham et al. 2000). This interpretation is substantiated by a study of Knutson et al. (2004), which shows that healthy probands displayed reduced ventral striatal activation by a dopamine release induced by amphetamines with the same paradigm. The detected correlation between the severity of the negative symptomatology and the reduced activation in the ventral striatum substantiates for the first time a close relationship between the schizophrenic negative symptomatology and a dysfunction of the reward system. This reinforces the hypothesis that a dysfunction of the reward system contributes to the development of negative symptomatology.

The findings of the cross-sectional study concerning the influence of typical and atypical neuroleptics on the reward system can be summed up as follows: (1) Schizophrenic patients treated with typical but not with atypical neuroleptics displayed a reduced activity of the ventral striatum compared to healthy controls and (2) the severity of the negative symptomatology correlated with the reduced activity in the ventral striatum in schizophrenic patients treated with typical neuroleptics. The better effectiveness of atypical neuroleptics, which is known from clinical studies, on schizophrenic negative symptomatology (Leucht et al. 1999; Möller 2003) is potentially correlated with lower impairment of the reward system. Atypical neuroleptics, e.g., risperidone or olanzapine, cause a reduced blockade of striatal D2 receptors and are not bound so tightly to these receptors. Moreover, they interact with other transmitter systems, e.g., the serotonergic system by its effect on the 5-HT2A receptors. These mechanisms could be based on the better efficiency of atypical neuroleptics on the negative symptomatology of schizophrenia.

The following restrictions concerning the interpretation of the result have to be made: On the one hand, there was only a relative low number of cases examined. Furthermore, the exact mechanism of the fundamental dysfunction cannot be clarified by means of fMRI. The application of multimodal methods with a combination of dopamine PET and fMRI could possibly explain which mechanisms cause the dysfunction of the reward system in schizophrenic patients. In comparing typical and atypical neuroleptics, a design with an intra-subjective comparison seems to be more adequate. Therefore, a replication of the results in a longitudinal design is planned by examining patients under the treatment of typical neuroleptics and afterwards under the treatment of atypical neuroleptics. The first pilot data showed respective results. In order to replicate a correlation between a dysfunction of the reward system and the negative symptomatology, a comparison between schizophrenic patients with extreme negative symptomatology and patients without this symptomatology could be suitable.

Concerning the aforementioned conception of schizophrenic dysfunctions, our results reveal that not only dysfunctions of the prefrontal function but also subcortical structures, e.g., the ventral striatum (including the nucleus accumbens), are involved in the development of negative symptoms. In this regard, MRI connectivity analyses could give more information concerning the fundamental variances.

12.7 Outlook: Further New Aspects

COMT polymorphism: Patients with Met allele are subject to a higher dopamine synthesis, which correlates in a higher ventral striatal reward activity (Schmack et al. 2008). This was found in a great number of healthy volunteers. The next step would be to determine the COMT activity in schizophrenic persons, too, in order to further examine the mechanism of the dysregulated ventral striatal reward activity (please refer to COMT studies concerning working memory of the dorsal lateral prefrontal cortex in schizophrenic patients; Goldberg et al. 2003). Possibly, this could be an explanation not only for different activations in a phasic but also tonic functional state of the dopaminergic system in schizophrenia. If the Knutson paradigm primarily shows the tonic dopaminergic activity in the ventral striatum, the hypothesis could be substantiated that especially VAL/VAL patients with a schizophrenic disease show a low activity in the Knutson paradigm concerning the ventral striatum, whereas patients with schizophrenia with MET/MET show a higher activity. This would possibly result in a more progressive course concerning prodromal and negative symptomatology.

A second new finding is related to the variation of the reward feedback in the medial prefrontal cortex and the relevance for the state of delusion in schizophrenia (Schlagenhauf et al. 2009). Increasing attribution of increased salience in neutral or aversive stimuli may be associated with the dysfunction of neuronal processing of positive and negative reinforcement and may contribute to the formation of delusion in schizophrenia. The study group impressively found out that in 15 unmedicated patients with schizophrenia compared to matched test persons, there is a correlation between responses to negative outcome in reward trails (abolition of gain, avoidance of loss) with an increased activation of mPFC in patients with schizophrenia. On the other hand, the known result of a low activation in connection with the anticipation in the ventral striatum in schizophrenic patients could be found. The following correlation could be made: increase of the level of delusion states in schizophrenia in correlation with the decrease of the activation in the medial prefrontal cortex caused by the response to missing rewards (loss avoidance) matching the brain development hypothesis (glutamatergic projection fibers from mPFC resulting in phasic dopamine increase in the ventral striatum) (Fig. 12.5). These findings substantiate impairment and a

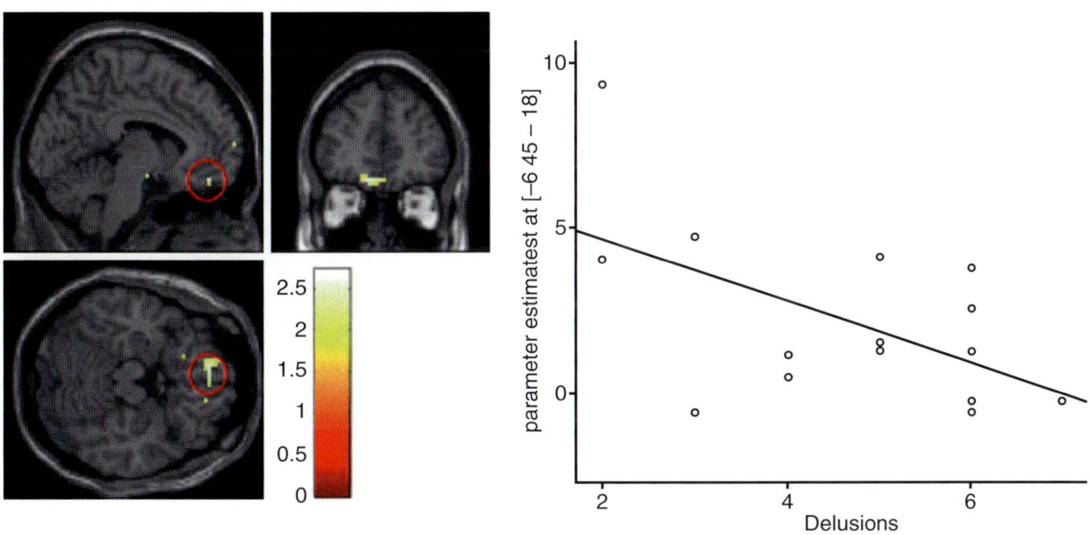

Fig. 12.5 Correlation between positive and negative symptom scales (subitem delusion) and the activation of feedback response to avoidance of loss in the medial prefrontal cortex in unmedicated schizophrenic patients: lower activation of the response to negative outcome with increase of the severity of delusion states (Taken from Schlagenhauf et al. (2009))

reduced functional connectivity between the ventral striatum and the medial prefrontal cortex during the processing of reward and avoidance of loss of reward in unmedicated patients with schizophrenia. Obviously, there is a connection between the occurrence of delusion states and the neuronal processing of negative outcome (nonoccurrence of gain anticipation).

References

Abi-Dargham A, Rodenhiser J, Printz D, Zea-Ponce Y, Gil R, Kegeles LS, Weiss R, Cooper TB, Mann JJ, Van Heertum RL, Gorman JM, Laruelle M (2000) Increased baseline occupancy of D2 receptors by dopamine in schizophrenia. Proc Natl Acad Sci U S A 97:8104–8109

Chau DT, Roth RM, Green AI (2004) The neural circuitry of reward and its relevance to psychiatric disorders. Curr Psychiatry Rep 6:391–399

Goldberg TE, Egan MF, Gscheidle T, Coppola R, Weickert T, Kolachana BS, Goldman D, Weinberger DR (2003) Executive subprocesses in working memory: relationship to catechol-O-methyltransferase Val158Met genotype and schizophrenia. Arch Gen Psychiatry 60(9): 889–896

Goldstein RZ, Volkow ND (2002) Drug addiction and its underlying neurobiological basis: neuroimaging evidence for the involvement of the frontal cortex. Am J Psychiatry 159:1642–1652

Green AI (2005) Schizophrenia and comorbid substance use disorder: effects of antipsychotics. J Clin Psychiatry 66(Suppl 6):21–26

Gründer G, Müller MJ, Andreas J, Heydari N, Wetzel H, Schlösser R, Schlegel S, Nickel O, Eissner D, Benkert O (1999) Occupancy of striatal D(2)-like dopamine receptors after treatment with the sigma ligand EMD 57445, a putative atypical antipsychotic. Psychopharmacology (Berl) 146(1):81–86

Heinz A, Weinberger DR (2000) Schizophrenia: the neurodevelopmental hypothesis. Current concepts in psychiatry (Psychiatrie der Gegenwart). Springer, Berlin/Heidelberg/New York, pp 89–104

Heinz A, Knable MB, Coppola R, Gorey JG, Jones DW, Lee KS, Weinberger DR (1998) Psychomotor slowing, negative symptoms and dopamine receptor availability–an IBZM SPECT study in neuroleptic-treated and drug-free schizophrenic patients. Schizophr Res 31:19–26

Heinz A, Romero B, Gallinat J, Juckel G, Weinberger DR (2003) Molecular brain imaging and the neurobiology and genetics of schizophrenia. Pharmacopsychiatry 36(Suppl 3):S152–S157

Heinz A, Braus DF, Romero B, Gallinat J, Puls I, Juckel G, Weinberger DR (2004) Genetic and pharmacological effects on prefrontal cortical function in schizophrenia. Nervenarzt 75:845–856

Hietala J, Någren K, Lehikoinen P, Ruotsalainen U, Syvälahti E (1999) Measurement of striatal D2 dopamine receptor density and affinity with [11C]-raclopride in vivo: a test-retest analysis. J Cereb Blood Flow Metab 19(2):210–217

Juckel G, Sass L, Heinz A (2003) Anhedonia, self-experience in schizophrenia, and implications for treatment. Pharmacopsychiatry 36(Suppl 3):S176–S180

Juckel G, Schlagenhauf F, Koslowski M, Wustenberg T, Villringer A, Knutson B, Wrase J, Heinz A (2006a) Dysfunction of ventral striatal reward prediction in schizophrenia. Neuroimage 29:409–416

Juckel G, Schlagenhauf F, Koslowski M, Filonov D, Wustenberg T, Villringer A, Knutson B, Kienast T, Gallinat J, Wrase J, Heinz A (2006b) Dysfunction of ventral striatal reward prediction in schizophrenic patients treated with typical, not atypical, neuroleptics. Psychopharmacology (Berl) 187:222–228

Juckel G, Friedel E, Koslowski M, Witthaus H, Özgürdal S, Gudlowski Y, Knutson B, Wrase J, Brüne M, Heinz A, Schlagenhauf F (2012) Ventral striatal activation during expectancy of reward in subjects with ultra-high risk for psychosis. Neuropsychobiology 66:50–56

Knutson B, Adams CM, Fong GW, Hommer D (2001) Anticipation of increasing monetary reward selectively recruits nucleus accumbens. J Neurosci 21:RC159

Knutson B, Bjork JM, Fong GW, Hommer D, Mattay VS, Weinberger DR (2004) Amphetamine modulates human incentive processing. Neuron 43:261–269

Laruelle M, bi-Dargham A, van Dyck CH, Gil R, D'Souza CD, Erdos J, McCance E, Rosenblatt W, Fingado C, Zoghbi SS, Baldwin RM, Seibyl JP, Krystal JH, Charney DS, Innis RB (1996) Single photon emission computerized tomography imaging of amphetamine-induced dopamine release in drug-free schizophrenic subjects. Proc Natl Acad Sci U S A 93: 9235–9240

Leucht S, Pitschel-Walz G, Abraham D, Kissling W (1999) Efficacy and extrapyramidal side-effects of the new antipsychotics olanzapine, quetiapine, risperidone, and sertindole compared to conventional antipsychotics and placebo. A meta-analysis of randomized controlled trials. Schizophr Res 35:51–68

Marenco S, Weinberger DR (2000) The neurodevelopmental hypothesis of schizophrenia: following a trail of evidence from cradle to grave. Dev Psychopathol 12:501–527

Meltzer HY (2013) Update on typical and atypical antipsychotic drugs. Annu Rev Med 64:393–406

Meyer-Lindenberg A, Miletich RS, Kohn PD, Esposito G, Carson RE, Quarantelli M, Weinberger DR, Berman KF (2002) Reduced prefrontal activity predicts exaggerated striatal dopaminergic function in schizophrenia. Nat Neurosci 5:267–271

Möller HJ (2003) Management of the negative symptoms of schizophrenia: new treatment options. CNS Drugs 17:793–823

O'Doherty JP, Dayan P, Friston K, Critchley H, Dolan RJ (2003) Temporal difference models and reward-related learning in the human brain. Neuron 38(2):329–337

Reith J, Benkelfat C, Sherwin A, Yasuhara Y, Kuwabara H, Andermann F, Bachneff S, Cumming P, Diksic M, Dyve SE, Etienne P, Evans AC, Lal S, Shevell M, Savard G, Wong DF, Chouinard G, Gjedde A (1994) Elevated dopa decarboxylase activity in living brain of patients with psychosis. Proc Natl Acad Sci U S A 91:11651–11654

Robbins TW, Everitt BJ (1996) Neurobehavioural mechanisms of reward and motivation. Curr Opin Neurobiol 6:228–236

Schlagenhauf F, Juckel G, Koslowski M, Kahnt T, Knutson B, Dembler T, Kienast T, Gallinat J, Wrase J, Heinz A (2008) Reward system activation in schizophrenic patients switched from typical neuroleptics to olanzapine. Psychopharmacology (Berl) 196(4):673–684

Schlagenhauf F, Sterzer P, Schmack K, Ballmaier M, Rapp M, Wrase J, Juckel G, Gallinat J, Heinz A (2009) Reward feedback alterations in unmedicated schizophrenia patients: relevance for delusions. Biol Psychiatry 65(12):1032–1039

Schlagenhauf F, Dinges M, Beck A, Wüstenberg T, Friedel E, Dembler T, Sarkar R, Wrase J, Gallinat J, Juckel G, Heinz A (2010) Switching schizophrenia patients from typical neuroleptics to aripiprazole: effects on working memory dependent functional activation. Schizophr Res 118(1–3):189–200

Schmack K, Schlagenhauf F, Sterzer P, Wrase J, Beck A, Dembler T, Kalus P, Puls I, Sander T, Heinz A, Gallinat J (2008) Catechol-O-methyltransferase val158met genotype influences neural processing of reward anticipation. Neuroimage 42(4):1631–1638

Schultz W (2002) Getting formal with dopamine and reward. Neuron 36:241–263

Schultz W, Apicella P, Ljungberg T (1993) Responses of monkey dopamine neurons to reward and conditioned stimuli during successive steps of learning a delayed response task. J Neurosci 13:900–913

Weinberger DR (1987) Implications of normal brain development for the pathogenesis of schizophrenia. Arch Gen Psychiatry 44:660–669

Wise RA (1982) Neuroleptics and operant-behavior - the anhedonia hypothesis. Behav Brain Sci 5:39–53

Part III

Disorders

Neuroimaging in Schizophrenia

13

Irina Falkenberg, Tilo Kircher, and Axel Krug

Abbreviations

ACC	Anterior cingulate cortex
ALE	Activation likelihood estimation
BA	Brodmann area
BD	Bipolar disorder
CS	Chronic schizophrenia
CT	Computed tomography
DTI	Diffusion tensor imaging
FA	Fractional anisotropy
FES	First episode of schizophrenia
FHR	Familial risk of schizophrenia
GM	Gray matter
MDD	Major depressive disorder
ROI	Regions of interest
SDM	Signed differential mapping
ToM	Theory of Mind
UHR	Ultrahigh risk
VBM	Voxel-based morphometry
WM	White matter

13.1 Introduction

Schizophrenia is a severe psychiatric disorder affecting 1 % of the population, and it is one of the leading causes of disability worldwide (Murray and Lopez 1997). It usually begins in late adolescence or early adulthood and is characterized by positive psychotic symptoms, such as delusions and hallucinations and disorganized speech, as well as negative symptoms such as emotional blunting and loss of motivation. These symptoms are accompanied by cognitive impairments, particularly impairments in memory and executive functions (Kurtz 2005), as well as social and occupational dysfunction. In recent years, psychiatric research has allowed clinicians to assess and to identify individuals in the early phases of schizophrenia, before the onset of frank disease. Individuals are deemed at increased clinical risk of developing schizophrenia because of the presence of an intermediate illness phenotype, such as the presence of subclinical psychotic symptoms (Yung et al. 1998). This state of increased clinical risk, also termed "ultrahigh-risk" (UHR) stage, is associated with a high risk of 36 % of transition to schizophrenia within the first 3 years of clinical presentation according to a recent meta-analysis (Fusar-Poli et al. 2012a, b).

More than 30 years ago, the demonstration of enlarged ventricles via computed tomography (CT) confirmed earlier pneumoencephalographic (Huber 1955, 1961) and neuropathological findings, paving the way for more interest in the biological basis of schizophrenia (Johnstone et al. 1976). Since then a host of structural and functional neuroimaging studies have demonstrated abnormalities in brain areas relevant to the disease, such as the hippocampus and parahippocampal gyrus, cingulate gyrus, insula, thalamus,

I. Falkenberg (✉) • T. Kircher • A. Krug
Department of Psychiatry and Psychotherapy,
Philipps-University Marburg,
Rudolf-Bultmann-Str. 8, 35039 Marburg, Germany
e-mail: irina.falkenberg@med.uni-marburg.de

C. Mulert, M.E. Shenton (eds.), *MRI in Psychiatry*,
DOI 10.1007/978-3-642-54542-9_13, © Springer Berlin Heidelberg 2014

and prefrontal and temporal cortices (Vita et al. 2006). These findings have been extended to first-episode groups, high-risk groups, and schizotypal disorder and have been integrated in a stage-based disease model (Pantelis et al. 2005). This chapter reviews the evidence for changes in the structure and function of the brain in patients with chronic schizophrenia (CS), patients in their first episode of schizophrenia (FES), and subjects at high risk of schizophrenia for clinical reasons (UHR) during the genesis of the disorder. We will highlight the structural and functional abnormalities that have been found in these groups and whether any similar or lesser abnormalities are also observed in subjects at increased familial risk of schizophrenia (familial high risk (FHR)) and healthy subjects carrying risk genes of schizophrenia.

13.2 Gray Matter Abnormalities in Different Stages of Schizophrenia

We will restrict our review of the available evidence mainly to meta-analyses. There are two procedures for imaging meta-analysis. The first pools volumetric studies of target regions of interest (ROI) using traditional meta-analytic techniques, where mean values for regional volumes reported in each study are used to calculate a standardized weighted mean difference (Cohen's d or Hedge's g), representing the magnitude of alterations reported within a defined anatomical region. The second approach is a voxel-based meta-analysis, which combines the results of voxel-based morphometry (VBM) studies. VBM studies use statistical maps to identify voxels with a significant probability of gray or white matter composition differences between patient and control groups. A voxel-based meta-analysis estimates a distribution map of anatomical localization, revealing locations of concurrence across the entire set of included studies. There are several techniques available for voxel-based meta-analysis, including activation likelihood estimation (ALE; Turkeltaub et al. 2002) and signed differential mapping (SDM; Radua and Mataix-Cols 2012). These techniques differ in analytic approach but both purport to combine significant voxels from multiple studies across the whole brain. These volumetric and voxel-based meta-analytic procedures assess distinct constructs of structural abnormality, and to date, reviews of findings from each technique have been reported separately.

13.2.1 Regional Distribution of Gray Matter Changes in Schizophrenia?

There is one mega-analysis of meta-analyses available, summarizing 32 published meta-analyses quantitatively (Shepherd et al. 2012). ROI meta-analyses with patient numbers between 447 (Baiano et al. 2007) and 2,771 (Sun et al. 2009) as well as VBM meta-analyses ranging from 408 (Fusar-Poli et al. 2012a, b) to 5,839 (Bora et al. 2011) patients, all compared to healthy control subjects, were included. The authors of this mega-analysis tried to consider the comparability of the information presented in duplicate systematic reviews and used a quality rating of included reviews, applying the AMSTAR checklist (Shea et al. 2009, 2007) and PRISMA guidelines (Moher et al. 2009). Across both volumetric and voxel-based meta-analyses, high-quality evidence identified regions of reduced gray matter in both first-episode and chronic illness, particularly in the frontal lobe gyri and cingulate cortex, thalamus, insula, postcentral gyrus, and medial temporal regions, as well as increased volumes of ventricles and cavum septum pellucidum. The findings of this meta-review are summarized in Fig. 13.1. They mainly confirm the results of previous meta-analyses.

13.2.2 Causes of Volume Changes in Schizophrenia

There is ongoing discussion as to whether volume changes in schizophrenia are caused by early developmental insults (genetic or environmental) and remain static over time or are caused

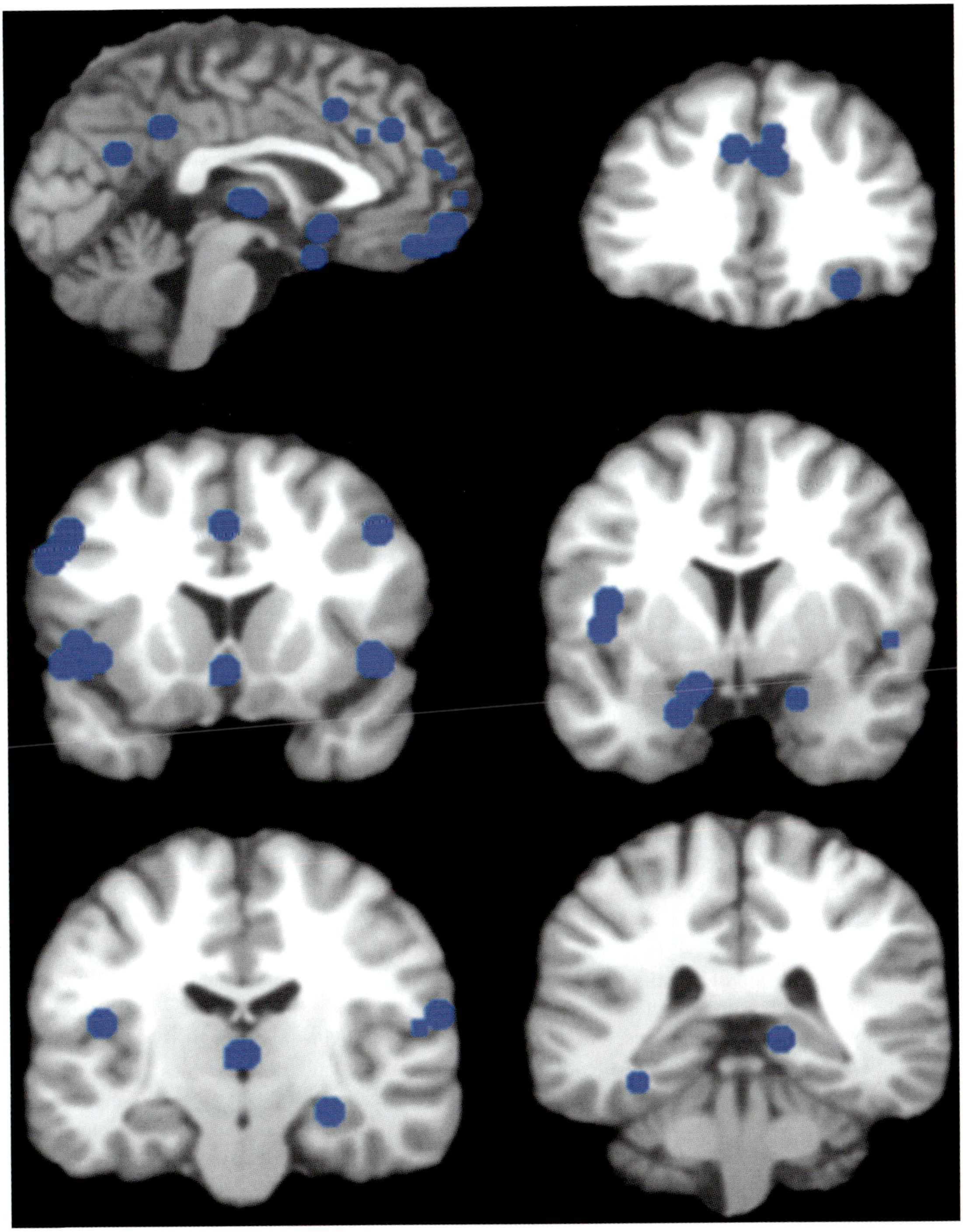

Fig. 13.1 Distribution of foci with reduced gray matter density in chronic schizophrenia vs. healthy controls (Shepherd et al. 2012)

by a progressive loss due to the disease process. This question is best addressed by investigating subjects at genetic risk (i.e., unaffected first-degree relatives), individuals in the putatively prodromal state of the illness, and patients over the course of several years of illness.

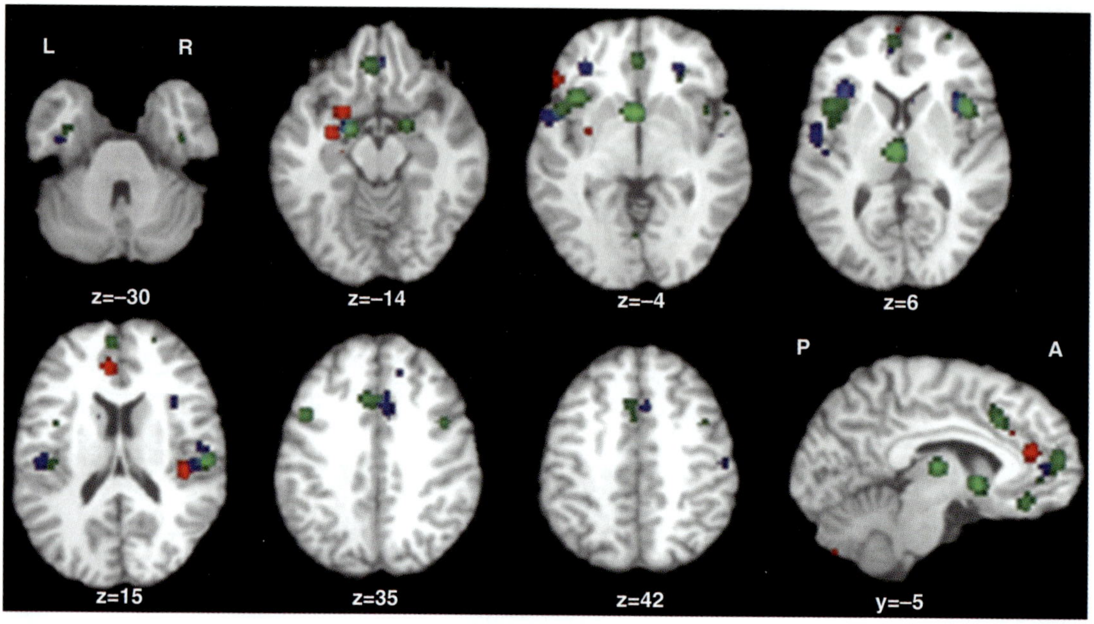

Fig. 13.2 Gray matter volume reductions in high-risk group (*red*), first-episode group (*blue*), and chronic patients (*green*) compared with healthy subjects (Chan et al. 2011)

First, we review studies on unaffected relatives of patients bearing familial liability to the disorder (familial high-risk (FHR) individuals). In the VBM meta-analysis by Chan et al. (2011), eight studies comparing first-degree relatives of patients and FES patients with control subjects were included with a total of 566 relatives. They found significantly reduced gray matter in the bilateral anterior cingulate gyrus (Brodmann area [BA] 32/24), right insula (BA 13), left amygdala, left subcallosal gyrus (BA 34), and left inferior frontal gyrus (BA 47) in FHR individuals relative to the healthy control group. Of these regions, the left amygdala, subcallosal gyrus, and inferior frontal gyrus were also smaller in the FHR as compared to the FES group (see Fig. 13.2).

In a similar meta-analysis, 16 VBM studies involving 733 FHR subjects (relatives of patients with schizophrenia), 563 healthy controls, and 474 patients were analyzed using the signed differential mapping (SDM) technique (Palaniyappan et al. 2012). A significant gray matter reduction in the lentiform nucleus, amygdala/parahippocampal gyrus, and medial prefrontal cortex was found in association with the genetic diathesis. Gray matter reduction in bilateral insula, inferior frontal gyrus, superior temporal gyrus, and the anterior cingulate was present in association with the disease expression. The neuroanatomical changes associated with the genetic diathesis to develop schizophrenia appear to be somewhat different in this meta-analysis from those contributing to the clinical expression of the illness. A notable feature of this analysis is that most VBM studies (five out of nine) do not find structural differences in FHR individuals when compared to healthy controls. This suggests that the anatomical correlates of genetic diathesis are either weak or inconsistently identified using the VBM approach. Some support to the latter notion comes from a head-to-head comparison of VBM approach and surface-based morphometric approaches (Palaniyappan and Liddle 2012), which suggests that subtle surface anatomical changes that may be important for the pathophysiology of schizophrenia may be missed when using VBM.

These two VBM meta-analytic results (Palaniyappan et al. 2012; Chan et al. 2011) are consistent with Fusar-Poli et al. (2011a, b, 2012a, b) who undertook two meta-analyses of the VBM studies on the genetic diathesis of schizophrenia.

The left parahippocampal gyrus emerged as the most significant locus with gray matter reduction in FHR compared to controls (Fusar-Poli et al. 2011a, b, 2012a, b). In addition, the anterior cingulate cortex was found to be linked to genetic diathesis (Fusar-Poli et al. 2011a, b, 2012a, b).

13.2.3 Structural Changes in Subjects at Ultrahigh Risk for Schizophrenia

Studies using voxel-based morphometry (VBM) aim to identify neuroanatomical correlates of increased vulnerability to schizophrenia and to predict consecutive transition to the full-blown disorder. MRI findings, mostly from cross-sectional studies, have reported structural abnormalities in the prepsychotic phase resembling those that have been described in patients with schizophrenia. Specifically, studies in UHR samples have shown that abnormalities in frontal, temporal, and limbic regions may predate illness onset (Borgwardt et al. 2007; Meisenzahl et al. 2008; Pantelis et al. 2003). However, as these individual studies have used different methodological approaches, relatively small sample sizes, and have produced divergent findings, they are insufficient for the characterization of brain regions associated with increased vulnerability to schizophrenia. Furthermore, other factors such as short-term use of antipsychotic medication, differences in ascertainment strategies for the UHR syndrome, and the fact that the syndrome is a dynamic condition with symptoms varying considerably over time may modulate findings. To adequately address such confounding factors and to examine the neurobiological predictors of transition from the high-risk state to the full-blown disorder, Fusar-Poli et al. (2011a, b) have conducted an activation likelihood estimation (ALE) meta-analysis of 19 VBM studies of UHR subjects. When comparing subjects at increased risk for schizophrenia (clinical risk and familial risk) to healthy controls, the authors found that at-risk subjects showed reduced gray matter volume in the temporal (right superior temporal gyrus), parietal (left precuneus), and limbic regions (bilateral parahippocampal/hippocampal regions, bilateral anterior cingulate) and in the prefrontal cortex bilaterally (left medial frontal gyrus, right middle frontal gyrus). As none of the at-risk subjects were psychotic, these findings could be interpreted as neuroanatomical correlates of an increased risk for schizophrenia. Structural alterations in these regions have been found in psychotic disorders (Tan et al. 2009; Boyer et al. 2007; Baiano et al. 2007; Sun et al. 2009; Cavanna and Trimble 2006) and all of the above named regions have also been found to be abnormal in fMRI studies of high-risk subjects (Fusar-Poli et al. 2007), indicating there may be a shared pathophysiological mechanism of functional and structural abnormalities in the high-risk group. However, clinical and familial risks of schizophrenia are not directly comparable, as different transition rates in the two groups suggest (20–40 % risk over 2 years for UHR cohorts vs. 5–9 % lifetime risk in FHR cohorts (Cannon et al. 2008; Knowles and Sharma 2004; Yung et al. 2003)). Thus, UHR subjects showed reduced gray matter in the bilateral anterior cingulate and increased gray matter in the left hippocampus and insula and the right superior temporal gyrus when compared to FHR subjects (Fusar-Poli et al. 2011a, b). These differences in gray matter changes may reflect the differential influences of genetic vs. psychopathological factors in the prepsychotic phase possibly underlying the differential rates of transition. A potential neurobiological predictor of schizophrenia transitions may be represented by gray matter reductions in the right inferior frontal gyrus and superior temporal gyrus, which are present in UHR subjects who go on to make a transition but not in those who don't (Fusar-Poli et al. 2011a). These findings suggest that although structural abnormalities identified by VBM studies show similar results in at-risk subjects and subjects with schizophrenia, they may be differentially associated with risk of transition.

We will next address the question as to whether the volume alterations seen in manifest schizophrenia patients are static or progress over the course of illness. Four ROI volumetric reviews assessed the temporal evolution of the

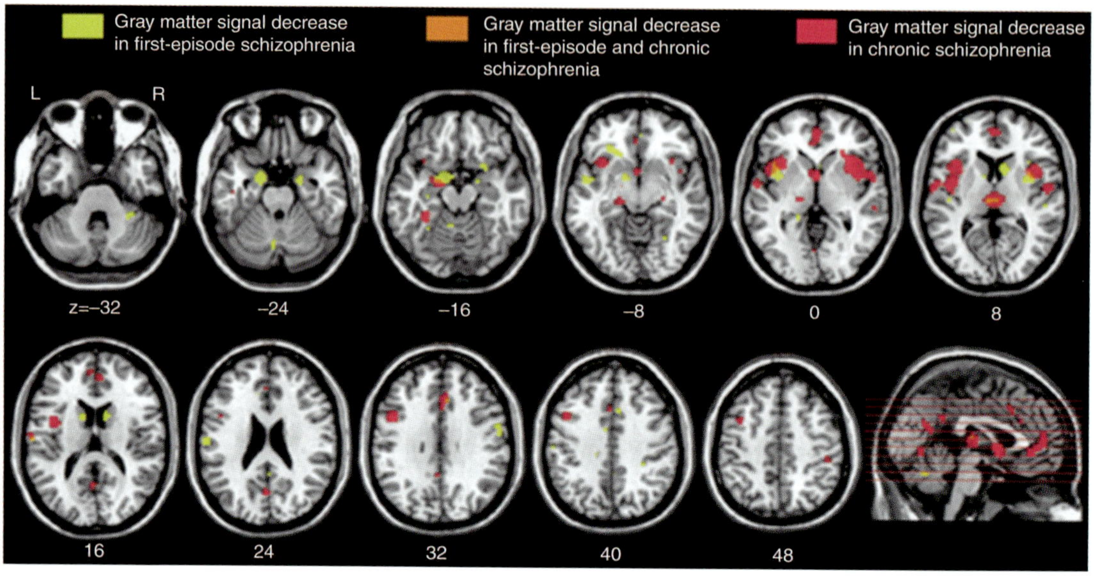

Fig. 13.3 Regional overlap of gray matter changes in first-episode and chronic schizophrenia (Ellison-Wright et al. 2008)

reported brain alterations (Hulshoff-Pol and Kahn 2008). The most comprehensive of these (Olabi et al. 2011) found a significantly greater degree of change over time in whole brain gray matter ($d = -0.520$), frontal lobe ($d = -0.340$), and left caudate ($d = -0.336$) in people with schizophrenia, though no progressive changes were reported in temporal, parietal, or occipital lobe gray matter, hippocampus, amygdala, or cerebellum over a range of 1–10 years. Significant progressive volume increases were reported in bilateral lateral ventricles ($d = 0.530$), with a trend toward significance for the third ventricle ($d = 0.180$) (Olabi et al. 2011). The difference between patients and control subjects in annualized percentage volume change was –0.07 % for the whole-brain volume and –0.59 % for the whole-brain gray matter (Olabi et al. 2011). An additional meta-analysis has assessed progressive changes in ventricular volume, reporting moderate- to high-quality evidence for increased lateral ventricular volume over time ($g = 0.449$), with no significant difference if the baseline measurement was in FES ($g = 0.491$) or CS ($g = 0.407$) (Kempton et al. 2010). Interscan duration ranged similarly between 1 and 10 years. In summary, there is clear evidence for progressive local (and global) volume losses across time in patients over and above these seen in healthy control subjects.

Three voxel-based meta-analyses performed subtraction analyses between the significant voxels identified in FES and CS cross-sectionally (Chan et al. 2011; Ellison-Wright et al. 2008) and largely support many of the distinctions identified by indirect comparisons across illness duration. While both groups showed alterations in the inferior frontal lobe, cingulate gyri, insula, and cerebellum (Fig. 13.3), differential changes were also reported. Relative to FES, CS were characterized by reduced GM in the medial frontal lobe, DLPFC, postcentral gyrus, uncus, temporal lobe gyri (inferior, middle, fusiform), and parahippocampus. Relative to CS, FES were characterized by reductions in the basal ganglia (caudate and putamen), temporal lobe gyri (superior and transverse), amygdala, and precentral gyrus.

In summary, there are structural volume losses in localized brain areas in patients with schizophrenia that are already present in subjects at increased clinical and familial risk for schizophrenia as well as in patients with a first psychotic episode. Brain abnormalities at first presentation include similar regions as in the risk groups, but additional fronto-striatal-temporal pathology emerges. Finally, when the chronic stage of the illness has been reached, the gray matter volume reduction has progressed further not only within the same regions but also involving more prefrontal cortical and thalamic loci. This corticothalamic pathology maps the neurochemical circuitry systems implicated in schizophrenia, namely, corticothalamic loop systems, regulated through a complex interplay of glutamate, c-aminobutyric acid, and dopamine neurotransmission. The changes across groups of subjects (UHR, FHR, FEP, and CS) and course of time are not associated with a unique and clear-cut clinical picture, thus reflecting heterogeneous etiology and clinical course of the disorder. Most likely, the changes are not linear over time and may fluctuate with phylogenesis and the course of disease. To conclude, volume changes both are of neurodevelopmental origin and progress further in the course of illness, with some, but not all, anatomical regions overlapping over time.

13.2.4 Volume Changes in Schizophrenia and in Other "Functional Psychoses"

Before we address the question of whether volume changes are unique to schizophrenia or are also present in the other "functional psychoses" such as bipolar disorder (BD) and major depressive disorder (MDD), we have to keep in mind that the three disorders, schizophrenia, BD, and MDD, are most likely not separate disease entities but mere clinical conventions. They share many symptoms; diagnoses can switch in the course of illness; they share a number of genetic and environmental risk factors; all three lead to cognitive deficits, particularly in the chronic, severe patients; and some atypical antipsychotics are efficient in acute- and long-term treatment in all three disorders. We therefore can expect some overlap in structural changes across the three disorders. It is of note that there are no meta-analyses and no original structural MRI studies, where the three disorders are compared directly, so we will consider only the comparison of two disorders at a time.

There are three meta-analyses comparing the findings of schizophrenia and BD (Ellison-Wright and Bullmore 2010; Yu et al. 2010; Bora et al. 2011). Since in the schizophrenia studies male subjects predominate, as compared to BD samples, and the brains of women and men are different, Bora et al. (2012a) controlled for gender effect. They found gray matter differences between gender-balanced schizophrenia and BD patients only in the right dorsomedial frontal cortex and left dorsolateral prefrontal cortex (smaller in schizophrenia; see Fig. 13.4). If the samples are not controlled for gender, the differences become more apparent (Ellison-Wright and Bullmore 2010; Yu et al. 2010; see Fig. 13.5).

There is no meta-analysis comparing schizophrenia and MDD directly, because there are no original studies comparing these groups. In the two VBM meta-analysis with MDD vs. control subjects (Du et al. 2012; Bora et al. 2012b), consistent gray matter reductions in MDD patients were identified in the bilateral subgenual anterior cingulate cortex (ACC) in both studies. Additionally only in the Du et al. meta-analysis (2012) were there differences in the right middle and inferior frontal gyrus, right hippocampus, and left thalamus. These areas are also affected in schizophrenia and BD, but probably to a lesser extent. Direct comparisons between the three disorders are urgently needed. We would hypothesize that the affected regions roughly overlap and correlate with disease severity (functional outcome) and neurocognitive dysfunctions across the three diagnostic groups.

13.2.5 Effects of Antipsychotic Medication on Volume Changes

Medication might alter the natural course of volume changes, i.e., volume differences between patients and controls could be due to neurodevelopmental insults, the natural course of illness, and/or antipsychotics. Although this is of great importance, it is only very rarely addressed in MR imaging studies, because of the difficulty in controlling for medication. Thus, the available

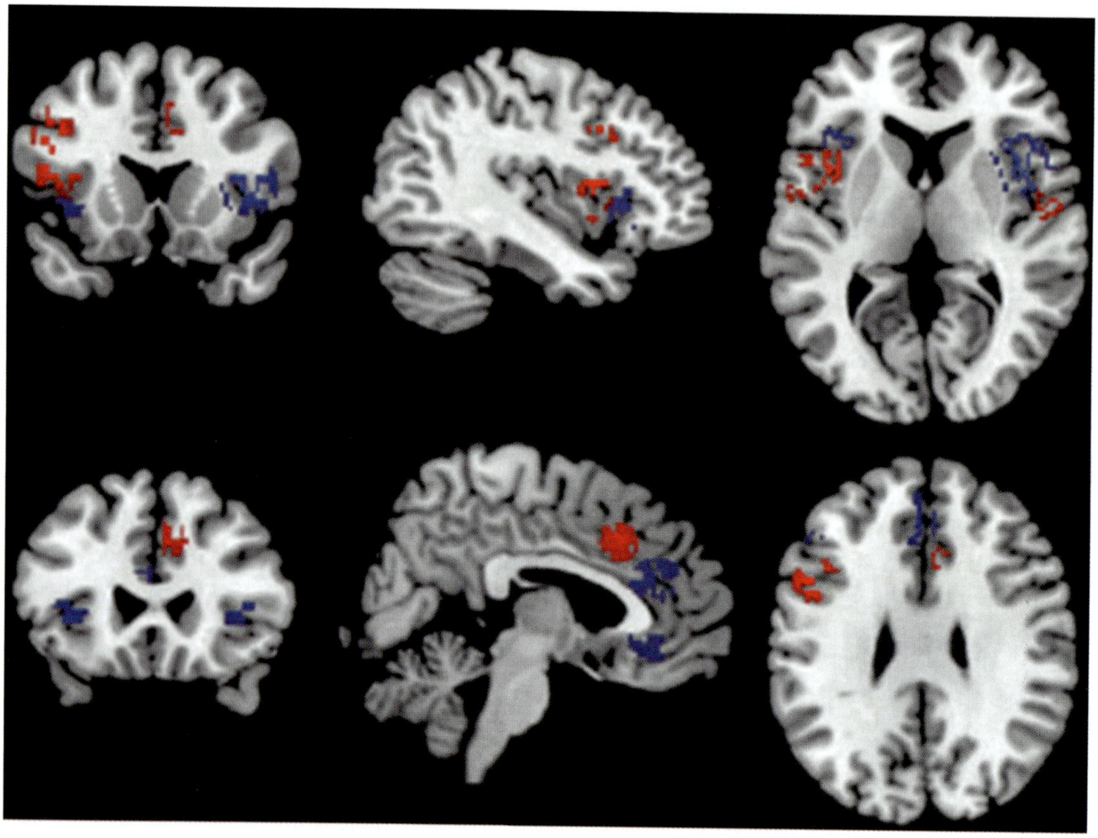

Fig. 13.4 Gray matter reductions in bipolar disorder (*blue*) and gender-balanced samples of schizophrenia (*red*) (Bora et al. 2011)

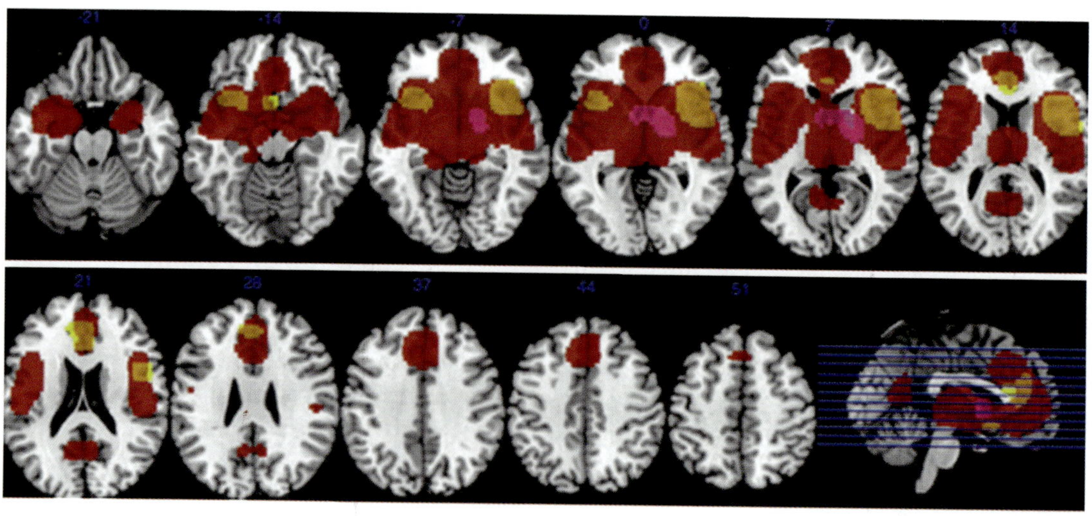

Fig 13.5 Gray matter decreases in subjects with bipolar disorder relative to controls (*yellow*). Regions of gray matter decreases (*red*) and increases (*purple*) in subjects with schizophrenia relative to controls (Ellison-Wright and Bullmore 2010)

data is extremely limited, populations are usually small, and findings must be interpreted with caution. Four reviews have qualitatively compared the structural integrity of various brain regions in treatment-naive patients, following short-term treatment administration and after long-term treatment (Moncrieff and Leo 2010; Navari and Dazzan 2009; Scherk and Falkai 2006; Smieskova et al. 2009). Notably, these reviews are of lower quality, predominantly because they report highly inconsistent data, often from insufficiently controlled studies or studies with small samples, and they use heterogeneous outcome measures or drug exposures. Furthermore, these reviews are presented in a narrative format that is susceptible to reporting bias and unable to be qualitatively assessed according to GRADE criteria. The available evidence is particularly conflicting for comparisons across medicated and unmedicated patients in early psychosis. Most consistent evidence supports increased basal ganglia volume following treatment with typical antipsychotics. However, atypical medications have been associated with both increases and reductions of basal ganglia volume, and other inconsistent findings of increased and decreased subcortical regions (including the hippocampus) have been reported for patients receiving typical antipsychotics. In long-term antipsychotic-medicated patients, effects of switching neuroleptics have been examined longitudinally. Switching from typical to atypical medication reduced basal ganglia and thalamus volume, but no difference was reported in ventricular or whole brain volume. Some evidence pointed to other antipsychotic class-specific changes, for example, longitudinal changes following typical antipsychotic administration but not following atypical administration. Nonetheless, over time, the differences between the two types of medication become less clear (Smieskova et al. 2009). Linear associations reported between reductions of regional brain volume and increasing antipsychotic exposures could represent a modulatory effect of the medication on brain structure but could also reflect a more severe course of illness requiring higher cumulative medication doses (Navari and Dazzan 2009). Two more recent original studies that had not been included in the above meta-analyses also reported conflicting results, with one study showing volume increases (van Haren et al. 2011) and the other decreases (Ho et al. 2011).

For first-episode patients, a voxel-based meta-analysis (Fusar-Poli et al. 2011a, b) compared medication-naive patients with healthy controls and reported GM reductions of superior temporal and insular cortices and cerebellum. However, another voxel-based meta-analysis explicitly considered the effect of medication status in FES (Leung et al. 2011), where reductions in insula, anterior cingulate cortex, middle and inferior frontal and precentral gyri, uncus/amygdala, and superior temporal gyrus were found in both medicated and unmedicated FES patients compared to healthy controls. Medicated FES patients were additionally characterized by reductions in the superior frontal gyri, inferior temporal gyrus, PCC, claustrum, cerebellum, and caudate. Additional reductions in unmedicated FES patients were located in medial frontal gyri, thalamus, and parahippocampus.

In summary, the evidence of antipsychotic medication on MRI volume changes is unclear. Volume increases and decreases probably interact in a complicated way with the individual course of illness. In general, it is yet not clear what kind of volume changes, decreases or increases, are potentially "good" or "bad" for the individual patient.

13.3 White Matter Changes Through the Course of Schizophrenia

13.3.1 White Matter Changes in Chronic Schizophrenia

White matter structure and integrity can be investigated with diffusion tensor imaging (DTI) and VBM using magnetic resonance imaging. Investigations in schizophrenia have identified compromised white matter integrity (Kubicki et al. 2007). Ellison-Wright and Bullmore (2009) conducted an ALE analysis of fractional anisotropy changes identified by DTI and identified two large clusters of reductions in patients with schizophrenia compared to controls, in the deep frontal and deep temporal white matter. In an ALE analysis of whole brain VBM studies, two small clusters of reduced white matter in the frontal lobe, including left medial frontal gyrus and right dorsolateral prefrontal cortex, as well as small clusters of reduced white matter in the left and right internal capsule were identified (Di et al. 2009). A comparison between VBM and fractional anisotropy analyses in white matter using SDM meta-analysis suggested that fractional anisotropy may be more sensitive to identifying changes in white matter (Bora et al. 2011). The VBM comparisons identified white matter reductions in temporal white matter including the internal capsule, a finding that was mirrored and extended to the genu of the corpus callosum and medial frontal regions bilaterally (Bora et al. 2011). Evidence from a recent meta-analysis of longitudinal volumetric studies in chronic schizophrenia suggests that these white matter changes show a progressive course over a follow-up period between 1 and 10 years and that this progress affects the frontal, temporal, and parietal white matter volume (Olabi et al. 2011).

13.3.2 Diffusion Tensor Imaging (DTI) in Patients with First-Episode Schizophrenia

There are subtle anomalies of white matter brain morphology in patients with schizophrenia which are already present at the onset of the disease, i.e., in first-episode patients, but have also been found in patients in their first episode of BD (Kempton et al. 2008; Pfeifer et al. 2008). A recent meta-analysis aimed to compare gray and white matter volume reductions, as detected by MRI, in first-episode patients with schizophrenia or BD to address the question of neuroanatomical specificity of these changes (De Peri et al. 2012). When first-episode BD and schizophrenia patients were conjunctly compared with healthy controls, a significant reduction of gray and white matter volume was found in the first-episode patients. The meta-analysis of studies comparing patients with first-episode schizophrenia and BD to healthy controls separately revealed gray matter volume deficits to be more prominent in first-episode schizophrenia patients, whereas white matter volume reduction was more pronounced in first-episode bipolar patients. However, as sample sizes were notably larger in the case of first-episode schizophrenia as compared to first-episode BD, it remains unclear whether this difference may be attributable to differences in statistical power between the meta-analyses of schizophrenia or BD and healthy controls.

Studies investigating white matter integrity as inferred by DTI mainly report abnormalities in the frontal, frontotemporal, and fronto-limbic connections, with tracts including the superior longitudinal fasciculus (Federspiel et al. 2006; Karlsgodt et al. 2009, 2008; Pérez-Iglesias et al. 2010), uncinate fasciculus (Kawashima et al. 2009; Luck et al. 2011; Price et al. 2008), and corpus callosum (Cheung et al. 2008; Dekker et al. 2010; Federspiel et al. 2006; Pérez-Iglesias et al. 2010). These anomalies seem to be associated with the exposure to antipsychotic medication and duration of treatment (Peters et al. 2008; Szeszko et al. 2008; White et al. 2011). However, all DTI studies of antipsychotic-naive patients with first-episode schizophrenia, to date, have reported widespread as opposed to localized FA changes, suggestive of extensive structural disconnectivity associated with illness onset, before the initiation of antipsychotic treatment (Cheung et al. 2008, 2011; Pérez-Iglesias et al. 2010).

13.3.3 Diffusion Tensor Imaging (DTI) in Ultrahigh-Risk Subjects

Schizophrenia has been found to involve cortico-cortical disconnectivity, both in terms of functional connectivity between different brain regions (McGuire and Frith 1996) and in terms of neuroanatomical alterations underlying the functional changes, particularly in white matter tracts connecting the frontal and temporal lobes (Ellison-Wright and Bullmore 2009). In line with the neurodevelopmental hypothesis of schizophrenia, reduced white matter volume and integrity have also been found in early and subclinical stages of the disorder, i.e., in first-episode patients (Gasparotti et al. 2009) and subjects at ultrahigh risk (Carletti et al. 2012). Most DTI studies comparing UHR subjects to healthy controls identified white matter abnormalities in the frontal (Bloemen et al. 2010; Karlsgodt et al. 2009; Peters et al. 2009) and temporal connections in UHR individuals (Bloemen et al. 2010; Jacobson et al. 2010). A recent multimodal MRI/EEG study, using VBM to identify volumetric changes in a sample of 39 UHR subjects compared to 41 healthy control subjects, found WM volume changes in the right frontal, temporal, and parietal regions that were associated with P300 amplitude in the two groups (Fusar-Poli et al. 2011a, b). Of the 39 UHR subjects, 26 % developed a psychotic illness within the follow-up period of 2 years. The subgroup in whom psychosis subsequently developed had a smaller volume of white matter underlying the left precuneus and the right middle temporal gyrus and increased volume in the white matter underlying the right middle frontal gyrus as compared to those who did not develop psychosis. A second study looking at longitudinal outcome in UHR subjects using DTI also reported temporal WM changes, with both increased (left medial temporal lobe) and decreased (left superior temporal lobe) fractional anisotropy (FA) being found in subjects who converted to psychosis compared to those who did not (Bloemen et al. 2010), indicating that the medial temporal lobe has an important role in the progression from a high-risk state to frank psychosis. Despite the small number of studies using DTI in UHR subjects, these findings of WM alterations in frontal and temporal regions in UHR relative to control subjects suggest that these regions might be indicators of conferred risk of developing schizophrenia. This hypothesis is further supported by a recent study investigating the time course of WM changes by using DTI to study UHR subjects before and after the onset of illness and comparing WM DTI properties of UHR subjects ($n=32$) to those of patients with FEP ($n=15$) and healthy controls ($n=32$) (Carletti et al. 2012). At baseline, FA was lowest in FEP subjects, highest in healthy controls, and intermediate in the UHR group in two clusters. The first cluster comprised voxels in areas corresponding to the splenium and body of the corpus callosum, the left inferior and superior longitudinal fasciculus, and the left inferior fronto-occipital fasciculus. The second cluster included the right external capsule, the retrolenticular part of the right internal capsule, and the right posterior corona radiata. A follow-up scan was performed after 28 months, a period in which eight of the 32 UHR individuals had developed psychosis. As opposed to the findings of both decreased and increased FA in the temporal lobe in the prospective DTI study by Bloemen et al. (2010) mentioned above, this longitudinal analysis revealed a progressive reduction in FA in the left frontal WM in those UHR subjects who developed schizophrenia that was not evident in UHR subjects who did not make a transition. This difference in findings may be due to methodological differences between the two studies (e.g., prospective vs. longitudinal approach, age differences between the UHR samples, potentially confounding effects of medication). In summary, the available data suggest that the UHR stage is associated with WM changes in areas that have also been implicated in FES and chronic schizophrenia. There is now also first evidence that the onset of psychosis in UHR subjects may be associated with a longitudinal progression of abnormalities in the left frontal WM.

13.4 fMRI Findings in Schizophrenia

13.4.1 fMRI in Chronic Schizophrenia

fMRI has been extensively used to study the neurobiological basis of schizophrenia and a wide variety of paradigms has been applied. In the following section, we focus on fMRI findings of

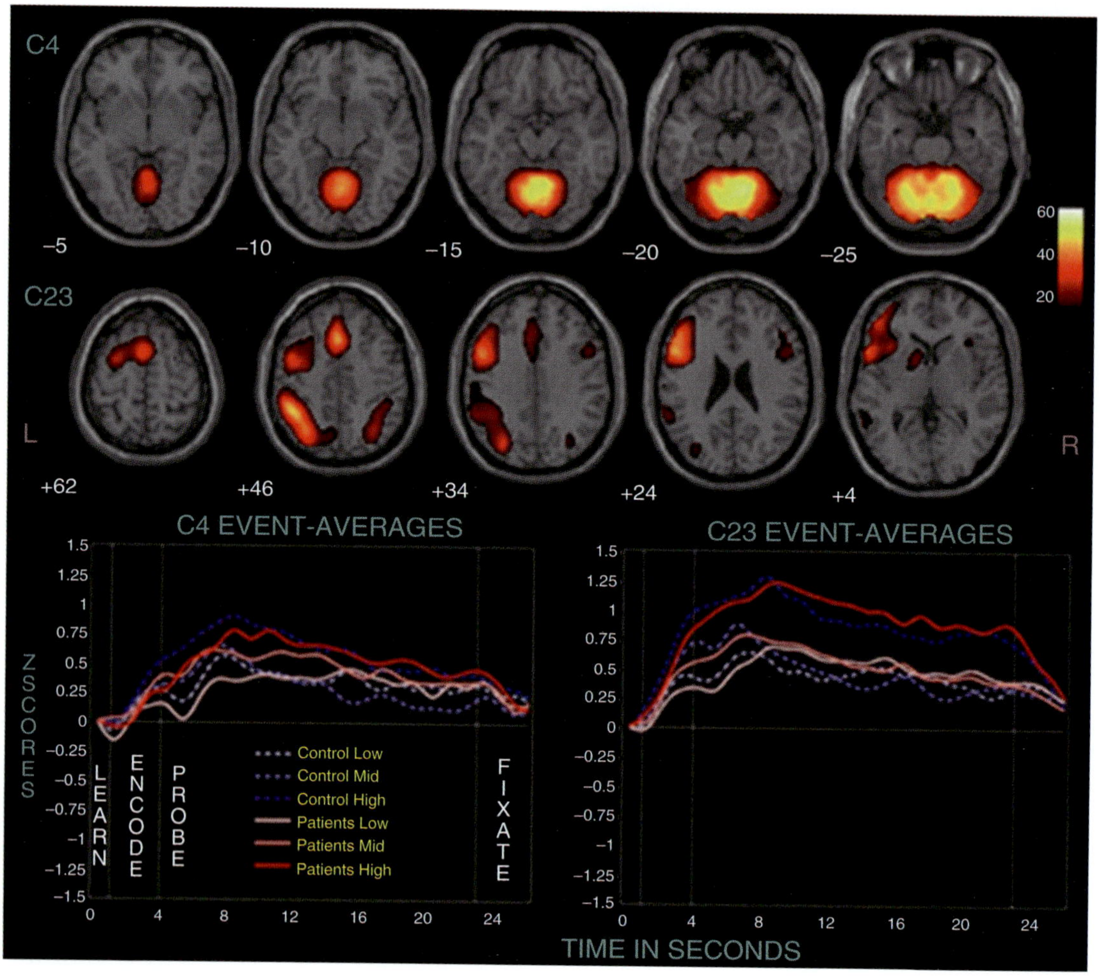

Fig. 13.6 Two positively modulated components (C4 and C23). Displayed are selected slices from each component thresholded at $p < 1.0 \times 10^{-13}$, FDR, along with their event-related averages (Kim et al. 2009)

working memory (WM), executive functioning, and social cognition, as these are among the domains most severely affected in schizophrenia. We summarize the available evidence from studies across different stages of the illness, focusing on those that did not pool patients in different stages of illness, but studied UHR, FES, and CS samples separately.

13.4.1.1 Working Memory

During WM tasks, several studies found marked differences in neural activation across several modalities of presentation. In a multicentric study, several components (as identified by ICA) differed between healthy subjects and patients with chronic schizophrenia ($n = 130$ and 115, respectively) (Kim et al. 2009). One component (C23) consisted of the ACC, left prefrontal cortex, and left inferior parietal cortex (see Fig. 13.6). Other areas displaying group differences were the superior and medial frontal cortex, the parahippocampal cortex, and parts of the temporal lobe.

In a similar study, differences in temporal profiles of WM-related components were found in subjects with CS and controls (Meda et al. 2009). This was particularly pronounced for left-lateralized networks. In another study, reduced functional connectivity of the left Sylvian-parietal-temporal area and left anterior insula

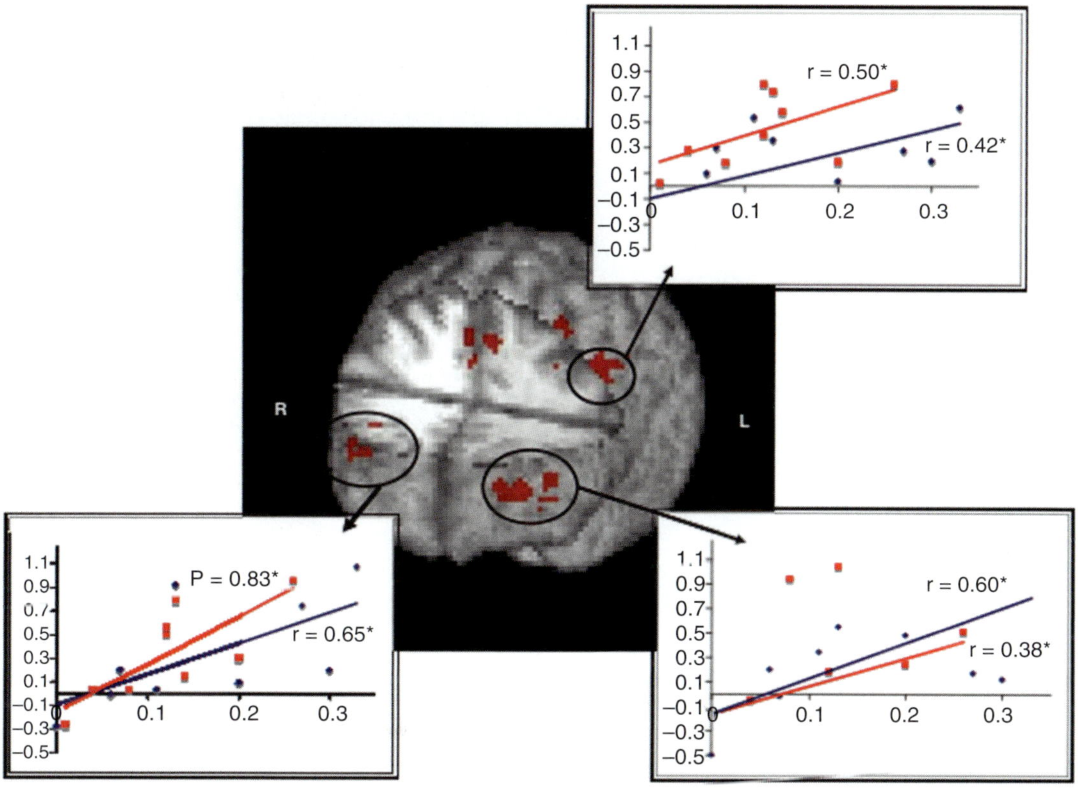

Fig. 13.7 Regions of overlap in the three-way interaction (load by time by group) between both the word and picture WM tasks showing activity increases in bilateral frontopolar regions, anterior cingulate cortex, and left dorsolateral/dorsal prefrontal cortex. Significant correlations between improved behavioral performance (*x*-axis, percentage of increase) and increased functional activation (*y*-axis, percentage of signal change) are also shown. Word 2-back is represented in *red* and picture 2-back in *blue* (Haut et al. 2010)

during auditory WM encoding was found in 14 patients with CS compared to healthy controls (Hashimoto et al. 2010). This reduction in functional connectivity was not observed for visual stimuli.

As WM and its underlying neural correlates is the most severely impaired domain in schizophrenia, it has been the target of interventions, e.g., cognitive remediation training. In one such study, the bilateral frontopolar cortex and ACC were found to show signs of plasticity induced by remediation training (Haut et al. 2010) (see Fig. 13.7).

To summarize, the available evidence on brain activation dysfunctions in CS during WM suggests a widespread network comprising the ACC as well as lateral prefrontal and parietal areas is affected in patients.

13.4.1.2 Executive Functions

In an exemplary study on executive function, the neural correlates of discrimination of novel or old visual stimuli, four groups were investigated (healthy, UHR, FES, CS; Morey et al. 2005) (see Fig. 13.8). It could be shown that especially the anterior cingulate, as well as additional prefrontal areas, is involved in this task. These areas show decrements in all groups compared to healthy subjects.

To summarize, the available evidence on brain activation dysfunctions in CS during executive function suggests that medial and lateral prefrontal cortices are affected in patients.

13.4.1.3 Social Cognition

In chronic schizophrenia, processing of emotional faces is impaired, and patients already show

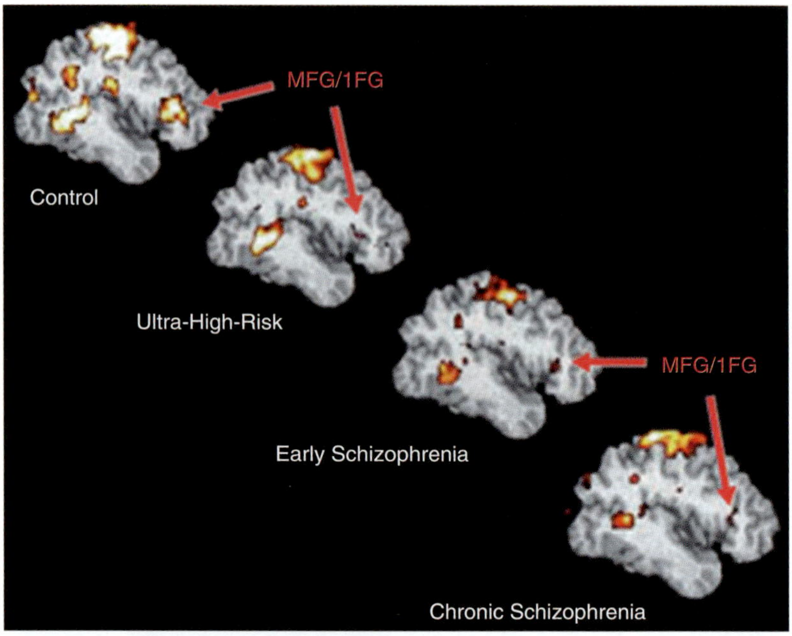

Fig. 13.8 Group-averaged, target-related prefrontal activation in control, ultrahigh-risk, early, and chronic groups. *IFG* inferior frontal gyrus, *MFG* middle frontal gyrus (Morey et al. 2005)

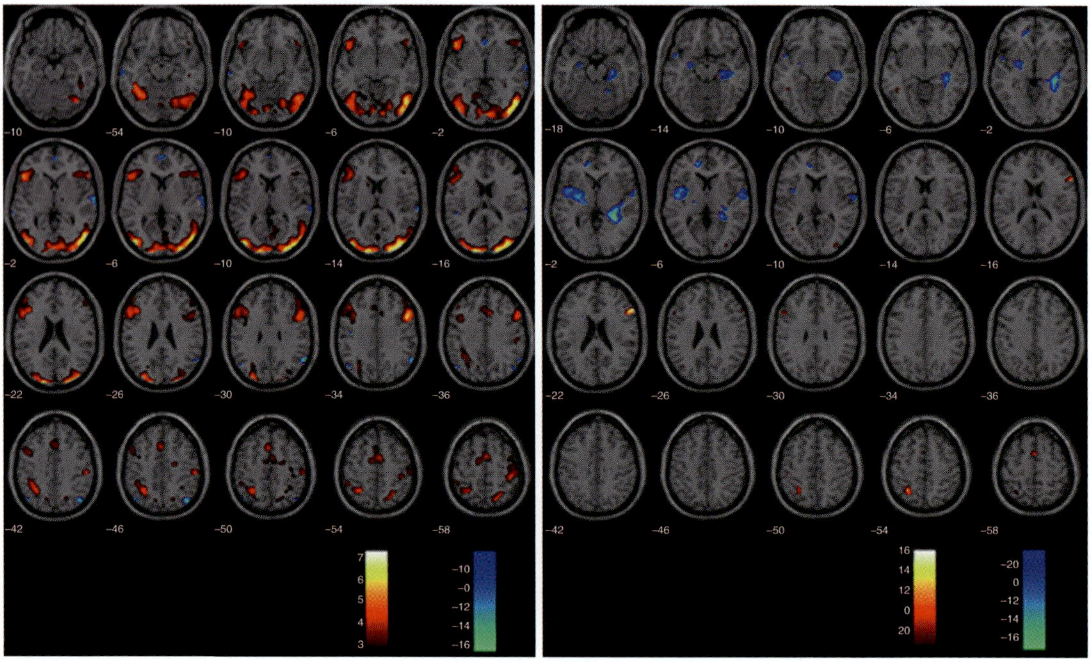

Fig 13.9 Processing of sad vs. neutral faces (SPM2, repeated measures ANOVA, $p<0.005$ uncorrected). Illustrated are the contrasts sad > neutral (*red/yellow*) and neutral > sad (*blue/green*) in controls on the *left* and in UHR subjects, respectively, on the *right* (Seiferth et al. 2008)

difficulties in processing neutral faces (Surguladze et al. 2006) (see Fig. 13.9). In a study with 11 chronic patients, compared to healthy controls, face-processing areas such as the fusiform gyrus and the parahippocampal gyrus were hyperactivated in patients while processing neutral faces.

In a social appraisal task (Taylor et al. 2011), 22 patients with CS depicted greater activations for social vs. gender judgments of negative faces than healthy subjects. These hyperactivations were found in the ACC, superior medial gyrus, inferior temporal and parietal cortex, as well as precuneus, which were also correlated with poor social cognition and poor social adjustment. As can be seen from these two studies, in CS, face processing is already disturbed for neutral faces and is also characterized by abnormal recruitment of networks involved in social appraisal.

To summarize, the available evidence on brain activation dysfunctions in CS during social cognition suggests that the medial temporal cortex (including the hippocampus) and prefrontal (including ACC) and parietal areas (including the precuneus) are affected in patients.

13.4.2 fMRI in the Ultrahigh-Risk Stage

13.4.2.1 Social Cognition

Although research has highlighted that early social functioning may be associated with an increased risk of developing a psychotic disorder (Malmberg et al. 1998; Cannon et al. 1997), there is only limited published literature on social cognitive performance in UHR populations in general and on the role of neural correlates in particular. Social cognitive domains that have been primarily investigated thus far using fMRI in UHR populations include emotion processing and Theory of Mind (ToM).

13.4.2.2 Emotion Processing

To date, only one study has explicitly investigated the neural correlates of emotion processing in UHR individuals. Seiferth et al. (2008) used an fMRI facial emotion discrimination paradigm to study 12 UHR individuals and 12 healthy controls. No group differences were evident at the behavioral level, but emotion discrimination was associated with greater activation in the right lingual and fusiform gyri and in the left middle occipital gyrus in the UHR subjects compared with controls. The UHR group also showed more activation than controls in the inferior and superior frontal gyri, the cuneus, the thalamus, and the hippocampus when viewing neutral relative to emotional faces (Fig. 13.9). The latter findings suggest that there is a greater neural response to stimuli which would not normally be regarded as emotionally "salient," consistent with the notion that the aberrant assignment of salience underlies the formation of psychotic symptoms (Kapur 2003) and evidence that motivational salience processing is altered in UHR subjects (Roiser et al. 2013).

Pauly et al. (2010) studied emotional experience and the interaction between emotion and cognitive performance in 12 UHR subjects and controls. During the induction of negative emotion by olfactory stimulation, UHR subjects exhibited decreased activation in the right insula. When unpleasant olfactory stimulation was given while the subjects were performing a WM task, the UHR group showed more activation than controls in the cerebellum. The differential cerebellar activation was interpreted as a compensatory response to the increased demands of the WM task in the presence of a distracting emotion induction. Altered cerebellar activation was not observed when the same emotion-cognition interaction task was studied in adolescent-onset schizophrenia (Pauly et al. 2008), although differential activation during olfactory emotion induction has previously been described in the insula in patients with schizophrenia (Crespo-Facorro et al. 2001). In the above studies, the UHR samples were not followed up clinically, so it is unclear if the findings were generic to all UHR subjects or specific to subjects who later develop psychosis.

13.4.2.3 Theory of Mind (ToM)

Evidence of both verbal and visual ToM abnormalities lying intermediately between those of first-episode patients and healthy controls has been provided by a recent meta-analysis of behavioral ToM studies (Bora and Pantelis 2013). To date, only one study investigated patterns of brain activation during ToM task performance in UHR subjects, chronic schizophrenia patients, and healthy controls (Brüne et al. 2011). Brain activation in regions associated with the ToM network such as the prefrontal cortex, posterior cingulate, and the temporoparietal cortex was most pronounced in UHR subjects as compared to chronic patients and, in part, also as compared to healthy controls in the absence of behavioral differences. This finding was interpreted as being indicative of a compensatory overactivation of brain regions involved in empathic responses during mental state attribution in UHR subjects. However, as the sample size in the UHR group was relatively small ($n=10$ vs. 22 chronic patients and 26 healthy controls) and only one UHR subject subsequently developed psychosis, it is unclear how differences in group sizes may have influenced the results and whether greater activation in ToM-related brain regions is specifically related to the development of schizophrenia.

To summarize, the available evidence on brain activation dysfunctions in UHR during social cognition suggests that the temporo-occipital (including the hippocampus, insula, lingual and fusiform gyrus), temporoparietal, prefrontal, subcortical, and cerebellar regions are affected in UHR subjects.

13.4.2.4 Working Memory and Executive Function

WM impairment is the most consistently observed cognitive deficit exhibited by patients with schizophrenia (cf. Forbes et al. 2009). Prefrontal dysfunction appears to underlie many of the cognitive impairments seen in schizophrenia, particularly deficits in WM function (Lee and Park 2005; Glahn et al. 2005). Neurocognitive deficits are also a core characteristic of the UHR syndrome (Brewer et al. 2005; Lencz et al. 2006; Eastvold et al. 2007; Nieman et al. 2007), particularly affecting WM domains (Nieman et al. 2007). UHR subjects can be distinguished from healthy controls specifically on the basis of WM performance (Smith et al. 2006; Pflueger et al. 2007) and their baseline WM functioning can predict the longitudinal development of psychosis (Pukrop et al. 2007). To identify the neurofunctional correlates of neurocognitive impairment (WM, executive function) in the UHR stage, functional magnetic imaging has been employed and abnormalities in the prefrontal and temporal lobes during cognitive functioning have been shown that are qualitatively similar though less marked than those in subjects with established schizophrenia (Broome et al. 2010, 2009). PFC dysfunction during cognitive task performance has been shown to be greatest in those UHR subjects who subsequently go on to develop psychosis (Allen et al. 2012). Such alterations are not attributable to effects of the illness or its treatment and may represent core neurobiological markers of an increased vulnerability to psychosis.

To summarize, the available evidence on brain activation dysfunctions in the UHR stage during WM and executive function suggests that the PFC is affected in UHR subjects.

13.4.3 fMRI in First-Episode Schizophrenia

13.4.3.1 Working Memory

In first-episode schizophrenia, WM deficits become more apparent than in UHR, and neural correlates of this impairment have been investigated in several studies. They comprise mainly frontal, but also temporal and parietal, areas (Broome et al. 2009) (see Fig. 13.10).

In another study where patients ($n=11$) and healthy subjects had to maintain and manipulate letters, first-episode patients exhibited lower activations mainly in the middle frontal gyrus compared to healthy controls, while activations of the bilateral inferior frontal gyrus were greater in patients (Tan et al. 2005). In a multicentric study (Schneider et al. 2007), a classical letter version of the n-back task was carried out by healthy subjects and first-episode schizophrenia. While patients performed worse in several behavioral measures of the task, in the 2-back >0-back contrast, it was found that first-episode patients relative to controls depicted hypoactivations in the precuneus while at the same time exhibiting right-lateralized hyperactivations in the ventrolateral prefrontal cortex and insula.

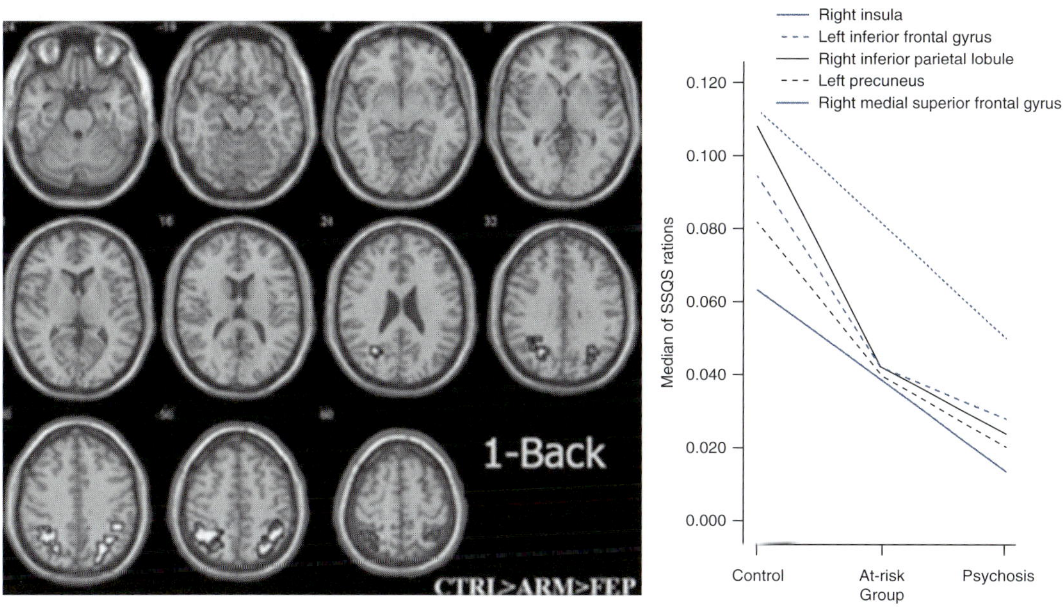

Fig 13.10 Group differences in cluster activation during the 2-back condition of the N-back task. Differential activation during the 2-back condition was greatest in controls, weakest in the psychosis group, and intermediate in the at-risk group in the lateral prefrontal, insular, and parietal cortex and in the precuneus. The left side of the brain is shown on the *left* of the figure (voxel $p<0.05$, cluster $p<0.01$). *SSQRs* sum of squares of deviations due to the residuals (Broome et al. 2009)

As can be seen from these studies, patients experiencing their first episode already exhibit behavioral deficits in WM while at the same time aberrant neural activations are found, mainly in the insula, precuneus, and bilateral prefrontal cortex.

13.4.3.2 Executive Functions

Executive functioning such as task switching and problem solving is among the domains in which first-episode patients display deficits. It could be shown that in a "Tower of London" task, first-episode patients ($n=10$) did not show performance deficits but displayed differences in right prefrontal and temporal activation compared to healthy subjects during the task (Rasser et al. 2005).

Another domain in which patients with schizophrenia exhibit deficits, and which is most often categorized as executive function, is verbal fluency. In a study by Broome et al. (2009), see Fig. 13.11, verbal fluency was also investigated in subjects at elevated risk and first episode. As can be seen from these selected studies, depending on the specific task, neural inefficiencies are found in frontal and parietal brain regions.

To summarize, the available evidence on brain activation dysfunctions in FE during executive functions shows that prefrontal and temporal cortices are affected in patients.

13.4.3.3 Social Cognition

Social cognition is an important domain as it might translate directly into employment status and number of social contacts (leading to greater support). One of the fundamental underlying prerequisites to engage in most kinds of social contact is facial processing which could be shown to be impaired in the disorder. These deficits arise from difficulties in facial affect recognition and discrimination. While some studies found difficulties in assessing negative affect in general, others found this impairment to be more pronounced in a specific emotion such as fear. So far, the relationship of this impairment

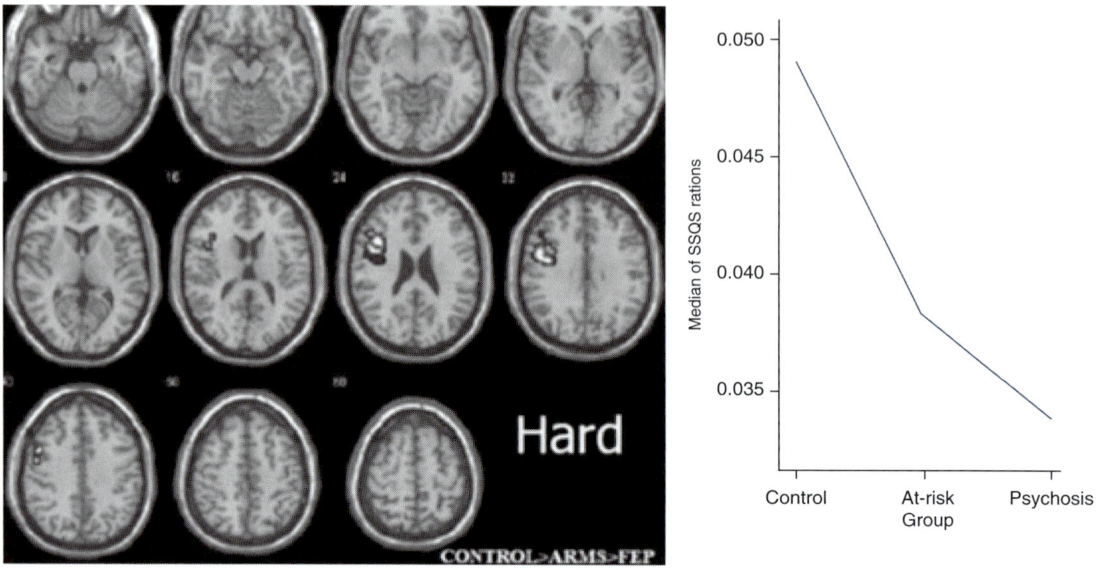

Fig. 13.11 When the task demands were high, there was a differential engagement of dorsolateral prefrontal cortex with activation greatest in the control group, weakest in the psychosis group, and intermediate in the at-risk group. However, on the same version of the task, there was differential engagement of the left anterior insula. When task demands were high, activation in this region was greatest in the psychosis group, weakest in the controls, and intermediate in the at-risk group. The left side of the brain is shown on the *left* of the figure (voxel $p<0.05$, cluster $p<0.01$). *SSQRs* sum of squares of deviations due to the residuals (Broome et al. 2009)

and current psychopathology is unclear as some studies found correlations with positive and negative symptoms or overall severity of symptoms (Marwick and Hall 2008).

In a study investigating social cognition in recovered first-episode schizophrenia, it could be shown that improvement in insight and social functioning was correlated with positive signal change in the left medial prefrontal cortex (Lee et al. 2006) (see Fig. 13.12).

Given the importance of social cognition as a good predictor of general social outcome, this domain could be an important target for therapeutic interventions in the disorder (Marwick and Hall 2008).

13.5 Imaging Genetics

For the past 10 years, researchers have employed fMRI investigations to study the effects of genetic variation on brain activation of candidate genes for schizophrenia. The tasks studied are typically selected to match cognitive domains known to be impaired in schizophrenia, such as WM, verbal fluency, or episodic memory. However, this approach has not been limited to cognitive domains but has been carried out for morphometric studies and DTI studies as well (e.g., Baune et al. 2012; Kim et al. 2009; Nickl-Jockschat et al. 2012). As most of the first identified variants were common variants with a frequency of >10 % in the general population, most studies comprised of healthy subjects only. This approach circumvents effects of medication, current psychopathology, or course of the disorder. As some of those variants were identified in hypotheses-free approaches such as association studies, the effects of those variants were poorly understood, and it was hoped that imaging genetics could shed light on their effects on brain and behavior. The rationale for this approach is depicted in Fig. 13.13 (Rasetti and Weinberger 2011).

A recent review of several fMR imaging genetic studies found strong evidence for

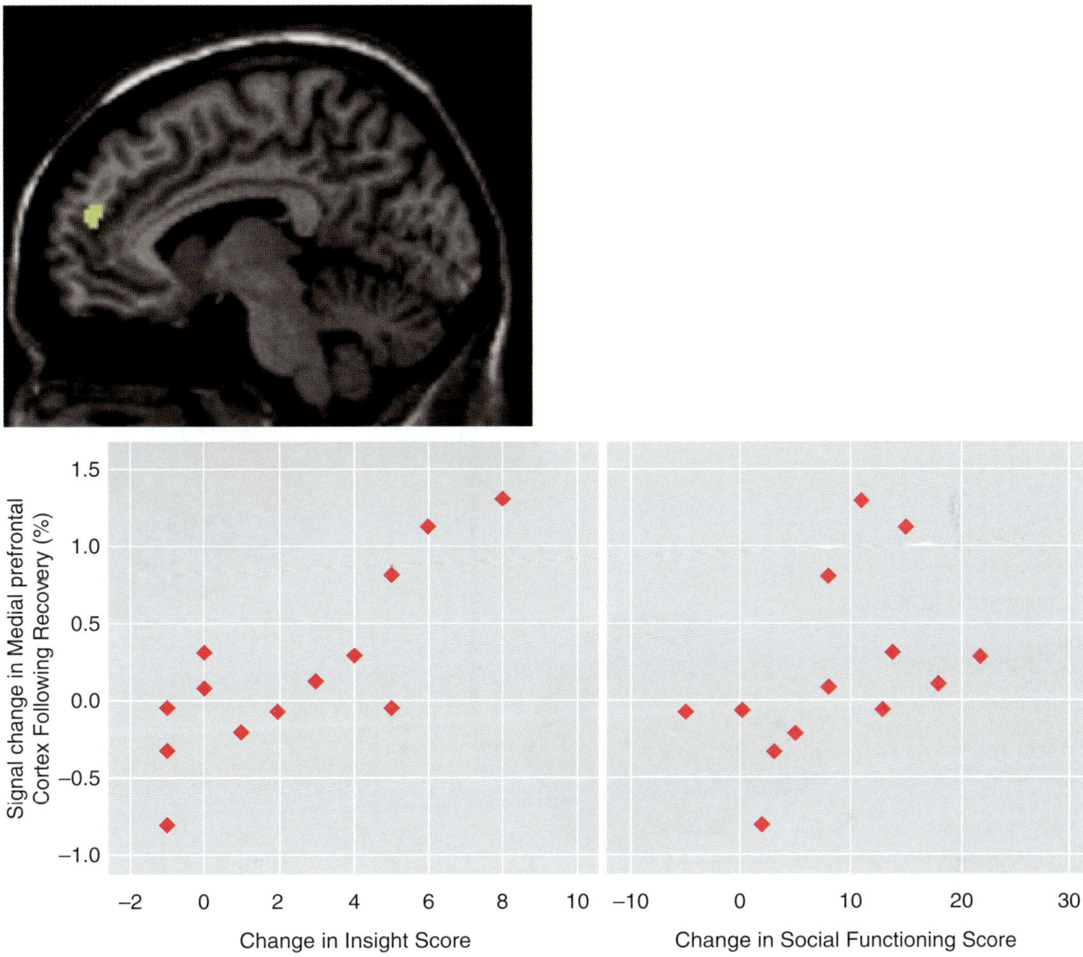

Fig. 13.12 Left medial prefrontal cortex (peak activation coordinates: $x=-6$, $y=51$, $z=16$; 32 voxels) activation common to empathic and forgivability judgments in patients with schizophrenia following recovery from an acute episode. The graphs show that the difference between mean blood-oxygen-level-dependent signal changes in the peak activation before and after recovery in patients was significantly correlated with improvement in insight ($r=0.81$, $p<0.001$) and was related at a less than significant level to improvement in social functioning scores ($r=0.51$, $p=0.06$) (Lee et al. 2006)

modulation of neural circuits by genetic variation such as NRG1 (e.g., Krug et al. 2010), DTNBP1, G72, COMT (e.g., Krug et al. 2009), etc. (Rasetti and Weinberger 2011).

As Fig. 13.14 shows, imaging genetics also employs cohorts of affected individuals. Adding patients to these models, it can be tested if variation in these susceptibility genes interacts with neural circuits in patients in a different way than in healthy subjects. So far, this approach is of great importance in assessing the impact of newly identified genetic variation that leads to elevated risk for the disorder. Further methodological refinement of this research area, such as multimodal imaging, multicentric imaging, computational neuroscience methods, machine-learning algorithms, etc., will certainly add to the understanding of genetically modulated neural networks in healthy subjects and schizophrenia.

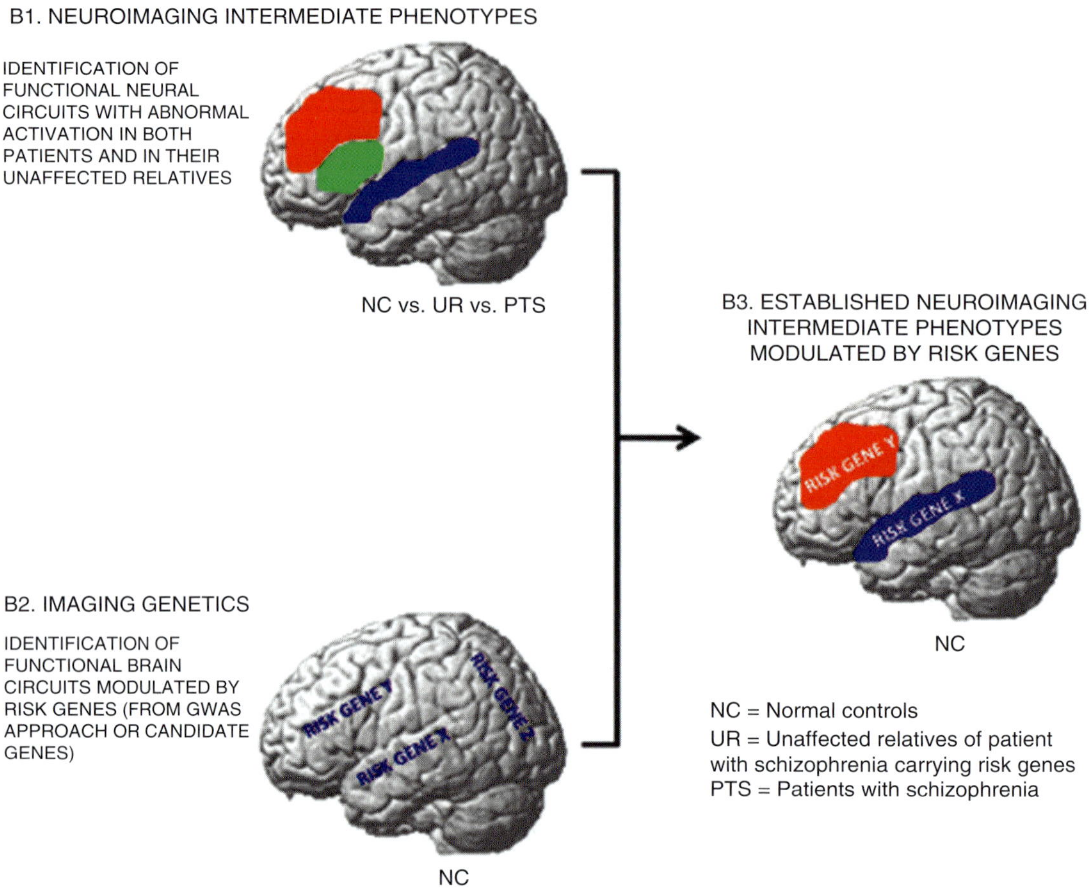

Fig. 13.13 Genetic risk on vulnerable brain circuits. *B1*. Identification of neuroimaging intermediate phenotypes – which are alterations in neural circuit functions in patients with psychiatric disorders as well as in high-genetic-risk subjects (i.e., unaffected relatives). *B2*. Imaging genetics defines neural systems that are modulated by genetic variations, including genetic variations that have been associated with increased risk for psychiatric disorders. *B3*. To increase the probability that the observed biological modulation by the risk genetic variation is the mechanism through which that gene increases the risk for a psychiatric disorder, it is important to demonstrate that the gene modulates a neuroimaging intermediate phenotype (From Rasetti and Weinberger (2011))

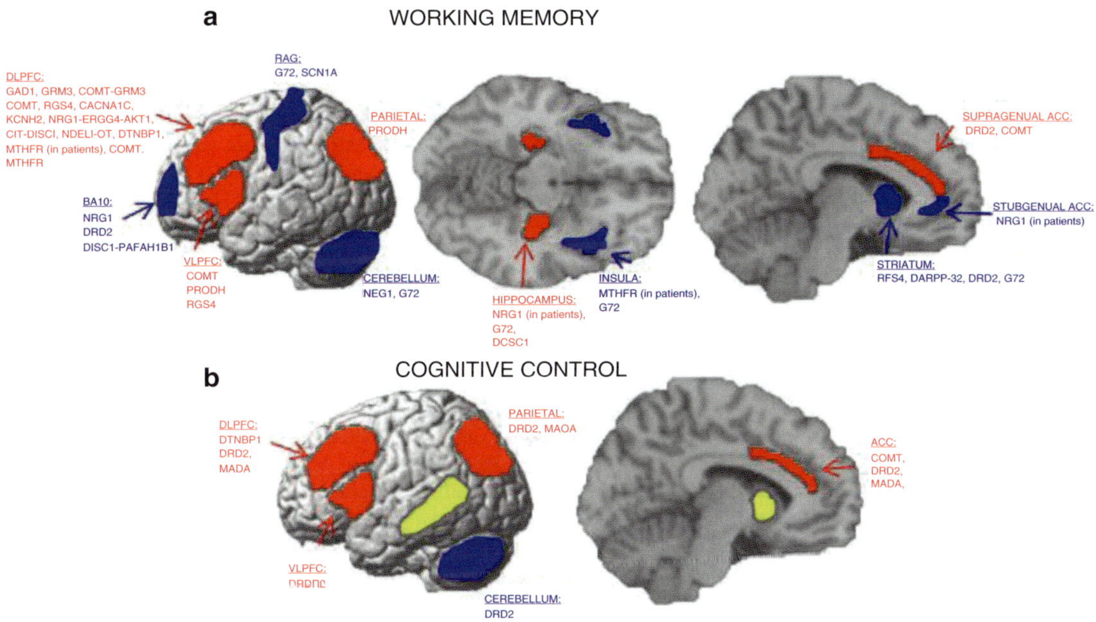

Fig. 13.14 Genetic modulation on vulnerable circuits. (**a**) WM. Most brain areas reported altered in patients and their healthy relatives during WM task are also modulated by a number of risk genes explored with the same paradigm (*red fields with square dots*) (DLPFC, VLPFC, ACC, parietal cortex, and HF). Many other effects of genes during WM paradigms have not been shown to be intermediate phenotypes (striatum, basal ganglia, subgenual ACC, insula, BA10, BA 4/6, cerebellum) (*blue fields*). (**b**) Cognitive control circuit. Several brain areas within the cognitive control circuit have been reported to be modulated by risk genes during cognitive control processing (PFC, especially ACC, superior temporal gyrus, parietal cortex, and cerebellum). Among these, only the PFC (DLPFC, VLPFC, and ACC) and parietal cortex have been consistently reported being altered in patients with schizophrenia and their unaffected relatives with cognitive control paradigms (*red fields with square dots*). The striatum and middle temporal gyrus (BA 21) have been reported altered in patients with schizophrenia and their healthy relatives during cognitive control, although none of the risk genes studied so far have shown modulation of these regions (*yellow fields with solid line*)

Conclusion

Neuroimaging studies support the presence of structural and functional alterations in schizophrenia, from its genetic risk, its putative prodromal phase to the chronic disorder. Abnormalities, even if subtle in magnitude and variable in extent across studies, are now reported in essentially every brain region. However, one core pathophysiological abnormality that is underlying schizophrenia has yet to be identified. This reflects the heterogeneity in etiology and course of the purely clinical category of schizophrenia. Therefore, it should not be believed that neuroimaging or any other "marker" will identify a pathognomonic abnormality of the disorder as it is presently defined by its symptoms (and course). Future studies should therefore be aided by employing more homogeneous classifications of psychotic disorders (e.g., using symptom-, cognitive-, genetic-, environmental risk-based approaches), comparing them with other diagnoses (e.g., using a dimensional approach), and relating imaging findings with other biological data. It is time for a novel, neurobiologically based characterization of affective and schizophrenic psychoses.

References

Allen P, Luigjes J, Howes OD, Egerton A, Hirao K, Valli I, Kambeitz J, Fusar-Poli P, Broome M, McGuire P (2012) Transition to psychosis associated with

prefrontal and subcortical dysfunction in ultra high-risk individuals. Schizophr Bull 38(6):1268–1276
Baiano M, David A, Versace A, Churchill R, Balestrieri M, Brambilla P (2007) Anterior cingulate volumes in schizophrenia: a systematic review and a meta-analysis of MRI studies. Schizophr Res 93(1–3):1–12
Baune BT, Konrad C, Grotegerd D, Suslow T, Birosova E, Ohrmann P, Bauer J, Arolt V, Heindel W, Domschke K, Schöning S, Rauch AV, Uhlmann C, Kugel H, Dannlowski U (2012) Interleukin-6 gene (IL-6): a possible role in brain morphology in the healthy adult brain. J Neuroinflammation 9:125
Bloemen OJN, de Koning MB, Schmitz N, Nieman DH, Becker HE, de Haan L, Dingemans P, Linszen DH, van Amelsvoort TAMJ (2010) White-matter markers for psychosis in a prospective ultra-high-risk cohort. Psychol Med 40(8):1297–1304
Bora E, Pantelis C (2013) Theory of mind impairments in first-episode psychosis, individuals at ultra-high risk for psychosis and in first-degree relatives of schizophrenia: systematic review and meta-analysis. Schizophr Res 144(1–3):31–36
Bora E, Fornito A, Radua J, Walterfang M, Seal M, Wood SJ, Yücel M, Velakoulis D, Pantelis C (2011) Neuroanatomical abnormalities in schizophrenia: a multimodal voxelwise meta-analysis and meta-regression analysis. Schizophr Res 127(1–3):46–57
Bora E, Fornito A, Yücel M, Pantelis C (2012a) The effects of gender on grey matter abnormalities in major psychoses: a comparative voxelwise meta-analysis of schizophrenia and bipolar disorder. Psychol Med 42(2):295–307. doi: 10.1017/S0033291711001450, Epub 2011
Bora E, Fornito A, Pantelis C, Yücel M (2012b) Gray matter abnormalities in major depressive disorder: a meta-analysis of voxel based morphometry studies. J Affect Disord 138(1–2):9–18
Borgwardt SJ, Riecher-Rössler A, Dazzan P, Chitnis X, Aston J, Drewe M, Gschwandtner U, Haller S, Pflüger M, Rechsteiner E, D'Souza M, Stieglitz R-D, Radü E-W, McGuire PK (2007) Regional gray matter volume abnormalities in the at risk mental state. Biol Psychiatry 61(10):1148–1156
Boyer P, Phillips JL, Rousseau FL, Ilivitsky S (2007) Hippocampal abnormalities and memory deficits: new evidence of a strong pathophysiological link in schizophrenia. Brain Res Rev 54(1):92–112
Brewer WJ, Francey SM, Wood SJ, Jackson HJ, Pantelis C, Phillips LJ, Yung AR, Anderson VA, McGorry PD (2005) Memory impairments identified in people at ultra-high risk for psychosis who later develop first-episode psychosis. Am J Psychiatry 162(1):71
Broome MR, Matthiasson P, Fusar-Poli P, Woolley JB, Johns LC, Tabraham P, Bramon E, Valmaggia L, Williams SCR, Brammer MJ, Chitnis X, McGuire PK (2009) Neural correlates of executive function and working memory in the "at-risk mental state". Br J Psychiatry 194(1):25–33
Broome MR, Fusar-Poli P, Matthiasson P, Woolley JB, Valmaggia L, Johns LC, Tabraham P, Bramon E, Williams SCR, Brammer MJ, Chitnis X, Zelaya F, McGuire PK (2010) Neural correlates of visuospatial working memory in the "at-risk mental state". Psychol Med 40(12):1987–1999
Brüne M, Ozgürdal S, Ansorge N, von Reventlow HG, Peters S, Nicolas V, Tegenthoff M, Juckel G, Lissek S (2011) An fMRI study of "theory of mind" in at-risk states of psychosis: comparison with manifest schizophrenia and healthy controls. Neuroimage 55(1):329–337
Cannon M, Jones P, Gilvarry C, Rifkin L, McKenzie K, Foerster A, Murray RM (1997) Premorbid social functioning in schizophrenia and bipolar disorder: similarities and differences. Am J Psychiatry 154(11):1544–1550
Cannon TD, Cadenhead K, Cornblatt B, Woods SW, Addington J, Walker E, Seidman LJ, Perkins D, Tsuang M, McGlashan T, Heinssen R (2008) Prediction of psychosis in youth at high clinical risk: a multisite longitudinal study in North America. Arch Gen Psychiatry 65(1):28
Carletti F, Woolley JB, Bhattacharyya S, Perez-Iglesias R, Poli PF, Valmaggia L, Broome MR, Bramon E, Johns L, Giampietro V, Williams SCR, Barker GJ, McGuire PK (2012) Alterations in white matter evident before the onset of psychosis. Schizophr Bull 38(6):1170–1179
Cavanna AE, Trimble MR (2006) The precuneus: a review of its functional anatomy and behavioural correlates. Brain 129(Pt 3):564–583
Chan RCK, Di X, McAlonan GM, Gong Q (2011) Brain anatomical abnormalities in high-risk individuals, first-episode, and chronic schizophrenia: an activation likelihood estimation meta-analysis of illness progression. Schizophr Bull 37(1):177–188
Cheung V, Cheung C, McAlonan GM, Deng Y, Wong JG, Yip L, Tai KS, Khong PL, Sham P, Chua SE (2008) A diffusion tensor imaging study of structural dysconnectivity in never-medicated, first-episode schizophrenia. Psychol Med 38(6):877–885
Cheung V, Chiu CPY, Law CW, Cheung C, Hui CLM, Chan KKS, Sham PC, Deng MY, Tai KS, Khong P-L, McAlonan GM, Chua S-E, Chen E (2011) Positive symptoms and white matter microstructure in never-medicated first episode schizophrenia. Psychol Med 41(8):1709–1719
Crespo-Facorro B, Paradiso S, Andreasen NC, O'Leary DS, Watkins GL, Ponto LL, Hichwa RD (2001) Neural mechanisms of anhedonia in schizophrenia. JAMA: J Am Med Assoc 286(4):427
De Peri L, Crescini A, Deste G, Fusar-Poli P, Sacchetti E, Vita A (2012) Brain structural abnormalities at the onset of schizophrenia and bipolar disorder: a meta-analysis of controlled magnetic resonance imaging studies. Curr Pharm Des 18(4):486–494
Dekker N, Schmitz N, Peters BD, van Amelsvoort TA, Linszen DH, de Haan L (2010) Cannabis use and callosal white matter structure and integrity in recent-onset schizophrenia. Psychiatry Res 181(1):51–56
Di X, Chan RCK, Gong Q-Y (2009) White matter reduction in patients with schizophrenia as revealed by

voxel-based morphometry: an activation likelihood estimation meta-analysis. Prog Neuropsychopharmacol Biol Psychiatry 33(8):1390–1394
Du M-Y, Wu Q-Z, Yue Q, Li J, Liao Y, Kuang W-H, Huang X-Q, Chan RCK, Mechelli A, Gong Q-Y (2012) Voxelwise meta-analysis of gray matter reduction in major depressive disorder. Prog Neuropsychopharmacol Biol Psychiatry 36(1):11–16
Eastvold AD, Heaton RK, Cadenhead KS (2007) Neurocognitive deficits in the (putative) prodrome and first episode of psychosis. Schizophr Res 93(1–3):266–277
Ellison-Wright I, Bullmore E (2009) Meta-analysis of diffusion tensor imaging studies in schizophrenia. Schizophr Res 108(1–3):3–10
Ellison-Wright I, Bullmore E (2010) Anatomy of bipolar disorder and schizophrenia: a meta-analysis. Schizophr Res 117(1):1–12
Ellison-Wright I, Glahn DC, Laird AR, Thelen SM, Bullmore E (2008) The anatomy of first-episode and chronic schizophrenia: an anatomical likelihood estimation meta-analysis. Am J Psychiatry 165(8):1015–1023
Federspiel A, Begré S, Kiefer C, Schroth G, Strik WK, Dierks T (2006) Alterations of white matter connectivity in first episode schizophrenia. Neurobiol Dis 22(3):702–709
Forbes NF, Carrick LA, McIntosh AM, Lawrie SM (2009) Working memory in schizophrenia: a meta-analysis. Psychol Med 39(6):889–905
Fusar-Poli P, Perez J, Broome M, Borgwardt S, Placentino A, Caverzasi E, Cortesi M, Veggiotti P, Politi P, Barale F, McGuire P (2007) Neurofunctional correlates of vulnerability to psychosis: a systematic review and meta-analysis. Neurosci Biobehav Rev 31(4):465–484
Fusar-Poli P, Borgwardt S, Crescini A, Deste G, Kempton MJ, Lawrie S, Mc Guire P, Sacchetti E (2011a) Neuroanatomy of vulnerability to psychosis: a voxel-based meta-analysis. Neurosci Biobehav Rev 35(5):1175–1185
Fusar-Poli P, Crossley N, Woolley J, Carletti F, Perez-Iglesias R, Broome M, Johns L, Tabraham P, Bramon E, McGuire P (2011b) White matter alterations related to P300 abnormalities in individuals at high risk for psychosis: an MRI-EEG study. J Psychiatry Neurosci 36(4):239–248
Fusar-Poli P, Bonoldi I, Yung AR, Borgwardt S, Kempton MJ, Valmaggia L, Barale F, Caverzasi E, McGuire P (2012a) Predicting psychosis: meta-analysis of transition outcomes in individuals at high clinical risk. Arch Gen Psychiatry 69(3):220–229
Fusar-Poli P, Radua J, McGuire P, Borgwardt S (2012b) Neuroanatomical maps of psychosis onset: voxel-wise meta-analysis of antipsychotic-naive VBM studies. Schizophr Bull 38(6):1297–1307
Gasparotti R, Valsecchi P, Carletti F, Galluzzo A, Liserre R, Cesana B, Sacchetti E (2009) Reduced fractional anisotropy of corpus callosum in first-contact, antipsychotic drug-naive patients with schizophrenia. Schizophr Res 108(1–3):41–48
Glahn DC, Ragland JD, Abramoff A, Barrett J, Laird AR, Bearden CE, Velligan DI (2005) Beyond hypofrontality: a quantitative meta-analysis of functional neuroimaging studies of working memory in schizophrenia. Hum Brain Mapp 25(1):60–69
Hashimoto R, Lee K, Preus A, McCarley RW, Wible CG (2010) An fMRI study of functional abnormalities in the verbal working memory system and the relationship to clinical symptoms in chronic schizophrenia. Cereb Cortex 20(1):46–60
Haut KM, Lim KO, MacDonald A 3rd (2010) Prefrontal cortical changes following cognitive training in patients with chronic schizophrenia: effects of practice, generalization, and specificity. Neuropsychopharmacology 35(9):1850–1859
Ho B-C, Andreasen NC, Ziebell S, Pierson R, Magnotta V (2011) Long-term antipsychotic treatment and brain volumes: a longitudinal study of first-episode schizophrenia. Arch Gen Psychiatry 68(2):128–137
Huber G (1955) The pneumoencephalogram at the onset of schizophrenic disease. Archiv für Psychiatrie und Nervenkrankheiten, vereinigt mit Zeitschrift für die gesamte Neurologie und Psychiatrie 193(4):406–426
Huber G (1961) Clinical and neuroradiological research on chronic schizophrenics. Nervenarzt 32:7–15
Hulshoff Pol HE, Kahn RS (2008) What happens after the first episode? A review of progressive brain changes in chronically ill patients with schizophrenia. Schizophr Bull 34(2):354–366
Jacobson S, Kelleher I, Harley M, Murtagh A, Clarke M, Blanchard M, Connolly C, O'Hanlon E, Garavan H, Cannon M (2010) Structural and functional brain correlates of subclinical psychotic symptoms in 11–13year old schoolchildren. Neuroimage 49(2):1875–1885
Johnstone EC, Crow TJ, Frith CD, Husband J, Kreel L (1976) Cerebral ventricular size and cognitive impairment in chronic schizophrenia. Lancet 2(7992):924–926
Kapur S (2003) Psychosis as a state of aberrant salience: a framework linking biology, phenomenology, and pharmacology in schizophrenia. Am J Psychiatry 160(1):13
Karlsgodt KH, van Erp TGM, Poldrack RA, Bearden CE, Nuechterlein KH, Cannon TD (2008) Diffusion tensor imaging of the superior longitudinal fasciculus and working memory in recent-onset schizophrenia. Biol Psychiatry 63(5):512–518
Karlsgodt KH, Niendam TA, Bearden CE, Cannon TD (2009) White matter integrity and prediction of social and role functioning in subjects at ultra-high risk for psychosis. Biol Psychiatry 66(6):562–569
Kawashima T, Nakamura M, Bouix S, Kubicki M, Salisbury DF, Westin C-F, McCarley RW, Shenton ME (2009) Uncinate fasciculus abnormalities in recent onset schizophrenia and affective psychosis: a diffusion tensor imaging study. Schizophr Res 110(1–3):119–126
Kempton MJ, Geddes JR, Ettinger U, Williams SCR, Grasby PM (2008) Meta-analysis, database, and meta-regression of 98 structural imaging studies in bipolar disorder. Arch Gen Psychiatry 65(9):1017–1032

Kempton MJ, Stahl D, Williams SCR, DeLisi LE (2010) Progressive lateral ventricular enlargement in schizophrenia: a meta-analysis of longitudinal MRI studies. Schizophr Res 120(1–3):54–62

Kim DI, Manoach DS, Mathalon DH, Turner JA, Mannell M, Brown GG, Ford JM, Gollub RL, White T, Wible C, Belger A, Bockholt HJ, Clark VP, Lauriello J, O'Leary D, Mueller BA, Lim KO, Andreasen N, Potkin SG, Calhoun VD (2009) Dysregulation of working memory and default-mode networks in schizophrenia using independent component analysis, an fBIRN and MCIC study. Hum Brain Mapp 30(11):3795–3811

Knowles L, Sharma T (2004) Identifying vulnerability markers in prodromal patients: a step in the right direction for schizophrenia prevention. CNS Spectr 9:595–603

Krug A, Markov V, Sheldrick A, Krach S, Jansen A, Zerres K, Eggermann T, Stöcker T, Shah NJ, Kircher T (2009) The effect of the COMT val(158)met polymorphism on neural correlates of semantic verbal fluency. Eur Arch Psychiatry Clin Neurosci 259(8):459–465

Krug A, Markov V, Krach S, Jansen A, Zerres K, Eggermann T, Stöcker T, Shah NJ, Nöthen MM, Treutlein J, Rietschel M, Kircher T (2010) The effect of Neuregulin 1 on neural correlates of episodic memory encoding and retrieval. Neuroimage 53(3): 985–991

Kubicki M, McCarley R, Westin C-F, Park H-J, Maier S, Kikinis R, Jolesz FA, Shenton ME (2007) A review of diffusion tensor imaging studies in schizophrenia. J Psychiatr Res 41(1–2):15–30

Kurtz MM (2005) Neurocognitive impairment across the lifespan in schizophrenia: an update. Schizophr Res 74(1):15–26

Lee J, Park S (2005) Working memory impairments in schizophrenia: a meta-analysis. J Abnorm Psychol 114(4):599–611

Lee K-H, Brown WH, Egleston PN, Green RDJ, Farrow TFD, Hunter MD, Parks RW, Wilkinson ID, Spence SA, Woodruff PWR (2006) A functional magnetic resonance imaging study of social cognition in schizophrenia during an acute episode and after recovery. Am J Psychiatry 163(11):1926–1933

Lencz T, Smith CW, McLaughlin D, Auther A, Nakayama E, Hovey L, Cornblatt BA (2006) Generalized and specific neurocognitive deficits in prodromal schizophrenia. Biol Psychiatry 59(9):863–871

Leung M, Cheung C, Yu K, Yip B, Sham P, Li Q, Chua S, McAlonan G (2011) Gray matter in first-episode schizophrenia before and after antipsychotic drug treatment. Anatomical likelihood estimation meta-analyses with sample size weighting. Schizophr Bull 37(1):199–211

Luck D, Buchy L, Czechowska Y, Bodnar M, Pike GB, Campbell JSW, Achim A, Malla A, Joober R, Lepage M (2011) Fronto-temporal disconnectivity and clinical short-term outcome in first episode psychosis: a DTI-tractography study. J Psychiatr Res 45(3):369–377

Malmberg A, Lewis G, David A, Allebeck P (1998) Premorbid adjustment and personality in people with schizophrenia. Br J Psychiatry 172:308–313, discussion 314–315

Marwick K, Hall J (2008) Social cognition in schizophrenia: a review of face processing. Br Med Bull 88(1):43–58

McGuire PK, Frith CD (1996) Disordered functional connectivity in schizophrenia. Psychol Med 26(4):663–667

Meda SA, Stevens MC, Folley BS, Calhoun VD, Pearlson GD (2009) Evidence for anomalous network connectivity during working memory encoding in schizophrenia: an ICA based analysis. PLoS ONE 4(11):e7911

Meisenzahl EM, Koutsouleris N, Gaser C, Bottlender R, Schmitt GJE, McGuire P, Decker P, Burgermeister B, Born C, Reiser M, Möller H-J (2008) Structural brain alterations in subjects at high-risk of psychosis: a voxel-based morphometric study. Schizophr Res 102(1–3):150–162

Moher D, Liberati A, Tetzlaff J, Altman DG (2009) Preferred reporting items for systematic reviews and meta-analyses: the PRISMA statement. BMJ 339:b2535, Clinical research ed

Moncrieff J, Leo J (2010) A systematic review of the effects of antipsychotic drugs on brain volume. Psychol Med 40(9):1409–1422

Morey RA, Inan S, Mitchell TV, Perkins DO, Lieberman JA, Belger A (2005) Imaging frontostriatal function in ultra-high-risk, early, and chronic schizophrenia during executive processing. Arch Gen Psychiatry 62(3):254–262

Murray CJ, Lopez AD (1997) Alternative projections of mortality and disability by cause 1990–2020: Global Burden of Disease Study. Lancet 349(9064):1498–1504

Navari S, Dazzan P (2009) Do antipsychotic drugs affect brain structure? A systematic and critical review of MRI findings. Psychol Med 39(11):1763–1777

Nickl-Jockschat T, Stöcker T, Markov V, Krug A, Huang R, Schneider F, Habel U, Zerres K, Nöthen MM, Treutlein J, Rietschel M, Shah NJ, Kircher T (2012) The impact of a dysbindin schizophrenia susceptibility variant on fiber tract integrity in healthy individuals: a TBSS-based diffusion tensor imaging study. Neuroimage 60(2):847–853

Nieman D, Becker H, van de Fliert R, Plat N, Bour L, Koelman H, Klaassen M, Dingemans P, Niessen M, Linszen D (2007) Antisaccade task performance in patients at ultra high risk for developing psychosis. Schizophr Res 95(1–3):54–60

Olabi B, Ellison-Wright I, McIntosh AM, Wood SJ, Bullmore E, Lawrie SM (2011) Are there progressive brain changes in schizophrenia? A meta-analysis of structural magnetic resonance imaging studies. Biol Psychiatry 70(1):88–96

Palaniyappan L, Liddle PF (2012) Differential effects of surface area, gyrification and cortical thickness on voxel based morphometric deficits in schizophrenia. Neuroimage 60(1):693–699

Palaniyappan L, Balain V, Liddle PF (2012) The neuroanatomy of psychotic diathesis: a meta-analytic review. J Psychiatr Res 46(10):1249–1256

Pantelis C, Velakoulis D, McGorry PD, Wood SJ, Suckling J, Phillips LJ, Yung AR, Bullmore ET, Brewer W, Soulsby B, Desmond P, McGuire PK (2003) Neuroanatomical abnormalities before and after onset of psychosis: a cross-sectional and longitudinal MRI comparison. Lancet 361(9354):281–288

Pantelis C, Yücel M, Wood SJ, Velakoulis D, Sun D, Berger G, Stuart GW, Yung A, Phillips L, McGorry PD (2005) Structural brain imaging evidence for multiple pathological processes at different stages of brain development in schizophrenia. Schizophr Bull 31(3):672–696

Pauly K, Seiferth NY, Kellermann T, Backes V, Vloet TD, Shah NJ, Schneider F, Habel U, Kircher TT (2008) Cerebral dysfunctions of emotion–cognition interactions in adolescent-onset schizophrenia. J Am Acad Child Adolesc Psychiatry 47(11):1299–1310

Pauly K, Seiferth NY, Kellermann T, Ruhrmann S, Daumann B, Backes V, Klosterkötter J, Shah NJ, Schneider F, Kircher T, Habel U (2010) The interaction of working memory and emotion in persons clinically at risk for psychosis: an fMRI pilot study. Schizophr Res 120(1–3):167–176

Pérez-Iglesias R, Tordesillas-Gutiérrez D, Barker GJ, McGuire PK, Roiz-Santiañez R, Mata I, de Lucas EM, Quintana F, Vazquez-Barquero JL, Crespo-Facorro B (2010) White matter defects in first episode psychosis patients: a voxelwise analysis of diffusion tensor imaging. Neuroimage 49(1):199–204

Peters BD, de Haan L, Dekker N, Blaas J, Becker HE, Dingemans PM, Akkerman EM, Majoie CB, van Amelsvoort T, den Heeten GJ, Linszen DH (2008) White matter fibertracking in first-episode schizophrenia, schizoaffective patients and subjects at ultra-high risk of psychosis. Neuropsychobiology 58(1):19–28

Peters BD, Schmitz N, Dingemans PM, van Amelsvoort TA, Linszen DH, de Haan L, Majoie CB, den Heeten GJ (2009) Preliminary evidence for reduced frontal white matter integrity in subjects at ultra-high-risk for psychosis. Schizophr Res 111(1–3):192–193

Pfeifer JC, Welge J, Strakowski SM, Adler CM, DelBello MP (2008) Meta-analysis of amygdala volumes in children and adolescents with bipolar disorder. J Am Acad Child Adolesc Psychiatry 47(11):1289–1298

Pflueger MO, Gschwandtner U, Stieglitz R-D, Riecher-Rössler A (2007) Neuropsychological deficits in individuals with an at risk mental state for psychosis - working memory as a potential trait marker. Schizophr Res 97(1–3):14–24

Price G, Cercignani M, Parker GJM, Altmann DR, Barnes TRE, Barker GJ, Joyce EM, Ron MA (2008) White matter tracts in first-episode psychosis: a DTI tractography study of the uncinate fasciculus. Neuroimage 39(3):949–955

Pukrop R, Ruhrmann S, Schultze-Lutter F, Bechdolf A, Brockhaus-Dumke A, Klosterkötter J (2007) Neurocognitive indicators for a conversion to psychosis: comparison of patients in a potentially initial prodromal state who did or did not convert to a psychosis. Schizophr Res 92(1–3):116–125

Radua J, Mataix-Cols D (2012) Meta-analytic methods for neuroimaging data explained. Biol Mood Anxiety Disord 2(1):6

Rasetti R, Weinberger DR (2011) Intermediate phenotypes in psychiatric disorders. Curr Opin Genet Dev 21(3):340–348

Rasser PE, Johnston P, Lagopoulos J, Ward PB, Schall U, Thienel R, Bender S, Toga AW, Thompson PM (2005) Functional MRI BOLD response to tower of London performance of first-episode schizophrenia patients using cortical pattern matching. Neuroimage 26(3):941–951

Roiser JP, Howes OD, Chaddock CA, Joyce EM, McGuire P (2013) Neural and behavioral correlates of aberrant salience in individuals at risk for psychosis. Schizophr Bull 39(6):1328–1336

Scherk H, Falkai P (2006) Effects of antipsychotics on brain structure. Curr Opin Psychiatry 19(2):145–150

Schneider F, Habel U, Reske M, Kellermann T, Stöcker T, Shah NJ, Zilles K, Braus DF, Schmitt A, Schlösser R, Wagner M, Frommann I, Kircher T, Rapp A, Meisenzahl E, Ufer S, Ruhrmann S, Thienel R, Sauer H, Henn FA, Gaebel W (2007) Neural correlates of working memory dysfunction in first-episode schizophrenia patients: an fMRI multi-center study. Schizophr Res 89(1–3):198–210

Seiferth NY, Pauly K, Habel U, Kellermann T, Jon Shah N, Ruhrmann S, Klosterkötter J, Schneider F, Kircher T (2008) Increased neural response related to neutral faces in individuals at risk for psychosis. Neuroimage 40(1):289–297

Shea BJ, Grimshaw JM, Wells GA, Boers M, Andersson N, Hamel C, Porter AC, Tugwell P, Moher D, Bouter LM (2007) Development of AMSTAR: a measurement tool to assess the methodological quality of systematic reviews. BMC Med Res Methodol 7:10

Shea BJ, Hamel C, Wells GA, Bouter LM, Kristjansson E, Grimshaw J, Henry DA, Boers M (2009) AMSTAR is a reliable and valid measurement tool to assess the methodological quality of systematic reviews. J Clin Epidemiol 62(10):1013–1020

Shepherd AM, Laurens KR, Matheson SL, Carr VJ, Green MJ (2012) Systematic meta-review and quality assessment of the structural brain alterations in schizophrenia. Neurosci Biobehav Rev 36(4):1342–1356

Smieskova R, Fusar-Poli P, Allen P, Bendfeldt K, Stieglitz RD, Drewe J, Radue EW, McGuire PK, Riecher-Rössler A, Borgwardt SJ (2009) The effects of antipsychotics on the brain: what have we learnt from structural imaging of schizophrenia?–a systematic review. Curr Pharm Des 15(22):2535–2549

Smith CW, Park S, Cornblatt B (2006) Spatial working memory deficits in adolescents at clinical high risk for schizophrenia. Schizophr Res 81(2–3):211–215

Sun J, Maller JJ, Guo L, Fitzgerald PB (2009) Superior temporal gyrus volume change in schizophrenia: a review on region of interest volumetric studies. Brain Res Rev 61(1):14–32

Surguladze S, Russell T, Kucharska-Pietura K, Travis MJ, Giampietro V, David AS, Phillips ML (2006) A reversal of the normal pattern of parahippocampal response to

neutral and fearful faces is associated with reality distortion in schizophrenia. Biol Psychiatry 60(5):423–431
Szeszko PR, Robinson DG, Ashtari M, Vogel J, Betensky J, Sevy S, Ardekani BA, Lencz T, Malhotra AK, McCormack J, Miller R, Lim KO, Gunduz-Bruce H, Kane JM, Bilder RM (2008) Clinical and neuropsychological correlates of white matter abnormalities in recent onset schizophrenia. Neuropsychopharmacology 33(5):976–984
Tan H-Y, Choo W-C, Fones CSL, Chee MWL (2005) fMRI study of maintenance and manipulation processes within working memory in first-episode schizophrenia. Am J Psychiatry 162(10):1849–1858
Tan H-Y, Callicott JH, Weinberger DR (2009) Prefrontal cognitive systems in schizophrenia: towards human genetic brain mechanisms. Cogn Neuropsychiatr 14(4–5):277–298
Taylor SF, Chen AC, Tso IF, Liberzon I, Welsh RC (2011) Social appraisal in chronic psychosis: role of medial frontal and occipital networks. J Psychiatr Res 45(4):526–538
Turkeltaub PE, Eden GF, Jones KM, Zeffiro TA (2002) Meta-analysis of the functional neuroanatomy of single-word reading: method and validation. Neuroimage 16(3 Pt 1):765–780
Van Haren NEM, Schnack HG, Cahn W, van den Heuvel MP, Lepage C, Collins L, Evans AC, Hulshoff Pol HE, Kahn RS (2011) Changes in cortical thickness during the course of illness in schizophrenia. Arch Gen Psychiatry 68(9):871–880
Vita A, De Peri L, Silenzi C, Dieci M (2006) Brain morphology in first-episode schizophrenia: a meta-analysis of quantitative magnetic resonance imaging studies. Schizophr Res 82(1):75–88
White T, Magnotta VA, Bockholt HJ, Williams S, Wallace S, Ehrlich S, Mueller BA, Ho B-C, Jung RE, Clark VP, Lauriello J, Bustillo JR, Schulz SC, Gollub RL, Andreasen NC, Calhoun VD, Lim KO (2011) Global white matter abnormalities in schizophrenia: a multisite diffusion tensor imaging study. Schizophr Bull 37(1):222–232
Yu K, Cheung C, Leung M, Li Q, Chua S, McAlonan G (2010) Are bipolar disorder and schizophrenia neuroanatomically distinct? An anatomical likelihood meta-analysis. Front Hum Neurosci 4:189
Yung AR, Phillips LJ, McGorry PD, McFarlane CA, Francey S, Harrigan S, Patton GC, Jackson HJ (1998) Prediction of psychosis. A step towards indicated prevention of schizophrenia. Br J Psychiatry Suppl 172(33):14–20
Yung AR, Phillips LJ, Yuen HP, Francey SM, McFarlane CA, Hallgren M, McGorry PD (2003) Psychosis prediction: 12-month follow up of a high-risk. Schizophr Res 60(1):21–32

14 MR Neuroimaging of Depression

Thomas Frodl

Abbreviations

ACC	Anterior cingulate cortex
BDNF	Brain-derived neurotrophic factor
BOLD	Blood oxygen level dependent
CBT	Cognitive behavioral therapy
DG	Dentate gyrus
dmPFC	Dorsomedial PFC
DTI	Diffusion tensor imaging
FA	Fractional anisotropy
fMRI	Functional magnetic resonance imaging
FOF	Fronto-occipital fasciculus
GC	Glucocorticoids
GILZ	Glucocorticoid-induced leucine zipper
HPA	Hypothalamic–pituitary–adrenal axis
MDD	Major depressive disorder
pCC	Posterior cingulate cortex
PET	Positron emission tomography
PFC	Prefrontal cortex
SEM	Structural equation modeling
SFG	Superior frontal gyrus
sgACC	Subgenual ACC
SLF	Superior longitudinal fasciculus
SMA	Supplementary motor area

T. Frodl, MD, MA
Department of Psychiatry, University of Dublin, Trinity College, Dublin, Ireland

Department of Psychiatry, University of Regensburg, Regensburg, Germany
e-mail: frodlt@tcd.ie

14.1 Introduction

Until only a few years ago, the adult brain was considered to be an organ with a fixed structure, unable to remodel or repair itself. However, recent research shows that both structural and physiological changes occur in the adult nervous system, some arising as a result of the individual's interaction with the surrounding environment and some from internal adaptation, also interacting with genetic factors. The best-characterized and most studied examples of neuroplasticity are the molecular and cellular adaptations underlying learning and memory (Feng et al. 2001; Eichenbaum and Harris 2000; Lisman 1999). In this context, a major breakthrough was the finding that, in *Aplysia californica*, long-term memory requires new protein synthesis (Kandel 2001). In human subjects also, studies provide evidence for structural brain changes associated with learning. In healthy young adults, for example, working memory capacity, traditionally considered to be constant, was found to be increased after a 5-week training program (Olesen et al. 2004). Consistent with this, increased hippocampal and parietal gray matter brain volumes have been detected in young adults after participation in intensive learning programs (Draganski et al. 2006).

On the other hand, chronic social stress has been shown to induce glucocorticoid-mediated pyramidal dendritic retraction in the hippocampus and changes in dendritic arborization in the prefrontal cortex (PFC) (Woolley et al. 1990; Magarinos et al. 1996; Wellman 2001; Kole et al. 2004), which

C. Mulert, M.E. Shenton (eds.), *MRI in Psychiatry*,
DOI 10.1007/978-3-642-54542-9_14, © Springer Berlin Heidelberg 2014

might be associated with the behavioral manifestations of stress-related disorders like major depressive disorder (MDD) (Macqueen and Frodl 2011). There is mounting evidence that specific neuronal circuits, particularly in the developing brain, are damaged by environmental stress inducing changes in the hypothalamic–pituitary–adrenal axis (HPA) and inflammatory pathways (Krishnan and Nestler 2008). Experimental studies have shown that stress or cortisol administration may lead to depressive-like states and atrophy of neurons in the hippocampus (Duman 2002) and that therapy with antidepressants reverses these changes (Santarelli et al. 2003). Moreover, chronic hypercortisolism has been shown to enhance tryptophan breakdown in the brain and induce neurodegenerative changes (Capuron and Miller 2011). Other research has shown that both physiological and psychological stress can induce increased production of proinflammatory mediators that can stimulate tryptophan catabolism in the brain (Myint et al. 2012), with consequences on neurotransmitter metabolism, neuroendocrine function, synaptic transmission, and neurocircuits that regulate mood, motor activity, motivation, anxiety, and alarm (Capuron and Miller 2011).

Major depressive disorder (MDD) is a disease affecting brain structure and function and where the ultimate goal of therapy is in remission. Thus, neuroimaging is a method that could be helpful for diagnosis, therapy evaluation, and therapy guidance. In this chapter, the aim is firstly to describe the main imaging findings obtained with structural MRI, diffusion tensor imaging, and functional MRI techniques and secondly to describe the current knowledge about how neuroimaging might be helpful to evaluate changes during therapy and for therapy guidance.

14.2 Structural Imaging in Major Depressive Disorder

14.2.1 Overview

Studies in humans strengthen the evidence that MDD is associated with structural changes. Many structural imaging studies have reported that the hippocampus is small in patients with MDD. A recent meta-analysis of hippocampal volumes in patients with MDD confirmed that patients had hippocampal volumes that were approximately 4–6 % smaller than matched control subjects in the left and right hippocampus. The analysis included 1,167 patients and 1,088 control subjects, across a wide range of ages from pediatric to geriatric populations (McKinnon et al. 2009). Conclusions from this meta-analysis were consistent with the findings of earlier meta-analyses of hippocampal volume in patients with MDD (Campbell et al. 2004; Videbech and Ravnkilde 2004). A recent meta-analysis over 226 studies found areas affected in MDD to be the hippocampus, basal ganglia, and orbitofrontal cortex, in particular the gyrus rectus (Kempton et al. 2011). The abovementioned associations between glucocorticoids and stress and neuronal damage in the hippocampus indicate that the neurotoxic effects of glucocorticoids (GC) on the hippocampus can be visualized in terms of overall volume changes. Evidence from neuroimaging, neuropathological, and lesion analysis studies further implicates limbic–cortical–striatal–pallidal–thalamic circuits, including the prefrontal cortex, amygdala, ventromedial striatum, mediodorsal and midline thalamic nuclei, and ventral pallidum, in the pathophysiology of mood disorders (Miller et al. 2010). In line with the glucocorticoid cascade theory are some longitudinal imaging studies. In a longitudinal study on 30 patients with MDD and 30 healthy controls, it was demonstrated that a negative clinical outcome with more relapses and a chronic course during a 3-year follow-up was associated with hippocampal, amygdala, anterior cingulate cortex, and dorsomedial prefrontal cortex volume decline (Frodl et al. 2008d). In a long-term follow-up study, it was evaluated whether any possible difference in hippocampal volume and brain structure between depressed inpatients and healthy controls at inclusion disappeared over an 11-year period when the patients were in remission. At baseline, patients had smaller volumes in the right and left superior and middle temporal gyri, medulla, and body of the right hippocampus. At follow-up, there were no significant local brain differences between patients and controls.

In a group of 19 patients and 19 controls who were investigated at baseline and follow-up, no significant hippocampal volume differences were detected in this study (Ahdidan et al. 2011). While the cross-sectional parts of the study are well powered, for the longitudinal part of the study, a larger sample would have been desirable, since a sample of 19 seems not to have enough power to detect small changes in brain structure. Moreover, another study showed that during successful treatment brain structures like the left inferior frontal cortex, right fusiform gyrus, and right cerebellum increased in size (Lai and Hsu 2011). On the other hand, smaller hippocampal volumes were also found to be predictive of a poor clinical outcome in 1- and 3-year follow-up studies and also for response to a course of antidepressant therapy (Frodl et al. 2004a; Frodl et al. 2008a; MacQueen and Frodl 2011). Therefore, a predisposition to depression might be associated with smaller hippocampal volumes, which might further decline during the course of a chronic depression, but normalize during remission. However, this is speculative at this stage and needs to be investigated in studies with longitudinal designs.

14.2.2 Causes for Structural Changes

Numerous factors like environmental life stressors, alcohol use, genetic influences, and age have been shown to influence brain structure. Here the focus will be on the main factors relevant for MDD. Firstly, life stressors were found to be associated with structural changes in patients with MDD. It was reported that patients with MDD with a history of emotional neglect during childhood had reduced left hippocampal white matter compared to those without a history of emotional neglect, but no significant differences were detected in the whole hippocampus. An effect of childhood adversity on hippocampal structure was also shown in 35 healthy controls and 22 unaffected first-degree relatives of patients with MDD. Moreover, in 30 patients with MDD this association between early life adversity and hippocampal volumes was modulated by parental history of depression (Rao et al. 2010). Smaller left hippocampal volumes in subjects with childhood maltreatment were later replicated by a further study in patients with MDD (Vythilingam et al. 2002). The finding that childhood maltreatment was associated with smaller hippocampal head volumes also in 20 unaffected first-degree relatives of patients with MDD is very interesting (Carballedo et al. 2012). There is an anterior–posterior gradient in the proportional volume of each subfield in the head, body, and tail of the hippocampus. Higher proportions of the CA1–3 and subiculum are found in the hippocampal head, whereas the hippocampal body includes the greatest proportion of the dentate gyrus (DG) (Malykhin et al. 2010).

In line with an impact from life stressors are ideas of an association between inflammation, the stress system, and structural brain changes. Preclinical and clinical MRI research suggests that neuroinflammation in MDD might be associated with structural and functional anomalies in various regions of the CNS. Recently, it was confirmed that IL-6 and CRP had a significant effect on the left and right hippocampus in 40 patients with MDD, independently of demographic variables. Moreover, there was an association between IL-6 and hippocampal volumes in healthy controls. Smaller hippocampal volumes were found in MDD patients with lower expression of glucocorticoid-induced leucine zipper (GILZ) mRNA or SGK-1 mRNA, respectively, as markers of reduced activation of the glucocorticoid system compared to those with higher serum and glucocorticoid-induced kinase (GILZ) mRNA expression (Frodl et al. 2012a). However, these findings need replication in a larger sample in order to confirm the relationships between inflammation, stress hormone system, and brain structure/function. Lower BDNF expression, increased IL-6 expression, and increased cortisol levels all significantly and independently predicted a smaller left hippocampal volume in first-episode psychosis patients (Mondelli et al. 2011).

A recent review including 20 studies about the association between cortisol measurements and brain structure in particular hippocampal volumes showed that smaller hippocampal volumes are associated with increased cortisol secretion during the day. Since MDD is associated with alterations in the HPA axis, changes in cortisol system may indeed be a reason for alterations in brain structure. However, to date there are only a few studies to determine how cortisol changes are associated with brain structure (Frodl and O'Keane 2013).

Hippocampal size has a moderate heritability (Lyons et al. 2001; Gilbertson et al. 2002). An association between heritability and brain size is supported by twin studies (Bartley et al. 1997) and a wide range of imaging genetic studies (Frodl et al. 2008a, b, c, d). A recent study examined hippocampal volumes in 112 unaffected mono- and dizygotic co-twins of twins affected or not affected by depression (Baare et al. 2010). The co-twins of affected twins had smaller hippocampal volumes than co-twins of unaffected twins; interestingly the effect was most pronounced among dizygotic twins. The investigators acknowledge, however, that the small hippocampal volumes in the unaffected dizygotic twins of affected co-twins may have resulted from shared environmental factors. These results are also consistent with another twin study in which twins with high trait anxiety and depression had gray matter reductions in the left posterior HC when compared to a less anxious co-twin (de Geus et al. 2007).

Alternative phenotypic markers for genetic association studies that are more closely related to the underlying neurobiology of the disease are increasingly being used for genetic association studies. This partly derives from a need to address problems of clinical heterogeneity in psychiatric disorders and partly from the need to delineate the functional consequences of identified risk variants at the level of brain structure and function. Imaging genetics facilitates an elucidation of the impact of genes at the level of the brain that can then be extended to the pathophysiology of the disease.

Elimination of 5-HT from the synaptic cleft is mediated by a single protein, the 5-HT transporter (5-HTT), which determines the size and duration of the serotonergic responses (Lesch and Mossner 1998). The promoter region of 5-HTT has a polymorphism that results in allelic variation of functional 5-HTT expression (Lesch et al. 1996). The long (l) allele is associated with production of more 5-HTT transcription than the short (s) allele and hence more functional 5-HT uptake than the s allele (Lesch et al. 1996). The sallele of the 5-HTT polymorphism is associated with anxiety, depression, and aggression-related personality traits in some reports (Lesch and Mossner 1998; Lesch et al. 1996). The importance of variation in 5-HTTLPR to hippocampal volume in patients with MDD has been noted in three published reports that utilized diallelic analysis and two reports using triallelic analysis (Frodl et al. 2004a, b, 2008a, b, c, d; Taylor et al. 2005).

Brain-derived neurotrophic factor (BDNF) regulates neuronal survival, migration, phenotypic differentiation, axonal and dendritic growth, and synapse formation (Huang and Reichardt 2001). BDNF is also a key regulator of synaptic plasticity and behavior (Lu 2003) and may be important for memory acquisition and consolidation (Tyler et al. 2002). A single-nucleotide polymorphism in the pro-domain of BDNF converts the 66th amino acid valine into methionine (Val66Met). This Val66Met polymorphism affects dendritic trafficking and synaptic localization of BDNF and impairs its secretion. The Val66Met SNP is associated with deficits in short-term episodic memory (Egan et al. 2003). Healthy Met-BDNF carriers have relatively small hippocampal volumes (Bueller et al. 2006; Pezawas et al. 2004); an effect of the Met-BDNF allele on hippocampal volume is also apparent in patients with MDD (Frodl et al. 2007a, b), although one recent report did not observe this association, but rather found larger differences for patients with the Val phenotype (Gonul et al. 2011). A study in healthy volunteers reported that subjects carrying the Met-BDNF allele have smaller hippocampal volumes when they have more subthreshold symptoms of depression and a higher extent of neuroticism (Joffe et al. 2009).

Gray matter volumes in temporal lobe regions including the HC were associated with glycogen synthase kinase-3 (GSK3) polymorphisms in a large sample of depressed patients and healthy controls (Inkster et al. 2009). The underlying neurobiological background of this finding is unknown; it is interesting, however, that GSK3 activity might be associated with therapeutic effects of antidepressants and lithium (Inkster et al. 2009). Thus, these first genetic imaging studies provide evidence that genetic polymorphisms are relevant for neurobiological mechanisms in MDD and for hippocampal volumes. These studies emphasize the potential importance of genetic influences on hippocampal volumes.

Genetic and environmental stress factors need to be considered together. Research demonstrates that genetic and environmental factors interact and increase the vulnerability for major depression as well as the impact on brain structure and function. In MDD samples there was an interaction between early life adversity and a genetic variant in the promoter region of the serotonin transporter (5-HTTLPR) (Frodl et al. 2010b) and the BDNF Val66Met polymorphism (Carballedo et al. 2013) on hippocampal volumes. Moreover, the tentative protective effect of allele of a variant in the BICC gene seemed to have been blocked by early life adversity showing structural changes in the hippocampus and functional brain alterations in patients with MDD (Bermingham et al. 2012). Evidence for the importance of gene by stress interactions derives also from a study in healthy controls using magnetic resonance imaging, genotyping, and self-reported life stress. It could be shown that life stress affected, as a function of serotonin transporter genotype, gray matter structural features as well as functional connectivity of the amygdala and hippocampus with a wide network of other regions (Canli et al. 2006). Another recent study in healthy controls could demonstrate that Met-allele carriers of the BDNF polymorphism have smaller hippocampal volumes when they have elevations in stress as well as when they show more neuroticism (Joffe et al. 2009). In a recent study of 89 healthy participants, significant interactions between BDNF genotype and early life stress were detected on hippocampal and amygdala volumes, heart rate, and working memory. With structural equation modeling (SEM), the explicit pathways were investigated through which BDNF genotype and early life stress interact to produce effects on brain structure, body arousal, and emotional stability and in turn predict alterations in symptoms and cognition. SEM suggested that the combination of Met carrier status and exposure to early life stress predicted reduced HC volumes and associated lateral prefrontal cortex volumes and, in turn, higher depression (Gatt et al. 2009).

Therefore, there is evidence that genetic polymorphisms and early life stress may contribute to hippocampal volumes prior to onset of illness. Repeated episodes of illness may further contribute to loss of hippocampal volume via stress toxicity, reduced neurotrophic factors, and excess in glutamatergic neurotransmission on hippocampal neurons. Speculatively, the changes in the hippocampus may then contribute to treatment resistance or chronicity and may further increase the vulnerability for the disease (Fig. 14.1).

14.3 Functional Imaging in MDD

fMRI studies examining neural responses to emotional stimuli in patients with major depression (MDD) indicated increased responses in the amygdala, anterior cingulate cortex (ACC), fusiform gyrus, putamen, and prefrontal cortical regions (Fu et al. 2004; Surguladze et al. 2005; Frodl et al. 2007a, b). Since many researchers assume that the depressive syndrome might arise from abnormal interactions between brain regions, functional neuroimaging studies have examined the connectivity of the neural network. With respect to connectivity a study in 15 unmedicated patients with major depression and 15 healthy volunteers found decreased correlations between ACC and limbic regions, which is consistent with the hypothesis that decreased cortical regulation of limbic activation in response to negative stimuli may be present in depression (Anand et al. 2005). Again the amygdala was negatively coupled with the ACC but also positively coupled bilaterally with medial temporal and ventral occipital regions in 19 unmedicated patients with major depression and 19 healthy volunteers (Chen et al. 2008). Studies on functional connectivity in patients with MDD receiving antidepressant medication achieved varied

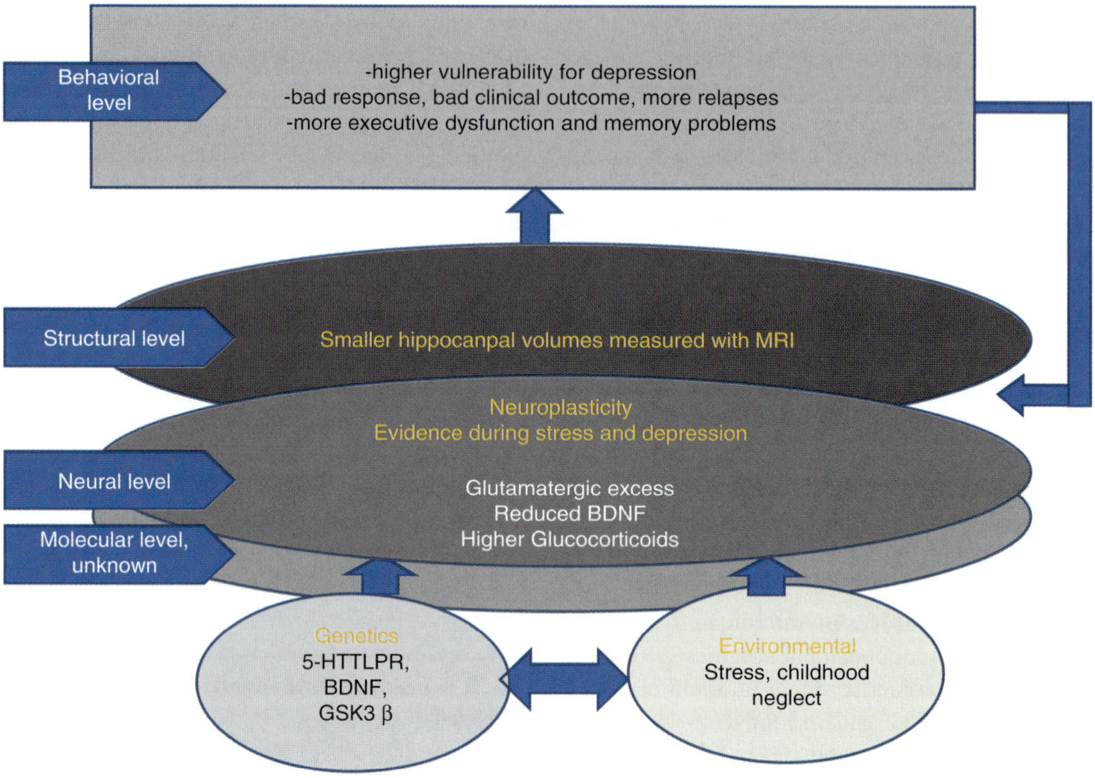

Fig. 14.1 Overview of the factors influencing small HC volumes and the consequence of HC volume changes in MDD: there is evidence that genetic polymorphisms and early life stress may contribute to HC volumes prior to onset of illness. Repeated episodes of illness may further contribute to loss of HC volume via stress toxicity, reduced neurotrophic factors, and excess in glutamatergic neurotransmission on hippocampal neurons. The underlying molecular basis is an ongoing matter of research. Speculatively, the changes in the HC may then contribute to treatment resistance or chronicity and may further increase the vulnerability for the disease. Consistent with this are studies reporting that small HC volumes predict poor short- and long-term responsiveness to treatment and a higher vulnerability. A more chronic course of disease and relapses seem to have further effects on neural connectivity and might result in further structural changes (Reprinted courtesy of Molecular Psychiatry, with permission (MacQueen and Frodl 2011))

results. The results indicated that a neural network consisting of the cingulate region, prefrontal cortical regions, amygdala, and subcortical regions may play key roles in MDD: compared to healthy controls, patients with depression showed increased functional connectivity between the amygdala, hippocampus, and caudate–putamen regions during emotion processing (Hamilton and Gotlib 2008) but significantly reduced amygdala–prefrontal connectivity (Dannlowski et al. 2009). Uncoupling of the prefrontal cortex and gyrus cinguli was found in 14 patients with MDD and 14 healthy controls during a verbal working memory task (Vasic et al. 2008). Resting-state fMRI showed that subgenual cingulate and thalamic functional connectivity was significantly increased in 20 patients with MDD than in 20 healthy controls (Greicius et al. 2009). In 25 drug-free patients with MDD compared to 15 healthy controls, dorsal anterior cingulate cortex, precuneus, and cerebellum activity showed less connectivity with the OFC in patients than in controls, while functional connectivity between the OFC and the right dorsolateral prefrontal cortex (DLPFC), right inferior frontal operculum, and left motor areas was increased in patients than in healthy controls (Frodl et al. 2010a, b, c).

Interest is growing in the use of resting-state fMRI, which does not require the use of a task and which has become a popular means of complementing the results of task-based fMRI studies. Resting-state fMRI allows for the examination of large-scale neural systems that exhibit spontaneous synchronous fluctuations during goal-directed and non-goal-directed behavior (Castellanos and Proal 2012). These low-frequency (<0.1 Hz) spontaneous fluctuations in blood oxygenated level dependent (BOLD) signal correlate with interactions between adjacent and nonadjacent brain areas that form spatially distributed networks of brain function (Raichle et al. 2001). Thus, functional connectivity is the observed correlation in spontaneous neural activity between brain areas at rest (Deco and Corbetta 2011). Several resting-state studies have found increased resting-state functional connectivity in the cognitive control network (Zhou et al. 2010; Sheline et al. 2010), increased default mode network (DMN) connectivity (Zhou et al. 2010; Grimm et al. 2009; Sheline et al. 2010), as well as increased functional connectivity in the affective network (Sheline et al. 2010).

14.4 Diffusion Imaging in MDD

Structural neuroimaging studies in MDD show volume reductions, and functional imaging studies indicate dysconnectivity in limbic and frontal brain regions. Whether white matter fiber bundles between limbic and frontal brain regions are altered can be investigated with diffusion tensor imaging (DTI). Recent DTI studies have suggested that there is a strong correlation between depression and reduced fractional anisotropy (FA) with the nature of this relationship representing a topic of great interest. A study comparing 13 patients with late-life depression to matched healthy controls found a reduction in FA in both the frontal and temporal lobes of depressed patients (Nobuhara et al. 2006). In addition, an inverse relationship was established between FA values and symptom severity (Nobuhara et al. 2006). Another recent study conducted in MDD patients using whole-brain DTI analysis found reduced FA in the left striatum, right cingulate cortex, and posterior body of the corpus callosum, areas of the brain believed to play an important role in emotional regulation (Kieseppä et al. 2010). Importantly, reductions in FA have also been associated with early life adversity in the form of disrupted maternal–infant attachment and correlate with an increased risk of both anxiety and depression (Coplan et al. 2010). A study comparing 12 maternally deprived adult male macaques to nine normally reared controls found significant reductions in FA in the anterior limb of the internal capsule in the maternally deprived macaques (Coplan et al. 2010). This is another brain region important in emotional regulation and is involved in the medial and the basolateral limbic circuits (Coplan et al. 2010). Thus, disruption of this region may alter functional connectivity between the frontal and temporal lobes conferring an increased risk of MDD (Coplan et al. 2010). Another study demonstrating the microstructural implications of early life adversity found significantly reduced FA within the genu of the corpus callosum among those exposed to high levels of early life stress (Paul et al. 2008). DTI represents the forefront of neuroimaging techniques in the characterization of microstructural alterations occurring in the brain both as an antecedent to and as a consequence of depression (Frodl et al. 2012b). These findings appear to be heterogeneous, and therefore, we conducted a meta-analysis over seven available studies including 188 patients with MDD and 221 healthy control patients with depression which showed decreased white matter FA values in the superior longitudinal fasciculus (SLF) and increased FA values in the fronto-occipital fasciculus (FOF) compared to controls. In conclusion, the meta-analysis revealed a significant reduction in FA values in the left SLF, which may ultimately play an important role in the pathology of depression. More research in larger samples is needed to particularly track changes during the disease course using DTI.

14.5 Monitoring Treatment Effects

Functional magnetic resonance imaging (fMRI) is becoming established as a method of visualizing the action of drugs on animal and human brain; in this context it is called pharmacoMRI or phMRI. Longitudinal functional imaging studies (without a comparison treatment arm) that investigated changes in the brains of patients with MDD after treatment with an antidepressant medication have found different results. Only a few studies have tried to elucidate medication effects on the structure of the brain. Increases in hippocampal volume following treatment with the antidepressants sertraline and paroxetine have been reported in patients with post-traumatic stress disorder (PTSD) (Vermetten et al. 2003; Bremner and Vermetten 2004). In MDD, the results are less clear, with some studies failing to find an association between antidepressant therapy and hippocampal volumes (Vythilingam et al. 2004). One study, however, found a trend toward increases in hippocampal volumes in patients who had declines in cortisol levels under pharmacotherapy with either amitriptyline or paroxetine (Colla et al. 2007). In a longitudinal 3-year follow-up study, it was found that continuous antidepressant treatment was associated with an increase in hippocampal volumes (Frodl et al. 2008a, b, c, d). In another longitudinal 8-year follow-up study, 31 remitted former inpatients with moderate to severe MDD showed no differences in hippocampal volume compared to 37 healthy comparison subjects (Hviid et al. 2010). However, a more recent study in 15 young patients with MDD has shown an association between antidepressant treatments and increased posterior hippocampal volumes correlating with improved performance in neuropsychological tests (Schermuly et al. 2011). Moreover, the antidepressant duloxetine increased gray matter volume in the left inferior frontal cortex, right fusiform gyrus, and right cerebellar areas after a 6-week therapy with 60 mg duloxetine in 15 patients with MDD and panic disorder compared to healthy controls (Lai and Hsu 2011). Studies in larger samples with controlled designs will be necessary to elucidate the effect of antidepressants and psychotherapy on brain structure.

fMRI showed evidence that altered brain function normalizes during effective treatment against depression. In 19 patients with major depressive disorder (MDD) treated with fluoxetine for 8 weeks, BOLD responses decreased significantly not only in the basal ganglia and thalamus regions but also in the amygdala, ACC, insula, precentral and postcentral gyrus, and inferior parietal lobule (Fu et al. 2004). Moreover, in this sample, treatment with fluoxetine was associated with a significant increase in functional coupling between the amygdala and subgenual ACC (Chen et al. 2008). Effects on amygdala activation were also found in other studies; for example, exaggerated left amygdala activation in 11 patients with major depression during a face-matching paradigm decreased to a normal activation level following treatment with the SSRI sertraline (Sheline et al. 2001). This effect may have been a direct effect of the antidepressant: after 21 days of treatment with escitalopram, 13 healthy volunteers without depression showed less activation in the amygdala when shown fearful faces than when shown control shapes (Arce et al. 2008). Only one earlier study in 12 patients with MDD found significantly increased activation in the left insular cortex after 2 weeks of treatment with venlafaxine and in the left anterior cingulate cortex after 8 weeks of treatment (Davidson et al. 2003). Thus, the majority of studies of SSRIs found decreased BOLD responses in cortical and subcortical brain regions. Patients receiving mirtazapine showed a significantly different pattern of changes: these patients showed more BOLD responses after 4 weeks' treatment than before in the left and right MCC and the left and right supplementary motor area (SMA), which indicates that mirtazapine had some stimulating effects in these brain areas (Frodl et al. 2011). A recent pharmacological MRI study in 45 healthy male volunteers who were randomly allocated to receive mirtazapine or placebo in a double-blind fashion found increased activation in the right orbitofrontal cortex after a single oral dose of mirtazapine (Vollm et al. 2006).

14.6 Predicting Therapy Response

Interestingly, the neural correlates of emotional processing in MDD may have predictive value in determining which patients will respond to treatment. The first evidence came from positron emission tomography (PET) which suggests that the subgenual ACC (sgACC) may be relevant in determining biomarkers for treatment response.

Differential metabolism in the sgACC predicts response to antidepressant treatment (Mayberg et al. 1997; Wu et al. 1999), an observation which has led to the utilization of this area as a target for deep brain stimulation in chronically treatment-resistant depression (Mayberg et al. 2005). Other regions are also interesting with respect to treatment response. Decreased metabolism in the insular cortex was found by Mayberg et al. (1999) and Kennedy et al. (2001) to be associated with response to treatment in patients with MDD. In addition, such functional alterations are not limited to the use of psychotropic medications for the treatment of MDD. Response to cognitive behavioral therapy (CBT) has been linked to metabolic increases in the hippocampus and poster cingulate (pCC) (BA 24) and decreases in the dorsal (BA 9/46), ventral (BA 47/11), and medial (BA 9/10/11) frontal cortices (Goldapple et al. 2004). Moreover, amygdala hyperactivation and ACC hypoactivation during fMRI predicted response to CBT (Siegle et al. 2006; Fu et al. 2004).

Moreover, decreases in glucose metabolism in ventral regions of the PFC (Brody et al. 1999; Kennedy et al. 2001) and increases in the temporal cortex (Buchsbaum et al. 1997; Brody et al. 2001) have been previously associated with the response to selective serotonin reuptake inhibitors (SSRIs). Pre- vs. posttreatment changes in the ventrolateral prefrontal and temporal cortex, posterior cingulate (BA 29), and putamen have also been reported with non-SSRI antidepressant pharmacotherapy (Davies et al. 2003; Martin et al. 2001). Furthermore, the caudate nucleus is discussed to be a trait marker of depression vulnerability, and caudate activation is elevated even in recovered depressed patients (Norbury et al. 2010). Before treatment, patients with MDD responding better to pharmacological treatment showed greater activation in the dorsomedial PFC (dmPFC), posterior cingulate cortex (pCC), and superior frontal gyrus (SFG) when viewing of negative emotional pictures compared with the resting condition. Activations in the caudate nucleus and insula contrasted for emotional compared to neutral stimuli were also associated with successful treatment (Samson et al. 2011). In a meta-analysis a difference in the activation of a limbic–cortical path of the hippocampus–OFC–ACC–lateral PFC network was found between responders and nonresponders to paroxetine and a combination of various medications. The pathway OFC–ACC–lateral PFC was also characteristic for responders treated with various antidepressants (Seminowicz et al. 2004). Indeed good OFC connectivity observed before treatment was shown to be associated with response to antidepressant (Lisiecka et al. 2011).

Responders may, therefore, be more likely to exhibit compensatory functional hyperactivation during fMRI tasks, and thus, responders might be likely to be identified by functional MRI. However, this link and the underlying neurochemical changes of functional hyperactivation require further exploration.

Conclusion

Improved MRI techniques and fMRI tasks that make possible the study of altered brain function in depression will likely be able to provide useful information on therapy effects and may, in the future, be able to predict therapy response. More longitudinal research during the disease course is needed to achieve these aims.

References

Ahdidan J, Hviid LB, Chakravarty MM, Ravnkilde B, Rosenberg R, Rodell A, Stodkilde-Jorgensen H, Videbech P (2011) Longitudinal MR study of brain structure and hippocampus volume in major depressive disorder. Acta Psychiatr Scand 123(3): 211–219

Anand A, Li Y, Wang Y, Wu J, Gao S, Bukhari L, Mathews VP, Kalnin A, Lowe MJ (2005) Activity and connectivity of brain mood regulating circuit in depression: a functional magnetic resonance study. Biol Psychiatry 57(10):1079–1088

Arce E, Simmons AN, Lovero KL, Stein MB, Paulus MP (2008) Escitalopram effects on insula and amygdala BOLD activation during emotional processing. Psychopharmacology (Berl) 196(4):661–672

Baare WF, Vinberg M, Knudsen GM, Paulson OB, Langkilde AR, Jernigan TL, Kessing LV (2010) Hippocampal volume changes in healthy subjects at risk of unipolar depression. J Psychiatr Res 44(10): 655–662

Bartley AJ, Jones DW, Weinberger DR (1997) Genetic variability of human brain size and cortical gyral patterns. Brain 120(Pt 2):257–269

Bermingham R, Carballedo A, Lisiecka D, Fagan A, Morris D, Fahey C, Donohoe G, Meaney J, Gill M, Frodl T (2012) Effect of genetic variant in BICC1 on functional and structural brain changes in depression. Neuropsychopharmacology 37(13):2855–2862

Bremner JD, Vermetten E (2004) Neuroanatomical changes associated with pharmacotherapy in posttraumatic stress disorder. Ann N Y Acad Sci 1032: 154–157

Brody AL, Saxena S, Silverman DH, Alborzian S, Fairbanks LA, Phelps ME, Huang SC, Wu HM, Maidment K, Baxter LR Jr (1999) Brain metabolic changes in major depressive disorder from pre- to post-treatment with paroxetine. Psychiatry Res 91(3): 127–139

Brody AL, Saxena S, Mandelkern MA, Fairbanks LA, Ho ML, Baxter LR (2001) Brain metabolic changes associated with symptom factor improvement in major depressive disorder. Biol Psychiatry 50(3):171–178

Buchsbaum MS, Wu J, Siegel BV, Hackett E, Trenary M, Abel L, Reynolds C (1997) Effect of sertraline on regional metabolic rate in patients with affective disorder. Biol Psychiatry 41(1):15–22

Bueller JA, Aftab M, Sen S, Gomez-Hassan D, Burmeister M, Zubieta JK (2006) BDNF Val66Met allele is associated with reduced hippocampal volume in healthy subjects. Biol Psychiatry 59(9):812–815

Campbell S, Marriott M, Nahmias C, MacQueen GM (2004) Lower hippocampal volume in patients suffering from depression: a meta-analysis. Am J Psychiatry 161(4):598–607

Canli T, Qiu M, Omura K, Congdon E, Haas BW, Amin Z, Herrmann MJ, Constable RT, Lesch KP (2006) Neural correlates of epigenesis. Proc Natl Acad Sci U S A 103(43):16033–16038

Capuron L, Miller AH (2011) Immune system to brain signaling: neuropsychopharmacological implications. Pharmacol Ther 130(2):226–238. [Review]

Carballedo A, Lisiecka D, Fagan A, Saleh K, Ferguson Y, Connolly G, Meaney J, Frodl T (2012) Early life adversity is associated with brain changes in subjects at family risk for depression. World J Biol Psychiatry 13:569–578

Carballedo A, Morris D, Zill P, Fahey C, Reinhold E, Meisenzahl E, Bondy B, Gill M, Moller HJ, Frodl T (2013) Brain-derived neurotrophic factor Val66Met polymorphism and early life adversity affect hippocampal volume. Am J Med Genet B Neuropsychiatr Genet 162B(2):183–190

Castellanos FX, Proal E (2012) Large-scale brain systems in ADHD: beyond the prefrontal–striatal model. Trends Cogn Sci 16(1):17–26

Chen CH, Suckling J, Ooi C, Fu CH, Williams SC, Walsh ND, Mitterschiffthaler MT, Pich EM, Bullmore E (2008) Functional coupling of the amygdala in depressed patients treated with antidepressant medication. Neuropsychopharmacology 33(8):1909–1918

Colla M, Kronenberg G, Deuschle M, Meichel K, Hagen T, Bohrer M, Heuser I (2007) Hippocampal volume reduction and HPA-system activity in major depression. J Psychiatr Res 41(7):553–560

Coplan J, Abdallah C, Tang C, Mathew S, Martinez J, Hof P, Smith E, Dwork A, Perera T, Pantol G (2011) The role of early life stress in development of the anterior limb of the internal capsule in non-human primates. Neurosci Lett 480:93–96

Dannlowski U, Ohrmann P, Konrad C, Domschke K, Bauer J, Kugel H, Hohoff C, Schoning S, Kersting A, Baune BT, Mortensen LS, Arolt V, Zwitserlood P, Deckert J, Heindel W, Suslow T (2009) Reduced amygdala-prefrontal coupling in major depression: association with MAOA genotype and illness severity. Int J Neuropsychopharmacol 12(1):11–22

Davidson RJ, Irwin W, Anderle MJ, Kalin NH (2003) The neural substrates of affective processing in depressed patients treated with venlafaxine. Am J Psychiatry 160(1):64–75

Davies J, Lloyd KR, Jones IK, Barnes A, Pilowsky LS (2003) Changes in regional cerebral blood flow with venlafaxine in the treatment of major depression. Am J Psychiatry 160(2):374–376

de Geus EJ, van't Ent D, Wolfensberger SP, Heutink P, Hoogendijk WJ, Boomsma DI, Veltman DJ (2007) Intrapair differences in hippocampal volume in monozygotic twins discordant for the risk for anxiety and depression. Biol Psychiatry 61(9):1062–1071

Deco G, Corbetta M (2011) The dynamical balance of the brain at rest. Neuroscientist 17(1):107–123

Draganski B, Gaser C, Kempermann G, Kuhn HG, Winkler J, Buchel C, May A (2006) Temporal and spatial dynamics of brain structure changes during extensive learning. J Neurosci 26(23):6314–6317

Duman RS (2002) Pathophysiology of depression: the concept of synaptic plasticity. Eur Psychiatry 17(Suppl 3):306–310

Egan MF, Kojima M, Callicott JH, Goldberg TE, Kolachana BS, Bertolino A, Zaitsev E, Gold B, Goldman D, Dean M, Lu B, Weinberger DR (2003) The BDNF val66met polymorphism affects activity-dependent secretion of BDNF and human memory and hippocampal function. Cell 112(2):257–269

Eichenbaum H, Harris K (2000) Toying with memory in the hippocampus. Nat Neurosci 3(3):205–206

Feng R, Rampon C, Tang YP, Shrom D, Jin J, Kyin M, Sopher B, Miller MW, Ware CB, Martin GM, Kim SH, Langdon RB, Sisodia SS, Tsien JZ (2001) Deficient neurogenesis in forebrain-specific presenilin-1 knockout mice is associated with reduced clearance of hippocampal memory traces. Neuron 32(5):911–926

Frodl T, O'Keane V (2013) How does the brain deal with cumulative stress? A review with focus on developmental stress, HPA axis function and hippocampal structure in humans. Neurobiol Dis 52:24–37

Frodl T, Meisenzahl EM, Zetzsche T, Hohne T, Banac S, Schorr C, Jager M, Leinsinger G, Bottlender R, Reiser M, Moller HJ (2004a) Hippocampal and amygdala changes in patients with major depressive disorder and healthy controls during a 1-year follow-up. J Clin Psychiatry 65(4):492–499

Frodl T, Meisenzahl EM, Zill P, Baghai T, Rujescu D, Leinsinger G, Bottlender R, Schule C, Zwanzger P, Engel RR, Rupprecht R, Bondy B, Reiser M, Moller HJ (2004b) Reduced hippocampal volumes associated with the long variant of the serotonin transporter polymorphism in major depression. Arch Gen Psychiatry 61(2):177–183

Frodl T, Scheuerecker J, Albrecht J, Kleemann AM, Muller-Schunk S, Koutsouleris N, Moller HJ, Bruckmann H, Wiesmann M, Meisenzahl E (2007a) Neuronal correlates of emotional processing in patients with major depression. World J Biol Psychiatry 26:1–7

Frodl T, Schule C, Schmitt G, Born C, Baghai T, Zill P, Bottlender R, Rupprecht R, Bondy B, Reiser M, Moller HJ, Meisenzahl EM (2007b) Association of the brain-derived neurotrophic factor Val66Met polymorphism with reduced hippocampal volumes in major depression. Arch Gen Psychiatry 64(4):410–416

Frodl T, Jager M, Smajstrlova I, Born C, Bottlender R, Palladino T, Reiser M, Moller HJ, Meisenzahl EM (2008a) Effect of hippocampal and amygdala volumes on clinical outcomes in major depression: a 3-year prospective magnetic resonance imaging study. J Psychiatry Neurosci 33(5):423–430

Frodl T, Moller HJ, Meisenzahl E (2008b) Neuroimaging genetics: new perspectives in research on major depression? Acta Psychiatr Scand 118(5):363–372

Frodl T, Zill P, Baghai T, Schule C, Rupprecht R, Zetzsche T, Bondy B, Reiser M, Moller HJ, Meisenzahl EM (2008c) Reduced hippocampal volumes associated with the long variant of the tri- and diallelic serotonin transporter polymorphism in major depression. Am J Med Genet B Neuropsychiatr Genet 147B(7):1003–1007

Frodl TS, Koutsouleris N, Bottlender R, Born C, Jäger M, Scupin I, Reiser M, Möller HJ, Meisenzahl EM (2008d) Depression-related variation in brain morphology over 3 years: effects of stress? Arch Gen Psychiatry 65(10):1156–1165

Frodl T, Bokde AL, Scheuerecker J, Lisiecka D, Schoepf V, Hampel H, Moller HJ, Bruckmann H, Wiesmann M, Meisenzahl E (2010a) Functional connectivity bias of the orbitofrontal cortex in drug-free patients with major depression. Biol Psychiatry 67(2):161–167

Frodl T, Reinhold E, Koutsouleris N, Donohoe G, Bondy B, Reiser M, Moller HJ, Meisenzahl EM (2010b) Childhood stress, serotonin transporter gene and brain structures in major depression. Neuropsychopharmacology 35(6):1383–1390

Frodl T, Reinhold E, Koutsouleris N, Reiser M, Meisenzahl EM (2010c) Interaction of childhood stress with hippocampus and prefrontal cortex volume reduction in major depression. J Psychiatr Res 44(13):799–807

Frodl T, Scheuerecker J, Schoepf V, Linn J, Koutsouleris N, Bokde AL, Hampel H, Moller HJ, Bruckmann H, Wiesmann M, Meisenzahl E (2011) Different effects of mirtazapine and venlafaxine on brain activation: an open randomized controlled fMRI study. J Clin Psychiatry 72(4):448–457

Frodl T, Carballedo A, Hughes MM, Saleh K, Fagan A, Skokauskas N, McLoughlin DM, Meaney J, O'Keane V, Connor TJ (2012a) Reduced expression of glucocorticoid-inducible genes GILZ and SGK-1: high IL-6 levels are associated with reduced hippocampal volumes in major depressive disorder. Transl Psychiatry 2:e88

Frodl T, Carballedo A, Fagan AJ, Lisiecka D, Ferguson Y, Meaney JF (2012b) Effects of early-life adversity on white matter diffusivity changes in patients at risk for major depression. J Psychiatry Neurosci. 37(1):37–45

Fu CH, Williams SC, Cleare AJ, Brammer MJ, Walsh ND, Kim J, Andrew CM, Pich EM, Williams PM, Reed LJ, Mitterschiffthaler MT, Suckling J, Bullmore ET (2004) Attenuation of the neural response to sad faces in major depression by antidepressant treatment: a prospective, event-related functional magnetic resonance imaging study. Arch Gen Psychiatry 61(9):877–889

Gatt JM, Nemeroff CB, Dobson-Stone C, Paul RH, Bryant RA, Schofield PR, Gordon E, Kemp AH, Williams LM (2009) Interactions between BDNF Val66Met polymorphism and early life stress predict brain and arousal pathways to syndromal depression and anxiety. Mol Psychiatry 14(7):681–695

Gilbertson MW, Shenton ME, Ciszewski A, Kasai K, Lasko NB, Orr SP, Pitman RK (2002) Smaller hippocampal volume predicts pathologic vulnerability to psychological trauma. Nat Neurosci 5(11):1242–1247

Goldapple K, Segal Z, Garson C, Lau M, Bieling P, Kennedy S, Mayberg H (2004) Modulation of cortical-limbic pathways in major depression: treatment-specific effects of cognitive behavior therapy. Arch Gen Psychiatry 61(1):34–41

Gonul AS, Kitis O, Eker MC, Eker OD, Ozan E, Coburn K (2011) Association of the brain-derived neurotrophic factor Val66Met polymorphism with hippocampus volumes in drug-free depressed patients. World J Biol Psychiatry 12(2):110–118

Greicius MD, Supekar K, Menon V, Dougherty RF (2009) Resting-state functional connectivity reflects structural connectivity in the default mode network. Cereb Cortex 19(1):72–78

Grimm S, Boesiger P, Beck J, Schuepbach D, Bermpohl F, Walter M, Ernst J, Hell D, Boeker H, Northoff G (2009) Altered negative BOLD responses in the default-mode network during emotion processing in depressed subjects. Neuropsychopharmacology 34(4):932–943

Hamilton JP, Gotlib IH (2008) Neural substrates of increased memory sensitivity for negative stimuli in major depression. Biol Psychiatry 63(12):1155–1162

Huang EJ, Reichardt LF (2001) Neurotrophins: roles in neuronal development and function. Annu Rev Neurosci 24:677–736

Hviid LB, Ravnkilde B, Ahdidan J, Rosenberg R, Stodkilde-Jorgensen H, Videbech P (2010) Hippocampal visuospatial function and volume in remitted depressed patients: an 8-year follow-up study. J Affect Disord 125(1–3):177–183

Inkster B, Nichols TE, Saemann PG, Auer DP, Holsboer F, Muglia P, Matthews PM (2009) Association of GSK3beta polymorphisms with brain structural changes in major depressive disorder. Arch Gen Psychiatry 66(7):721–728

Joffe RT, Gatt JM, Kemp AH, Grieve S, Dobson-Stone C, Kuan SA, Schofield PR, Gordon E, Williams LM (2009) Brain derived neurotrophic factor Val66Met polymorphism, the five factor model of personality and hippocampal volume: implications for depressive illness. Hum Brain Mapp 30(4):1246–1256

Kandel ER (2001) The molecular biology of memory storage: a dialogue between genes and synapses. Science 294(5544):1030–1038

Kempton MJ, Salvador Z, Munafo MR, Geddes JR, Simmons A, Frangou S, Williams SC (2011) Structural neuroimaging studies in major depressive disorder. Meta-analysis and comparison with bipolar disorder. Arch Gen Psychiatry 68(7):675–690

Kennedy SH, Evans KR, Kruger S, Mayberg HS, Meyer JH, McCann S, Arifuzzman AI, Houle S, Vaccarino FJ (2001) Changes in regional brain glucose metabolism measured with positron emission tomography after paroxetine treatment of major depression. Am J Psychiatry 158(6):899–905

Kieseppä T, Eerola M, Mäntylä R, Neuvonen T, Poutanen V, Luoma K, Tuulio-Henriksson A, Jylhä P, Mantere O (2010) Major depressive disorder and white matter abnormalities: a diffusion tensor imaging study with tract-based spatial statistics. J Affect Disord 120(1-3): 240–244

Kole MH, Czeh B, Fuchs E (2004) Homeostatic maintenance in excitability of tree shrew hippocampal CA3 pyramidal neurons after chronic stress. Hippocampus 14(6):742–751

Krishnan V, Nestler EJ (2008) The molecular neurobiology of depression. Nature 455(7215):894–902. [Research Support, N.I.H., Extramural Review]

Lai CH, Hsu YY (2011) A subtle grey-matter increase in first-episode, drug-naive major depressive disorder with panic disorder after 6 weeks' duloxetine therapy. Int J Neuropsychopharmacol 14(2):225–235

Lesch KP, Mossner R (1998) Genetically driven variation in serotonin uptake: is there a link to affective spectrum, neurodevelopmental, and neurodegenerative disorders? Biol Psychiatry 44(3):179–192

Lesch KP, Bengel D, Heils A, Sabol SZ, Greenberg BD, Petri S, Benjamin J, Muller CR, Hamer DH, Murphy DL (1996) Association of anxiety-related traits with a polymorphism in the serotonin transporter gene regulatory region. Science 274(5292):1527–1531

Lisiecka D, Meisenzahl E, Scheuerecker J, Schoepf V, Whitty P, Chaney A, Moeller HJ, Wiesmann M, Frodl T (2011) Neural correlates of treatment outcome in major depression. Int J Neuropsychopharmacol 14(4): 521–534

Lisman JE (1999) Relating hippocampal circuitry to function: recall of memory sequences by reciprocal dentate-CA3 interactions. Neuron 22(2):233–242

Lu B (2003) Pro-region of neurotrophins: role in synaptic modulation. Neuron 39(5):735–738

Lyons DM, Yang C, Sawyer-Glover AM, Moseley ME, Schatzberg AF (2001) Early life stress and inherited variation in monkey hippocampal volumes. Arch Gen Psychiatry 58(12):1145–1151

Macqueen G, Frodl T (2011) The hippocampus in major depression: evidence for the convergence of the bench and bedside in psychiatric research? Mol Psychiatry 16(3):252–264

Magarinos AM, McEwen BS, Flugge G, Fuchs E (1996) Chronic psychosocial stress causes apical dendritic atrophy of hippocampal CA3 pyramidal neurons in subordinate tree shrews. J Neurosci 16(10):3534–3540

Malykhin NV, Lebel RM, Coupland NJ, Wilman AH, Carter R (2010) In vivo quantification of hippocampal subfields using 4.7 T fast spin echo imaging. Neuroimage 49(2):1224–1230

Martin SD, Martin E, Rai SS, Richardson MA, Royall R (2001) Brain blood flow changes in depressed patients treated with interpersonal psychotherapy or venlafaxine hydrochloride: preliminary findings. Arch Gen Psychiatry 58(7):641–648

Mayberg HS, Brannan SK, Mahurin RK, Jerabek PA, Brickman JS, Tekell JL, Silva JA, McGinnis S, Glass TG, Martin CC, Fox PT (1997) Cingulate function in depression: a potential predictor of treatment response. Neuroreport 8(4):1057–1061

Mayberg HS, Liotti M, Brannan SK, McGinnis S, Mahurin RK, Jerabek PA, Silva JA, Tekell JL, Martin CC, Lancaster JL, Fox PT (1999) Reciprocal limbic-cortical function and negative mood: converging PET findings in depression and normal sadness. Am J Psychiatry 156(5):675–682

Mayberg HS, Lozano AM, Voon V, McNeely HE, Seminowicz D, Hamani C, Schwalb JM, Kennedy SH (2005) Deep brain stimulation for treatment-resistant depression. Neuron 45(5):651–660

McKinnon MC, Yucel K, Nazarov A, MacQueen GM (2009) A meta-analysis examining clinical predictors of hippocampal volume in patients with major depressive disorder. J Psychiatry Neurosci 34(1):41–54

Miller EJ, Saint Marie LR, Breier MR, Swerdlow NR (2010) Pathways from the ventral hippocampus and caudal amygdala to forebrain regions that regulate sensorimotor gating in the rat. Neuroscience 165(2):601–611

Mondelli V, Cattaneo A, Belvederi Murri M, Di Forti M, Handley R, Hepgul N, Miorelli A, Navari S, Papadopoulos AS, Aitchison KJ, Morgan C, Murray RM, Dazzan P, Pariante CM (2011) Stress and inflammation reduce brain-derived neurotrophic factor expression in first-episode psychosis: a pathway to smaller hippocampal volume. J Clin Psychiatry 72(12):1677–1684

Myint AM, Schwarz MJ, Muller N (2012) The role of the kynurenine metabolism in major depression. J Neural Transm 119(2):245–251 [Research Support, Non-U.S. Gov't]

Nobuhara K, Okugawa G, Sugimoto T, Minami T, Tamagaki C, Takase K, Saito Y, Sawada S, Kinoshita T (2006) Frontal white matter anisotropy and symptom severity of late-life depression: a magnetic resonance diffusion tensor imaging study. J Neurol Neurosurg Psychiatry 77(1):120

Norbury R, Selvaraj S, Taylor MJ, Harmer C, Cowen PJ (2010) Increased neural response to fear in patients recovered from depression: a 3T functional magnetic resonance imaging study. Psychol Med 40(3): 425–432

Olesen PJ, Westerberg H, Klingberg T (2004) Increased prefrontal and parietal activity after training of working memory. Nat Neurosci 7(1):75–79

Paul R, Henry L, Grieve S, Guilmette T, Niaura R, Bryant R, Bruce S, Williams L, Richard C, Cohen R (2008) The relationship between early life stress and microstructural integrity of the corpus callosum in a nonclinical population. Neuropsychiatr Dis Treat 4(1):193

Pezawas L, Verchinski BA, Mattay VS, Callicott JH, Kolachana BS, Straub RE, Egan MF, Meyer-Lindenberg A, Weinberger DR (2004) The brain-derived neurotrophic factor val66met polymorphism and variation in human cortical morphology. J Neurosci 24(45):10099–10102

Raichle ME, MacLeod AM, Snyder AZ, Powers WJ, Gusnard DA, Shulman GL (2001) A default mode of brain function. Proc Natl Acad Sci U S A 98(2):676–682

Rao U, Chen LA, Bidesi AS, Shad MU, Thomas MA, Hammen CL (2010) Hippocampal changes associated with early-life adversity and vulnerability to depression. Biol Psychiatry 67(4):357–364

Samson AC, Meisenzahl E, Scheuerecker J, Rose E, Schoepf V, Wiesmann M, Frodl T (2011) Brain activation predicts treatment improvement in patients with major depressive disorder. J Psychiatr Res 45(9):1214–1222

Santarelli L, Saxe M, Gross C, Surget A, Battaglia F, Dulawa S, Weisstaub N, Lee J, Duman R, Arancio O, Belzung C, Hen R (2003) Requirement of hippocampal neurogenesis for the behavioral effects of antidepressants. Science 301(5634):805–809

Schermuly I, Wolf D, Lieb K, Stoeter P, Fellgiebel A (2011) State dependent posterior hippocampal volume increases in patients with major depressive disorder. J Affect Disord 135(1–3):405–409

Seminowicz DA, Mayberg HS, McIntosh AR, Goldapple K, Kennedy S, Segal Z, Rafi-Tari S (2004) Limbic-frontal circuitry in major depression: a path modeling metanalysis. Neuroimage 22(1):409–418

Sheline YI, Barch DM, Donnelly JM, Ollinger JM, Snyder AZ, Mintun MA (2001) Increased amygdala response to masked emotional faces in depressed subjects resolves with antidepressant treatment: an fMRI study. Biol Psychiatry 50(9):651–658

Sheline YI, Price JL, Yan Z, Mintun MA (2010) Resting-state functional MRI in depression unmasks increased connectivity between networks via the dorsal nexus. Proc Natl Acad Sci U S A 107(24):11020–11025

Siegle GJ, Carter CS, Thase ME (2006) Use of FMRI to predict recovery from unipolar depression with cognitive behavior therapy. Am J Psychiatry 163(4):735–738

Surguladze S, Brammer MJ, Keedwell P, Giampietro V, Young AW, Travis MJ, Williams SC, Phillips ML (2005) A differential pattern of neural response toward sad versus happy facial expressions in major depressive disorder. Biol Psychiatry 57(3):201–209

Taylor WD, Steffens DC, Payne ME, MacFall JR, Marchuk DA, Svenson IK, Krishnan KR (2005) Influence of serotonin transporter promoter region polymorphisms on hippocampal volumes in late-life depression. Arch Gen Psychiatry 62(5):537–544

Tyler WJ, Alonso M, Bramham CR, Pozzo-Miller LD (2002) From acquisition to consolidation: on the role of brain-derived neurotrophic factor signaling in hippocampal-dependent learning. Learn Mem 9(5):224–237

Vasic N, Walter H, Sambataro F, Wolf RC (2008) Aberrant functional connectivity of dorsolateral prefrontal and cingulate networks in patients with major depression during working memory processing. Psychol Med 10:1–11

Vermetten E, Vythilingam M, Southwick SM, Charney DS, Bremner JD (2003) Long-term treatment with paroxetine increases verbal declarative memory and hippocampal volume in posttraumatic stress disorder. Biol Psychiatry 54(7):693–702

Videbech P, Ravnkilde B (2004) Hippocampal volume and depression: a meta-analysis of MRI studies. Am J Psychiatry 161(11):1957–1966

Vollm B, Richardson P, McKie S, Elliott R, Deakin JF, Anderson IM (2006) Serotonergic modulation of neuronal responses to behavioural inhibition and reinforcing stimuli: an fMRI study in healthy volunteers. Eur J Neurosci 23(2):552–560

Vythilingam M, Heim C, Newport J, Miller AH, Anderson E, Bronen R, Brummer M, Staib L, Vermetten E, Charney DS, Nemeroff CB, Bremner JD (2002) Childhood trauma associated with smaller hippocampal volume in women with major depression. Am J Psychiatry 159(12):2072–2080

Vythilingam M, Vermetten E, Anderson GM, Luckenbaugh D, Anderson ER, Snow J, Staib LH, Charney DS, Bremner JD (2004) Hippocampal volume, memory, and cortisol status in major depressive disorder: effects of treatment. Biol Psychiatry 56(2):101–112

Wellman CL (2001) Dendritic reorganization in pyramidal neurons in medial prefrontal cortex after chronic corticosterone administration. J Neurobiol 49(3):245–253

Woolley CS, Gould E, Frankfurt M, McEwen BS (1990) Naturally occurring fluctuation in dendritic spine density on adult hippocampal pyramidal neurons. J Neurosci 10(12):4035–4039

Wu J, Buchsbaum MS, Gillin JC, Tang C, Cadwell S, Wiegand M, Najafi A, Klein E, Hazen K, Bunney WE Jr, Fallon JH, Keator D (1999) Prediction of antidepressant effects of sleep deprivation by metabolic rates in the ventral anterior cingulate and medial prefrontal cortex. Am J Psychiatry 156(8):1149–1158

Zhou Y, Yu C, Zheng H, Liu Y, Song M, Qin W, Li K, Jiang T (2010) Increased neural resources recruitment in the intrinsic organization in major depression. J Affect Disord 121(3):220–230

Functional Neuroimaging of Anxiety Disorders

15

Victor I. Spoormaker, Eric Vermetten, Michael Czisch, and Frank H. Wilhelm

Abbreviations

BOLD	Blood oxygen level dependent
CCK-4	Cholecystokinin tetrapeptide
CS−	Safety stimulus
CS+	Conditioned stimulus
dACC	Dorsal anterior cingulate cortex
dmPFC	Dorsomedial prefrontal cortex
EMG	Electromyography
ITC	Intercalated cells
OCD	Obsessive-compulsive disorder
vmPFC	Ventromedial prefrontal cortex
PTSD	Posttraumatic stress disorder
SCR	Skin conductance response
SNRI	Serotonin-norepinephrine reuptake inhibitor

V.I. Spoormaker, PhD (✉) • M. Czisch
Max Planck Institute of Psychiatry, Munich, Germany
e-mail: spoormaker@mpipsykl.mpg.de

E. Vermetten, MD, PhD
Department of Psychiatry, Leiden University Medical Center, Leiden, The Netherlands

F.H. Wilhelm, PhD
Clinical Psychology, Psychotherapy & Health Psychology, Institute for Psychology, University of Salzburg, Salzburg, Austria

15.1 Introduction: Structural and Functional Imaging Tools of Anxiety Disorders

Anxiety disorders are the most prevalent group of psychiatric disorders with a 12-month prevalence of 14 %, affecting roughly 60 million people in the 27 countries of the European Union (Wittchen et al. 2011). Although the direct healthcare and indirect costs are lower in anxiety disorders compared to mood or psychotic disorders, the sheer amount of patients makes it one of the top five costliest brain disorders in Europe, surpassing an estimated 70 billion Euro a year (Gustavsson et al. 2011). Furthermore, prevalence estimates in the United States are comparable or even higher (Kessler et al. 2012).

In this chapter, we will address the role of neuroimaging markers in anxiety disorders, with a specific focus on posttraumatic stress disorder (PTSD), panic disorder, specific phobias, and social anxiety disorder. Moreover, we will briefly discuss obsessive-compulsive disorder (OCD). The development of neuroimaging methodology for anxiety disorders can be illustrated by the progress made in PTSD, a disorder that may develop after exposure to a potentially traumatic event and is characterized by intrusive memories, avoidance behavior, numbing, and hyperarousal. Before the widespread availability of functional magnetic resonance imaging (fMRI), studies addressed structural differences between PTSD patients and controls and observed volumetric differences in hippocampus (Bremner et al. 1995, 1997). When fMRI was

C. Mulert, M.E. Shenton (eds.), *MRI in Psychiatry*,
DOI 10.1007/978-3-642-54542-9_15, © Springer Berlin Heidelberg 2014

technically suited to assess functional parameters of brain activity, the attention shifted from hippocampus to amygdala and the role of the prefrontal cortex. Research started to address the functioning of brain regions in PTSD by means of combining functional neuroimaging with symptom provocation tasks (e.g., with trauma scripts or stimuli), general affective tasks, and functional connectivity during the so-called resting state (Francati et al. 2007; Liberzon and Sripada 2008). These studies revealed intriguing altered functional patterns of activity and connectivity in the amygdala, the hippocampus, and the medial prefrontal cortex (mPFC) in PTSD (Francati et al. 2007; Liberzon and Sripada 2008; Vermetten and Lanius 2012). In more recent years, functional neuroimaging studies have increasingly focused on the underlying processes proposed to be involved in the etiology of PTSD, such as fear conditioning and extinction (Pitman et al. 2012). Importantly, these neuroimaging findings are starting to be integrated with data from genetics, endocrinology, and immunology to provide an integrative biological framework for PTSD in which prefrontal cortex dysregulation in conjunction with fear-conditioned amygdala response may eventually come to play a key role (Vermetten and Lanius 2012).

In this chapter, we would like to stress the utility of functional imaging tools for anxiety disorders, given that anatomical markers have revealed rather limited effects in most anxiety disorders (with the exception of OCD). For instance, one meta-analysis on volumetric differences in PTSD observed effect sizes *d* of 0.07–0.14 for the amygdala and 0.28–0.29 for the hippocampus (Karl et al. 2006). Some methodological issues deserve mentioning: the amount of positive reports on reduced hippocampus sizes in PTSD (about half of all published volumetric studies in the meta-analysis by Karl et al.) is rather high given the small effect and sample sizes (Ioannidis 2011), suggesting that there may be an analysis or reporting bias in the literature. Reduced hippocampus size may not be specific to PTSD since it is also observed in depression – yet, it is interesting that the volumetric differences are predominantly found in depressive patients with early childhood trauma (Vythilingam et al. 2002). Furthermore, the relevance of reduced hippocampus size may lie in altered hippocampal *functioning*, i.e., hippocampus size is, for instance, relevant for declarative memory *performance* (Starkman et al. 2003). Impaired functioning of the hippocampus may therefore be more closely related to anxiety symptoms and behavior. Although its use as a biomarker for PTSD or other anxiety disorders may be limited, reduced hippocampus sizes are of basic scientific interest considering whether they constitute a premorbid risk factor for PTSD or are a consequence of PTSD, or both – as indicated by longitudinal studies (Pitman et al. 2012). The more sensitive and specific endophenotype is, however, *dysfunctional* neural circuitry that is directly correlated to anxiety symptoms and behavior.

Functional imaging tools are as effective as the task that elicits (or modulates) brain activity to probe for abnormal functioning in specific circuitry. In other words, functional imaging tools depend on customized tasks to challenge the brain, as this may constitute a primary marker for whether brain functioning "breaks down" during stress. For task optimization, the related field of psychophysiology can be informative as many of the primary readouts are closely related to relevant neural circuitry (e.g., limbic and more downstream regions). A meta-analysis of psychophysiological effects in PTSD concluded that effect sizes of relevant physiological measures were larger for tasks involving trauma cues than for tasks involving trauma-unrelated stimuli, such as presentation of unexpected acoustic startle sounds, and that any task showed better effects than resting-state measurements (Pole 2007). For anxiety disorders, where a stimulus, context, event, or memory elicits specific fear (e.g., phobias, panic disorder, PTSD), symptom provocation by presentation of reminder cues will likely probe the neural fear circuitry to a much larger extent (and with more experimental control) than resting-state tasks. A promising alternative is to directly examine the underlying process that has been proposed to be responsible for pathological anxiety: fear conditioning and extinction (see Lissek et al. 2005).

15.2 Experimental Tasks for Fear-Related Anxiety Disorders: Fear Conditioning and Extinction

As mentioned above, one major advantage of functional imaging tasks for anxiety disorders is that much can be learned from 80 years of experience in experimental psychology. Utilizing skin conductance response (SCR) or eyeblink startle electromyography (EMG) as physiological readouts, psychophysiological studies have provided a wealth of data on relevant functional tasks and dependent variables in anxiety disorders (Lissek et al. 2005). Although functional imaging can be viewed as adding a new dependent variable to the existing functional task and psychophysiological readouts, functional imaging research is only rarely tapping into this existing body of knowledge. In our opinion, functional imaging studies should be firmly rooted in experimental psycho(physio)logy, so that the neural correlates can be adequately labeled and interpreted.

Fear conditioning is a robust experimental task that has been studied over eighty years in human subjects, ever since pioneering studies observed that pairing an electric shock (unconditioned stimulus, US) with a light caused an SCR at subsequent light presentation (Steckle 1933; Switzer 1934; Rodnick 1937), which therefore became a conditioned stimulus (CS+) and the onset SCR an anticipatory, conditioned response (see Fig. 15.1, upper panel). This provided a well controllable translational model of classical conditioning, first reported in dogs by Pavlov (Pavlov 1927), and a template for further variations. The introduction of a safety stimulus (CS−) into the task allowed examination of the *differential* conditioned response, to control for individual variability in SCR intensity and frequency. After initial fear conditioning, the mere exposure of CS+ no longer followed by shocks

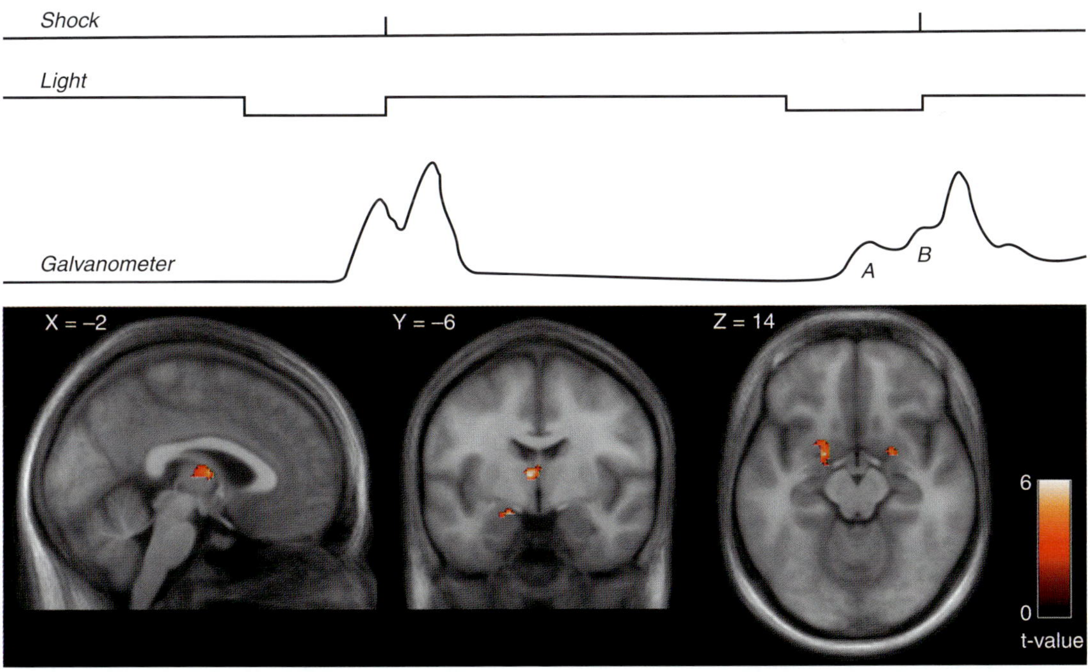

Fig. 15.1 *Upper panel*: example of an early human fear-conditioning task describing anticipatory skin conductance responses (SCR) to a conditioned stimulus (CS+), in the figure at locations *A* and *B* (From Rodnick (1937)). *Lower panels*: activity correlated with a parametric modulation of the amplitude of anticipatory SCR to CS+ during the course of fear conditioning (decreasing slope), showing clusters of activity in the bilateral amygdala and thalamus ($N=34$, $p_{FDR}<0.025$) (Reproduced by kind permission of Elsevier from Spoormaker et al. (2011))

constituted *fear extinction*, which as a laboratory model shares overlap with exposure-based therapy (Rauch et al. 2006). Variations in CS+ shapes provided a means to probe for *fear generalization,* which is a relevant process for several anxiety disorders (e.g., generalization of trauma- or panic-attack reminders is typically reported in PTSD and panic disorder, whereas a specific phobia may be characterized by a non-generalized fear to a specific object or context, Lissek 2012).

Using fMRI, limbic and paralimbic regions such as the amygdala, hippocampus, dorsal anterior cingulate cortex (dACC), and ventromedial prefrontal cortex (vmPFC), as well as the insula and thalamus, have been repeatedly demonstrated to subserve fear conditioning and extinction in healthy individuals (for a systematic review, see Sehlmeyer et al. 2009). Already early fMRI studies on human fear conditioning revealed activity in the dACC and bilateral insula associated with CS+ (Büchel et al. 1998; LaBar et al. 1998), whereas the first human fMRI fear extinction study showed an increase in activity in the vmPFC during fear extinction (Phelps et al. 2004). In analyses addressing integration of SCR and fMRI data, the onset SCR to CS+ as described above appears specifically correlated to amygdala and thalamus activity to CS+ during fear conditioning (e.g., Spoormaker et al. 2011), see Fig. 15.1, lower panels. A quickly growing body of research has started to address the overlap in neurocognitive mechanisms underlying fear conditioning and extinction in preclinical and clinical studies, leading authoritative reviews to conclude that this is a successful model for translating data from animal models to human subjects and psychiatric patients (Milad and Quirk 2012; Pitman et al. 2012).

15.2.1 Functional Imaging of Fear Conditioning and Extinction: Relevance for PTSD

Fear conditioning and extinction has been increasingly recognized as a promising human and animal model for studying the neural circuitry involved in anxiety disorders, most notably PTSD (Rauch et al. 2006; Pape and Pare 2010). Besides forming the core of theoretical models on the etiology of normal and pathological anxiety (Lissek et al. 2005), multiple independent research groups have now provided the critical evidence for impaired extinction in PTSD and other anxiety disorders (Lissek et al. 2005; Blechert et al. 2007; Wessa and Flor 2007; Milad et al. 2009).

Crucially, the amygdala and dACC show abnormal fMRI activity in PTSD patients during extinction and at recall of extinction (Milad et al. 2009), whereas hypoactivity was observed in both the vmPFC and hippocampus. This pattern of activation and deactivation has also been reported by a recent meta-analysis of 26 functional neuroimaging studies of PTSD employing various tasks (from symptom provocation to fear extinction) that found that the dACC and amygdala were the most strongly activated regions in PTSD, whereas the vmPFC and the inferior frontal gyri were the most strongly deactivated regions in PTSD (Hayes et al. 2012). Moreover, a decrease in vmPFC activity was associated with increased amygdala activity (Hayes et al. 2012), in line with the inhibitory connections from the vmPFC to amygdala (LeDoux 2000). Neurocircuitry models of PTSD postulate that this failure of the vmPFC and hippocampus to downregulate a hyperactive amygdala is the core feature characterizing PTSD, which is postulated to be responsible for increased fear responses, impaired fear extinction learning and retrieval, and impairments in general emotion regulation and attentional biases towards threat (Rauch et al. 2006). In a recent comprehensive review on biological studies of PTSD (Pitman et al. 2012), a main conclusion was that the amygdala was hyperactive in PTSD in response to trauma-related stimuli as well as generic threat stimuli, and during fear conditioning, compared to control subjects. Here it is important to note, however, that amygdala deactivation has been observed in PTSD as well (Etkin and Wager 2007). Such amygdala deactivation has been proposed to be characteristic of dissociative subtype of PTSD, with limbic overmodulation rather than undermodulation as a core feature (Lanius et al. 2010a), which results in increased frontal cortical activity and decreased amygdala activity to traumatic script challenge. Patients responding with this pattern are typically characterized by early trauma histories (Lanius et al. 2010b). This would be an interesting area for further research on the use of neuroimaging tools in PTSD patient stratification.

The dACC showed increased activity during fear conditioning and recall of extinction in PTSD compared to trauma-exposed healthy controls; the vmPFC, in contrast, was concluded to have decreased activity during both trauma-related and trauma-unrelated affective stimuli. Hippocampus involvement may be task dependent and is likely related to "deficits in recognizing safe contexts" (Pitman et al. 2012). Context dependencies of CS+ (e.g., fear conditioning in context A and fear extinction in context B, manipulated by presenting different background colors during both phases) involve the hippocampus (Milad et al. 2007) and are impaired in PTSD patients compared to trauma-exposed controls (Milad et al. 2009). The hippocampus is involved at a more basic level in pattern completion or separation to CS+, which would activate the vmPFC in turn (Lissek 2012). The emotional circuitry sensitive to PTSD extends to other regions involved in affective processing, such as the insula, but this has also been observed in other anxiety disorders and may not be specific to PTSD (Etkin and Wager 2007).

15.2.2 Comparison of PTSD with Other Anxiety Disorders

Which brain activity is specific to PTSD? This question was addressed by Etkin and Wager (2007) in a meta-analysis of imaging studies that compares PTSD with specific phobias, social anxiety disorder, and fear conditioning in healthy subjects. Most functional imaging (positron emission tomography, PET; and functional magnetic resonance imaging, fMRI) studies included in the meta-analysis used symptom provocation: i.e., 10 out of 15 imaging studies on PTSD used visual- or script-based trauma cues, five out of eight imaging studies on social phobia used negative facial expressions (the other addressing public speaking and speech anticipation), and six out of seven imaging studies on specific phobias used images, words, or videos of the phobic object/animal/situation.

Results revealed that all disorders (patients > control subjects) showed hyperactivity in the amygdala and insula and that this pattern of activity was analogous to fear conditioning in healthy subjects. This suggests overlapping fear circuitry involved in both symptom provocation in patients and fear conditioning in controls, relevant across disorders. However, only PTSD further showed hypoactivity in the ventro- and dorsomedial prefrontal cortex (dmPFC), anterior hippocampus, and parahippocampal gyrus, among others, that was *specific* to this disorder. According to the meta-analysis, results of social and specific phobias were strongly overlapping, whereas the pattern of activation and deactivation in PTSD showed more deviation and complexity. In PTSD, hypoactivity in medial prefrontal regions tended to co-occur, whereas a pattern of hypoactivity in the medial prefrontal and hyperactivity in the amygdala was also observed in this meta-analysis. Interestingly, hyperactivity in the amygdala tended to occur more frequently in phobias than in PTSD (Etkin and Wager 2007), with potential relevance for PTSD subtype stratification (Lanius et al. 2010a, b) as mentioned above.

Due to this specificity, hypoactivity in medial prefrontal and hippocampal areas in PTSD may be of special interest. A logical prediction is that hypoactivity in these regions is correlated with extinction and/or generalization impairments. Future research has yet to address this question; however, studies have already shown that mPFC activity correlates with subjectively reported PTSD symptom severity (Shin et al. 2005). Intriguingly and consistent with this, symptom severity improvements after cognitive-behavioral therapy for PTSD have been found to be related to an increase in mPFC and a decrease in amygdala activity (Felmingham et al. 2007). Although the hypoactivity of the vmPFC and hyperactivity of the amygdala in PTSD are unequivocal, there

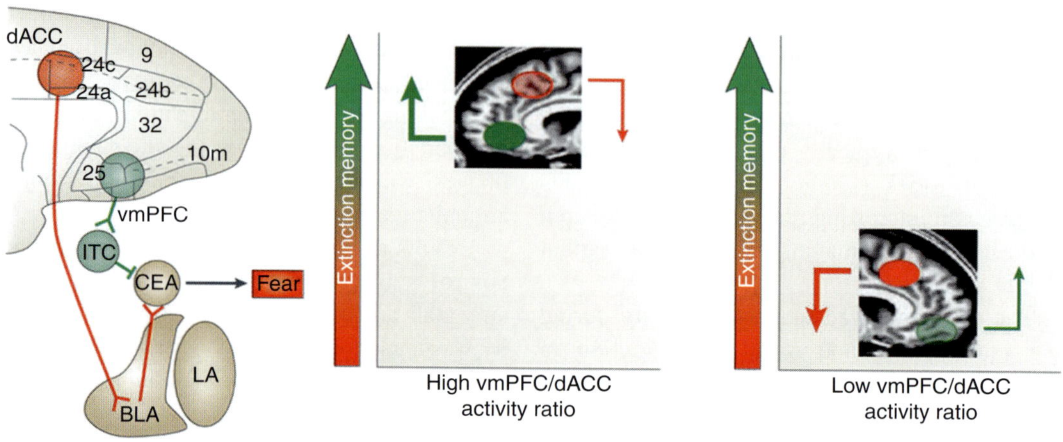

Fig. 15.2 Functional interactions among fear circuitry regions. A recent review on the biology of PTSD described the relevance of dACC excitatory connections to the basolateral amygdala (*BLA*), vmPFC inhibitory connections to the central nucleus of the amygdala (CEA) through intercalated cells (ITC), and a high vmPFC/dACC activity ratio for successful fear memory encoding and retrieval (*left*). As compared to stronger inhibitory influence of the vmPFC and lower excitatory input of the dACC over the amygdala under healthy conditions (*middle*), excess dACC activity during fear conditioning and extinction results in hyperactivity in the amygdala and reduced vmPFC activity (*right*), which further implies a failure to downregulate the amygdala, causing functional impairments in extinction memory encoding and/or retrieval (Reproduced by kind permission of Macmillan Publishers from Pitman et al. (2012))

is some lack of clarity regarding whether or not the dmPFC (dACC) is hyper- or hypoactivated in PTSD: Etkin and Wager (2007) found clear hypoactivity in PTSD, while other studies have shown hyperactivity of the dACC during fear conditioning in PTSD (Milad et al. 2009) and positive correlations between dACC and amygdala activity (Pitman et al. 2012). It is possible though that hyperactivity is confined to the more posterior dACC (middle cingulate cortex) that extends into the supplementary motor area, which has been found to consistently activate during fear conditioning (Sehlmeyer et al. 2009). More rostral portions of the dACC and ACC extending into the ventral parts may then be more hypoactivated in PTSD and involved in extinction learning and recall (Phelps et al. 2004; Etkin and Wager 2007; Milad et al. 2009). For a brief summary, see Fig. 15.2.

15.2.3 Fear Conditioning: Functional and Imaging Limitations

Although fear conditioning and extinction constitutes one of the most robust paradigms in experimental psychophysiology, allowing disturbing symptom provocation tasks to be replaced by less stressful and better controlled experimental stimuli, the task has too received critique for exclusively focusing on physiological (and subjective) components of emotion while not addressing the behavioral component of (pathological) anxiety, i.e., avoidance and safety behavior (Beckers et al. 2013). New tasks combining fear conditioning with exploration and avoidance behavior in virtual reality environments are a promising start into this understudied dimension (Grillon et al. 2006). Moreover, PTSD is primarily characterized by distressing intrusive memories in the form of flashbacks and nightmares, and electrical shocks as US do not tap into (dys)function of such memory aspects. Tasks exploiting conditioning with highly intrusive movie scenes as US (Wegerer et al. 2013) may provide a promising and clinically useful extension of the typical fear-conditioning task. Furthermore, stimuli in fear-conditioning tasks are typically visual or auditory; more research attention could be devoted to alternative stimuli given the promise of olfactory trauma reminders in neuroimaging studies of PTSD (Vermetten et al. 2007).

It is possibly one of the great paradoxes in functional MR imaging of anxiety disorders that the brain regions that are most relevant, such as the amygdala, vmPFC, subgenual ACC, and orbitofrontal cortex, are among the regions that are most strongly affected by magnetic susceptibility-induced artifacts and consequently signal loss (Schwarzbauer et al. 2010). Such artifacts are caused by inhomogeneities of the magnetic field and arise especially at and in proximity of brain/air borders, due to the anatomy itself. fMRI is especially sensitive to local variations of field homogeneity, as the primary source of the image contrast is its sensitivity to small changes in the magnetic properties of hemoglobin, depending on the blood oxygenation (leading to the so-called blood oxygen level dependent – BOLD – effect, which is reflected by changes of the T2* relaxation rates of neighboring nuclear spins). As a result, fMRI image acquisition optimization is desirable to recover signal from these essential regions.

15.2.4 Panic Disorder

Panic disorder is arguably an anxiety disorder most closely related to the specific phobias and PTSD since in panic disorder, patients have learned to fear upcoming panic attacks (US) often triggered by interoceptive stimuli (CS+) thought to prequel a following attack. Agoraphobia, a fear of open or public spaces, is a frequent comorbid disorder in panic disorder. Psychophysiological research has shown impairments in extinction learning (Michael et al. 2007) and overgeneralization in panic disorder patients to various deviating versions of CS+ (Lissek et al. 2010). No imaging data have yet been published on these tasks; however, clear predictions of neural correlates of overgeneralization would point to the hippocampus given its role in pattern completion versus separation and the mPFC in relation to impaired extinction learning (Lissek 2012). Interestingly, an initial study showed increased midbrain activity in panic patients during fear conditioning (Lueken et al. 2013), which was paralleled by impaired differential learning in panic disorder patients (manifest in subjective ratings); however, no psychophysiological measures were recorded. Another fear conditioning study without psychophysiological measures also observed increased midbrain activity in panic disorder patients in response to safety cues, coactivating with clusters of activity in the ventral striatum, the anterior medial temporal lobe, and the subgenual ACC (Tuescher et al. 2011).

In general, it is of note that panic disorder has been studied to a lesser degree than PTSD, and the imaging studies that have been conducted to date, employing general affective tasks (e.g., emotional stroop, facial recognition, negative pictures), have shown mixed results (see Dresler et al. 2013, for a review). fMRI studies of affective tasks ($N=9$) were inconclusive, but abnormal activity in the ACC (either increased or decreased) occurred across tasks (Dresler et al. 2013). Hyperactivity of the amygdala has not been consistently found during these tasks; however, a few case studies of spontaneous panic attacks during functional imaging sessions reported increased amygdala and insula activity (Dresler et al. 2013). Another review has implicated the amygdala, insula, mPFC and hippocampus in panic disorder (de Carvalho et al. 2010); yet as no meta-analyses have yet been conducted, a formal statistical evaluation of such potentially relevant effects is precluded.

For panic disorder, promising alternative models have been tested that may eventually be combined with fear conditioning to achieve a comprehensive view on this disorder. Hyper- and hypocapnia (elevated and reduced CO_2 levels in the blood) have been experimentally induced in panic disorder (Martinez et al. 1998). This resulted in differences in heart rate and blood pressure between panic disorder patients and controls (Martinez et al. 1998), the correlates of which in fMRI need to be more closely examined. Pharmacological models of panic attacks have also been reported, such as with panicogenic neuropeptide cholecystokinin tetrapeptide (CCK-4), showing increased amygdala activation compared to placebo after CCK-4 administration in healthy subjects, as well as in the vmPFC, lateral prefrontal regions, brainstem, and cerebellum, among others (Eser et al. 2009). The reported anticipatory anxiety associated with the dACC in this study is also worth mentioning. CCK-4 led to an increase in subjective panic symptoms and heart rate, which may interfere with the experimental readout in fMRI studies – also here closer examination of potential interference of autonomic responses to panicogenic stimuli with the BOLD signal appears of interest, although the specific, localized increase in limbic and paralimbic areas after CKK-4 seems to go beyond potential unspecific baseline differences.

One methodological issue that is relevant for panic disorder but also for other anxiety disorders is that anxiety typically correlates with respiratory alterations that may produce significant changes in cerebral blood flow independent of task-related neural activation (Giardino et al. 2007). Measuring arterial carbon dioxide tension (or at the very least, breathing patterns) may be a useful way to control for variance in fMRI data of nonneuronal origin.

15.2.5 Specific Phobias, Social Anxiety Disorder, and Generalized Anxiety Disorder

A recent meta-analysis on specific phobias has confirmed the initial conclusions by Etkin and Wager (2007) and reported that phobic stimuli were associated with increased activity in the amygdala, insula, and pallidum and that these regions and the cerebellum and thalamus were also more activated in phobic patients relative to healthy controls in response to phobic stimuli (Ipser et al. 2013). Differences between patients and controls in the ACC were related to general affective processing. Noteworthy is that this meta-analysis also observed reduced activity in the dACC, thalamus, insula, and more lateral prefrontal regions after cognitive-behavioral treatment ($N=3$) (Ipser et al. 2013). The same group performed a meta-analysis on social anxiety disorder as a distinct and epidemiologically relevant anxiety disorder, which also revealed increased activity in the amygdala, pallidum, and ACC in response to social stimuli (e.g., faces, linguistic stimuli), as well as in the hippocampal complex (perirhinal cortex). These conclusions are still tentative as the amount of included fMRI studies was rather small ($N=7$); however, the reported regions fit well with the regions activated in fear-conditioning tasks (Sehlmeyer et al. 2009), and the absence of hypoactivity in the mPFC could have functional relevance for the specificity of the phobic fear (Lissek 2012) versus the overgeneralization in, for instance, PTSD.

Generalized anxiety disorder, an anxiety disorder characterized by excessive and diffuse anxiety and worrying rather than specific fear, has shown less consistent neural correlates than the fear-related anxiety disorders. An increased amygdala response to all – aversive and neutral – stimuli has been observed in generalized anxiety disorder (Nitschke et al. 2009), although another study reported decreased amygdala activity in response to fearful faces in patients compared to healthy volunteers (Blair et al. 2008), and further studies did not observe any group differences between patients and controls during affective stimulus processing (Whalen et al. 2008; Etkin et al. 2010). It may be that diffuse anxiety may be associated with different neural correlates than amygdala-mediated fear and a role for the involvement of the bed nucleus of the stria terminalis has been proposed (Davis 1998) and observed in generalized anxiety disorder patients in response to uncertainty (Yassa et al. 2012). Moreover, a few studies point to abnormal patterns of (de)activation in the dorsomedial and dorsolateral prefrontal cortices associated with emotional (dys)regulation (Blair et al. 2012; Ball et al. 2012), which differed from healthy controls but not from patients with generalized social phobia (Blair et al. 2012). Further research is necessary to replicate these promising observations.

15.3 Obsessive-Compulsive Disorder

The situation is different in obsessive-compulsive disorder (OCD), which has received much attention in neuroimaging and is rather discrepant from the other anxiety disorders, both in symptomatology and imaging findings. OCD is characterized by obsessive thoughts and images, with which persons cope by means of ritualistic and stereotyped overt or covert behavior. The disorder has been removed from the section of anxiety disorders in the fifth version of the diagnostic and statistical manual of mental disorders (DSM-5) and now has a separate chapter: Obsessive-Compulsive and Related Disorders. This also makes sense from an imaging perspective: instead of dysfunctional neural circuitry primarily in limbic and medial prefrontal areas, both anatomical (Radua and Mataix-Cols 2009; Peng et al. 2012) and functional (Menzies et al. 2008; Rotge et al. 2008) meta-analyses have implicated the striatum and particularly cortico-striato-thalamic loops in OCD, with additional differential activity in the orbito- and prefrontal cortices (Menzies et al. 2008; Rotge et al. 2008). Changes in striatal volumes have not always been found (Rotge et al. 2009), although heterogeneity in the meta-analyses may be due to methodological differences (Ferreira and Busatto 2010).

The meta-analysis by Menzies et al. (2008) on a variety of functional tasks in OCD revealed increased activity in the orbitofrontal cortex and caudate nucleus and medial and lateral prefrontal cortex. Lateral orbitofrontal and medial prefrontal cortex abnormalities were proposed to be related to the intrusive obsessions, whereas striatal abnormalities have been proposed to be related to the stereotyped behavior. A recent meta-analysis on brain structure abnormalities has pointed out that the dmPFC and vmPFC/orbitofrontal cortex have reduced volumes in OCD, whereas the caudate nucleus shows increased volumes (Peng et al. 2012); see Fig. 15.3.

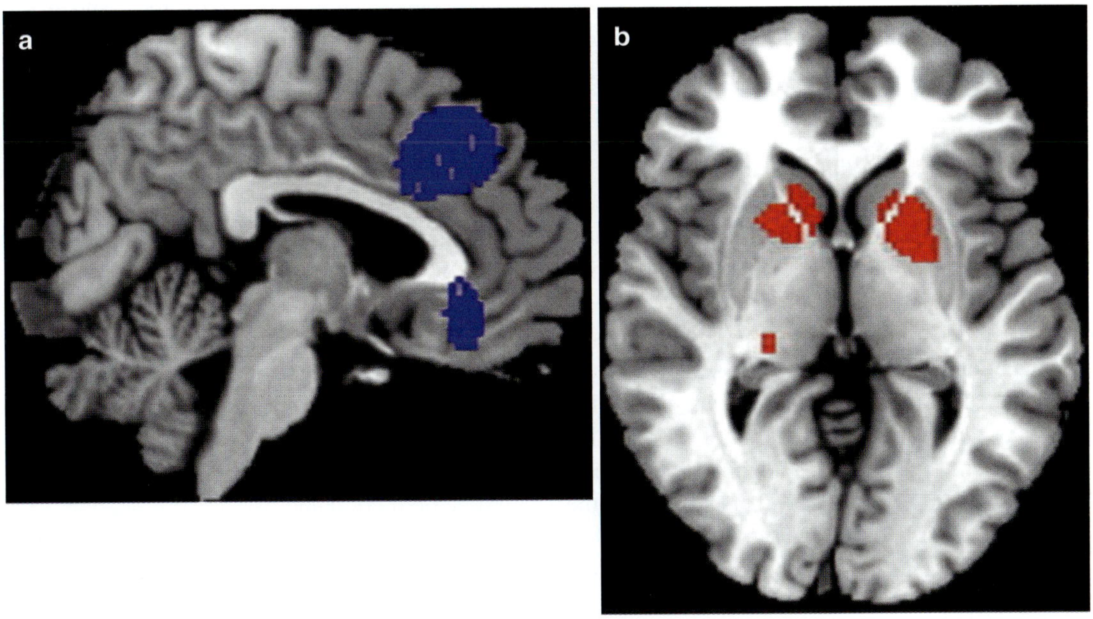

Fig. 15.3 Gray matter differences between patients with obsessive-compulsive disorder and healthy controls, with decreased gray matter volumes in patients depicted in *blue* (**a**) and increased volumes in *red* (**b**) (Reproduced by kind permission from Elsevier from Peng et al. (2012))

15.4 Outlook: Imaging Markers for Prediction of Treatment Outcome and Tracking of Treatment Success

A promising application for functional imaging may be to stratify patients and track treatment progress in an objective manner. Exposure-based cognitive-behavioral therapy is one of the most successful treatments for anxiety disorders and shows effects in a majority of patients. However, a minority does not respond to this therapeutic approach and would possibly benefit more from pharmacological treatment. Imaging tasks related to fear conditioning and extinction could help to predict who might benefit from exposure- (and self-regulation) based therapies and who may better start with pharmacological or combined treatment. Research into this field has started to develop, with a recent fMRI study of 39 social anxiety disorder patients as a timely example: this study observed that activity in the occipitotemporal cortex before treatment was positively correlated to cognitive-behavioral treatment outcome 12 weeks later (Doehrmann et al. 2013). Other studies with a similar setup focused on treatment response in generalized anxiety disorder and noted that dACC activity predicted response to the serotonin-norepinephrine reuptake inhibitor (SNRI) venlafaxine (Whalen et al. 2008; Nitschke et al. 2009). Intriguingly, an early study reported that long-term treatment with the selective serotonin reuptake inhibitor (SSRI) paroxetine resulted in an increase in hippocampal volumes – and verbal declarative memory performance – in PTSD patients (Vermetten et al. 2003).

Moreover, a study in PTSD patients found amygdala and vmPFC activity to be predictive of treatment response to cognitive-behavioral therapy (Bryant et al. 2008). This relates to the previously described finding that symptom severity improvements after cognitive-behavioral therapy for PTSD were related to an increase in the mPFC and a decrease in amygdala activity (Felmingham et al. 2007). Analogous findings were observed in a meta-analysis on specific phobia, with reduced activity in the dACC, thalamus, insula, and more lateral prefrontal regions after cognitive-behavioral treatment (Ipser et al. 2013). However, findings of this meta-analysis are tentative due to small group sizes, typical noninclusion of patient control groups, not reporting the group × time interactions, or employing no or inadequate multiple test corrections in the included studies.

Conclusion

Functional neuroimaging has proved to be a relevant tool for anxiety disorders, including PTSD, specific phobias, social anxiety disorder, and panic disorder. Fear conditioning and extinction tasks have become the dominant paradigm for the etiology (and treatment) of these anxiety disorders and have been successfully combined with functional neuroimaging tools, extending decades of psychophysiological research. Most neuroimaging attention to date has been devoted to PTSD, where hypoactivity in the vmPFC, dmPFC, and hippocampal complex may be a specific feature compared to phobias. In contrast, hyperactivity in amygdala seems to be a general feature shared by most anxiety disorders. Particularly for specific phobias, generalized anxiety disorder, and panic disorder, more replications are required. In contrast, the literature on OCD is more mature and converges towards both functional and structural abnormalities in cortico-striato-thalamic circuitry including the orbitofrontal cortex, albeit with considerable heterogeneity. Methodological challenges lie in the optimization of fMRI acquisition protocols for limbic and medial prefrontal circuitry, the control for respiration-induced vascular changes, and the expansion of fear-conditioning tasks with intrusion-relevant unconditioned stimuli and measurable avoidance behavior. Potential applications seem promising for treatment outcome prediction and patient stratification.

References

Ball TM, Ramsawh HJ, Campbell-Sills L, Paulus MP, Stein MB (2012) Prefrontal dysfunction during emotion regulation in generalized anxiety and panic disorders. Psychol Med 31:1–12

Beckers T, Krypotos AM, Boddez Y, Effting M, Kindt M (2013) What's wrong with fear conditioning? Biol Psychol 92:90–96

Blair K, Shaywitz J, Smith BW, Rhodes R, Geraci M, Jones M, McCaffrey D, Vythilingam M, Finger E, Mondillo K, Jacobs M, Charney DS, Blair RJ, Drevets WC, Pine DS (2008) Response to emotional expressions in generalized social phobia and generalized anxiety disorder: evidence for separate disorders. Am J Psychiatry 165:1193–1202

Blair KS, Geraci M, Smith BW, Hollon N, DeVido J, Otero M, Blair JR, Pine DS (2012) Reduced dorsal anterior cingulate cortical activity during emotional regulation and top-down attentional control in generalized social phobia, generalized anxiety disorder, and comorbid generalized social phobia/generalized anxiety disorder. Biol Psychiatry 72:476–482

Blechert J, Michael T, Vriends N, Margraf J, Wilhelm FH (2007) Fear conditioning in posttraumatic stress disorder: evidence for delayed extinction of autonomic, experiential, and behavioural responses. Behav Res Ther 45(9):2019–2033

Bremner JD, Randall P, Scott TM, Bronen RA, Seibyl JP, Southwick SM, Delaney RC, McCarthy G, Charney DS, Innis RB (1995) MRI-based measurement of hippocampal volume in patients with combat-related posttraumatic stress disorder. Am J Psychiatry 152:973–981

Bremner JD, Randall P, Vermetten E, Staib L, Bronen RA, Mazure C, Capelli S, McCarthy G, Innis RB, Charney DS (1997) Magnetic resonance imaging-based measurement of hippocampal volume in posttraumatic stress disorder related to childhood physical and sexual abuse–a preliminary report. Biol Psychiatry 41:23–32

Bryant RA, Felmingham K, Kemp A, Das P, Hughes G, Peduto A, Williams L (2008) Amygdala and ventral anterior cingulate activation predicts treatment response to cognitive behaviour therapy for posttraumatic stress disorder. Psychol Med 38:555–561

Büchel C, Morris J, Dolan RJ, Friston KJ (1998) Brain systems mediating aversive conditioning: an event-related fMRI study. Neuron 20:947–957

Davis M (1998) Are different parts of the extended amygdala involved in fear versus anxiety? Biol Psychiatry 44:1239–1247

de Carvalho MR, Dias GP, Cosci F, de-Melo-Neto VL, Bevilaqua MC, Gardino PF, Nardi AE (2010) Current findings of fMRI in panic disorder: contributions for the fear neurocircuitry and CBT effects. Expert Rev Neurother 10:291–303

Doehrmann O, Ghosh SS, Polli FE, Reynolds GO, Horn F, Keshavan A, Triantafyllou C, Saygin ZM, Whitfield-Gabrieli S, Hofmann SG, Pollack M, Gabrieli JD (2013) Predicting treatment response in social anxiety disorder from functional magnetic resonance imaging. JAMA Psychiatry 70:87–97

Dresler T, Guhn A, Tupak SV, Ehlis AC, Herrmann MJ, Fallgatter AJ, Deckert J, Domschke K (2013) Revise the revised? New dimensions of the neuroanatomical hypothesis of panic disorder. J Neural Transm 120:3–29

Eser D, Leicht G, Lutz J, Wenninger S, Kirsch V, Schüle C, Karch S, Baghai T, Pogarell O, Born C, Rupprecht R, Mulert C (2009) Functional neuroanatomy of CCK-4-induced panic attacks in healthy volunteers. Hum Brain Mapp 30:511–522

Etkin A, Wager TD (2007) Functional neuroimaging of anxiety: a meta-analysis of emotional processing in PTSD, social anxiety disorder, and specific phobia. Am J Psychiatry 164:1476–1488

Etkin A, Prater KE, Hoeft F, Menon V, Schatzberg AF (2010) Failure of anterior cingulate activation and connectivity with the amygdala during implicit regulation of emotional processing in generalized anxiety disorder. Am J Psychiatry 167:545–554

Felmingham K, Kemp A, Williams L, Das P, Hughes G, Peduto A, Bryant R (2007) Changes in anterior cingulate and amygdala after cognitive behavior therapy of posttraumatic stress disorder. Psychol Sci 18:127–129

Ferreira LK, Busatto GF (2010) Heterogeneity of coordinate-based meta-analyses of neuroimaging data: an example from studies in OCD. Br J Psychiatry 197:76–77

Francati V, Vermetten E, Bremner JD (2007) Functional neuroimaging studies in posttraumatic stress disorder: review of current methods and findings. Depress Anxiety 24:202–218

Giardino ND, Friedman SD, Dager SR (2007) Anxiety, respiration, and cerebral blood flow: implications for functional brain imaging. Compr Psychiatry 48:103–112

Grillon C, Baas JMP, Cornwell B, Johnson L (2006) Context conditioning and behavioral avoidance in a virtual reality environment: effect of predictability. Biol Psychiatry 60:752–759

Gustavsson A, Svensson M, Jacobi F, Allgulander C, Alonso J, Beghi E, Dodel R, Ekman M, Faravelli C, Fratiglioni L, Gannon B, Jones DH, Jennum P, Jordanova A, Jönsson L, Karampampa K, Knapp M, Kobelt G, Kurth T, Lieb R, Linde M, Ljungcrantz C, Maercker A, Melin B, Moscarelli M, Musayev A, Norwood F, Preisig M, Pugliatti M, Rehm J, Salvador-Carulla L, Schlehofer B, Simon R, Steinhausen HC, Stovner LJ, Vallat JM, Van den Bergh P, van Os J, Vos P, Xu W, Wittchen HU, Jönsson B, Olesen J, CDBE2010Study Group (2011) Cost of disorders of the brain in Europe 2010. Eur Neuropsychopharmacol 21:718–779

Hayes JP, Hayes SM, Mikedis AM (2012) Quantitative meta-analysis of neural activity in posttraumatic stress disorder. Biol Mood Anxiety Disord 2:9

Ioannidis JP (2011) Excess significance bias in the literature on brain volume abnormalities. Arch Gen Psychiatry 68:773–780

Ipser JC, Singh L, Stein DJ (2013) Meta-analysis of functional brain imaging in specific phobia. Psychiatry Clin Neurosci 67:311–322

Karl A, Schaefer M, Malta LS, Dörfel D, Rohleder N, Werner A (2006) A meta-analysis of structural brain abnormalities in PTSD. Neurosci Biobehav Rev 30:1004–1031

Kessler RC, Petukhova M, Sampson NA, Zaslavsky AM, Wittchen HU (2012) Twelve-month and lifetime prevalence and lifetime morbid risk of anxiety and mood disorders in the United States. Int J Methods Psychiatr Res 21:169–184

LaBar KS, Gatenby JC, Gore JC, LeDoux JE, Phelps EA (1998) Human amygdala activation during conditioned fear acquisition and extinction: a mixed-trial fMRI study. Neuron 20:937–945

Lanius RA, Vermetten E, Loewenstein RJ, Brand B, Schmahl C, Bremner JD, Spiegel D (2010a) Emotion modulation in PTSD: clinical and neurobiological evidence for a dissociative subtype. Am J Psychiatry 167:640–647

Lanius RA, Frewen PA, Vermetten E, Yehuda R (2010b) Fear conditioning and early life vulnerabilities: two distinct pathways of emotional dysregulation and brain dysfunction in PTSD. Eur J Psychotraumatol 1:5467

LeDoux JE (2000) Emotion circuits in the brain. Ann Rev Neurosci 23:155–184

Liberzon I, Sripada CS (2008) The functional neuroanatomy of PTSD: a critical review. Prog Brain Res 167:151–169

Lissek S (2012) Toward an account of clinical anxiety predicated on basic, neurally mapped mechanisms of Pavlovian fear-learning: the case for conditioned overgeneralization. Depress Anxiety 29:257–263

Lissek S, Powers AS, McClure EB, Phelps EA, Woldehawariat G, Grillon C, Pine DS (2005) Classical fear conditioning in the anxiety disorders: a meta-analysis. Behav Res Ther 43:1391–1424

Lissek S, Rabin S, Heller RE, Lukenbaugh D, Geraci M, Pine DS, Grillon C (2010) Overgeneralization of conditioned fear as a pathogenic marker of panic disorder. Am J Psychiatry 167:47–55

Lueken U, Straube B, Reinhardt I, Maslowski NI, Wittchen HU, Ströhle A, Wittmann A, Pfleiderer B, Konrad C, Ewert A, Uhlmann C, Arolt V, Jansen A, Kircher T (2013) Altered top-down and bottom-up processing of fear conditioning in panic disorder with agoraphobia. Psychol Med 23:1–14

Martinez JM, Coplan JD, Browne ST, Goetz R, Welkowitz LA, Papp LA, Klein DF, Gorman JM (1998) Hemodynamic response to respiratory challenges in panic disorder. J Psychosom Res 44:153–161

Menzies L, Chamberlain SR, Laird AR, Thelen SM, Sahakian BJ, Bullmore ET (2008) Integrating evidence from neuroimaging and neuropsychological studies of obsessive-compulsive disorder: the orbitofronto-striatal model revisited. Neurosci Biobehav Rev 32:525–549

Michael T, Blechert J, Vriends N, Margraf J, Wilhelm FH (2007) Fear conditioning in panic disorder: enhanced resistance to extinction. J Abnorm Psychol 116:612–617

Milad MR, Quirk GJ (2012) Fear extinction as a model for translational neuroscience: ten years of progress. Annu Rev Psychol 63:129–151

Milad MR, Wright CI, Orr SP, Pitman RK, Quirk GJ, Rauch SL (2007) Recall of fear extinction in humans activates the ventromedial prefrontal cortex and hippocampus in concert. Biol Psychiatry 62:446–454

Milad MR, Pitman RK, Ellis CB, Gold AL, Shin LM, Lasko NB, Zeidan MA, Handwerger K, Orr SP, Rauch SL (2009) Neurobiological basis of failure to recall extinction memory in posttraumatic stress disorder. Biol Psychiatry 66:1075–1082

Nitschke JB, Sarinopoulos I, Oathes DJ, Johnstone T, Whalen PJ, Davidson RJ, Kalin NH (2009) Anticipatory activation in the amygdala and anterior cingulate in generalized anxiety disorder and prediction of treatment response. Am J Psychiatry 166:302–310

Pape HC, Pare D (2010) Plastic synaptic networks of the amygdala for the acquisition, expression, and extinction of conditioned fear. Physiol Rev 90:419–463

Pavlov IP (1927) Conditioned reflexes. Oxford University Press, London

Peng Z, Lui SS, Cheung EF, Jin Z, Miao G, Jing J, Chan RC (2012) Brain structural abnormalities in obsessive-compulsive disorder: converging evidence from white matter and grey matter. Asian J Psychiatry 5:290–296

Phelps EA, Delgado MR, Nearing KI, LeDoux JE (2004) Extinction learning in humans: role of the amygdala and vmPFC. Neuron 43:897–905

Pitman RK, Rasmusson AM, Koenen KC, Shin LM, Orr SP, Gilbertson MW, Milad MR, Liberzon I (2012) Biological studies of post-traumatic stress disorder. Nat Rev Neurosci 13:769–787

Pole N (2007) The psychophysiology of posttraumatic stress disorder: a meta-analysis. Psychol Bull 133:725–746

Radua J, Mataix-Cols D (2009) Voxel-wise meta-analysis of grey matter changes in obsessive-compulsive disorder. Br J Psychiatry 195:393–402

Rauch SL, Shin LM, Phelps EA (2006) Neurocircuitry models of posttraumatic stress disorder and extinction: human neuroimaging research–past, present, and future. Biol Psychiatry 60:376–382

Rodnick EH (1937) Characteristics of delayed and trace conditioned responses. J Exp Psychol 20:409–425

Rotge JY, Guehl D, Dilharreguy B, Cuny E, Tignol J, Bioulac B, Allard M, Burbaud P, Aouizerate B (2008) Provocation of obsessive-compulsive symptoms: a quantitative voxel-based meta-analysis of functional neuroimaging studies. J Psychiatry Neurosci 33:405–412

Rotge JY, Guehl D, Dilharreguy B, Tignol J, Bioulac B, Allard M, Burbaud P, Aouizerate B (2009) Meta-analysis of brain volume changes in obsessive-compulsive disorder. Biol Psychiatry 65:75–83

Schwarzbauer C, Mildner T, Heinke W, Brett M, Deichmann R (2010) Dual echo EPI–the method of choice for fMRI in the presence of magnetic field inhomogeneities? Neuroimage 49:316–326

Sehlmeyer C, Schöning S, Zwitserlood P, Pfleiderer B, Kircher T, Arolt V, Konrad C (2009) Human fear conditioning and extinction in neuroimaging: a systematic review. PLoS One 4:e5865

Shin LM, Wright CI, Cannistraro PA, Wedig MM, McMullin K, Martis B, Macklin ML, Lasko NB,

Cavanagh SR, Krangel TS, Orr SP, Pitman RK, Whalen PJ, Rauch SL (2005) A functional magnetic resonance imaging study of amygdala and medial prefrontal cortex responses to overtly presented fearful faces in posttraumatic stress disorder. Arch Gen Psychiatry 62:273–281

Spoormaker VI, Andrade KC, Schröter MS, Sturm A, Goya-Maldonado R, Sämann PG, Czisch M (2011) The neural correlates of negative prediction error signaling in human fear conditioning. Neuroimage 54:2250–2256

Starkman MN, Giordani B, Gebarski SS, Schteingart DE (2003) Improvement in learning associated with increase in hippocampal formation volume. Biol Psychiatry 53(3):233–238

Steckle LC (1933) A trace conditioning of the galvanic reflex. Gen Psychol 9:475–480

Switzer SA (1934) Anticipatory and inhibitory characteristics of delayed conditioned reactions. J Exp Psychol 17:603–620

Tuescher O, Protopopescu X, Pan H, Cloitre M, Butler T, Goldstein M, Root JC, Engelien A, Furman D, Silverman M, Yang Y, Gorman J, LeDoux J, Silbersweig D, Stern E (2011) Differential activity of subgenual cingulate and brainstem in panic disorder and PTSD. J Anxiety Disord 25:251–257

Vermetten E, Lanius RA (2012) Biological and clinical framework for posttraumatic stress disorder. In: Schlaepfer TE, Nemeroff CB (eds) Neurobiology of psychiatric disorders, vol 106 (3rd series), Handbook of clinical neurology. Elsevier, London, pp 291–342

Vermetten E, Vythilingam M, Southwick SM, Charney DS, Bremner JD (2003) Long-term treatment with paroxetine increases verbal declarative memory and hippocampal volume in posttraumatic stress disorder. Biol Psychiatry 54:693–702

Vermetten E, Schmahl C, Southwick SM, Bremner JD (2007) Positron tomographic emission study of olfactory induced emotional recall in veterans with and without combat-related posttraumatic stress disorder. Psychopharmacol Bull 40:8–30

Vythilingam M, Heim C, Newport J, Miller AH, Anderson E, Bronen R, Brummer M, Staib L, Vermetten E, Charney DS, Nemeroff CB, Bremner JD (2002) Childhood trauma associated with smaller hippocampal volume in women with major depression. Am J Psychiatry 159:2072–2080

Wegerer M, Blechert J, Kerschbaum H, Wilhelm FH (2013) Relationship between fear conditionability and aversive memories: evidence from a novel conditioned-intrusion paradigm. PLoS ONE 8:e79025. doi:10.1371/journal.pone.0079025

Wessa M, Flor H (2007) Failure of extinction of fear responses in posttraumatic stress disorder: evidence from second-order conditioning. Am J Psychiatry 164:1684–1692

Whalen PJ, Johnstone T, Somerville LH, Nitschke JB, Polis S, Alexander AL, Davidson RJ, Kalin NH (2008) A functional magnetic resonance imaging predictor of treatment response to venlafaxine in generalized anxiety disorder. Biol Psychiatry 63:858–863

Wittchen HU, Jacobi F, Rehm J, Gustavsson A, Svensson M, Jönsson B, Olesen J, Allgulander C, Alonso J, Faravelli C, Fratiglioni L, Jennum P, Lieb R, Maercker A, van Os J, Preisig M, Salvador-Carulla L, Simon R, Steinhausen HC (2011) The size and burden of mental disorders and other disorders of the brain in Europe 2010. Eur Neuropsychopharmacol 21:655–679

Yassa MA, Hazlett RL, Stark CE, Hoehn-Saric R (2012) Functional MRI of the amygdala and bed nucleus of the stria terminalis during conditions of uncertainty in generalized anxiety disorder. J Psychiatr Res 46:1045–1052

The Structural and Functional Neuroanatomy of Bipolar Disorder

16

Sophia Frangou

Abbreviations

ACC	Anterior cingulate cortex
BA	Brodmann area
BD	Bipolar disorder
DLPFC	Dorsolateral prefrontal cortex
DTI	Diffusion tensor imaging
FA	Fractional anisotropy
fMRI	Functional magnetic resonance imaging
GAF	Global Assessment of Functioning
MDD	Major depressive disorder
PFC	Prefrontal cortex
ROI	Region of interest
sMRI	Structural magnetic resonance imaging
SZ	Schizophrenia
VBM	Voxel brain morphology
VLPFC	Lateral PFC
VMPFC	Medial PFC
VPFC	The ventral PFC
WMH	White matter hyperintensities

S. Frangou, MD, MSc, PhD, FRCPsych
Department of Psychiatry,
Icahn School of Medicine at Mount Sinai,
1425 Madison Ave, 1230,
New York, NY 10029, USA
e-mail: sophia.frangou@mssm.edu

16.1 Introduction

Bipolar disorder (BD) is characterised by emotional dysregulation associated with abnormal depressive or manic affect (APA 1994). The last two decades have witnessed a concerted effort to identify neural underpinnings of BD and to develop a neurobiological model for the disorder within the wider context of affective neuroscience. In this respect, neuroimaging studies have begun to yield consistent and biologically meaningful information on the impact of BD on neural networks involved in affect regulation. The remainder of this chapter presents an overview and critical synthesis of the findings so far and highlights areas for future research.

16.2 Neural Networks for Emotional Regulation

The regulation and integration of emotion with cognitive and visceral functions engage distributed neural networks. Key nodes within these networks include the ventral and dorsal prefrontal cortex (PFC), the anterior cingulate cortex (ACC), the amygdala and the parahippocampal gyrus and the insula. These regions are heavily interconnected and also connected with other brain structures, particularly the thalamus, hypothalamus and striatum.

The ventral PFC (VPFC) (Brodmann Area (BA) 10, 47, 11) is traditionally described as having a lateral (VLPFC) and a medial (VMPFC)

C. Mulert, M.E. Shenton (eds.), *MRI in Psychiatry*,
DOI 10.1007/978-3-642-54542-9_16, © Springer Berlin Heidelberg 2014

division based on anatomical and functional considerations (Gläscher et al. 2012). Switching or inhibiting responses when contingencies change engages the VLPFC (Robins 2007), while the ability to use motivational or reward value information to make strategically appropriate choices depends on an intact VMPFC (Bechara and Van Der Linden 2005). The ACC (BA25, 32, 33, 24) possesses extensive cortico-cortical connections with both the VLPFC and the dorsolateral prefrontal cortex (DLPFC) (Paus 2001). The dorsal ACC in particular operates closely with the lateral PFC and with the motor cortices in tasks that require increased monitoring and implementation of performance adjustments (Gläscher et al. 2012). The subgenual and perigenual ACC are extensively connected with the VMPFC, the basolateral complex of the amygdala, the hypothalamus and the periaqueductal grey and thus provide cortical control over autonomic and endocrine function (Ongür et al. 1998). The parahippocampal gyrus and the amygdala often coactivate during emotional processing (Fusar-Poli et al. 2009), but their engagement is thought to subserve partially segregated functions. That is, amygdala engagement may signal salience or ambiguity (Gerber et al. 2008; Santos et al. 2011), while parahippocampal involvement may reflect context appraisal (Gerdes et al. 2010). The insula is considered important in a variety of conditions involving the primary and multimodal integrative processing of interoceptive stimuli, supporting both a coherent representation and a conscious awareness of feelings toward stimuli or events (Craig 2011).

16.3 Brain Morphology in BD

Structural magnetic resonance imaging (sMRI) studies have consistently identified morphological differences between BD patients and healthy individuals. There are now several reviews and meta-analyses that have synthesised this extensive primary sMRI literature (McDonald et al. 2004; Kempton et al. 2008, 2011; Arnone et al. 2009; Beyer et al. 2009; Vita et al. 2009; Ellison-Wright and Bullmore 2010; Bora et al. 2010; Selvaraj et al. 2012). Ventricular enlargement of approximately 20 % is a consistent although nonspecific finding (McDonald et al. 2004; Arnone et al. 2009; Kempton et al. 2008, 2011). Morphological changes in white and grey matter are also reliably present and more informative in terms of linking brain morphology to disease expression (Ellison-Wright and Bullmore 2010; Bora et al. 2010; Kempton et al. 2011; Selvaraj et al. 2012).

16.3.1 White Matter

There is considerable evidence for white matter involvement in BD (Mahon et al. 2010). Deep white matter hyperintensities (WMH) are more prevalent in BD patients than healthy individuals (odds ratios 3.5, 95 %; CI 2.2, 4.5) regardless of age (Beyer et al. 2009). Increased rates of WMH are not disease specific as they have been observed in major depressive disorder (MDD) and in schizophrenia (SZ) (Beyer et al. 2009). Their presence in BD, however, may be of prognostic significance; a higher number of WMH has been linked to poor outcome (Moore et al. 2001) and to treatment resistance (Regenold et al. 2008). White matter volume is also reduced both globally (Kempton et al. 2008; Vita et al. 2009) and regionally, particularly in the corpus callosum (Arnone et al. 2008; Walterfang et al. 2009). These findings are of clinical relevance. Global white matter volume reduction is probably one of the earliest brain structural abnormalities in BD and can be detected in first-episode patients (Vita et al. 2009). Moreover, white matter integrity is a significant predictor of outcome as shown by Forcada and colleagues (2011). They investigated the association between patients' functional status, assessed with the Global Assessment of Functioning (GAF; APA 1994), with clinical, cognitive and brain structural variables. They found that IQ, global white matter volume and a predominantly depressive illness course were independently associated with functional outcome and accounted for 53 % of the variance in patients' GAF scores (Fig. 16.1). Regional white matter abnormalities in the genu of the corpus

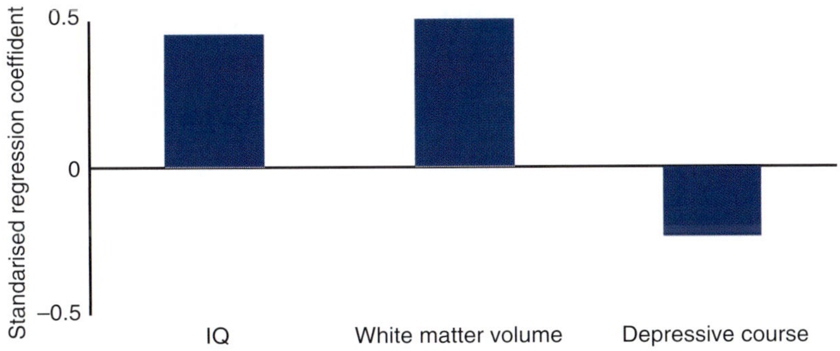

Fig. 16.1 Predictors of global function in bipolar disorder

callosum are also clinically important as have been linked to increased motor and non-planning impulsivity as well as suicidality in BD (Matsuo et al. 2010).

The application of diffusion tensor imaging (DTI) techniques has revealed that white matter deficits in BD are also present at the microstructural level (Brambilla et al. 2009; Vederine et al. 2011). DTI infers information about the integrity of white matter tracts based on measures of fractional anisotropy (FA) and diffusivity. Reduced FA in BD patients compared to healthy individuals has been most consistently described in the white matter surrounding the amygdala/parahippocampal gyrus and the white matter of the VPFC extending to the ventral ACC, corona radiata and genu of the corpus callosum (Benedetti et al. 2011; Vederine et al. 2011). Importantly, reduced FA in the medial VPFC and ventral ACC has been associated with increased impulsivity and suicidality (Mahon et al. 2012).

16.3.2 Grey Matter

Grey matter changes in BD are generally subtle as their average effect size is 0.5 or less (Kempton et al. 2008). There is little evidence for global grey matter abnormalities especially in first-episode patients (Vita et al. 2009). Regional grey matter volume reductions have been consistently reported in BD patients compared to healthy individuals in the VPFC, insula and ACC (Bora et al. 2010; Ellison-Wright and Bullmore 2010; Selvaraj et al. 2012). Despite initial reports implicating the amygdala, meta-analyses of sMRI data from voxel brain morphology (VBM) studies in adult BD patients have not found consistent alterations in this structure. Similarly, meta-analyses of region-of-interest (ROI) studies report minimal differences in amygdala volume between adult BD patients and healthy individuals (Hedges' g −0.04 to −0.07) (Kempton et al. 2008). However, amygdala volume reduction has emerged as a consistent finding in paediatric BD (Terry et al. 2009). Evidence from longitudinal studies indicate that the smaller size of the amygdala early in life in BD may be accounted for by delayed or abnormal developmental trajectories in patients compared to healthy youth (Bitter et al. 2011).

Grey matter reductions in the VPFC and the ACC represent a robust correlate of disease expression in BD and are present regardless the age of onset. Studies that have examined the relevance of these findings to other disorders have reported overlap but also differences. Grey matter volume reductions in SZ are generally more extensive and more severe, but there is substantial overlap between the two disorders in the insula and ACC (Ellison-Wright and Bullmore 2010). Patients with SZ also appear to have greater ventricular enlargement and smaller amygdala volumes than those with BD (Arnone et al. 2009). A substantial overlap has also been found in brain structural changes in BD and MDD (Kempton et al. 2011). However, MDD patients may have smaller hippocampal and basal ganglia volume and larger corpus callosum areas compared to BD patients (Kempton et al. 2011).

16.4 Brain Function in BD

16.4.1 Resting State Studies

Functional magnetic resonance imaging (fMRI) techniques have been adapted to allow examination of resting state data and subsequent connectivity analyses of resting state neural networks. The first report was from Anand and colleagues (2009) who found decreased connectivity between the pregenual ACC and the thalamus, basal ganglia and amygdala when comparing resting state data from unmedicated symptomatic (manic=6, depressed=5) BD patients to that from healthy individuals. Chepenik and colleagues (2010) focused on VMPFC connectivity based on resting state data from 15 BD patients with variable symptomatology compared to 10 healthy individuals. Patients had reduced negative connectivity between the VMPFC and the amygdala and increased positive connectivity between the VMFC and the ipsilateral VLPFC and ACC. These findings were largely replicated by Chai et al. (2011) in 14 acutely manic and psychotic BD patients. In contrast to healthy individuals, patients showed positive VMPFC-amygdala and VMPFC-VLPFC connectivity. Two further studies have established that abnormal VPFC-amygdala connectivity persists in euthymic BD patients (Anticevic et al. 2013; Torrisi et al. 2013) while a further study confirmed the relevance of these connectivity finding to disease expression for BD as opposed to SZ (Mamah et al. 2013; Meda et al. 2012).

16.4.2 Activation Studies

Functional neuroimaging studies in BD have employed a variety of activation paradigms. Three recent quantitative meta-analyses have usefully summarised the literature in terms of the neural regions implicated across paradigms (Chen et al. 2011) and with regard to specificity compared to MDD (Delvecchio et al. 2012) and SZ (Delvecchio et al. 2013). Chen and colleagues reviewed 65 fMRI studies comparing BD patients to healthy individuals and categorised tasks into cognitive and emotional based on the absence or presence of emotive content. During cognitive task performance, BD patients showed abnormally decreased activation in the VLPFC, the putamen and lingual gyrus. Abnormally decreased activation in the VLPFC in patients was also present during emotional tasks where additionally patients demonstrated abnormally increased engagement in the parahippocampal gyrus extending to the amygdala and in the putamen and pallidum. Delvecchio et al. (2012) further demonstrated that increased activation of the parahippocampal gyrus/amygdala during emotional tasks was a shared feature of BD and MDD. However, diagnosis-specific differences were also observed as decreased VLPFC engagement and increased responsiveness in the thalamus and basal ganglia were associated with BD and not MDD. Comparison between BD and SZ during emotional processing tasks revealed significant diagnostic differences; SZ was associated with underactivation both in medial temporal and in prefrontal regions but greater engagement in posterior associative visual cortices (Delvecchio et al. 2013). This finding suggests that the origin of emotional processing abnormalities in BD probably relates to dysfunctional emotional regulation while abnormal visual integration may be the core underlying deficit in SZ.

16.5 Key Moderating Factors

Sex does not appear to have a major moderator role in brain morphology in BD (Jogia et al. 2012). Age and duration of illness are highly collinear. As there is a paucity of longitudinal studies in BD, the effects of these two factors cannot be easily disentangled. Sarnicola et al. (2009) did not find differential age-related changes in brain morphology in BD patients and healthy individuals over an age span of 40 years. However, multi-episode (as opposed to first episode) BD patients may have smaller VPFC and larger striatum and amygdala (Bora et al. 2010). The latter observations may reflect chronic exposure to medication and particularly antipsychotics and lithium. The association between antipsychotics and increased striatal volume is well replicated and is considered to reflect diagnosis-independent microstructural changes in response to long-term reduction in dopamine neurotransmission (Navari and Dazzan 2009). Moore et al. (2000) were the first to report that treatment with lithium increased global grey matter volume in BD patients. This observation has been robustly replicated with regard to global (Bearden et al. 2007; Lyoo et al. 2010) and regional grey matter volume increases particularly in the amygdala, rostral ACC and PFC (Bearden et al. 2007; Germaná et al. 2010; Moore et al. 2009). This brain volume "response" to lithium is of therapeutic significance as it has been linked to symptomatic improvement (Lyoo et al. 2010; Moore et al. 2009). Although the precise mechanism underlying lithium-induced brain structural changes in BD remains to be determined, the drug's primary target, glycogen synthase kinase-3, is involved in pathways with neurotrophic, neurogenetic and neuroprotective functions (Machado-Vieira et al. 2009). Clinical symptoms at the time of scanning may also impact neuroimaging results. This is mostly the case for functional rather than anatomical studies. In general, acute symptoms, either depressive or manic, appear to increase fMRI signal in subcortical regions (hippocampus, amygdala) and reduce PFC engagement (Chen et al. 2011; Delvecchio et al. 2013).

16.6 Translational Potential of Neuroimaging for BD

The findings described above demonstrate the contribution of neuroimaging to our understanding of the neural correlates in BD. However, their impact on clinical practice has been negligible. This is primarily because conventional data analysis computes differences between groups either within predefined anatomical regions of interest (ROI) or throughout the brain using univariate mass analyses such as voxel-based morphometry (VBM). Both types of analyses are not able to differentiate between groups where mean differences are small to moderate. The ROI approach is statistically more powerful but restricted to regions that can be reliably anatomically defined. VBM requires correction for multiple comparisons that limits its statistical power to detect subtle changes. Additionally, neither approach utilises information about the spatial distribution of case-control differences. Current research interest is focusing on maximising the translational value of neuroimaging by developing new computational tools for data analysis that could assist with the clinical evaluation of patients.

Calhoun and colleagues (2008) used independent component analysis to extract mode images of brain activity during resting state and during performance on a selective attention task in patients with BD or SZ and healthy individuals. A multistage classification algorithm was then applied to concatenate each participant's mode images into a single image. A mean image was computed for each group, and the Euclidean distance between an individual's brain image and each group's images was computed. Using this approach, the authors were able to classify participants according to diagnostic group with an average sensitivity of 90 % and specificity of 95 %.

Another computational approach involves the use of supervised machine learning algorithms to predict diagnostic classification. Costafreda and colleagues (2011) applied support vector machine (SVM) to functional imaging data derived from patients with BD or SZ and healthy individuals while performing a verbal fluency task. Classification accuracy for SZ was 92 % (91 % sensitivity and 92 % specificity) and for BD 79 % (56 % sensitivity and 89 % specificity); misclassification occurred mostly because of the wrong categorization of BD patients as healthy individuals. Rocha-Rego and colleagues (2013) applied Gaussian Process Classifiers (GPCs) to sMRI data to evaluate the feasibility of using pattern recognition techniques for the diagnostic differentiation of patients with BD from healthy individuals. They chose to focus on sMRI because of its high acceptability by patients and wide availability in clinical settings. GPCs represent a significant advance as they combine equivalent predictive performance to SVM with the additional benefit of probabilistic classification (Marquand et al. 2010). Grey and white matter classifiers correctly assigned patients to the appropriate diagnostic category

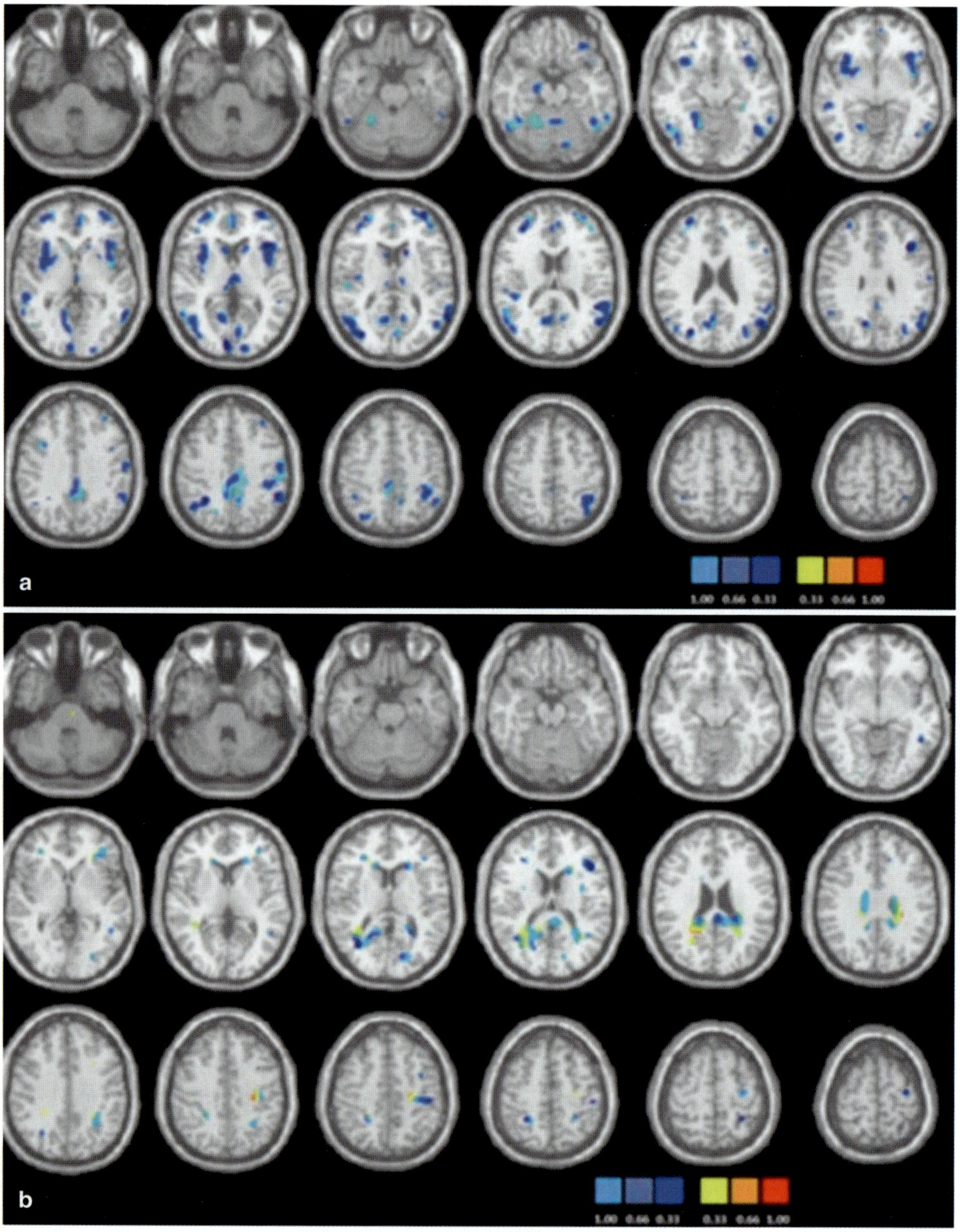

Fig. 16.2 Discrimination maps for grey matter (**a**) and white matter (**b**) classification between patients with bipolar disorder and healthy individuals

with respective accuracy levels of 73 % (sensitivity 69 %, specificity 77 %) and 69 % (sensitivity 69 %, specificity 69 %). Grey matter discriminative clusters were localised within cortical and subcortical structures implicated in BD, including the VPFC, ACC, parahippocampal gyrus, insula, thalamus and striatum (Fig. 16.2a). White matter discriminative clusters were identified mainly within the cingulum, occipital regions and the genu of the corpus callosum (Fig. 16.2b).

16.7 Concluding Remarks

This chapter summarised the effect of disease expression for BD on brain structure and function. There are four key messages. First, the data demonstrate the central role of VPFC dysfunction in BD. Postmortem studies in BD report neuropathological abnormalities in the VPFC leading to regional reductions in the number and density of pyramidal cells and PV interneurons (Cotter et al. 2005; Pantazopoulos et al. 2007). The mechanisms involved are not established, but multiple lines of evidence implicate reduced expression of neurotrophins (Berk et al. 2011), abnormalities in oxidative energy generation (Berk et al. 2011) and mitochondrial dysfunction resulting in altered Ca+2 regulation and PV-interneuron reduction (Berk et al. 2011; Powell et al. 2012). Second, the evidence suggests that medial temporal lobe regions (amygdala/parahippocampal gyrus), although not structurally compromised, are functionally abnormal in terms of neural responsiveness and connectivity, both consistent with loss of regulatory input. Third, additional abnormalities are present in subcortical regions, particularly the insula, thalamus and basal ganglia, suggesting that mechanisms relating to BD impact on the representation and awareness of feelings toward stimuli or events (insula) (Craig 2011), directing and maintaining attention toward emotional stimuli (thalamus) (Pessoa and Adolphs 2010), and on response selection (basal ganglia) (Robins 2007). Fourth, increased computational sophistication may allow in the near future the use of neuroimaging as a clinical evaluation tool. The data summarised here also highlight a major knowledge gap. That is, although we are able to define the "biological phenotype" of BD in increasing detail and precision, we are still behind in delineating the pathophysiological mechanisms that contribute to this phenotype.

References

American Psychiatric Association (1994) Diagnostic and statistical manual of mental disorders: DSM-IV. American Psychiatric Press, Washington, DC

Anand A, Li Y, Wang Y, Lowe MJ et al (2009) Resting state corticolimbic connectivity abnormalities in unmedicated bipolar disorder and unipolar depression. Psychiatry Res 171:189–198

Anticevic A, Brumbaugh MS, Winkler AM et al (2013) Global prefrontal and fronto-amygdala dysconnectivity in bipolar I disorder with psychosis history. Biol Psychiatry 73:565–573

Arnone D, McIntosh AM, Chandra P et al (2008) Meta-analysis of magnetic resonance imaging studies of the corpus callosum in bipolar disorder. Acta Psychiatr Scand 118:357–362

Arnone D, Cavanagh J, Gerber D et al (2009) Magnetic resonance imaging studies in bipolar disorder and schizophrenia: meta-analysis. Br J Psychiatry 195:194–201

Bearden CE, Thompson PM, Dalwani M et al (2007) Greater cortical gray matter density in lithium-treated patients with bipolar disorder. Biol Psychiatry 62:7–16

Bechara A, Van Der Linden M (2005) Decision-making and impulse control after frontal lobe injuries. Curr Opin Neurol 18:734–739

Benedetti F, Absinta M, Rocca MA et al (2011) Tract-specific white matter structural disruption in patients with bipolar disorder. Bipolar Disord 13:414–424

Berk M, Kapczinski F, Andreazza AC et al (2011) Pathways underlying neuroprogression in bipolar disorder: focus on inflammation, oxidative stress and neurotrophic factors. Neurosci Biobehav Rev 35:804–817

Beyer JL, Young R, Kuchibhatla M et al (2009) Hyperintense MRI lesions in bipolar disorder: a meta-analysis and review. Int Rev Psychiatry 21:394–409

Bitter SM, Mills NP, Adler CM et al (2011) Progression of amygdala volumetric abnormalities in adolescents after their first manic episode. J Am Acad Child Adolesc Psychiatry 50:1017–1026

Bora E, Fornito A, Yücel M et al (2010) Voxelwise meta-analysis of gray matter abnormalities in bipolar disorder. Biol Psychiatry 67:1097–1105

Brambilla P, Bellani M, Yeh PH et al (2009) White matter connectivity in bipolar disorder. Int Rev Psychiatry 21:380–386

Calhoun VD, Maciejewski PK, Pearlson GD et al (2008) Temporal lobe and "default" hemodynamic brain modes discriminate between schizophrenia and bipolar disorder. Hum Brain Mapp 29:1265–1275

Chai XJ, Whitfield-Gabrieli S, Shinn AK et al (2011) Abnormal medial prefrontal cortex resting-state connectivity in bipolar disorder and schizophrenia. Neuropsychopharmacology 36:2009–2017

Chen CH, Suckling J, Lennox BR et al (2011) A quantitative meta-analysis of fMRI studies in bipolar disorder. Bipolar Disord 13:1–15

Chepenik LG, Raffo M, Hampson M et al (2010) Functional connectivity between ventral prefrontal

cortex and amygdala at low frequency in the resting state in bipolar disorder. Psychiatry Res 182:207–210
Costafreda SG, Fu CH, Picchioni M et al (2011) Pattern of neural responses to verbal fluency shows diagnostic specificity for schizophrenia and bipolar disorder. BMC Psychiatry. doi:10.1186/1471-244X-11-18
Cotter D, Hudson L, Landau S (2005) Evidence for orbitofrontal pathology in bipolar disorder and major depression, but not in schizophrenia. Bipolar Disord 7:358–369
Craig AD (2011) Significance of the insula for the evolution of human awareness of feelings from the body. Ann NY Acad Sci 1225:72–82
Delvecchio G, Fossati P, Boyer P et al (2012) Common and distinct neural correlates of emotional processing in bipolar disorder and major depressive disorder: a voxel-based meta-analysis of functional magnetic resonance imaging studies. Eur Neuropsychopharmacol 22:100–113
Delvecchio G, Sugranyes G, Frangou S (2013) Evidence of diagnostic specificity in the neural correlates of facial affect processing in bipolar disorder and schizophrenia: a meta-analysis of functional imaging studies. Psychol Med 43:553–569
Ellison-Wright I, Bullmore E (2010) Anatomy of bipolar disorder and schizophrenia: a meta-analysis. Schizophr Res 117:1–12
Forcada I, Papachristou E, Mur M et al (2011) The impact of general intellectual ability and white matter volume on the functional outcome of patients with bipolar disorder and their relatives. J Affect Disord 130:413–420
Fusar-Poli P, Placentino A, Carletti F et al (2009) Functional atlas of emotional faces processing: a voxel-based meta-analysis of 105 functional magnetic resonance imaging studies. J Psychiatry Neurosci 34:418–432
Gerber AJ, Posner J, Gorman D et al (2008) An affective circumplex model of neural systems subserving valence, arousal, and cognitive overlay during the appraisal of emotional faces. Neuropsychologia 46:2129–2139
Gerdes AB, Wieser MJ, Muhlberger A et al (2010) Brain activations to emotional pictures are differentially associated with valence and arousal ratings. Front Hum Neurosci 4:175
Germaná C, Kempton MJ, Sarnicola A et al (2010) The effects of lithium and anticonvulsants on brain structure in bipolar disorder. Acta Psychiatr Scand 122:481–487
Gläscher J, Adolphs R, Damasio H et al (2012) Lesion mapping of cognitive control and value-based decision making in the prefrontal cortex. Proc Natl Acad Sci U S A 109:14681–14686
Jogia J, Dima D, Frangou S (2012) Sex differences in bipolar disorder: a review of neuroimaging findings and new evidence. Bipolar Disord 14:461–471
Kempton MJ, Geddes JR, Ettinger U et al (2008) Meta-analysis, database, and meta-regression of 98 structural imaging studies in bipolar disorder. Arch Gen Psychiatry 65:1017–1032
Kempton MJ, Salvador Z, Munafò MR et al (2011) Structural neuroimaging studies in major depressive disorder. Meta-analysis and comparison with bipolar disorder. Arch Gen Psychiatry 68:675–690
Lyoo IK, Dager SR, Kim JE et al (2010) Lithium-induced gray matter volume increase as a neural correlate of treatment response in bipolar disorder: a longitudinal brain imaging study. Neuropsychopharmacology 35:1743–1750
Machado-Vieira R, Manji HK, Zarate CA Jr (2009) The role of lithium in the treatment of bipolar disorder: convergent evidence for neurotrophic effects as a unifying hypothesis. Bipolar Disord 11(Suppl 2):92–109
Mahon K, Burdick KE, Szeszko PR (2010) A role for white matter abnormalities in the pathophysiology of bipolar disorder. Neurosci Biobehav Rev 34:533–554
Mahon K, Burdick KE, Wu J et al (2012) Relationship between suicidality and impulsivity in bipolar I disorder: a diffusion tensor imaging study. Bipolar Disord 14:80–89
Mamah D, Barch DM, Repovš G (2013) Resting state functional connectivity of five neural networks in bipolar disorder and schizophrenia. J Affect Disord 150:601–609
Marquand A, Howard M, Brammer M et al (2010) Quantitative prediction of subjective pain intensity from whole-brain fMRI data using Gaussian processes. Neuroimage 49:2178–2189
Matsuo K, Nielsen N, Nicoletti MA et al (2010) Anterior genu corpus callosum and impulsivity in suicidal patients with bipolar disorder. Neurosci Lett 469:75–80
McDonald C, Zanelli J, Rabe-Hesketh S et al (2004) Meta-analysis of magnetic resonance imaging brain morphometry studies in bipolar disorder. Biol Psychiatry 56:411–417
Meda SA, Gill A, Stevens MC, Lorenzoni RP et al (2012) Differences in resting-state functional magnetic resonance imaging functional network connectivity between schizophrenia and psychotic bipolar probands and their unaffected first-degree relatives. Biol Psychiatry 71:881–889
Moore GJ, Bebchuk JM, Wilds IB et al (2000) Lithium-induced increase in human brain grey matter. Lancet 356(9237):1241–1242
Moore PB, Shepherd DJ, Eccleston D et al (2001) Cerebral white matter lesions in bipolar affective disorder: relationship to outcome. Br J Psychiatry 178:172–176
Moore GJ, Cortese BM, Glitz DA et al (2009) A longitudinal study of the effects of lithium treatment on prefrontal and subgenual prefrontal gray matter volume in treatment-responsive bipolar disorder patients. J Clin Psychiatry 70:699–705
Navari S, Dazzan P (2009) Do antipsychotic drugs affect brain structure? A systematic and critical review of MRI findings. Psychol Med 39:1763–1777
Ongür D, An X, Price JL (1998) Prefrontal cortical projections to the hypothalamus in macaque monkeys. J Comp Neurol 401:480–505

Pantazopoulos H, Lange N, Baldessarini RJ et al (2007) Parvalbumin neurons in the entorhinal cortex of subjects diagnosed with bipolar disorder or schizophrenia. Biol Psychiatry 61:640–652
Paus T (2001) Primate anterior cingulate cortex: where motor control, drive and cognition interface. Nat Rev Neurosci 2:417–424
Pessoa L, Adolphs R (2010) Emotion processing and the amygdala: from a 'low road' to 'many roads' of evaluating biological significance. Nat Rev Neurosci 11:773–783
Powell SB, Sejnowski TJ, Behrens MM (2012) Behavioral and neurochemical consequences of cortical oxidative stress on parvalbumin-interneuron maturation in rodent models of schizophrenia. Neuropharmacology 62:1322–1331
Regenold WT, Hisley KC, Phatak P et al (2008) Relationship of cerebrospinal fluid glucose metabolites to MRI deep white matter hyperintensities and treatment resistance in bipolar disorder patients. Bipolar Disord 10:753–764
Robins TW (2007) Shifting and stopping: fronto-striatal substrates, neurochemical modulation and clinical implications. Philos Trans R Soc B Biol Sci 362: 917–932
Rocha-Rego V, Jogia J, Marquand AF et al (2013) Examination of the predictive value of structural magnetic resonance scans in bipolar disorder: a pattern classification approach. Psychol Med. doi:10.1017/S0033291713001013
Santos A, Mier D, Kirsch P, Meyer-Lindenberg A (2011) Evidence for a general face salience signal in human amygdala. Neuroimage 54:3111–3116
Sarnicola A, Kempton M, Germanà C et al (2009) No differential effect of age on brain matter volume and cognition in bipolar patients and healthy individuals. Bipolar Disord 11:316–322
Selvaraj S, Arnone D, Job D et al (2012) Grey matter differences in bipolar disorder: a meta-analysis of voxel-based morphometry studies. Bipolar Disord 14:135–145
Terry J, Lopez-Larson M, Frazier JA (2009) Magnetic resonance imaging studies in early onset bipolar disorder: an updated review. Child Adolesc Psychiatr Clin N Am 18:421–439
Torrisi S, Moody TD, Vizueta N, Thomason ME (2013) Differences in resting corticolimbic functional connectivity in bipolar I euthymia. Bipolar Disord 15:156–166
Vederine FE, Wessa M, Leboyer M et al (2011) A meta-analysis of whole-brain diffusion tensor imaging studies in bipolar disorder. Prog Neuropsychopharmacol Biol Psychiatry 35:1820–1826
Vita A, De Peri L, Sacchetti E (2009) Gray matter, white matter, brain, and intracranial volumes in first-episode bipolar disorder: a meta-analysis of magnetic resonance imaging studies. Bipolar Disord 11:807–814
Walterfang M, Wood AG, Barton S et al (2009) Corpus callosum size and shape alterations in individuals with bipolar disorder and their first-degree relatives. Prog Neuropsychopharmacol Biol Psychiatry 33:1050–1057

Structural and Functional Brain Imaging in Borderline, Antisocial, and Narcissistic Personality Disorder

17

Lars Schulze and Stefan Roepke

Abbreviations

ACC	Anterior cingulate cortex
BOLD	Blood oxygenation level dependant
BPD	Borderline personality disorder
CSP	Cavum septum pellucidum
DTI	Diffusion tensor imaging
GM	Gray matter
NPD	Narcissistic personality disorder
OFC	Orbitofrontal cortex
PCL-R	Psychopathy Checklist Revised
PET	Positron emission tomography
PFC	Prefrontal cortex (PFC)
PPI	Psychopathic Personality Inventory
PTSD	Posttraumatic stress disorder
SPECT	Single-photon emission computed tomography
VBM	Voxel-based morphometry

L. Schulze (✉)
Department of Clinical Psychology and Psychotherapy, Freie Universität Berlin, Berlin, Germany
e-mail: lars.schulze@fu-berlin.de

S. Roepke (✉)
Department of Psychiatry and Psychotherapy, Charité Universitätsmedizin Berlin, Berlin, Germany
e-mail: stefan.roepke@charite.de

17.1 Individuals with Borderline Personality Disorder

17.1.1 Introduction

Borderline personality disorder (BPD) is a serious mental disorder that affects up to 3 % of the population in western nations (Lenzenweger et al. 2007; Trull et al. 2010) and is characterized by a pervasive pattern of instability in affect, behavior, interpersonal relationships, and self-identity (DSM-IV-TR 2000). Dissociation, impulsive aggression, self-injury, and chronic suicidal tendencies are the most prominent clinical symptoms (Leichsenring et al. 2011), resulting in pronounced impairments of psychosocial functioning, even in comparison to patients with mood or other personality disorders (e.g., Ansell et al. 2007; Skodol et al. 2002). Due to severe dysfunctions in their daily lives, BPD patients engage in extensive utilization of mental health resources (Ansell et al. 2007; Bender et al. 2001; Zanarini et al. 2001) and are usually overrepresented in psychiatric in- and outpatient samples (Grilo et al. 1998; Korzekwa et al. 2008; Marinangeli et al. 2000). The course of BPD symptomatology was traditionally considered chronic and intractable. However, this pessimistic outlook has fundamentally changed following recent investigations into the long-term course of BPD symptomatology as well as research demonstrating the effectiveness of specialized psychotherapeutic options (e.g., Bateman and Fonagy 2008; Giesen-Bloo et al. 2006; Linehan

C. Mulert, M.E. Shenton (eds.), *MRI in Psychiatry*,
DOI 10.1007/978-3-642-54542-9_17, © Springer Berlin Heidelberg 2014

et al. 2006). The 10-year course of BPD, for instance, is characterized by high rates of remissions and low rates of relapse. However, BPD patients experience persistent disturbance in social functioning (Gunderson et al. 2011).

Factor analytic studies identified three major domains of psychopathological impairments in BPD (Sanislow et al. 2002; Skodol et al. 2002): affective dysregulation, behavioral dysregulation (i.e., impulsivity), and disturbed relatedness.[1] Although disturbance in interpersonal relations is highly characteristic of personality disorders in general, disturbed relatedness in BPD is uniquely characterized by turbulent relationships and an excessive fear of perceived or real abandonment. This specific interactional style was, accordingly, proposed to best discriminate BPD from other personality disorders (Gunderson et al. 1995). Disturbed relatedness is presumably a consequence of presence of traits such as marked rejection sensitivity and pronounced difficulties in recognizing others' intentions and feelings (Gunderson 2007). Impaired mentalization capabilities are consistently observed in BPD (Bland et al. 2004; Levine et al. 1997; Preissler et al. 2010; Ritter et al. 2011; Unoka et al. 2011). Impairments in mentalizing capabilities are particularly prominent in more naturalistic settings (for a review, see Roepke et al. 2012), e.g., when patients have to integrate facial and prosodic information (Minzenberg et al. 2006) or are required to rapidly categorize facial expressions (Dyck et al. 2009). Along with biases in the evaluation of others (e.g., Arntz and Haaf 2012; Arntz and Veen 2001; Barnow et al. 2009), these aberrations contribute significantly to impaired functioning in interpersonal contexts (for reviews, see Dinsdale and Crespi 2013; Domes et al. 2009). Experience sampling studies provide further evidence that perceived rejection triggers anger and aggression (Berenson et al. 2011), states of tension, as well as the urge to engage in self-injurious behavior (Stiglmayr et al. 2005).

Negative emotional states are much more frequent and longer lasting in BPD patients (Ebner-Priemer et al. 2007; Stiglmayr et al. 2005, 2001), who show marked emotional reactions (even in response to presumably neutral stimuli) along with an aberrant variability in affective states (for a review, see Rosenthal et al. 2008). Correspondingly, the diagnostic criterion of affective dysregulation is the most frequent and stable criterion in BPD (Glenn and Klonsky 2009; Gunderson et al. 2011; McGlashan et al. 2005). Abnormalities in the processing and regulation of emotions are therefore widely considered to be "at the core of borderline pathology" (Stiglmayr et al. 2005, p. 372). Importantly, most clinical symptoms in BPD are directly related to affective dysregulation. Self-injurious behavior in BPD, for instance, is reportedly used to escape from undesired or extreme emotions (Chapman et al. 2006; Kleindienst et al. 2008) and, thus, might represent an efficient, but highly dysfunctional means of regulating aversive emotional states (Klonsky 2007; Niedtfeld and Schmahl 2009). Furthermore, affective instability was found to be related to feelings of emptiness, interpersonal problems, and identity disturbances in BPD (e.g., Kleindienst et al. 2008; Klonsky 2007; Koenigsberg et al. 2001; Tragesser et al. 2007).

Impulsive behaviors in BPD such as unsafe sexual practices, gambling, reckless driving, substance abuse, or disordered eating (e.g., Soloff et al. 2000; Trull et al. 2000) might also reflect dysfunctional attempts to cope with negative emotionality (Brown et al. 2002; Yen et al. 2004). Several studies, however, provided evidence that affective *and* behavioral dysregulation are distinct phenotypic traits of BPD (Tragesser and Robinson 2009; Zanarini et al. 2005). That is, impulse control difficulties predicted unique features of BPD (even when controlling for affective instability). Thus, it was argued that impulsivity must be considered independently from the affective domain to better understand the range of BPD symptoms (for a discussion, see Tragesser and Robinson 2009).

The prominent psychopathological impairments of BPD provided impetus for increased

[1] Alternative findings suggest that unifactorial models best account for variance in BPD psychopathology (e.g., Clifton and Pilkonis 2007; Johansen et al. 2004).

research investigating the neurobiological correlates of BPD symptomatology. The following sections will focus on summarizing available findings from the structural and functional neuroimaging literature (for a review of PET studies, see Lis et al. 2007). First, we will start with a short overview of structural brain abnormalities. Then, findings of functional neuroimaging will be presented and are organized with respect to the major psychopathological domains of BPD: processing and regulation of emotions, self-injurious behavior and pain processing, as well as interpersonal disturbances.

17.1.2 Structural Neuroimaging in BPD

Previous studies on brain structure in BPD used manual tracing methods, which allow the precise detection of small volume differences in a set of a priori-defined brain regions. The majority of these studies were interested in structural properties of the limbic system and found significantly smaller gray matter volume in the amygdala and hippocampus (Driessen et al. 2000; Rusch et al. 2003; Schmahl et al. 2003; Tebartz van Elst et al. 2003), although other studies failed to find group differences in amygdala volume (Brambilla et al. 2004; Zetzsche et al. 2006). Discrepancies regarding structural findings of the limbic system were argued to be partially due to different comorbidities in the investigated BPD samples. Recent meta-analyses aimed to clarify this point and found smaller gray matter volumes in the bilateral amygdala and hippocampus of about 13 and 11 %, respectively (Nunes et al. 2009; Ruocco et al. 2012). Interestingly, abnormalities of gray matter volume in these regions were independent of comorbid disorders or treatment experience. Accordingly, gray matter abnormalities of the limbic system were argued to serve as "candidate endophenotypes" of BPD (Ruocco et al. 2012). This interpretation, however, should be treated with caution as it is based on only a few studies with small sample sizes. Smaller gray matter volumes of the limbic system are, for example, also common in patients with posttraumatic stress disorder (PTSD; for a meta-analysis, see Karl et al. 2006) or in women with a history of traumatization (Dannlowski et al. 2012). Both conditions are highly prevalent in BPD (Lieb et al. 2004) and might contribute to abnormalities in the limbic system of BPD patients. Up until now, just one study evaluated the effects of co-occurring PTSD in BPD patients (Schmahl et al. 2009). In this study, smaller hippocampal volumes were only found in BPD patients with PTSD, thereby highlighting the necessity to disentangle the respective disorders' effects on structural brain abnormalities.

Apart from the limbic system, structural brain abnormalities were also evaluated in the frontal lobe of BPD patients (e.g., Hazlett et al. 2005; Sala et al. 2011; Tebartz van Elst et al. 2003). Particularly the advent of voxel-based morphometry (VBM) in recent years allowed an unbiased quantification of whole-brain structural properties rather than just of a few selected brain regions (Mechelli et al. 2005). Available VBM studies illustrated group differences in gray matter in the anterior cingulate cortex, the dorsolateral prefrontal cortex, as well as the orbitofrontal cortex (e.g., Brunner et al. 2010; Sato et al. 2012; Soloff et al. 2012; Vollm et al. 2009) while replicating smaller gray matter volume of the limbic system (Kuhlmann et al. 2012; Rusch et al. 2003). Prefrontal abnormalities of BPD patients were also related to impulsivity and suicidality (Sala et al. 2011; Soloff et al. 2008, 2012; Vollm et al. 2009).

While the investigation of structural white matter integrity in BPD is still in its infancy, available results highlight abnormalities in anterior parts of the corpus callosum and the bilateral orbitofrontal cortex (Carrasco et al. 2012) as well as in the anterior cingulate cortex, which are characterized by diminished interhemispheric structural connectivity (Rusch et al. 2010). Reduced integrity of inferior frontal white matter circuits was more specifically related to the BPD symptomatology, such as impaired affect regulation and enhanced impulsivity (Grant et al. 2007; Rusch et al. 2007).

In sum, findings of previous morphometric studies point predominantly toward abnormalities in limbic and prefrontal brain regions in BPD, especially in the amygdala and hippocampus, as well as in parts of the anterior cingulate cortex, dorsolateral prefrontal cortex, and orbitofrontal cortex. Structural properties of these regions might serve as biological markers to discriminate patients with BPD from healthy controls on the single-subject level (Sato et al. 2012).

17.1.3 Functional Neuroimaging in BPD

17.1.3.1 Processing and Regulation of Emotions

BPD is characterized by a pervasive pattern of emotional instability that is directly related to many aspects of patients' psychopathology, e.g., self-injury, interpersonal disturbance, and identity disturbance. Given the pivotal role of emotional instability in BPD symptomatology, this endophenotype is a major target for current treatment approaches, which aim to provide patients with techniques supporting the identification and regulation of negative emotional states. These include mentalization-based therapy (Bateman and Fonagy 2008) and dialectical behavior therapy (for a meta-analysis see Kliem et al. 2010). Emotional instability is theorized to be the result of two factors: (a) an enhanced emotional reactivity and (b) an inability to control intense negative emotional states (Koenigsberg 2010; Minzenberg et al. 2008; Posner et al. 2003). The review of available neuroimaging findings will be accordingly subdivided, although one should keep in mind that the line between emotion processing and emotion regulation is blurry at best as the generation of emotional responses is already subject to (implicit) modulations by regulatory-oriented processes (Gross 1998, 2002).

In accordance with two-factor models, BPD patients described more intense responses to emotionally evocative stimuli and exhibited greater instability of emotions in both laboratory and naturalistic settings (Ebner-Priemer et al. 2007; Herpertz et al. 1997; Rosenthal et al. 2008; Stiglmayr et al. 2005). Experimental studies provided further support for impaired inhibition of negative stimuli (Domes et al. 2006; Silbersweig et al. 2007), deficits in the disengagement of attentional resources from negative facial expression (von Ceumern-Lindenstjerna et al. 2010), and a hypervigilance to emotional cues (Arntz et al. 2000), especially in the presence of BPD-specific cues (Sieswerda et al. 2007).

At the neural level, most studies find support for abnormalities in the emotion processing circuitry of patients with BPD (for a detailed description of brain regions underlying emotion processing, see, e.g., Kober et al. 2008; Murphy et al. 2003; Tsuchiya and Adolphs 2007). An initial study by Herpertz et al. (2001), for instance, presented patients with aversive social scenes and found a link between patients' experience of intense emotions and enhanced activations of the bilateral amygdala. Stronger neural activity was also found in the bilateral fusiform gyrus, which might reflect an increased vigilance for negative emotional stimuli, as the visual cortex has dense connections to the amygdala (Amaral et al. 2003). Subsequent work found enhanced reactivity of the left amygdala in BPD in response to pictures of facial expressions of emotions, regardless of their specific valence (Donegan et al. 2003), whereas another study demonstrated a more specific pattern of limbic hyperarousal. In this study, enhanced activity of the right amygdala in BPD was only observed in response to fearful compared to neutral faces, whereas attenuated activation of the bilateral amygdala was found during the presentation of angry facial expressions (Minzenberg et al. 2007).

Overall, available evidence in BPD is consistent with the conclusion that limbic and paralimbic hyperreactivity holds particularly for negative and neutral stimuli (e.g., Jacob et al. 2012; Niedtfeld et al. 2010; Schulze et al. 2011), whereas the existing results regarding the processing of positive stimuli are to date still inconclusive. Koenigsberg and colleagues (2009b) found that compared to healthy controls, BPD patients exhibited heightened activity of the limbic system only in response to negative stimuli; they failed to find group differences for positive images. They also observed greater activation of the dorsolateral prefrontal cortex during the presentation of negative stimuli in healthy controls. In contrast, Hazlett et al. (2012), as well as Donegan et al. (2003), found greater amygdala activity in BPD patients across all picture types, including positive stimuli. Additionally, recent findings suggest abnormalities in emotion processing might affect the functioning of the reward circuitry in BPD (i.e., the ventral striatum, the pregenual anterior cingulate, and the ventromedial prefrontal cortex) and the functioning of the cortical midline structures during the anticipation of reward (Enzi et al. 2013).

Although most studies primarily investigated the processing of facial expressions or social scenes, heightened negative emotionality in patients with BPD has been demonstrated across a wide range of different stimuli. For instance, Jacob et al. (2012) reported stronger left-lateralized activation of the amygdala, along with attenuated activation of the subgenual anterior cingulate during an auditory anger induction procedure. Similarly, BPD patients showed stronger activation of the amygdala and insular region while recalling unresolved autobiographic life events (Beblo et al. 2006) or reading negatively valenced words (Silbersweig et al. 2007). Exaggerated and temporally prolonged amygdala responses were also observed in BPD patients during a procedure in which patients were presented with simple squares paired with painful electrodermal stimulation while being scanned (Kamphausen et al. 2013).

However, as outlined above, heightened emotional response represents only one facet of emotional instability in BPD. Impairment in the cognitive control of intense emotional states is argued to be equally important for understanding the disturbance patients experience in their daily lives, as the ability to regulate emotions has important effects on physical and mental health (Davidson 2000; Eftekhari et al. 2009; Gross and John 2003). As a consequence, studies of neural mechanisms associated with emotional regulation have rapidly increased in the last years. These studies have examined several different forms of cognitive control, ranging from attentional control processes, such as (in)voluntary shifts of attentional focus or limited allocation of resources necessary for the processing of external stimuli, to cognitive change via reappraisal (for reviews, see Ochsner and Gross 2005; Ochsner et al. 2012). The impact of cognitive reappraisal on neural activity in the limbic and paralimbic regions is observed across a complex prefrontal network involving multiple neural regions; however, the impact is particularly observable in the dorso- and ventrolateral, orbitofrontal, and anterior cingulate regions (e.g., Banks et al. 2007; Domes et al. 2010; Johnstone et al. 2007; Ochsner et al. 2004; Wager et al. 2008). The variability of prefrontal localizations associated with cognitive reappraisal might be attributable to temporal aspects of the reappraisal process, which first involves implementation and subsequently involves maintenance and monitoring of the effects of reappraisal on emotional responses (Kalisch 2009).

Recent studies of BPD patients investigated the effects of two different strategies of cognitive reappraisal: psychological distancing and reinterpretation of emotional situations. In one of the first neuroimaging studies of this phenomenon conducted by Koenigsberg et al. (2009a), patients were asked to decrease their emotional response to negative visual scenes using psychological distancing strategies. Compared to healthy controls, BPD patients showed difficulties engaging the dorsal anterior cingulate cortex and inferior parietal lobe during conditions of psychological distancing while showing a paradoxical increase of neural activity in the right amygdala. In another study conducted by Schulze et al. (2011), a delayed reappraisal paradigm was used to more clearly distinguish between the initial emotional reactivity to a negative stimulus and the subsequent modulation of emotional responses by cognitive reappraisal (Jackson et al. 2000). Statistical analyses during the initial viewing phase replicated previous results of limbic and paralimbic hyperreactivity during the presentation of negative and neutral stimuli in BPD. With regard to the regulatory phase, BPD patients showed attenuated activations of the left (more lateral) orbitofrontal cortex, along with attenuated reduction of activity in the bilateral insula during attempts to voluntarily decrease their emotional reactions. The orbitofrontal cortex is thought to play an important role in successful reappraisal via altering and updating the context-sensitive relevance of stimuli (Ochsner et al. 2004; Rolls 2000), suggesting specific deficits in BPD patients' ability to implement new appraisals for the presented stimuli. Overall, the results by Schulze et al. (2011) support two-factor models of emotional instability in BPD, i.e., enhanced emotional responding and deficits in cognitive regulation of emotions. Interestingly, neurofunctional abnormalities in regions associated with the processing or regulation of emotions were only prominent when participants were instructed to attenuate negative emotions, whereas no group differences were found when participants attempted to increase their emotions. Additionally, recent work aimed to disentangle the specific effects of BPD from those of traumatization on the neural correlates of cognitive reappraisal (Lang et al. 2012). In this study, healthy participants without any history of traumatization showed an initial increase in prefrontal activation (e.g., dorsolateral and dorsomedial prefrontal regions, anterior cingulate cortex), whereas trauma-exposed BPD patients and a comparison group of healthy individuals with traumatic experiences showed initial deactivations in those regions. Intriguingly, the group differences in the initial phase of emotion regulation might also indicate specific impairments in the implementation and generation of alternative appraisals (Kalisch 2009). In addition, the results highlight the necessity to more adequately address individual differences in BPD, e.g., trauma exposure, PTSD, or treatment experience.

Several studies examined attentional control (i.e., behavioral inhibition) of BPD patients in the context of negative stimuli (e.g., Domes et al. 2006; Fertuck et al. 2006; Rentrop et al. 2008). At the neural level, deficits in the inhibition of behavioral responses in a linguistic Go/No-Go task, for instance, were related to relatively decreased activity in the ventromedial prefrontal cortex (including orbitofrontal and subgenual anterior cingulate regions) compared to healthy controls (Silbersweig et al. 2007). Attenuated activations of the anterior cingulate cortex during response inhibition were also observed in an emotional Stroop paradigm (Wingenfeld et al. 2009). While these studies assessed inhibition and negative emotions in general, recent work also addressed how performance on an impulse control task is modulated by more specific emotional states (Jacob et al. 2012). Listening to an anger-inducing story, BPD patients showed a stronger activation of the right amygdala and the nucleus subthalamicus, whereas healthy controls showed a stronger recruitment of the subgenual anterior cingulate cortex. The subsequent performance on a Go/No-Go task was accompanied by attenuated activation of the left inferior frontal cortex in BPD.

Finally, two studies investigated the influence of emotional distraction on working memory performance. In both studies, BPD patients showed longer reaction times during emotional distraction along with enhanced activations of the amygdala compared to healthy controls (Krause-Utz et al. 2012; Prehn et al. 2013b). Moreover, limbic and paralimbic activity was negatively correlated with current states of dissociation, suggesting that dissociation has a dampening effect on emotional reactivity (Krause-Utz et al. 2012). Dissociation is considered a regulatory strategy to cope with overwhelming emotions, presumably via prefrontal inhibition of limbic regions (Lanius et al. 2010; Sierra and Berrios 1998). While dissociation represents a highly efficient way to shut down the emotional system (Ebner-Priemer et al. 2005, 2009), such states also prevent patients from learning to implement new strategies in emotionally arousing situations and represent a major challenge for psychotherapeutic treatment of BPD (Kleindienst et al. 2011).

The study by Prehn et al. (2013b) further assessed the role of working memory load on emotion processing. Despite heightened distractibility by highly negative stimuli in general, healthy controls and criminal offenders with BPD showed equivalent declines of limbic activity associated with manipulations affecting working memory load. In other words, the engagement in a secondary cognitively challenging task downregulates amygdala activity, presumably because these tasks draw resources necessary for the processing of emotional stimuli (see also Kanske et al. 2011; Van Dillen et al. 2009). Hence, attentional deployment strategies might be particularly useful for attenuating emotional reactions in BPD.

To summarize, findings of functional neuroimaging studies support two-factor conceptualizations of emotional instability by illustrating abnormalities in the processing *and* regulation of emotions. Comparing negative emotionality between patients with BPD and healthy controls, enhanced activation of the amygdala and insular region, along with attenuated activation of the rostral part of the anterior cingulate cortex, was consistently reported. As stated initially, the boundary between processing and regulation of emotions is blurry at best (Gross 1998, 2002), and quite congruently, self-reports of emotional regulation capabilities correlate positively with neural activity in limbic and paralimbic regions (Niedtfeld et al. 2010; Schulze et al. 2011). Preliminary results also suggest dialectical behavior therapy—focusing particularly on improving emotion regulation—attenuates neural activity of limbic regions in response to emotionally arousing stimuli (Schnell and Herpertz 2007), providing a promising starting point for assessing the effects of specific treatment modules on affective hyperarousal in BPD. With regard to the cognitive control of emotions, BPD seems to be characterized by an attenuated functioning of the dorsal part of anterior cingulate cortex and medial and lateral parts of the orbitofrontal cortex and deficient activation of the dorsolateral prefrontal cortex. Therefore, dysfunctional behaviors, such as self-injury or dissociation, might be best understood as an attempt to initiate inhibitory control of intense emotions via increased activation of the prefrontal regions, e.g., dorsolateral prefrontal cortex (Krause-Utz et al. 2012; Niedtfeld et al. 2012; see also next paragraph).

17.1.3.2 Self-Injury and Pain Processing

Deliberate self-injurious behavior, such as cutting or burning, is one of the most prominent clinical symptoms observed in BPD patients, with rates as high as 90 % observed in some studies (Skodol et al. 2002; Zanarini et al. 2008). Patients frequently state that they use non-suicidal self-injurious behavior to escape from undesired or extreme negative emotions (Chapman et al. 2006; Kleindienst et al. 2008). They also experience an immediate relief of tension, a decrease of dissociative symptoms, and elevations of mood (Herpertz 1995; Kemperman et al. 1997; Reitz et al. 2012). Self-injury was therefore proposed to represent a dysfunctional attempt to regulate negative emotional states directly corresponding to emotional instability in BPD (Klonsky 2007; Welch et al. 2008).

Patients with BPD display a reduced sensitivity to painful stimulation across different nociceptive modalities (e.g., Cardenas-Morales et al. 2011; Magerl et al. 2012; Russ et al. 1992; Schmahl et al. 2004), which is further decreased under conditions of elevated stress or states of dissociation (Bohus et al. 2000; Ludascher et al. 2007). Conversely, discontinuation of intentional self-injurious behavior results in a normalization of pain perception (Ludascher et al. 2009). The reduced sensitivity to pain, however, is not attributable to impairments in the sensory-discriminative component of pain processing, as BPD patients did not exhibit abnormal performance in the spatial discrimination of nociceptive stimuli or in the detection of painful stimulation (Ludascher et al. 2007; Schmahl et al. 2004). Therefore, BPD patients' reduced sensitivity might be better explained by abnormalities in the affective-motivational component of pain sensation, which in turn can be modulated by the context and cognitive appraisals of pain. Importantly, recent work by Magerl et al. (2012) found that endogenous antinociception in BPD is best predicted by the recency of self-injurious behavior, whereas psychometric measures of BPD psychopathology were unrelated to pain sensitivity. Thus, the authors suggested reduced pain sensitivity might not be a consequence of BPD psychopathology per se, but rather the result of neurofunctional rearrangement of brain circuits involved in the processing of pain due to BPD-related self-injurious behavior.

Neural models of pain sensation identified two anatomically distinct neural pathways that might be implicated in BPD patients' self-injurious behavior—sensory discriminative and affective motivational (for a review, see Treede et al. 1999). The sensory-discriminative system, necessary for stimulus localization and perception of pain quality and intensity, projects from the *lateral* thalamic nuclei to the primary and secondary somatosensory cortices. In contrast, the affective-motivational pathway, which underlies evaluation of pain and subsequent emotional or behavioral reactions, projects from the *medial* thalamic nuclei to the anterior cingulate cortex and insula, which is in turn reciprocally connected to the brainstem and the amygdala (Augustine 1996; Barbas et al. 2003).

Neurobiological correlates of altered pain processing in BPD were initially investigated by Schmahl and colleagues (2006). In this study, heat stimuli of two different intensities were delivered via a thermode. Participants either received a standardized temperature of 43 °C or an individual temperature adjusted for equal subjective pain in all individuals. Here, BPD patients again showed much higher pain thresholds than healthy controls. In response to individually adjusted heat stimuli, patients showed neural deactivation in the perigenual anterior cingulate cortex and the amygdala, while greater BOLD responses were found in the dorsolateral prefrontal cortex. Therefore, antinociceptive mechanisms might be attributable to increased top-down control closely associated with deactivations of the limbic system, particularly in patients with comorbid posttraumatic stress disorder (Kraus et al. 2009). Subsequent work combined thermal sensory stimulation with the prior induction of negative affect to further elucidate the potential role of pain with respect to regulation of affective states in BPD (Niedtfeld et al. 2010). In line with previous work, the BPD sample showed enhanced activations of the amygdala, insula, and anterior cingulate cortex in response to negative (but also to neutral) stimuli. The subsequent administration of heat stimulation again served to suppress neural activation of the amygdala and anterior cingulate cortex. Interestingly, there was only limited support for a BPD-specific role of pain in affect regulation, as both groups showed comparably attenuated activity in the amygdala. As a result, the authors proposed a general mechanism to explain the soothing effects of pain. In particular, they argued that painful sensory stimulation might draw resources from the processing of negative emotions, similar to cognitive strategies of "attentional deployment" (Gross 2002; Ochsner and Gross 2005). Additional connectivity analyses indicated that inhibitory influence on limbic regions during painful stimulation is particularly attributable to activation of the medial and dorsolateral parts of the prefrontal cortex (Niedtfeld et al. 2012).

In a different study, participants were presented with an audiotape describing an act of self-injury comprising different aspects of the situation, such as the respective trigger, related cognitive and emotional reactions, the act of self-harming behavior, and the subsequent relaxation (Kraus et al. 2010). While listening to emotional reactions, BPD patients displayed attenuated activation of the orbitofrontal cortex in comparison to healthy controls, presumably due to impaired ability to inhibit or modulate the elicited emotions. The description of the self-injurious act itself was again accompanied by decreased activity of the mid-cingulate cortex in BPD.

Alternative accounts of self-injury highlight the importance of the endogenous opioid system (for reviews, see Bandelow et al. 2010; New and Stanley 2010; Stanley and Siever 2010). The human opioid system consists of β-endorphin, met-enkephalin, and dynorphin, with all three neurotransmitters broadly involved in antinociception and stress-induced analgesic states (Fields 2004; Flor et al. 2002). More specifically, it is suggested that several forms of dysfunctional behavior in BPD (e.g., self-injury, risk-seeking behavior, promiscuity, substance abuse) might reflect attempts to raise relatively low levels of opioids (Bandelow et al. 2010; New and Stanley 2010). Reduced availability of dynorphin might also explain feelings of chronic emptiness and dysphoria (Land et al. 2008; for a discussion, see Stanley and Siever 2010). An initial investigation into this phenomenon found lower levels of cerebrospinal fluid β-endorphins and met-enkephalins in individuals with a history of self-injury, mainly patients with BPD (Stanley et al. 2010). These findings were paralleled by a PET study measuring μ-opioid receptor availability during neutral and dysphoric states (Prossin et al. 2010). At baseline, greater μ-opioid receptor availability in BPD was found in the orbitofrontal cortex, caudate nucleus, nucleus accumbens, and amygdala, presumably reflecting lower levels of endogenous opioids. The induction of sad affect led to a greater activation of the opioid system in the anterior cingulate cortex, orbitofrontal cortex, ventral pallidum, and amygdala, indicating a compensatory response to basal levels of μ-opioid in BPD.

To summarize, theoretical conceptualizations propose that self-injury is a dysfunctional, but highly efficient strategy for the regulation of negative emotional states. Higher pain thresholds in BPD are not attributable to impairment in the sensory-discriminative component of pain processing, but rather to abnormalities in affective-motivational components. The results of imaging studies further bolster such conceptualizations. In response to pain, patients with BPD show enhanced activation of the prefrontal cortex, along with attenuated activation of the anterior cingulate cortex and limbic system, whereas no functional abnormalities are found in somatosensory cortices. Therefore, the soothing effects of self-harm might be best explained by an increased top-down control of limbic regions through medial and dorsolateral prefrontal regions. Further emphasis must be placed on understanding associations between self-injury and the opioid system. Given that recent findings indicate that BPD patients exhibit baseline deficits in endogenous opioids, this area might hold some promise as a target for pharmacotherapy in BPD.

17.1.3.3 Interpersonal Disturbances

Disturbed relatedness was acknowledged in earliest descriptions of BPD (Kernberg 1967), and patients' difficulties in interpersonal contexts are currently considered the most unique and discriminant feature of BPD psychopathology (Gunderson et al. 1995). Up until the present, studies in this area have focused predominantly on abnormalities in the perception and recognition of others' mental and emotional states as a social-cognitive factor contributing to disturbance in BPD patients' social interactions (for a review of social cognition in BPD, see Roepke et al. 2012). Consistent with this notion, most behavioral studies found that BPD patients exhibit impairment in basic emotion recognition (e.g., Bland et al. 2004; Levine et al. 1997; Merkl et al. 2010; Ritter et al. 2011; Unoka et al. 2011) as well as a bias to interpret ambiguous stimuli more negatively (Domes et al. 2008; Meyer et al. 2004; Wagner and Linehan 1999), although a number of studies also reported superior attribution and detection of mental states in BPD (Fertuck et al. 2009; Lynch et al. 2006; Schulze et al. 2013a). Deficits in the recognition of emotions were particularly prominent in more complex and naturalistic settings (for a discussion of the role of ecological validity, see Roepke et al. 2012), e.g., when subjects have to integrate facial and prosodic information (Minzenberg et al. 2006), are required to rapidly discriminate facial expressions (Dyck et al. 2009), or have to judge video-based social interactions (Preissler et al. 2010; Sharp et al. 2011). In sum, these studies point predominantly toward impairments in BPD patients' capacity to infer the emotional states of other individuals.

However, cognitive empathy represents only one facet of social cognition (Decety and Meyer 2008; Singer 2006); a second component represents the emotional response to the observed emotional state of another person, i.e., emotional empathy (Eisenberg and Miller 1987; Mehrabian and Epstein 1972). The few experimental studies available on the subject found that BPD patients exhibit a reduced capacity to experience empathic concern for other people's observed distress (Dziobek et al. 2011; Ritter et al. 2011). In combination with aforementioned findings of deficits in cognitive empathy, these findings suggest that misperceptions of others' emotional states might lead to dysfunctional emotional responses in social interactions (New et al. 2008).

Consequently, recent neuroimaging studies have aimed to identify functional brain abnormalities associated with impaired cognitive empathy in BPD. An initial study by Guitart-Masip et al. (2009) investigated patients' brain responses while discriminating emotional facial expressions from neutral expressions. Compared to healthy controls, BPD patients made more mistakes in discriminating between negative and neutral facial expressions. More specifically, patients less accurately discriminated between disgusted and neutral expressions and showed a trend for similar difficulties in distinguishing between fearful and neutral faces. These behavioral impairments in BPD were accompanied by greater activation of the left inferior and middle temporal cortex (including parts of the superior temporal gyrus), possibly reflecting abnormalities in the perceptual processing of visual stimuli.

Subsequently, Dziobek et al. (2011) used a more ecologically valid paradigm to measure functional neural correlates of social cognition in BPD. The Multifaceted Empathy Test (MET; Dziobek et al. 2008) depicts people in emotionally charged situations and allows a separate assessment of cognitive and emotional aspects of state empathy. Participants are required to either infer the mental states of the depicted person (cognitive empathy) or to rate their level of empathic concern for the displayed individual (emotional empathy). Behaviorally, patients showed impairments in cognitive and emotional aspects of empathy compared to healthy controls. At the neural level, both groups activated a similar neural network during empathic processes, which comprised the temporal pole, the temporoparietal junction, the orbitofrontal cortex, and the superior temporal gyrus for cognitive empathy (for a review, see Frith and Frith 2005) and, in addition, the insula and medial prefrontal cortex for emotional facets of empathy. Subsequent group comparisons revealed significantly reduced activation of the left superior temporal sulcus and gyrus of BPD patients while inferring others' mental states. Moreover, this reduction was related to symptoms of intrusion in the BPD group. In contrast, enhanced activation of the right mid-insula was found when patients were emotionally attuned to another person. This brain abnormality was additionally associated with emotional arousal, as assessed by skin conductance responses.

Abnormalities in the left superior temporal sulcus of BPD patients were also found by Mier et al. and by Frick et al. (2012). In the former study, neural responses were measured while participants performed basic or more complex social-cognitive tasks. Healthy controls showed stronger activation of the superior temporal sulcus in response to greater demands on social-cognitive processes; such a pattern, however, was not seen in BPD patients. Hypo-activation of this brain region was consequently particularly prominent when patients performed complex social-cognitive tasks, i.e., attributing others' intentions. Furthermore, the authors found that BPD patients exhibited hyperresponsiveness of the limbic system independent of the specific social-cognitive process investigated. In line with these findings, Frick and colleagues (2012) also observed attenuated activation of the superior temporal sulcus and enhanced activation of the right amygdala during a mental state discrimination task (based on the eye region of individuals only). In this study, patients with BPD were more accurate in their ability to discern others' intentions than nonclinical controls (see also Scott et al. 2011).

Taken together, available studies illustrate abnormalities in BPD patients' ability to infer others' mental and emotional states, with traumatization being a negative predictor of social-cognitive abilities in BPD (Dyck et al. 2009; Preissler et al. 2010). Neuroimaging studies of mentalization found consistently altered functioning of the superior temporal sulcus in BPD. The superior temporal sulcus might be particularly vulnerable to the impact of traumatic experiences, as this brain region matures rather late in adolescence (Paus 2005). Alternative accounts propose that emotional hyperreactivity in concert with deficient functioning of the prefrontal cortex might interfere with social-cognitive processes (for a discussion, see Domes et al. 2009). The data provided by Dziobek et al. (2011) support the hypothesis that emotional arousal interferes with empathic processes in BPD. Thus, abnormalities related to the processing and regulation of emotions might be linked to altered social-cognitive processes in BPD patients.

As illustrated above, up until now, studies investigating social cognition in BPD primarily focused on the ability of BPD patients to accurately recognize others' emotions and intentions, i.e., cognitive empathy. Such abilities, however, only represent a small part of successful interpersonal interactions, which are characterized by an ongoing exchange of social signals between individuals (cf. Dziobek 2012). For that reason, recent studies adopted multiplayer economic exchange games to model interactional behavior in BPD (see also Seres et al. 2009). In multi-round trust games without any feedback about the back transfer of the other partner, BPD patients transferred smaller amounts of monetary units compared to nonclinical controls or patients with major depression (Unoka et al. 2009). Impaired trust during these sequential interactions was also related to individual difficulties in interpersonal relationships.

Despite the prominence of interpersonal disturbances in BPD, there is to date only one study available that examined the underlying neurofunctional correlates. King-Casas and colleagues (2008) used a multi-round economic exchange game with functional neuroimaging to investigate interactional behavior in patients with BPD. Their findings suggested that interpersonal disturbances in BPD are related to difficulties in maintaining cooperation with a healthy partner as well as an impaired capacity to restore broken cooperation. Neurologically, abnormal activation patterns of the bilateral anterior insula differentiated healthy controls from individuals with BPD. While activation of the insula in healthy controls tracked changes in cooperative behavior, in patients with BPD, activity of the anterior insula was unrelated to the magnitude of offers received (input) in BPD. Importantly, no insular abnormalities were found to be related to repayment of partners (output). Thus, the attenuated activity observed in the anterior insula during input led the authors to propose that "anomalous [perception of] social 'input' … (rather than 'output') is a potential mechanism of interpersonal dysfunction" (King-Casas and Chiu 2012; p. 121), particularly with regard to disturbed perception of social norms and an inability to recognize cues signaling waning trust of partners.

17.1.4 Summary

To recapitulate, structural brain imaging findings consistently illustrate smaller gray matter volume in the amygdala and hippocampus of BPD patients. Furthermore, available evidence suggests that BPD patients exhibit structural abnormalities in prefrontal brain regions, such as the anterior cingulate cortex, dorsolateral prefrontal cortex, and orbitofrontal regions.

Functional neuroimaging studies also support frontolimbic abnormalities across a variety of tasks. Processing of emotionally arousing stimuli, for instance, was associated with enhanced activation of the amygdala. BPD patients' attempts to cognitively control affective states were found to be less effective than those of healthy controls, which is presumably the result of attenuated functioning of prefrontal regions, specifically the dorsolateral prefrontal cortex, as well as medial and lateral regions of the orbitofrontal cortex. Interestingly, dysfunctional behaviors, such as self-injury or dissociation, were found to increase activation of prefrontal regions. Consequently, the soothing effects of self-harm on intense affective states might be best explained by an increased top-down control of limbic regions by medial and dorsolateral prefrontal regions.

Finally, several studies addressed interpersonal disturbances in BPD patients and investigated neural processes associated with inferring others' mental and emotional states. Findings highlight abnormal functioning of the superior temporal sulcus and additionally suggest that patients' heightened emotional arousal interferes with social-cognitive processes.

17.2 Individuals with Antisocial Behavior and Psychopathic Traits

17.2.1 Introduction

Empirical research on antisocial behavior has identified a number of genetic, environmental, psychological, and social pathways that potentially lead to these behaviors (Holmes et al.

2001; Moffitt 2005; Murray and Farrington 2010; Raine 2002; Vermeiren et al. 2002). Additionally, a growing body of evidence has linked antisocial behavior to functional and structural brain abnormalities. Any attempt to summarize this literature has several significant limitations. Most important of which is the lack of one common and coherent underlying psychopathological construct. Several studies are based on one of two official classification systems, the DSM and the ICD, which already show little overlap in constructs that capture antisocial behavior. For example, antisocial personality disorder in DSM-IV-TR and dissocial personality disorder in ICD-10 were found to be least concordant, when compared to all other personality disorders (Ottosson et al. 2002). However, studies in child or adolescent populations are mainly based on a different DSM-IV-TR construct; conduct disorder, often with an additional specifier for callous; and unemotional traits (Viding et al. 2012). Other studies focus on very specific aspects of antisocial behavior, e.g., violence (Raine et al. 1997; Volkow et al. 1995), aggressive and impulsive behavior (Dolan et al. 2002), or pathological lying (Yang et al. 2005). The most developed construct capturing antisocial behavior is psychopathy (Hare 2006). Psychopathy is characterized by a cluster of interpersonal, affective, and behavioral characteristics including impulsivity, callousness, and persistent antisocial behavior with profound lack of guilt or remorse (Hare 2006). A further limitation associated with summarizing the current literature is the diverse characteristics of study samples used to research antisocial behaviors. Included populations range from university or community samples with self-reported psychopathic traits (e.g., assessed with the Psychopathic Personality Inventory, PPI; Lilienfeld and Andrews 1996) to nonclinical temporary agency employees to incarcerated men with full criteria psychopathy (e.g., assessed with Psychopathy Checklist-Revised, PCL-R, Hare 2003). Further, study populations vary widely in sample size, gender ratio, education level, cognitive abilities, age, and comorbid conditions, e.g., duration of substance use disorder. A further limitation is that a number of publications refer to the same study population.

Possibly due to these limitations, functional and structural brain findings do not present a consistent picture underlining antisocial behavior and trait psychopathy (Koenigs et al. 2011; featured review). Although most fMRI studies assessed the processing of social or emotional information (e.g., within decision making, conditioning, or reward designs), differences in activation patterns are widespread across all four lobes of the cortex and subcortical structures (Koenigs et al. 2011; review). Also, structural brain findings have not yet provided robust replicable data. Because of these inconsistencies in current data, in this chapter, we focus on specific brain structures and functional networks that are theory driven and have been associated with antisocial behavior and trait psychopathy, presenting promising starting points for both understanding the neural mechanisms underlying these constructs and guiding future research.

17.2.2 Amygdala

Very early studies in individuals with psychopathic traits revealed that these patients reported decreased fear and showed decreased autonomic responses to aversive stimuli, results which both point toward involvement of the amygdala (Lykken 1957; Patrick et al. 1993). Indeed, one of the most robust functional MRI findings across a variety of tasks is attenuated activity of the amygdala. Thus, during an affective memory task, criminal psychopaths ($N=8$, PCL-R ≥ 28) from a maximum security prison showed significantly less affect-related activity in the amygdala compared to controls (Kiehl et al. 2001). In an aversive delay conditioning paradigm, criminal psychopaths ($N=4$, mean PCL-R $=25$) showed attenuated amygdala response compared to control groups (Veit et al. 2002). While performing an affect recognition task within a group of male college students ($N=20$), those with high PPI scores (mean split) displayed lower amygdala activation compared to the low PPI group. In later studies, lower amygdala activation has been confirmed in a number of conditions, e.g., by a group of prison inmates ($N=16$, PCL-R ≥ 30) viewing pictures depicting moral norm violations (Harenski et al. 2010), during acquisition in a fear-conditioning paradigm in a group of 10 criminal psychopaths (mean PCL-R = 25, Birbaumer et al. 2005), and by a group of schizophrenic patients with trait psychopathy from a secure psychiatric inpatient facility (group composition determined by a mean split of the PCL screening version score, Hart et al. 1995) while viewing fearful faces (Dolan and Fullam 2009). Further, in nonclinical samples, psychopathy score (PPI) correlated negatively with amygdala activation during a prisoner's dilemma task (Rilling et al. 2007) and while viewing pictures of aversive stimuli ($N=10$ females, psychopathy assessed with the PPI; Harenski et al. 2009). Lower amygdala activity during passive avoidance learning was also found in youth ($N=15$) with psychopathic traits (PCL youth version (YV) ≥ 20, Finger et al. 2011). Glenn et al. (2009) found that reduced activity in the amygdala during an emotional moral decision-making task correlated with a higher psychopathy score in a sample of 17 community participants with varying degrees of psychopathy (PCL-R = 7.4–32). Also in youth with callous unemotional traits ($N=12$, PCL:YV ≥ 20), amygdala activation was reduced compared to control groups while processing fearful faces (Marsh et al. 2008). Further, Jones et al. (2009) found reduced amygdala activation while viewing fearful faces in boys ($N=17$) with callous unemotional traits (assessed with the antisocial process screening device, Frick and Hare (2001)) compared to controls.

Nevertheless, there are also contradictory findings. Müller et al. (2003), for instance, found increased amygdala activation in criminal psychopaths from a high-security psychiatric facility ($N=6$, PCL-R > 30) while viewing pictures with negative content. Convergent evidence for involvement of the amygdala in antisocial behavior is provided by studies looking for structural alterations in the brains of antisocial individuals. For instance, reduced amygdala volume has been found in unsuccessful psychopaths (defined as PCL-R ≥ 23 and history of prosecution for criminal acts, $N=16$; Yang et al. 2010) recruited from temporary employment agencies compared to controls. An additional study with 27 individuals recruited from temporary employment agencies with high psychopathy scores (PCL-R ≥ 23) also confirmed amygdala volume reduction in the high psychopathy group compared to a control group; however, this reduction was more associated with affective and interpersonal facets of psychopathy than the impulsive and behavioral symptoms (Yang and Raine 2009). Further, this study assessed amygdala deformation, which was most prominent in basolateral, lateral, cortical, and central nuclei in participants with psychopathy (Yang and Raine 2009). Finally, in a study including a large number of male prison inmates ($N=296$), reduced volume in the amygdala was associated with psychopathy score (measured with the PCL-R; Ermer et al. 2012).

In contrast to the results of reduced amygdala volume, a study of 26 violent offenders from forensic psychiatric hospitals meeting criteria for antisocial and dissocial personality disorder revealed larger amygdala volume compared to a control group (Boccardi et al. 2011). Volume reduction in psychopaths was only found in the basolateral amygdala (a region with reciprocal connections to the orbitofrontal cortex—OFC). The central and lateral nuclei, in contrast, were enlarged (presenting components of the threat circuit that is involved in fear conditioning, Boccardi et al. 2011). A limitation of these results is that all offenders in this sample met the criteria for substance abuse (Boccardi et al. 2011).

In sum, most functional MRI studies point toward a deficit in amygdala activation in trait psychopathy compared to controls, Further, amygdala volume seems to be reduced, at least in some subnuclei, in individuals with trait psychopathy. However, to date there is a lack of studies combining functional and structural analyses.

17.2.3 Prefrontal Cortex

Early findings of antisocial behavior in individuals with prefrontal brain damage pointed to the potential involvement of this brain structure

(Anderson et al. 1999; Blumer and Benson 1975; Harlow 1868). Initial fMRI data confirmed this hypothesis and revealed that the orbitofrontal and ventromedial (OFC/vm) and prefrontal cortex (PFC) were particularly implicated in psychopathy (for review see Anderson and Kiehl 2012). In particular, the aforementioned study by Rilling et al. (2007) revealed lower OFC activity in participants with higher psychopathy scores when choosing to cooperate in a prisoner's dilemma task. Also, the previously mentioned Finger et al. (2011) study showed reduced OFC activity in youth with callous and unemotional traits in response to early stimulus-reinforcement exposure and to reward in a passive avoidance learning task. Criminal offenders (*N*=10) with dissocial personality disorder (PCL-R>28) in a forensic setting also showed decreased prefrontal cortex activation while engaged in a task designed to assess the influence of emotion on cognitive processes (Muller et al. 2008). Results from a study in which ten male psychopathic patients (mean PCL:SV=16) from a forensic setting and healthy controls engaged in a competitive reinforcement fMRI task revealed that psychopathic patients had lower activation in the mPFC during retaliation, but increased activation in the mPFC when seeing an opponent being punished compared to controls (Veit et al. 2010). Further, reduced activity of the mPFC has been observed when psychopaths (22 men with history of severe criminal offense and PCL-R>20) are engaged in moral reasoning compared to controls (Pujol et al. 2012). In addition, one study of ten nonclinical female adults (psychopathy measured with PPI) found a negative correlation between psychopathy score and mPFC activation while engaged in fMRI tasks presenting pictures of moral norm violation (Harenski et al. 2009). In line with these findings, lack of vmPFC activation while engaged in a moral norm violation paradigm has been found in 16 male prison inmates with psychopathy (PCL-R≥30) compared to controls (Harenski et al. 2010).

Decreased prefrontal activity was also found during a decision-making task in 12 offenders with antisocial personality disorder and emotional hypo-reaction recruited from a high-security forensic facility (Prehn et al. 2013a). Finally, Birbaumer et al. (2005) also found reduced OFC activation in schizophrenic patients with trait psychopathy during the acquisition phase in a fear-conditioning paradigm. Notably, not all fMRI experiments revealed reduced PFC activation in individuals with antisocial behavior and trait psychopathy. For instance, an fMRI study by Sommer et al. (2010) using a theory of mind task revealed increased activation of OFC and mPFC in 14 criminal patients from a forensic setting with antisocial personality disorder and PCL-R score>28 compared to controls.

Additionally, structural data support the argument that volume reduction in PFC structures is linked to antisocial behavior. In a study by Raine et al. (2000), 21 men with antisocial personality disorder recruited from temporary employment agencies showed reduced gray matter volume compared to controls. Boccardi et al. (2011) found a reduction in OFC gray matter. Tiihonen et al. (2008) also found OFC gray matter atrophy in 26 forensic patients fulfilling criteria for antisocial and dissocial personality disorder. Also, the study by de Oliveira-Souza et al. (2008) found gray matter reduction in the OFC in patients with antisocial personality disorder compared to controls. Further, volume reduction in individuals with antisocial behavior and trait psychopathy has been found in the anterior frontopolar region of the PFC (de Oliveira-Souza et al. 2008; Gregory et al. 2012; Tiihonen et al. 2008). Assessing cortical thickness in 27 nonclinical psychopaths (recruited from temporary employment agencies, PCL-R≥23) revealed a higher inverse correlation between cortical thickness in the OFC and response perseveration in psychopaths compared to controls (Yang et al. 2011). A recent meta-analysis of 43 structural and functional brain imaging studies (also including PET and SPECT studies) of antisocial, violent, and psychopathic individuals revealed prefrontal structural and functional reduction, especially in the right OFC and left dlPFC compared to controls (Yang and Raine 2009).

Despite the large number of consistent findings, a study by Schiffer et al. (2011) revealed that reduced gray matter volumes in the PFC/OFC were more associated with substance use disorder than with violent behavior. Consistent with this finding, a study by Laakso et al. (2002) found that volume reduction in the dlPFC, OFC, and mPFC in 24 forensic patients with antisocial personality disorder was better explained by duration of alcohol consumption than by trait psychopathy.

17.2.4 Frontolimbic Circuitry

Based on the distinction between reactive and instrumental aggression (Berkowitz 1993), one prominent model (Blair 2010) integrates amygdala and PFC findings in individuals with antisocial behavior and trait psychopathy. Within the framework of this model, reactive aggression follows perceived frustrating or threatening events and is defined as an unplanned and enraged attack against these triggers. Instrumental aggression, on the other hand, is more goal directed and purposeful. Instrumental aggression is thus conceived as a premeditated means of obtaining a goal, other than harming the victim, and being proactive rather than reactive. Psychopathic individuals can show both reactive and instrumental aggression (Anderson et al. 1999). Individuals who predominately engage in reactive aggression are hypothesized to exhibit increased responsiveness of the amygdala-hypothalamus-periaqueductal gray threat system, which might be accompanied by reduced frontal regulatory activity. Findings in patients with predominately reactive aggression confirmed the hypothesis of increased amygdala activity in response to threat stimuli compared to controls (e.g., individuals with impulsive aggression in Coccaro et al. 2007; spouse abusers in Lee et al. 2008). Also, individuals with borderline personality disorder (BPD) show more reactive aggression, and fMRI studies of BPD patients confirm the expected corresponding hyperresponsiveness of the amygdala in this population (Donegan et al. 2003; Herpertz et al. 2001; Koenigsberg et al. 2009a; Minzenberg et al. 2007). In line with the theory of reactive and instrumental aggression, Prehn et al. (2013b) found increased amygdala activity in a study investigating the influence of emotional stimuli on working memory in 15 male criminal offenders from a forensic setting with antisocial personality and comorbid BPD, all meeting the affective instability and lack of anger control criteria. However, it must be pointed out that the cited fMRI studies in individuals with predominately reactive aggression did not consistently show reduced frontal regulatory activity.

Individuals who engage in predominantly instrumental aggression are hypothesized to show specific impairment of the amygdala-dependent stimulus-reinforcement learning and impairment of the prefrontal cortex-related representation of reinforcement expectancies (for a review, see Crowe and Blair 2008). As outlined above, a number of fMRI studies confirmed reduced amygdala and OFC/vmPFC activation in individuals with psychopathic traits or antisocial behavior. Nevertheless, some inconsistent findings must be noted (e.g., lack of differences between psychopaths and controls in OFC activation while viewing emotional faces, Finger et al. 2008; Gordon et al. 2004; Jones et al. 2009; Marsh et al. 2011, 2008). An fMRI study by Marsh et al. (2011) showed a reduced activation in the amygdala and the OFC in adolescents with psychopathic traits (PCL:YV ≥ 20) when making judgments about legal actions, in addition to a reduced functional connectivity between these two regions.

As mentioned previously, structural findings also point toward involvement of the amygdala and OFC in antisocial behavior and trait psychopathy. In addition to these findings, a study using diffusion tensor MRI (Craig et al. 2009) found abnormal structural connectivity between the amygdala and OFC (uncinate fasciculus) in nine forensic patients with psychopathy (PCL-R ≥ 25) compared to controls. These results were confirmed by Motzkin et al. (2011), who found reduced structural integrity (DTI) and reduced functional connectivity in the right uncinate fasciculus in 14 male inmates from a correctional institution (PCL-R ≥ 30) compared to controls.

In sum, a number of functional and structural MRI studies are in favor of Blair's theory of impaired amygdala-dependent stimulus-reinforcement learning and impairment of prefrontal cortex-related representation of reinforcement expectancies in individuals with psychopathic traits and antisocial behavior.

17.2.5 Paralimbic Structures

Another prominent theoretical model of trait psychopathy is grounded in the study of cytoarchitectonics of subcortical brain structures, which extends the model of amygdala and PFC dysfunction to paralimbic structures, specifically, the parahippocampal gyrus, the temporal pole, the insula, and the anterior and posterior cingulate cortex (Kiehl 2006). These regions serve as the transition between subcortical structures and higher neocortical regions (Mesulam 2000) and might be involved in deviant information processing in trait psychopathy. Although most functional and structural MRI findings in these brain regions are contradicted by negative results from other studies (Blair 2010), growing empirical evidence supports the hypothesis of the involvement of paralimbic brain structures in antisocial behavior and trait psychopathy (Anderson and Kiehl 2012).

Thus, functional MRI analyses revealed abnormal activation patterns in the superior temporal cortex in individuals with antisocial behavior and psychopathic traits compared to controls in a number of studies (Finger et al. 2008; Jones et al. 2009; Kiehl et al. 2001; Marsh et al. 2008). Also, abnormal activation in the posterior cingulate cortex has been repeatedly found in these populations compared to control groups (Birbaumer et al. 2005; Finger et al. 2008; Glenn et al. 2009; Kiehl et al. 2001; Marsh et al. 2008; Rilling et al. 2007). Further, some evidence is indicative of abnormal functioning in the parahippocampal gyrus (Kiehl et al. 2001; Marsh et al. 2008), the anterior cingulate cortex (ACC Birbaumer et al. 2005; Kiehl et al. 2001; Prehn et al. 2013a), and the insular region (Birbaumer et al. 2005).

Also, findings from structural MRI studies indicate involvement of paralimbic regions in antisocial behavior and trait psychopathy. For example, volume reduction in the hippocampus (Boccardi et al. 2010; Laakso et al. 2001) and the parahippocampal gyrus (Boccardi et al. 2011) in individuals with trait psychopathy and antisocial behavior has been found compared to controls. Also, abnormal morphological features of the hippocampus have been found to correlate with trait psychopathy (Boccardi et al. 2010). Further, volume reduction in the anterior temporal cortex has been found in individuals with trait psychopathy and antisocial behavior compared to controls (Ermer et al. 2012; Yang et al. 2011). A study including 17 convicted criminals from high-security forensic facilities (PRL-R > 28) also showed gray matter reduction in the frontal and temporal brain regions, especially in the temporal pole, compared to controls (Muller et al. 2008).

In addition, a study of 17 violent offenders with antisocial personality disorder and PCL-R $\geq$ 25 revealed gray matter volume reduction in the temporal poles (Gregory et al. 2012). In a study comparing 21 male psychopaths from a correctional facility (PCL-R $\geq$30) to controls, psychopaths were found to have thinner cortex in the insular regions (Ly et al. 2012). Also, a voxel-based morphometry (VBM) analysis by de Oliveira-Souza et al. (2008) found gray matter reduction in the insular region.

Further, volume reduction in the ACC (Boccardi et al. 2011) and dorsal (d)ACC (Ly et al. 2012) has been found in psychopathic individuals. Finally, the very robust data of Ermer et al. (2012) confirmed an association between trait psychopathy and decreased gray matter in the posterior cingulate, parahippocampal regions, and temporal poles, in addition to the amygdala and OFC.

In sum, abnormalities in paralimbic structures were less reliably found across studies as the OFC and amygdala findings. Nevertheless, the meta-analysis performed by Yang and Raine (2009) confirmed structural abnormalities in the right ACC. Further, most of these paralimbic regions show structural connectivity to the OFC (Budhani et al. 2007) and the amygdala (Price 2003); thus, independent involvement of these regions in antisocial behavior and trait psychopathy still must be elucidated in future studies.

17.2.6 Developmental Hypothesis

Early structural brain studies point toward a neurodevelopmental deficit in individuals with antisocial behavior and trait psychopathy. Also, these early studies are supported by functional MRI studies of individuals with antisocial behavior and psychopathic traits which indicate that abnormalities are already present in childhood (e.g., Finger et al. 2008).

For instance, a study of 15 individuals with antisocial personality disorder recruited from temporary employment agencies (PCL-R $\geq$ 23) found that individuals with antisocial personality disorder had increased volume of the corpus callosum (which might be the result of an early arrest of axonal pruning) compared to controls (Raine et al. 2003). Further, the observed larger volumes in posterior brain regions in individuals with ASP might reflect disruption of brain maturation, and thus, a deficit in neurodevelopmental processes (for a discussion, see Tiihonen et al. 2008). Additional evidence stems from the observed presence of cavum septum pellucidum (CSP), which is considered to be a marker for fetal neural maldevelopment. Individuals ($N = 18$) recruited from temporary employment agencies with CSP had significantly higher levels of antisocial personality disorder (Raine et al. 2010). These results were recently replicated in a sample of 32 youths with conduct disorder and oppositional defiant disorder (White et al. 2013).

In sum, early functional and structural brain data point toward a neurodevelopmental deficit in individuals with antisocial behavior and trait psychopathy. Nevertheless, this hypothesis needs to be tested in further studies.

17.2.7 Summary

Based on findings from the reviewed studies, antisocial behavior and psychopathic traits are associated with altered brain structure and function. Frontolimbic circuits including the OFC/vmPFC and the amygdala are characterized by volume reduction and hypofunctionality. Since these structures are typically involved in emotion processing and in the mental representation of behavioral outcomes and punishment, reduced activation in these structures might represent a neural correlate of the predominately instrumental aggression in individuals with trait psychopathy. Additionally, cytoarchitectonically related paralimbic structures might also be involved in trait psychopathy, as decreased volume and hypofunctionality of these structures imply. Further, fMRI activation patterns particularly differentiate individuals with trait psychopathy from individuals with predominately impulsive aggression, e.g., patients with BPD, which are characterized by increased activation of the amygdala. Finally, brain structural data point toward a neurodevelopmental component in antisocial behavior and trait psychopathy.

17.3 Individuals with Narcissistic Personality Disorder

17.3.1 Introduction

The defining features of narcissistic personality disorder (NPD) are "a pervasive pattern of grandiosity, need for admiration, and lack of empathy" (DSM-IV-TR 2000). Previous studies on NPD revealed that the disorder has a median prevalence of about 1 % in the general population (Pincus and Lukowitsky 2010) and is associated with severe impairments in psychosocial functioning (Miller et al. 2007; Stinson et al. 2008), a high comorbidity rate with affective as well as substance use disorders (Ritter et al. 2010; Stinson et al. 2008), and an increased rate of suicidal behavior (Blasco-Fontecilla et al. 2009; Ronningstam et al. 2008). However, despite the disorder's impact on the mental health system, there is to date a fundamental lack of empirical research on NPD, a fact which nearly provoked elimination of NPD as a distinct personality disorder in the upcoming DSM-5 (for discussions, see Miller et al. 2010; Ronningstam 2011). Although NPD will be retained in the DSM 5, the uncertainty regarding diagnostic classification of this disorder highlighted a pressing need for additional research examining the psychological and neurobiological factors associated with the clinical presentation of NPD (Miller and Campbell 2010).

17.3.2 Neuroimaging in NPD

The few presently available studies focused on the most characteristic feature of NPD patients: "lack of empathy." Impairment in the empathic processes is considered a hallmark of NPD and has played a long-standing role in theoretical conceptualizations of this mental disorder (e.g., Akhtar and Thomson 1982; Blais et al. 1997; Kernberg 1970; Ronningstam 2010). Empathy is commonly proposed to be best understood in terms of a multidimensional model comprising cognitive and emotional aspects (Blair 2005; Decety and Meyer 2008; Singer 2006). Cognitive empathy refers to the ability to infer mental states of another person (i.e., perspective taking, social cognition; Baron-Cohen and Wheelwright 2004), whereas emotional empathy refers to the emotional response of an individual when observing the emotional state of another person (Eisenberg and Miller 1987; Mehrabian and Epstein 1972). While most experts agree that impairment of the empathic process is a central feature of NPD, it was only recently that an experimental investigation by our research group more thoroughly investigated empathic capabilities in this population (Ritter et al. 2011). More specifically, Ritter et al. investigated cognitive and emotional aspects of empathy in a sample of NPD patients, a nonclinical and a psychopathological control group. The experimental assessment of empathy (Dziobek et al. 2006, 2008) illustrated that NPD patients are neither characterized by a general lack of empathy nor by impaired abilities in cognitive empathy, but rather by distinct and specific impairment in emotional empathy (Ritter et al. 2011). In other words, NPD patients are able to cognitively understand and represent the mental state of others but are impaired in their ability to mirror or emotionally respond to the observed emotional state of another person.

At the neural level, the anterior insular cortex is particularly implicated in one's ability to be attuned to another person's emotional response (e.g., de Greck et al. 2012; Dziobek et al. 2011) and was shown to be part of a neural network underlying empathy (for meta-analyses, see Fan et al. 2011a; Lamm et al. 2011). Meta-analytic evidence suggests that a core neural network, comprising the bilateral anterior insula, the anterior and median parts of the cingulate cortex, and the supplementary motor area, is particularly implicated in empathic processes (Fan et al. 2011a). Complementary evidence regarding the neural representation of empathy was provided by volumetric studies. For instance, a loss of GM volume in the ventromedial prefrontal cortex of individuals with schizophrenia was shown to significantly contribute to deficits in theory of mind skills (i.e., cognitive empathy; Hooker et al. 2011). A lack of emotional empathy, in contrast, was related to smaller GM volume of the bilateral anterior insula in adolescents with conduct disorder (Sterzer et al. 2007).

The prominent role of the insular region as a potential neural candidate for the observed impairments in emotional empathy of NPD patients was highlighted in a study investigating nonclinical individuals either high or low in narcissistic traits. Here, while being asked to emotionally empathize with others, healthy participants high in self-reported narcissism showed an attenuated activation in the right anterior insula, in the dorsolateral prefrontal and posterior cingulate cortex, as well as in the right premotor cortex (Fan et al. 2011b). Subsequent analyses confirmed that only the functional abnormalities in the anterior insula were unequivocally related to empathic processes, whereas abnormalities in the other brain regions might be related to more general functions, such as face perception or evaluation.

Recently, we aimed to provide initial insight into structural brain abnormalities in NPD by using voxel-based morphometry to compare local GM volume between patients with NPD and healthy controls (Schulze et al. 2013b). The results revealed smaller GM volume in the left anterior insula in NPD patients. Subsequent regression analyses, comprising scores of cognitive and emotional empathy, underlined the crucial role of the left anterior insula in emotionally empathic responding. Complementary whole-brain analyses provided additional support for gray matter abnormalities in a fronto-paralimbic network comprising the rostral and medial cingulate cortex as well as the dorsolateral and medial parts of the prefrontal cortex. Interestingly, these brain regions spatially overlap with the neural circuitry commonly implicated in the representation of empathy (for meta-analyses, see Fan et al. 2011a; Lamm et al. 2011).

17.3.3 Summary

The studies presented above represent the only neuroimaging work available in narcissistic individuals. As such, these findings need to be considered as a preliminary, albeit promising, starting point for the investigation of neurobiological processes in pathological narcissism. Nevertheless, both studies consistently suggest that abnormalities in the anterior insula represent an especially promising neural candidate for NPD patients' impaired ability to mirror or emotionally respond to the observed emotional states of others.

References

Akhtar S, Thomson JA Jr (1982) Overview: narcissistic personality disorder. Am J Psychiatry 139:12–20

Amaral DG, Behniea H, Kelly JL (2003) Topographic organization of projections from the amygdala to the visual cortex in the macaque monkey. Neuroscience 118:1099–1120

Anderson NE, Kiehl KA (2012) The psychopath magnetized: insights from brain imaging. Trends Cogn Sci 16:52–60

Anderson SW, Bechara A, Damasio H, Tranel D, Damasio AR (1999) Impairment of social and moral behavior related to early damage in human prefrontal cortex. Nat Neurosci 2:1032–1037

Ansell EB, Sanislow CA, McGlashan TH, Grilo CM (2007) Psychosocial impairment and treatment utilization by patients with borderline personality disorder, other personality disorders, mood and anxiety disorders, and a healthy comparison group. Compr Psychiatry 48:329–336

Arntz A, Haaf JT (2012) Social cognition in borderline personality disorder: evidence for dichotomous thinking but no evidence for less complex attributions. Behav Res Ther 50:707–718

Arntz A, Veen G (2001) Evaluations of others by borderline patients. J Nerv Ment Dis 189:513–521

Arntz A, Appels C, Sieswerda S (2000) Hypervigilance in borderline disorder: a test with the emotional Stroop paradigm. J Pers Disord 14:366–373

Augustine JR (1996) Circuitry and functional aspects of the insular lobe in primates including humans. Brain Res Brain Res Rev 22:229–244

Bandelow B, Schmahl C, Falkai P, Wedekind D (2010) Borderline personality disorder: a dysregulation of the endogenous opioid system? Psychol Rev 117:623–636

Banks SJ, Eddy KT, Angstadt M, Nathan PJ, Phan KL (2007) Amygdala-frontal connectivity during emotion regulation. Soc Cogn Affect Neurosci 2:303–312

Barbas H, Saha S, Rempel-Clower N, Ghashghaei T (2003) Serial pathways from primate prefrontal cortex to autonomic areas may influence emotional expression. BMC Neurosci 4:25

Barnow S, Stopsack M, Grabe HJ, Meinke C, Spitzer C, Kronmuller K, Sieswerda S (2009) Interpersonal evaluation bias in borderline personality disorder. Behav Res Ther 47:359–365

Baron-Cohen S, Wheelwright S (2004) The empathy quotient: an investigation of adults with Asperger syndrome or high functioning autism, and normal sex differences. J Autism Dev Disord 34:163–175

Bateman A, Fonagy P (2008) 8-year follow up of patients treated for borderline personality disorder: mentalization-based treatment versus treatment as usual. Am J Psychiatry 165:631–638

Beblo T, Driessen M, Mertens M, Wingenfeld K, Piefke M, Rullkoetter N, Silva-Saavedra A, Mensebach C, Reddemann L, Rau H, Markowitsch HJ, Wulff H, Lange W, Berea C, Ollech I, Woermann FG (2006) Functional MRI correlates of the recall of unresolved life events in borderline personality disorder. Psychol Med 36:845–856

Bender DS, Dolan RT, Skodol AE, Sanislow CA, Dyck IR, McGlashan TH, Shea MT, Zanarini MC, Oldham JM, Gunderson JG (2001) Treatment utilization by patients with personality disorders. Am J Psychiatry 158:295–302

Berenson KR, Downey G, Rafaeli E, Coifman KG, Paquin NL (2011) The rejection-rage contingency in borderline personality disorder. J Abnorm Psychol 120:681–690

Berkowitz L (1993) Aggression: its causes, consequences and control. McGraw-Hill, New York

Birbaumer N, Veit R, Lotze M, Erb M, Hermann C, Grodd W, Flor H (2005) Deficient fear conditioning in psychopathy: a functional magnetic resonance imaging study. Arch Gen Psychiatry 62:799–805

Blair RJ (2005) Responding to the emotions of others: dissociating forms of empathy through the study of typical and psychiatric populations. Conscious Cogn 14:698–718

Blair RJ (2010) Neuroimaging of psychopathy and antisocial behavior: a targeted review. Curr Psychiatry Rep 12:76–82

Blais MA, Hilsenroth MJ, Castlebury FD (1997) Content validity of the DSM-IV borderline and narcissistic personality disorder criteria sets. Compr Psychiatry 38:31–37

Bland AR, Williams CA, Scharer K, Manning S (2004) Emotion processing in borderline personality disorders. Issues Ment Health Nurs 25:655–672

Blasco-Fontecilla H, Baca-Garcia E, Dervic K, Perez-Rodriguez MM, Lopez-Castroman J, Saiz-Ruiz J, Oquendo MA (2009) Specific features of suicidal behavior in patients with narcissistic personality disorder. J Clin Psychiatry 70:1583–1587

Blumer D, Benson DF (1975) Personality changes with frontal and temporal lesion. In: Benson DF, Blumer D (eds) Psychiatric aspects of neurological disease. Stratton, New York, pp 151–169

Boccardi M, Ganzola R, Rossi R, Sabattoli F, Laakso MP, Repo-Tiihonen E, Vaurio O, Kononen M, Aronen HJ, Thompson PM, Frisoni GB, Tiihonen J (2010) Abnormal hippocampal shape in offenders with psychopathy. Hum Brain Mapp 31:438–447

Boccardi M, Frisoni GB, Hare RD, Cavedo E, Najt P, Pievani M, Rasser PE, Laakso MP, Aronen HJ, Repo-Tiihonen E, Vaurio O, Thompson PM, Tiihonen J (2011) Cortex and amygdala morphology in psychopathy. Psychiatry Res 193:85–92

Bohus M, Limberger M, Ebner U, Glocker FX, Schwarz B, Wernz M, Lieb K (2000) Pain perception during self-reported distress and calmness in patients with borderline personality disorder and self-mutilating behavior. Psychiatry Res 95:251–260

Brambilla P, Soloff PH, Sala M, Nicoletti MA, Keshavan MS, Soares JC (2004) Anatomical MRI study of borderline personality disorder patients. Psychiatry Res 131:125–133

Brown MZ, Comtois KA, Linehan MM (2002) Reasons for suicide attempts and nonsuicidal self-injury in women with borderline personality disorder. J Abnorm Psychol 111:198–202

Brunner R, Henze R, Parzer P, Kramer J, Feigl N, Lutz K, Essig M, Resch F, Stieltjes B (2010) Reduced prefrontal and orbitofrontal gray matter in female adolescents with borderline personality disorder: is it disorder specific? Neuroimage 49:114–120

Budhani S, Marsh AA, Pine DS, Blair RJ (2007) Neural correlates of response reversal: considering acquisition. Neuroimage 34:1754–1765

Cardenas-Morales L, Fladung AK, Kammer T, Schmahl C, Plener PL, Connemann BJ, Schonfeldt-Lecuona C (2011) Exploring the affective component of pain perception during aversive stimulation in borderline personality disorder. Psychiatry Res 186:458–460

Carrasco JL, Tajima-Pozo K, Diaz-Marsa M, Casado A, Lopez-Ibor JJ, Arrazola J, Yus M (2012) Microstructural white matter damage at orbitofrontal areas in borderline personality disorder. J Affect Disord 139:149–153

Chapman AL, Gratz KL, Brown MZ (2006) Solving the puzzle of deliberate self-harm: the experiential avoidance model. Behav Res Ther 44:371–394

Clifton A, Pilkonis PA (2007) Evidence for a single latent class of diagnostic and statistical manual of mental disorders borderline personality pathology. Compr Psychiatry 48:70–78

Coccaro EF, McCloskey MS, Fitzgerald DA, Phan KL (2007) Amygdala and orbitofrontal reactivity to social threat in individuals with impulsive aggression. Biol Psychiatry 62:168–178

Craig MC, Catani M, Deeley Q, Latham R, Daly E, Kanaan R, Picchioni M, McGuire PK, Fahy T, Murphy DG (2009) Altered connections on the road to psychopathy. Mol Psychiatry 14:946–953

Crowe SL, Blair RJ (2008) The development of antisocial behavior: what can we learn from functional neuroimaging studies? Dev Psychopathol 20:1145–1159

Dannlowski U, Stuhrmann A, Beutelmann V, Zwanzger P, Lenzen T, Grotegerd D, Domschke K, Hohoff C, Ohrmann P, Bauer J, Lindner C, Postert C, Konrad C, Arolt V, Heindel W, Suslow T, Kugel H (2012) Limbic scars: long-term consequences of childhood maltreatment revealed by functional and structural magnetic resonance imaging. Biol Psychiatry 71:286–293

Davidson RJ (2000) Affective style, psychopathology, and resilience: brain mechanisms and plasticity. Am Psychol 55:1196–1214

de Greck M, Wang G, Yang X, Wang X, Northoff G, Han S (2012) Neural substrates underlying intentional empathy. Soc Cogn Affect Neurosci 7:135–144

de Oliveira-Souza R, Hare RD, Bramati IE, Garrido GJ, Azevedo Ignacio F, Tovar-Moll F, Moll J (2008) Psychopathy as a disorder of the moral brain: fronto-temporo-limbic grey matter reductions demonstrated by voxel-based morphometry. Neuroimage 40:1202–1213

Decety J, Meyer M (2008) From emotion resonance to empathic understanding: a social developmental neuroscience account. Dev Psychopathol 20:1053–1080

Dinsdale N, Crespi BJ (2013) The borderline empathy paradox: evidence and conceptual models for empathic enhancements in borderline personality disorder. J Pers Disord 27:172–195

Dolan MC, Fullam RS (2009) Psychopathy and functional magnetic resonance imaging blood oxygenation level-dependent responses to emotional faces in violent patients with schizophrenia. Biol Psychiatry 66:570–577

Dolan MC, Deakin JF, Roberts N, Anderson IM (2002) Quantitative frontal and temporal structural MRI studies in personality-disordered offenders and control subjects. Psychiatry Res 116:133–149

Domes G, Winter B, Schnell K, Vohs K, Fast K, Herpertz SC (2006) The influence of emotions on inhibitory functioning in borderline personality disorder. Psychol Med 36:1163–1172

Domes G, Czieschnek D, Weidler F, Berger C, Fast K, Herpertz SC (2008) Recognition of facial affect in borderline personality disorder. J Pers Disord 22:135–147

Domes G, Schulze L, Herpertz SC (2009) Emotion recognition in borderline personality disorder-a review of the literature. J Pers Disord 23:6–19

Domes G, Schulze L, Bottger M, Grossmann A, Hauenstein K, Wirtz PH, Heinrichs M, Herpertz SC

(2010) The neural correlates of sex differences in emotional reactivity and emotion regulation. Hum Brain Mapp 31:758–769

Donegan NH, Sanislow CA, Blumberg HP, Fulbright RK, Lacadie C, Skudlarski P, Gore JC, Olson IR, McGlashan TH, Wexler BE (2003) Amygdala hyperreactivity in borderline personality disorder: implications for emotional dysregulation. Biol Psychiatry 54:1284–1293

Driessen M, Herrmann J, Stahl K, Zwaan M, Meier S, Hill A, Osterheider M, Petersen D (2000) Magnetic resonance imaging volumes of the hippocampus and the amygdala in women with borderline personality disorder and early traumatization. Arch Gen Psychiatry 57:1115–1122

DSM-IV-TR (2000) Diagnostic and statistical manual of mental disorders: DSM-IV-TR, 4th edn. American Psychiatric Association, Washington, DC (Text Revision ed)

Dyck M, Habel U, Slodczyk J, Schlummer J, Backes V, Schneider F, Reske M (2009) Negative bias in fast emotion discrimination in borderline personality disorder. Psychol Med 39:855–864

Dziobek I (2012) Towards a more ecologically valid assessment of empathy. Emot Rev 4:18–19

Dziobek I, Fleck S, Kalbe E, Rogers K, Hassenstab J, Brand M, Kessler J, Woike JK, Wolf OT, Convit A (2006) Introducing MASC: a movie for the assessment of social cognition. J Autism Dev Disord 36:623–636

Dziobek I, Rogers K, Fleck S, Bahnemann M, Heekeren HR, Wolf OT, Convit A (2008) Dissociation of cognitive and emotional empathy in adults with Asperger syndrome using the Multifaceted Empathy Test (MET). J Autism Dev Disord 38:464–473

Dziobek I, Preissler S, Grozdanovic Z, Heuser I, Heekeren HR, Roepke S (2011) Neuronal correlates of altered empathy and social cognition in borderline personality disorder. Neuroimage 57:539–548

Ebner-Priemer UW, Badeck S, Beckmann C, Wagner A, Feige B, Weiss I, Lieb K, Bohus M (2005) Affective dysregulation and dissociative experience in female patients with borderline personality disorder: a startle response study. J Psychiatr Res 39:85–92

Ebner-Priemer UW, Welch SS, Grossman P, Reisch T, Linehan MM, Bohus M (2007) Psychophysiological ambulatory assessment of affective dysregulation in borderline personality disorder. Psychiatry Res 150:265–275

Ebner-Priemer UW, Mauchnik J, Kleindienst N, Schmahl C, Peper M, Rosenthal MZ, Flor H, Bohus M (2009) Emotional learning during dissociative states in borderline personality disorder. J Psychiatry Neurosci 34:214–222

Eftekhari A, Zoellner LA, Vigil SA (2009) Patterns of emotion regulation and psychopathology. Anxiety Stress Coping 22:571–586

Eisenberg N, Miller PA (1987) The relation of empathy to prosocial and related behaviors. Psychol Bull 101:91–119

Enzi B, Doering S, Faber C, Hinrichs J, Bahmer J, Northoff G (2013) Reduced deactivation in reward circuitry and midline structures during emotion processing in borderline personality disorder. World J Biol Psychiatry 14:45–56

Ermer E, Cope LM, Nyalakanti PK, Calhoun VD, Kiehl KA (2012) Aberrant paralimbic gray matter in criminal psychopathy. J Abnorm Psychol 121:649–658

Fan Y, Duncan NW, de Greck M, Northoff G (2011a) Is there a core neural network in empathy? An fMRI based quantitative meta-analysis. Neurosci Biobehav Rev 35:903–911

Fan Y, Wonneberger C, Enzi B, de Greck M, Ulrich C, Tempelmann C, Bogerts B, Doering S, Northoff G (2011b) The narcissistic self and its psychological and neural correlates: an exploratory fMRI study. Psychol Med 41:1641–1650

Fertuck EA, Lenzenweger MF, Clarkin JF, Hoermann S, Stanley B (2006) Executive neurocognition, memory systems, and borderline personality disorder. Clin Psychol Rev 26:346–375

Fertuck EA, Jekal A, Song I, Wyman B, Morris MC, Wilson ST, Brodsky BS, Stanley B (2009) Enhanced 'reading the mind in the eyes' in borderline personality disorder compared to healthy controls. Psychol Med 39:1979–1988

Fields H (2004) State-dependent opioid control of pain. Nat Rev Neurosci 5:565–575

Finger EC, Marsh AA, Mitchell DG, Reid ME, Sims C, Budhani S, Kosson DS, Chen G, Towbin KE, Leibenluft E, Pine DS, Blair JR (2008) Abnormal ventromedial prefrontal cortex function in children with psychopathic traits during reversal learning. Arch Gen Psychiatry 65:586–594

Finger EC, Marsh AA, Blair KS, Reid ME, Sims C, Ng P, Pine DS, Blair RJ (2011) Disrupted reinforcement signaling in the orbitofrontal cortex and caudate in youths with conduct disorder or oppositional defiant disorder and a high level of psychopathic traits. Am J Psychiatry 168:152–162

Flor H, Birbaumer N, Schulz R, Grusser SM, Mucha RF (2002) Pavlovian conditioning of opioid and nonopioid pain inhibitory mechanisms in humans. Eur J Pain 6:395–402

Frick PJ, Hare RD (2001) The Antisocial Process Screening Device (ASPD). Multi Health Systems, Toronto

Frick C, Lang S, Kotchoubey B, Sieswerda S, Dinu-Biringer R, Berger M, Veser S, Essig M, Barnow S (2012) Hypersensitivity in borderline personality disorder during mindreading. PLoS One 7:e41650

Frith C, Frith U (2005) Theory of mind. Curr Biol 15:R644–R646

Giesen-Bloo J, van Dyck R, Spinhoven P, van Tilburg W, Dirksen C, van Asselt T, Kremers I, Nadort M, Arntz A (2006) Outpatient psychotherapy for borderline personality disorder: randomized trial of schema-focused therapy vs transference-focused psychotherapy. Arch Gen Psychiatry 63:649–658

Glenn CR, Klonsky ED (2009) Emotion dysregulation as a core feature of borderline personality disorder. J Pers Disord 23:20–28

Glenn AL, Raine A, Schug RA (2009) The neural correlates of moral decision-making in psychopathy. Mol Psychiatry 14:5–6

Gordon HL, Baird AA, End A (2004) Functional differences among those high and low on a trait measure of psychopathy. Biol Psychiatry 56:516–521

Grant JE, Correia S, Brennan-Krohn T, Malloy PF, Laidlaw DH, Schulz SC (2007) Frontal white matter integrity in borderline personality disorder with self-injurious behavior. J Neuropsychiatry Clin Neurosci 19:383–390

Gregory S, Ffytche D, Simmons A, Kumari V, Howard M, Hodgins S, Blackwood N (2012) The antisocial brain: psychopathy matters. Arch Gen Psychiatry 69:962–972

Grilo CM, McGlashan TH, Quinlan DM, Walker ML, Greenfeld D, Edell WS (1998) Frequency of personality disorders in two age cohorts of psychiatric inpatients. Am J Psychiatry 155:140–142

Gross JJ (1998) Antecedent- and response-focused emotion regulation: divergent consequences for experience, expression, and physiology. J Pers Soc Psychol 74:224–237

Gross JJ (2002) Emotion regulation: affective, cognitive, and social consequences. Psychophysiology 39:281–291

Gross JJ, John OP (2003) Individual differences in two emotion regulation processes: implications for affect, relationships, and well-being. J Pers Soc Psychol 85:348–362

Guitart-Masip M, Pascual JC, Carmona S, Hoekzema E, Berge D, Perez V, Soler J, Soliva JC, Rovira M, Bulbena A, Vilarroya O (2009) Neural correlates of impaired emotional discrimination in borderline personality disorder: an fMRI study. Prog Neuropsychopharmacol Biol Psychiatry 33:1537–1545

Gunderson JG (2007) Disturbed relationships as a phenotype for borderline personality disorder. Am J Psychiatry 164:1637–1640

Gunderson JG, Zanarini MC, Kisiel C (1995) Borderline personality disorder. In: Widiger TA, Frances A, Pincus H, Ross R, First M, Davis W (eds) DSM-IV sourcebook. American Psychiatric Press, Washington, DC, pp 717–732

Gunderson JG, Stout RL, McGlashan TH, Shea MT, Morey LC, Grilo CM, Zanarini MC, Yen S, Markowitz JC, Sanislow C, Ansell E, Pinto A, Skodol AE (2011) Ten-year course of borderline personality disorder: psychopathology and function from the collaborative longitudinal personality disorders study. Arch Gen Psychiatry 68:827–837

Hare RD (2003) The Hare psychopathy checklist-revised, 2nd edn. Multi-Health Systems, Toronto

Hare RD (2006) Psychopathy: a clinical and forensic overview. Psychiatr Clin North Am 29:709–724

Harenski CL, Kim SH, Hamann S (2009) Neuroticism and psychopathy predict brain activation during moral and nonmoral emotion regulation. Cogn Affect Behav Neurosci 9:1–15

Harenski CL, Harenski KA, Shane MS, Kiehl KA (2010) Aberrant neural processing of moral violations in criminal psychopaths. J Abnorm Psychol 119:863–874

Harlow JM (1868) Recovery from the passage of an iron bar through the head. Publ Mass Med Soc 2:327–347

Hart SD, Cox DN, Hare RD (1995) The Hare Psychopathy Checklist — Revised Screening Version (PCL:SV). Toronto: Multi-Health Systems

Hazlett EA, New AS, Newmark R, Haznedar MM, Lo JN, Speiser LJ, Chen AD, Mitropoulou V, Minzenberg M, Siever LJ, Buchsbaum MS (2005) Reduced anterior and posterior cingulate gray matter in borderline personality disorder. Biol Psychiatry 58:614–623

Hazlett EA, Zhang J, New AS, Zelmanova Y, Goldstein KE, Haznedar MM, Meyerson D, Goodman M, Siever LJ, Chu KW (2012) Potentiated amygdala response to repeated emotional pictures in borderline personality disorder. Biol Psychiatry 72:448–456

Herpertz SC (1995) Self-injurious behaviour. Psychopathological and nosological characteristics in subtypes of self-injurers. Acta Psychiatr Scand 91:57–68

Herpertz SC, Gretzer A, Steinmeyer EM, Muehlbauer V, Schuerkens A, Sass H (1997) Affective instability and impulsivity in personality disorder. Results of an experimental study. J Affect Disord 44:31–37

Herpertz SC, Dietrich TM, Wenning B, Krings T, Erberich SG, Willmes K, Thron A, Sass H (2001) Evidence of abnormal amygdala functioning in borderline personality disorder: a functional MRI study. Biol Psychiatry 50:292–298

Holmes SE, Slaughter JR, Kashani J (2001) Risk factors in childhood that lead to the development of conduct disorder and antisocial personality disorder. Child Psychiatry Hum Dev 31:183–193

Hooker CI, Bruce L, Lincoln SH, Fisher M, Vinogradov S (2011) Theory of mind skills are related to gray matter volume in the ventromedial prefrontal cortex in schizophrenia. Biol Psychiatry 70:1169–1178

Jackson DC, Malmstadt JR, Larson CL, Davidson RJ (2000) Suppression and enhancement of emotional responses to unpleasant pictures. Psychophysiology 37:515–522

Jacob GA, Zvonik K, Kamphausen S, Sebastian A, Maier S, Philipsen A, Tebartz van Elst L, Lieb K, Tuscher O (2012) Emotional modulation of motor response inhibition in women with borderline personality disorder: an fMRI study. J Psychiatry Neurosci 37:120029

Johansen M, Karterud S, Pedersen G, Gude T, Falkum E (2004) An investigation of the prototype validity of the borderline DSM-IV construct. Acta Psychiatr Scand 109:289–298

Johnstone T, van Reekum CM, Urry HL, Kalin NH, Davidson RJ (2007) Failure to regulate: counterproductive recruitment of top-down prefrontal-subcortical circuitry in major depression. J Neurosci 27:8877–8884

Jones AP, Laurens KR, Herba CM, Barker GJ, Viding E (2009) Amygdala hypoactivity to fearful faces in boys with conduct problems and callous-unemotional traits. Am J Psychiatry 166:95–102

Kalisch R (2009) The functional neuroanatomy of reappraisal: time matters. Neurosci Biobehav Rev 33:1215–1226

Kamphausen S, Schroder P, Maier S, Bader K, Feige B, Kaller CP, Glauche V, Ohlendorf S, Elst LT, Kloppel S,

Jacob GA, Silbersweig D, Lieb K, Tuscher O (2013) Medial prefrontal dysfunction and prolonged amygdala response during instructed fear processing in borderline personality disorder. World J Biol Psychiatry 14:307–318
Kanske P, Heissler J, Schonfelder S, Bongers A, Wessa M (2011) How to regulate emotion? Neural networks for reappraisal and distraction. Cereb Cortex 21:1379–1388
Karl A, Schaefer M, Malta LS, Dorfel D, Rohleder N, Werner A (2006) A meta-analysis of structural brain abnormalities in PTSD. Neurosci Biobehav Rev 30:1004–1031
Kemperman I, Russ MJ, Shearin E (1997) Self-injurious behavior and mood regulation in borderline patients. J Pers Disord 11:146–157
Kernberg OF (1967) Borderline personality organization. J Am Psychoanal Assoc 15:641–685
Kernberg OF (1970) Factors in psychoanalytic treatment of narcissistic personalities. J Am Psychoanal Assoc 18:51–85
Kiehl KA (2006) A cognitive neuroscience perspective on psychopathy: evidence for paralimbic system dysfunction. Psychiatry Res 142:107–128
Kiehl KA, Smith AM, Hare RD, Mendrek A, Forster BB, Brink J, Liddle PF (2001) Limbic abnormalities in affective processing by criminal psychopaths as revealed by functional magnetic resonance imaging. Biol Psychiatry 50:677–684
King-Casas B, Chiu PH (2012) Understanding interpersonal function in psychiatric illness through multiplayer economic games. Biol Psychiatry 72:119–125
King-Casas B, Sharp C, Lomax-Bream L, Lohrenz T, Fonagy P, Montague PR (2008) The rupture and repair of cooperation in borderline personality disorder. Science 321:806–810
Kleindienst N, Bohus M, Ludascher P, Limberger MF, Kuenkele K, Ebner-Priemer UW, Chapman AL, Reicherzer M, Stieglitz RD, Schmahl C (2008) Motives for nonsuicidal self-injury among women with borderline personality disorder. J Nerv Ment Dis 196:230–236
Kleindienst N, Limberger MF, Ebner-Priemer UW, Keibel-Mauchnik J, Dyer A, Berger M, Schmahl C, Bohus M (2011) Dissociation predicts poor response to dialectial behavioral therapy in female patients with borderline personality disorder. J Pers Disord 25:432–447
Kliem S, Kroger C, Kosfelder J (2010) Dialectical behavior therapy for borderline personality disorder: a meta-analysis using mixed-effects modeling. J Consult Clin Psychol 78:936–951
Klonsky ED (2007) The functions of deliberate self-injury: a review of the evidence. Clin Psychol Rev 27:226–239
Kober H, Barrett LF, Joseph J, Bliss-Moreau E, Lindquist K, Wager TD (2008) Functional grouping and cortical-subcortical interactions in emotion: a meta-analysis of neuroimaging studies. Neuroimage 42:998–1031
Koenigs M, Baskin-Sommers A, Zeier J, Newman JP (2011) Investigating the neural correlates of psychopathy: a critical review. Mol Psychiatry 16:792–799
Koenigsberg HW (2010) Affective instability: toward an integration of neuroscience and psychological perspectives. J Pers Disord 24:60–82
Koenigsberg HW, Harvey PD, Mitropoulou V, New AS, Goodman M, Silverman J, Serby M, Schopick F, Siever LJ (2001) Are the interpersonal and identity disturbances in the borderline personality disorder criteria linked to the traits of affective instability and impulsivity? J Pers Disord 15:358–370
Koenigsberg HW, Fan J, Ochsner KN, Liu X, Guise KG, Pizzarello S, Dorantes C, Guerreri S, Tecuta L, Goodman M, New A, Siever LJ (2009a) Neural correlates of the use of psychological distancing to regulate responses to negative social cues: a study of patients with borderline personality disorder. Biol Psychiatry 66:854–863
Koenigsberg HW, Siever LJ, Lee H, Pizzarello S, New AS, Goodman M, Cheng H, Flory J, Prohovnik I (2009b) Neural correlates of emotion processing in borderline personality disorder. Psychiatry Res 172:192–199
Korzekwa MI, Dell PF, Links PS, Thabane L, Webb SP (2008) Estimating the prevalence of borderline personality disorder in psychiatric outpatients using a two-phase procedure. Compr Psychiatry 49:380–386
Kraus A, Esposito F, Seifritz E, Di Salle F, Ruf M, Valerius G, Ludaescher P, Bohus M, Schmahl C (2009) Amygdala deactivation as a neural correlate of pain processing in patients with borderline personality disorder and co-occurrent posttraumatic stress disorder. Biol Psychiatry 65:819–822
Kraus A, Valerius G, Seifritz E, Ruf M, Bremner JD, Bohus M, Schmahl C (2010) Script-driven imagery of self-injurious behavior in patients with borderline personality disorder: a pilot FMRI study. Acta Psychiatr Scand 121:41–51
Krause-Utz A, Oei NY, Niedtfeld I, Bohus M, Spinhoven P, Schmahl C, Elzinga BM (2012) Influence of emotional distraction on working memory performance in borderline personality disorder. Psychol Med 42:2181–2192
Kuhlmann A, Bertsch K, Schmidinger I, Thomann PA, Herpertz SC (2012) Morphometric differences in central stress-regulating structures between women with and without borderline personality disorder. J Psychiatry Neurosci 37:120039
Laakso MP, Vaurio O, Koivisto E, Savolainen L, Eronen M, Aronen HJ, Hakola P, Repo E, Soininen H, Tiihonen J (2001) Psychopathy and the posterior hippocampus. Behav Brain Res 118:187–193
Laakso MP, Gunning-Dixon F, Vaurio O, Repo-Tiihonen E, Soininen H, Tiihonen J (2002) Prefrontal volumes in habitually violent subjects with antisocial personality disorder and type 2 alcoholism. Psychiatry Res 114:95–102
Lamm C, Decety J, Singer T (2011) Meta-analytic evidence for common and distinct neural networks associated with directly experienced pain and empathy for pain. Neuroimage 54:2492–2502
Land BB, Bruchas MR, Lemos JC, Xu M, Melief EJ, Chavkin C (2008) The dysphoric component of stress is encoded by activation of the dynorphin kappa-opioid system. J Neurosci 28:407–414
Lang S, Kotchoubey B, Frick C, Spitzer C, Grabe HJ, Barnow S (2012) Cognitive reappraisal in trauma-

exposed women with borderline personality disorder. Neuroimage 59:1727–1734

Lanius RA, Vermetten E, Loewenstein RJ, Brand B, Schmahl C, Bremner JD, Spiegel D (2010) Emotion modulation in PTSD: clinical and neurobiological evidence for a dissociative subtype. Am J Psychiatry 167:640–647

Lee TM, Chan SC, Raine A (2008) Strong limbic and weak frontal activation to aggressive stimuli in spouse abusers. Mol Psychiatry 13:655–656

Leichsenring F, Leibing E, Kruse J, New AS, Leweke F (2011) Borderline personality disorder. Lancet 377:74–84

Lenzenweger MF, Lane MC, Loranger AW, Kessler RC (2007) DSM-IV personality disorders in the National Comorbidity Survey Replication. Biol Psychiatry 62:553–564

Levine D, Marziali E, Hood J (1997) Emotion processing in borderline personality disorders. J Nerv Ment Dis 185:240–246

Lieb K, Zanarini MC, Schmahl C, Linehan MM, Bohus M (2004) Borderline personality disorder. Lancet 364:453–461

Lilienfeld SO, Andrews BP (1996) Development and preliminary validation of a self-report measure of psychopathic personality traits in noncriminal populations. J Pers Assess 66:488–524

Linehan MM, Comtois KA, Murray AM, Brown MZ, Gallop RJ, Heard HL, Korslund KE, Tutek DA, Reynolds SK, Lindenboim N (2006) Two-year randomized controlled trial and follow-up of dialectical behavior therapy vs therapy by experts for suicidal behaviors and borderline personality disorder. Arch Gen Psychiatry 63:757–766

Lis E, Greenfield B, Henry M, Guile JM, Dougherty G (2007) Neuroimaging and genetics of borderline personality disorder: a review. J Psychiatry Neurosci 32:162–173

Ludascher P, Bohus M, Lieb K, Philipsen A, Jochims A, Schmahl C (2007) Elevated pain thresholds correlate with dissociation and aversive arousal in patients with borderline personality disorder. Psychiatry Res 149:291–296

Ludascher P, Greffrath W, Schmahl C, Kleindienst N, Kraus A, Baumgartner U, Magerl W, Treede RD, Bohus M (2009) A cross-sectional investigation of discontinuation of self-injury and normalizing pain perception in patients with borderline personality disorder. Acta Psychiatr Scand 120:62–70

Ly M, Motzkin JC, Philippi CL, Kirk GR, Newman JP, Kiehl KA, Koenigs M (2012) Cortical thinning in psychopathy. Am J Psychiatry 169:743–749

Lykken DT (1957) A study of anxiety in the sociopathic personality. J Abnorm Psychol 55:6–10

Lynch TR, Rosenthal MZ, Kosson DS, Cheavens JS, Lejuez CW, Blair RJ (2006) Heightened sensitivity to facial expressions of emotion in borderline personality disorder. Emotion 6:647–655

Magerl W, Burkart D, Fernandez A, Schmidt LG, Treede RD (2012) Persistent antinociception through repeated self-injury in patients with borderline personality disorder. Pain 153:575–584

Marinangeli MG, Butti G, Scinto A, Di Cicco L, Petruzzi C, Daneluzzo E, Rossi A (2000) Patterns of comorbidity among DSM-III-R personality disorders. Psychopathology 33:69–74

Marsh AA, Finger EC, Mitchell DG, Reid ME, Sims C, Kosson DS, Towbin KE, Leibenluft E, Pine DS, Blair RJ (2008) Reduced amygdala response to fearful expressions in children and adolescents with callous-unemotional traits and disruptive behavior disorders. Am J Psychiatry 165:712–720

Marsh AA, Finger EC, Fowler KA, Jurkowitz IT, Schechter JC, Yu HH, Pine DS, Blair RJ (2011) Reduced amygdala-orbitofrontal connectivity during moral judgments in youths with disruptive behavior disorders and psychopathic traits. Psychiatry Res 194:279–286

McGlashan TH, Grilo CM, Sanislow CA, Ralevski E, Morey LC, Gunderson JG, Skodol AE, Shea MT, Zanarini MC, Bender D, Stout RL, Yen S, Pagano M (2005) Two-year prevalence and stability of individual DSM-IV criteria for schizotypal, borderline, avoidant, and obsessive-compulsive personality disorders: toward a hybrid model of axis II disorders. Am J Psychiatry 162:883–889

Mechelli A, Price CJ, Friston KJ, Ashburner J (2005) Voxel-based morphometry of the human brain: methods and applications. Curr Med Imaging Rev 1:105–113

Mehrabian A, Epstein N (1972) A measure of emotional empathy. J Pers 40:525–543

Merkl A, Ammelburg N, Aust S, Roepke S, Reinecker H, Trahms L, Heuser I, Sander T (2010) Processing of visual stimuli in borderline personality disorder: a combined behavioural and magnetoencephalographic study. Int J Psychophysiol 78:257–264

Mesulam M (2000) Brain, mind, and the evolution of connectivity. Brain Cogn 42:4–6

Meyer B, Pilkonis PA, Beevers CG (2004) What's in a (neutral) face? Personality disorders, attachment styles, and the appraisal of ambiguous social cues. J Pers Disord 18:320–336

Miller JD, Campbell WK (2010) The case for using research on trait narcissism as a building block for understanding narcissistic personality disorder. Personal Disord Theory Res Treat 1:180–191

Miller JD, Campbell WK, Pilkonis PA (2007) Narcissistic personality disorder: relations with distress and functional impairment. Compr Psychiatry 48:170–177

Miller JD, Widiger TA, Campbell WK (2010) Narcissistic personality disorder and the DSM-V. J Abnorm Psychol 119:640–649

Minzenberg MJ, Poole JH, Vinogradov S (2006) Social-emotion recognition in borderline personality disorder. Compr Psychiatry 47:468–474

Minzenberg MJ, Fan J, New AS, Tang CY, Siever LJ (2007) Fronto-limbic dysfunction in response to facial emotion in borderline personality disorder: an event-related fMRI study. Psychiatry Res 155:231–243

Minzenberg MJ, Poole JH, Vinogradov S (2008) A neurocognitive model of borderline personality disorder:

effects of childhood sexual abuse and relationship to adult social attachment disturbance. Dev Psychopathol 20:341–368

Moffitt TE (2005) Genetic and environmental influences on antisocial behaviors: evidence from behavioral-genetic research. Adv Genet 55:41–104

Motzkin JC, Newman JP, Kiehl KA, Koenigs M (2011) Reduced prefrontal connectivity in psychopathy. J Neurosci 31:17348–17357

Müller JL, Sommer M, Wagner V, Lange K, Taschler H, Roder CH, Schuierer G, Klein HE, Hajak G (2003) Abnormalities in emotion processing within cortical and subcortical regions in criminal psychopaths: evidence from a functional magnetic resonance imaging study using pictures with emotional content. Biol Psychiatry 54:152–162

Muller JL, Ganssbauer S, Sommer M, Dohnel K, Weber T, Schmidt-Wilcke T, Hajak G (2008) Gray matter changes in right superior temporal gyrus in criminal psychopaths. Evidence from voxel-based morphometry. Psychiatry Res 163:213–222

Murphy FC, Nimmo Smith I, Lawrence AD (2003) Functional neuroanatomy of emotions: a meta-analysis. Cogn Affect Behav Neurosci 3:207–233

Murray J, Farrington DP (2010) Risk factors for conduct disorder and delinquency: key findings from longitudinal studies. Can J Psychiatry 55:633–642

New AS, Stanley B (2010) An opioid deficit in borderline personality disorder: self-cutting, substance abuse, and social dysfunction. Am J Psychiatry 167:882–885

New AS, Goodman M, Triebwasser J, Siever LJ (2008) Recent advances in the biological study of personality disorders. Psychiatr Clin North Am 31:441–461, vii

Niedtfeld I, Schmahl C (2009) Emotion regulation and pain in borderline personality disorder. Curr Psychiatry Rev 5:48–54

Niedtfeld I, Schulze L, Kirsch P, Herpertz SC, Bohus M, Schmahl C (2010) Affect regulation and pain in borderline personality disorder: a possible link to the understanding of self-injury. Biol Psychiatry 68:383–391

Niedtfeld I, Kirsch P, Schulze L, Herpertz SC, Bohus M, Schmahl C (2012) Functional connectivity of pain-mediated affect regulation in borderline personality disorder. PLoS One 7:e33293

Nunes PM, Wenzel A, Borges KT, Porto CR, Caminha RM, de Oliveira IR (2009) Volumes of the hippocampus and amygdala in patients with borderline personality disorder: a meta-analysis. J Pers Disord 23:333–345

Ochsner KN, Gross JJ (2005) The cognitive control of emotion. Trends Cogn Sci 9:242–249

Ochsner KN, Ray RD, Cooper JC, Robertson ER, Chopra S, Gabrieli JD, Gross JJ (2004) For better or for worse: neural systems supporting the cognitive down- and up-regulation of negative emotion. Neuroimage 23:483–499

Ochsner KN, Silvers JA, Buhle JT (2012) Functional imaging studies of emotion regulation: a synthetic review and evolving model of the cognitive control of emotion. Ann N Y Acad Sci 1251:E1–E24

Ottosson H, Ekselius L, Grann M, Kullgren G (2002) Cross-system concordance of personality disorder diagnoses of DSM-IV and diagnostic criteria for research of ICD-10. J Pers Disord 16:283–292

Patrick CJ, Bradley MM, Lang PJ (1993) Emotion in the criminal psychopath: startle reflex modulation. J Abnorm Psychol 102:82–92

Paus T (2005) Mapping brain maturation and cognitive development during adolescence. Trends Cogn Sci 9:60–68

Pincus AL, Lukowitsky MR (2010) Pathological narcissism and narcissistic personality disorder. Annu Rev Clin Psychol 6:421–446

Posner MI, Rothbart MK, Vizueta N, Thomas KM, Levy KN, Fossella J, Silbersweig D, Stern E, Clarkin J, Kernberg O (2003) An approach to the psychobiology of personality disorders. Dev Psychopathol 15:1093–1106

Prehn K, Schlagenhauf F, Schulze L, Berger C, Vohs K, Fleischer M, Hauenstein K, Keiper P, Domes G, Herpertz SC (2013a) Neural correlates of risk taking in violent criminal offenders characterized by emotional hypo- and hyper-reactivity. Soc Neurosci 8:136–147

Prehn K, Schulze L, Rossmann S, Berger C, Vohs K, Fleischer M, Hauenstein K, Keiper P, Domes G, Herpertz SC (2013b) Effects of emotional stimuli on working memory processes in male criminal offenders with borderline and antisocial personality disorder. World J Biol Psychiatry 14:71–78

Preissler S, Dziobek I, Ritter K, Heekeren HR, Roepke S (2010) Social cognition in borderline personality disorder: evidence for disturbed recognition of the emotions, thoughts, and intentions of others. Front Behav Neurosci 4:182

Price JL (2003) Comparative aspects of amygdala connectivity. Ann N Y Acad Sci 985:50–58

Prossin AR, Love TM, Koeppe RA, Zubieta JK, Silk KR (2010) Dysregulation of regional endogenous opioid function in borderline personality disorder. Am J Psychiatry 167:925–933

Pujol J, Batalla I, Contreras-Rodriguez O, Harrison BJ, Pera V, Hernandez-Ribas R, Real E, Bosa L, Soriano-Mas C, Deus J, Lopez-Sola M, Pifarre J, Menchon JM, Cardoner N (2012) Breakdown in the brain network subserving moral judgment in criminal psychopathy. Soc Cogn Affect Neurosci 7:917–923

Raine A (2002) Biosocial studies of antisocial and violent behavior in children and adults: a review. J Abnorm Child Psychol 30:311–326

Raine A, Buchsbaum M, LaCasse L (1997) Brain abnormalities in murderers indicated by positron emission tomography. Biol Psychiatry 42:495–508

Raine A, Lencz T, Bihrle S, LaCasse L, Colletti P (2000) Reduced prefrontal gray matter volume and reduced autonomic activity in antisocial personality disorder. Arch Gen Psychiatry 57:119–127

Raine A, Lencz T, Taylor K, Hellige JB, Bihrle S, Lacasse L, Lee M, Ishikawa S, Colletti P (2003) Corpus callosum abnormalities in psychopathic antisocial individuals. Arch Gen Psychiatry 60:1134–1142

Raine A, Lee L, Yang Y, Colletti P (2010) Neurodevelopmental marker for limbic maldevelopment in antisocial personality disorder and psychopathy. Br J Psychiatry 197:186–192

Reitz S, Krause-Utz A, Pogatzki-Zahn EM, Ebner-Priemer U, Bohus M, Schmahl C (2012) Stress regulation and incision in borderline personality disorder–a pilot study modeling cutting behavior. J Pers Disord 26:605–615

Rentrop M, Backenstrass M, Jaentsch B, Kaiser S, Roth A, Unger J, Weisbrod M, Renneberg B (2008) Response inhibition in borderline personality disorder: performance in a Go/Nogo task. Psychopathology 41:50–57

Rilling JK, Glenn AL, Jairam MR, Pagnoni G, Goldsmith DR, Elfenbein HA, Lilienfeld SO (2007) Neural correlates of social cooperation and non-cooperation as a function of psychopathy. Biol Psychiatry 61:1260–1271

Ritter K, Roepke S, Merkl A, Heuser I, Fydrich T, Lammers CH (2010) Comorbidity in patients with narcissistic personality disorder in comparison to patients with borderline personality disorder. Psychother Psychosom Med Psychol 60:14–24

Ritter K, Dziobek I, Preissler S, Ruter A, Vater A, Fydrich T, Lammers CH, Heekeren HR, Roepke S (2011) Lack of empathy in patients with narcissistic personality disorder. Psychiatry Res 187:241–247

Roepke S, Vater A, Preissler S, Heekeren H, Dziobek I (2012) Social cognition in borderline personality disorder. Front Neurosci 6

Rolls ET (2000) The orbitofrontal cortex and reward. Cereb Cortex 10:284–294

Ronningstam EF (2010) Narcissistic personality disorder: a current review. Curr Psychiatry Rep 12:68–75

Ronningstam EF (2011) Narcissistic personality disorder in DSM-V–in support of retaining a significant diagnosis. J Pers Disord 25:248–259

Ronningstam EF, Weinberg I, Maltsberger JT (2008) Eleven deaths of Mr. K.: contributing factors to suicide in narcissistic personalities. Psychiatry 71:169–182

Rosenthal MZ, Gratz KL, Kosson DS, Cheavens JS, Lejuez CW, Lynch TR (2008) Borderline personality disorder and emotional responding: a review of the research literature. Clin Psychol Rev 28:75–91

Ruocco AC, Amirthavasagam S, Zakzanis KK (2012) Amygdala and hippocampal volume reductions as candidate endophenotypes for borderline personality disorder: a meta-analysis of magnetic resonance imaging studies. Psychiatry Res 201:245–252

Rusch N, van Elst LT, Ludaescher P, Wilke M, Huppertz HJ, Thiel T, Schmahl C, Bohus M, Lieb K, Hesslinger B, Hennig J, Ebert D (2003) A voxel-based morphometric MRI study in female patients with borderline personality disorder. Neuroimage 20:385–392

Rusch N, Weber M, Il'yasov KA, Lieb K, Ebert D, Hennig J, van Elst LT (2007) Inferior frontal white matter microstructure and patterns of psychopathology in women with borderline personality disorder and comorbid attention-deficit hyperactivity disorder. Neuroimage 35:738–747

Rusch N, Bracht T, Kreher BW, Schnell S, Glauche V, Il'yasov KA, Ebert D, Lieb K, Hennig J, Saur D, van Elst LT (2010) Reduced interhemispheric structural connectivity between anterior cingulate cortices in borderline personality disorder. Psychiatry Res 181:151–154

Russ MJ, Roth SD, Lerman A, Kakuma T, Harrison K, Shindledecker RD, Hull J, Mattis S (1992) Pain perception in self-injurious patients with borderline personality disorder. Biol Psychiatry 32:501–511

Sala M, Caverzasi E, Lazzaretti M, Morandotti N, De Vidovich G, Marraffini E, Gambini F, Isola M, De Bona M, Rambaldelli G, d'Allio G, Barale F, Zappoli F, Brambilla P (2011) Dorsolateral prefrontal cortex and hippocampus sustain impulsivity and aggressiveness in borderline personality disorder. J Affect Disord 131:417–421

Sanislow CA, Grilo CM, Morey LC, Bender DS, Skodol AE, Gunderson JG, Shea MT, Stout RL, Zanarini MC, McGlashan TH (2002) Confirmatory factor analysis of DSM-IV criteria for borderline personality disorder: findings from the collaborative longitudinal personality disorders study. Am J Psychiatry 159:284–290

Sato JR, de Araujo Filho GM, de Araujo TB, Bressan RA, de Oliveira PP, Jackowski AP (2012) Can neuroimaging be used as a support to diagnosis of borderline personality disorder? An approach based on computational neuroanatomy and machine learning. J Psychiatr Res 46:1126–1132

Schiffer B, Muller BW, Scherbaum N, Hodgins S, Forsting M, Wiltfang J, Gizewski ER, Leygraf N (2011) Disentangling structural brain alterations associated with violent behavior from those associated with substance use disorders. Arch Gen Psychiatry 68:1039–1049

Schmahl C, Vermetten E, Elzinga BM, Douglas Bremner J (2003) Magnetic resonance imaging of hippocampal and amygdala volume in women with childhood abuse and borderline personality disorder. Psychiatry Res 122:193–198

Schmahl C, Greffrath W, Baumgartner U, Schlereth T, Magerl W, Philipsen A, Lieb K, Bohus M, Treede RD (2004) Differential nociceptive deficits in patients with borderline personality disorder and self-injurious behavior: laser-evoked potentials, spatial discrimination of noxious stimuli, and pain ratings. Pain 110:470–479

Schmahl C, Bohus M, Esposito F, Treede RD, Di Salle F, Greffrath W, Ludaescher P, Jochims A, Lieb K, Scheffler K, Hennig J, Seifritz E (2006) Neural correlates of antinociception in borderline personality disorder. Arch Gen Psychiatry 63:659–667

Schmahl C, Berne K, Krause A, Kleindienst N, Valerius G, Vermetten E, Bohus M (2009) Hippocampus and amygdala volumes in patients with borderline personality disorder with or without posttraumatic stress disorder. J Psychiatry Neurosci 34:289–295

Schnell K, Herpertz SC (2007) Effects of dialectic-behavioral-therapy on the neural correlates of affective hyperarousal in borderline personality disorder. J Psychiatr Res 41:837–847

Schulze L, Domes G, Kruger A, Berger C, Fleischer M, Prehn K, Schmahl C, Grossmann A, Hauenstein K, Herpertz SC (2011) Neuronal correlates of cognitive reappraisal in borderline patients with affective instability. Biol Psychiatry 69:564–573

Schulze L, Domes G, Koppen D, Herpertz SC (2013a) Enhanced detection of emotional facial expressions in borderline personality disorder. Psychopathology 46:217–224

Schulze L, Dziobek I, Vater A, Heekeren HR, Bajbouj M, Renneberg B, Heuser I, Roepke S (2013b) Gray matter abnormalities in patients with narcissistic personality disorder. J Psychiatr Res 47:1363–1369

Scott LN, Levy KN, Adams RB Jr, Stevenson MT (2011) Mental state decoding abilities in young adults with borderline personality disorder traits. Personal Disord 2:98–112

Seres I, Unoka Z, Keri S (2009) The broken trust and cooperation in borderline personality disorder. Neuroreport 20:388–392

Sharp C, Pane H, Ha C, Venta A, Patel AB, Sturek J, Fonagy P (2011) Theory of mind and emotion regulation difficulties in adolescents with borderline traits. J Am Acad Child Adolesc Psychiatry 50:563–573.e1

Sierra M, Berrios GE (1998) Depersonalization: neurobiological perspectives. Biol Psychiatry 44:898–908

Sieswerda S, Arntz A, Mertens I, Vertommen S (2007) Hypervigilance in patients with borderline personality disorder: specificity, automaticity, and predictors. Behav Res Ther 45:1011–1024

Silbersweig D, Clarkin JF, Goldstein M, Kernberg OF, Tuescher O, Levy KN, Brendel G, Pan H, Beutel M, Pavony MT, Epstein J, Lenzenweger MF, Thomas KM, Posner MI, Stern E (2007) Failure of frontolimbic inhibitory function in the context of negative emotion in borderline personality disorder. Am J Psychiatry 164:1832–1841

Singer T (2006) The neuronal basis and ontogeny of empathy and mind reading: review of literature and implications for future research. Neurosci Biobehav Rev 30:855–863

Skodol AE, Gunderson JG, Pfohl B, Widiger TA, Livesley WJ, Siever LJ (2002) The borderline diagnosis I: psychopathology, comorbidity, and personality structure. Biol Psychiatry 51:936–950

Soloff PH, Lynch KG, Kelly TM, Malone KM, Mann JJ (2000) Characteristics of suicide attempts of patients with major depressive episode and borderline personality disorder: a comparative study. Am J Psychiatry 157:601–608

Soloff PH, Nutche J, Goradia D, Diwadkar V (2008) Structural brain abnormalities in borderline personality disorder: a voxel-based morphometry study. Psychiatry Res 164:223–236

Soloff PH, Pruitt P, Sharma M, Radwan J, White R, Diwadkar VA (2012) Structural brain abnormalities and suicidal behavior in borderline personality disorder. J Psychiatr Res 46:516–525

Sommer M, Sodian B, Dohnel K, Schwerdtner J, Meinhardt J, Hajak G (2010) In psychopathic patients emotion attribution modulates activity in outcome-related brain areas. Psychiatry Res 182:88–95

Stanley B, Siever LJ (2010) The interpersonal dimension of borderline personality disorder: toward a neuropeptide model. Am J Psychiatry 167:24–39

Stanley B, Sher L, Wilson S, Ekman R, Huang YY, Mann JJ (2010) Non-suicidal self-injurious behavior, endogenous opioids and monoamine neurotransmitters. J Affect Disord 124:134–140

Sterzer P, Stadler C, Poustka F, Kleinschmidt A (2007) A structural neural deficit in adolescents with conduct disorder and its association with lack of empathy. Neuroimage 37:335–342

Stiglmayr CE, Shapiro DA, Stieglitz RD, Limberger MF, Bohus M (2001) Experience of aversive tension and dissociation in female patients with borderline personality disorder – a controlled study. J Psychiatr Res 35:111–118

Stiglmayr CE, Grathwol T, Linehan MM, Ihorst G, Fahrenberg J, Bohus M (2005) Aversive tension in patients with borderline personality disorder: a computer-based controlled field study. Acta Psychiatr Scand 111:372–379

Stinson FS, Dawson DA, Goldstein RB, Chou SP, Huang B, Smith SM, Ruan WJ, Pulay AJ, Saha TD, Pickering RP, Grant BF (2008) Prevalence, correlates, disability, and comorbidity of DSM-IV narcissistic personality disorder: results from the wave 2 national epidemiologic survey on alcohol and related conditions. J Clin Psychiatry 69:1033–1045

Tebartz van Elst L, Hesslinger B, Thiel T, Geiger E, Haegele K, Lemieux L, Lieb K, Bohus M, Hennig J, Ebert D (2003) Frontolimbic brain abnormalities in patients with borderline personality disorder: a volumetric magnetic resonance imaging study. Biol Psychiatry 54:163–171

Tiihonen J, Rossi R, Laakso MP, Hodgins S, Testa C, Perez J, Repo-Tiihonen E, Vaurio O, Soininen H, Aronen HJ, Kononen M, Thompson PM, Frisoni GB (2008) Brain anatomy of persistent violent offenders: more rather than less. Psychiatry Res 163:201–212

Tragesser SL, Robinson RJ (2009) The role of affective instability and UPPS impulsivity in borderline personality disorder features. J Pers Disord 23:370–383

Tragesser SL, Solhan M, Schwartz-Mette R, Trull TJ (2007) The role of affective instability and impulsivity in predicting future BPD features. J Pers Disord 21:603–614

Treede RD, Kenshalo DR, Gracely RH, Jones AK (1999) The cortical representation of pain. Pain 79:105–111

Trull TJ, Sher KJ, Minks-Brown C, Durbin J, Burr R (2000) Borderline personality disorder and substance use disorders: a review and integration. Clin Psychol Rev 20:235–253

Trull TJ, Jahng S, Tomko RL, Wood PK, Sher KJ (2010) Revised NESARC personality disorder diagnoses: gender, prevalence, and comorbidity with substance dependence disorders. J Pers Disord 24:412–426

Tsuchiya N, Adolphs R (2007) Emotion and consciousness. Trends Cogn Sci 11:158–167

Unoka Z, Seres I, Aspan N, Bodi N, Keri S (2009) Trust game reveals restricted interpersonal transactions in patients with borderline personality disorder. J Pers Disord 23:399–409

Unoka Z, Fogd D, Fuzy M, Csukly G (2011) Misreading the facial signs: specific impairments and error patterns in recognition of facial emotions with negative valence in borderline personality disorder. Psychiatry Res 189:419–425

Van Dillen LF, Heslenfeld DJ, Koole SL (2009) Tuning down the emotional brain: an fMRI study of the effects of cognitive load on the processing of affective images. Neuroimage 45:1212–1219

Veit R, Flor H, Erb M, Hermann C, Lotze M, Grodd W, Birbaumer N (2002) Brain circuits involved in emotional learning in antisocial behavior and social phobia in humans. Neurosci Lett 328:233–236

Veit R, Lotze M, Sewing S, Missenhardt H, Gaber T, Birbaumer N (2010) Aberrant social and cerebral responding in a competitive reaction time paradigm in criminal psychopaths. Neuroimage 49:3365–3372

Vermeiren R, Deboutte D, Ruchkin V, Schwab-Stone M (2002) Antisocial behaviour and mental health. Findings from three communities. Eur Child Adolesc Psychiatry 11:168–175

Viding E, Fontaine NM, McCrory EJ (2012) Antisocial behaviour in children with and without callous-unemotional traits. J R Soc Med 105:195–200

Volkow ND, Tancredi LR, Grant C, Gillespie H, Valentine A, Mullani N, Wang GJ, Hollister L (1995) Brain glucose metabolism in violent psychiatric patients: a preliminary study. Psychiatry Res 61:243–253

Vollm BA, Zhao L, Richardson P, Clark L, Deakin JF, Williams S, Dolan MC (2009) A voxel-based morphometric MRI study in men with borderline personality disorder: preliminary findings. Crim Behav Ment Health 19:64–72

von Ceumern-Lindenstjerna IA, Brunner R, Parzer P, Mundt C, Fiedler P, Resch F (2010) Attentional bias in later stages of emotional information processing in female adolescents with borderline personality disorder. Psychopathology 43:25–32

Wager TD, Davidson ML, Hughes BL, Lindquist MA, Ochsner KN (2008) Prefrontal-subcortical pathways mediating successful emotion regulation. Neuron 59:1037–1050

Wagner AW, Linehan MM (1999) Facial expression recognition ability among women with borderline personality disorder: implications for emotion regulation? J Pers Disord 13:329–344

Welch SS, Linehan MM, Sylvers P, Chittams J, Rizvi SL (2008) Emotional responses to self-injury imagery among adults with borderline personality disorder. J Consult Clin Psychol 76:45–51

White SF, Brislin S, Sinclair S, Fowler KA, Pope K, Blair RJ (2013) The relationship between large cavum septum pellucidum and antisocial behavior, callous-unemotional traits and psychopathy in adolescents. J Child Psychol Psychiatry 54:575–581

Wingenfeld K, Rullkoetter N, Mensebach C, Beblo T, Mertens M, Kreisel S, Toepper M, Driessen M, Woermann FG (2009) Neural correlates of the individual emotional Stroop in borderline personality disorder. Psychoneuroendocrinology 34:571–586

Yang Y, Raine A (2009) Prefrontal structural and functional brain imaging findings in antisocial, violent, and psychopathic individuals: a meta-analysis. Psychiatry Res 174:81–88

Yang Y, Raine A, Lencz T, Bihrle S, Lacasse L, Colletti P (2005) Prefrontal white matter in pathological liars. Br J Psychiatry 187:320–325

Yang Y, Raine A, Colletti P, Toga AW, Narr KL (2010) Morphological alterations in the prefrontal cortex and the amygdala in unsuccessful psychopaths. J Abnorm Psychol 119:546–554

Yang Y, Raine A, Colletti P, Toga AW, Narr KL (2011) Abnormal structural correlates of response perseveration in individuals with psychopathy. J Neuropsychiatry Clin Neurosci 23:107–110

Yen S, Shea MT, Sanislow CA, Grilo CM, Skodol AE, Gunderson JG, McGlashan TH, Zanarini MC, Morey LC (2004) Borderline personality disorder criteria associated with prospectively observed suicidal behavior. Am J Psychiatry 161:1296–1298

Zanarini MC, Frankenburg FR, Khera GS, Bleichmar J (2001) Treatment histories of borderline inpatients. Compr Psychiatry 42:144–150

Zanarini MC, Frankenburg FR, Hennen J, Reich DB, Silk KR (2005) Psychosocial functioning of borderline patients and axis II comparison subjects followed prospectively for six years. J Pers Disord 19:19–29

Zanarini MC, Frankenburg FR, Reich DB, Fitzmaurice G, Weinberg I, Gunderson JG (2008) The 10-year course of physically self-destructive acts reported by borderline patients and axis II comparison subjects. Acta Psychiatr Scand 117:177–184

Zetzsche T, Frodl T, Preuss UW, Schmitt G, Seifert D, Leinsinger G, Born C, Reiser M, Moller HJ, Meisenzahl EM (2006) Amygdala volume and depressive symptoms in patients with borderline personality disorder. Biol Psychiatry 60:302–310

18 The Use of Magnetic Resonance Imaging (MRI) in Eating Disorders

Valentina Cardi, Masashi Suda, and Janet Treasure

Abbreviations

ACC	Anterior cingulate cortex
AN	Anorexia nervosa
B/P	Binge eating and purging
BED	Binge eating disorder
BMI	Body mass index
BN	Bulimia nervosa
CSF	Cerebrospinal fluid
CT	Computed tomography
DLPFC	Dorsolateral PFC
ED	Eating disorder
EDNOS	Eating Disorder Not Otherwise Specified
Glx	Glutamate/glutamine
GM	Gray matter
HC	Healthy control
LH	Left-handed
mI	Myoinositol
NAA	N-acetylaspartate
NR	Not reported
OFC	Orbitofrontal cortex
PFC	Prefrontal cortex
r	Recovered
-R	Restricting
RH	Right-handed
SPECT	Single-photon emission computed tomography
SSRI	Selective serotonin reuptake inhibitor
VBM	Voxel-based morphometry
WM	White matter

V. Cardi, PhD (✉) • M. Suda, MD, PhD
J. Treasure, PhD, FRCP, FRCPsych
Department of Psychological Medicine, Institute of Psychiatry, King's College London, London, UK
e-mail: valentina.cardi@googlemail.com; sudamasashi7@gmail.com; janet.tresure@iop.kcl.ac.uk

18.1 Introduction

Eating disorders (EDs) are psychiatric disorders characterized by eating behavior problems. Although they are currently classified into three subgroups, anorexia nervosa (AN), bulimia nervosa (BN), and eating disorder not otherwise specified (EDNOS), changes are planned for the DSM-5 and ICD to include binge eating disorder (BED) as a separate category and to reduce the size of the nonspecified group.

In this chapter we review brain imaging studies according to the type of methodologies used, such as structural and functional.

18.2 Structural Studies

Brain morphometric studies using computed tomography (CT) were conducted in AN in the 1980s. Most of these studies reported diffuse brain atrophy in AN, such as enlarged ventricle and fissures. From the mid-1990s, magnetic resonance imaging (MRI) studies using manual or semiautomated measurements of structures of interest have been reported. Early MRI studies in people with an ED indicated that AN may be

Valentina Cardi and Masashi Suda are joint first authors

C. Mulert, M.E. Shenton (eds.), *MRI in Psychiatry*,
DOI 10.1007/978-3-642-54542-9_18, © Springer Berlin Heidelberg 2014

associated with (partially reversible) structural brain changes, for example, increased cerebrospinal fluid (CSF) volumes and reduced white matter (WM) and gray matter (GM) volumes (Katzman et al. 1997, 2001; Kingston et al. 1996; Kohn et al. 1997; Lambe et al. 1997).

Studies that examine specific regional changes in areas of particular interest including the hippocampus (Connan et al. 2006), the hippocampus–amygdala formation (Giordano et al. 2001), and the anterior cingulate cortex (McCormick et al. 2008) have been conducted. The development of automated techniques such as voxel-based morphometry (VBM) (Ashburner and Friston 2000) allows measurements to be conducted for large groups of people without the need for time-consuming manual measurements or subjective visual assessments (Whitwell 2009). A recent systematic review of these VBM studies in EDs (Van den Eynde et al. 2012) reported on ten studies in a total of 236 people with a current or past ED and 257 healthy controls as shown in Table 18.1. Sample heterogeneity prohibited a meta-analytic approach. The findings do not unequivocally indicate gray or white matter volume abnormalities in people with an ED. Nevertheless, these preliminary data suggest that, compared with healthy controls, people with AN have decreased gray matter in a range of brain regions and that those with BN have increased gray matter volumes in frontal and ventral striatal areas (Van den Eynde et al. 2012). Research in the recovery phase and longitudinal studies suggests that potential brain tissue abnormalities may recover with clinical improvement. Overall, as the available data are inconclusive, further efforts in this field are warranted.

18.3 Functional Studies

In this section, we review functional MRI studies according to the different aspects of the eating disorders psychopathology investigated. Eating disorders involve changes in eating behavior and usually, but not always, are associated with overvalued ideas about shape and weight. Additional features include abnormalities in cognitive style, emotional regulation, social functioning, reward sensitivity, and interoceptive awareness (Treasure et al. 2011).

18.3.1 Eating

Abnormal eating behaviors lie at the heart of all forms of EDs. This includes both over and under control of eating, sometimes associated with the interruption of the process of digestion by spitting/vomiting. Table 18.2 summarizes neuroimaging findings on functional activations to food stimuli in people with EDs.

18.3.1.1 Eating in Anorexia Nervosa

A recent review of neuroimaging studies involving food-related stimuli in EDs concludes that there are altered activations in the parietal and temporal cortices, anterior and subgenual cingulate cortex, and frontal cortex in AN (van Kuyck et al. 2009). A meta-analysis of functional MRI studies using coordinate-based meta-analysis methodology reports increased activation of medial frontal and caudate regions and reduced activation in parietal areas in AN during exposure to food cues (Zhu et al. 2012). In restricting AN, food stimuli are associated with the activity in the brain network responsible for the identification of emotional significance of the stimuli, affective states, and autonomic regulation (e.g., bottom-up processes), including the right amygdala (Joos et al. 2011b), ventral striatum (Wagner et al. 2008), orbitofrontal cortex (Uher et al. 2004), and insular cortex (Uher et al. 2004; Gizewski et al. 2010; Schienle et al. 2009; Vocks et al. 2010). Also, food stimuli are associated with abnormalities in brain areas involved in regulatory processes (e.g., top-down), comprising dorsal regions of the anterior cingulate cortex, posterior cingulate cortex (Gizewski et al. 2010), prefrontal cortex (Pietrini et al. 2011), medial prefrontal cortex (Uher et al. 2004), and dorsolateral prefrontal cortex (Brooks et al. 2011b). Inconsistent findings have been shown in relation to the activity of the cingulate cortex, which was found to be reduced in some studies (Gizewski et al. 2010; Pietrini et al. 2011; Joos et al. 2011a)

Table 18.1 Structural imaging studies in anorexia nervosa, bulimia nervosa, and binge eating disorder

Study	Participants	Age (years)	BMI (kg/m²)	Duration of illness/recovery	Medication	Additional relevant information (e.g., specific exclusion criteria, report on ED severity measure)
Boghi et al. (2011)	AN-R: $n=21$	AN-R: 29 ± 10.1	AN-R: 15.5 ± 1.8	Illness (in years):	All patients on SSRI treatment	Exclusion criterion: past or current alcohol or substance abuse
	"Short" duration: $n=10$	HC: 30.8 ± 8.7	HC: 21.9 ± 1.5	AN-R: 11.3 ± 12.1		No measure of ED severity reported
	"Long" duration: $n=11$			"Short": 1.9 ± 1.3		
	HC: $n=27$ (Age-matched)			"Long": 19.8 ± 11.1		
Brooks et al. (2011a)	AN: $n=14$	AN: 26 ± 7.1	AN: 15.6 ± 1.5	Illness (in years):	NR	ED severity measure: Eating Disorder Examination Questionnaire
	"AN-R: $n=8$"	"AN-R: 26 ± 10.9"	"AN-R: 15.1 ± 1.9"	AN: 9.2 ± 7.1		
	"AN-BP: $n=6$"	"AN-BP: 27 ± 9.7"	"AN-BP: 16.2 ± 1.1"	"AN-R: 9.2 ± 11.6"		
	HC: $n=21$ (Age-matched)	HC: 26 ± 9.6	HC: 21.4 ± 2.3	"AN-BP: 9.2 ± 8.2"		
Castro Fornieles et al. (2009)	AN: $n=12$	AN: 14.5 ± 1.5	AN (baseline): 14.8 ± 2.0	NR	1 at baseline and 3 at follow-up on fluoxetine or fluvoxamine	One male in each group
	(9 AN-R; 3 AN-BP)	HC: 14.6 ± 3.2	AN (follow-up): 18.8 ± 0.4			ED severity measure: Eating Attitudes Test
	HC: $n=9$		HC: not reported			
Gaudio et al. (2011)	AN-R: $n=16$	AN-R: 15.2 ± 1.7	AN-R: 14.2 ± 1.4	Illness (in months):	All on clomipramine or SSRI (duration of treatment: 16.4 ± 6.4 days)	Exclusion criterion: current or past other DSM-IV-TR disorders
	HC: $n=16$	HC: 15.1 ± 1.5	HC: 20.2 ± 1.6	AN-R 5.3 ± 3.2	10 on haloperidol (0.5–2.0 mg/day; mean dose: 1.3 ± 0.5 mg/day; mean duration of treatment: 9.2 ± 6.4 days)	ED severity measure: Eating Attitudes Test

(continued)

Table 18.1 (continued)

Study	Participants	Age (years)	BMI (kg/m^2)	Duration of illness/ recovery	Medication	Additional relevant information (e.g., specific exclusion criteria, report on ED severity measure)
Joos et al. (2010)	AN-R: n = 12	AN-R: 25.0 ± 4.8	AN-R: 16.0 ± 1.2	Illness (in years):	AN-R:1 on sertraline 75 mg/day	AN-R group: 1 participant had "inconstant bulimic phases" and 1 used laxatives
	BN: n = 17	BN: 24.5 ± 4.8	BN: 21.1 ± 2.5	AN-R: 4.7 ± 3.6 years		ED severity measure: Eating Disorder Inventory
	HC: n = 18 (Age-matched)	HC: 26.9 ± 5.7	HC: 21.2 ± 2.0	BN: 7.5 ± 5.7 years		
Muhlau et al. (2007)	rAN-R: n = 22	rAN-R: median 22.3 (range: 18.4–40.8)	rAN-R: median 19.5 (range: 17.0–22.8)	Recovery (in months):	NR	Exclusion criteria: lifetime history of post-traumatic stress disorder, manic episodes, schizophrenia, obsessive compulsive disorder, substance use disorders, or borderline personality disorder. Major depressive disorder occurring outside episodes of low weight was also an exclusion criterion. During episodes of low weight at least 16 rAN-R, participants had met the criteria for depressive disorder at least once
	HC: n = 37	HC: median 23.8 (range: 18.3–40.2)	HC: 20.1 median 22.3 (range: 18.3–24.8)	rAN-R: median 15.5 (range: 6–60)		Lifetime lowest BMI: median 13.5 (range: 10.0–16.1) Duration of AN (years): median 5 (range: 1–23)
Roberto et al. (2011)	AN: n = 32	AN: 26.9 ± 6.4	AN (baseline): 16.0 ± 1.6	Illness (in years):	No medication	Exclusion criteria: "axis I disorder other than major depression" and "history of suicide attempt or other self-injurious behavior within the previous 6 months"
	(14 AN-R; 18 AN-BP) HC: n = 21	HC: 25.0 ± 3.2	AN (follow-up): 20.0 ± 0.6 HC (baseline): 20.8 ± 1.2 HC (follow-up): 20.6 ± 1.2	AN: 10.2 ± 6.2		Amenorrhea was not considered as a criterion for AN

Schafer et al. (2010)	BN-P: $n=14$ BED: $n=17$ HC: $n=19$	BN: 23.1±3.8 BED: 26.4±6.4 HC: 22.3±2.6	BN: 22.1±2.5 BED: 32.2±4.0 HC: 21.7±1.4	Illness (in years): BN: 7.3±3.6 BED: 6.8±4.0	No medication	Exclusion criterion: "clinically relevant depression" ED severity measure: Eating Disorder Inventory
Suchan et al. (2010)	AN: $n=15$ HC: $n=15$	AN: 26.8±8.4 HC: 29.5±8.2	AN: 16.0±1.3 HC: 22.0±2.1	Illness (in years): AN: 5.5±5	NR	No measure of ED severity reported
Wagner et al. (2006)	rAN-R: $n=14$ rAN-BP: $n=16$ rBN: $n=10$ HC: $n=31$	rANR: 23.7±5.3 rANBP: 27.4±7.2 rBN: 24.0±6.1 HC: 26.8±7.3	rANR: 21.2±2.0 rANBP: 21.2±1.5 rBN: 23.1±2.4 HC: 21.9±2.0	Recovery (in months): rAN-R: 28.7±20.4 rAN-BP: 39.5±52.7 rBN: 29.8±18.1	No medication	Comorbid axis I and II assessed but not included in report Lifetime lowest BMI: rANR: 14.1±1.4 rANBP: 14.8±2.0 rBN: 19.2±2.1 HC: 20.1±1.4

Age and body mass index (BMI) are reported as mean ± standard deviation

AN anorexia nervosa, *BN* bulimia nervosa, *BED* binge eating disorder, *HC* healthy control, *ED* eating disorder, *r* recovered, *RH* right-handed, *LH* left-handed, *NR* not reported, *BMI* body mass index, *SSRI* selective serotonin reuptake inhibitor

Table 18.2 Summary of neuroimaging findings on functional activations to food stimuli in people with eating disorders

Study	Task	Sample	Main findings	Interpretation
Bohon and Stice (2011)	Gustatory paradigm (palatable vs. neutral food)	BN vs. subthreshold BN vs. HCs	BN < HCs: right prefrontal gyrus (anticipatory and consummatory), left middle frontal gyrus, left thalamus, right posterior insula (consummatory)	People with bulimic symptoms may experience less activation in gustatory and reward regions during anticipation and receipt of palatable foods
Brooks et al. (2011b)	Food vs. nonfood pictures	AN vs. BN vs. HCs	HC > AN, BN: right insular cortex, right superior temporal gyrus, left side cerebellum, left caudate body AN > BN, HC: left visual cortex, cerebellum, right DLPFC, right precuneus BN > AN, HC: right visual cortex, right insula, left prefrontal gyrus BN < HC: bilateral superior temporal gyrus, insular cortex, left visual cortex BN < AN: right parietal lobe, left dorsal posterior cingulate BN > AN: right caudate, right superior temporal gyrus, left supplementary motor area	AN and BN: activation of cognitive control AN and BN: sustained visual attention to food stimuli (possible substrate for attentional bias) BN: greater reward sensitivity
Burger and Stice (2011)	Gustatory paradigm (palatable food) and food pictures	HCs: high vs. low dietary restraint	High dietary restraint > low dietary restraint (food receipt): OFC, DLPFC	Individuals who report high dietary restraint have a hyperresponsivity in reward-related brain regions when food intake is occurring
Frank et al. (2011)	Reward learning paradigm (association learning between conditioned visual and unconditioned taste stimuli)	BN vs. HCs	BN < HCs: insula, ventral putamen, amygdala, OFC	Altered temporal learning in BN, which could be due to episodic excessive food stimulation which results in desensitization of dopamine circuits
Gizewski et al. (2010)	Food vs. nonfood pictures	AN-R vs. HCs	AN = HCs (hunger): insula (anterior insula in AN; posterior insula in BN)	Food stimuli more emotionally arousing to AN, but more physically stimulating to HCs
Joos et al. (2011a, b)	Food vs. nonfood pictures	AN-R vs. HCs	AN R > HCs: right amygdala AN R < HCs: cingulate cortex	AN: negative feedback loop of emotional processing (dysfunction of the top-down processes of the dorsal stream)
Pietrini et al. (2011)	Review	AN-R vs. AN-b/p	AN-R: rest < symptom provocation: frontal cortex; cingulate cortex AN-B/P: rest < symptom provocation: frontal, parietal, and cingulate	Possible disturbance of a network involving frontal, parietal, and cingulate metabolism at rest in AN that normalize after recovery

Santel et al. (2006)	Food vs. nonfood pictures	AN vs. HCs	AN < HCs (satiety): left inferior parietal cortex AN < HCs (hunger): right visual occipital cortex	Decreased food-related somatosensory processing in AN during satiety. Attentional mechanisms during hunger might facilitate restricted eating
Schienle et al. (2009)	Food vs. nonfood pictures	BED vs. BN vs. HCs vs. overweight	Food > nonfood: OFC, ACC, insula BED > overweight, HCs, BN: medial OFC BN > overweight, HCs, BED: ACC, insula	Differential brain activations in reward circuitry associated with food in patients suffering from BED and BN
Stice et al. (2010)	Palatable food vs. unpalatable food vs. neutral food pictures	HCs (from lean to obese)	Palatable food < unpalatable/neutral food: frontal operculum, lateral OFC, striatum predicted > BMI for those with DRD2 TaqIA A1 allele or DRD4-7R allele	Responsivity of reward circuitry to food increases risk for future weight gain, with the moderating effect of genes that impact dopamine signaling capacity
Stice et al. (2011)	Gustatory paradigm (receipt and anticipated receipt of palatable vs. neutral food stimuli)	HCs: high vs. low risk for obesity	High > low-risk obesity (palatable food receipt): caudate, parietal operculum, frontal operculum No differences in response to anticipated food reward	Youth at risk for obesity show elevated reward circuitry responsivity in general, coupled with elevated somatosensory region responsivity to food
Uher et al. (2003)	Food vs. nonfood pictures	Recovered AN-R vs. AN-R vs. HCs	REC > HCs (food): medial prefrontal cortex, dorsal anterior cingulate, cerebellum	Frontal lobe reactivity could be evaluated as a candidate factor predictive of outcome in AN
			REC < HCs (food): left parietal lobule and visual occipital cortex REC, HCs > AN (food): right lateral PFC, apical PFC, dorsal ACC AN > REC, HCs (food): superior medial PFC REC, HCs > AN: apical PFC, bilateral DLPFC, medial paracentral cortex AN, REC > HCs: medial PFC, cerebellum	The brain response in people recovered from AN is a combination of the responses seen in ill patients (medial frontal) and those in HCs (apical and lateral prefrontal)
Uher et al. (2004)	Food vs. nonfood pictures	AN vs. BN vs. HCs	HCs > EDs (food): left lateral PFC, left parietal cortex, bilateral visual cortex, cerebellum EDs > HCs (food): left medial orbitofrontal and anterior cingulate cortices ED < HCs (food): lateral PFC, inferior parietal lobule, cerebellum BN < HCs (food): lateral and apical PFC	Abnormal focus of food-related activity in the medial prefrontal region identified in a large number of ED patients

(continued)

Table 18.2 (continued)

Study	Task	Sample	Main findings	Interpretation
van Kuyck et al. (2009)	Review	AN-R	Parietal cortex: < Anterior and subgenual cingulate cortex: < rest; > symptom provocation Temporal lobe: symptom provocation: > insula; amygdala	No specific imaging biomarker or pattern for AN discovered
Vocks et al. (2011)	Gustatory paradigm (chocolate vs. hunger vs. satiety)	AN vs. HCs	AN > HCs (chocolate + hunger): amygdala and left medial temporal gyrus HC > AN (chocolate + hunger): right medial frontal gyrus	AN: fear response to high-calorie food HC: high reward anticipation
Wagner et al. (2008)	Gustatory paradigm (sugar vs. water)	Recovered AN-R vs. HCs	Recovered AN < HCs: insula, dorsal and middle caudate, dorsal and ventral putamen, anterior cingulate AN = HCs: anterioventral striatum, amygdala, and OFC	Individuals recovered from anorexia have disturbances of gustatory processing

AN anorexia nervosa, *BN* bulimia nervosa, *BED* binge eating disorder, *HCs* healthy control, *ED* eating disorder, *-R* restricting, *B/P* binge eating and purging, *PFC* prefrontal cortex, *DLPFC* dorsolateral PFC, *OFC* orbitofrontal cortex, *ACC* anterior cingulate cortex and increased in others (Uher et al. 2004). The overall conclusion is that there is heightened cognitive control and low reward associated with food in patients with AN. Amygdala hyperreactivity has been found in restrictive AN during symptom provocation using visual stimuli (Pietrini et al. 2011) and a gustatory paradigm (Gizewski et al. 2010), suggesting a heightened fear response. Furthermore, people with AN show lower activations in the parietal cortex when hungry (Uher et al. 2004; Santel et al. 2006) and during symptom provocation (Pietrini et al. 2011), possibly reflecting a decreased food-related somatosensory processing (Santel et al. 2006). The increased left visual cortex response to food suggests sustained attention towards food stimuli (Uher et al. 2004; Brooks et al. 2011b). A bilateral reduction of the frontal, parietal, and cingulate function at rest has been found in the binge/purge subtype of AN (Pietrini et al. 2011).

18.3.1.2 Eating in Bulimia Nervosa

Palatable "binge" foods (Small et al. 2001; de Araujo et al. 2008) activate reward-related brain areas (Wang et al. 2004). The areas in the brain linked to reward processing include the mesolimbic (dopaminergic/opioid) system, which extends from the ventral tegmental area of the midbrain to the nucleus accumbens (NA) in the striatum and includes the prefrontal cortex, amygdala, and hippocampus (Lowe et al. 2009). In particular, neuroimaging data suggest that the insula and frontal operculum are involved in food craving and anticipated reward from food and that, further, the orbitofrontal cortex, amygdala, and striatum encode the reward value for food (Gottfried et al. 2003; Small et al. 2001). Evidence shows that the activation of this system is powerful enough to promote food consumption beyond physiological satiation (Berthoud 2004) and that its overactivation also promotes heightened reward sensitivity and overconsumption (Kelley 2004; Berthoud and Morrison 2008). A recent study (Burger and Stice 2012) reports that individuals with high versus low dietary restraint scores show hyperresponsivity to food stimuli in regions associated with food reward. Recent data also suggests that brain regions belonging to the prefrontal cortex may regulate both excitatory and inhibitory reactions to appetitive stimuli (Berthoud and Morrison 2008). Several strands of evidence implicate the right hemisphere of the prefrontal cortex as being most directly related to the cognitive control of food intake (Wang et al. 2004; Miller and Cohen 2001; Uher and Treasure 2005) and show that left-sided prefrontal asymmetry correlates with measures of disinhibition, hunger, and appetitive responsivity in overweight individuals (Ochner et al. 2009).

In support of the reward hypersensitivity theory, women with bulimic symptoms have higher activation of somatosensory (Brooks et al. 2011b), motor (Brooks et al. 2011b; Marsh et al. 2009; Geliebter et al. 2006), and reward-associated brain areas to food stimuli (Brooks et al. 2011b; Marsh et al. 2009; Uher et al. 2004; Schienle et al. 2009) in comparison to healthy controls. Women recovered from BN continue to show an aberrant pattern of activation in the striatum in response to reward (Wagner et al. 2010) and a reduced pattern of activation in prefrontal regions to the taste of glucose (Frank et al. 2006).

18.3.1.3 Conclusion

In conclusion, functional MRI studies investigating brain activations to food stimuli in EDs seem to suggest a low sensitivity to reward associated with food in AN and an increased sensitivity in BN, with abnormalities persisting also after recovery.

18.3.2 Body Image

Body dissatisfaction is important in the development, maintenance, and relapse of EDs. Cognitive processing of weight- and shape-related stimuli has been found to differ in EDs from those of healthy controls. Several neuroimaging studies using functional MRI (Uher et al. 2005; Wagner et al. 2003) or SPECT (Beato Fernandez et al. 2009; Rodriguez Cano et al. 2009) found that people with EDs have abnormal activations in the body-shape processing network comprising the lateral occipitotemporal gyrus (including the extrastriate body area and the fusiform body area). Uher and colleagues (Uher et al. 2005) reported that females with AN showed reduced brain activation of the fusiform gyrus compared to healthy females when perceiving drawings of body-shape cartoons varying in weight and asked to judge their acceptability. Others have shown that viewing self-portraits (photographs) is associated with reduced brain activation of the prefrontal cortex, insula, precuneus (Sachdev et al. 2008), and the parietal lobe (Vocks et al. 2010). Other fMRI studies in EDs, which employed self-portraits with a judgment made of satisfaction and size, reported increased activation of the insula, prefrontal cortex, and precuneus (Mohr et al. 2010), while a distorted self-image has been shown to produce amygdala activation (Miyake et al. 2010). Furthermore, a self-"ideal" comparison (fashion models vs. ideal homes) produced increased activation of the insula and premotor cortex and reduced activation of the anterior cingulate cortex (Friederich et al. 2010). People with AN have also shown a different pattern of activation in the ventral striatum when asked to imagine having a body form depicted across a range of body size (thin to fat); in women with AN, activation was higher during the processing of underweight stimuli compared with normal weight stimuli, and the reverse pattern was observed in healthy females (Fladung et al. 2010). Nevertheless, differences in the type of stimuli, the instructions, the processing of the images, and the overall context may account for the variability between studies.

18.3.3 Reward and Self-Regulatory Circuit

The extreme control or loss of control in relationship to food is suggestive of abnormalities in reward and drive mechanisms. The brain responses to more abstract forms of reward such as those associated with winning and losing in a game were attenuated in people with an ED after recovery (Wagner et al. 2007, 2010). This may suggest that in AN, generalized anhedonia (asceticism) is possibly a trait vulnerability. Also, hyperactivation within a network of frontal brain regions associated with executive function and cognitive control including the dorsolateral prefrontal cortex, anterior cingulated cortex, presupplementary motor cortex, and anterior insular cortex (Uher et al. 2004) has been found. Activation in this system, especially activation of the right dorsolateral prefrontal cortex, is most marked in the restricting form of AN (Brooks et al. 2011b) and is probably due to high restraint. People suffering from the binge/purge subtype of AN show decreased dorsolateral prefrontal activity in contrast to restrictive AN (Pietrini et al. 2011), maybe suggesting less cognitive control. Those who have recovered from AN have a similar pattern of brain activation to those who are currently unwell but with additional activation in the prefrontal areas and anterior cingulate regions (Uher et al. 2003). These may represent the new, non-fear learning or extinction circuits that have been developed as part of the recovery process.

Functional MRI studies in BN suggest both hyper- and hypo-activation of the brain circuits associated with food reward, including the amygdala, insula, striatum, and orbitofrontal cortex. Abnormalities within prefrontal regions have been shown, suggesting higher excitatory and lower inhibitory reactions to appetitive stimuli. Aberrant pattern of activations within the hedonic (Wagner et al. 2010) and regulatory (Frank et al. 2006) areas have been found also in people recovered from BN.

18.3.4 Social Cognition

People with EDs have high levels of social anxiety and experience social and interpersonal difficulties (Tiller et al. 1997; Zucker et al. 2007); social isolation is a key maintaining factor (McKnight and Boughton 2009). In functional MRI studies related to social cognition, reduced activation of social cognitive brain areas such as the temporoparietal junction, the inferior frontal gyrus, the fusiform gyrus, the medial prefrontal cortex, and the temporal poles was found when using a theory of mind task involving shapes depicting emotional interactions (McAdams and Krawczyk 2011). Social emotions (anger and disgust) have been associated with reduced activation in the precuneus in females with BN (Ashworth et al. 2011). Cues eliciting negative self-beliefs were associated with decreased limbic activation in AN (Miyake et al. 2012), and self-representation reduced activation in the inferior parietal lobe in AN (Vocks et al. 2010). Although the detected brain regions have differed among the studies, social cognitive abnormalities in EDs may be related to reduced brain activation in neuroanatomy of social cognition.

18.4 Functional Connectivity

Recently, three papers reported the characteristics of functional connectivity in EDs (Favaro et al. 2012; Kim et al. 2012). Favaro and colleagues examined the functional connectivity of networks involved in visuospatial and somatosensory processing in 16 recovered and 29 acute AN patients using resting-state functional connectivity. As a result, both AN groups showed areas of decreased connectivity in the ventral visual network, a network involved in the "what?" pathway of visual perception. Only the acute AN group, but not the recovered AN group, displayed increased co-activation in the left parietal cortex, encompassing the somatosensory cortex, in an area implicated in long-term multimodal spatial memory and representation, even in the absence of visual information (Favaro et al. 2012). This may indicate the failure of the integration process between visual and somatosensory perceptual information (Favaro et al. 2012). On the other hand, Cowdrey and colleagues (2012) reported increased temporal correlation (coherence) in the default mode network which is thought to be involved in self-referential processing in recovered AN in their resting-state functional connectivity study (Cowdrey et al. 2014). In that paper, the recovered AN participants showed increased temporal coherence between the default mode network and the precuneus and the dorsolateral prefrontal cortex/inferior frontal gyrus compared to healthy controls. The authors speculated that these findings support the view that dysfunction in resting-state functional connectivity in regions involved in self-referential processing and cognitive control might be a vulnerability marker for the development of AN. In the study from Kim and colleagues (Kim et al. 2012), the left anterior insula was defined as a region of interest; seed-based functional connectivity and effective connectivity were performed in both AN and BN participants. The results showed that the left anterior insula had significant interactions with the right insula and the right inferior frontal gyrus in the AN group, whereas it demonstrated significant interactions with the medial orbitofrontal cortex in the BN sample. The authors concluded that the distinct patterns of functional and effective connectivity of the anterior insula may contribute to the different clinical features of AN and BN.

18.5 Magnetic Resonance Spectroscopy

Magnetic resonance spectroscopy has been also applied in EDs, although the findings are still controversial and contradictory. Castro-Fornieles et al. (2007) reported that N-acetylaspartate (NAA), glutamate/glutamine (Glx), and myoinositol (mI) levels were low in the frontal gray matter of adolescents with acute AN. They also found NAA levels to be restored after short-term weight recovery (Castro Fornieles et al. 2007) and total choline (Cho) and creatine (Cr) levels restoring later (Castro Fornieles et al. 2009). Blasel et al. (2012) reported different findings, with increased NAA level found in the frontal cortex and Glx, pooled creatine (tCr), and choline (tCho) in both frontal and parietal cortices in acute AN patients compared to healthy controls (Ohrmann et al. 2004). Although Ohrmann et al. (2004) described a reduction of Glx in the anterior cingulate cortex and reduced levels of mI and Cr in the dorsolateral prefrontal cortex in AN (Ohrmann et al. 2004), Joos et al. reported no differences in NAA, Cho, Cr, mI, Glu, and Glx in the frontal cortex and anterior cingulate cortex when comparing 17 adult female patients with EDs (10 with BN, 7 with AN) and 14 healthy controls (Joos et al. 2011a). Studies with more power and similar protocols may resolve these discrepancies in outcomes.

18.6 Summary

People with EDs show abnormalities in brain structure, functional connectivity, and neural activations to symptoms-related cues. Overall, decreased gray matter in a range of brain regions in AN and increased gray matter volumes in frontal and ventral striatal areas in BN have been reported. Studies using functional magnetic resonance describe aberrant patterns of response to body shape and food cues and more general abnormal reward sensitivity with differences between diagnoses. People with AN show higher cognitive control and low sensitivity to reward, whereas people with BN seem to have higher reward sensitivity. Reduced brain activation in neuroanatomy of social cognition has also been reported, as well as dysfunctions in resting-state functional connectivity in regions involved in self-referential processing.

Overall, neuroimaging studies in EDs have often been underpowered and have used a variety of different probes, which tap into various aspects of brain function; therefore, it is difficult to synthesize the results. The next phase of research will hopefully rectify these problems by focusing on refining and exploring some of the theoretical models that have been developed to explain the evolution and maintenance of EDs.

References

Ashburner J, Friston KJ (2000) Voxel-based morphometry–the methods. Neuroimage 11:805–821

Ashworth F, Pringle A, Norbury R, Harmer CJ, Cowen PJ, Cooper MJ (2011) Neural response to angry and disgusted facial expressions in bulimia nervosa. Psychol Med 41:2375–2384

Beato Fernandez L, Rodriguez Cano T, Garcia Vilches I, Garcia Vicente A, Poblete Garcia V, Castrejon AS, Toro J (2009) Changes in regional cerebral blood flow after body image exposure in eating disorders. Psychiatry Res 171:129–137

Berthoud HR (2004) Neural control of appetite: cross-talk between homeostatic and non-homeostatic systems. Appetite 43:315–317

Berthoud HR, Morrison C (2008) The brain, appetite, and obesity. Annu Rev Psychol 59:55–92

Blasel S, Pilatus U, Magerkurth J, von Stauffenberg M, Vronski D, Mueller M, Woeckel L, Hattingen E (2012) Metabolic gray matter changes of adolescents with anorexia nervosa in combined MR proton and phosphorus spectroscopy. Neuroradiology 54: 753–764

Boghi A, Sterpone S, Sales S, D'Agata F, Bradac GB, Zullo G, Munno D (2011) In vivo evidence of global and focal brain alterations in anorexia nervosa. Psychiatry Res 192:154–159

Bohon C, Stice E (2011) Reward abnormalities among women with full and subthreshold bulimia nervosa: a functional magnetic resonance imaging study. Int J Eat Disord 44:585–595

Brooks SJ, Barker GJ, O'Daly OG, Brammer M, Williams SC, Benedict C, Schioth HB, Treasure J, Campbell IC (2011a) Restraint of appetite and reduced regional brain volumes in anorexia nervosa: a voxel-based morphometric study. BMC Psychiatry 11:179

Brooks SJ, O'Daly OG, Uher R, Friederich HC, Giampietro V, Brammer M, Williams SC, Schioth HB, Treasure J, Campbell IC (2011b) Differential neural responses to food images in women with bulimia versus anorexia nervosa. PLoS One 6:e22259

Burger K, Stice E (2011) Relation of dietary restraint scores to activation of reward-related brain regions in response to food intake, anticipated intake, and food pictures. Neuroimage 55:233–239

Burger KS, Stice E (2012) Frequent ice cream consumption is associated with reduced striatal response to receipt of an ice cream-based milkshake. Am J Clin Nutr 95:810–817

Castro Fornieles J, Bargallo N, Lazaro L, Andres S, Falcon C, Plana MT, Junque C (2007) Adolescent anorexia nervosa: cross-sectional and follow-up frontal gray matter disturbances detected with proton magnetic resonance spectroscopy. J Psychiatr Res 41:952–958

Castro Fornieles J, Bargallo N, Lazaro L, Andres S, Falcon C, Plana MT, Junque C (2009) A cross-sectional and follow-up voxel-based morphometric MRI study in adolescent anorexia nervosa. J Psychiatr Res 43:331–340

Connan F, Murphy F, Connor SE, Rich P, Murphy T, Bara Carill N, Landau S, Krljes S, Ng V, Williams S, Morris RG, Campbell IC, Treasure J (2006) Hippocampal volume and cognitive function in anorexia nervosa. Psychiatry Res 146:117–125

Cowdrey FA, Filippini N, Park RJ, Smith SM, McCabe C (2014) Increased resting state functional connectivity in the default mode network in recovered anorexia nervosa. Hum Brain Mapp 35(2):483–491. doi: 10.1002/hbm.22202 de Araujo IE, Oliveira Maia AJ, Sotnikova TD, Gainetdinov RR, Caron MG, Nicolelis MA, Simon SA (2008) Food reward in the absence of taste receptor signaling. Neuron 57:930–941

Favaro A, Santonastaso P, Manara R, Bosello R, Bommarito G, Tenconi E, Di Salle F (2012) Disruption of visuospatial and somatosensory functional connectivity in anorexia nervosa. Biol Psychiatry. 72(10):864–870.

Fladung AK, Gron G, Grammer K, Herrnberger B, Schilly E, Grasteit S, Wolf RC, Walter H, von Wietersheim J (2010) A neural signature of anorexia nervosa in the ventral striatal reward system. Am J Psychiatry 167:206–212

Frank G, Reynolds J, Shott M, O'Reilly R (2011) Altered temporal difference learning in bulimia nervosa. Biol Psychatry 70:728–735

Frank GK, Wagner A, Achenbach S, McConaha C, Skovira K, Aizenstein H, Carter CS, Kaye WH (2006) Altered brain activity in women recovered from bulimic-type eating disorders after a glucose challenge: a pilot study. Int J Eat Disord 39:76–79

Friederich HC, Brooks S, Uher R, Campbell IC, Giampietro V, Brammer M, Williams SC, Herzog W, Treasure J (2010) Neural correlates of body dissatisfaction in anorexia nervosa. Neuropsychologia 48:2878–2885

Gaudio S, Nocchi F, Franchin T, Genovese E, Cannata V, Longo D, Fariello G (2011) Gray matter decrease distribution in the early stages of anorexia nervosa restrictive type in adolescents. Psychiatry Res 191:24–30

Geliebter A, Ladell T, Logan M, Schneider T, Sharafi M, Hirsch J (2006) Responsivity to food stimuli in obese and lean binge eaters using functional MRI. Appetite 46:31–35

Giordano GD, Renzetti P, Parodi RC, Foppiani L, Zandrino F, Giordano G, Sardanelli F (2001) Volume measurement with magnetic resonance imaging of hippocampus-amygdala formation in patients with anorexia nervosa. J Endocrinol Invest 24:510–514

Gizewski ER, Rosenberger C, de Greiff A, Moll A, Senf W, Wanke I, Forsting M, Herpertz S (2010) Influence of satiety and subjective valence rating on cerebral activation patterns in response to visual stimulation with high-calorie stimuli among restrictive anorectic and control women. Neuropsychobiology 62:182–192

Gottfried JA, O'Doherty J, Dolan RJ (2003) Encoding predictive reward value in human amygdala and orbitofrontal cortex. Science 301:1104–1107

Joos A, Kloppel S, Hartmann A, Glauche V, Tuscher O, Perlov E, Saum B, Freyer T, Zeeck A, Tebartz van Elst L (2010) Voxel-based morphometry in eating disorders: correlation of psychopathology with grey matter volume. Psychiatry Res 182:146–151

Joos AA, Perlov E, Buchert M, Hartmann A, Saum B, Glauche V, Freyer T, Weber Fahr W, Zeeck A, Tebartz van Elst L (2011a) Magnetic resonance spectroscopy of the anterior cingulate cortex in eating disorders. Psychiatry Res 191:196–200

Joos AA, Saum B, van Elst LT, Perlov E, Glauche V, Hartmann A, Freyer T, Tuscher O, Zeeck A (2011b) Amygdala hyperreactivity in restrictive anorexia nervosa. Psychiatry Res 191:189–195

Katzman DK, Zipursky RB, Lambe EK, Mikulis DJ (1997) A longitudinal magnetic resonance imaging study of brain changes in adolescents with anorexia nervosa. Arch Pediatr Adolesc Med 151:793–797

Katzman DK, Christensen B, Young AR, Zipursky RB (2001) Starving the brain: structural abnormalities and cognitive impairment in adolescents with anorexia nervosa. Semin Clin Neuropsychiatry 6:146–152

Kelley AE (2004) Ventral striatal control of appetitive motivation: role in ingestive behavior and reward-related learning. Neurosci Biobehav Rev 27:765–776

Kim KR, Ku J, Lee JH, Lee H, Jung YC (2012) Functional and effective connectivity of anterior insula in anorexia nervosa and bulimia nervosa. Neurosci Lett 521:152–157

Kingston K, Szmukler G, Andrewes D, Tress B, Desmond P (1996) Neuropsychological and structural brain changes in anorexia nervosa before and after refeeding. Psychol Med 26:15–28

Kohn MR, Ashtari M, Golden NH, Schebendach J, Patel M, Jacobson MS, Shenker IR (1997) Structural brain changes and malnutrition in anorexia nervosa. Ann N Y Acad Sci 817:398–399

Lambe EK, Katzman DK, Mikulis DJ, Kennedy SH, Zipursky RB (1997) Cerebral gray matter volume deficits after weight recovery from anorexia nervosa. Arch Gen Psychiatry 54:537–542

Lowe MR, van Steenburgh J, Ochner C, Coletta M (2009) Neural correlates of individual differences related to appetite. Physiol Behav 97:561–571

Marsh R, Steinglass JE, Gerber AJ, Graziano O'Leary K, Wang Z, Murphy D, Walsh BT, Peterson BS (2009) Deficient activity in the neural systems that mediate self-regulatory control in bulimia nervosa. Arch Gen Psychiatry 66:51–63

McAdams CJ, Krawczyk DC (2011) Impaired neural processing of social attribution in anorexia nervosa. Psychiatry Res 194:54–63

McCormick LM, Keel PK, Brumm MC, Bowers W, Swayze V, Andersen A, Andreasen N (2008) Implications of starvation-induced change in right dorsal anterior cingulate volume in anorexia nervosa. Int J Eat Disord 41:602–610

McKnight R, Boughton N (2009) A patient's journey. Anorexia nervosa. BMJ 339:b3800

Miller EK, Cohen JD (2001) An integrative theory of prefrontal cortex function. Annu Rev Neurosci 24:167–202

Miyake Y, Okamoto Y, Onoda K, Kurosaki M, Shirao N, Okamoto Y, Yamawaki S (2010) Brain activation during the perception of distorted body images in eating disorders. Psychiatry Res 181:183–192

Miyake Y, Okamoto Y, Onoda K, Shirao N, Okamoto Y, Yamawaki S (2012) Brain activation during the perception of stressful word stimuli concerning interpersonal relationships in anorexia nervosa patients with high degrees of alexithymia in an fMRI paradigm. Psychiatry Res 201:113–119

Mohr HM, Zimmermann J, Roder C, Lenz C, Overbeck G, Grabhorn R (2010) Separating two components of body image in anorexia nervosa using fMRI. Psychol Med 40:1519–1529

Muhlau M, Gaser C, Ilg R, Conrad B, Leibl C, Cebulla MH, Backmund H, Gerlinghoff M, Lommer P, Schnebel A, Wohlschlager AM, Zimmer C, Nunnemann S (2007) Gray matter decrease of the anterior cingulate cortex in anorexia nervosa. Am J Psychiatry 164:1850–1857

Ochner CN, Green D, van Steenburgh J, Kounios J, Lowe MR (2009) Asymmetric prefrontal cortex activation in relation to markers of overeating in obese humans. Appetite 53:44–49

Ohrmann P, Kersting A, Suslow T, Lalee Mentzel J, Donges US, Fiebich M, Arolt V, Heindel W, Pfleiderer B (2004) Proton magnetic resonance spectroscopy in anorexia nervosa: correlations with cognition. Neuroreport 15:549–553

Pietrini F, Castellini G, Ricca V, Polito C, Pupi A, Faravelli C (2011) Functional neuroimaging in anorexia nervosa: a clinical approach. Eur Psychiatry 26:176–182

Roberto CA, Mayer LE, Brickman AM, Barnes A, Muraskin J, Yeung LK, Steffener J, Sy M, Hirsch J, Stern Y, Walsh BT (2011) Brain tissue volume changes following weight gain in adults with anorexia nervosa. Int J Eat Disord 44:406–411

Rodriguez Cano T, Beato Fernandez L, Garcia Vilches I, Garcia Vicente A, Poblete Garcia V, Soriano Castrejon A (2009) Regional cerebral blood flow patterns of change following the own body image exposure in eating disorders: a longitudinal study. Eur Psychiatry 24:275–281

Sachdev P, Mondraty N, Wen W, Gulliford K (2008) Brains of anorexia nervosa patients process self-images differently from non-self-images: an fMRI study. Neuropsychologia 46:2161–2168

Santel S, Baving L, Krauel K, Munte TF, Rotte M (2006) Hunger and satiety in anorexia nervosa: fMRI during cognitive processing of food pictures. Brain Res 1114:138–148

Schafer A, Vaitl D, Schienle A (2010) Regional grey matter volume abnormalities in bulimia nervosa and binge-eating disorder. Neuroimage 50:639–643

Schienle A, Schafer A, Hermann A, Vaitl D (2009) Binge-eating disorder: reward sensitivity and brain activation to images of food. Biol Psychiatry 65:654–661

Small DM, Zatorre RJ, Dagher A, Evans AC, Jones-Gotman M (2001) Changes in brain activity related to eating chocolate: from pleasure to aversion. Brain 124:1720–1733

Stice E, Yokum S, Bohon C, Marti N, Smolen A (2010) Reward circuitry responsivity to food predicts future increases in body mass: moderating effects of DRD2 and DRD4. Neuroimage 50:1618–1625

Stice E, Yokum S, Burger KS, Epstein LH, Small DM (2011) Youth at risk for obesity show greater activation of striatal and somatosensory regions to food. J Neurosci 23:4360–4366

Suchan B, Busch M, Schulte D, Gronemeyer D, Herpertz S, Vocks S (2010) Reduction of gray matter density in the extrastriate body area in women with anorexia nervosa. Behav Brain Res 206:63–67

Tiller JM, Sloane G, Schmidt U, Troop N, Power M, Treasure JL (1997) Social support in patients with anorexia nervosa and bulimia nervosa. Int J Eat Disord 21:31–38

Treasure J, Cardi V, Kan C (2011) Eating in eating disorders. Eur Eat Disord Rev 20:e42–e49

Uher R, Treasure J (2005) Brain lesions and eating disorders. J Neurol Neurosurg Psychiatry 76:852–857

Uher R, Brammer MJ, Murphy T, Campbell IC, Ng VW, Williams SC, Treasure J (2003) Recovery and chronicity in anorexia nervosa: brain activity associated with differential outcomes. Biol Psychiatry 54:934–942

Uher R, Murphy T, Brammer MJ, Dalgleish T, Phillips ML, Ng VW, Andrew CM, Williams SC, Campbell IC, Treasure J (2004) Medial prefrontal cortex activity associated with symptom provocation in eating disorders. Am J Psychiatry 161:1238–1246

Uher R, Murphy T, Friederich HC, Dalgleish T, Brammer MJ, Giampietro V, Phillips ML, Andrew CM, Ng VW, Williams SC, Campbell IC, Treasure J (2005) Functional neuroanatomy of body shape perception in healthy and eating-disordered women. Biol Psychiatry 58:990–997

Van den Eynde F, Suda M, Broadbent H, Guillaume S, Van den Eynde M, Steiger H, Israel M, Berlim M, Giampietro V, Simmons A, Treasure J, Campbell I, Schmidt U (2012) Structural magnetic resonance imaging in eating disorders: a systematic review of voxel-based morphometry studies. Eur Eat Disord Rev 20:94–105

van Kuyck K, Gerard N, Van Laere K, Casteels C, Pieters G, Gabriels L, Nuttin B (2009) Towards a neurocircuitry in anorexia nervosa: evidence from functional neuroimaging studies. J Psychiatr Res 43:1133–1145

Vocks S, Busch M, Gronemeyer D, Schulte D, Herpertz S, Suchan B (2010) Neural correlates of viewing photographs of one's own body and another woman's body in anorexia and bulimia nervosa: an fMRI study. J Psychiatry Neurosci 35:163–176

Vocks S, Herpertz S, Rosenberger C, Senf W, Gizewski ER (2011) Effects of gustatory stimulation on brain activity during hunger and satiety in females with restricting-type anorexia nervosa: an fMRI study. J Psychiatr Res 45:395–403

Wagner A, Ruf M, Braus DF, Schmidt MH (2003) Neuronal activity changes and body image distortion in anorexia nervosa. Neuroreport 14:2193–2197

Wagner A, Greer P, Bailer UF, Frank GK, Henry SE, Putnam K, Meltzer CC, Ziolko SK, Hoge J, McConaha C, Kaye WH (2006) Normal brain tissue volumes after long-term recovery in anorexia and bulimia nervosa. Biol Psychiatry 59:291–293

Wagner A, Aizenstein H, Venkatraman VK, Fudge J, May JC, Mazurkewicz L, Frank GK, Bailer UF, Fischer L, Nguyen V, Carter C, Putnam K, Kaye WH (2007) Altered reward processing in women recovered from anorexia nervosa. Am J Psychiatry 164:1842–1849

Wagner A, Aizenstein H, Mazurkewicz L, Fudge J, Frank GK, Putnam K, Bailer UF, Fischer L, Kaye WH (2008) Altered insula response to taste stimuli in individuals recovered from restricting-type anorexia nervosa. Neuropsychopharmacology 33:513–523

Wagner A, Aizenstein H, Venkatraman VK, Bischoff Grethe A, Fudge J, May JC, Frank GK, Bailer UF, Fischer L, Putnam K, Kaye WH (2010) Altered striatal response to reward in bulimia nervosa after recovery. Int J Eat Disord 43:289–294

Wang GJ, Volkow ND, Telang F, Jayne M, Ma J, Rao M, Zhu W, Wong CT, Pappas NR, Geliebter A, Fowler JS (2004) Exposure to appetitive food stimuli markedly activates the human brain. Neuroimage 21:1790–1797

Whitwell JL (2009) Voxel-based morphometry: an automated technique for assessing structural changes in the brain. J Neurosci 29:9661–9664

Zhu Y, Hu X, Wang J, Chen J, Guo Q, Li C, Enck P (2012) Processing of food, body and emotional stimuli in anorexia nervosa: a systematic review and meta-analysis of functional magnetic resonance imaging studies. Eur Eat Disord Rev 20(6):439–450. doi: 10.1002/erv.2197. 4 Sept 2012

Zucker NL, Losh M, Bulik CM, LaBar KS, Piven J, Pelphrey KA (2007) Anorexia nervosa and autism spectrum disorders: guided investigation of social cognitive endophenotypes. Psychol Bull 133: 976–1006

Drug Addiction

19

Katrin Charlet, Anne Beck, and Andreas Heinz

Abbreviations

^{1}H-MRS	Proton magnetic resonance spectroscopy
5-HT	Serotonin
Cho	Choline
Cr	Creatine
DA	Dopamine
DRD_2	Dopamine D_2 receptor
DSM-IV	Diagnostic and Statistical Manual of Mental Disorders
DTI	Diffusion tensor imaging
fMRI	Functional magnetic resonance imaging
GABA	Gamma-aminobutyric acid
MRI	Magnetic resonance imaging
NAA	N-acetyl-aspartate
NMDA	N-methyl-D-aspartate
PET	Positron-emission tomography
PFC	Prefrontal cortex
SPECT	Single photon emission computed tomography

K. Charlet (✉) • A. Beck • A. Heinz
Department of Psychiatry and Psychotherapy, Charité Universitätsmedizin Berlin,
Campus Mitte, Berlin, Germany
e-mail: katrin.charlet@charite.de;
andreas.heinz@charite.de

19.1 Introduction

Magnetic resonance imaging (MRI) allows the noninvasive investigation of neuronal systems that are involved in basic and complex processes of everyday life. In this context, clinical studies of the last decades have used different MRI methods and revealed core systems that play pivotal roles in specific mental processes and their disorders, such as decision-making, emotion processing, and executive functioning. This variety of MRI methods, which have been introduced earlier in this book, enables us to achieve manifold insights into specific brain mechanisms involved in the development and maintenance of psychiatric disorders, i.e., alcohol dependence, and thus help to understand the neurobiological correlates of mental diseases. One day this may contribute to provide a personalized and more precise treatment of patients.

One of the most common psychiatric diseases worldwide is substance dependence, and its individual development is assumed to result from an interaction of various psychological, genetic, social, and substance-inherent psychotropic conditions. Maintenance of addictive behavior is assumed to be mediated by neuroadaptive processes in different neurotransmitter systems, such as dopaminergic, glutamatergic, serotonergic, and opioidergic neurotransmission. The interaction of such neurotransmitter systems with genetic and environmental factors helps to explain core aspects of addictive behavior, and therefore the research on neurobiological

C. Mulert, M.E. Shenton (eds.), *MRI in Psychiatry*,
DOI 10.1007/978-3-642-54542-9_19, © Springer Berlin Heidelberg 2014

correlates of drug dependence is necessary to create specific concepts and approaches for psychotherapeutic and pharmacological therapy.

19.2 Substance/Alcohol Dependence

19.2.1 Neuronal Mechanisms of the Aberrant Dopaminergic Reward System in Alcoholism

It is assumed that alcohol abuse and dependence is crucially mediated by an aberrant functioning of the mesocorticolimbic *reward system*. This dopaminergically modulated reward circuit consists of the following major neuronal components: the ventral tegmental area and the substantia nigra with their dopaminergic (DA) projections to the ventral (including the nucleus accumbens) and dorsal striatum, amygdala, olfactory tubercle, and frontal and limbic cortex regions (Charlet et al. 2013a; Ikemoto 2007). The pioneering study by Schultz et al. (1997) revealed that DA neurons are engaged in reward-processing and reward-depending learning: different types of appetitive stimuli induce phasic DA release and thus elicit orientating behavior. Changes of DA discharge were seen for unexpected rewards and for Pavlovian conditioned stimuli, which predict upcoming reward, but failed to appear at the moment of receiving the predicted reward (Schultz et al. 1997). Therefore, DA neurotransmission appears to function as a signaling network, indicating the occurrence of salient stimuli and so-called prediction errors (i.e., unexpected rewards; Schultz et al. 1997; for review, see Charlet et al. 2013a). Alcohol and other drugs of abuse release dopamine and signal larger-than-expected reward and thus reinforce consecutive drug intake (Di Chiara 2002). Compared to natural reinforcers, such as food, sleep, and sex, the rewarding effects of, for example, alcohol on DA release does not habituate (Di Chiara and Bassareo 2007). Since alcohol discharges up to two times as much DA as natural reinforcers usually do (Di Chiara and Imperato 1988), it was suggested that drugs of abuse "hijack" the reward system, which now preferentially responds to drug-associated reinforcement at the expense of natural nondrug reinforcement (Gardner 2005). Indeed, a reduced functional activation (measured with fMRI) of the ventral striatum – the core region of the described reward system – was found in alcohol-dependent patients who were confronted with monetary cues that indicated the availability of reward (Wrase et al. 2007). Moreover, the same patients displayed an increased activation of the ventral striatum when confronted with alcohol-associated stimuli. This activation pattern was also correlated with the severity of alcohol craving in alcohol-dependent individuals compared to healthy controls (Wrase et al. 2007). Reduced brain activation to naturally reward-indicating stimuli may thus interfere with the patients' motivation to experience other than drug-related rewarding situations. Thus, the repeated and continuous use of drugs shapes the reward system and determines the development and maintenance of substance dependence.

Alcohol-dependent behavior is assumed to involve dissociable aspects of reward, namely, *wanting* and *linking* of a positive reinforce, which on a neurobiological level can mainly be assigned to the dopaminergic neurotransmitter system and the opioidergic neurotransmitter system, respectively (Berridge et al. 2009; Charlet et al. 2013a; Heinz et al. 2009b). Since pharmacological DA blockage caused motivational apathy rather than anhedonia, DA is attributed to the *wanting* aspects of drug dependence (Boileau et al. 2003; Di Chiara 2002; Di Chiara and Bassareo 2007; Heinz et al. 2009a). Contrary, *liking* is referred to the drugs' pleasurable effects. According to Robinson and Berridge (1993), *liking* is associated with opioidergic neurotransmission, with one hedonic hotspot for opioidergic enhancement in the nucleus accumbens (Berridge 2009; Berridge and Kringelbach 2008; Robinson and Berridge 1993). This hypothesis is supported by the observation that hedonic effects of acute alcohol intake can be blocked by naltrexone (a mu-opiate-receptor antagonist), reflecting directly rewarding effects of alcohol due to endorphin release and endophinergic stimulation of mu-opiate receptors in the ventral striatum

(Volpicelli et al. 1995). A functional MRI study by Myrick et al. (2008) confirmed that functional activation in the ventral striatum elicited by alcohol-associated vs. neutral cues can be blocked by naltrexone alone or in combination with ondansetron, an antagonist of the serotonergic 5-HT3-receptors. Naltrexone is known to reduce relapse rates in alcohol-dependent patients by blocking the effects of alcohol and of alcohol-associated stimuli on endorphin release within the ventral striatum, hypothetically mediated by endorphinergic regulation of dopamine release.

19.2.2 Imaging Compensatory Neuroadaptive Processes in Alcohol Dependence

Beyond the described acute alcohol effects, chronic alcohol abuse triggers compensatory neuroadaptations in order to preserve homeostatic regulation, which can be explored, e.g., with the help of positron-emission tomography (PET) and functional magnetic resonance tomography (fMRI). For instance, mu-opiate receptors were found to be upregulated in the ventral striatum of detoxified chronic alcoholics, which was associated with the intensity of alcohol craving (Heinz et al. 2005a). Furthermore, neuroimaging studies using PET indicated that detoxified alcohol-dependent patients display a reduced DA synthesis capacity, an impaired DA release after stimulation, and a downregulation of DA D_2 receptors in the ventral striatum (Heinz et al. 1996; Volkow et al. 1996). Interestingly, the degree of DA D_2 receptor (DRD_2) downregulation and of DA synthesis capacity in the ventral striatum was linked with an increase in craving for alcohol (Heinz et al. 1996). Compensatory downregulation of striatal DA receptors following chronic alcohol intake appears to facilitate the development of tolerance towards high amounts of consumed alcohol; however, a vicious cycle can thus be generated in which alcohol intake is increased in order to generate the wanted psychotropic effects in spite of counteradaptive neuroadaptations in neurotransmitter systems activated by ethanol intake. Such a compensatory downregulation of striatal DA receptors seems to persist during early abstinence, and the severity of neuronal adaptations predicts relapse (Heinz et al. 1996, 2005b). As a consequence of neuroadaptations in, for example, DA neurotransmission following chronic alcohol intake, the DA signaling network seems to be biased towards preferential processing of the "strong" DA activating effects of alcohol and alcohol-associated stimuli, at the expense of weaker, "natural" reinforcers such as social interactions or non-drug-related food intake. This bias of the reward system towards drug effects and cues appears to cause impairment of learning processes associated with comparably "weak," non-drug-related reinforcers in alcohol-dependent subjects (Park et al. 2010). Additionally, intact DA neurotransmission is involved in the extinction or "unlearning" of operant responses that are no longer rewarded: in a multimodal imaging study combining PET and fMRI, ventral striatal dopamine D_2 receptor downregulation was associated with increased processing of alcohol-associated cues in the anterior cingulate cortex and medial prefrontal cortex (PFC), as well as with the severity of alcohol craving in detoxified alcohol-dependent patients (Heinz et al. 2004). This indicates that in early abstinence, drug-related stimuli are still salient and bias attention allocation in alcohol-dependent subjects and this bias is directly related to the degree of neuroadaptation in ventral striatal dopaminergic neurotransmission. Further fMRI studies showed that such an increased prefrontal drug cue-induced activation is associated with a higher prospective relapse risk (Beck et al. 2012; Grüsser et al. 2004). The degree of DRD_2 downregulation even directly predicted treatment outcome: in subsequent alcohol abstainers, D_2 receptor sensitivity recovered quickly during the first week of abstinence, while in patients who would relapse during the 6-month observation period, DA receptor sensitivity displayed delayed recovery during early abstinence (Heinz et al. 1996).

Besides neuronal findings of dysfunctional brain activation, most likely due to alcohols' neurotoxic effects, possible protective brain mechanisms have been revealed in another study by

Heinz and coworkers (2007): detoxified alcohol-dependent patients displayed heightened response to affectively positive stimuli in the anterior cingulate cortex, PFC, striatum, and thalamus, and this thalamic activation elicited by nondrug reinforcers was inversely correlated with subsequent drinking days and the amount of consumed alcohol in a 3-month observation phase after detoxification. In line, Charlet et al. (2013b) found enhanced functional activation in the anterior cingulate cortex and medial frontal gyrus during an aversive face-cue-comparison task in recently detoxified alcohol-dependent patients compared to controls, which in turn was associated with less (previous) lifetime alcohol intake, longer abstinence, and less subsequent binge drinking in patients during the 6-month follow-up period. Also, further findings showed that in contrast to prospective relapsers, alcohol-dependent patients who abstained from alcohol more than 7 months after detoxification showed increased brain activation in executive behavioral control brain areas, such as rostral and ventrolateral PFC and premotor cortex, when mastering high cognitive load in the n-back working memory task (Charlet et al. 2013c).

19.2.3 Serotonergic Mediation of Alcohol's Negative Reinforcement Effects

Another monoaminergic neurotransmitter which has been widely studied in the context of alcohol dependence is serotonin (5-HT). While rewarding effects of alcohol are attributed to positive reinforcement pathways via activation of the dopaminergic and opioidergic system, negative reinforcement effects of alcohol (avoidance or release of aversive states through repeated drug intake) are supposed to be mediated with alcohol-induced increases in 5-HT neurotransmission, which appears to reduce negative mood states (Heinz et al. 2001). In accordance with this hypothesis, persistence of negative mood states after detoxification was correlated with impaired 5-HT reuptake assessed with single photon emission computed tomography (SPECT), i.e., a decrease in 5-HT transporter availability in the brainstem (raphe) area, a center of origin of serotonergic projections (Heinz et al. 1998b). Serotonin dysfunction also plays a role in the disposition towards excessive alcohol intake: in rhesus monkeys suffering from early social isolation stress, a trait-like decrease in 5-HT turnover rates (measured as the availability of the serotonin metabolite 5-hydroxyindoleacetic acid in the cerebrospinal fluid) and a relative elevation of 5-HT transporter availability were found, which was directly correlated with the amount of alcohol intake when offered in a free-choice paradigm (Heinz et al. 1998a, 2003a). Likewise, a genetically driven relative increase in 5-HT transporter availability in humans (assessed with SPECT) was correlated with decreased sensitivity towards acute aversive effects of alcohol and excessive alcohol intake in adolescents and young adults, obviously because they lack a warning sign that they had been drinking "too much" (Heinz et al. 2000; Schuckit and Smith 1996). Indeed, humans and nonhuman primates who show a low response to acute alcohol intoxication consume more alcohol when given the choice (Heinz et al. 2003a; Hinckers et al. 2006; Schuckit et al. 1999). 5-HT dysfunction was further correlated with aggressiveness (Heinz et al. 1998a) and related to withdrawal-emerged anxiety and depressive symptoms in alcohol-dependent individuals (Heinz et al. 2001, 1998b).

19.2.4 Spectroscopic Insights of Glutamatergic and GABAergic Neuroadaptations

Beyond its effects on monoaminergic systems, alcohol stimulates gamma-aminobutyric acid (GABA) neurons and inhibits glutamatergic neurotransmission, which can lead to a counteradaptive downregulation of GABA receptors and an upregulation of glutamatergic NMDA receptors during chronic alcohol intake (Koob 2008). In case of interrupted alcohol consumption, e.g., during detoxification, alcohol-associated activation of GABAergic and inhibition of glutamatergic receptors (NMDA) suddenly stop, and the result is neuronal excitation, which can be manifested as alcohol-related epileptic seizures or aversive withdrawal symptoms (Heinz et al. 2009a; Krystal et al. 2006; Tsai et al. 1995) that may directly promote relapse. One pilot study suggested that withdrawal symptoms can also appear as a conditioned reaction, predating alcohol relapse during abstinence (Heinz et al. 2003b). Studies measuring the neuronal concentrations of GABA in alcohol-dependent individuals showed that GABA concentration remains low even after detoxification, suggesting that beyond the effects of chronic alcohol intake, a trait-like alteration in GABAergic neurotransmission may characterize alcohol-dependent patients (Krystal et al. 2006). Further, glutamatergic imbalance in chronically alcohol-abusing individuals as well as in an animal model of alcohol dependence was reported in a combined 3 T (humans) and 9.4 T (animals) proton magnetic resonance spectroscopy (^{1}H-MRS) study by Hermann et al. (2012): during acute alcohol withdrawal, significantly increased glutamate levels in prefrontocortical regions were found in treatment-seeking alcohol-dependent patients as well as in rats displaying signs of alcohol dependence compared to control subjects. Another MRS investigation in young alcohol-dependent men also observed an increased glutamate to creatine ratio in the anterior cingulate cortex (Lee et al. 2007). Such potentially compensatory, counteradaptive neuroadaptations (in this case, to alcohol's impairment of glutamatergic neurotransmission) are not immediately reversed once alcohol intake is stopped, for example, during acute detoxification. Instead, different durations of convalescence were found: while downregulated $GABA_A$ receptors and reduced GABA concentrations were seen during the first weeks of alcohol detoxification and are suggested to persist long into abstinence (Abi-Dargham et al. 1998; Krystal et al. 2006), a rather fast recovery progress was observed with respect to glutamatergic alterations in metabolite concentrations within 2–3 weeks of abstinence (Hermann et al. 2012). Pharmacological treatment with medications such as acamprosate can modulate glutamatergic transmission, and these neurobiological effects may be particularly strong in subjects suffering strongly from conditioned withdrawal symptoms due to an imbalance between excitatory and inhibitory neurotransmitter systems. Although it is presently unknown how acamprosate exactly works, a recent study using ^{1}H-MRS demonstrated a significant suppression of central glutamate during 4 weeks of acamprosate medication in alcohol-dependent patients following detoxification (Umhau et al. 2010).

19.2.5 Structural Magnetic Resonance Imaging of Alcohol's Neurotoxic Effects

Another well-investigated aspect of alcohol dependence is structural brain alteration caused by the neurotoxic effects of chronic alcohol intake. Here, widespread brain atrophy is a well-known phenomenon with considerable individual variation (Oscar-Berman and Marinkovic 2007). About 50–70 % of all alcohol-dependent patients are reported to suffer from some degree of ventricular enlargement and brain atrophy, which (partially) recovers when alcohol consumption is terminated (Carlen et al. 1978; Schroth et al. 1988). As Kril et al. (1997) reported, white matter volume correlated negatively with daily alcohol intake. In their diffusion tensor imaging (DTI) study, Harris and collaborates (2008) found microstructure deficits in white matter tracts connecting prefrontal and limbic regions in abstinent alcohol-dependent men. Another DTI investigation showed that age-related alterations of white matter integrity in terms of fiber tracks in the corpus callosum are increased in alcohol-dependent patients compared to age-matched controls (Pfefferbaum et al. 2006).

Studies using multimodal imaging showed that volume reductions in the frontal lobes are prominent in alcohol-dependent patients (Moselhy et al. 2001). Further morphological studies reported cerebellar atrophy and a general widening of the cerebral sulci, which exceeds comparable effects of age (Pfefferbaum et al. 1996; Sullivan 2000). Neuropathological and neuroimaging studies support the hypothesized model of "premature aging" due to neurotoxic effects of alcohol (Oscar-Berman and Marinkovic 2003). Indeed, older compared to younger alcohol-dependent patients display greater than expected reductions in size or blood flow in the cerebral cortex (Di Sclafani et al. 1995; Harris et al. 1999; Pfefferbaum et al. 1997), in the hippocampus (Laakso et al. 2000; Sullivan et al. 1995), in the cerebellum (Harris et al. 1999; Sullivan 2000), and in the corpus callosum (Pfefferbaum et al. 1996, 2006; Schulte et al. 2005).

It has long been discussed whether alcohol-related atrophy and recovery during abstinence are associated with fluctuations of water content or with some degree of neuronal regeneration, e.g., dendritic arborization. ^{1}H MRS is a useful tool to investigate the neuronal basis of atrophy and its recovery, because it can directly measure substances such as N-acetyl-aspartate (NAA), an indicator of neuronal integrity, and choline (Cho), which is found in neuronal membranes (Ross and Michaelis 1994; Vion-Dury et al. 1994). In this context, MRS studies observed reduced NAA concentrations in the frontal cortex (Vion-Dury et al. 1994) and cerebellum (Jagannathan et al. 1996; Seitz et al. 1999) in alcohol-dependent subjects in comparison with age-matched controls. Similarly, choline concentrations were decreased, referring to alterations or dysfunctions of cell membranes (Ende et al. 2005; Seitz et al. 1999). However, these first studies are bases on ratios of NAA and Cho compared with creatine (Cr) instead on absolute metabolite concentrations, assuming that creatine remains stable and can be used for normalizing ^{1}H MRS data, which has recently been disproved (Licata and Renshaw 2010). Nevertheless, measurements of (absolute) concentrations of the abovementioned metabolites as well as water content do not suggest that brain atrophy and its regeneration can be explained by dehydrating states during chronic alcohol intake (Agartz et al. 1991; Besson et al. 1989; Mann et al. 1993; Smith et al. 1988; Schroth et al. 1988). Instead, neuronal regeneration in prolonged abstinence seems to be mediated by a significant increase in Cho, potentially reflecting increased dendritic arborization and other forms of neuronal membrane regeneration (Bartsch et al. 2007; Bendszus et al. 2001; Durazzo et al. 2006; Ende et al. 2005; Martin et al. 1995).

Neurotoxic effects of chronic alcohol intake are particularly pronounced in women and older patients (Mann et al. 1992; Pfefferbaum et al. 1997). Compared with age-matched alcohol-dependent men, alcohol-dependent women display comparable brain atrophy, although the overall amount of lifetime alcohol intake is lower in these women and they had been suffering from alcohol dependence for a shorter duration of time (Mann et al. 2005; Schuckit et al. 1998). Indeed, once women and men are matched with respect to the amount of lifetime alcohol intake, women display increased brain atrophy in the corpus callosum, hippocampus, and cortical brain areas (Hommer et al. 1996, 2001; Agartz et al. 1991). Interestingly, women may show an increased ability to recover from alcohol-associated brain atrophy when remaining abstinent after detoxification (Mann et al. 2005). An overview of the reported MRI studies on alcohol dependence is given in Table 19.1.

19.3 Drug Dependence Other than Alcoholism

All psychotropic drugs of abuse are characterized by a dysfunctional responsiveness of the brain reward system and share adaptational mechanisms also found in alcohol dependence. Hence, drug and alcohol dependence is characterized by increased tolerance to excessive drug intake due to homeostatic counteradaptations such as reductions in neuroreceptors stimulated by drugs of abuse, and withdrawal symptoms appear as an expression of impaired homeostasis between excitatory and inhibitory neurotransmitter systems once drug intake is suddenly stopped during withdrawal (Koob 2003; Heinz et al. 2009a). Likewise, all drugs of abuse stimulate DA release, and this activating effect does not appear to habituate, resulting in counteradaptive downregulations of DA neurotransmission and impaired effects of non-drug rewards, thus biasing the individual towards urging for the relative strong effects of drug consumption (Heinz et al. 2009a, b).

Table 19.1 Human neuroimaging studies in alcohol dependence

MRI method	Authors/study	Major findings
fMRI	Gruesser et al. (2004)	Heightened cue-induced activation in the ACC, striatum, and mPFC predicted subsequent alcohol intake in alcoholics vs. controls
fMRI	Heinz et al. (2007)	Limbic brain activation towards affectively positive stimuli correlated inversely with prospective relapse in alcoholics vs. controls
fMRI	Wrase et al. (2007)	Reduced striatal functional activation during monetary gain anticipation in alcoholics vs. controls, which correlated negatively with craving Increased striatal activation towards alcohol-associated stimuli in alcoholics vs. controls, which correlated positively with craving severity
fMRI	Myrick et al. (2008)	Anti-craving medication naltrexone and ondansetron reduced alcohol cue-induced striatal activation compared to placebo treatment in alcoholics The combination treatment decreased craving while viewing alcohol cues in alcoholics
fMRI	Park et al. (2010)	Dysfunctional frontostriatal connectivity predicted abnormal decision-making and higher alcohol craving in male alcohol-dependent patients vs. healthy controls
fMRI + structural MRI	Beck et al. (2012)	Increased alcohol cue-induced mPFC activation in relapsers vs. abstainers mPFC gray matter loss pronounced in relapsers vs. abstainers
fMRI + structural MRI	Charlet et al. (2013b)	Increased ACC activation towards aversive facial stimuli correlated with less previous lifetime alcohol intake and better treatment outcome in alcoholics vs. healthy controls Gray matter alterations explained most of the reduced brain activation found in alcoholics
fMRI + structural MRI	Charlet et al. (2013c)	Increased PFC and premotor activations during high working memory load predict low relapse risk in alcoholics
Structural MRI	Schroth et al. (1988)	Reduction in CSF volume after 5 weeks of alcohol abstinence, but no white matter changes
Structural MRI	Di Sclafani et al. (1995)	Age-related ventricular enlargement in alcoholics vs. controls
Structural MRI	Sullivan et al. (1995)	Age-related gray matter volume loss in the anterior hippocampus and temporal cortex; white matter loss and ventricular enlargement in alcoholics vs. controls
Structural MRI	Hommer et al. (1996)	Smaller corpus callosum area among alcoholic women compared to healthy women and to alcoholic men
Structural MRI	Pfefferbaum et al. (1996)	Reduction of the callosal area in chronic alcoholics vs. controls
Structural MRI	Pfefferbaum et al. (1997)	More severe volume deficits in the cortical gray and white matter, sulcal and ventricular enlargement in older vs. younger alcoholics, e.g., in (pre)frontal regions
Structural MRI	Laakso et al. (2000)	Smaller right hippocampal volumes in late-onset type 1 alcoholics and early-onset type 2 alcoholics
Structural MRI	Hommer et al. (2001)	More pronounced reductions of gray and white matter volumes in alcoholic women vs. alcoholic men
DTI	Schulte et al. (2005)	Disruption of corpus callosum microstructure contributed to disturbance in interhemispheric processing in alcoholics
DTI	Pfefferbaum et al. (2006)	Age-related macro- and microstructural white matter disruption of the corpus callosum in alcoholics
DTI	Harris et al. (2008)	Diminished white matter density in the right frontal and cingulum in alcoholics, which was related to working memory performance
PET	Volkow et al. (1996)	Significant reductions in DA D_2 receptors (postsynaptic marker) but not in dopamine transporter availability (presynaptic marker) in alcoholics vs. nonalcoholics

(continued)

Table 19.1 (continued)

MRI method	Authors/study	Major findings
PET	Heinz et al. (2005b)	Low striatal DA synthesis capacity and low DA $D_{2/3}$ receptor availability was linked with high levels of craving in alcoholics Alcohol craving correlated positively with relapse
PET	Heinz et al. (2005a)	Elevated mu-opiate receptors availability in the ventral striatum, including the NAc in abstinent alcoholics vs. healthy controls, which was associated with higher craving Remained elevated even after 5 weeks of remeasurement
PET + fMRI	Heinz et al. (2004)	Less DA D_2 receptor availability was linked with alcohol craving and increased drug cue-induced striatal activation in detoxified alcoholics vs. controls
SPECT	Heinz et al. (1998b)	Reduced 5-HTT transporter availability in the brainstem area was linked to increased anxiety, depression, and lifetime alcohol consumption in alcoholics
SPECT	Abi-Dargham et al. (1998)	Decreased $GABA_A$ receptor density in the ACC, PFC, and cerebellum in abstinent alcoholics
SPECT	Harris et al. (1999)	Decline in cortical/cerebellar perfusion ratio with aging and age at last drink in alcoholics vs. controls
SPECT	Heinz et al. (2000)	Reduced raphe 5-HTT availability in ll-homozygous alcoholics correlated with higher alcohol consumption
^{1}H-MRS	Martin et al. (1995)	Increases of cerebellar Cho/NAA ratio during the first 3–4 weeks of abstinence; distinct reductions in the Cho/NAA ratio was associated with relapse
^{1}H-MRS	Jagannathan et al. (1996)	Reduced NAA/Cr and NAA/Cho ratios in the frontal lobe, cerebellum, and thalamus in alcoholics vs. controls
^{1}H-MRS	Ende et al. (2005)	Reduced Cho in the frontal white matter and cerebellum, which increased within 3 months of abstinence; trend towards reduced NAA in frontal white matter
^{1}H-MRS	Durazzo et al. (2006)	Partial recovery of regional NAA and Cho in frontal and parietal regions after 1 month of abstinence, which was inversely affected by chronic smoking
^{1}H-MRS	Umhau et al. (2010)	Reduction of the glutamate to Cr ratio across 4 weeks of acamprosate medication in alcoholics
^{1}H-MRS	Hermann et al. (2012)	Increased glutamate absolute concentrations during acute alcohol withdrawal in prefrontocortical regions in alcoholics vs. controls, which normalized within 2 weeks
^{1}H-MRS + structural MRI	Seitz et al. (1999)	Reductions of cerebellar tissue and cerebellar NAA/Cr and CH/Cr ratio in alcoholics vs. controls
^{1}H-MRS + structural MRI	Bendszus et al. (2001)	Normalization of acute withdrawal-related decreased NAA/Cr levels in the frontal lobe and cerebellum and decreased cerebellar CH/Cr ratio in alcoholics after 5 weeks of abstinence Reduction in the CSF volume within 5 weeks of abstinence in alcoholics
^{1}H-MRS + structural MRI	Bartsch et al. (2007)	Global volumetric brain recovery within 6–7 weeks of sobriety in alcoholics but not in controls, which was partially correlated with increasing cerebellar Cho and frontomesial NAA levels
^{1}H-MRS + structural MRI	Lee et al. (2007)	Increased glutamate to Cr ratio in the ACC was positively related to alcohol consumption during the last month in young abstinent alcoholics No group differences on gray matter volume found in alcoholics and controls

^{1}H-MRS proton magnetic resonance spectroscopy, *5-HTT* serotonin, *ACC* anterior cingulate cortex, *Cho* choline, *Cr* creatine, *CSF* cerebrospinal fluid, *DA* dopamine, *DTI* diffusion tensor imaging, *fMRI* functional magnetic resonance imaging, *GABA* gamma-aminobutyric acid, *mPFC* medial prefrontal cortex, *MRI* magnetic resonance imaging, *NAA* N-acetyl-aspartate, *Nac* nucleus accumbens, *PET* positron-emission tomography, *SPECT* single photon emission computed tomography

19.3.1 Heroin/Opiates

Opiates suppress the inhibitory effect of GABAergic neurons on dopaminergic neurons in the ventral tegmental area, resulting in an increased activity of DA neurotransmission in the ventral striatum (Kreek 2008). Opiates also strongly impact on opiate receptors and inhibit noradrenergic neurons in the locus coeruleus (Kreek 2008). The thus suppressed noradrenergic activity results in clinically manifest symptoms of opiate intoxication such as bradycardia, drowsiness, and slow respiration. This opioidergic inhibition of noradrenergic neurons in the locus coeruleus results in counteradaptive upregulation of noradrenergic neurotransmission, resulting in a new homeostatic balance between noradrenergic neuroadaptation and opioidergic inhibition as long as opiate intake is maintained on a relatively regular basis. However, when opioidergic inhibition of noradrenergic neurotransmission is suddenly interrupted during abstinence, noradrenergic hyperactivity occurs with typical physical and affective withdrawal symptoms, e.g., tachycardia, transpiration, restlessness, sleep disorders, and anxiety (Kosten and George 2002). Because of heroin's short half-life period of 6 h, dependent subjects need to consume heroin several times a day in order to avoid aversive withdrawal symptoms, thus negatively reinforcing drug intake via the avoidance of a highly unpleasant state. Interestingly, withdrawal symptoms can also occur as a consequence of context-conditioned learning: Siegel (1983) observed that counteradaptive mechanisms lead to withdrawal symptoms as soon as rats used to opiate application in a certain context (cage) are left in this environment without opiate supply. Context-dependent cues thus seem to activate, e.g., noradrenergic upregulation, resulting in hyper-excitation once opiate application does not occur as expected. If on the other hand usual opiate doses were applied in an unfamiliar context, e.g., an unknown cage, animals used to a certain opiate dose displayed symptoms of opiate intoxication and even died, indicating that in the unknown context, their organisms failed to actively prepare for renewed intoxication via context-conditioned neuroadaptational processes (Siegel 1983). In humans, comparable conditioned effects of chronic drug intake may manifest as perception biases towards drug-associated stimuli and memory including related emotions and modification of (aversive) emotional states. For instance, Lubman and colleagues observed that long-term heroin users vs. healthy controls displayed altered processing of drug-related (e.g., pictures of drug preparation and injection) and pleasant pictures (e.g., pictures of food, erotic nudes): positive pictures, contrary to drug-related pictures, elicited less physiological responses and lower self-reported arousal in drug-dependent subjects. Further, low valence ratings of pleasant pictures predicted consistently regular heroin use (at least once a week) at a median of a 6-month follow-up assessment (Lubman et al. 2009). Like in alcohol dependence, prolonged abstinence allows for neurofunctional recovery, hypothetically reflected in decreased salience of drug-associated cues, thereby, reducing the risk of relapses. Indeed, an fMRI study indicated that following long-term compared to short heroin abstinence, subjects displayed decreased neural responses to heroin-related cues in brain regions subserving visual sensory processing, attention, memory, and action planning (Lou et al. 2012).

19.3.2 Cocaine and Other Psychostimulants

The strong effects of cocaine on the DA system can be demonstrated by the fact that the consumption of this substance discharges about 400 % of the dopamine doses normally released by natural reinforcers, such as food or sexual activities (Hurd et al. 1990). These massive effects lead to a counteradaptive reduction in (striatal) DA D_2 receptor availability in cocaine-dependent subjects (Volkow et al. 1993), and this striatal downregulation predicted reduced functional responses to a non-drug-related reinforcer (monetary reward) in the thalamus, a brain region that is implicated in DA-modulated conditioned responses and reward expectation (Asensio et al. 2010). However, in this study also a low DA D_2 receptor availability in the dorsal and ventral striatum was found to be correlated with high responsiveness of the dorsal superior medial frontal gyrus (Brodmann area 6/8/32), an area that is crucially involved in behavioral monitoring. Asensio and coworkers assumed that cocaine-addicted individuals who displayed lower ventral striatal DA D_2 receptor availability are also the ones who possibly need to compensate particularly for their reduced sensitivity to monetary reinforcement (Goldstein et al. 2007) or for their increased impulsivity (Asensio et al. 2010). Other fMRI-based study revealed reduced brain activation elicited by naturally reinforcing visual stimuli, e.g., pictures with erotic content, while cocaine-associated cues which triggered craving elicited increased functional responses throughout the brain including prefrontal, limbic, and parietal regions (Garavan et al. 2000). It has been hypothesized that prefrontal brain activation elicited by drug-related stimuli impairs executive function and behavior control (Beck et al. 2012). Indeed, a prospective fMRI study by Paulus et al. (2005) observed that decreased activation of the PFC during a decision-making task was associated with the subsequent relapse risk in methamphetamine-dependent patients.

19.4 Non-substance-related Addiction

Another class of addictions are the so-called non-substance-related or behavioral addictions, e.g., pathological gambling. Referring to ICD-10, pathological gambling is subsumed in the chapter "Disorders of Adult Personality and Behavior" in the category "habit and impulse disorders" (Dilling et al. 2000), whereas in DSM-IV, pathological gambling is classified as an "impulse-control disorder" (Saß et al. 2003).

Looking at the criteria, pathological gambling and substance-related disorders share a variety of symptoms (Grüsser and Thalemann 2006; Potenza 2006):

- An appetitive urge or strong craving state prior to the engagement in the behavior
- Preoccupation with gambling (e.g., preoccupied with reliving past gambling experiences, handicapping or planning the next venture, or thinking of ways to get money with which to gamble)
- Need to gamble with increasing amounts of money in order to achieve the desired excitement
- Repeated unsuccessful efforts to control, cut back, or stop gambling
- Continued engagement in the behavior despite adverse consequences: the person has jeopardized or lost a significant relationship, job, or educational or career opportunity because of gambling

Also imaging data suggest a closer relationship between pathological gambling and substance use disorders than between pathological gambling and obsessive-compulsive disorder (Potenza 2008). For example, neurobiological studies investigated cue reactivity processes in behavioral addictions, i.e., when subjects are confronted with stimuli closely related to gambling. Crockford and associates (2005) observed that pathological gamblers exhibited significantly greater activity in the right dorsolateral prefrontal cortex (DLPFC), including the inferior and medial frontal gyri, the right parahippocampal gyrus, and the left occipital cortex, including the fusiform gyrus, during visual presentations of gambling-related video alternating with video of nature scenes. They moreover reported a significant increase in craving for gambling after the study.

Other studies suggested that there is also a reduction in the sensitivity of the reward system in behavioral addictions that resembles reward system dysfunction in substance dependence: Reuter and colleagues (2005) used fMRI to investigate pathological gamblers and healthy volunteers during a game. They observed a reduction of ventral striatal and ventromedial prefrontal activation in pathological gamblers, which was negatively correlated with gambling severity, linking hypoactivation of these areas to disease severity (Reuter et al. 2005). In another study among pathological gamblers, de Ruiter et al. (2009) showed a diminished reward and punishment sensitivity as indicated by hypoactivation of the ventrolateral PFC when money was gained and lost.

As described earlier, alcohol dependence is associated with smaller gray matter volumes in a variety of cortical and subcortical brain regions. Comparably van Holst et al. (2012) hypothesized that there might be structural abnormalities in pathological gamblers, because the same cognitive impairments have been found in patients suffering from problem gambling behavior and from substance dependence. However, in contrast to alcohol, gambling behavior does not expose the brain to toxic agents. Using voxel-based morphometry, van Holst et al. assessed gray matter volumes in treatment-seeking problem gamblers, subjects with alcohol use disorders, and healthy volunteers. They replicated previous findings of smaller gray matter volumes in subjects with alcohol use disorders, but did not observe significant morphological brain abnormalities in problem gamblers. Moreover, Joutsa et al. (2011) investigated possible changes in regional brain gray and white matter volumes as well as white matter integrity in pathological gamblers compared to healthy controls; again there was no volumetric difference in gray matter as well as in white matter between pathological gamblers and controls. However, in pathological gamblers, Joutsa and colleagues observed lower white matter integrity in multiple brain regions including the corpus callosum, the cingulum, the superior and inferior longitudinal fascicle, the anterior thalamic radiation, and the inferior fronto-occipital fascicle, which require replication in further studies (e.g., Lane et al. 2010).

19.5 Summary

Altogether, in vivo imaging using magnetic resonance techniques has enabled researchers to reveal major mechanisms involved in the development and maintenance of drug dependence and related behavioral addictions. Here, especially, the mechanism of unphysiologically high DA release by drugs of abuse motivates the individual to preferentially respond to drug-associated reinforcement at the expense of natural nondrug reinforcement, which can be observed in various fMRI studies testing brain responses towards drug-associated vs. monetary or emotional cues. Further mechanisms of maintaining the drug dependence involve neuroadaptive processes in different neurotransmitter systems. Thus, investigations using PET, SPECT, ^{1}H-MRS, and fMRI revealed (1) upregulated mu-opiate receptors; (2) reduced DA synthesis capacity, impaired DA release, and downregulated DA D2 receptors; (3) decreased 5-HT transporter availability; (4) downregulated GABA receptors; and (5) upregulated NMDA receptors in chronic alcohol-dependent patients, which were associated with craving, negative mood states, withdrawal symptoms, relapse risk, aberrant cue-induced brain activation, and impaired learning. However, despite findings of dysfunctional brain processes in addictive behavioral disorders, possible protective brain mechanisms were also observed which predicted better treatment outcome. Furthermore, findings of structural brain alterations caused by neurotoxic effects of chronic drug use indicate a premature aging phenomenon. While women and older individuals seem to be more vulnerable despite of less overall drug intake and shorter dependence duration, structural MR studies showed that particularly women may have an increased ability to recover from drug-associated brain atrophy. Thus, such MRI-based insights into the neurobiological correlates of addictive behaviors may not only help to understand these mental diseases. One day they may contribute to provide a personalized and more precise treatment of affected patients.

References

Abi-Dargham A et al (1998) Alterations of benzodiazepine receptors in type II alcoholic subjects measured with SPECT and [123I]iomazenil. Am J Psychiatry 155(11):1550–1555

Agartz I et al (1991) T1 and T2 relaxation time estimates and brain measures during withdrawal in alcoholic men. Drug Alcohol Depend 29(2):157–169

Asensio S et al (2010) Striatal dopamine D2 receptor availability predicts the thalamic and medial prefrontal responses to reward in cocaine abusers three years later. Synapse (New York, NY) 64(5):397–402

Bartsch A et al (2007) Manifestations of early brain recovery associated with abstinence from alcoholism. Brain 130(Pt 1):36–47

Beck A et al (2012) Relapse in alcohol dependence: the impact of brain structure, brain function and brain connectivity. Arch Gen Psychiatry 69:842–852

Bendszus M et al (2001) Sequential MR imaging and proton MR spectroscopy in patients who underwent recent detoxification for chronic alcoholism: correlation with clinical and neuropsychological data. AJNR Am J Neuroradiol 22(10):1926–1932

Berridge K (2009) Wanting and liking: observations from the neuroscience and psychology laboratory. Inquiry (Oslo, Norway) 52(4):378

Berridge K, Kringelbach M (2008) Affective neuroscience of pleasure: reward in humans and animals. Psychopharmacology (Berl) 199:457–480

Berridge K, Robinson T, Aldridge J (2009) Dissecting components of reward: "liking", "wanting", and learning. Curr Opin Pharmacol 9:65–73

Besson J et al (1989) Magnetic resonance imaging in Alzheimer's disease, multi-infarct dementia, alcoholic dementia and Korsakoff's psychosis. Acta Psychiatr Scand 80(5):451–458

Boileau I et al (2003) Alcohol promotes dopamine release in the human nucleus accumbens. Synapse 49:226–231

Carlen P et al (1978) Reversible cerebral atrophy in recently abstinent chronic alcoholics measured by computed tomography scans. Science (New York, NY) 200(4345):1076–1078

Charlet K, Beck A, Heinz A (2013a) The dopamine system in mediating alcohol effects in humans. Curr Top Behav Neurosci 13:461–488

Charlet K et al (2013b) Neural activation during processing of aversive faces predicts treatment outcome in alcoholism. Addict Biol. 2013 Mar 7. doi:10.1111/adb.12045.(Epub ahead of print)

Charlet K et al (2013c) Increased neural activity during high working memory load predicts low relapse risk in alcohol dependence. Addict Biol. 2013 Oct 22. doi:10.1111/adb.12103. (Epub ahead of print)

Crockford D et al (2005) Cue-induced brain activity in pathological gamblers. Biol Psychiatry 58(10): 787–795

de Ruiter M et al (2009) Response perseveration and ventral prefrontal sensitivity to reward and punishment in male problem gamblers and smokers. Neuropsychopharmacology 34(4):1027–1038

Di Chiara G (2002) Nucleus accumbens shell and core dopamine: differential role in behavior and addiction. Behav Brain Res 137:75–114

Di Chiara G, Bassareo V (2007) Reward system and addiction: what dopamine does and doesn't do. Curr Opin Pharmacol 7(1):69–76

Di Chiara G, Imperato A (1988) Drugs abused by humans preferentially increase synaptic dopamine concentrations in the mesolimbic system of freely moving rats. Proc Natl Acad Sci U S A 85(14):5274–5278

Di Sclafani V et al (1995) Brain atrophy and cognitive function in older abstinent alcoholic men. Alcohol Clin Exp Res 19(5):1121–1126

Dilling H, Mombour W, Schmidt M (2000) Internationale Klassifikation psychischer Störungen: ICD-10, Kapitel V (F) klinisch-diagnostische Leitlinien. Weltgesundheitsorganisation. Huber, Bern

Durazzo T et al (2006) Brain metabolite concentrations and neurocognition during short-term recovery from alcohol dependence: preliminary evidence of the effects of concurrent chronic cigarette smoking. Alcohol Clin Exp Res 30(3):539–551

Ende G et al (2005) Monitoring the effects of chronic alcohol consumption and abstinence on brain metabolism: a longitudinal proton magnetic resonance spectroscopy study. Biol Psychiatry 58(12):974–980

Garavan H et al (2000) Cue-induced cocaine craving: neuroanatomical specificity for drug users and drug stimuli. Am J Psychiatry 157(11):1789–1798

Gardner E (2005) Endocannabinoid signaling system and brain reward: emphasis on dopamine. Pharmacol Biochem Behav 81(2):263–284

Goldstein R et al (2007) Is decreased prefrontal cortical sensitivity to monetary reward associated with impaired motivation and self-control in cocaine addiction? Am J Psychiatry 164(1):43–51

Grüsser S, Thalemann C (2006) Verhaltenssucht – Diagnostik, Therapie, Forschung. Huber, Bern

Grüsser S et al (2004) Cue-induced activation of the striatum and medial prefrontal cortex is associated with subsequent relapse in abstinent alcoholics. Psychopharmacology (Berl) 175(3):296–302

Harris G et al (1999) Hypoperfusion of cerebellum and aging effects on cerebral cortex blood flow in abstinent alcoholics: a SPECT study. Alcohol Clin Exp Res 23(7):1219–1227

Harris G et al (2008) Frontal white matter and cingulum diffusion tensor imaging deficits in alcoholism. Alcohol Clin Exp Res 32(6):1001–1013

Heinz A et al (1996) Psychopathological and behavioral correlates of dopaminergic sensitivity in alcohol-dependent patients. Arch Gen Psychiatry 53(12):1123–1128

Heinz A, Higley J et al (1998a) In vivo association between alcohol intoxication, aggression, and serotonin transporter availability in nonhuman primates. Am J Psychiatry 155(8):1023–1028

Heinz A, Ragan P et al (1998b) Reduced serotonin transporters in alcoholism. Am J Psychiatry 155:1544–1549

Heinz A et al (2000) A relationship between serotonin transporter genotype and in vivo protein expression and alcohol neurotoxicity. Biol Psychiatry 47(7):643–649

Heinz A, Mann K, Weinberger D (2001) Serotonergic dysfunction, negative mood states, and response to alcohol. Alcohol Clin Exp Res 25:487–495

Heinz A, Jones D et al (2003a) Serotonin transporter availability correlates with alcohol intake in nonhuman primates. Mol Psychiatry 8(2):231–234

Heinz A, Löber S et al (2003b) Reward craving and withdrawal relief craving: assessment of different motivational pathways to alcohol intake. Alcohol Alcohol 38(1):35–39

Heinz A et al (2004) Correlation between dopamine D-2 receptors in the ventral striatum and central processing of alcohol cues and craving. Am J Psychiatry 161: 1783–1789

Heinz A, Reimold M et al (2005a) Correlation of stable elevations in striatal mu-opioid receptor availability in detoxified alcoholic patients with alcohol craving. Arch Gen Psychiatry 62:57–64

Heinz A, Siessmeier T et al (2005b) Correlation of alcohol craving with striatal dopamine synthesis capacity and D-2/3 receptor availability: a combined [18F] DOPA and [18F]DMFP PET study in detoxified alcoholic patients. Am J Psychiatry 162:1515–1520

Heinz A et al (2007) Brain activation elicited by affectively positive stimuli is associated with a lower risk of relapse in detoxified alcoholic subjects. Alcohol Clin Exp Res 31(7):1138–1147

Heinz A, Beck A, Grüsser S et al (2009a) Identifying the neural circuitry of alcohol craving and relapse vulnerability. Addict Biol 14(1):108–118

Heinz A, Beck A, Wrase J et al (2009b) Neurotransmitter systems in alcohol dependence. Pharmacopsychiatry 42(Suppl 1):S95–S101

Hermann D et al (2012) Translational magnetic resonance spectroscopy reveals excessive central glutamate levels during alcohol withdrawal in humans and rats. Biol Psychiatry 71:1015–1021

Hinckers A et al (2006) Low level of response to alcohol as associated with serotonin transporter genotype and high alcohol intake in adolescents. Biol Psychiatry 60(3):282–287

Hommer D et al (1996) Decreased corpus callosum size among alcoholic women. Arch Neurol 53(4): 359–363

Hommer D et al (2001) Evidence for a gender-related effect of alcoholism on brain volumes. Am J Psychiatry 158(2):198–204

Hurd Y et al (1990) The influence of cocaine self-administration on in vivo dopamine and acetylcholine neurotransmission in rat caudate-putamen. Neurosci Lett 109(1–2):227–233

Ikemoto S (2007) Dopamine reward circuitry: two projection systems from the ventral midbrain to the nucleus accumbens-olfactory tubercle complex. Brain Res Rev 56(1):27–78

Jagannathan N, Desai N, Raghunathan P (1996) Brain metabolite changes in alcoholism: an in vivo proton magnetic resonance spectroscopy (MRS) study. Magn Reson Imaging 14(5):553–557

Joutsa J et al (2011) Extensive abnormality of brain white matter integrity in pathological gambling. Psychiatry Res 194(3):340–346

Koob GF (2003) Alcoholism: allostasis and beyond. Alcohol Clin Exp Res 27(2):232–243

Koob GF (2008) Hedonic homeostatic dysregulation as a driver of drug-seeking behavior. Drug Discov Today Dis Model 5(4):207–215

Kosten T, George T (2002) The neurobiology of opioid dependence: implications for treatment. Sci Pract Perspect 1(1):13–20

Kreek M (2008) Neurobiology of opiates and opioids. In: Galanter M, Kleber HD (eds) Textbook of substance abuse treatment, 4th edn. American Psychiatric Publishing, Washington, pp 247–264

Kril J et al (1997) The cerebral cortex is damaged in chronic alcoholics. Neuroscience 79(4):983–998

Krystal J et al (2006) Gamma-aminobutyric acid type A receptors and alcoholism: intoxication, dependence, vulnerability, and treatment. Arch Gen Psychiatry 63(9):957–968

Laakso M et al (2000) A volumetric MRI study of the hippocampus in type 1 and 2 alcoholism. Behav Brain Res 109(2):177–186

Lane S et al (2010) Diffusion tensor imaging and decision making in cocaine dependence. PLoS ONE 5(7): e11591

Lee E et al (2007) Alteration of brain metabolites in young alcoholics without structural changes. Neuroreport 18(14):1511–1514

Licata S, Renshaw P (2010) Neurochemistry of drug action: insights from proton magnetic resonance spectroscopic imaging and their relevance to addiction. Ann N Y Acad Sci 1187:148–171

Lou M et al (2012) Cue-elicited craving in heroin addicts at different abstinent time: an fMRI pilot study. Subst Use Misuse 47:631–639

Lubman D et al (2009) Responsiveness to drug cues and natural rewards in opiate addiction: associations with later heroin use. Arch Gen Psychiatry 66(2):205–212

Mann K et al (1992) Do women develop alcoholic brain damage more readily than men? Alcohol Clin Exp Res 16(6):1052–1056

Mann K et al (1993) The reversibility of alcoholic brain damage is not due to rehydration: a CT study. Addiction (Abingdon, England) 88(5):649–653

Mann K et al (2005) Neuroimaging of gender differences in alcohol dependence: are women more vulnerable? Alcohol Clin Exp Res 29(5):896–901

Martin P et al (1995) Brain proton magnetic resonance spectroscopy studies in recently abstinent alcoholics. Alcohol Clin Exp Res 19(4):1078–1082

Moselhy H, Georgiou G, Kahn A (2001) Frontal lobe changes in alcoholism: a review of the literature. Alcohol Alcohol 36(5):357–368

Myrick H et al (2008) Effect of naltrexone and ondansetron on alcohol cue-induced activation of the ventral striatum in alcohol-dependent people. Arch Gen Psychiatry 65(4):466–475

Oscar-Berman M, Marinkovic K (2003) Alcoholism and the brain: an overview. Alcohol Res Health 27(2): 125–133

Oscar-Berman M, Marinkovic K (2007) Alcohol: effects on neurobehavioral functions and the brain. Neuropsychol Rev 17:239–257

Park SQ et al (2010) Prefrontal cortex fails to learn from reward prediction errors in alcohol dependence. J Neurosci 30(22):7749–7753

Paulus M, Tapert S, Schuckit M (2005) Neural activation patterns of methamphetamine-dependent subjects during decision making predict relapse. Arch Gen Psychiatry 62(7):761–768

Pfefferbaum A et al (1996) Thinning of the corpus callosum in older alcoholic men: a magnetic resonance imaging study. Alcohol Clin Exp Res 20(4):752–757

Pfefferbaum A et al (1997) Frontal lobe volume loss observed with magnetic resonance imaging in older chronic alcoholics. Alcohol Clin Exp Res 21(3): 521–529

Pfefferbaum A, Adalsteinsson E, Sullivan E (2006) Dysmorphology and microstructural degradation of the corpus callosum: interaction of age and alcoholism. Neurobiol Aging 27(7):994–1009

Potenza M (2006) Should addictive disorders include non-substance-related conditions? Addiction (Abingdon, England) 101(Suppl):142–151

Potenza M (2008) Review. The neurobiology of pathological gambling and drug addiction: an overview and new findings. Philos Trans R Soc B-Biol Sci 363(1507): 3181–3189

Reuter J et al (2005) Pathological gambling is linked to reduced activation of the mesolimbic reward system. Nat Neurosci 8(2):147–148

Robinson T, Berridge K (1993) The neural basis of drug craving - an incentive-sensitization theory of addiction. Brain Res Brain Res Rev 18(3):247–291

Ross B, Michaelis T (1994) Clinical applications of magnetic resonance spectroscopy. Magn Reson Q 10(4):191–247

Saß H et al (2003) Diagnostisches und Statistisches Manual Psychischer Störungen DSM-IV-TR. Hogrefe, Göttingen

Schroth G et al (1988) Reversible brain shrinkage in abstinent alcoholics, measured by MRI. Neuroradiology 30(5):385–389

Schuckit M, Smith T (1996) An 8-year follow-up of 450 sons of alcoholic and control subjects. Arch Gen Psychiatry 53(3):202–210

Schuckit M et al (1998) The clinical course of alcohol-related problems in alcohol dependent and nonalcohol dependent drinking women and men. J Stud Alcohol 59(5):581–590

Schuckit M et al (1999) Selective genotyping for the role of 5-HT2A, 5-HT2C, and GABA alpha 6 receptors and the serotonin transporter in the level of response to alcohol: a pilot study. Biol Psychiatry 45(5):647–651

Schulte T et al (2005) Corpus callosal microstructural integrity influences interhemispheric processing: a diffusion tensor imaging study. Cereb Cortex 15(9):1384–1392

Schultz W, Dayan P, Montague P (1997) A neural substrate of prediction and reward. Science 275:1593–1599

Seitz D et al (1999) Localized proton magnetic resonance spectroscopy of the cerebellum in detoxifying alcoholics. Alcohol Clin Exp Res 23(1):158–163

Siegel S (1983) No title. In classical conditioning, drug tolerance and drug dependence. In: Israel I, Glaser FB et al (eds) Research advances in alcohol and drug problems. Plenum Press, New York, pp 207–246

Smith M et al (1988) Brain hydration during alcohol withdrawal in alcoholics measured by magnetic resonance imaging. Drug Alcohol Depend 21(1):25–28

Sullivan E (2000) Neuropsychological vulnerability to alcoholism: evidence from neuroimaging studies. In: Noronha A, Eckardt M, Warren K (eds) Review of NIAAA's neuroscience and behavioral research. NIAAA, Bethesda, pp 473–508, Monograph No. 34

Sullivan E et al (1995) Anterior hippocampal volume deficits in nonamnesic, aging chronic alcoholics. Alcohol Clin Exp Res 19(1):110–122

Tsai G, Gastfriend D, Coyle J (1995) The glutamatergic basis of human alcoholism. Am J Psychiatry 152:332–340

Umhau J et al (2010) Effect of acamprosate on magnetic resonance spectroscopy measures of central glutamate in detoxified alcohol-dependent individuals: a randomized controlled experimental medicine study. Arch Gen Psychiatry 67(10):1069–1077

van Holst R et al (2012) A voxel-based morphometry study comparing problem gamblers, alcohol abusers, and healthy controls. Drug Alcohol Depend 124:142–148

Vion-Dury J et al (1994) What might be the impact on neurology of the analysis of brain metabolism by in vivo magnetic resonance spectroscopy? J Neurol 241(6):354–371

Volkow N et al (1993) Decreased dopamine D2 receptor availability is associated with reduced frontal metabolism in cocaine abusers. Synapse 14(2):169–177

Volkow N et al (1996) Decreases in dopamine receptors but not in dopamine transporters in alcoholics. Alcohol Clin Exp Res 20:1594–1598

Volpicelli J et al (1995) Effect of naltrexone on alcohol "high" in alcoholics. Am J Psychiatry 152(4): 613–615

Wrase J et al (2007) Dysfunction of reward processing correlates with alcohol craving in detoxified alcoholics. Neuroimage 35(2):787–794

Functional and Structural MRI in Alzheimer's Disease: A Multimodal Approach

20

Michel J. Grothe, Arun L.W. Bokde, and Stefan J. Teipel

Abbreviations

AD	Alzheimer's Disease
ADNI	Alzheimer's Disease Neuroimaging Initiative
APOE4	ε4 Allele of the APOE Gene (Encoding the Apolipoprotein E)
APP	Amyloid Precursor Protein
AxD	Axial Diffusivity
CC	Corpus Callosum
CSF	Cerebrospinal Fluid
DMN	Default Mode Network
DTI	Diffusion Tensor Imaging
EDSD	European DTI Study in Dementia
FA	Fractional Anisotropy
ICA	Independent Component Analysis
ICN	Intrinsic Connectivity Network
MCI	Mild Cognitive Impairment
MD	Mean Diffusivity
MRI	Magnetic Resonance Imaging
MTL	Medial Temporal Lobe
PET	Positron Emission Tomography
RaD	Radial Diffusivity
ROI	Region of Interest
TRD	Task-Related Deactivation
VBM	Voxel-Based Morphometry

M.J. Grothe (✉)
German Center for Neurodegenerative Diseases (DZNE), Rostock/Greifswald, Gehlsheimer Str. 20, Rostock 18147, Germany
e-mail: michel.grothe@dzne.de

A.L.W. Bokde, PhD
Discipline of Psychiatry, School of Medicine and Trinity College Institute of Neuroscience, Trinity College Dublin, Adelaide and Meath Hospital incorporating National Children's Hospital, Dublin, Ireland

S.J. Teipel
German Center for Neurodegenerative Diseases (DZNE), Rostock/Greifswald, Gehlsheimer Str. 20, Rostock 18147, Germany

Department of Psychosomatic Medicine, University Rostock, Rostock, Germany

20.1 Introduction

Alzheimer's disease (AD) is a progressive neurodegenerative brain disorder and the most common cause of dementia in the elderly. Histopathologic hallmarks of AD are extracellular deposits of aggregated amyloid-β protein, called senile plaques, and intracellular fibrils of hyperphosphorylated tau protein, called neurofibrillary tangles. These pathologic events are accompanied by altered synapse function, decreased cellular metabolism, and neurodegeneration. Based on the seminal autopsy studies by Braak and Braak in the early 1990s, AD is regarded as a system-specific brain disease that affects discrete neuronal systems in a fairly consistent temporospatial pattern while other brain regions are widely spared or only affected in the ultimate stages of the disease (Braak and Braak 1995).

Over the last years, several neuroimaging tools have become available that make it possible to study AD-related structural and functional brain

C. Mulert, M.E. Shenton (eds.), *MRI in Psychiatry*,
DOI 10.1007/978-3-642-54542-9_20, © Springer Berlin Heidelberg 2014

changes in the living patient. Compared to postmortem autopsy studies, which are mostly limited to advanced, and thus fatal, stages of AD, in vivo neuroimaging studies allow for bigger sample sizes and significantly facilitate the study of early pathological changes of the disease. Neuroimaging studies further permit longitudinal study designs, enabling the assessment of progressive AD pathology in individual patients followed over time. Hence, the role of neuroimaging in the investigation of dementia has gained increasing importance for research into disease mechanisms and in the differential neurobiological effects of genetic, cardiovascular, and lifestyle factors known to influence the development and clinical progression of AD. Neuroimaging is also being explored as a tool for early and differential diagnosis of AD and for the assessment of neurobiological effects of medical treatments in clinical trials. Several recent in vivo neuroimaging studies have confirmed and extended the initial findings by Braak and Braak on the temporospatial patterns of system-specific brain changes associated with AD. Regionally specific structural and functional deteriorations were found to be related to impairments in specific cognitive domains that characterize different disease stages of progressing AD pathology. Furthermore, longitudinal studies of initially healthy elderly subjects as well as studies of neuropsychologically, genetically, or molecularly defined risk models for AD have demonstrated that abnormal structural and functional brain changes precede the onset of dementia by several years.

This increased knowledge about presymptomatic and disease-specific brain changes has motivated the ongoing development of imaging-derived biomarkers, which best represent neurobiological changes associated with different stages of disease pathogenesis, for potential use as diagnostic markers or as secondary outcomes in clinical trials (Teipel et al. 2013). To date, diagnosis of AD is based on clinical history, psychiatric and neurological examination, neuropsychological testing, and supportive measures including structural imaging and blood tests to exclude other causes of dementia (McKhann et al. 1984). However, diagnostic criteria for AD are currently being revised to take into account the wealth of scientific knowledge on disease mechanisms and disease progression that has been acquired over the last years since diagnostic criteria were first published in 1984 (Jack et al. 2011). Main differences to the original criteria include the incorporation of positive biomarkers as supportive features for diagnosis as well as the consideration of different stages of disease, from asymptomatic preclinical stages over a symptomatic predementia phase to clinical AD dementia. The inclusion of biomarkers from structural, functional, and molecular imaging aims both to improve specificity for the diagnosis of AD dementia (McKhann et al. 2011) and to reach a diagnosis of AD before the onset of manifest dementia. These criteria are primarily intended for use in research settings but may also impact the clinical diagnosis of early stages of AD in the future. An early diagnosis of AD will become fundamental at the time disease-modifying and preventive treatments become available. However, the current development of such treatment strategies also depends on the availability of accurate biomarkers for the selection of at-risk samples for clinical trials and to serve as surrogate markers of treatment effects that are sensitive for the underlying pathology.

Development and improvement of neuroimaging techniques is a growing field of research. There are now several methods for in vivo assessment of brain gray matter atrophy (structural MRI), integrity of white matter tracts (diffusion-weighted MRI), and altered brain activation and functional connectivity (functional MRI). In the following sections, we will provide an overview of the role of these principal MRI-based imaging modalities for current AD research. We will describe how these MRI modalities have furthered our understanding of AD pathogenesis, each through the assessment of different structural or functional correlates of the underlying pathology, and how these findings can be translated into potential applications in clinical research and clinical practice by the development of stage-specific disease biomarkers.

20.2 Structural MRI

20.2.1 The Temporospatial Course of Progressing Neurodegeneration in AD

The neurodegeneration in AD leads to a marked reduction of brain tissue in patients with AD dementia which is clearly visible on MRI scans and primarily presents as signal loss in the medial temporal lobe (MTL) and widening of cerebrospinal fluid (CSF) spaces. Already in the 1990s, increasing evidence from postmortem autopsy studies pointed to a temporospatial pattern of progressive AD pathology that manifests at first in the peri- and entorhinal cortices of the MTL and extends sequentially to include other allocortical regions followed by neocortical association areas of the temporal, parietal, and frontal lobes. Primary sensory-motor areas are relatively spared or only affected in highly advanced stages of the disease (Braak and Braak 1995). Based on the consistency of these observations, a neuropathological staging scheme of AD severity into six stages of cortical involvement was proposed. Importantly, this staging scheme predicted that the first stages of AD pathology, the "transentorhinal" stages I and II, correspond to clinically silent cases, while the "limbic" (III and IV) and "neocortical" (V and VI) stages represent incipient and fully developed AD dementia, respectively (Fig. 20.1).

20.2.1.1 Manual Volumetry

Given the prominence of MTL atrophy on MRI scans, which corresponds to postmortem evidence of earliest and most severe neurofibrillary degeneration in this region, first attempts to quantify AD-related atrophy on in vivo MRI scans focused on volumetric measurements of MTL structures, most notably the hippocampus (Seab et al. 1988; Killiany et al. 1993). Manually measured hippocampus volumes in patients with AD dementia showed a consistent reduction of up to 40 % compared to healthy controls of the same age, allowing for highly accurate group separations on the basis of the MRI scan only. Hippocampal volume as measured by MRI scans post-mortem or in end stages of disease was further shown to account for a high proportion of 50–80 % of variance in measures of neuronal loss and neurofibrillary pathology at autopsy (Bobinski et al. 2000; Csernansky et al. 2004). While subregions of the MTL showed comparably high degrees of atrophy even in mild stages of AD (Lehericy et al. 1994; Teipel et al. 2006), neocortical involvement was more consistently detected in advanced stages of AD dementia and showed a pattern of declining atrophy severity from temporal over parietal to frontal and occipital lobes (Rusinek et al. 1991; Teipel et al. 2003a).

The Braak staging scheme of progressive neurofibrillary degeneration predicts earliest structural changes to precede the onset of dementia by several years, and thus, AD-typical atrophy patterns on MRI should also be detectable in asymptomatic or very mildly impaired subjects at high risk of developing AD. One of those risk groups is given by subjects with a diagnosis of amnestic mild cognitive impairment (MCI), which is defined as a relatively isolated memory impairment that is greater as what is expected by age alone, but is not severe enough to interfere with activities of daily living and does not suffice for a diagnosis of dementia (Petersen et al. 1999). Thus, patients diagnosed with MCI exhibit an increased risk for developing AD with incidence rates of 10–15 % per year, compared to 1–2 % per year for healthy unimpaired subjects of the same age (Gauthier et al. 2006). The syndrome is seen as a boundary or transitional stage between normal aging and dementia and often reflects prodromal AD (Guillozet et al. 2003). When comparing temporal lobe volumes between healthy elderly subjects, individuals with a diagnosis of MCI, and AD patients, a significant reduction of hippocampus volume was detectable in the MCI (−14 %) and the AD group (−22 %), but atrophy of the temporal neocortex was only found in the AD group (Convit et al. 1997). Interestingly, in a clinical follow-up study on subjects with MCI, neocortical temporal lobe volume was already significantly reduced in those subjects that later declined to AD, which significantly distinguished them from the MCI subjects that remained stable (Convit et al. 2000).

Especially informative for the study of the earliest structural changes in the course of AD is MRI data acquired in elderly individuals who were completely asymptomatic at the time of the MRI scan and later developed cognitive impairments leading to diagnoses of MCI or AD. Using such data, a few studies showed that volumetric reduction of the anterior MTL, including the hippocampus, the amygdala, and the entorhinal cortex, precedes the onset of cognitive deterioration by several years (Kaye et al. 1997; den Heijer et al. 2006; Martin et al. 2010). Thus, volumes of the amygdala and the hippocampus were found to be on average 5 % smaller in clinically declining subjects compared to stable controls already 6 years before the diagnosis of dementia was made.

20.2.1.2 Automated Data-Driven Methods

Manual volumetric methods are associated with several drawbacks that limit their utility for the assessment of brain atrophy in neurodegenerative diseases. Most notably, manual delineation of regions of interest (ROIs) is rater dependent and relatively labor intensive, and it is limited to brain structures that provide clearly delineable borders defined by anatomical landmarks. Since the beginning of this century, automated methods are being developed to determine regionally specific changes in brain structure using hypothesis- and rater-independent approaches (Ashburner and Friston 2000; Fischl et al. 2004). The most commonly used automated morphometry approach is voxel-based morphometry (VBM). In its original implementation, the method is based on a low-dimensional spatial transformation of brain scans into a common reference space to control for global differences in brain size and shape. After segmentation of the scans into brain tissue and CSF spaces, remaining local differences in gray matter volumes, which are not modeled by the global spatial normalization, are the parameters of interest that drive a voxel-based univariate statistic. This approach has been criticized because the concept of global vs. regional effects cannot be operationalized and modeled effects will depend on the characteristics of the normalization algorithm. Furthermore, the accuracy of normalization may vary with neurobiological differences and thus effects in VBM may be driven by group differences in normalization accuracy rather than the neurobiological differences themselves (Bookstein 2001). These arguments are a matter of debate (Ashburner and Friston 2001) and are still not finally resolved. From a pragmatic perspective, VBM has widely been used and results from these analyses have proven to be reproducible across different scanners and processing approaches (Ewers et al. 2006) and to spatially agree with effects from other imaging techniques (Hutton et al. 2009) and autopsy studies (Whitwell et al. 2008).

When applied to demented AD patients, VBM analyses demonstrate the expected pattern of global cortical atrophy that is most pronounced in medial and lateral temporoparietal cortices with a relative sparing of the sensorimotor cortex, the occipital poles, and the cerebellum (Baron et al. 2001; Karas et al. 2003). The spatial pattern of progressive neuropathology in AD as suggested by the postmortem autopsy studies was impressively demonstrated using VBM and serial MRI of MCI subjects that progressed to AD (Whitwell et al. 2007). Hence, early atrophic changes in subjects with MCI, approximately 3 years prior to the clinical diagnosis of AD, were found to be confined to the MTL. At a later stage, but still with the diagnosis of MCI, atrophy had spread to include lateral temporal and temporoparietal association areas of the neocortex. At the time the diagnosis of AD was made, atrophy had become more severe and widespread and then also involved areas of the prefrontal cortex, probably reflecting neurofibrillary tangle pathology of the neocortical Braak stages V and VI (Whitwell et al. 2008). Accordingly, cross-sectional VBM studies on MCI consistently demonstrate atrophy of the MTL, whereas the degree of involvement of neocortical association areas varies across studies, probably due to the heterogeneity of MCI as a transitional state between normal aging and AD (Ries et al. 2008) (Fig. 20.1). Thus, MCI subjects with an imminent conversion to AD dementia show greater reductions in MTL volume but especially greater atrophy of the temporoparietal

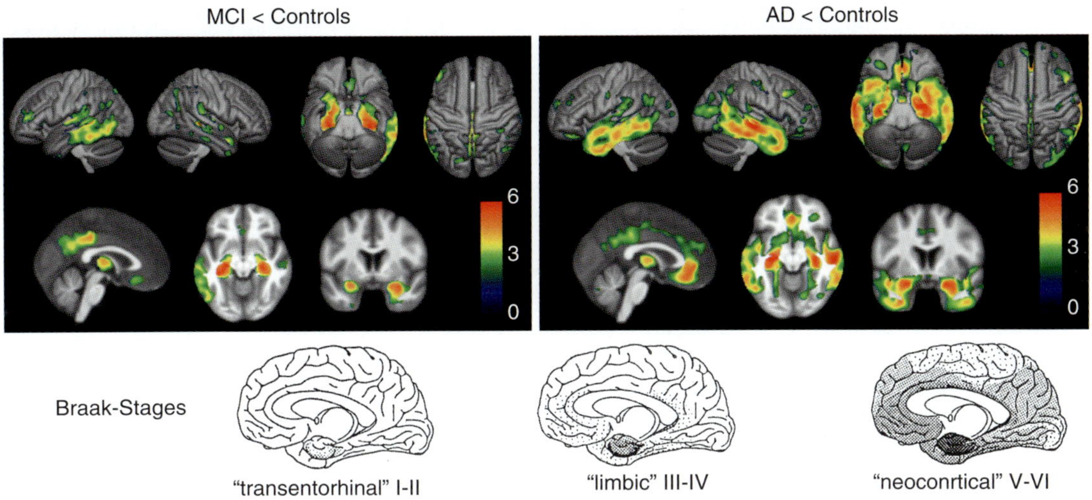

Fig. 20.1 *Top panel*: voxel-wise maps of gray matter volume reductions in a group of MCI subjects ($N=69$) and a group of AD patients ($N=28$), respectively, compared to the same group of healthy age-matched controls ($N=95$); statistical threshold set at $p < 0.01$, false discovery rate corrected (Based on data published in Grothe et al. (2012); MRI images retrieved from publicly available database: www.oasis-brains.org). *Bottom panel*: drawings of the Braak staging scheme of neurofibrillary pathology in AD (Adapted from Braak et al. (1993) with permission to reprint granted by S. Karger AG, Basel). Note that voxel-wise pattern of gray matter atrophy in MCI and AD resemble the "limbic" and "neocortical" stages of neurofibrillary degeneration, respectively, as outlined by Braak et al.

neocortex and the posterior cingulate/precuneus when compared to MCI subjects that remain stable during clinical follow-up (Chetelat et al. 2005; Risacher et al. 2009). Using serial MRI, it could be further shown that most of these regions showed not only lower volumes at baseline but also higher longitudinal rates of volume decline as MCI subjects progress to a diagnosis of AD (Chetelat et al. 2005; Risacher et al. 2010; Grothe et al. 2013).

Unbiased VBM studies could also complement manual volumetry findings of earliest structural changes in the (anterior) MTL in cognitively normal subjects that are destined to develop AD, by providing evidence for concomitant atrophy outside the MTL, including the basal forebrain (Hall et al. 2008; Grothe et al. 2013) and the posterior cingulate/precuneus (Tondelli et al. 2012). Analysis of serial MRI data over a follow-up of up to 10 consecutive years further revealed that age-related volume loss is apparent and widespread even in cognitively stable elderly, but the regional brain changes associated with a conversion to MCI differ significantly in location and magnitude from the pattern associated with normal brain aging (Driscoll et al. 2009).

20.2.1.3 Advanced Image Analysis

Based on improving hard- and software resources, increasingly sophisticated computational methods for image processing and analysis are being developed that allow the automated segmentation and detailed subfield analysis of the hippocampus and other ROIs in large imaging datasets (Van Leemput et al. 2009; Wang et al. 2009b; Leung et al. 2010; Heckemann et al. 2011). Modern computational methods demonstrate high anatomic accuracy of the segmented volumes when compared to manual volumetry techniques (Klein et al. 2009; Leung et al. 2010) and even allow for the detailed structural analysis of brain regions that are less amenable to manual delineation. The overarching principle behind the majority of these approaches is the use of high-dimensional image deformation algorithms. Different from the classic VBM analysis, these deformation-based approaches employ hundreds of thousands of nonlinear parameters to warp the individual brain images into the reference space, yielding a highly exact match to the template image and thereby eliminating spatial differences among the individual scans. The information on interindividual spatial variability then resides entirely in the deformation functions themselves, rendering the differentiation in global and regional effects obsolete. This opens the way for the high-resolution study of anatomical details by direct extraction of volumetric information from the individual deformation fields or by automated segmentation of the individual brains into structural or functional ROIs based on inverse warping of detailed atlases in the template space.

Subregional analysis of hippocampal shape revealed a selective atrophy of the CA1 and subicular subfields at preclinical and predementia stages of AD, while atrophy was found to aggravate and to spread to CA2-3 subfields as subjects progressed to clinically manifest AD (Csernansky et al. 2005; Apostolova et al. 2010; Hanseeuw et al. 2011; Achterberg et al. 2013). This selective involvement of hippocampal subregions is in good accordance with patterns of neuronal cell loss described by detailed histopathology and a similarly coherent pattern of selective atrophy could also be demonstrated for the subnuclei of the amygdala (Cavedo et al. 2011).

While MRI research on structural brain atrophy in AD generally focuses on the MTL and neocortical regions, the involvement of subcortical nuclei across disease stages is less thoroughly explored. This is surprising given the prevalent neuropathological evidence of selective involvement of specific subcortical areas, most notably the monoaminergic brainstem nuclei (Lyness et al. 2003; Grinberg et al. 2011; Braak and Del Tredici 2012) and the cholinergic neurons of the basal forebrain (Bartus et al. 1982; Mesulam et al. 2004), and may be partly explained by technical issues associated with subcortical in vivo volumetry. The complex anatomy of the basal forebrain and the indistinguishability of cholinergic cells on normal MRI scans place important constraints on the in vivo structural analysis of cholinergic degeneration in AD (Huckman 1995; Hanyu et al. 2002). However, the recent development of cytoarchitectonic maps of the cholinergic nuclei in MRI standard space, based on combined histology and postmortem MRI (Teipel et al. 2005; Zaborszky et al. 2008; Kilimann et al. 2014), now renders the cholinergic basal forebrain amenable to in vivo volumetry by the use of advanced computational methods for automated regional volume extraction. Using such an approach the selective involvement of different cholinergic nuclei across disease stages could be studied in vivo. A progressive and differential decline in volume of the cholinergic subnuclei was found across the spectrum from normal aging over MCI to AD, resulting in severe atrophy of the nucleus basalis of Meynert in patients with mild to moderate AD which was comparable in magnitude to atrophy of the hippocampus (Teipel et al. 2011a; Grothe et al. 2012, 2013) (Fig. 20.2). Interestingly, this MRI-based in vivo marker of cholinergic basal forebrain atrophy was also found to reproduce some of the major pathologic features of cholinergic degeneration in AD as reported from neuropathological autopsy studies, including correlations with specific cognitive deficits as well as distinct associations with amyloid deposition and neurodegeneration in the cortical projection sites of the cholinergic nuclei (Perry et al. 1978; Cullen et al. 1997; Grothe et al. 2010, 2014; Kilimann et al. 2014). Recently, novel MRI-based imaging techniques sensitive for subtle tissue abnormalities within the brainstem have been described (Shibata et al. 2008; Lambert et al. 2013), but these methods still await application in the study of AD-related brainstem pathology.

20.2.1.4 The Cognitive Correlates of Regional Brain Atrophy

Given the progressive and regionally heterogeneous brain atrophy in AD, question arises regarding the functional significance of regional volume loss for the expression of cognitive impairments. The focal reduction of MTL volumes in predementia stages of AD with relatively isolated memory impairment suggests an important role of MTL atrophy for the emergence of memory deficits. This notion has been substantiated in early MRI studies by the demonstration of associations between manually measured MTL volumes and scores on tests of declarative memory in AD patients (Deweer et al. 1995), subjects with MCI (Killiany et al. 2002), and even within the normal variance among healthy elderly individuals (Hackert et al. 2002). Subsequent studies employed specific neuropsychological tests for

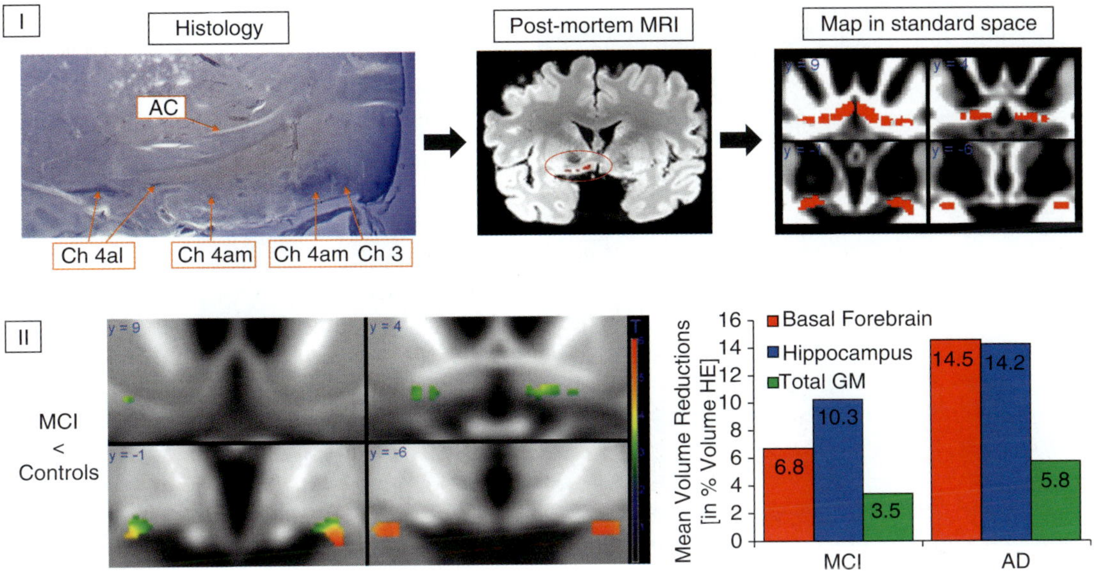

Fig. 20.2 (*I*) Generation of a cytoarchitectonic map of basal forebrain cholinergic nuclei in MRI standard space. Cholinergic cells are stained in histologic slices, and cholinergic nuclei are manually transferred to corresponding sections of a postmortem MRI, which is then automatically normalized into MRI standard space as defined by the Montreal Neurological Institute (MNI). *AC* anterior commissure, *CH3* cholinergic cells corresponding to the vertical limb of the diagonal band of Broca, *CH4 am*, *al* cholinergic cells corresponding to the anterior medial and lateral parts of the nucleus basalis of Meynert (Based on data published in Teipel et al. (2005)). (*II*) *Left*: voxel-wise reductions in gray matter volume within the basal forebrain region of interest in MCI subjects compared to healthy controls. *Right*: group means of extracted basal forebrain, hippocampus, and total gray matter (*GM*) volumes in groups of MCI subjects and AD patients plotted as percent volume reduction compared to a healthy elderly (*HE*) reference group (Based on data published in Grothe et al. (2012))

different aspects of memory function in combination with fine-grained structural analysis on high-resolution MRI scans, to provide increasingly detailed evidence for functional specialization and hemispheric differentiation among MTL structures (de Toledo-Morrell et al. 2000; Rosen et al. 2003; Chen et al. 2010; Mueller et al. 2011) (Fig. 20.3).

In addition to the detailed examination of MTL atrophy in memory dysfunction, unbiased automated analyses have been used to search the whole brain for structural correlates of neuropsychological deficits in AD, including deficits in different aspects of memory, executive function, language function, and visuospatial ability (Chang et al. 2010; Grothe et al. 2010; Cardenas et al. 2011; Wolk and Dickerson 2011; Gross et al. 2012; McDonald et al. 2012). While the central role of MTL atrophy for declarative memory deficits in AD could be confirmed by these studies, the data also indicated that additional neocortical and subcortical changes were related to further neuropsychological deficits that are commonly expressed by AD patients. For example, executive function deficits were found to be more strongly associated with anterior frontal cortex atrophy, whereas deficits in naming and verbal fluency were found to be more strongly associated with atrophy of the left anterior temporal and frontotemporal lobes, respectively. The combined investigation of volumetric and neuropsychological data also showed that cognitive deficits as assessed by standard neuropsychological tests do not usually map to discrete cerebral sites but rather reflect atrophic processes in widespread and partially overlapping networks. For example, memory deficits were also found to be associated with atrophy of brain regions outside the MTL, including the lateral temporal lobe, the thalamus, the medial prefrontal cortex, and the posterior cingulate cortex (Fig. 20.3). Hence, these

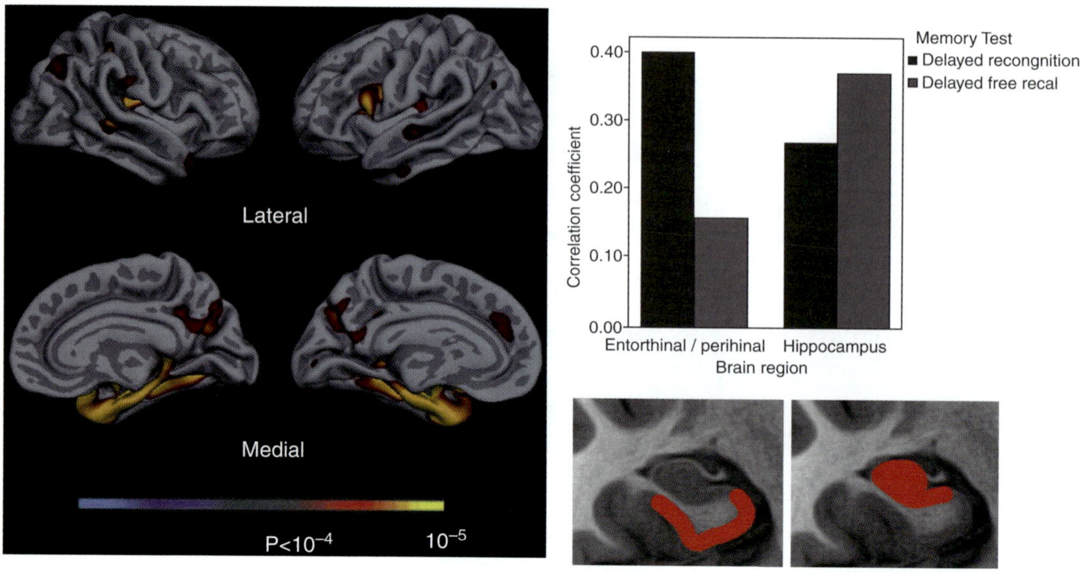

Fig. 20.3 *Left*: exploratory analysis of reductions in cortical thickness associated with impaired delayed recognition memory in AD patients. *Right*: Pearson correlations for hippocampus and perirhinal/entorhinal cortex (rostral MTL) volumes with delayed recall and recognition discrimination performance, controlled for age, education, and gender. Hippocampal atrophy is more strongly related to impaired delayed recall while perirhinal/entorhinal cortex atrophy is more strongly associated with impaired recognition memory performance. The *bottom section* of the figure represents the "idealized" anatomic localization of the two regions being compared (Adapted from Wolk and Dickerson (2011) with permission to reprint granted by Elsevier)

studies lend further experimental support to the notion that cognition depends on integrated networks of cortical and subcortical brain structures that differentially contribute to different aspects of the cognitive function, which they bring forth in concert. Associations between brain volumes and cognitive test performances are also found in other neurodegenerative diseases (Grossman et al. 2004; Luks et al. 2010) and even in the process of normal brain aging (Fjell and Walhovd 2010). Hence, age- and disease-related neurodegeneration is increasingly employed as a natural lesion model for the study of structure-cognition relationships in the human brain, thereby complementing findings from functional imaging studies of altered neuronal activity in response to cognitive tasks (see Sect. 20.4).

20.2.2 Risk and Resilience

Over the last years, large epidemiological studies have identified several factors that increase an individual's risk for late-life cognitive decline and AD dementia (McDowell 2001). Among these risk factors are disadvantageous or predisposing combinations of specific genes (St George-Hyslop 2000), molecular signs of AD pathology (Rabinovici and Jagust 2009; Blennow et al. 2010), and lifestyle factors (Daviglus et al. 2011) that are mainly related to cardiovascular risk and obesity. On the other hand, higher education and socioeconomic status, specific diets, as well as maintained physical, social, and cognitive activity through old age have been found to reduce the risk of cognitive decline. Investigation of the effects that these genetic, molecular, and lifestyle factors exert on brain structure and function will further our understanding of the pathogenesis of AD.

20.2.2.1 Genetic Risk and Predisposition

Genetic mutations that affect the cellular processing of the amyloid precursor protein (APP) lead to increased production of neurotoxic amyloid-β proteins and cause the relatively rare autosomal dominantly inherited form of AD (familial AD). Mutation carriers are essentially certain to develop the disease and thus provide an opportunity to study brain changes that predate symptoms or clinical diagnosis. Consistently, mutation carriers were found to show accelerated atrophy of the hippocampus as well as of the posterior cingulate cortex and the precuneus already several years prior to diagnosis (Ridha et al. 2006; Knight et al. 2011). Similar to the genetic mutations in familial AD, the triplication of the APP gene on chromosome 21 in Down's syndrome (trisomy 21) also leads inevitably to an increased amyloid-β deposition as well as AD-like neurodegenerative changes and cognitive impairments as subjects age (Mann 1988). Thus, MRI-derived volumes of the MTL were found to be significantly reduced in demented subjects with Down's syndrome, but increased age-related reductions in AD-typical allo- and neocortical areas could also be demonstrated in nondemented individuals, providing another interesting model for presymptomatic brain changes in AD (Teipel et al. 2003b).

Apart from genetic mutations that inevitably lead to familial AD, a familial history of sporadic AD significantly increases the risk of AD in the offspring, pointing to disadvantageous and predisposing genotypes. The best investigated genetic risk factor for sporadic AD is the ε4 allele of the apolipoprotein E (APOE4) (Corder et al. 1993). Apolipoprotein E is believed to be an important promoter of axonal growth and synaptogenesis during neuronal growth, repair, and regeneration (Mahley and Rall 2000). It was also found to promote the proteolytic degradation of amyloid deposits (Jiang et al. 2008; Castellano et al. 2011) which may be dysfunctional in the APOE4 isoform, possibly explaining the increased amyloid deposition with age in APOE4 carriers (Reiman et al. 2009; Morris et al. 2010). In accordance with the increased risk for developing AD, healthy elderly individuals that posses the ε4 allele have higher rates of cognitive decline than noncarriers (Caselli et al. 2007) which is accompanied by increased rates of temporal lobe atrophy, mapping mainly to the MTL and to the entorhinal cortex and subiculum in particular (Honea et al. 2009; Donix et al. 2010) (Fig. 20.4). Structural MRI studies further showed that the dose-dependent and region-specific effect of the ε4 allele on brain structure leads to distinguishable AD phenotypes (Wolk and Dickerson 2010; Morgen et al. 2013). Thus, among AD patients, ε4 carriers exhibit greater atrophy of the temporal lobes, whereas noncarriers have greater frontoparietal atrophy. In line with the differential patterns of atrophy, carriers were found to be more impaired on episodic memory tests, whereas noncarriers performed worse on tests of working memory and executive function.

In contrast to the ε4 allele, the APOE ε2 allele is associated with a reduced risk of AD and lower amyloid deposition compared to the most common ε3 allele (Corder et al. 1994; Morris et al. 2010). Consistent with the supposed protective effect against AD pathology, healthy elderly carriers of the ε2 allele were found to exhibit increased thickness of the temporal cortex (Fan et al. 2010) as well as bigger hippocampus volumes (Hostage et al. 2013) and slower rates of hippocampal atrophy compared to ε3 carriers (Chiang et al. 2010).

20.2.2.2 Increased Biomarker Levels of Molecular AD Pathology

Postmortem autopsy studies indicate that a considerable proportion of cognitively normal subjects of advanced age harbor AD pathology in the form of amyloid plaques and neurofibrillary tangles (Knopman et al. 2003). These findings could be reproduced using in vivo CSF-based and novel positron emission tomography (PET)-based biomarkers of molecular AD pathology. According to these studies, approximately 20–30 % of the normal elderly population aged 70 years and above harbor clinically silent amyloid pathology, which is associated with an increased risk of future cognitive decline and conversion to AD dementia (Rabinovici and Jagust 2009; Blennow et al. 2010).

In vivo MRI studies are now beginning to map the effects of abnormal levels of molecular AD pathology to structural brain changes. Thus, cross-sectional MRI studies found that an increased cerebral amyloid load in cognitively

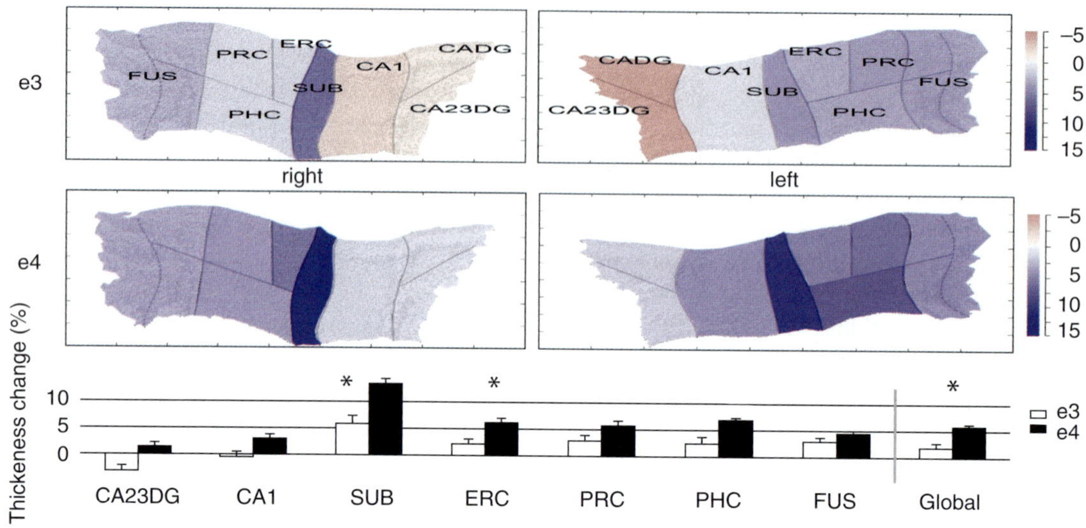

Fig. 20.4 Increased atrophy rates of hippocampal subfields in cognitively normal APOE4 carriers compared to noncarriers revealed by a cortical unfolding technique. The *color-coded* flat maps visualize cortical thickness change over a 2-year follow-up period in APOE4 carriers and noncarriers for both hemispheres separately. The graph below provides average (across both hemispheres) cortical thickness change for all subregions over time (mean ± SE). Significant between-group differences are indicated (*, $p<0.05$, two tailed). APOE4 carriers compared to noncarriers showed significantly greater cortical thinning in SUB, ERC, and Global. *CADG* anterior CA fields and dentate gyrus, *CA23DG* CA fields 2, 3, and dentate gyrus, *CA1* CA field 1, *SUB* subiculum, *ERC* entorhinal cortex, *PRC* perirhinal cortex, *PHC* parahippocampal cortex, *FUS* fusiform cortex, *Global* average thickness across all subregions, *e3* non-APOE 4 carriers, *e4* APOE 4 subjects (From Donix et al. (2010) with permission to reprint granted by Elsevier)

normal elderly, as assessed by CSF biomarkers (Tosun et al. 2010; Sabuncu et al. 2011) or amyloid imaging (Dickerson et al. 2009; Storandt et al. 2009; Becker et al. 2011), is associated with lower regional brain volumes in sites of early AD-related neurodegeneration, particularly the hippocampus, the temporoparietal neocortex, and the posterior cingulate/precuneus (Fig. 20.5). These findings could be further substantiated using serial MRI data, where those individuals with abnormally elevated levels of amyloid deposition were found to exhibit an accelerated decline of MTL and posteromedial parietal volumes compared to amyloid-negative controls (Scheinin et al. 2009; Chetelat et al. 2012; Ewers et al. 2012; Knopman et al. 2013). Moreover, subtle cognitive deficits in nondemented subjects with elevated amyloid burden were shown to be mediated by these structural brain changes (Ewers et al. 2012).

However, the relationship between amyloid pathology and changes in brain volume appears to be complex and disease stage specific. Hence, other studies did not find accelerated brain atrophy in cognitively normal amyloid-positive individuals during a 10-year interval preceding the amyloid scan (Driscoll et al. 2011), or did only find associations between amyloid load and reduced brain volume when limiting the sample to individuals with subjective cognitive impairment or to those with a high amyloid burden (Chetelat et al. 2010; Fjell et al. 2010; Grothe et al. 2014). Furthermore, in patients with clinically manifest AD, global amyloid burden is not or only weakly related to regional brain volume (Chetelat et al. 2010) or rates of longitudinal volume decline (Scheinin et al. 2009), possibly reflecting saturated cerebral amyloid deposition in the face of ongoing neuronal degeneration at the clinical stage of AD (Jack et al. 2009; Buchhave et al. 2012). On the other hand, elevated CSF markers of tau protein, indicative of neurofibrillary pathology, were found to be more robustly associated with in vivo measures of

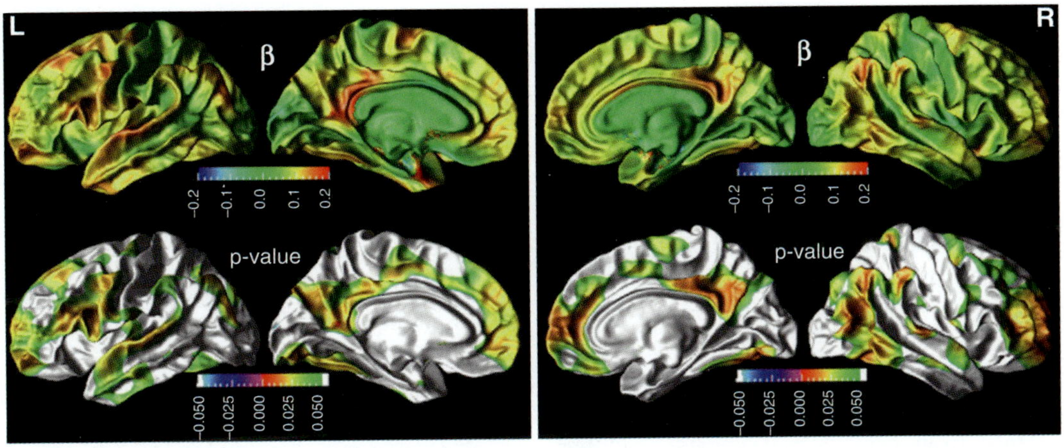

Fig. 20.5 Association between cerebrospinal fluid (CSF) beta amyloid (Aβ)$_{1-42}$ concentrations and cortical thickness in a group of cognitively normal elderly subjects. *Top row*: cortical maps of regression coefficients for Aβ$_{1-42}$. *Bottom row*: false discovery rate corrected *p*-value maps (From Tosun et al. (2010) with permission to reprint granted by Elsevier)

brain atrophy across MCI and AD groups, especially regarding MTL atrophy (Tosun et al. 2010; Desikan et al. 2011; de Souza et al. 2012). Together, these data agree with a proposed sequence of pathological events in AD that begins with cerebral amyloid deposition several years to decades before the first symptoms appear. Over the years, the primary amyloid-related molecular pathology would lead to structural brain changes, probably mediated by amyloid-induced neurofibrillary pathology, which finally trigger and parallel cognitive decline. In this respect, the combination of molecular and structural imaging can be used to test hypotheses on the pathogenetic sequence of AD development in vivo (see Box 20.1: The Amyloid-Cascade Model of AD).

20.2.2.3 Cardiovascular Risk and Lifestyle

Several cardiovascular risk factors have been associated with an increased risk for cognitive decline and AD (Daviglus et al. 2011). This increased risk is suggested to be mediated by the negative effects that these factors exert on brain structure and function, which is now being increasingly investigated by in vivo imaging studies. Thus, several MRI studies could confirm that risk conditions such as smoking (Almeida et al. 2011), insulin resistance (Rasgon et al. 2011), diabetes mellitus (Chen et al. 2012), obesity (Jagust et al. 2005), as well as abnormal plasma levels of cholesterol (Ward et al. 2010) and homocysteine (den Heijer et al. 2003) are associated with atrophic brain changes in otherwise healthy and cognitively normal individuals. However, conditions appear to be differentially related to regional brain changes and do not necessarily map to regions that are most vulnerable in AD. For example, although smoking was found to be associated with lower brain volume in the posteromedial parietal cortex, significant effects were also found in regions less affected in AD, including the cerebellum and the frontal lobes, and no negative effects on hippocampal volume were found. In contrast, specific effects on hippocampal volume could be demonstrated for obesity as well as high levels of plasma homocysteine in nondemented older adults. Importantly, much of these risk conditions can be targeted by a change in lifestyle, dietary supplementation, or medical intervention, providing the possibility to actively reduce an individual's risk for brain atrophy and cognitive decline. Thus, in vivo imaging studies have provided first evidence that successful smoking cessation, physical exercise training, as well as homocysteine lowering by B-vitamin intake may reduce or delay brain atrophy and cognitive decline in the elderly

Box 20.1: The Amyloid-Cascade Hypothesis

The amyloid-cascade hypothesis of AD states that cerebral amyloid deposition, as caused by altered processing of the membrane-bound amyloid precursor protein (APP), is the primary etiologic event in AD pathogenesis. Over the years, clinically silent amyloid deposition would initiate downstream pathologic events, such as the formation of intracellular neurofibrillary tangles, that ultimately lead to neuronal dysfunction, cell loss, and cognitive decline (Hardy and Selkoe 2002).

This model is based on substantial evidence from genetic, pathological, and biochemical studies:

- All AD patients have amyloid plaques
- Aβ induces hyperphosphorylation of tau protein, neurodegeneration, and inflammatory processes in vitro
- All known genetic predispositions for AD involve proteins in amyloid processing
- Triplication of APP in Down's syndrome leads to life-long Aβ deposition that inevitably results in AD
- Transgenic mouse models expressing mutant human APP gene reproduces major features of AD

More recently, in vivo imaging- and CSF biomarkers could lend further support to the proposed sequence of pathologic events in the course of AD:

- Amyloid biomarkers are the first to become abnormal but are relatively decoupled from atrophy and cognitive decline at the time clinical symptoms become manifest
- Tau biomarkers are more consistently associated with imaging markers of neuronal dysfunction and atrophy
- Atrophy on MRI both precedes and parallels cognitive decline

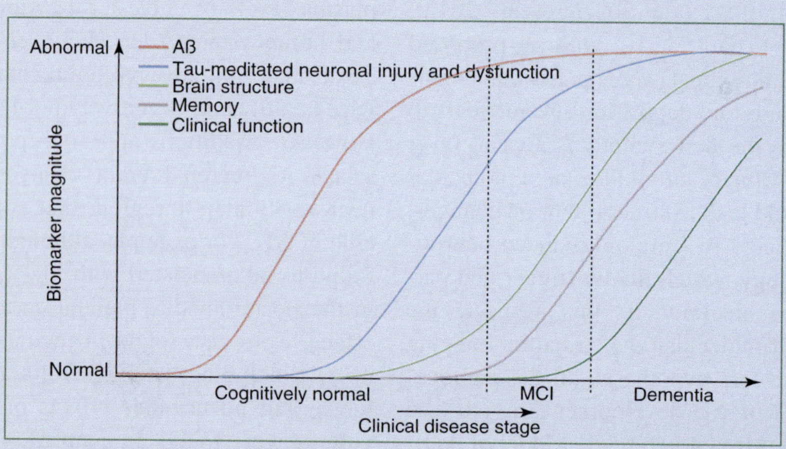

From Jack et al. (2010) with permission to reprint granted by Elsevier

Although the initial biomarker evidence in favor of the amyloid-cascade hypothesis is compelling, this generic model of AD pathogenesis will certainly have to be modified and refined in the future, to account for inter-individual variability in the dynamics of pathologic changes and their relation to the expression of clinical symptoms (Jack et al. 2013; Chetelat 2013).

(Smith et al. 2010; Almeida et al. 2011; Erickson et al. 2011).

Already in the 1990s, epidemiological studies suggested that the incidence of AD is decreased in subjects with higher education. Interestingly, higher education did not reduce the risk for neurodegenerative AD pathology, but rather appeared to mitigate the impact of pathology on the clinical expression of dementia (Brayne et al. 2010). In accordance with that, MRI studies in AD showed that at a given level of cognitive impairment, patients with a higher education had more severe brain atrophy, suggesting that these subjects exhibit some kind of cognitive reserve which allows them to compensate more severe cerebral pathology (Perneczky et al. 2009). A similar resilience against AD pathology was also observed in patients with larger head circumference, which may be related to bigger brain size and thus higher brain reserve (Perneczky et al. 2010; Guo et al. 2013). In addition, healthy elderly subjects with higher education were found to have regionally increased cortical gray matter compared to less educated healthy subjects (Liu et al. 2012; Arenaza-Urquijo et al. 2013). Together, these data suggest that the increased resilience to AD pathology, or cognitive reserve, in higher educated individuals may be partly mediated by education-related increases in brain size, that is, by a higher brain reserve.

20.2.3 Imaging-Based Diagnosis

As a direct clinical application, atrophy patterns on MRI can be employed as positive markers of AD for diagnostic purposes as well as for the tracking of disease progression. In clinical trials on potential disease-modifying or preventive drugs, imaging-derived markers of early AD pathology may serve to enrich and stratify at-risk populations and may at the same time be used as surrogate markers of disease-modifying treatment effects. The progressive character of AD pathology that exhibits a long initial phase of asymptomatic brain changes may provide a possibility to detect the disease before the onset of manifest dementia and thus at a time when potential preventive therapies will be most effective. To improve early diagnosis of AD at predementia stages, there is an ongoing search for sensitive and specific imaging markers that best discriminate between healthy elderly, MCI subjects, and AD patients and that may predict a future conversion to AD in asymptomatic elderly subjects or subjects with MCI.

Given the severe atrophy of the MTL in AD, even simple visual rating scales of MTL atrophy can distinguish AD patients from healthy controls with high diagnostic accuracies of approximately 85 % (Scheltens et al. 1992). However, the visual ratings are not sensitive enough to detect subtle changes in MTL volume over time, suggesting that this approach may not be usefully employed as a secondary endpoint in clinical trials (Ridha et al. 2007). In contrast, manually measured volumes of the MTL allow for a more sensitive detection of longitudinal atrophy, demonstrating hippocampus atrophy rates of 3–8 % in AD, with an acceleration of rates from MCI to clinically manifest AD, while healthy controls exhibit atrophy rates around 1 % in later life (Raz et al. 2004; Barnes et al. 2009). In cross-sectional settings, volumes of the hippocampus and the entorhinal cortex can separate AD patients and MCI subjects from healthy controls with accuracies ranging from 70 % for early stages of MCI to complete group separation for advanced stages of AD dementia (Seab et al. 1988; Du et al. 2001; Teipel et al. 2006). Furthermore, both hippocampus and entorhinal cortex volume predict future conversion to AD in individuals with MCI at accuracies around 80–85 %, whereas the predictive value of entorhinal cortex volume seems to be slightly superior over hippocampus volume (Jack et al. 1999; Killiany et al. 2002; deToledo-Morrell et al. 2004).

Due to the laborious nature of manual volumetry methods, most of these studies have only assessed the diagnostic accuracy of select MTL structures in relatively small samples of monocentric data, limiting the heuristic value of the findings and their generalizability to clinical multicenter settings. Novel automated volumetry techniques now enable the automated segmentation of the hippocampus and other ROIs in large datasets in a rater-independent and timely manner. Automated segmentations of the hippocampus exhibit high anatomic accuracy when compared to manual delineations (Klein et al. 2009; Leung et al. 2010) and provide comparable diagnostic power (Colliot et al. 2008; Mak et al. 2011). They further demonstrate high sensitivity to change over 1-year intervals and may even be more reliable than manual volumetry approaches for the assessment of longitudinal atrophy rates (van de Pol et al. 2007). When tested for their diagnostic potential in large multicenter data sets, such as provided by the *Alzheimer's Disease*

Neuroimaging Initiative (ADNI), automatically segmented volumes of the hippocampus achieve accuracies of around 70 and 80 % for separating MCI subjects and AD patients, respectively, from healthy controls (Calvini et al. 2009; Chupin et al. 2009).

Based on the increasing awareness of early cholinergic degeneration in the course of AD, which also appears to be related to the severity of cortical amyloid pathology (Mesulam et al. 2004; Potter et al. 2011; Grothe et al. 2014), MRI-based volumetry of the cholinergic basal forebrain has emerged as a promising diagnostic marker for AD. Thus, the diagnostic accuracy of in vivo basal forebrain volumetry for the detection of early and predementia stages of AD was found to be comparable to the accuracy of hippocampus volume (Kilimann et al. 2014), but basal forebrain atrophy was also a significant predictor of amyloid pathology among predemented subjects, whereas hippocampus atrophy was not (Teipel et al. 2014a).

Diagnostic models based on single regional markers do not account for the AD-typical pattern of progressive atrophy that spreads from the MTL/basal forebrain to the temporoparietal neocortex early in the course of the disease. Modern image-processing software enables the automated parcellation of the brain into several neuroanatomic ROIs that can then be tested separately and in combination for their diagnostic potential using logistic regression or discriminant analysis. Using such an approach, it could be shown that a combined structural marker of the hippocampus, the entorhinal cortex, and the temporoparietal junction could increase diagnostic accuracy for the separation of MCI from controls in a multicenter setting to 90 % (Desikan et al. 2009). Optimal combinations of automatically extracted regional markers for distinguishing between AD patients and controls could even yield complete group separations (Lerch et al. 2008; Desikan et al. 2009). Accordingly, composite markers that combined AD-typical regional measures, including the MTL, temporoparietal association areas, and medial parietal regions, showed promising accuracies of around 75 % for the prediction of imminent conversion to AD in multicenter data of subjects with MCI (Bakkour et al. 2009; Querbes et al. 2009). Atrophy across several a priori cortical predilection sites for AD-related neurodegeneration was even found to be a significant predictor of subsequent memory decline and the development of AD dementia in cognitively normal elderly subjects (Dickerson et al. 2011). An extension of logistic regression and discriminant analysis approaches for the construction of diagnostic models is given by multivariate pattern classification techniques, including principal component analysis and machine learning algorithms such as support vector machine (Teipel et al. 2007a; Davatzikos et al. 2008; Plant et al. 2010). These techniques make optimal use of the data by extracting and recognizing diagnosis-specific regional patterns of atrophy from preprocessed imaging data in an unsupervised manner and can distinguish between AD patients, MCI subjects, and controls with high diagnostic accuracy. When trained to recognize spatial patterns of brain atrophy which best distinguish between AD patients and controls, pattern classifiers can predict an imminent conversion to AD dementia in MCI subjects with accuracies around 80 % (Misra et al. 2009) and may even be used to assess an individual's risk for future cognitive decline in asymptomatic elderly subjects (Davatzikos et al. 2009). Machine learning algorithms are already now being developed for the software of scanner consoles as radiological expert systems based on anatomical MRI data. In the future, these algorithms may be expanded to integrate diverse information from a multitude of image and biomarker modalities to improve the automated detection of prodromal AD stages based on high-dimensional pattern recognition (Davatzikos et al. 2011; Liu et al. 2013).

20.3 Diffusion Tensor Imaging

20.3.1 White Matter Pathology in AD

Although AD is primarily considered a brain gray matter disease, histopathological studies have also detected pathologic alterations of the subcortical white matter. The origin of these white matter lesions is a matter of ongoing debate. While secondary (Wallerian) degeneration due to cortical atrophy is the most likely cause, other origins such as direct amyloid-myelin interactions, inflammatory processes, and hypoperfusional/hypoxic events are also discussed (Englund et al. 1988; Xu et al. 2001; Roher et al. 2002). In the "retrogenesis" model of white matter degeneration, myelin breakdown is hypothesized to specifically affect the late-myelinating neurons of higher-order association cortices due to their smaller axonal diameter and lower oligodendrocyte-to-axon ratio, rendering them more vulnerable to age- and disease-related insults. Thus, the theory posits a progressive demyelination pattern that recapitulates the developmental process of myelination in reverse (Bartzokis 2004). Viewed from the point of axonal degeneration, the functional deficits in AD may also be described in terms of a cortical disconnection syndrome that is characterized by impaired neuronal communication between widely distributed neural networks that subserve higher cognitive function (Delbeuck et al. 2003). Diffusion tensor imaging (DTI) now allows studying the individual anatomy and microstructural integrity of white matter fiber tracts in vivo and provides the opportunity to investigate the role of white matter pathology and cortical disconnection in the pathogenesis of AD in unprecedented detail.

20.3.1.1 Regional Pattern of White Matter Changes in the Course of Progressing AD Pathology

By examining diffusion indices of microstructural integrity within ROIs placed in normal appearing lobar white matter, it could be shown that AD patients exhibit increased mean diffusivity (MD) and decreased fractional anisotropy (FA) in the corpus callosum (CC) as well as in the frontal, temporal, and parietal lobes when compared to controls. In contrast, microstructural integrity of the occipital lobe and the internal capsule was found to be widely preserved (Bozzali et al. 2002). In the prodromal MCI stage, similar alterations of white matter integrity were found in proximity to cortical regions that are typically affected early in the course of AD, particularly the temporal and parietal white matter, but not in frontal regions or the CC (Fellgiebel et al. 2004; Huang and Auchus 2007). However, the increased contrast for white matter compartments in diffusion-weighted data also allows the structural identification and quantification of specific fiber tracts of interest as opposed to relatively unspecific lobar white matter ROIs. Accordingly, reduced fiber integrity in AD was found for widespread intracortical projecting fibers, including the splenium of the CC, the superior longitudinal fasciculus, and the cingulum bundle, but not for the pyramidal tract (Rose et al. 2000). In MCI subjects tract-specific analysis revealed disruptions of white matter integrity mainly in (limbic) fiber tracts with direct connections to MTL structures, including the posterior and parahippocampal cingulum, the perforant path, the fornix, and the uncinate fasciculus (Fellgiebel et al. 2005; Kalus et al. 2006; Zhang et al. 2007; Sexton et al. 2010). Especially the disruption of the cingulum bundle due to MTL atrophy is believed to contribute to early functional deficits in the course of AD pathogenesis by disconnecting the posterior cingulate cortex from the MTL (Villain et al. 2010; Bozzali et al. 2012).

Similar to the study of gray matter changes, automated voxel-based as well as atlas-based parcellation methods for the analysis of DTI data have been developed that greatly facilitate the comprehensive assessment of microstructural white matter changes throughout the whole brain in AD (Medina et al. 2006; Teipel et al. 2007b; Zhuang et al. 2010; Huang et al. 2012). Despite considerable methodical heterogeneity, leading to partly discrepant findings, results from automated and manual studies converge on a pattern of microstructural white matter changes in AD that begin and are most severe in limbic tracts, including the fornix and posterior and parahippocampal fibers of the cingulum, and sequentially extend to include more lateral temporoparietal association fibers, commissural fibers of the splenium, and finally long-ranging association tracts involving the frontal white matter (Fig. 20.6). Importantly, first microstructural alterations of the limbic tracts were already detectable in presymptomatic subjects approximately 2 years

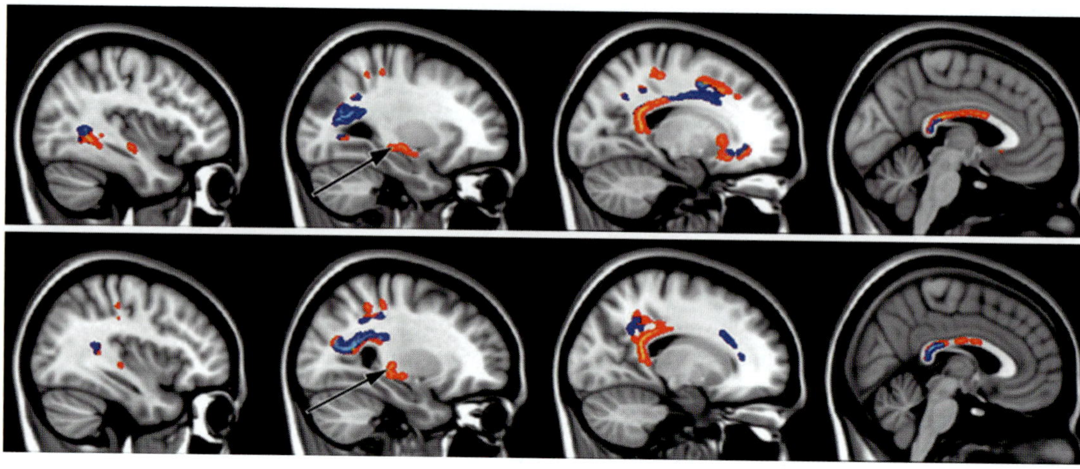

Fig. 20.6 Sagittal sections illustrating significant fractional anisotropy (FA) reduction in MCI subjects compared with controls as revealed by tract-based spatial statistics (TBSS) analysis ($p<0.05$, corrected for multiple comparisons). *Light blue* represents white matter regions with significant FA reduction overlapping with white matter (WM) lesions. Areas of significant FA decrease without the involvement of WM lesions are marked in *red-yellow*. *Black arrows* show the location of the crus of fornix (From Zhuang et al. (2010) with permission to reprint granted by Elsevier)

before they developed cognitive deficits and received a diagnosis of MCI (Zhuang et al. 2012). On the other hand, diffusivity increases in frontal and parietal white matter, but not in temporal white matter, distinguish between MCI subjects that imminently convert to AD dementia and those that remain stable (Scola et al. 2010). Extracortical projecting fiber tracts, in contrast, are relatively preserved until advanced stages of the disease (Teipel et al. 2014c). The functional consequences of these microstructural white matter changes appear to be widely consistent with a priori models of the representation of cognitive function across neuronal networks in the human brain. Thus, fiber tract disruptions of the fornix and the parahippocampal white matter, including the cingulum and the perforant path, have been shown to be associated with impaired episodic memory function in AD patients as well as in individuals with MCI (Huang and Auchus 2007; Fellgiebel et al. 2008; Sexton et al. 2010). Executive function deficits, on the other hand, were found to map to white matter disruptions in frontal and anterior cingulate regions (Huang and Auchus 2007; Chen et al. 2009; Grambaite et al. 2011b). Although disruptions of most fiber tracts are highly correlated to gray matter atrophy in the respective cortical projection sites (Sydykova et al. 2007; Avants et al. 2010) (Fig. 20.7), microstructural fiber alterations have been found to contribute independently to cognitive impairments, highlighting the functional significance of cortical disconnection in addition to local cortical lesions (Delbeuck et al. 2003; Grambaite et al. 2011a; Bozzali et al. 2012).

Although the current model of progressive white matter changes and associated cognitive impairments in the course of AD is mainly derived from cross-sectional studies on at-risk populations, first results from longitudinal follow-up studies largely confirm the extrapolated pattern of white matter degeneration spreading from limbic tracts over temporoparietal association fibers to the frontal white matter (Scola et al. 2010; Teipel et al. 2010b; Mielke et al. 2012; Zhuang et al. 2012; Douaud et al. 2013).

20.3.1.2 Comprehensive Analysis of the Diffusion Tensor

While the majority of studies so far concentrated on FA or MD as markers of microstructural tissue alteration, more recent studies suggest that simultaneous assessment of the full range of

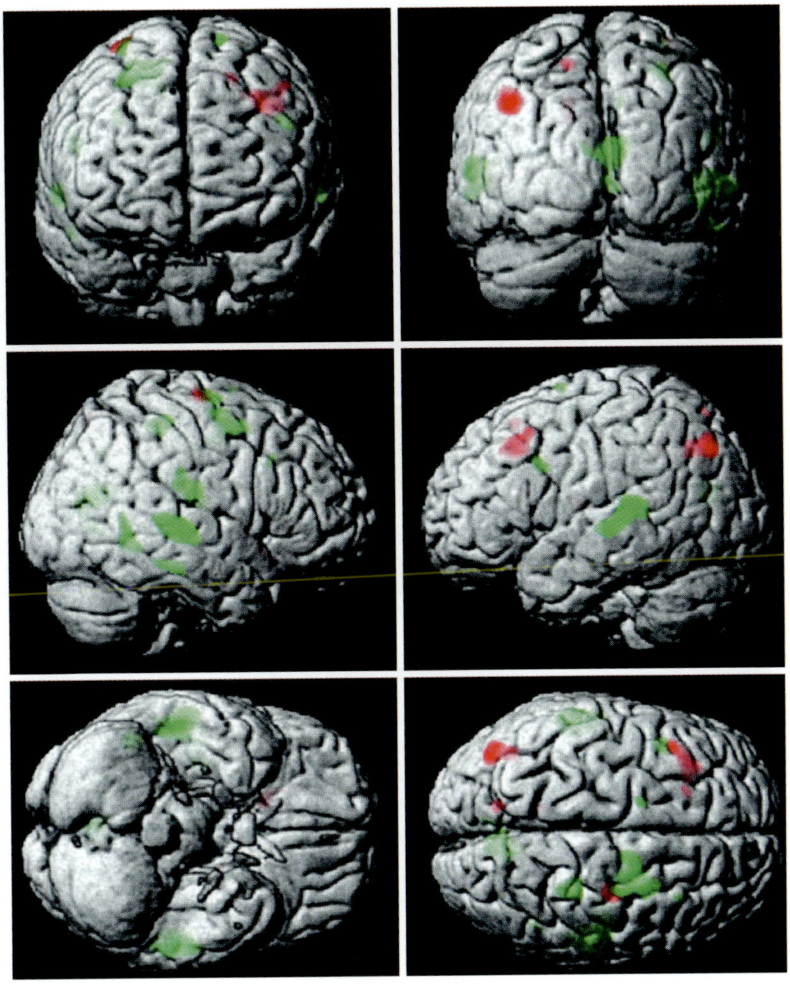

Fig. 20.7 Voxel-wise correlation between cortical gray matter volume and fractional anisotropy (*FA*) value in the anterior (*red*) and posterior (*green*) corpus callosum in patients with AD. Cluster extension set at ≥50 contiguous voxels passing the significance threshold of $p<0.001$ (From Sydykova et al. (2007) with permission to reprint granted by Oxford University Press)

tensor-derived diffusion indices, including FA, MD, as well as axial (AxD) and radial diffusivity (RaD), may provide more detailed information about the specifics of white matter damage in AD. Thus, increases in RaD have been specifically associated with myelin degeneration, whereas changes in AxD are suspected to reflect direct axonal damage (Song et al. 2003). In general, absolute diffusivities (i.e., MD, AxD, and RaD) have been found to be more sensitive markers of AD-related white matter pathology than FA, particularly in early and prodromal stages of the disease (Acosta-Cabronero et al. 2010; Bosch et al. 2012) (Fig. 20.8). More specifically, AD-related microstructural changes in limbic tracts were found to be characterized by increased RaD but relatively preserved AxD, whereas an opposite pattern was found for projection tracts. Degeneration of commissural and association tracts, on the other hand, was characterized by increases in RaD as well as in AxD, thus indicating differing processes of tissue disruption among fiber populations in AD (Huang et al. 2012). A differential contribution of axial and radial diffusivities was also found for AD-related tissue disruption in CC subregions (Di Paola et al. 2010). An increase in AxD in combination with unchanged FA values within the splenium was interpreted by the authors as indicative of Wallerian degeneration secondary to temporoparietal cortex atrophy, whereas a specific increase in RaD in combination with preserved AxD was suggested to reflect demyelination of frontal fibers in the course of retrogenesis.

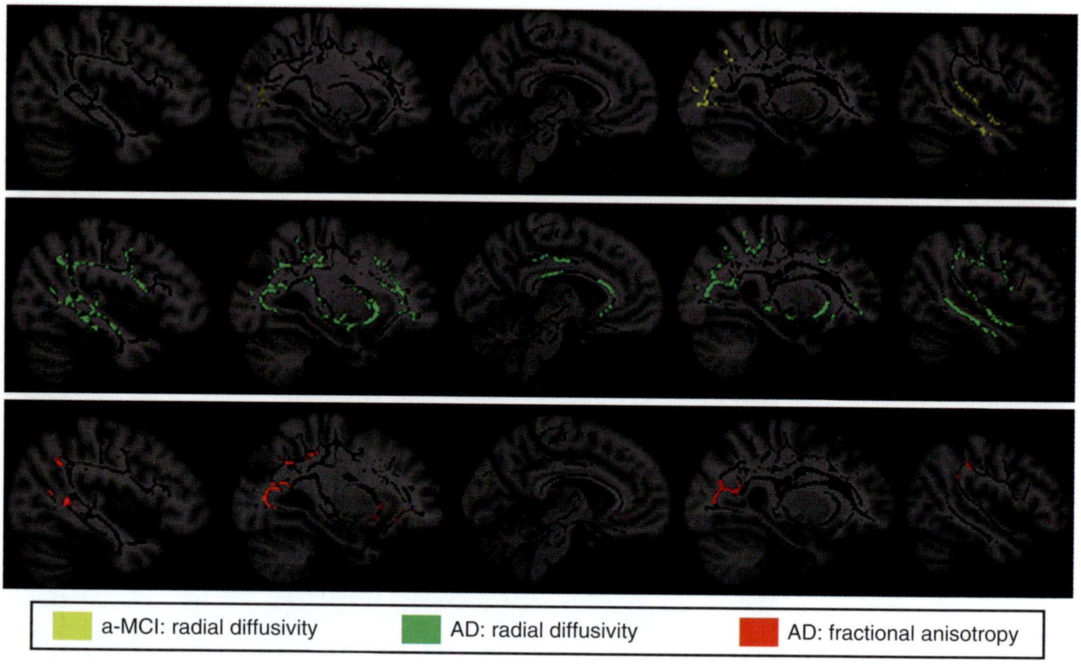

Fig. 20.8 White matter areas showing significant radial diffusivity increases in amnestic MCI (a-MCI) and AD patients compared to healthy elderly controls as assessed by tract-based spatial statistics (TBSS) analysis ($p < 0.05$, corrected for multiple comparisons). Note that regions displaying significant radial diffusivity increases among a-MCI show anatomical overlap with those of reduced FA among demented patients (From Bosch et al. (2012) with permission to reprint granted by Elsevier)

When compared with young adults, both healthy elderly subjects and AD patients demonstrated widespread FA reductions and increases in RaD that were characterized by a gradient of decreasing lesion severity from late-myelinating frontal to early-myelinating posterior fibers (Head et al. 2004; Damoiseaux et al. 2009; Gao et al. 2011). Together these data suggest that retrogenesis may account to a large extent for age-related white matter disruptions, most prominently affecting frontal regions, whereas AD pathology may accelerate this process slightly and additionally affects posterior fibers through Wallerian degeneration secondary to severe temporoparietal atrophy.

20.3.1.3 Fiber Tractography

Apart from deriving scalar indices of fiber integrity from the diffusion tensor, diffusion-weighted data can also be used to automatically reconstruct individual fiber tracts from selected seed points based on the inter-voxel continuation of the principal diffusion direction. This tractography approach can be used to derive indices of structural connectivity between cortical regions, such as the number of streamlines that reached a cortical target region from a distant seed region or the average microstructural integrity of the reconstructed fiber connection. Thus, by sending fiber tracking streamlines from hippocampus and posterior cingulate seeds to the rest of the cortex, it could be shown that the number of reconstructed fiber tracts was significantly reduced for both seeds in AD and for the hippocampus seed in MCI. Moreover, reduced numbers of hippocampal fibers were also shown to correlate with episodic memory performance, indicating that global structural disconnection of the hippocampus already occurs in prodromal stages of AD and contributes to early memory deficits (Zhou et al. 2008). An intriguing new approach to the study of cortical disconnection is given by the reconstruction of individual whole-brain connectivity networks and subsequent analysis of their topological organization using graph-theoretical analysis (He and Evans 2010). Using white matter tractography-derived connectivity information to build cortical connectivity networks, it could be shown that presence of AD alters discrete topological network metrics, mostly related to increased shortest path lengths and less efficient network wiring (Lo et al. 2010; Shao et al. 2012). Analysis of abnormal systematic brain integrity in AD may be an interesting alternative to regional lesion analysis as it inherently accounts for the structural and functional network architecture of the human brain. A major shortcoming of current DTI-based fiber tracking algorithms is that they cannot reliably resolve crossing and touching fiber bundles due to their indirect reconstruction of the principal fiber direction through the Gaussian parameterization of diffusion in the diffusion tensor. Recently developed model-free generalizations of DTI, such as diffusion spectrum imaging, allow a more accurate reconstruction of crossing fiber tracts (Wedeen et al. 2008) and may be used in the future to study age- and AD-related structural connectivity changes in greater detail (Teipel et al. 2014b).

20.3.2 Genes, Molecular Pathology, and Lifestyle Factors

Compared with the study of cortical gray matter, the effects that the various genetic, molecular, cardiovascular, and lifestyle factors of risk and resilience exert on the brain's white matter integrity are less thoroughly explored.

20.3.2.1 Influence of Susceptibility Genes

Studies that investigated white matter integrity in relation to carrier status of familial AD mutations found decreased FA values in preclinical carriers compared to noncarriers in the fornix as well as in the parahippocampal and orbitofrontal white matter. FA reductions of the fornix were already marked in completely presymptomatic mutation carriers with preserved hippocampal volume and may thus represent a very early pathologic event in AD (Ringman et al. 2007; Ryan et al. 2013). Similarly, a family history of sporadic AD and possession of the APOE4 allele have been shown to be associated with reduced white matter integrity even in asymptomatic healthy elderly subjects without signs of cortical atrophy (Gold et al. 2010). Interestingly, the effects of family history and APOE status were found to be partly independent, indicating that further unknown genetic risk factors may affect brain structure independent from, and in addition to, APOE status (Bendlin et al. 2010; Xiong et al. 2011). In group comparisons with healthy elderly controls homozygous for the APOE ε3 allele, APOE4 effects mainly mapped to posterior AD-susceptible fiber tracts, including the parahippocampal white matter and the splenium of the CC (Persson et al. 2006; Honea et al. 2009; Kljajevic et al. 2013). However, possession of the APOE4 allele was also shown to dose dependently increase the negative effects of advanced age on white matter microstructure in frontal and posterior regions (Ryan et al. 2011). Accordingly, asymptomatic APOE4 carriers also demonstrated an accelerated age-related loss of graph-theoretical measures of local cortical interconnectivity, mainly affecting posteromedial and lateral parietal as well as orbitofrontal cortices (Brown et al. 2011). Based on the observation that reduced fiber integrity can also be observed in young adult carriers compared to noncarriers (Heise et al. 2011) and that the physiological role of the apolipoprotein E is associated with both white matter development and repair (Mahley and Rall 2000), it is currently not clear whether APOE4 effects on white matter microstructure are primarily of neurodegenerative or developmental nature.

20.3.2.2 Markers of Molecular Pathology

Although AD-related changes in DTI indices of white matter integrity are still little explored in relation to molecular biomarkers of amyloid pathology, there is now initial evidence that increased levels of tau protein in the CSF are related to microstructural fiber disruption. The microtubuli associated tau protein is essential for the maintenance of axonal integrity. Increased levels of CSF tau protein have been

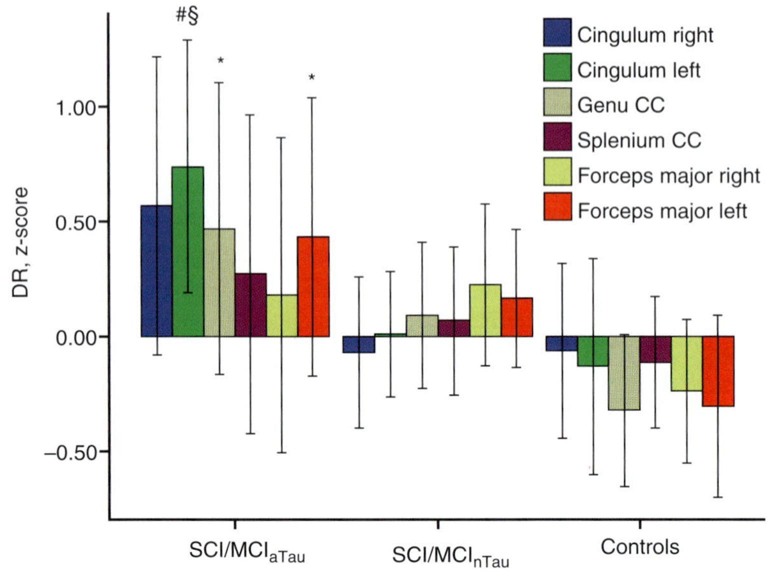

Fig. 20.9 Standardized residuals (*z*-scores) of radial diffusivity after correcting for age, sex, volume of white matter hyperintensities, and different MRI scanner. Subjects with subjective cognitive impairment (*SCI*) or MCI and abnormal CSF tau protein levels (*SCI/MCI$_{aTau}$*) had significantly higher radial diffusivity (*DR*) in genu corpus callosum, left forceps major, and left posterior cingulum compared to controls and higher DR in the left posterior cingulum compared to subjects with SCI or MCI and normal CSF tau protein levels SCI/MCI$_{nTau}$ (Mann-Whitney *U*-test). *Statistically significant vs. controls ($p<0.05$, uncorrected). #Statistically significant vs. controls ($p<0.01$, uncorrected). §Statistically significant vs. SCI/MCI$_{nTau}$ ($p<0.05$, uncorrected). *DR* radial diffusivity (Adapted from Stenset et al. (2011) with permission to reprint granted by Elsevier)

found in several conditions with severe axonal damage such as head trauma, AD, or large cerebral infarcts. In accordance with the pathophysiological significance of increased CSF tau levels, MCI subjects with increased levels of CSF tau had significantly lower FA and increased RaD in the posterior cingulum fibers compared to those with normal levels (Stenset et al. 2011) (Fig. 20.9). Increased CSF tau was also found to be associated with microstructural tissue disruption in a group of asymptomatic middle-aged adults with a family history of AD (Bendlin et al. 2012). Surprisingly, no association could be found between tau levels and AxD, which had been proposed as a marker of direct axonal damage as opposed to damage to the myelin sheath in previous studies. There is, however, evidence for a stage-specific alteration of AxD in the pathogenesis of AD, where AxD is decreased in early stages of disease, suggesting functional alterations of intra-axonal transport, but increased with loss of gross axonal integrity in more advanced stages. These initial findings lend substantial support to the utility of CSF tau as a molecular marker of axonal degeneration induced by neurofibrillary pathology and warrant further exploration of the relationship between biomarkers of molecular AD pathology and diffusion indices of microstructural fiber integrity.

20.3.2.3 Cardiovascular Risk and Lifestyle

Vascular risk was found to worsen age-related white matter changes by exacerbating frontal fiber disruptions and further driving the expansion of white matter damage from anterior to posterior regions (Kennedy and Raz 2009). When compared to the region-specific effects of AD, vascular risk factors were found to contribute independently to reductions in white matter integrity and with a different region-specific injury pattern (Lee et al. 2009). This has been demonstrated in detail for CC fiber disruptions, where vascular risk was associated with reduced FA in most parts of the CC, but FA reductions in the genu and especially the splenium were predominantly driven by AD diagnosis (Lee et al. 2010). Interestingly, white matter disruptions in obese but otherwise healthy adults were found to map more selectively to AD-susceptible regions, especially the fornix and the splenium of the CC (Stanek et al. 2011; Xu et al. 2013).

Although possible positive effects on white matter integrity through changes in lifestyle have not yet been demonstrated in longitudinal trials, there is initial evidence from cross-sectional studies that smoking cessation and aerobic fitness may partly reverse the negative effects on white matter integrity of cardiovascular risk factors (Gons et al. 2011, 2013; Tseng et al. 2013). In addition, cognitive training programs in older adults have been shown to benefit white matter integrity, where increases in FA correlated with memory improvement in the training group (Engvig et al. 2012). DTI can also contribute to further our understanding of cognitive reserve in high-educated individuals (Teipel et al. 2009; Arenaza-Urquijo et al. 2011). Thus, it could be shown that AD patients with higher education showed reduced fiber integrity in limbic and association fibers when compared to similarly impaired patients with less education, indicating that in high-educated subjects more fiber disruptions are necessary to induce the same level of cognitive impairment. In healthy elderly subjects on the other hand, higher education was associated with greater white matter integrity in similar fiber systems. Thus, the cognitive reserve capacity of high-educated individuals may be mediated not only by bigger brain sizes or selectively increased gray matter volumes but also by education-induced increases in cortical network wiring which is probably associated with facilitated intracortical communication.

20.3.3 DTI as a Diagnostic Marker

The group differences in microstructural integrity of select white matter tracts may also be used to aid early diagnosis of AD at the individual level. Several studies have examined the diagnostic potential of averaged measures of microstructural integrity from ROIs in lobar white matter as well as in delineated tracts of interest, most notably the fornix, cingulum, or CC. In line with the findings from intergroup comparisons, reduced microstructural integrity of the parahippocampal and posterior cingulum, the fornix, and the splenium of the CC were the most reliable predictors of AD and MCI. However, the reported diagnostic accuracies differed widely between studies, particularly for separating MCI subjects from healthy controls, and ranged from hardly significant group separations (Johnson et al. 2010) to accuracies between 70 and 80 % (Chua et al. 2009; Wang et al. 2009a; Zhuang et al. 2010). These discrepancies may be related to small and heterogeneous study samples but also to differences in image processing strategies and the employed diffusion metrics. In a meta-analytic study, diffusion changes in limbic regions yielded effects sizes at predementia stages of AD that were in the range of effect sizes for MTL volumes (Clerx et al. 2012). However, given that microstructural fiber tract alterations and regional gray matter atrophy carry complementary information of the disease process, highest diagnostic accuracies were generally found when diffusivity markers and macroscopic atrophy markers were combined (Zhang et al. 2007; Wang et al. 2009a).

Recent longitudinal studies suggest that especially diffusivity changes of the fornix appear to be good predictors of imminent memory decline, both at the preclinical (Zhuang et al. 2013) and predementia stages of AD (Mielke et al. 2012). White matter abnormalities further appear to track cognitive deterioration at the dementia stage of AD (Acosta-Cabronero et al. 2012), and mean diffusivity of the posterior cingulum was found to be a more sensitive indicator of dementia severity than hippocampal volume (Nakata et al. 2009). Moreover, in one study diffusion indices of microstructural tissue integrity were found to be better predictors of conversion to AD in MCI subjects when compared to volumetric measures across several white and gray matter ROIs (Scola et al. 2010).

A promising application of DTI as early marker of AD pathology is also given by the assessment of diffusion abnormalities in gray matter regions, most notably the hippocampus, as

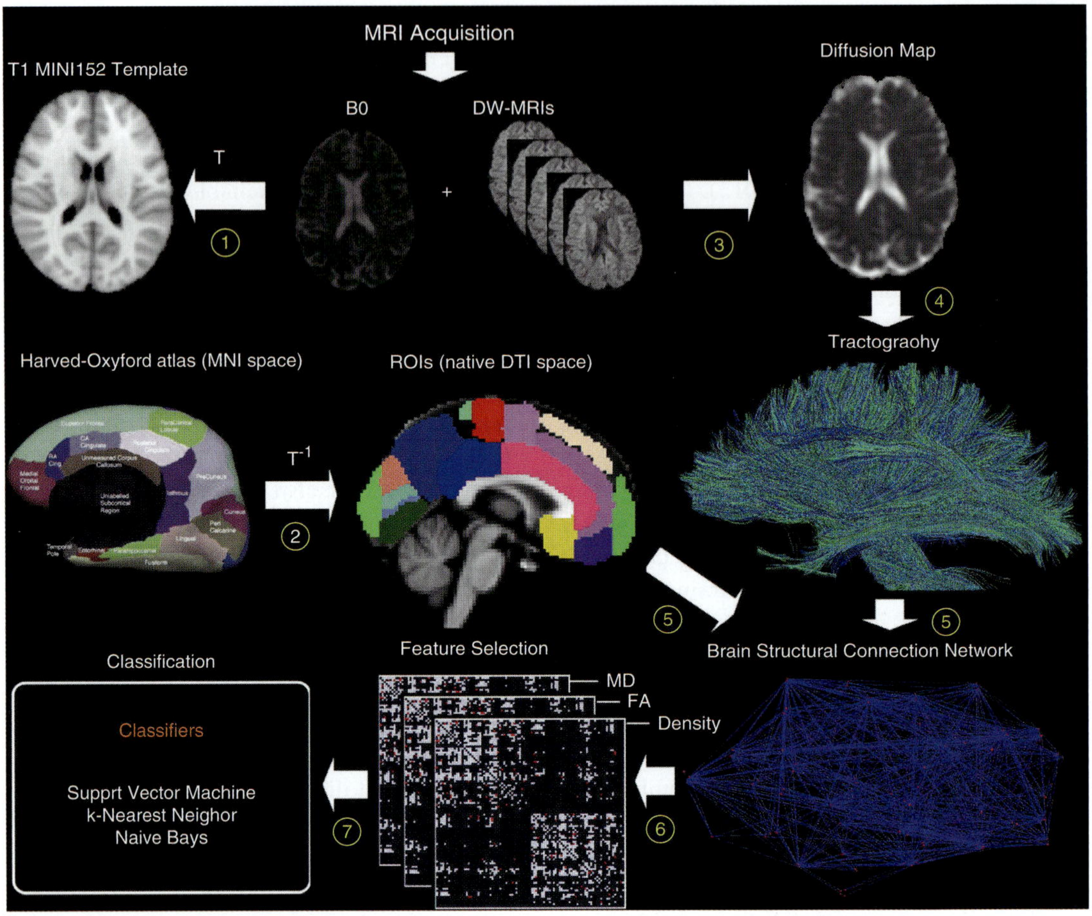

Fig. 20.10 Flowchart of diffusion-weighted image analysis for automated classification based on structural connectivity networks. Individual diffusion-weighted images are normalized to the MRI standard space (*1*) and automatically parcellated based on inverse transformation of an anatomical atlas in standard space (*2*). A voxel-wise diffusion tensor model is used to derive maps of local properties of water diffusion (e.g., fractional anisotropy (*FA*) or mean diffusivity (*MD*)) (*3*). Whole-brain tractography is performed providing an estimate of axonal trajectories across the entire white matter (*4*). Individual structural connectivity networks (*ISCNs*) are constructed by combining the output of both cortical parcellation and diffusion tractography for each individual subject (*5*). The most distinctive connections of ISCNs among groups are identified by a feature selection criterion for different attributes of fiber density, FA, and MD (*6*). ISCNs of patients with MCI and mild AD, respectively, and healthy control subjects are classified by three different pattern recognition algorithms (*7*) (From Shao et al. (2012) with permission to reprint granted by Elsevier)

they may detect microstructural tissue disruption before volumetric loss occurs. Accordingly, at the predementia stage of AD, mean diffusivity of the hippocampus was found to yield higher diagnostic accuracies than hippocampal volume (Fellgiebel and Yakushev 2011; Clerx et al. 2012).

Recently, multivariate approaches which take into consideration the covariance structure of the DTI measures across the whole brain are being explored for their potential to improve diagnosis of AD. Thus, mean diffusivity maps of the whole brain derived from principal component analysis showed equally high diagnostic accuracy as maps of structural volumetric changes for separating AD patients from controls (Friese et al. 2010). When feeding various tract-based measures of diffusion abnormalities in a classifier based on a machine learning algorithm (support vector machine), optimal classification accuracies above 90 % were achieved for separating MCI subjects from controls as well as for separating stable MCI subjects from those that imminently progressed to AD (Haller et al. 2010). Similarly high accuracies for the imaging-based diagnosis of MCI were achieved when feeding a support vector machine classifier with network metrics derived from DTI-based cortical connectivity networks (Shao et al. 2012) (Fig. 20.10).

It has to be noted, however, that all of these studies used monocentric data to estimate diagnostic accuracies, and DTI-derived measures of tissue integrity have only recently entered the state of multicenter trials. Within the framework of the *European DTI Study in Dementia* (EDSD), a clinical and physical phantom study suggested an about 50 % higher variability of multicenter acquired DTI data compared to classical anatomical MRI scans, significantly limiting its applicability in the wider clinical setting (Teipel et al. 2011b). The variability was further found to differ between diffusion indices and fiber tracts which should be taken into account for the design of DTI-based diagnostic markers which, in the end, are being developed for application in clinical settings. Based on a subsample of the EDSD multicenter data, posterior cingulate fiber tracking yielded a diagnostic accuracy of about 70 % for markers of fiber tract integrity with a considerable variation across centers (Fischer et al. 2012). Voxel-based analysis of FA and MD in the extended EDSD cohort with 137 AD patients and 143 controls yielded about 82 % accuracy in group separation across nine different sites (Teipel et al. 2012). Currently, multivariate machine learning algorithms are being developed for classification of multicenter DTI data (Dyrba et al. 2013). Prospective protocols for acquisition of multicenter DTI data for the study of neurodegenerative disorders have been suggested (Turner et al. 2011) and will be evaluated in future studies.

20.4 Functional MRI

20.4.1 Progression of Brain Dysfunction in AD

Based on the saliency of memory deficits in AD as well as evidence from neuropathological and structural imaging studies pointing to the MTL as the site of earliest and most pronounced AD pathology, functional neuroimaging studies in AD have focused on the neural correlates of memory deficits with particular emphasis on the function of the hippocampus and the extended memory network. However, neural changes in response to other cognitive and perceptual challenges are also being studied. Of particular interest have been the functional compensatory mechanisms that evolve in the face of increasing pathology within particular neuronal systems to maintain the function of initially spared cognitive domains. More recently, the discovery of intrinsic brain activity during resting baseline conditions has shifted the focus of functional imaging studies from task-related activity changes in perceptual or cognitive networks to the study of dysfunctional suppression of baseline activity and disrupted network connectivity at rest.

20.4.1.1 Task-Based fMRI

Memory Function

In order to examine the earliest cognitive impairments related to AD, a significant number of functional imaging studies have focused on memory paradigms and hippocampal function. Thus, studies have utilized a paradigm where participants have to encode a single stimulus at a time, be it an image or word. The activation pattern in the hippocampus were found to change along a linear continuum, where the activation level in the hippocampus decreased from healthy subjects to MCI subjects to patients with clinical AD. Thus, when investigating the activation differences in the medial temporal lobe between an AD patient group and a healthy control group during encoding of images, it was found that AD patients showed reduced activation of the left hippocampus and the bilateral parahippocampal gyri (Rombouts et al. 2000). Another memory encoding study with images as stimuli reported lower hippocampal activation in both an AD group and an MCI group when compared to a control group of age-matched healthy subjects (Machulda et al. 2003). Similarly, compared to age-matched healthy controls, MCI subjects were found to show decreased activation in the right hippocampus during encoding of line drawings (Johnson et al. 2006a).

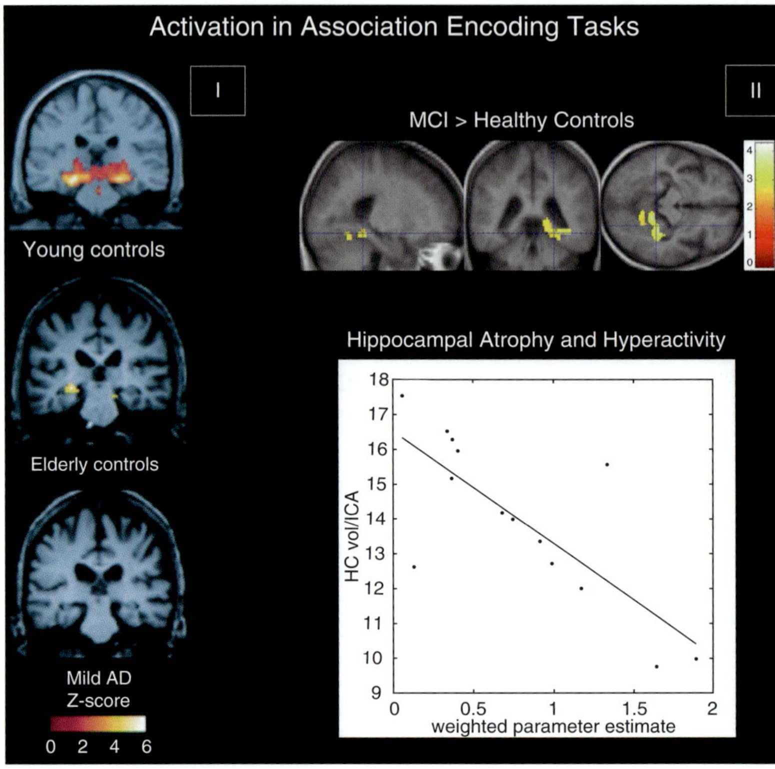

Fig. 20.11 (*I*) Within-group, random effects average activation maps for the encoding of novel face-name pairs compared to repeated (familiar) face-name pairs for young controls, elderly controls, and patients with mild AD, shown on a representative structural image from each group in the coronal plane at the level of the hippocampal formation (Adapted from Sperling et al. (2003) with permission to reprint granted by BMJ Publishing Group Ltd). (*II*) *Top*: between-group statistical activation map showing the increased left parahippocampal and fusiform activation ($p<0.05$, corrected; crosshair position −36, −40, −12) in MCI when compared to controls during associative encoding of novel picture-word pairs (contrasted to resting baseline). *Color bar* presents *T* values. *Bottom*: scatter plot illustrating the inverse relationship between hippocampal volume and parahippocampal activation in MCI. The hippocampal volumes (*HC vol.*; mm^3) on the *Y*-axis were normalized by the intracranial area (*ICA*; mm^2). The weighted parameter estimate on the *X*-axis is an estimate of the magnitude of parahippocampal activation in the encoding-baseline comparison (Adapted from Hamalainen et al. (2007) with permission to reprint granted by Elsevier)

Another approach to investigating memory function is given by associative encoding tasks, which involve the pairing of two non-semantically related stimuli such as a face with a name. In one associative memory study investigating changes with normal aging and in early-stage AD, there was significant activation of the hippocampus in both young and older healthy subjects when encoding novel face-name pairs compared to a well-learned face-name pair, whereas patients with mild AD failed to show such activation (Sperling et al. 2003) (Fig. 20.11). However, in contrast to previous findings of a linear decrease of hippocampal activation from healthy aging over MCI to AD, several studies reported increased MTL activation in MCI subjects during associative memory tasks compared to both healthy elderly controls and AD patients (Dickerson et al. 2005; Hamalainen et al. 2007). This hyperactivity was suggested to represent a compensatory mechanism for beginning MTL atrophy, as hippocampal volume and parahippocampal activity were negatively correlated in MCI subjects, but not in controls or patients with AD (Hamalainen et al. 2007) (Fig. 20.11). In line with the notion of a nonlinear continuum of hippocampal activity changes due to early compensatory processes, another study found hyperactivation in the hippocampus only in early-stage MCI subjects, whereas more impaired MCI subjects and AD patients demonstrated significantly reduced levels of activity compared to healthy controls (Celone et al. 2006). In addition, a longitudinal follow-up study found that healthy subjects with more rapid cognitive decline over a 2-year period had both the highest hippocampal activation at baseline and the greatest loss of hippocampal activation over follow-up, where the rate of activation loss correlated with the rate of cognitive decline (O'Brien et al. 2010).

Another cognitive domain that has received significant interest is working memory (see review (Bokde et al. 2009)). Thus, an altered verbal working memory process in AD patients was reported using an n-back verbal working memory task, which revealed a decreased prefrontal activation in the AD patient group but also an increased parietal activation compared to healthy controls, suggestive of a compensatory mechanism (Lim et al. 2008). In another working memory paradigm with letters as stimuli, it was reported that AD patients showed increased activation during the recognition phase in several brain areas, including the left hippocampal formation (Peters et al. 2009). Hence, AD patients may rely on the recruitment of alternative recognition mechanisms when performing short-term memory tasks. Similar alternative processing mechanisms during working memory performance have also been observed in MCI subjects (Bokde et al. 2010a). It was found that response time in the healthy control group was correlated to the left hippocampus during the encoding phase and to parietal and frontal areas during the recognition phase of the employed working memory paradigm. Contrary to the healthy control group, response time in the MCI group was strongly correlated to the inferior and middle temporal gyri during encoding, the middle frontal gyrus during the maintenance phase, and the hippocampus during the recognition phase.

Perception

Another area of investigation has been visual perception, where one study investigated the alterations in functional activation along the ventral and dorsal visual pathways using two different cognitive paradigms that were matched for performance (Bokde et al. 2008). The healthy control group selectively activated the ventral and dorsal pathways during the face and location matching tasks, respectively, while the MCI group did not show this functional separation. In addition, the MCI group had greater activation than the HC group in the left frontal lobe during the location matching task – a task that recruits the dorsal visual pathway. Similarly, in a cohort of AD patients and using the same tasks, it was reported that during the location-matching task the AD group recruited additional regions along the dorsal visual pathway, primarily in the parietal and frontal lobes (Bokde et al. 2010b). In the same paradigm, examination of the functional connectivity of the fusiform gyrus (a key area for processing of faces) in MCI subjects and healthy controls revealed that healthy controls showed higher positive linear correlation between the right middle fusiform gyrus and the visual cortex, the parietal lobes, and the right dorsolateral prefrontal cortex compared to the MCI group (Fig. 20.12), whereas the latter showed higher positive linear correlation with the left cuneus (Bokde et al. 2006). Hence, the putative presence of AD neuropathology in MCI affects the functional connectivity of the fusiform gyrus during processing of faces despite normal task performance. In addition, higher linear correlation in the MCI group in the parietal lobe may indicate the initial appearance of compensatory processes. Another study utilized an angle discrimination task to investigate changes in AD patients along the visual pathways and found the most pronounced differences compared to healthy controls in the superior parietal lobule (more activity in controls) and the occipital-temporal cortex (more activity in patients) (Prvulovic et al. 2002). Given that the differences in functional activation between the AD patients and controls were partly explained by the differences in superior parietal lobe atrophy, these results indicate

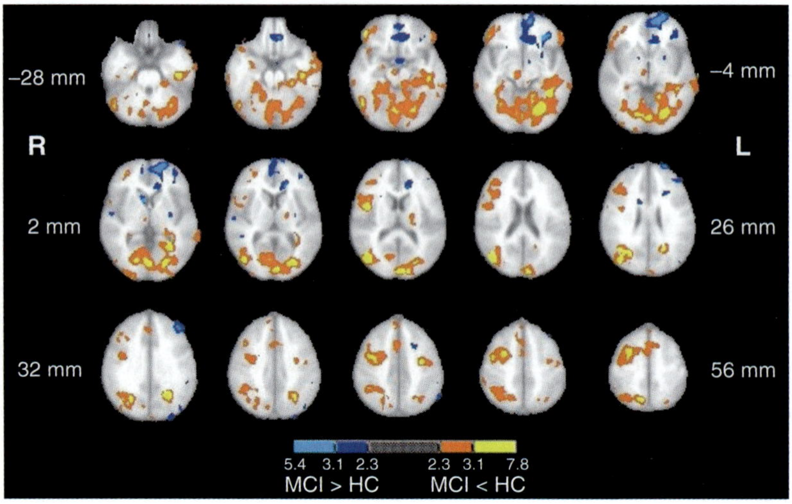

Fig. 20.12 Map of the regions showing statistically significant differences in functional connectivity of the fusiform gyrus during the visual discrimination task between healthy control and MCI groups (From Bokde et al. (2006) with permission to reprint granted by Oxford University Press)

that atrophy-related parietal dysfunction in mild to moderate AD may be compensated by recruitment of the ventral visual pathway.

Task-Related Deactivation

While cognitive tasks generally induce a significant increase of activity in brain areas subserving various aspects of the specific task at hand, there is also a network of brain regions that consistently reduces its activity from a baseline resting condition during performance of a wide range of cognitively demanding tasks. These task-related deactivations (TRDs) are suggestive of intrinsic brain activity attributable to a baseline state or "default mode" and led to the definition of the default mode network (DMN) (Raichle et al. 2001). The DMN includes the posterior cingulate cortex/precuneus, the dorsal and ventral medial prefrontal cortex, the lateral inferior parietal cortices, and the medial temporal lobes. It has been suggested to have a role in intrinsic thought processing, including autobiographical memory, prospection, and mind-wandering, all of which represent typically self-focused cognitive processes in the absence of tasks demanding external attentional focus and effort (Buckner et al. 2008). Thus, under normal conditions activity in the DMN is highly anticorrelated with activity in cortical areas associated with executive function and visuospatial attention which are involved in several forms of active task processing and are commonly referred to as "task-positive network" (Fox et al. 2005). The graded deactivation of the DMN with increasing attentional task demand is consistent with the hypothesis that there is a reallocation of cognitive resources from self-focused processes towards externally directed tasks (Harrison et al. 2011) and failure to deactivate the DMN has been linked to momentary lapses in attention and impaired cognitive control (Weissman et al. 2006; Persson et al. 2007). Consistent with the notion that memory tasks require attentional focus during stimulus presentation in the encoding phase as well as introspective processes during memory retrieval and recognition, dissociation of activity patterns among DMN components has been observed in functional studies with memory paradigms. Thus, while the hippocampus shows positive activation during both encoding and retrieval of successfully encoded novel items, the posteromedial node of the DMN deactivates during successful encoding and activates during retrieval (Vannini et al. 2011). Accordingly, reduced posteromedial deactivation during encoding has been found to predict subsequent retrieval errors (Daselaar et al. 2009), and normal age-related memory impairments are believed to be primarily related to a loss of medial parietal deactivation during encoding rather than hippocampal dysfunction (Miller et al. 2008a).

Several studies have used cognitively demanding tasks, such as semantic classification, verb

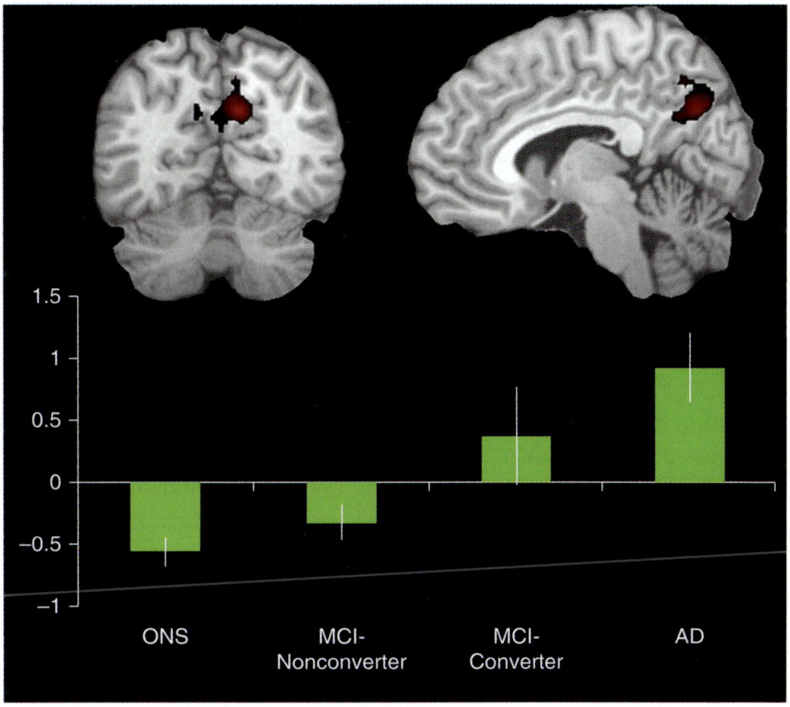

Fig. 20.13 Bar graph shows activation magnitude parameter estimates in a posteromedial region of interest during encoding of novel compared to familiar face-name pairs, demonstrating a continuum from control to MCI-nonconverter, to MCI-converter, and to AD. Differences between groups are statistically significant ($p<0.05$) with the exception of the control vs MCI-nonconverter comparison and the AD vs MCI-converter comparison. Note the overall pattern of negative activation magnitude in the control and MCI-nonconverter groups and positive activation magnitude in the AD and MCI-converter groups (Adapted from Petrella et al. (2007); open-access license, no further permission required)

generation, and working memory paradigms, to study dysfunctional TRD in the continuum from healthy aging to clinical AD (Lustig et al. 2003; Rombouts et al. 2005; Persson et al. 2007). Thus, it could be shown that compared to young adults, older healthy subjects had a smaller magnitude of TRD in the DMN, the differences being especially striking in tasks with higher cognitive demand. Moreover, task-related deactivations of the DMN were further reduced in AD patients of the same age, particularly within the posteromedial node, whereas TRDs in MCI subjects were found to be lower than in healthy controls but still higher than in AD. This continuum of reduced TRD from healthy aging to clinically manifest AD was also noted in a clinical follow-up study, where the magnitude of encoding-related deactivation in the posteromedial cortex was found to be highest in healthy elderly subjects, intermediate in longitudinally stable MCI subjects, and lowest in clinically declining MCI subjects and AD patients (Petrella et al. 2007, 2011) (Fig. 20.13). In the progressive MCI subjects and patients with AD, the sign of the posteromedial activation magnitude parameter estimates was even reversed to positive values, resulting in a paradoxical activation pattern. In the light of these findings and the respective roles for the posteromedial cortex and the hippocampus in memory function, the frequently observed hippocampal hyperactivity in prodromal AD stages (see above as well as in Sect. 20.4.2) may reflect a compensatory mechanism to achieve successful encoding despite reduced parietal deactivation (Miller et al. 2008a). In later stages of accumulating AD neuropathology, increasing hippocampal atrophy and dysfunction may hinder these early compensatory processes, leading to breakdown of the neuronal network subserving memory function.

20.4.1.2 Resting-State fMRI

Activations in task-based fMRI studies are usually performance dependent, which limits their applicability for the study of demented subjects. Resting-state fMRI provides a performance level-independent method to studying functional deficits in AD, thereby greatly facilitating data acquisition and minimizing performance-related variability of the functional data. Based on the coherence of spontaneous low-frequency signal fluctuations within functional brain systems, resting-state fMRI data can be used to extract functional estimates of so-called intrinsic connectivity networks (ICNs), which largely overlap with the various functional networks observed in task-based activation studies across multiple sensory and cognitive domains (Smith et al. 2009). Among the most widely used analysis techniques are functional connectivity analysis, a measure of interregional time-course correlation between regions of interest (Biswal et al. 1995), and independent component analysis (ICA), which automatically isolates networks of synchronized coactivation from the fMRI data in a purely data-driven approach (Beckmann et al. 2005). While the ICA approach has the advantage over functional connectivity analyses that the multiple resting-state networks can be detected without necessity for defining (arbitrary) a priori reference points in the data, it is also computationally more demanding and is limited to the study of coherent intra-network coactivation, disregarding potentially important interactions between functional networks.

Given the substantial overlap of the DMN with the predilection sites of AD neuropathology in the MTL, the posterior cingulate, and the temporoparietal neocortex, it is the earliest and most extensively studied functional network in AD (see also the section on TRD in the previous chapter). Thus, using ICA to extract individual representations of the DMN, it could be shown that AD patients exhibit decreased resting-state coactivation of the posterior cingulate and the hippocampus (Greicius et al. 2004). Subsequent ICA-based resting-state studies that investigated AD-related changes across several cognitive ICNs, in addition to the DMN, could confirm these initial findings and further highlighted the relative specificity of DMN disconnection in AD pathology (Agosta et al. 2012; Binnewijzend et al. 2012). While disrupted DMN coactivation in AD was the most consistent finding among these studies, increased resting-state coactivation in frontal networks was also observed, which was interpreted as a possible compensatory mechanism in an attempt to maintain cognitive efficiency. In addition, subtle DMN disruptions, particularly affecting coactivation of the posterior cingulate and the hippocampus, were already detectable in MCI subjects. This is in line with a previous ICA-based investigation of resting network disruption in MCI, which found reduced network-related activity in the DMN and also within its anticorrelated independent component representing the executive-attention network (Sorg et al. 2007). However, other ICA studies focusing on DMN disruption in the predementia phase of AD found that although the MCI subjects exhibited decreased coactivation in the posterior parietal and medial temporal lobes, other regions of the DMN, including the superior prefrontal cortex and the middle temporal lobe, showed increased resting activity compared to healthy controls (Qi et al. 2010; Damoiseaux et al. 2012). Similarly, in a longitudinal study on functional DMN disruption in MCI that rigorously controlled for regional gray matter atrophy, MCI subjects showed even increased posteromedial DMN coactivation at baseline. However, over clinical follow-up the MCI subjects showed strongly decreasing coactivation of the posterior and anterior-ventral midline nodes and increasing activity in superior frontal regions compared to age-matched controls. Furthermore, the longitudinal decline in resting posteromedial coactivation correlated strongly with declining episodic memory function as assessed by neuropsychological tests outside the scanner (Bai et al. 2011a). These findings indicate that during the preclinical phase of AD, both disruptive and compensative processes coexist within the DMN resting activity.

Complementary to the ICA-based studies on intrinsic network changes, functional connectivity analysis has been widely used to study functional disconnection in the course of AD. Reference points for voxel-based functional connectivity analyses have been typically chosen among prominent nodes of the DMN, including the posterior cingulate, hippocampus, and medial prefrontal cortex (Wang et al. 2006; Zhang et al. 2010; Bai et al. 2011b; Gili et al. 2011), but also within other ROIs, such as the dorsolateral prefrontal cortex (Liang et al. 2011), temporoparietal

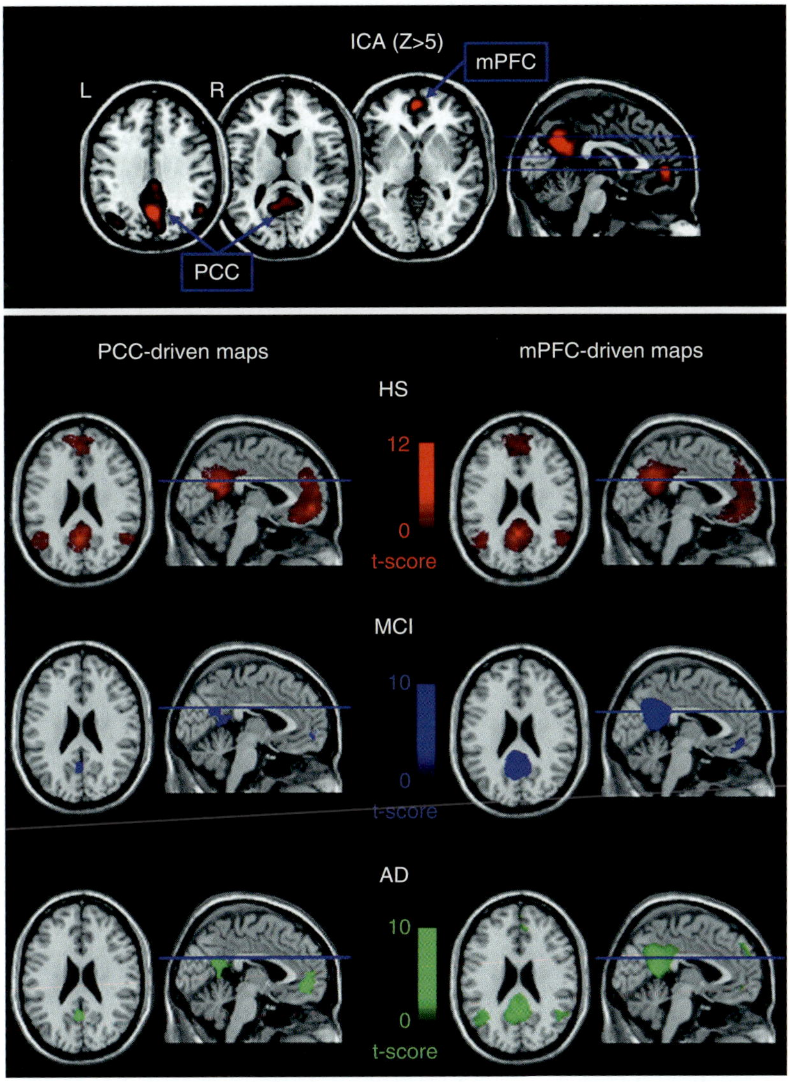

Fig. 20.14 *Top panel* illustrates the mean spatial independent component representing the default mode network. *Blue arrows* indicate the two regions (posterior cingulate (*PCC*) and medial prefrontal cortex (*mPFC*)) with the highest *Z* peak, which were selected as regions of interest for subsequent correlation analysis. *Bottom panel* shows, for each studied group (healthy subjects (*HS*), amnesic mild cognitive impairment (*a-MCI*), and Alzheimer's disease (*AD*) patients), results of correlation analysis obtained using the PCC (*left*) and the mPFC ROI (*right*) to produce connectivity maps. All functional results are superimposed on a high-resolution single subject T1-weighted volume template (From Gili et al. (2011) with permission to reprint granted by BMJ Publishing Group Ltd.)

junction (Liang et al. 2012), insula (Xie et al. 2012), or the thalamus (Wang et al. 2012). Despite some heterogeneity, which may be attributable to differing sample characteristics, image processing strategies, and seed point specification, the findings from these functional connectivity analyses clearly confirm a progressive disconnection among the DMN components in the course of AD (Fig. 20.14). The functional isolation of the posterior cingulate from its main interaction sites in the MTL and the medial prefrontal cortex appears to emerge particularly early in the disease process and was found to be related to worsening episodic memory function in MCI subjects (Bai et al. 2011b; Gili et al. 2011). However, while disconnection between DMN components is the most consistent finding across seed-based functional connectivity studies, there is also preliminary evidence of impaired connectivity among components of the executive-attention network, which deserve further investigation (Liang et al. 2011, 2012; Brier et al. 2012).

An alternative approach that was used to study connectivity changes over the entire brain, independent of a specific seed point, consisted in subdividing the brain into 116 anatomically defined regions and examining the functional connectivity across all pairs of ROIs (Wang et al. 2007).

It was found that AD patients had decreased positive correlations between the prefrontal and parietal lobes but also increased positive correlations within the prefrontal lobe, parietal lobe, and occipital lobe. Hence, the results are consistent with previous reports of a large-scale anterior-posterior disconnection across the brain and increased within-lobe functional connectivity in AD. Interestingly, it was also found that AD patients showed significantly lower negative correlations between components corresponding to the DMN and components of its intrinsically anticorrelated "task-positive network," which may be related to the decreased downregulation of DMN baseline activity in attentionally demanding tasks (Lustig et al. 2003; Kelly et al. 2008). Other studies used similar parcellations of the brain into anatomically defined ROIs (nodes) to reconstruct whole-brain functional networks based on appropriate thresholding of the functional connectivity matrix of pair-wise correlation coefficients (paths) (Supekar et al. 2008; Sanz-Arigita et al. 2010). Application of graph-theoretical analysis revealed that the so constructed functional brain networks in healthy controls showed small-world organization of brain activity, characterized by a high clustering coefficient and a low characteristic path length, indicative of strong local connectivity and an optimal network structure through efficient long-range information transfer. In contrast, the functional brain networks in AD showed loss of small-world properties, characterized by a significantly lower clustering coefficient or an increased characteristic path length, rendering the network metrics closer to the theoretical values of random networks and supporting the hypothesis of disrupted global information integration in AD (Fig. 20.15).

20.4.2 Genes, Molecular Pathology and Lifestyle Factors

20.4.2.1 Genetic Risk and Familial AD

In addition to the changes in brain activity seen in AD patients and neuropsychologically defined risk models of AD, such as MCI, there is also evidence of genotypic effects contributing to abnormal brain function. These effects can be exerted by fully penetrant mutations of familial AD, APOE genotype, or yet to be identified risk genes associated with a positive family history of sporadic AD.

One study in cognitively normal older healthy subjects found that carriers of the APOE4 allele showed higher activation in task-relevant areas during the encoding and recall of semantically unrelated words when compared to healthy noncarriers (Bookheimer et al. 2000). Again, the higher levels of brain activation in the APOE4 carriers were interpreted as being due to compensatory mechanisms. Other studies also reported results that are consistent with the hypothesis of increased activation due to compensatory mechanisms in APOE4 carriers, for example, during encoding of pictures (Bondi et al. 2005), during encoding of word pairs (Fleisher et al. 2005), and during working memory with an n-back task (Wishart et al. 2006); in these studies there were no regions with decreased activation in the APOE4 carriers compared to the noncarriers. However, this pattern of activation has not been a consistent finding. Thus, another study reported decreased hippocampal activation in APOE4 carriers compared to noncarriers during performance of an encoding task (Trivedi et al. 2006). Similarly, during a semantic categorization task, healthy subjects with the APOE4 allele had lower activation in the parietal cortex compared to noncarriers (Lind et al. 2006). The differences in activation between APOE4 carriers among the studies may be explained, in part, due to differences in the cognitive paradigms utilized and the age of the participants in the studies (ranging from mid-40s to 80s) but also due to the presence of other risk factors such as a positive family history of sporadic AD. Thus, in a task comparing new items to previously learned items, persons with a first-degree family history of AD had lower functional activation to novel items in the medial temporal lobe and fusiform gyrus bilaterally compared to controls without a family history of sporadic AD (Johnson et al. 2006b). In hippocampal areas, the group composed of subjects without family history but with APOE4

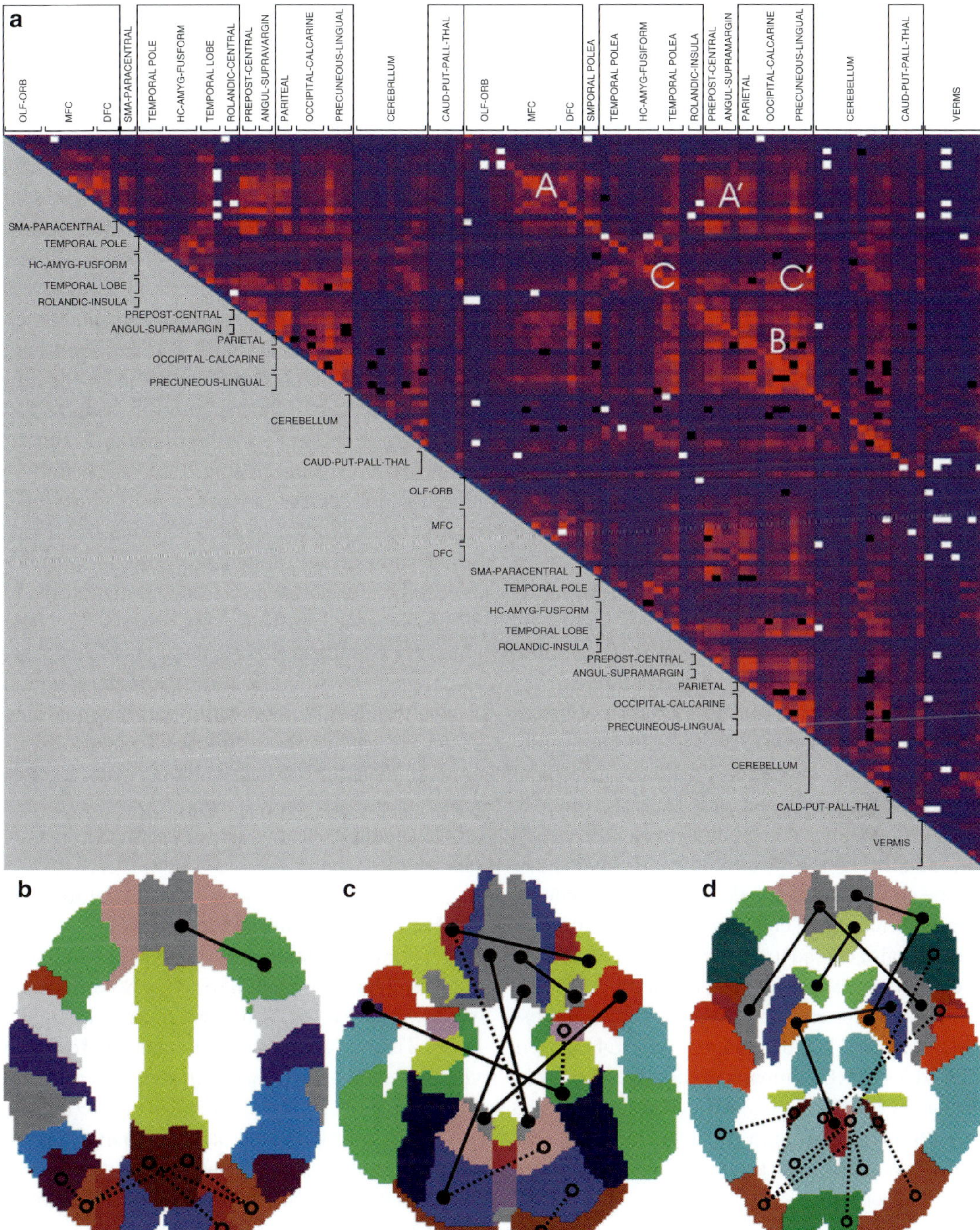

Fig. 20.15 Matrix of significant differences in functional connectivity between AD and controls (two-tail *t*-test, $p<0.05$ uncorrected). The *white* and *black dots* represent pairs of brain areas with increased and decreased synchronization in AD, respectively. (**b–d**) A subset of connectional differences corresponding to the matrix (**a**) are plotted at three superior to inferior levels through the brain template of the structural atlas: (**b**) = z53; (**c**) = z73; (**d**) = z111. *Lines* depict synchronization between pairs of regions: *solid lines* = enhanced synchronization; *dashed lines* = reduced synchronization. Note the pattern of generalized posterior (parietal and occipital) synchronization reductions and increased frontal synchronization (From Sanz-Arigita et al. (2010); open-access license, no further permission required)

genotype exhibited the greatest activation, and the group composed of subjects with positive family history and APOE4 genotype had the smallest level of activation. Furthermore, in another study it was found that family history and age exerted independent effects on brain activation patterns during episodic encoding, whereas APOE interacted with age such that APOE4 carriers exhibited age-related increases in activity in the hippocampus (Trivedi et al. 2008).

One study compared the brain activation patterns during an encoding task between cognitively healthy subjects with APOE4 genetic risk and subjects with fully penetrant familial Alzheimer's disease mutations, providing evidence that APOE4-associated effects on brain activity may be partially independent of those exerted by familial Alzheimer's disease mutations (Ringman et al. 2011). They found that the presymptomatic mutation carriers had lower activation in the anterior cingulate gyrus bilaterally and the left frontal pole compared to non-mutation carriers, whereas the APOE4 carriers had greater activation during the task compared to noncarriers of the APOE ε4 allele. These findings suggest that the increased activation seen in the APOE4 carriers may not be due solely to compensatory mechanisms but may indicate, in part, an independent physiological mechanism.

The changes in regional task-related activation seen in healthy subjects at genetic risk for sporadic AD also extend to a dysfunctional regulation of default mode activity. Thus, when performing a semantic categorization task known to suppress DMN activity, healthy elderly APOE4 carriers showed less TRD compared to age-matched noncarriers (Persson et al. 2008). This finding was later reproduced using a memory encoding task and it was shown that greater failure of posteromedial deactivation during encoding was related to worse delayed recall performance as assessed by neuropsychological tests (Pihlajamaki et al. 2010). Using an ICA approach to investigate DMN activity at rest, it could be demonstrated that healthy APOE4 carriers showed increased hippocampal DMN synchronization compared to noncarriers, which was supposed to reflect ε4-related failure in hippocampal decoupling, possibly resulting in an elevated hippocampal metabolic burden and increased risk of cognitive decline and AD (Westlye et al. 2011). However, functional connectivity analysis of resting-state data demonstrated decreased connectivity between a posteromedial seed and other DMN nodes in healthy APOE4 carriers compared to noncarriers (Machulda et al. 2011). Interestingly, a similar connectivity disruption could also be observed in healthy APOE4 carriers that showed no signs of presymptomatic amyloid pathology as assessed by molecular PET and CSF profiling, indicating that functional effects of APOE genotype may be partly independent of primary AD pathology (Sheline et al. 2010). This is further supported in a recent resting-state fMRI study in middle-aged healthy subjects which found that, relative to ε3 homozygotes, changes in functional connectivity associated with the ε4 risk allele were similar in direction and magnitude to those associated with the "protective" ε2 allele (Trachtenberg et al. 2012). Thus, APOE appears to have an intrinsic effect on the differentiation of functional networks in the brain, which may be related to the role of this gene in neurodevelopment (Mahley and Rall 2000), whereas differential interactions between AD pathology and APOE genotype may determine allele-specific risk profiles.

20.4.2.2 Amyloid and Functional Networks

An interesting association exists between cortical amyloid deposition and default activity in the brain. Thus, brain amyloid as measured with in vivo molecular PET was found to preferentially accumulate in prominent nodes of the DMN, including the prefrontal, lateral parietal, lateral temporal, and posterior cingulate cortices, with the intriguing exception of the medial temporal cortex (Buckner et al. 2005; Sperling et al. 2009) (Fig. 20.16). The regions with the highest amyloid accumulation were also found to be among the most highly interconnected regions in the brain (so-called cortical hubs), indicating that they may constitute critical way stations for global information processing (Buckner et al. 2009). Hence, the key regions for neural activity

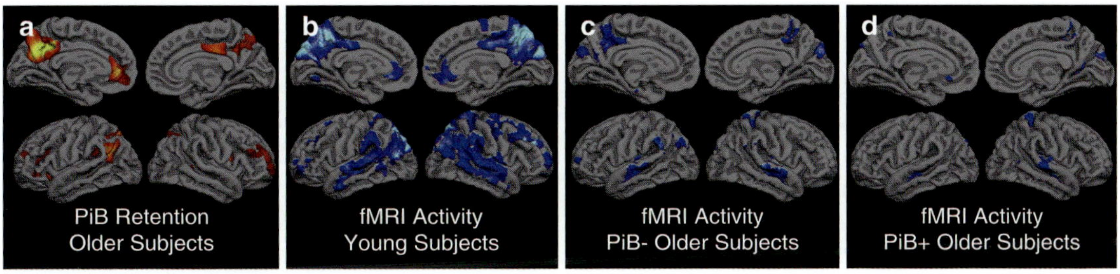

Fig. 20.16 Anatomic distribution of amyloid (PiB)-PET and fMRI default network activity. (**a**) Anatomic distribution of PiB retention (i.e., amyloid deposition) for the group-wise comparison of PiB+ ($n=13$) > PiB− ($n=22$) older subjects thresholded at $p<0.001$ FWE corrected for multiple comparisons. (**b–d**) voxel-wise one-sample t test of significant task-related decreases in fMRI activity (deactivation shown in *blue*) during successful encoding of face-name pairs (fixation > high confidence correct "hit" responses based on subsequent memory testing) in young subjects (**b**), PiB− older subjects (**c**), and PiB+ older subjects (**d**) (From Sperling et al. (2009) with permission to reprint granted by Elsevier)

integration in the brain also appear to be the most vulnerable to AD pathology and may thus serve as key points in the cascade leading to brain disease and resulting cognitive deficits in AD.

Supportive of this idea, combined functional MRI and amyloid-PET studies could show that cortical amyloid deposition affects brain function already long before the appearance of clinical symptoms in AD. Thus, cognitively healthy subjects in the presymptomatic phase of AD, as indicated by considerable cortical amyloid deposition, were found to show disrupted resting connectivity of the DMN (Hedden et al. 2009; Drzezga et al. 2011; Mormino et al. 2011) as well as impaired posteromedial deactivation during performance of a memory task (Sperling et al. 2009) (Fig. 20.16) when compared to age-matched healthy subjects without amyloid deposition. Similar to the findings in neuropsychologically or genetically defined models of prodromal AD, healthy elderly subjects with high amyloid load also showed presumptive compensatory mechanisms presenting as heightened MTL activation during episodic memory encoding (Mormino et al. 2012) as well as increased resting connectivity in dorsal prefrontal and lateral temporal cortices (Mormino et al. 2011).

20.4.2.3 Cognitive Reserve and Physical Exercise

The nonlinear pattern of activity changes from prodromal and at-risk stages to clinical AD point to the existence of potential compensatory processes that maintain cognitive function in the face of early pathologic alterations. Education has been identified as a possible mediator of increased resilience to AD pathology, a concept termed cognitive reserve. In line with this, higher cognitive reserve in healthy subjects was found to be associated with lower deactivations within the DMN and lower task-related activity, which was interpreted to reflect increased neural efficiency. In contrast, the amnestic MCI subjects and AD patients with higher cognitive reserve had greater activity in task-related brain areas and increased deactivations within the posterior cingulate and medial frontal regions compared to those with lower cognitive reserve. Thus, a greater reallocation of processing resources from the DMN to the neural network engaged in the experimental task could reflect increased reliance on compensatory resources to maintain cognitive function in these patients (Bosch et al. 2010). This is also consistent with previous findings in healthy adult subjects where DMN deactivation was strongly associated with subjectively experienced task difficulty (Harrison et al. 2011).

Apart from the protective effect of higher education, a recent resting-state fMRI study in healthy elderly individuals also provides interesting insights for the role of exercise training in active attenuation of age-related brain dysfunction (Voss et al. 2010). Thus, 1 year of aerobic exercise training (walking) effectively increased the functional connectivity of higher cognitive networks, including the DMN and the frontal-executive network. Most importantly, these exercise-induced connectivity changes were also shown to be behaviorally relevant in that increased functional connectivity was associated with improvements in executive function.

20.4.3 fMRI for the Diagnosis of AD

Neuroimaging studies are only beginning to test and quantify the prognostic value of functional MRI for the diagnosis of AD. Using task-based functional MRI with a memory paradigm, one study could show that the degree of hippocampal hyperactivity in MCI subjects significantly predicted cognitive decline and conversion to dementia over a clinical follow-up of 6 years (Miller et al. 2008b). Greicius and colleagues used the goodness of fit to a standard DMN template on individual representations of the DMN as extracted by ICA and found that AD patients and healthy elderly controls could be correctly categorized with a sensitivity of 85 % and a specificity of 77 % (Greicius et al. 2004). Similar goodness-of-fit scores of ICA-extracted DMN components from task-based functional MRI data were further shown to be predictive of cognitive decline and conversion to AD dementia in MCI subjects (Petrella et al. 2011). Quantification of longitudinal changes in functional connectivity strength between hippocampal subfields and the posterior cingulate cortex yielded diagnostic accuracies well over 80 % for the distinction between MCI and healthy controls as well as for discriminating stable MCI subjects from those that converted to AD over follow-up (Bai et al. 2011b). Another study used both ROI-based interconnectivity analysis and ICA-based analysis of the DMN for separating AD patients from controls (Koch et al. 2012). Although functional connectivity analysis appeared to be of slightly higher value for correct group classification compared to ICA, the combination of both approaches yielded the highest diagnostic accuracy of 97 % (sensitivity 100 %, specificity 95 %). However, all of these studies focused on the DMN as a whole or on connectivity measures between some of its key nodes, thereby disregarding potential connectivity changes across other cognitive networks. Other studies examined the diagnostic accuracy of functional connectivity changes in a more comprehensive manner by dividing the brain into several (90 or more) anatomical ROIs based on inverse atlas labeling. Pair-wise connectivity indices (Chen et al. 2011) or second-order network indices from graph-theoretical analysis (Supekar et al. 2008; Wee et al. 2012) could then be used to separate the diagnostic groups with considerable diagnostic accuracies between 80 and 95 %. Finally, the diagnostic value to separate asymptomatic subjects at high risk for AD (family history and possession of the APOE4 allele) from healthy low-risk controls has been directly compared between measures of altered functional activation during a cognitive task and disrupted DMN connectivity in the resting state (Fleisher et al. 2009). It was found that resting-state functional connectivity discriminated between the two risk groups with a significantly larger effect size compared with task-related functional changes (regional indices of activation and deactivation).

An issue for the use of the DMN as a potential marker for diagnosis of dementia is the stability of the DMN as assessed within individuals over time. Using ICA on resting-state data of 18 healthy young subjects the DMN was reproducible within a single imaging session as well as between imaging sessions 12 h and 1 week apart (Meindl et al. 2010). The reproducibility over time has also been shown for other frequently reported ICNs (Zuo et al. 2010) and using functional connectivity analysis instead of ICA (Shehzad et al. 2009). Finally, ICNs as detected by both functional connectivity analysis and ICA were also shown to be consistent across multi-center data (Biswal et al. 2010). Thus, functional MRI and particularly resting-state fMRI as a performance-independent measure provide high potential as imaging markers for early AD pathology. However, the precise implementation of various types of functional indices for the design of diagnostic markers has not yet been explored thoroughly and clinical follow-up studies to evaluate the prognostic value of these markers to predict future cognitive decline in predementia stages of AD are still scarce (Petrella et al. 2007; Miller et al. 2008b; Bai et al. 2011b).

20.5 Multimodal Integration

20.5.1 Multimodal Assessment of Brain Structure and Function

Most of the studies described in the previous chapters have focused on a single imaging modality to study specific aspects of structural or functional brain alterations in the course of AD. Although these modality-specific studies have revealed important insights into the pathogenesis of AD, integration of the isolated findings into a comprehensive framework of disease stage-specific brain changes is limited by the large variability in study populations, image processing strategies, and statistical methods. Concurrent assessment of multimodal markers of structural and functional brain integrity in the same study population significantly increases the heuristic value of imaging studies and enables the examination of interrelationships among these markers, which are assumed to reflect distinct but interrelated pathologic characteristics. Thus, multimodal imaging data may reveal new insights on disease mechanisms not amenable to single-modality studies. A particularly intriguing disease mechanism in early stage AD could be revealed by demonstrating that atrophy of the MTL was associated with both dysfunction of the posterior cingulate cortex and a disruption of cingulum fibers which connect these two regions (Villain et al. 2008; Miettinen et al. 2011; Yakushev et al. 2011; Bozoki et al. 2012). Longitudinal multimodal imaging data further allowed reconstruction of the sequential relationships among these pathologic events, such that an initial MTL atrophy leads to a disruption of the cingulum bundle, which in turn induces dysfunction in the disconnected posterior cingulate cortex (Villain et al. 2010).

Although most microstructural white matter alterations are correlated with atrophic changes in corresponding gray matter regions (Sydykova et al. 2007; Avants et al. 2010) (Fig. 20.7), they also provide complementary information of the disease process not obtainable with volumetric mapping (Canu et al. 2010), and combination of both imaging markers are more accurate in explaining the severity of cognitive deficits in MCI and AD (Walhovd et al. 2009; Bozzali et al. 2012). The unique contribution of DTI measurements over and above volumetric changes may be especially important for the detection of early AD pathology, given that AD-associated diffusion abnormalities in the MTL and the fornix were found to precede macroscopic abnormalities and may thus provide a more sensitive marker of early AD pathology (Fellgiebel and Yakushev 2011; Zhuang et al. 2013).

Currently, one of the most interesting multimodal approaches is the comparative assessment of structural and functional networks defined by multimodal sources of connectivity information. The most common imaging-derived metrics for network connectivity are direct structural connections derived from DTI-based fiber tract reconstruction and functional connectivity information as assessed by resting-state fMRI. However, connectivity information can also be derived from structural MRI data based on the observation that related cortical areas covary in size (gray matter volume or cortical thickness) across individuals, probably as a result of mutually trophic influences or common experience-related plasticity (Mechelli et al. 2005; He and Evans 2010). Multimodal imaging studies are only beginning to explore how network characteristics that rely on distinct forms of connectivity information relate to each other and how they are affected by disease. Using structural covariance analysis on networks defined by resting-state fMRI, it could be shown that several nodes within functionally connected networks also exhibit correlated gray matter volumes across healthy adult subjects (Seeley et al. 2009; Segall et al. 2012) (Fig. 20.17). On the other hand, only approximately 35–40 % of the cortical areas that show covarying cortical thickness exhibit direct fiber connections as derived from DTI-based tractography (Gong et al. 2012). The combination of resting-state fMRI with DTI data showed that in healthy adult subjects, the presence of a direct fiber connection is almost

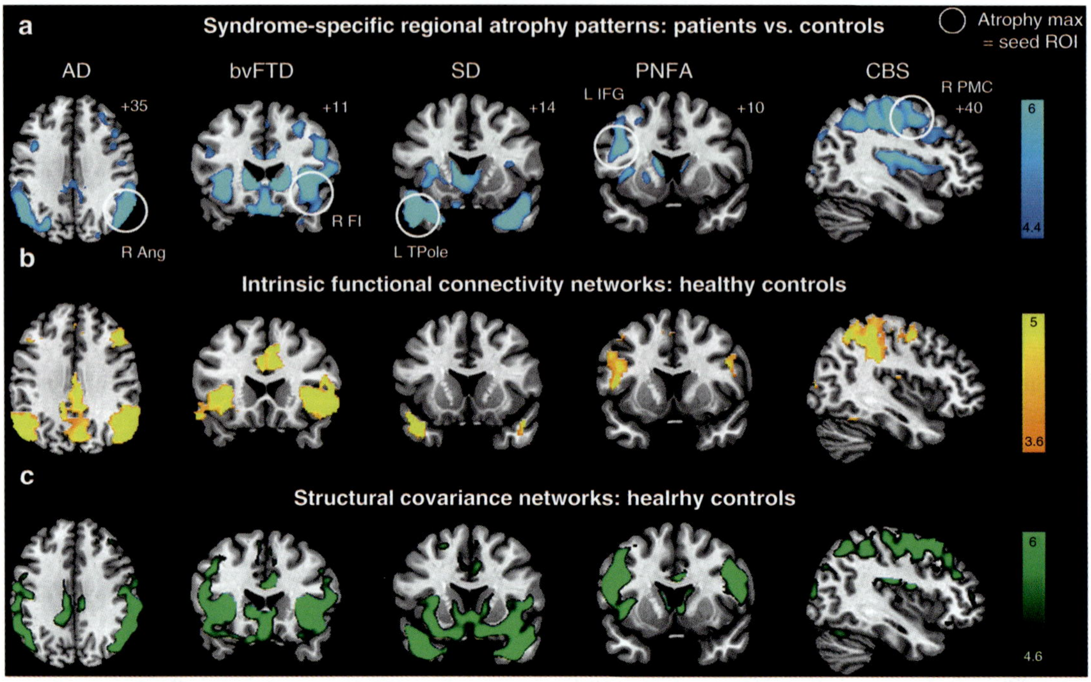

Fig. 20.17 Convergent syndromic atrophy, healthy intrinsic connectivity networks (ICNs), and healthy structural covariance patterns. (**a**) Five distinct clinical syndromes showed dissociable atrophy patterns, whose cortical maxima (*circled*) provided seed ROIs for functional connectivity and structural covariance analyses. (**b**) ICN mapping experiments identified five distinct networks anchored by the five syndromic atrophy seeds. (**c**) Healthy subjects further showed gray matter volume covariance patterns that recapitulated results shown in (**a**) and (**b**). For visualization purposes, results are shown at $p<0.00001$ uncorrected (**a** and **c**) and $p<0.001$ corrected height and extent thresholds (**b**). In (**a**–**c**), results are displayed on representative sections of the MNI template brain. *Color bars* indicate *t*-scores. In coronal and axial images, the *left side* of the image corresponds to the left side of the brain. *ANG* angular gyrus, *FI* frontoinsula, *IFGoper* inferior frontal gyrus, pars opercularis, *PMC* premotor cortex, *TPole* temporal pole (From Seeley et al. (2009) with permission to reprint granted by Elsevier)

always correlated with functional connectivity in the corresponding brain regions (Damoiseaux and Greicius 2009). However, the presence of functional connectivity between distinct brain regions is not necessarily implying the presence of a direct fiber connection, and the dependence on direct structural connections between network nodes varies among the different large-scale functional networks (Horn et al. 2013). Within the default mode network, it has been found that functional connectivity strength is largely predefined by the structural integrity of fiber tracts connecting the key nodes of this network. This finding has been replicated using both fiber tractography (van den Heuvel et al. 2008; Greicius et al. 2009) and multivariate analysis of anisotropy index maps (Teipel et al. 2010a) (Fig. 20.18).

Interestingly, it appears that neurodegenerative diseases selectively target specific functionally defined networks (Seeley et al. 2009) (Fig. 20.17). Thus, in AD the DMN has been shown to be selectively vulnerable to cortical amyloid deposition (Buckner et al. 2005), disruption of functional connectivity (Greicius et al. 2004; Sorg et al. 2007), and neurodegeneration (Seeley et al. 2009). On the other hand, cerebral pathologies other than AD were found to be associated with characteristic changes in distinct but specific cognitive networks, such as the "salience network" in behavioral variant frontotemporal dementia (Seeley et al. 2009; Zhou et al. 2010). When studying AD patients with resting-state fMRI and DTI, it can be shown that the decline of functional connectivity within the DMN is paralleled by a decline of microstructural tract integrity in fiber tracts connecting the key nodes of this network (Hahn et al. 2013; Likitjaroen et al. submitted) (Fig. 20.18). Combined assessment of structural and functional connectivity within the DMN of asymptomatic middle-aged APOE4 carriers indicated that abnormalities in functional network communication precede the breakdown of structural white matter connections in the pathogenesis of AD (Patel et al. 2013). Together, these data suggest that the declining functional connectivity in AD is accompanied by a decline in underlying fiber tract integrity where the combination of both markers can help to discriminate between two types of functional changes, (1) breakdown of functional and structural connectivity and (2) compensatory reallocation of neuronal networks.

20.5.2 Multimodal Diagnostic Markers

Given that each modality provides partially independent information of the underlying pathobiology of the disease, diagnostic markers that combine multimodal imaging data in multivariate statistical models capture a more complete picture of the pathologic profile of AD and may thus improve diagnostic accuracy. The recent incorporation of machine learning algorithms such as support vector machines for the analysis of multimodal imaging data allows the extraction of complex patterns of covariance between multidimensional datasets and enables the identification of binary outcomes based on a very efficient classification algorithm. Thus, recent studies that used combined information from regional atrophy and DTI-based changes in structural connectivity to train a support vector machine classifier could show first promising results for improved discrimination between AD patients and control subjects as well as between patients with MCI and healthy elderly controls (Cui et al. 2012; Dyrba et al. 2012). Another study used linear discriminant analysis to optimally combine multimodal structural MRI measurements (T1- and T2-weighted contrasts and DTI) for the distinction between AD patients and cognitively normal controls. These AD-specific multimodal MRI indices in combination with the discriminant function yielded a highly accurate model (around 90 % accuracy) for the distinction of patients with MCI that later converted to AD from those that remained stable over clinical follow-up (Oishi et al. 2011). Multiple linear discriminant analysis-based classifiers have also been used to integrate volumetric information from structural MRI and several metrics of functional connectivity derived from resting-state fMRI into a highly predictive classification model for AD (Dai et al. 2012). Consistent with a model of specific default mode network vulnerability in AD, the most discriminative features identified by this classifier involved several structural and functional indices from medial frontal, posterior cingulate, and MTL areas. Combining DTI and resting-state fMRI, recent studies used network metrics derived from both tractography-based structural networks and functional resting-state networks to train a support vector machine classifier for the automated diagnosis of MCI (Wee et al. 2012). It could be shown that this multimodality classification approach on graph-theoretical network metrics significantly outperformed the single-modality-based methods as well as the direct data fusion method, yielding a close to perfect diagnostic accuracy of approximately 95 % for the separation of MCI subjects from healthy controls. Network analysis of the specific fibers connecting the nodes of large-scale functional brain networks also proved useful for the separation of individual MCI patients converting to AD dementia from nonconverters (Hahn et al. 2013).

20.6 Summary/Conclusions

Throughout the last years, in vivo neuroimaging research has greatly deepened our understanding of the progressive character of AD pathogenesis. Based on in vivo assessments of neuropsychologically, genetically, and molecularly defined models of predementia AD, as well as longitudinal observations of initially healthy elderly individuals that later developed clinical signs of AD, neuroimaging studies could substantiate and expand earlier findings from autopsy studies pointing to a presymptomatic phase of AD pathogenesis. Together, the existing data supports a pathogenetic model of AD that begins with cortical amyloid deposition long before the first cognitive symptoms appear. Over the course of several years, the primary amyloid-related molecular pathology is

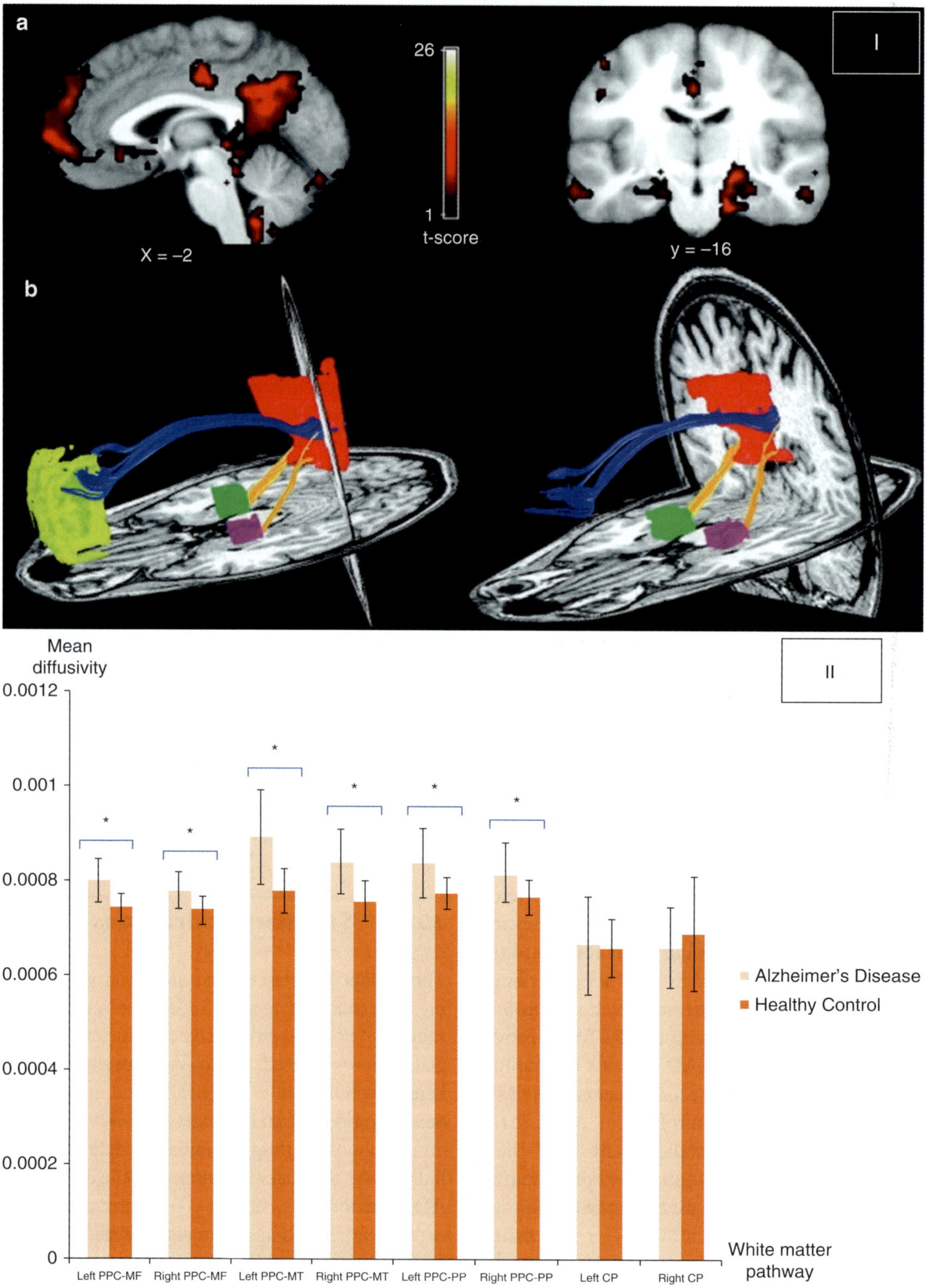
a
26
1
t-score
X = −2
y = −16
I
b
Mean diffusivity
0.0012
0.001
0.0008
0.0006
0.0004
0.0002
0
II
Left PPC-MF
Right PPC-MF
Left PPC-MT
Right PPC-MT
Left PPC-PP
Right PPC-PP
Left CP
Right CP
White matter pathway
Alzheimer's Disease
Healthy Control

assumed to initiate downstream pathologic events, leading to impaired synaptic function and neuronal degeneration which precede and parallel clinically detectable cognitive decline. Of note, the atrophic changes progressively affect only specific neuronal systems in an AD-specific pattern that begins in the anterior MTL at a time when subjects are still cognitively normal and gradually proceeds to affect association areas of the temporoparietal neocortex, the posterior cingulate/precuneus, and ultimately the prefrontal cortex. This neurodegenerative process is accompanied by the sequential emergence of cognitive deficits whose specific characteristics depend on the brain regions affected. While there is initial evidence that cortical atrophy is preceded by synaptic impairments and altered brain function, the pathological nature of axonal degeneration is still little explored and may vary among different neuronal systems. Furthermore, the atrophy pattern of subcortical neuron populations and specific neurotransmitter systems across disease stages and their contribution to cognitive deficits are only beginning to be investigated by neuroimaging studies.

Several genetic, molecular, and lifestyle factors of risk and resilience have been identified that modify the risk for, and the clinical course of, the disease. Assessment of the effects that these factors exert on brain structure and function and how they interact with each other provides important insights for our understanding of the pathogenesis of AD. This will help to develop new preventive strategies and at the same time provide the methodical background to assess the neurobiological mechanism of action of such preventive interventions.

The advances in imaging markers and biomarkers of AD have fueled the definition of new diagnostic categories, including the concepts of presymptomatic and predementia stages of AD. This stage model relies on the in vivo detection of brain pathology and serves to guide the development of diagnostic algorithms implementing different levels of biomarkers as surrogate markers for different stages of disease pathogenesis. In vivo imaging markers of diverse aspects of AD pathology have already been shown to provide clinically significant diagnostic accuracies in multicenter studies, even at early and predementia stages of the disease. Future diagnostic models will rely on the optimized feature selection properties of modern machine learning algorithms to integrate a broad range of disease-specific information into comprehensive classification models. Thus, distinct indices of structural and functional integrity from various neuroimaging modalities will be complemented by CSF-based markers of molecular AD pathology as well as by a multitude of genetic, neuropsychological, sociodemographic, cardiovascular, and lifestyle factors that have been shown to modulate an individual's risk for AD pathology or the expression of clinical symptoms. These models for the in vivo detection of an individual's risk for AD will first be of interest for risk stratification and sample selection in clinical trials aimed at modifying disease progression at different stages of the pathologic process. However, with the expected emergence of disease-modifying strategies in the midterm future, reliable diagnosis of presymptomatic AD stages will also become important in clinical practice.

Fig. 20.18 (*I*) Functional connectivity reflects structural connectivity in the default mode network (DMN). (**a**) Task-free, functional connectivity in the DMN is shown in a group of six subjects. The posterior cingulate/retrosplenial cortex (PCC/RSC) and medial prefrontal cortex (MPFC) clusters are best appreciated on the sagittal view. Prominent bilateral medial temporal lobe (MTL) clusters are seen on the coronal image. (**b**) DTI fiber tractography in a single subject demonstrates the cingulum bundle (*blue tracts*) connecting the PCC/RSC to the MPFC. The *yellow tracts* connect the bilateral MTL to the PCC/RSC. There were no tracts connecting the MPFC to the MTL (Adapted from Greicius et al. (2009) with permission to reprint granted by Oxford University Press). (*II*) Reconstructed fiber tracts connecting the key nodes of the DMN show increased mean diffusivity in AD patients compared to healthy controls. * = difference between groups significant at $p<0.001$. This effect is relatively specific for fibers connecting the DMN nodes as reconstructed fibers of the cerebellar peduncles did not show altered microstructural integrity in AD patients (Data from Likitjaroen et al. (submitted) with permission to reprint granted by the authors)

References

Achterberg HC, van der Lijn F, den Heijer T, Vernooij MW, Ikram MA, Niessen WJ et al (2013) Hippocampal shape is predictive for the development of dementia in a normal, elderly population. Hum Brain Mapp (in press) doi:10.1002/hbm.22333

Acosta-Cabronero J, Williams GB, Pengas G, Nestor PJ (2010) Absolute diffusivities define the landscape of white matter degeneration in Alzheimer's disease. Brain 133:529–539

Acosta-Cabronero J, Alley S, Williams GB, Pengas G, Nestor PJ (2012) Diffusion tensor metrics as biomarkers in Alzheimer's disease. PLoS One 7:e49072

Agosta F, Pievani M, Geroldi C, Copetti M, Frisoni GB, Filippi M (2012) Resting state fMRI in Alzheimer's disease: beyond the default mode network. Neurobiol Aging 33:1564–1578

Almeida OP, Garrido GJ, Alfonso H, Hulse G, Lautenschlager NT, Hankey GJ et al (2011) 24-month effect of smoking cessation on cognitive function and brain structure in later life. Neuroimage 55:1480–1489

Apostolova LG, Mosconi L, Thompson PM, Green AE, Hwang KS, Ramirez A et al (2010) Subregional hippocampal atrophy predicts Alzheimer's dementia in the cognitively normal. Neurobiol Aging 31:1077–1088

Arenaza-Urquijo EM, Bosch B, Sala-Llonch R, Sole-Padulles C, Junque C, Fernandez-Espejo D et al (2011) Specific anatomic associations between white matter integrity and cognitive reserve in normal and cognitively impaired elders. Am J Geriatr Psychiatry 19:33–42

Arenaza-Urquijo EM, Landeau B, La Joie R, Mevel K, Mezenge F, Perrotin A et al (2013) Relationships between years of education and gray matter volume, metabolism and functional connectivity in healthy elders. Neuroimage 83C:450–457

Ashburner J, Friston KJ (2000) Voxel-based morphometry–the methods. Neuroimage 11:805–821

Ashburner J, Friston KJ (2001) Why voxel-based morphometry should be used. Neuroimage 14:1238–1243

Avants BB, Cook PA, Ungar L, Gee JC, Grossman M (2010) Dementia induces correlated reductions in white matter integrity and cortical thickness: a multivariate neuroimaging study with sparse canonical correlation analysis. Neuroimage 50:1004–1016

Bai F, Watson DR, Shi Y, Wang Y, Yue C, YuhuanTeng et al (2011a) Specifically progressive deficits of brain functional marker in amnestic type mild cognitive impairment. PLoS One 6:e24271

Bai F, Xie C, Watson DR, Shi Y, Yuan Y, Wang Y et al (2011b) Aberrant hippocampal subregion networks associated with the classifications of aMCI subjects: a longitudinal resting-state study. PLoS One 6:e29288

Bakkour A, Morris JC, Dickerson BC (2009) The cortical signature of prodromal AD: regional thinning predicts mild AD dementia. Neurology 72:1048–1055

Barnes J, Bartlett JW, van de Pol LA, Loy CT, Scahill RI, Frost C et al (2009) A meta-analysis of hippocampal atrophy rates in Alzheimer's disease. Neurobiol Aging 30:1711–1723

Baron JC, Chetelat G, Desgranges B, Perchey G, Landeau B, de la Sayette V et al (2001) In vivo mapping of gray matter loss with voxel-based morphometry in mild Alzheimer's disease. Neuroimage 14:298–309

Bartus RT, Dean RL 3rd, Beer B, Lippa AS (1982) The cholinergic hypothesis of geriatric memory dysfunction. Science 217:408–414

Bartzokis G (2004) Age-related myelin breakdown: a developmental model of cognitive decline and Alzheimer's disease. Neurobiol Aging 25:5–18; author reply 49–62

Becker JA, Hedden T, Carmasin J, Maye J, Rentz DM, Putcha D et al (2011) Amyloid-beta associated cortical thinning in clinically normal elderly. Ann Neurol 69:1032–1042

Beckmann CF, DeLuca M, Devlin JT, Smith SM (2005) Investigations into resting-state connectivity using independent component analysis. Philos Trans R Soc Lond B Biol Sci 360:1001–1013

Bendlin BB, Ries ML, Canu E, Sodhi A, Lazar M, Alexander AL et al (2010) White matter is altered with parental family history of Alzheimer's disease. Alzheimers Dement 6:394–403

Bendlin BB, Carlsson CM, Johnson SC, Zetterberg H, Blennow K, Willette AA et al (2012) CSF T-Tau/Abeta42 predicts white matter microstructure in healthy adults at risk for Alzheimer's disease. PLoS One 7:e37720

Binnewijzend MA, Schoonheim MM, Sanz-Arigita E, Wink AM, van der Flier WM, Tolboom N et al (2012) Resting-state fMRI changes in Alzheimer's disease and mild cognitive impairment. Neurobiol Aging 33:2018–2028

Biswal B, Yetkin FZ, Haughton VM, Hyde JS (1995) Functional connectivity in the motor cortex of resting human brain using echo-planar MRI. Magn Reson Med 34:537–541

Biswal BB, Mennes M, Zuo XN, Gohel S, Kelly C, Smith SM et al (2010) Toward discovery science of human brain function. Proc Natl Acad Sci U S A 107:4734–4739

Blennow K, Hampel H, Weiner M, Zetterberg H (2010) Cerebrospinal fluid and plasma biomarkers in Alzheimer disease. Nat Rev Neurol 6:131–144

Bobinski M, de Leon MJ, Wegiel J, Desanti S, Convit A, Saint Louis LA et al (2000) The histological validation of post mortem magnetic resonance imaging-determined hippocampal volume in Alzheimer's disease. Neuroscience 95:721–725

Bokde AL, Lopez-Bayo P, Meindl T, Pechler S, Born C, Faltraco F et al (2006) Functional connectivity of the fusiform gyrus during a face-matching task in subjects with mild cognitive impairment. Brain 129:1113–1124

Bokde AL, Lopez-Bayo P, Born C, Dong W, Meindl T, Leinsinger G et al (2008) Functional abnormalities of the visual processing system in subjects with mild cognitive impairment: an fMRI study. Psychiatry Res 163:248–259

Bokde AL, Ewers M, Hampel H (2009) Assessing neuronal networks: understanding Alzheimer's disease. Prog Neurobiol 89:125–133

Bokde AL, Karmann M, Born C, Teipel SJ, Omerovic M, Ewers M et al (2010a) Altered brain activation during a verbal working memory task in subjects with amnestic mild cognitive impairment. J Alzheimers Dis 21:103–118

Bokde AL, Lopez-Bayo P, Born C, Ewers M, Meindl T, Teipel SJ et al (2010b) Alzheimer disease: functional abnormalities in the dorsal visual pathway. Radiology 254:219–226

Bondi MW, Houston WS, Eyler LT, Brown GG (2005) fMRI evidence of compensatory mechanisms in older adults at genetic risk for Alzheimer disease. Neurology 64:501–508

Bookheimer SY, Strojwas MH, Cohen MS, Saunders AM, Pericak-Vance MA, Mazziotta JC et al (2000) Patterns of brain activation in people at risk for Alzheimer's disease. N Engl J Med 343:450–456

Bookstein FL (2001) "Voxel-based morphometry" should not be used with imperfectly registered images. Neuroimage 14:1454–1462

Bosch B, Bartres-Faz D, Rami L, Arenaza-Urquijo EM, Fernandez-Espejo D, Junque C et al (2010) Cognitive reserve modulates task-induced activations and deactivations in healthy elders, amnestic mild cognitive impairment and mild Alzheimer's disease. Cortex 46:451–461

Bosch B, Arenaza-Urquijo EM, Rami L, Sala-Llonch R, Junque C, Sole-Padulles C et al (2012) Multiple DTI index analysis in normal aging, amnestic MCI and AD. Relationship with neuropsychological performance. Neurobiol Aging 33:61–74

Bozoki AC, Korolev IO, Davis NC, Hoisington LA, Berger KL (2012) Disruption of limbic white matter pathways in mild cognitive impairment and Alzheimer's disease: a DTI/FDG-PET study. Hum Brain Mapp 33:1792–1802

Bozzali M, Falini A, Franceschi M, Cercignani M, Zuffi M, Scotti G et al (2002) White matter damage in Alzheimer's disease assessed in vivo using diffusion tensor magnetic resonance imaging. J Neurol Neurosurg Psychiatry 72:742–746

Bozzali M, Giulietti G, Basile B, Serra L, Spano B, Perri R et al (2012) Damage to the cingulum contributes to Alzheimer's disease pathophysiology by deafferentation mechanism. Hum Brain Mapp 33:1295–1308

Braak H, Braak E (1995) Staging of Alzheimer's disease-related neurofibrillary changes. Neurobiol Aging 16:271–278; discussion 8–84

Braak H, Del Tredici K (2012) Where, when, and in what form does sporadic Alzheimer's disease begin? Curr Opin Neurol 25:708–714

Braak H, Braak E, Bohl J (1993) Staging of Alzheimer-related cortical destruction. Eur Neurol 33:403–408

Brayne C, Ince PG, Keage HA, McKeith IG, Matthews FE, Polvikoski T et al (2010) Education, the brain and dementia: neuroprotection or compensation? Brain 133:2210–2216

Brier MR, Thomas JB, Snyder AZ, Benzinger TL, Zhang D, Raichle ME et al (2012) Loss of intranetwork and internetwork resting state functional connections with Alzheimer's disease progression. J Neurosci 32:8890–8899

Brown JA, Terashima KH, Burggren AC, Ercoli LM, Miller KJ, Small GW et al (2011) Brain network local interconnectivity loss in aging APOE-4 allele carriers. Proc Natl Acad Sci U S A 108:20760–20765

Buchhave P, Minthon L, Zetterberg H, Wallin AK, Blennow K, Hansson O (2012) Cerebrospinal fluid levels of beta-amyloid 1-42, but not of tau, are fully changed already 5 to 10 years before the onset of Alzheimer dementia. Arch Gen Psychiatry 69:98–106

Buckner RL, Snyder AZ, Shannon BJ, LaRossa G, Sachs R, Fotenos AF et al (2005) Molecular, structural, and functional characterization of Alzheimer's disease: evidence for a relationship between default activity, amyloid, and memory. J Neurosci 25:7709–7717

Buckner RL, Andrews-Hanna JR, Schacter DL (2008) The brain's default network: anatomy, function, and relevance to disease. Ann N Y Acad Sci 1124:1–38

Buckner RL, Sepulcre J, Talukdar T, Krienen FM, Liu H, Hedden T et al (2009) Cortical hubs revealed by intrinsic functional connectivity: mapping, assessment of stability, and relation to Alzheimer's disease. J Neurosci 29:1860–1873

Calvini P, Chincarini A, Gemme G, Penco MA, Squarcia S, Nobili F et al (2009) Automatic analysis of medial temporal lobe atrophy from structural MRIs for the early assessment of Alzheimer disease. Med Phys 36:3737–3747

Canu E, McLaren DG, Fitzgerald ME, Bendlin BB, Zoccatelli G, Alessandrini F et al (2010) Microstructural diffusion changes are independent of macrostructural volume loss in moderate to severe Alzheimer's disease. J Alzheimers Dis 19:963–976

Cardenas VA, Chao LL, Studholme C, Yaffe K, Miller BL, Madison C et al (2011) Brain atrophy associated with baseline and longitudinal measures of cognition. Neurobiol Aging 32:572–580

Caselli RJ, Reiman EM, Locke DE, Hutton ML, Hentz JG, Hoffman-Snyder C et al (2007) Cognitive domain decline in healthy apolipoprotein E epsilon4 homozygotes before the diagnosis of mild cognitive impairment. Arch Neurol 64:1306–1311

Castellano JM, Kim J, Stewart FR, Jiang H, DeMattos RB, Patterson BW et al (2011) Human apoE isoforms differentially regulate brain amyloid-beta peptide clearance. Sci Transl Med 3:89ra57

Cavedo E, Boccardi M, Ganzola R, Canu E, Beltramello A, Caltagirone C et al (2011) Local amygdala structural differences with 3T MRI in patients with Alzheimer disease. Neurology 76:727–733

Celone KA, Calhoun VD, Dickerson BC, Atri A, Chua EF, Miller SL et al (2006) Alterations in memory networks in mild cognitive impairment and Alzheimer's disease: an independent component analysis. J Neurosci 26:10222–10231

Chang YL, Jacobson MW, Fennema-Notestine C, Hagler DJ Jr, Jennings RG, Dale AM et al (2010) Level of executive function influences verbal memory in amnestic mild cognitive impairment and predicts prefrontal and posterior cingulate thickness. Cereb Cortex 20:1305–1313

Chen TF, Chen YF, Cheng TW, Hua MS, Liu HM, Chiu MJ (2009) Executive dysfunction and periventricular diffusion tensor changes in amnesic mild cognitive impairment and early Alzheimer's disease. Hum Brain Mapp 30:3826–3836

Chen KH, Chuah LY, Sim SK, Chee MW (2010) Hippocampal region-specific contributions to memory performance in normal elderly. Brain Cogn 72:400–407

Chen G, Ward BD, Xie C, Li W, Wu Z, Jones JL et al (2011) Classification of Alzheimer disease, mild cognitive impairment, and normal cognitive status with large-scale network analysis based on resting-state functional MR imaging. Radiology 259:213–221

Chen Z, Li L, Sun J, Ma L (2012) Mapping the brain in type II diabetes: voxel-based morphometry using DARTEL. Eur J Radiol 81:1870–1876

Chetelat G (2013) Alzheimer disease: Abeta-independent processes-rethinking preclinical AD. Nat Rev Neurol 9:123–124

Chetelat G, Landeau B, Eustache F, Mezenge F, Viader F, de la Sayette V et al (2005) Using voxel-based morphometry to map the structural changes associated with rapid conversion in MCI: a longitudinal MRI study. Neuroimage 27:934–946

Chetelat G, Villemagne VL, Bourgeat P, Pike KE, Jones G, Ames D et al (2010) Relationship between atrophy and beta-amyloid deposition in Alzheimer disease. Ann Neurol 67:317–324

Chetelat G, Villemagne VL, Villain N, Jones G, Ellis KA, Ames D et al (2012) Accelerated cortical atrophy in cognitively normal elderly with high beta-amyloid deposition. Neurology 78:477–484

Chiang GC, Insel PS, Tosun D, Schuff N, Truran-Sacrey D, Raptentsetsang ST et al (2010) Hippocampal atrophy rates and CSF biomarkers in elderly APOE2 normal subjects. Neurology 75:1976–1981

Chua TC, Wen W, Chen X, Kochan N, Slavin MJ, Trollor JN et al (2009) Diffusion tensor imaging of the posterior cingulate is a useful biomarker of mild cognitive impairment. Am J Geriatr Psychiatry 17:602–613

Chupin M, Gerardin E, Cuingnet R, Boutet C, Lemieux L, Lehericy S et al (2009) Fully automatic hippocampus segmentation and classification in Alzheimer's disease and mild cognitive impairment applied on data from ADNI. Hippocampus 19:579–587

Clerx L, Visser PJ, Verhey F, Aalten P (2012) New MRI markers for Alzheimer's disease: a meta-analysis of diffusion tensor imaging and a comparison with medial temporal lobe measurements. J Alzheimers Dis 29:405–429

Colliot O, Chetelat G, Chupin M, Desgranges B, Magnin B, Benali H et al (2008) Discrimination between Alzheimer disease, mild cognitive impairment, and normal aging by using automated segmentation of the hippocampus. Radiology 248:194–201

Convit A, De Leon MJ, Tarshish C, De Santi S, Tsui W, Rusinek H et al (1997) Specific hippocampal volume reductions in individuals at risk for Alzheimer's disease. Neurobiol Aging 18:131–138

Convit A, de Asis J, de Leon MJ, Tarshish CY, De Santi S, Rusinek H (2000) Atrophy of the medial occipitotemporal, inferior, and middle temporal gyri in non-demented elderly predict decline to Alzheimer's disease. Neurobiol Aging 21:19–26

Corder EH, Saunders AM, Strittmatter WJ, Schmechel DE, Gaskell PC, Small GW et al (1993) Gene dose of apolipoprotein E type 4 allele and the risk of Alzheimer's disease in late onset families. Science 261:921–923

Corder EH, Saunders AM, Risch NJ, Strittmatter WJ, Schmechel DE, Gaskell PC Jr et al (1994) Protective effect of apolipoprotein E type 2 allele for late onset alzheimer disease. Nat Genet 7:180–184

Csernansky JG, Hamstra J, Wang L, McKeel D, Price JL, Gado M et al (2004) Correlations between antemortem hippocampal volume and postmortem neuropathology in AD subjects. Alzheimer Dis Assoc Disord 18:190–195

Csernansky JG, Wang L, Swank J, Miller JP, Gado M, McKeel D et al (2005) Preclinical detection of Alzheimer's disease: hippocampal shape and volume predict dementia onset in the elderly. Neuroimage 25:783–792

Cui Y, Wen W, Lipnicki DM, Beg MF, Jin JS, Luo S et al (2012) Automated detection of amnestic mild cognitive impairment in community-dwelling elderly adults: a combined spatial atrophy and white matter alteration approach. Neuroimage 59:1209–1217

Cullen KM, Halliday GM, Double KL, Brooks WS, Creasey H, Broe GA (1997) Cell loss in the nucleus basalis is related to regional cortical atrophy in Alzheimer's disease. Neuroscience 78:641–652

Dai Z, Yan C, Wang Z, Wang J, Xia M, Li K et al (2012) Discriminative analysis of early Alzheimer's disease using multi-modal imaging and multi-level characterization with multi-classifier (M3). Neuroimage 59:2187–2195

Damoiseaux JS, Greicius MD (2009) Greater than the sum of its parts: a review of studies combining structural connectivity and resting-state functional connectivity. Brain Struct Funct 213:525–533

Damoiseaux JS, Smith SM, Witter MP, Sanz-Arigita EJ, Barkhof F, Scheltens P et al (2009) White matter tract integrity in aging and Alzheimer's disease. Hum Brain Mapp 30:1051–1059

Damoiseaux JS, Prater KE, Miller BL, Greicius MD (2012) Functional connectivity tracks clinical deterioration in Alzheimer's disease. Neurobiol Aging 33(828):e19–e30

Daselaar SM, Prince SE, Dennis NA, Hayes SM, Kim H, Cabeza R (2009) Posterior midline and ventral parietal activity is associated with retrieval success and encoding failure. Front Hum Neurosci 3:13

Davatzikos C, Fan Y, Wu X, Shen D, Resnick SM (2008) Detection of prodromal Alzheimer's disease via pattern classification of magnetic resonance imaging. Neurobiol Aging 29:514–523

Davatzikos C, Xu F, An Y, Fan Y, Resnick SM (2009) Longitudinal progression of Alzheimer's-like patterns of atrophy in normal older adults: the SPARE-AD index. Brain 132:2026–2035

Davatzikos C, Bhatt P, Shaw LM, Batmanghelich KN, Trojanowski JQ (2011) Prediction of MCI to AD conversion, via MRI, CSF biomarkers, and pattern classification. Neurobiol Aging 32(2322):e19–e27

Daviglus ML, Plassman BL, Pirzada A, Bell CC, Bowen PE, Burke JR et al (2011) Risk factors and preventive interventions for Alzheimer disease: state of the science. Arch Neurol 68:1185–1190

de Souza LC, Chupin M, Lamari F, Jardel C, Leclercq D, Colliot O et al (2012) CSF tau markers are correlated with hippocampal volume in Alzheimer's disease. Neurobiol Aging 33:1253–1257

de Toledo-Morrell L, Dickerson B, Sullivan MP, Spanovic C, Wilson R, Bennett DA (2000) Hemispheric differences in hippocampal volume predict verbal and spatial memory performance in patients with Alzheimer's disease. Hippocampus 10:136–142

Delbeuck X, Van der Linden M, Collette F (2003) Alzheimer's disease as a disconnection syndrome? Neuropsychol Rev 13:79–92

den Heijer T, Vermeer SE, Clarke R, Oudkerk M, Koudstaal PJ, Hofman A et al (2003) Homocysteine and brain atrophy on MRI of non-demented elderly. Brain 126:170–175

den Heijer T, Geerlings MI, Hoebeek FE, Hofman A, Koudstaal PJ, Breteler MM (2006) Use of hippocampal and amygdalar volumes on magnetic resonance imaging to predict dementia in cognitively intact elderly people. Arch Gen Psychiatry 63:57–62

Desikan RS, Cabral HJ, Hess CP, Dillon WP, Glastonbury CM, Weiner MW et al (2009) Automated MRI measures identify individuals with mild cognitive impairment and Alzheimer's disease. Brain 132:2048–2057

Desikan RS, McEvoy LK, Thompson WK, Holland D, Roddey JC, Blennow K et al (2011) Amyloid-beta associated volume loss occurs only in the presence of phospho-tau. Ann Neurol 70:657–661

deToledo-Morrell L, Stoub TR, Bulgakova M, Wilson RS, Bennett DA, Leurgans S et al (2004) MRI-derived entorhinal volume is a good predictor of conversion from MCI to AD. Neurobiol Aging 25:1197–1203

Deweer B, Lehericy S, Pillon B, Baulac M, Chiras J, Marsault C et al (1995) Memory disorders in probable Alzheimer's disease: the role of hippocampal atrophy as shown with MRI. J Neurol Neurosurg Psychiatry 58:590–597

Di Paola M, Di Iulio F, Cherubini A, Blundo C, Casini AR, Sancesario G et al (2010) When, where, and how the corpus callosum changes in MCI and AD: a multimodal MRI study. Neurology 74:1136–1142

Dickerson BC, Salat DH, Greve DN, Chua EF, Rand-Giovannetti E, Rentz DM et al (2005) Increased hippocampal activation in mild cognitive impairment compared to normal aging and AD. Neurology 65:404–411

Dickerson BC, Bakkour A, Salat DH, Feczko E, Pacheco J, Greve DN et al (2009) The cortical signature of Alzheimer's disease: regionally specific cortical thinning relates to symptom severity in very mild to mild AD dementia and is detectable in asymptomatic amyloid-positive individuals. Cereb Cortex 19:497–510

Dickerson BC, Stoub TR, Shah RC, Sperling RA, Killiany RJ, Albert MS et al (2011) Alzheimer-signature MRI biomarker predicts AD dementia in cognitively normal adults. Neurology 76:1395–1402

Donix M, Burggren AC, Suthana NA, Siddarth P, Ekstrom AD, Krupa AK et al (2010) Longitudinal changes in medial temporal cortical thickness in normal subjects with the APOE-4 polymorphism. Neuroimage 53:37–43

Douaud G, Menke RA, Gass A, Monsch AU, Rao A, Whitcher B et al (2013) Brain microstructure reveals early abnormalities more than two years prior to clinical progression from mild cognitive impairment to Alzheimer's disease. J Neurosci 33:2147–2155

Driscoll I, Davatzikos C, An Y, Wu X, Shen D, Kraut M et al (2009) Longitudinal pattern of regional brain volume change differentiates normal aging from MCI. Neurology 72:1906–1913

Driscoll I, Zhou Y, An Y, Sojkova J, Davatzikos C, Kraut MA et al (2011) Lack of association between 11C-PiB and longitudinal brain atrophy in non-demented older individuals. Neurobiol Aging 32:2123–2130

Drzezga A, Becker JA, Van Dijk KR, Sreenivasan A, Talukdar T, Sullivan C et al (2011) Neuronal dysfunction and disconnection of cortical hubs in non-demented subjects with elevated amyloid burden. Brain 134:1635–1646

Du AT, Schuff N, Amend D, Laakso MP, Hsu YY, Jagust WJ et al (2001) Magnetic resonance imaging of the entorhinal cortex and hippocampus in mild cognitive impairment and Alzheimer's disease. J Neurol Neurosurg Psychiatry 71:441–447

Dyrba M, Ewers M, Wegrzyn M, Kilimann I, Plant C, Oswald A et al (2012) Combining DTI and MRI for the automated detection of Alzheimer's disease using a large European multicenter dataset. Lect Notes Comput Sci 7509:18–28

Dyrba M, Ewers M, Wegrzyn M, Kilimann I, Plant C, Oswald A et al (2013) Robust automated detection of microstructural white matter degeneration in Alzheimer's disease using machine learning classification of multicenter DTI data. PLoS One 8:e64925

Englund E, Brun A, Alling C (1988) White matter changes in dementia of Alzheimer's type. Biochemical and neuropathological correlates. Brain 111(Pt 6):1425–1439

Engvig A, Fjell AM, Westlye LT, Moberget T, Sundseth O, Larsen VA et al (2012) Memory training impacts short-term changes in aging white matter: a longitudinal diffusion tensor imaging study. Hum Brain Mapp 33:2390–2406

Erickson KI, Voss MW, Prakash RS, Basak C, Szabo A, Chaddock L et al (2011) Exercise training increases size of hippocampus and improves memory. Proc Natl Acad Sci U S A 108:3017–3022

Ewers M, Teipel SJ, Dietrich O, Schonberg SO, Jessen F, Heun R et al (2006) Multicenter assessment of reliability of cranial MRI. Neurobiol Aging 27:1051–1059

Ewers M, Insel P, Jagust WJ, Shaw L, Trojanowski JQ, Aisen P et al (2012) CSF biomarker and PIB-PET-derived beta-amyloid signature predicts metabolic, gray matter, and cognitive changes in nondemented subjects. Cereb Cortex 22:1993–2004

Fan M, Liu B, Zhou Y, Zhen X, Xu C, Jiang T (2010) Cortical thickness is associated with different apolipoprotein E genotypes in healthy elderly adults. Neurosci Lett 479:332–336

Fellgiebel A, Yakushev I (2011) Diffusion tensor imaging of the hippocampus in MCI and early Alzheimer's disease. J Alzheimers Dis 26(Suppl 3):257–262

Fellgiebel A, Wille P, Muller MJ, Winterer G, Scheurich A, Vucurevic G et al (2004) Ultrastructural hippocampal and white matter alterations in mild cognitive impairment: a diffusion tensor imaging study. Dement Geriatr Cogn Disord 18:101–108

Fellgiebel A, Muller MJ, Wille P, Dellani PR, Scheurich A, Schmidt LG et al (2005) Color-coded diffusion-tensor-imaging of posterior cingulate fiber tracts in mild cognitive impairment. Neurobiol Aging 26:1193–1198

Fellgiebel A, Schermuly I, Gerhard A, Keller I, Albrecht J, Weibrich C et al (2008) Functional relevant loss of long association fibre tracts integrity in early Alzheimer's disease. Neuropsychologia 46:1698–1706

Fischer F, Scheurich A, Wegrzyn M, Schermuly I, Bokde ALW, Klöppel S et al (2012) Automated tractography of the cingulate bundles in Alzheimer's disease: a multicenter DTI study. J Magn Reson Imaging 36:84–911

Fischl B, van der Kouwe A, Destrieux C, Halgren E, Segonne F, Salat DH et al (2004) Automatically parcellating the human cerebral cortex. Cereb Cortex 14:11–22

Fjell AM, Walhovd KB (2010) Structural brain changes in aging: courses, causes and cognitive consequences. Rev Neurosci 21:187–221

Fjell AM, Walhovd KB, Fennema-Notestine C, McEvoy LK, Hagler DJ, Holland D et al (2010) Brain atrophy in healthy aging is related to CSF levels of Abeta1-42. Cereb Cortex 20:2069–2079

Fleisher AS, Houston WS, Eyler LT, Frye S, Jenkins C, Thal LJ et al (2005) Identification of Alzheimer disease risk by functional magnetic resonance imaging. Arch Neurol 62:1881–1888

Fleisher AS, Sherzai A, Taylor C, Langbaum JB, Chen K, Buxton RB (2009) Resting-state BOLD networks versus task-associated functional MRI for distinguishing Alzheimer's disease risk groups. Neuroimage 47:1678–1690

Fox MD, Snyder AZ, Vincent JL, Corbetta M, Van Essen DC, Raichle ME (2005) The human brain is intrinsically organized into dynamic, anticorrelated functional networks. Proc Natl Acad Sci U S A 102:9673–9678

Friese U, Meindl T, Herpertz SC, Reiser MF, Hampel H, Teipel SJ (2010) Diagnostic utility of novel MRI-based biomarkers for Alzheimer's disease: diffusion tensor imaging and deformation-based morphometry. J Alzheimers Dis 20:477–490

Gao J, Cheung RT, Lee TM, Chu LW, Chan YS, Mak HK et al (2011) Possible retrogenesis observed with fiber tracking: an anteroposterior pattern of white matter disintegrity in normal aging and Alzheimer's disease. J Alzheimers Dis 26:47–58

Gauthier S, Reisberg B, Zaudig M, Petersen RC, Ritchie K, Broich K et al (2006) Mild cognitive impairment. Lancet 367:1262–1270

Gili T, Cercignani M, Serra L, Perri R, Giove F, Maraviglia B et al (2011) Regional brain atrophy and functional disconnection across Alzheimer's disease evolution. J Neurol Neurosurg Psychiatry 82:58–66

Gold BT, Powell DK, Andersen AH, Smith CD (2010) Alterations in multiple measures of white matter integrity in normal women at high risk for Alzheimer's disease. Neuroimage 52:1487–1494

Gong G, He Y, Chen ZJ, Evans AC (2012) Convergence and divergence of thickness correlations with diffusion connections across the human cerebral cortex. Neuroimage 59:1239–1248

Gons RA, van Norden AG, de Laat KF, van Oudheusden LJ, van Uden IW, Zwiers MP et al (2011) Cigarette smoking is associated with reduced microstructural integrity of cerebral white matter. Brain 134:2116–2124

Gons RA, Tuladhar AM, de Laat KF, van Norden AG, van Dijk EJ, Norris DG et al (2013) Physical activity is related to the structural integrity of cerebral white matter. Neurology 81:971–976

Grambaite R, Reinvang I, Selnes P, Fjell AM, Walhovd KB, Stenset V et al (2011a) Pre-dementia memory impairment is associated with white matter tract affection. J Int Neuropsychol Soc 17:143–153

Grambaite R, Selnes P, Reinvang I, Aarsland D, Hessen E, Gjerstad L et al (2011b) Executive dysfunction in mild cognitive impairment is associated with changes in frontal and cingulate white matter tracts. J Alzheimers Dis 27:453–462

Greicius MD, Srivastava G, Reiss AL, Menon V (2004) Default-mode network activity distinguishes Alzheimer's disease from healthy aging: evidence from functional MRI. Proc Natl Acad Sci U S A 101:4637–4642

Greicius MD, Supekar K, Menon V, Dougherty RF (2009) Resting-state functional connectivity reflects structural connectivity in the default mode network. Cereb Cortex 19:72–78

Grinberg LT, Rueb U, Heinsen H (2011) Brainstem: neglected locus in neurodegenerative diseases. Front Neurol 2:42

Gross AL, Manly JJ, Pa J, Johnson JK, Park LQ, Mitchell MB et al (2012) Cortical signatures of cognition and their relationship to Alzheimer's disease. Brain Imaging Behav 6:584–598

Grossman M, McMillan C, Moore P, Ding L, Glosser G, Work M et al (2004) What's in a name: voxel-based morphometric analyses of MRI and naming difficulty in Alzheimer's disease, frontotemporal dementia and corticobasal degeneration. Brain 127:628–649

Grothe M, Zaborszky L, Atienza M, Gil-Neciga E, Rodriguez-Romero R, Teipel SJ et al (2010) Reduction of basal forebrain cholinergic system parallels cognitive impairment in patients at high risk of developing Alzheimer's disease. Cereb Cortex 20:1685–1695

Grothe M, Heinsen H, Teipel SJ (2012) Atrophy of the cholinergic Basal forebrain over the adult age range and in early stages of Alzheimer's disease. Biol Psychiatry 71:805–813

Grothe M, Heinsen H, Teipel S (2013) Longitudinal measures of cholinergic forebrain atrophy in the transition from healthy aging to Alzheimer's disease. Neurobiol Aging 34:1210–1220

Grothe MJ, Ewers M, Krause B, Heinsen H, Teipel SJ (2014) Basal forebrain atrophy and cortical amyloid deposition in nondemented elderly subjects. Alzheimers Dement (in press) doi:10.1016/j.jalz.2013.09.011

Guillozet al, Weintraub S, Mash DC, Mesulam MM (2003) Neurofibrillary tangles, amyloid, and memory in aging and mild cognitive impairment. Arch Neurol 60:729–736

Guo LH, Alexopoulos P, Wagenpfeil S, Kurz A, Perneczky R (2013) Brain size and the compensation of Alzheimer's disease symptoms: a longitudinal cohort study. Alzheimers Dement 9:580–586

Hackert VH, den Heijer T, Oudkerk M, Koudstaal PJ, Hofman A, Breteler MM (2002) Hippocampal head size associated with verbal memory performance in nondemented elderly. Neuroimage 17:1365–1372

Hahn K, Myers N, Prigarin S, Rodenacker K, Kurz A, Forstl H et al (2013) Selectively and progressively disrupted structural connectivity of functional brain networks in Alzheimer's disease - revealed by a novel framework to analyze edge distributions of networks detecting disruptions with strong statistical evidence. Neuroimage 81:96–109

Hall AM, Moore RY, Lopez OL, Kuller L, Becker JT (2008) Basal forebrain atrophy is a presymptomatic marker for Alzheimer's disease. Alzheimers Dement 4:271–279

Haller S, Nguyen D, Rodriguez C, Emch J, Gold G, Bartsch A et al (2010) Individual prediction of cognitive decline in mild cognitive impairment using support vector machine-based analysis of diffusion tensor imaging data. J Alzheimers Dis 22:315–327

Hamalainen A, Pihlajamaki M, Tanila H, Hanninen T, Niskanen E, Tervo S et al (2007) Increased fMRI responses during encoding in mild cognitive impairment. Neurobiol Aging 28:1889–1903

Hanseeuw BJ, Van Leemput K, Kavec M, Grandin C, Seron X, Ivanoiu A (2011) Mild cognitive impairment: differential atrophy in the hippocampal subfields. AJNR Am J Neuroradiol 32:1658–1661

Hanyu H, Asano T, Sakurai H, Tanaka Y, Takasaki M, Abe K (2002) MR analysis of the substantia innominata in normal aging, Alzheimer disease, and other types of dementia. AJNR Am J Neuroradiol 23:27–32

Hardy J, Selkoe DJ (2002) The amyloid hypothesis of Alzheimer's disease: progress and problems on the road to therapeutics. Science 297:353–356

Harrison BJ, Pujol J, Contreras-Rodriguez O, Soriano-Mas C, Lopez-Sola M, Deus J et al (2011) Task-induced deactivation from rest extends beyond the default mode brain network. PLoS One 6:e22964

He Y, Evans A (2010) Graph theoretical modeling of brain connectivity. Curr Opin Neurol 23:341–350

Head D, Buckner RL, Shimony JS, Williams LE, Akbudak E, Conturo TE et al (2004) Differential vulnerability of anterior white matter in nondemented aging with minimal acceleration in dementia of the Alzheimer type: evidence from diffusion tensor imaging. Cereb Cortex 14:410–423

Heckemann RA, Keihaninejad S, Aljabar P, Gray KR, Nielsen C, Rueckert D et al (2011) Automatic morphometry in Alzheimer's disease and mild cognitive impairment. Neuroimage 56:2024–2037

Hedden T, Van Dijk KR, Becker JA, Mehta A, Sperling RA, Johnson KA et al (2009) Disruption of functional connectivity in clinically normal older adults harboring amyloid burden. J Neurosci 29:12686–12694

Heise V, Filippini N, Ebmeier KP, Mackay CE (2011) The APOE varepsilon4 allele modulates brain white matter integrity in healthy adults. Mol Psychiatry 16: 908–916

Honea RA, Vidoni E, Harsha A, Burns JM (2009) Impact of APOE on the healthy aging brain: a voxel-based MRI and DTI study. J Alzheimers Dis 18:553–564

Horn A, Ostwald D, Reisert M, Blankenburg F (2013) The structural-functional connectome and the default network of the human brain. Neuroimage (in press) doi:10.1016/j.neuroimage.2013.09.069

Hostage CA, Roy Choudhury K, Doraiswamy PM, Petrella JR (2013) Dissecting the gene dose-effects of the APOE epsilon4 and epsilon2 alleles on hippocampal volumes in aging and Alzheimer's disease. PLoS One 8:e54483

Huang J, Auchus AP (2007) Diffusion tensor imaging of normal appearing white matter and its correlation with cognitive functioning in mild cognitive impairment and Alzheimer's disease. Ann N Y Acad Sci 1097: 259–264

Huang H, Fan X, Weiner M, Martin-Cook K, Xiao G, Davis J et al (2012) Distinctive disruption patterns of white matter tracts in Alzheimer's disease with full diffusion tensor characterization. Neurobiol Aging 33:2029–2045

Huckman MS (1995) Where's the chicken? AJNR Am J Neuroradiol 16:2008–2009

Hutton C, Draganski B, Ashburner J, Weiskopf N (2009) A comparison between voxel-based cortical thickness and voxel-based morphometry in normal aging. Neuroimage 48:371–380

Jack CR Jr, Petersen RC, Xu YC, O'Brien PC, Smith GE, Ivnik RJ et al (1999) Prediction of AD with MRI-based hippocampal volume in mild cognitive impairment. Neurology 52:1397–1403

Jack CR Jr, Lowe VJ, Weigand SD, Wiste HJ, Senjem ML, Knopman DS et al (2009) Serial PIB and MRI in normal, mild cognitive impairment and Alzheimer's disease: implications for sequence of pathological events in Alzheimer's disease. Brain 132:1355–1365

Jack CR Jr, Knopman DS, Jagust WJ, Shaw LM, Aisen PS, Weiner MW et al (2010) Hypothetical model of dynamic biomarkers of the Alzheimer's pathological cascade. Lancet Neurol 9:119–128

Jack CR Jr, Albert MS, Knopman DS, McKhann GM, Sperling RA, Carrillo MC et al (2011) Introduction to the recommendations from the National Institute on

Aging-Alzheimer's Association workgroups on diagnostic guidelines for Alzheimer's disease. Alzheimers Dement 7:257–262

Jack CR Jr, Knopman DS, Jagust WJ, Petersen RC, Weiner MW, Aisen PS et al (2013) Tracking pathophysiological processes in Alzheimer's disease: an updated hypothetical model of dynamic biomarkers. Lancet Neurol 12:207–216

Jagust W, Harvey D, Mungas D, Haan M (2005) Central obesity and the aging brain. Arch Neurol 62: 1545–1548

Jiang Q, Lee CY, Mandrekar S, Wilkinson B, Cramer P, Zelcer N et al (2008) ApoE promotes the proteolytic degradation of Abeta. Neuron 58:681–693

Johnson SC, Schmitz TW, Moritz CH, Meyerand ME, Rowley HA, Alexander AL et al (2006a) Activation of brain regions vulnerable to Alzheimer's disease: the effect of mild cognitive impairment. Neurobiol Aging 27:1604–1612

Johnson SC, Schmitz TW, Trivedi MA, Ries ML, Torgerson BM, Carlsson CM et al (2006b) The influence of Alzheimer disease family history and apolipoprotein E epsilon4 on mesial temporal lobe activation. J Neurosci 26:6069–6076

Johnson DK, Barrow W, Anderson R, Harsha A, Honea R, Brooks WM et al (2010) Diagnostic utility of cerebral white matter integrity in early Alzheimer's disease. Int J Neurosci 120:544–550

Kalus P, Slotboom J, Gallinat J, Mahlberg R, Cattapan-Ludewig K, Wiest R et al (2006) Examining the gateway to the limbic system with diffusion tensor imaging: the perforant pathway in dementia. Neuroimage 30:713–720

Karas GB, Burton EJ, Rombouts SA, van Schijndel RA, O'Brien JT, Scheltens P et al (2003) A comprehensive study of gray matter loss in patients with Alzheimer's disease using optimized voxel-based morphometry. Neuroimage 18:895–907

Kaye JA, Swihart T, Howieson D, Dame A, Moore MM, Karnos T et al (1997) Volume loss of the hippocampus and temporal lobe in healthy elderly persons destined to develop dementia. Neurology 48:1297–1304

Kelly AM, Uddin LQ, Biswal BB, Castellanos FX, Milham MP (2008) Competition between functional brain networks mediates behavioral variability. Neuroimage 39:527–537

Kennedy KM, Raz N (2009) Pattern of normal age-related regional differences in white matter microstructure is modified by vascular risk. Brain Res 1297:41–56

Kilimann I, Grothe M, Heinsen H, Alho EJ, Grinberg L, Amaro Jr E, et al (2014) Subregional Basal Forebrain Atrophy in Alzheimer's Disease: A Multicenter Study. J Alzheimers Dis (in press)

Killiany RJ, Moss MB, Albert MS, Sandor T, Tieman J, Jolesz F (1993) Temporal lobe regions on magnetic resonance imaging identify patients with early Alzheimer's disease. Arch Neurol 50:949–954

Killiany RJ, Hyman BT, Gomez-Isla T, Moss MB, Kikinis R, Jolesz F et al (2002) MRI measures of entorhinal cortex vs hippocampus in preclinical AD. Neurology 58:1188–1196

Klein A, Andersson J, Ardekani BA, Ashburner J, Avants B, Chiang MC et al (2009) Evaluation of 14 nonlinear deformation algorithms applied to human brain MRI registration. Neuroimage 46:786–802

Kljajevic V, Meyer P, Holzmann C, Dyrba M, Kasper E, Bokde AL et al (2013) The epsilon4 genotype of apolipoprotein E and white matter integrity in Alzheimer's disease. Alzheimers Dement (in press) doi:10.1016/j.jalz.2013.02.008

Knight WD, Kim LG, Douiri A, Frost C, Rossor MN, Fox NC (2011) Acceleration of cortical thinning in familial Alzheimer's disease. Neurobiol Aging 32: 1765–1773

Knopman DS, Parisi JE, Salviati A, Floriach-Robert M, Boeve BF, Ivnik RJ et al (2003) Neuropathology of cognitively normal elderly. J Neuropathol Exp Neurol 62:1087–1095

Knopman DS, Jack CR, Wiste HJ, Weigand SD, Vemuri P, Lowe VJ et al (2013) Selective worsening of brain injury biomarker abnormalities in cognitively normal elderly persons with beta-Amyloidosis. JAMA Neurol 70:1030–1038

Koch W, Teipel S, Mueller S, Benninghoff J, Wagner M, Bokde AL et al (2012) Diagnostic power of default mode network resting state fMRI in the detection of Alzheimer's disease. Neurobiol Aging 33:466–478

Lambert C, Chowdhury R, Fitzgerald TH, Fleming SM, Lutti A, Hutton C et al (2013) Characterizing aging in the human brainstem using quantitative multimodal MRI analysis. Front Hum Neurosci 7:462

Lee DY, Fletcher E, Martinez O, Ortega M, Zozulya N, Kim J et al (2009) Regional pattern of white matter microstructural changes in normal aging, MCI, and AD. Neurology 73:1722–1728

Lee DY, Fletcher E, Martinez O, Zozulya N, Kim J, Tran J et al (2010) Vascular and degenerative processes differentially affect regional interhemispheric connections in normal aging, mild cognitive impairment, and Alzheimer disease. Stroke 41:1791–1797

Lehericy S, Baulac M, Chiras J, Pierot L, Martin N, Pillon B et al (1994) Amygdalohippocampal MR volume measurements in the early stages of Alzheimer disease. AJNR Am J Neuroradiol 15:929–937

Lerch JP, Pruessner J, Zijdenbos AP, Collins DL, Teipel SJ, Hampel H et al (2008) Automated cortical thickness measurements from MRI can accurately separate Alzheimer's patients from normal elderly controls. Neurobiol Aging 29:23–30

Leung KK, Barnes J, Ridgway GR, Bartlett JW, Clarkson MJ, Macdonald K et al (2010) Automated cross-sectional and longitudinal hippocampal volume measurement in mild cognitive impairment and Alzheimer's disease. Neuroimage 51:1345–1359

Liang P, Wang Z, Yang Y, Jia X, Li K (2011) Functional disconnection and compensation in mild cognitive impairment: evidence from DLPFC connectivity using resting-state fMRI. PLoS One 6:e22153

Liang P, Wang Z, Yang Y, Li K (2012) Three subsystems of the inferior parietal cortex are differently affected in mild cognitive impairment. J Alzheimers Dis 30: 475–487

Likitjaroen Y, Grothe M, Bauer A, Wegrzyn M, Hauenstein KH, Teipel SJ (submitted) Disrupted structural connectivity of the default mode network in Alzheimer's disease

Lim HK, Juh R, Pae CU, Lee BT, Yoo SS, Ryu SH et al (2008) Altered verbal working memory process in patients with Alzheimer's disease: an fMRI investigation. Neuropsychobiology 57:181–187

Lind J, Persson J, Ingvar M, Larsson A, Cruts M, Van Broeckhoven C et al (2006) Reduced functional brain activity response in cognitively intact apolipoprotein E epsilon4 carriers. Brain 129:1240–1248

Liu Y, Julkunen V, Paajanen T, Westman E, Wahlund LO, Aitken A et al (2012) Education increases reserve against Alzheimer's disease–evidence from structural MRI analysis. Neuroradiology 54:929–938

Liu Y, Mattila J, Ruiz MA, Paajanen T, Koikkalainen J, van Gils M et al (2013) Predicting AD conversion: comparison between prodromal AD guidelines and computer assisted PredictAD tool. PLoS One 8:e55246

Lo CY, Wang PN, Chou KH, Wang J, He Y, Lin CP (2010) Diffusion tensor tractography reveals abnormal topological organization in structural cortical networks in Alzheimer's disease. J Neurosci 30:16876–16885

Luks TL, Oliveira M, Possin KL, Bird A, Miller BL, Weiner MW et al (2010) Atrophy in two attention networks is associated with performance on a Flanker task in neurodegenerative disease. Neuropsychologia 48:165–170

Lustig C, Snyder AZ, Bhakta M, O'Brien KC, McAvoy M, Raichle ME et al (2003) Functional deactivations: change with age and dementia of the Alzheimer type. Proc Natl Acad Sci U S A 100:14504–14509

Lyness SA, Zarow C, Chui HC (2003) Neuron loss in key cholinergic and aminergic nuclei in Alzheimer disease: a meta-analysis. Neurobiol Aging 24:1–23

Machulda MM, Ward HA, Borowski B, Gunter JL, Cha RH, O'Brien PC et al (2003) Comparison of memory fMRI response among normal, MCI, and Alzheimer's patients. Neurology 61:500–506

Machulda MM, Jones DT, Vemuri P, McDade E, Avula R, Przybelski S et al (2011) Effect of APOE epsilon4 status on intrinsic network connectivity in cognitively normal elderly subjects. Arch Neurol 68:1131–1136

Mahley RW, Rall SC Jr (2000) Apolipoprotein E: far more than a lipid transport protein. Annu Rev Genomics Hum Genet 1:507–537

Mak HK, Zhang Z, Yau KK, Zhang L, Chan Q, Chu LW (2011) Efficacy of voxel-based morphometry with DARTEL and standard registration as imaging biomarkers in Alzheimer's disease patients and cognitively normal older adults at 3.0 Tesla MR imaging. J Alzheimers Dis 23:655–664

Mann DM (1988) Alzheimer's disease and Down's syndrome. Histopathology 13:125–137

Martin SB, Smith CD, Collins HR, Schmitt FA, Gold BT (2010) Evidence that volume of anterior medial temporal lobe is reduced in seniors destined for mild cognitive impairment. Neurobiol Aging 31:1099–1106

McDonald CR, Gharapetian L, McEvoy LK, Fennema-Notestine C, Hagler DJ Jr, Holland D et al (2012) Relationship between regional atrophy rates and cognitive decline in mild cognitive impairment. Neurobiol Aging 33:242–253

McDowell I (2001) Alzheimer's disease: insights from epidemiology. Aging (Milano) 13:143–162

McKhann G, Drachman D, Folstein M, Katzman R, Price D, Stadlan EM (1984) Clinical diagnosis of Alzheimer's disease: report of the NINCDS-ADRDA Work Group under the auspices of Department of Health and Human Services Task Force on Alzheimer's Disease. Neurology 34:939–944

McKhann GM, Knopman DS, Chertkow H, Hyman BT, Jack CR Jr, Kawas CH et al (2011) The diagnosis of dementia due to Alzheimer's disease: recommendations from the National Institute on Aging-Alzheimer's Association workgroups on diagnostic guidelines for Alzheimer's disease. Alzheimers Dement 7:263–269

Mechelli A, Friston KJ, Frackowiak RS, Price CJ (2005) Structural covariance in the human cortex. J Neurosci 25:8303–8310

Medina D, DeToledo-Morrell L, Urresta F, Gabrieli JD, Moseley M, Fleischman D et al (2006) White matter changes in mild cognitive impairment and AD: a diffusion tensor imaging study. Neurobiol Aging 27:663–672

Meindl T, Teipel S, Elmouden R, Mueller S, Koch W, Dietrich O et al (2010) Test-retest reproducibility of the default-mode network in healthy individuals. Hum Brain Mapp 31:237–246

Mesulam M, Shaw P, Mash D, Weintraub S (2004) Cholinergic nucleus basalis tauopathy emerges early in the aging-MCI-AD continuum. Ann Neurol 55: 815–828

Mielke MM, Okonkwo OC, Oishi K, Mori S, Tighe S, Miller MI et al (2012) Fornix integrity and hippocampal volume predict memory decline and progression to Alzheimer's disease. Alzheimers Dement 8:105–113

Miettinen PS, Pihlajamaki M, Jauhiainen AM, Niskanen E, Hanninen T, Vanninen R et al (2011) Structure and function of medial temporal and posteromedial cortices in early Alzheimer's disease. Eur J Neurosci 34:320–330

Miller SL, Celone K, DePeau K, Diamond E, Dickerson BC, Rentz D et al (2008a) Age-related memory impairment associated with loss of parietal deactivation but preserved hippocampal activation. Proc Natl Acad Sci U S A 105:2181–2186

Miller SL, Fenstermacher E, Bates J, Blacker D, Sperling RA, Dickerson BC (2008b) Hippocampal activation in adults with mild cognitive impairment predicts subsequent cognitive decline. J Neurol Neurosurg Psychiatry 79:630–635

Misra C, Fan Y, Davatzikos C (2009) Baseline and longitudinal patterns of brain atrophy in MCI patients, and their use in prediction of short-term conversion to AD: results from ADNI. Neuroimage 44:1415–1422

Morgen K, Frolich L, Tost H, Plichta MM, Kolsch H, Rakebrandt F et al (2013) APOE-dependent phenotypes in subjects with mild cognitive impairment converting to Alzheimer's disease. J Alzheimers Dis 37:389–401

Mormino EC, Smiljic A, Hayenga AO, Onami SH, Greicius MD, Rabinovici GD et al (2011) Relationships between beta-amyloid and functional connectivity in different components of the default mode network in aging. Cereb Cortex 21:2399–2407

Mormino EC, Brandel MG, Madison CM, Marks S, Baker SL, Jagust WJ (2012) Abeta deposition in aging is associated with increases in brain activation during successful memory encoding. Cereb Cortex 22:1813–1823

Morris JC, Roe CM, Xiong C, Fagan AM, Goate AM, Holtzman DM et al (2010) APOE predicts amyloid-beta but not tau Alzheimer pathology in cognitively normal aging. Ann Neurol 67:122–131

Mueller SG, Chao LL, Berman B, Weiner MW (2011) Evidence for functional specialization of hippocampal subfields detected by MR subfield volumetry on high resolution images at 4 T. Neuroimage 56:851–857

Nakata Y, Sato N, Nemoto K, Abe O, Shikakura S, Arima K et al (2009) Diffusion abnormality in the posterior cingulum and hippocampal volume: correlation with disease progression in Alzheimer's disease. Magn Reson Imaging 27:347–354

O'Brien JL, O'Keefe KM, LaViolette PS, DeLuca AN, Blacker D, Dickerson BC et al (2010) Longitudinal fMRI in elderly reveals loss of hippocampal activation with clinical decline. Neurology 74:1969–1976

Oishi K, Akhter K, Mielke M, Ceritoglu C, Zhang J, Jiang H et al (2011) Multi-modal MRI analysis with disease-specific spatial filtering: initial testing to predict mild cognitive impairment patients who convert to Alzheimer's disease. Front Neurol 2:54

Patel KT, Stevens MC, Pearlson GD, Winkler AM, Hawkins KA, Skudlarski P et al (2013) Default mode network activity and white matter integrity in healthy middle-aged ApoE4 carriers. Brain Imaging Behav 7:60–67

Perneczky R, Wagenpfeil S, Lunetta KL, Cupples LA, Green RC, DeCarli C et al (2009) Education attenuates the effect of medial temporal lobe atrophy on cognitive function in Alzheimer's disease: the MIRAGE study. J Alzheimers Dis 17:855–862

Perneczky R, Wagenpfeil S, Lunetta KL, Cupples LA, Green RC, Decarli C et al (2010) Head circumference, atrophy, and cognition: implications for brain reserve in Alzheimer disease. Neurology 75:137–142

Perry EK, Tomlinson BE, Blessed G, Bergmann K, Gibson PH, Perry RH (1978) Correlation of cholinergic abnormalities with senile plaques and mental test scores in senile dementia. Br Med J 2:1457–1459

Persson J, Lind J, Larsson A, Ingvar M, Cruts M, Van Broeckhoven C et al (2006) Altered brain white matter integrity in healthy carriers of the APOE epsilon4 allele: a risk for AD? Neurology 66:1029–1033

Persson J, Lustig C, Nelson JK, Reuter-Lorenz PA (2007) Age differences in deactivation: a link to cognitive control? J Cogn Neurosci 19:1021–1032

Persson J, Lind J, Larsson A, Ingvar M, Sleegers K, Van Broeckhoven C et al (2008) Altered deactivation in individuals with genetic risk for Alzheimer's disease. Neuropsychologia 46:1679–1687

Peters F, Collette F, Degueldre C, Sterpenich V, Majerus S, Salmon E (2009) The neural correlates of verbal short-term memory in Alzheimer's disease: an fMRI study. Brain 132:1833–1846

Petersen RC, Smith GE, Waring SC, Ivnik RJ, Tangalos EG, Kokmen E (1999) Mild cognitive impairment: clinical characterization and outcome. Arch Neurol 56:303–308

Petrella JR, Prince SE, Wang L, Hellegers C, Doraiswamy PM (2007) Prognostic value of posteromedial cortex deactivation in mild cognitive impairment. PLoS One 2:e1104

Petrella JR, Sheldon FC, Prince SE, Calhoun VD, Doraiswamy PM (2011) Default mode network connectivity in stable vs progressive mild cognitive impairment. Neurology 76:511–517

Pihlajamaki M, K OK, Bertram L, Tanzi RE, Dickerson BC, Blacker D et al (2010) Evidence of altered posteromedial cortical FMRI activity in subjects at risk for Alzheimer disease. Alzheimer Dis Assoc Disord 24:28–36

Plant C, Teipel SJ, Oswald A, Bohm C, Meindl T, Mourao-Miranda J et al (2010) Automated detection of brain atrophy patterns based on MRI for the prediction of Alzheimer's disease. Neuroimage 50:162–174

Potter PE, Rauschkolb PK, Pandya Y, Sue LI, Sabbagh MN, Walker DG et al (2011) Pre- and post-synaptic cortical cholinergic deficits are proportional to amyloid plaque presence and density at preclinical stages of Alzheimer's disease. Acta Neuropathol 122:49–60

Prvulovic D, Hubl D, Sack AT, Melillo L, Maurer K, Frolich L et al (2002) Functional imaging of visuospatial processing in Alzheimer's disease. Neuroimage 17:1403–1414

Qi Z, Wu X, Wang Z, Zhang N, Dong H, Yao L et al (2010) Impairment and compensation coexist in amnestic MCI default mode network. Neuroimage 50:48–55

Querbes O, Aubry F, Pariente J, Lotterie JA, Demonet JF, Duret V et al (2009) Early diagnosis of Alzheimer's disease using cortical thickness: impact of cognitive reserve. Brain 132:2036–2047

Rabinovici GD, Jagust WJ (2009) Amyloid imaging in aging and dementia: testing the amyloid hypothesis in vivo. Behav Neurol 21:117–128

Raichle ME, MacLeod AM, Snyder AZ, Powers WJ, Gusnard DA, Shulman GL (2001) A default mode of brain function. Proc Natl Acad Sci U S A 98:676–682

Rasgon NL, Kenna HA, Wroolie TE, Kelley R, Silverman D, Brooks J et al (2011) Insulin resistance and hippocampal volume in women at risk for Alzheimer's disease. Neurobiol Aging 32:1942–1948

Raz N, Rodrigue KM, Head D, Kennedy KM, Acker JD (2004) Differential aging of the medial temporal lobe: a study of a five-year change. Neurology 62:433–438

Reiman EM, Chen K, Liu X, Bandy D, Yu M, Lee W et al (2009) Fibrillar amyloid-beta burden in cognitively normal people at 3 levels of genetic risk for Alzheimer's disease. Proc Natl Acad Sci U S A 106:6820–6825

Ridha BH, Barnes J, Bartlett JW, Godbolt A, Pepple T, Rossor MN et al (2006) Tracking atrophy progression in familial Alzheimer's disease: a serial MRI study. Lancet Neurol 5:828–834

Ridha BH, Barnes J, van de Pol LA, Schott JM, Boyes RG, Siddique MM et al (2007) Application of automated medial temporal lobe atrophy scale to Alzheimer disease. Arch Neurol 64:849–854

Ries ML, Carlsson CM, Rowley HA, Sager MA, Gleason CE, Asthana S et al (2008) Magnetic resonance imaging characterization of brain structure and function in mild cognitive impairment: a review. J Am Geriatr Soc 56:920–934

Ringman JM, O'Neill J, Geschwind D, Medina L, Apostolova LG, Rodriguez Y et al (2007) Diffusion tensor imaging in preclinical and presymptomatic carriers of familial Alzheimer's disease mutations. Brain 130:1767–1776

Ringman JM, Medina LD, Braskie M, Rodriguez-Agudelo Y, Geschwind DH, Macias-Islas MA et al (2011) Effects of risk genes on BOLD activation in presymptomatic carriers of familial Alzheimer's disease mutations during a novelty encoding task. Cereb Cortex 21:877–883

Risacher SL, Saykin AJ, West JD, Shen L, Firpi HA, McDonald BC (2009) Baseline MRI predictors of conversion from MCI to probable AD in the ADNI cohort. Curr Alzheimer Res 6:347–361

Risacher SL, Shen L, West JD, Kim S, McDonald BC, Beckett LA et al (2010) Longitudinal MRI atrophy biomarkers: relationship to conversion in the ADNI cohort. Neurobiol Aging 31:1401–1418

Roher AE, Weiss N, Kokjohn TA, Kuo YM, Kalback W, Anthony J et al (2002) Increased A beta peptides and reduced cholesterol and myelin proteins characterize white matter degeneration in Alzheimer's disease. Biochemistry 41:11080–11090

Rombouts SA, Barkhof F, Veltman DJ, Machielsen WC, Witter MP, Bierlaagh MA et al (2000) Functional MR imaging in Alzheimer's disease during memory encoding. AJNR Am J Neuroradiol 21:1869–1875

Rombouts SA, Barkhof F, Goekoop R, Stam CJ, Scheltens P (2005) Altered resting state networks in mild cognitive impairment and mild Alzheimer's disease: an fMRI study. Hum Brain Mapp 26:231–239

Rose SE, Chen F, Chalk JB, Zelaya FO, Strugnell WE, Benson M et al (2000) Loss of connectivity in Alzheimer's disease: an evaluation of white matter tract integrity with colour coded MR diffusion tensor imaging. J Neurol Neurosurg Psychiatry 69:528–530

Rosen AC, Prull MW, Gabrieli JD, Stoub T, O'Hara R, Friedman L et al (2003) Differential associations between entorhinal and hippocampal volumes and memory performance in older adults. Behav Neurosci 117:1150–1160

Rusinek H, de Leon MJ, George AE, Stylopoulos LA, Chandra R, Smith G et al (1991) Alzheimer disease: measuring loss of cerebral gray matter with MR imaging. Radiology 178:109–114

Ryan L, Walther K, Bendlin BB, Lue LF, Walker DG, Glisky EL (2011) Age-related differences in white matter integrity and cognitive function are related to APOE status. Neuroimage 54:1565–1577

Ryan NS, Keihaninejad S, Shakespeare TJ, Lehmann M, Crutch SJ, Malone IB et al (2013) Magnetic resonance imaging evidence for presymptomatic change in thalamus and caudate in familial Alzheimer's disease. Brain 136:1399–1414

Sabuncu MR, Desikan RS, Sepulcre J, Yeo BT, Liu H, Schmansky NJ et al (2011) The dynamics of cortical and hippocampal atrophy in Alzheimer disease. Arch Neurol 68:1040–1048

Sanz-Arigita EJ, Schoonheim MM, Damoiseaux JS, Rombouts SA, Maris E, Barkhof F et al (2010) Loss of 'small-world' networks in Alzheimer's disease: graph analysis of FMRI resting-state functional connectivity. PLoS One 5:e13788

Scheinin NM, Aalto S, Koikkalainen J, Lotjonen J, Karrasch M, Kemppainen N et al (2009) Follow-up of [11C]PIB uptake and brain volume in patients with Alzheimer disease and controls. Neurology 73:1186–1192

Scheltens P, Leys D, Barkhof F, Huglo D, Weinstein HC, Vermersch P et al (1992) Atrophy of medial temporal lobes on MRI in "probable" Alzheimer's disease and normal ageing: diagnostic value and neuropsychological correlates. J Neurol Neurosurg Psychiatry 55:967–972

Scola E, Bozzali M, Agosta F, Magnani G, Franceschi M, Sormani MP et al (2010) A diffusion tensor MRI study of patients with MCI and AD with a 2-year clinical follow-up. J Neurol Neurosurg Psychiatry 81:798–805

Seab JP, Jagust WJ, Wong ST, Roos MS, Reed BR, Budinger TF (1988) Quantitative NMR measurements of hippocampal atrophy in Alzheimer's disease. Magn Reson Med 8:200–208

Seeley WW, Crawford RK, Zhou J, Miller BL, Greicius MD (2009) Neurodegenerative diseases target large-scale human brain networks. Neuron 62:42–52

Segall JM, Allen EA, Jung RE, Erhardt EB, Arja SK, Kiehl K et al (2012) Correspondence between structure and function in the human brain at rest. Front Neuroinform 6:10

Sexton CE, Mackay CE, Lonie JA, Bastin ME, Terriere E, O'Carroll RE et al (2010) MRI correlates of episodic memory in Alzheimer's disease, mild cognitive impairment, and healthy aging. Psychiatry Res 184:57–62

Shao J, Myers N, Yang Q, Feng J, Plant C, Bohm C et al (2012) Prediction of Alzheimer's disease using individual structural connectivity networks. Neurobiol Aging 33:2756–2765

Shehzad Z, Kelly AM, Reiss PT, Gee DG, Gotimer K, Uddin LQ et al (2009) The resting brain: unconstrained yet reliable. Cereb Cortex 19:2209–2229

Sheline YI, Morris JC, Snyder AZ, Price JL, Yan Z, D'Angelo G et al (2010) APOE4 allele disrupts resting state fMRI connectivity in the absence of amyloid plaques or decreased CSF Abeta42. J Neurosci 30:17035–17040

Shibata E, Sasaki M, Tohyama K, Otsuka K, Endoh J, Terayama Y et al (2008) Use of neuromelanin-sensitive MRI to distinguish schizophrenic and depressive patients and healthy individuals based on signal alterations in the substantia nigra and locus ceruleus. Biol Psychiatry 64:401–406

Smith SM, Fox PT, Miller KL, Glahn DC, Fox PM, Mackay CE et al (2009) Correspondence of the brain's functional architecture during activation and rest. Proc Natl Acad Sci U S A 106:13040–13045

Smith AD, Smith SM, de Jager CA, Whitbread P, Johnston C, Agacinski G et al (2010) Homocysteine-lowering by B vitamins slows the rate of accelerated brain atrophy in mild cognitive impairment: a randomized controlled trial. PLoS One 5:e12244

Song SK, Sun SW, Ju WK, Lin SJ, Cross AH, Neufeld AH (2003) Diffusion tensor imaging detects and differentiates axon and myelin degeneration in mouse optic nerve after retinal ischemia. Neuroimage 20:1714–1722

Sorg C, Riedl V, Muhlau M, Calhoun VD, Eichele T, Laer L et al (2007) Selective changes of resting-state networks in individuals at risk for Alzheimer's disease. Proc Natl Acad Sci U S A 104:18760–18765

Sperling RA, Bates JF, Chua EF, Cocchiarella AJ, Rentz DM, Rosen BR et al (2003) fMRI studies of associative encoding in young and elderly controls and mild Alzheimer's disease. J Neurol Neurosurg Psychiatry 74:44–50

Sperling RA, Laviolette PS, O'Keefe K, O'Brien J, Rentz DM, Pihlajamaki M et al (2009) Amyloid deposition is associated with impaired default network function in older persons without dementia. Neuron 63:178–188

St George-Hyslop PH (2000) Molecular genetics of Alzheimer's disease. Biol Psychiatry 47:183–199

Stanek KM, Grieve SM, Brickman AM, Korgaonkar MS, Paul RH, Cohen RA et al (2011) Obesity is associated with reduced white matter integrity in otherwise healthy adults. Obesity (Silver Spring) 19:500–504

Stenset V, Bjornerud A, Fjell AM, Walhovd KB, Hofoss D, Due-Tonnessen P et al (2011) Cingulum fiber diffusivity and CSF T-tau in patients with subjective and mild cognitive impairment. Neurobiol Aging 32: 581–589

Storandt M, Mintun MA, Head D, Morris JC (2009) Cognitive decline and brain volume loss as signatures of cerebral amyloid-beta peptide deposition identified with Pittsburgh compound B: cognitive decline associated with Abeta deposition. Arch Neurol 66:1476–1481

Supekar K, Menon V, Rubin D, Musen M, Greicius MD (2008) Network analysis of intrinsic functional brain connectivity in Alzheimer's disease. PLoS Comput Biol 4:e1000100

Sydykova D, Stahl R, Dietrich O, Ewers M, Reiser MF, Schoenberg SO et al (2007) Fiber connections between the cerebral cortex and the corpus callosum in Alzheimer's disease: a diffusion tensor imaging and voxel-based morphometry study. Cereb Cortex 17:2276–2282

Teipel SJ, Bayer W, Alexander GE, Bokde AL, Zebuhr Y, Teichberg D et al (2003a) Regional pattern of hippocampus and corpus callosum atrophy in Alzheimer's disease in relation to dementia severity: evidence for early neocortical degeneration. Neurobiol Aging 24:85–94

Teipel SJ, Schapiro MB, Alexander GE, Krasuski JS, Horwitz B, Hoehne C et al (2003b) Relation of corpus callosum and hippocampal size to age in nondemented adults with Down's syndrome. Am J Psychiatry 160:1870–1878

Teipel SJ, Flatz WH, Heinsen H, Bokde AL, Schoenberg SO, Stockel S et al (2005) Measurement of basal forebrain atrophy in Alzheimer's disease using MRI. Brain 128:2626–2644

Teipel SJ, Pruessner JC, Faltraco F, Born C, Rocha-Unold M, Evans A et al (2006) Comprehensive dissection of the medial temporal lobe in AD: measurement of hippocampus, amygdala, entorhinal, perirhinal and parahippocampal cortices using MRI. J Neurol 253:794–800

Teipel SJ, Born C, Ewers M, Bokde AL, Reiser MF, Moller HJ et al (2007a) Multivariate deformation-based analysis of brain atrophy to predict Alzheimer's disease in mild cognitive impairment. Neuroimage 38:13–24

Teipel SJ, Stahl R, Dietrich O, Schoenberg SO, Perneczky R, Bokde AL et al (2007b) Multivariate network analysis of fiber tract integrity in Alzheimer's disease. Neuroimage 34:985–995

Teipel SJ, Meindl T, Wagner M, Kohl T, Burger K, Reiser MF et al (2009) White matter microstructure in relation to education in aging and Alzheimer's disease. J Alzheimers Dis 17:571–583

Teipel SJ, Bokde AL, Meindl T, Amaro E Jr, Soldner J, Reiser MF et al (2010a) White matter microstructure underlying default mode network connectivity in the human brain. Neuroimage 49:2021–2032

Teipel SJ, Meindl T, Wagner M, Stieltjes B, Reuter S, Hauenstein KH et al (2010b) Longitudinal changes in fiber tract integrity in healthy aging and mild cognitive impairment: a DTI follow-up study. J Alzheimers Dis 22:507–522

Teipel SJ, Meindl T, Grinberg L, Grothe M, Cantero JL, Reiser MF et al (2011a) The cholinergic system in mild cognitive impairment and Alzheimer's disease: an in vivo MRI and DTI study. Hum Brain Mapp 32:1349–1362

Teipel SJ, Reuter S, Stieltjes B, Acosta-Cabronero J, Ernemann U, Fellgiebel A et al (2011b) Multicenter stability of diffusion tensor imaging measures: a European clinical and physical phantom study. Psychiatry Res 194:363–371

Teipel SJ, Wegrzyn M, Meindl T, Frisoni G, Bokde AL, Fellgiebel A et al (2012) Anatomical MRI and DTI in the diagnosis of Alzheimer's disease: a European multicenter study. J Alzheimers Dis 31(Suppl 3):S33–S47

Teipel S, Heinsen H, Amaro E Jr, Grinberg LT, Krause B, Grothe M (2014a) Cholinergic basal forebrain atrophy predicts amyloid burden in Alzheimer's disease. Neurobiol Aging 35(3):482–491

Teipel SJ, Grothe M, Lista S, Toschi N, Garaci FG, Hampel H (2013) Relevance of magnetic resonance imaging for early detection and diagnosis of Alzheimer disease. Med Clin North Am 97:399–424

Teipel SJ, Lerche M, Kilimann I, O'Brien K, Grothe M, Meyer P et al (2014b) Decline of fiber tract integrity over the adult age range – a diffusion spectrum imaging study. J Magn Reson Imaging (in press) doi:10.1002/jmri.24420

Teipel SJ, Grothe MJ, Filippi M, Fellgiebel A, Dyrba M, Frisoni GB, et al (2014c) Fractional Anisotropy Changes in Alzheimer's Disease Depend on the Underlying Fiber Tract Architecture: A Multiparametric DTI Study using Joint Independent Component Analysis. J Alzheimers Dis (in press)

Tondelli M, Wilcock GK, Nichelli P, De Jager CA, Jenkinson M, Zamboni G (2012) Structural MRI changes detectable up to ten years before clinical Alzheimer's disease. Neurobiol Aging 33(825):e25–e36

Tosun D, Schuff N, Truran-Sacrey D, Shaw LM, Trojanowski JQ, Aisen P et al (2010) Relations between brain tissue loss, CSF biomarkers, and the ApoE genetic profile: a longitudinal MRI study. Neurobiol Aging 31:1340–1354

Trachtenberg AJ, Filippini N, Ebmeier KP, Smith SM, Karpe F, Mackay CE (2012) The effects of APOE on the functional architecture of the resting brain. Neuroimage 59:565–572

Trivedi MA, Schmitz TW, Ries ML, Torgerson BM, Sager MA, Hermann BP et al (2006) Reduced hippocampal activation during episodic encoding in middle-aged individuals at genetic risk of Alzheimer's disease: a cross-sectional study. BMC Med 4:1

Trivedi MA, Schmitz TW, Ries ML, Hess TM, Fitzgerald ME, Atwood CS et al (2008) fMRI activation during episodic encoding and metacognitive appraisal across the lifespan: risk factors for Alzheimer's disease. Neuropsychologia 46:1667–1678

Tseng BY, Gundapuneedi T, Khan MA, Diaz-Arrastia R, Levine BD, Lu H et al (2013) White matter integrity in physically fit older adults. Neuroimage 82:510–516

Turner MR, Grosskreutz J, Kassubek J, Abrahams S, Agosta F, Benatar M et al (2011) Towards a neuroimaging biomarker for amyotrophic lateral sclerosis. Lancet Neurol 10:400–403

van de Pol LA, Barnes J, Scahill RI, Frost C, Lewis EB, Boyes RG et al (2007) Improved reliability of hippocampal atrophy rate measurement in mild cognitive impairment using fluid registration. Neuroimage 34: 1036–1041

van den Heuvel M, Mandl R, Luigjes J, Hulshoff Pol H (2008) Microstructural organization of the cingulum tract and the level of default mode functional connectivity. J Neurosci 28:10844–10851

Van Leemput K, Bakkour A, Benner T, Wiggins G, Wald LL, Augustinack J et al (2009) Automated segmentation of hippocampal subfields from ultra-high resolution in vivo MRI. Hippocampus 19:549–557

Vannini P, O'Brien J, O'Keefe K, Pihlajamaki M, Laviolette P, Sperling RA (2011) What goes down must come up: role of the posteromedial cortices in encoding and retrieval. Cereb Cortex 21:22–34

Villain N, Desgranges B, Viader F, de la Sayette V, Mezenge F, Landeau B et al (2008) Relationships between hippocampal atrophy, white matter disruption, and gray matter hypometabolism in Alzheimer's disease. J Neurosci 28:6174–6181

Villain N, Fouquet M, Baron JC, Mezenge F, Landeau B, de La Sayette V et al (2010) Sequential relationships between grey matter and white matter atrophy and brain metabolic abnormalities in early Alzheimer's disease. Brain 133:3301–3314

Voss MW, Prakash RS, Erickson KI, Basak C, Chaddock L, Kim JS et al (2010) Plasticity of brain networks in a randomized intervention trial of exercise training in older adults. Front Aging Neurosci 2

Walhovd KB, Fjell AM, Amlien I, Grambaite R, Stenset V, Bjornerud A et al (2009) Multimodal imaging in mild cognitive impairment: metabolism, morphometry and diffusion of the temporal-parietal memory network. Neuroimage 45:215–223

Wang L, Zang Y, He Y, Liang M, Zhang X, Tian L et al (2006) Changes in hippocampal connectivity in the early stages of Alzheimer's disease: evidence from resting state fMRI. Neuroimage 31:496–504

Wang K, Liang M, Wang L, Tian L, Zhang X, Li K et al (2007) Altered functional connectivity in early Alzheimer's disease: a resting-state fMRI study. Hum Brain Mapp 28:967–978

Wang L, Goldstein FC, Veledar E, Levey AI, Lah JJ, Meltzer CC et al (2009a) Alterations in cortical thickness and white matter integrity in mild cognitive impairment measured by whole-brain cortical thickness mapping and diffusion tensor imaging. AJNR Am J Neuroradiol 30:893–899

Wang L, Khan A, Csernansky JG, Fischl B, Miller MI, Morris JC et al (2009b) Fully automated, multi-stage hippocampus mapping in very mild Alzheimer disease. Hippocampus 19:541–548

Wang Z, Jia X, Liang P, Qi Z, Yang Y, Zhou W et al (2012) Changes in thalamus connectivity in mild cognitive impairment: evidence from resting state fMRI. Eur J Radiol 81:277–285

Ward MA, Bendlin BB, McLaren DG, Hess TM, Gallagher CL, Kastman EK et al (2010) Low HDL cholesterol is associated with lower gray matter volume in cognitively healthy adults. Front Aging Neurosci 2

Wedeen VJ, Wang RP, Schmahmann JD, Benner T, Tseng WY, Dai G et al (2008) Diffusion spectrum magnetic resonance imaging (DSI) tractography of crossing fibers. Neuroimage 41:1267–1277

Wee CY, Yap PT, Zhang D, Denny K, Browndyke JN, Potter GG et al (2012) Identification of MCI individuals using structural and functional connectivity networks. Neuroimage 59:2045–2056

Weissman DH, Roberts KC, Visscher KM, Woldorff MG (2006) The neural bases of momentary lapses in attention. Nat Neurosci 9:971–978

Westlye ET, Lundervold A, Rootwelt H, Lundervold AJ, Westlye LT (2011) Increased hippocampal default mode synchronization during rest in middle-aged and elderly APOE epsilon4 carriers: relationships with memory performance. J Neurosci 31: 7775–7783

Whitwell JL, Przybelski SA, Weigand SD, Knopman DS, Boeve BF, Petersen RC et al (2007) 3D maps from multiple MRI illustrate changing atrophy patterns as subjects progress from mild cognitive impairment to Alzheimer's disease. Brain 130:1777–1786

Whitwell JL, Josephs KA, Murray ME, Kantarci K, Przybelski SA, Weigand SD et al (2008) MRI correlates

of neurofibrillary tangle pathology at autopsy: a voxel-based morphometry study. Neurology 71:743–749
Wishart HA, Saykin AJ, Rabin LA, Santulli RB, Flashman LA, Guerin SJ et al (2006) Increased brain activation during working memory in cognitively intact adults with the APOE epsilon4 allele. Am J Psychiatry 163:1603–1610
Wolk DA, Dickerson BC (2010) Apolipoprotein E (APOE) genotype has dissociable effects on memory and attentional-executive network function in Alzheimer's disease. Proc Natl Acad Sci U S A 107:10256–10261
Wolk DA, Dickerson BC (2011) Fractionating verbal episodic memory in Alzheimer's disease. Neuroimage 54:1530–1539
Xie C, Bai F, Yu H, Shi Y, Yuan Y, Chen G et al (2012) Abnormal insula functional network is associated with episodic memory decline in amnestic mild cognitive impairment. Neuroimage 63:320–327
Xiong C, Roe CM, Buckles V, Fagan A, Holtzman D, Balota D et al (2011) Role of family history for Alzheimer biomarker abnormalities in the adult children study. Arch Neurol 68:1313–1319
Xu J, Chen S, Ahmed SH, Chen H, Ku G, Goldberg MP et al (2001) Amyloid-beta peptides are cytotoxic to oligodendrocytes. J Neurosci 21:RC118
Xu J, Li Y, Lin H, Sinha R, Potenza MN (2013) Body mass index correlates negatively with white matter integrity in the fornix and corpus callosum: a diffusion tensor imaging study. Hum Brain Mapp 34:1044–1052
Yakushev I, Schreckenberger M, Muller MJ, Schermuly I, Cumming P, Stoeter P et al (2011) Functional implications of hippocampal degeneration in early Alzheimer's disease: a combined DTI and PET study. Eur J Nucl Med Mol Imaging 38:2219–2227
Zaborszky L, Hoemke L, Mohlberg H, Schleicher A, Amunts K, Zilles K (2008) Stereotaxic probabilistic maps of the magnocellular cell groups in human basal forebrain. Neuroimage 42:1127–1141
Zhang Y, Schuff N, Jahng GH, Bayne W, Mori S, Schad L et al (2007) Diffusion tensor imaging of cingulum fibers in mild cognitive impairment and Alzheimer disease. Neurology 68:13–19
Zhang HY, Wang SJ, Liu B, Ma ZL, Yang M, Zhang ZJ et al (2010) Resting brain connectivity: changes during the progress of Alzheimer disease. Radiology 256:598–606
Zhou Y, Dougherty JH Jr, Hubner KF, Bai B, Cannon RL, Hutson RK (2008) Abnormal connectivity in the posterior cingulate and hippocampus in early Alzheimer's disease and mild cognitive impairment. Alzheimers Dement 4:265–270
Zhou J, Greicius MD, Gennatas ED, Growdon ME, Jang JY, Rabinovici GD et al (2010) Divergent network connectivity changes in behavioural variant frontotemporal dementia and Alzheimer's disease. Brain 133:1352–1367
Zhuang L, Wen W, Zhu W, Trollor J, Kochan N, Crawford J et al (2010) White matter integrity in mild cognitive impairment: a tract-based spatial statistics study. Neuroimage 53:16–25
Zhuang L, Sachdev PS, Trollor JN, Kochan NA, Reppermund S, Brodaty H et al (2012) Microstructural white matter changes in cognitively normal individuals at risk of amnestic MCI. Neurology 79:748–754
Zhuang L, Sachdev PS, Trollor JN, Reppermund S, Kochan NA, Brodaty H et al (2013) Microstructural white matter changes, not hippocampal atrophy, detect early amnestic mild cognitive impairment. PLoS One 8:e58887
Zuo XN, Kelly C, Adelstein JS, Klein DF, Castellanos FX, Milham MP (2010) Reliable intrinsic connectivity networks: test-retest evaluation using ICA and dual regression approach. Neuroimage 49:2163–2177

Index

C. Mulert, M.E. Shenton (eds.), *MRI in Psychiatry*,
DOI 10.1007/978-3-642-54542-9, © Springer Berlin Heidelberg 2014

W

Z

Printing and Binding: Stürtz GmbH, Würzburg